MW00988081

FLUID MECHANICS WITH ENGINEERING APPLICATIONS

FLUID MECHANICS WITH ENGINEERING APPLICATIONS

Eighth Edition

Robert L. Daugherty, A.B., M.E.

Late Professor Emeritus of Mechanical and Hydraulic Engineering
California Institute of Technology

Joseph B. Franzini, Ph.D.

Professor of Civil Engineering
Stanford University

E. John Finnemore, Ph.D.

Associate Professor of Civil Engineering
University of Santa Clara

McGraw-Hill, Inc.

New York St. Louis San Francisco Auckland Bogotá
Caracas Lisbon London Madrid Mexico Milan
Montreal New Delhi Paris San Juan Singapore
Sydney Tokyo Toronto

This book was set in Times Roman.
The editor was Kiran Verma;
the production supervisor was Marietta Breitwieser;
the cover was designed by Nadja Furlan-Lorbek.
Project supervision was done by Albert Harrison, Harley Editorial Services.
R. R. Donnelley & Sons Company was printer and binder.

FLUID MECHANICS WITH ENGINEERING APPLICATIONS

Copyright © 1985, 1977, 1965 by McGraw-Hill, Inc. All rights reserved. Copyright 1954 by McGraw-Hill, Inc. All rights reserved. Formerly published under the title of *Hydraulics*, copyright 1937, 1925, 1919, 1916 by McGraw-Hill, Inc. All rights reserved. Copyright renewed 1953, 1965 by R. L. Daugherty. Copyright renewed 1982 by Marguerite R. Daugherty. Printed in the United States of America. Except as permitted under the United States Copyright Act of 1976, no part of this publication may be reproduced or distributed in any form or by any means, or stored in a data base or retrieval system, without the prior written permission of the publisher.

9 10 11 12 DOC/DOC 9 9 8 7 6 5 4 3

ISBN 0-07-015441-4

Library of Congress Cataloging in Publication Data

Daugherty, Robert L. (Robert Long), date
Fluid mechanics with engineering applications.

Bibliography: p.
Includes index.
1. Fluid mechanics. I. Franzini, Joseph B.
II. Finnemore, E. John. III. Title.
TA357.D37 1985 532 84-10050
ISBN 0-07-015441-4

CONTENTS

PREFACE

This eighth edition of *Fluid Mechanics with Engineering Applications* has been written to serve as a textbook for a first course in fluid mechanics for engineering students. In most curricula this course comes in the junior year and the student will already have had courses in differential and integral calculus and engineering mechanics. The student may also have had a course in thermodynamics and possibly in differential equations. The coverage in this book is broad, so that it can be used in a number of ways for a second course in fluid mechanics if desired.

The basic approach to the presentation of fluid mechanics as an engineering subject that was developed by Professor Daugherty over his many years of teaching has been retained in this edition. The book has been in existence since 1916. The first edition was authored by Robert L. Daugherty and its title was *Hydraulics*. He revised the book four times. On the fourth revision (fifth edition) he was assisted by Dr. Alfred C. Ingersoll and the title of the book was changed to *Fluid Mechanics with Engineering Applications*. The fifth and sixth revisions (sixth and seventh editions) were entirely the work of Professor Franzini. In planning ahead, to maintain the perpetuity of this popular book, Professor Franzini enlisted the services of Professor E. John Finnemore to assist him with this eighth edition. Professor Finnemore made substantial contributions to the textual material and took primary responsibility for the preparation of the Solutions Manual. The plan is for him to take an increasing role in future revisions of the book.

Both of us (Franzini and Finnemore) feel it is most important that the engineering student clearly visualize the physical situation under consideration. Hence there is considerable emphasis on physical phenomena throughout the book. Numerous illustrative examples are given to indicate to the student how the basic principles of fluid mechanics can be applied to particular engineering problems. These examples also help to clarify the text. We feel that the subject matter is best learned by placing heavy emphasis on the development of basic principles, the assumptions made in the development of these principles, and their limits of applicability. The problems presented for assignment purposes have been carefully selected to provide the student with a thorough workout in the application of basic principles. Only through working numerous problems will the student experience the evolution so necessary to the learning process.

In this revision, even though English units (feet, slugs, seconds, pounds) are used as the primary system of units, the corresponding SI units are given in the text material. Some illustrative examples and problems are given in English units and others in SI units. Quite often illustrative examples or problems are given in English units followed by the corresponding SI unit, each in parentheses, rounded off to a value approximately equal to the English unit. Thus, where data are given in both systems of units, the user of the text can follow illustrative examples or work problems in SI units by employing the data given in parentheses. Every effort is made to ease the changeover from English units to SI units. A brief discussion of SI units is presented with the front matter. We encourage instructors to assign problems in both sets of units so that students become conversant with both. A substantial number of new problems in SI units have been added to this edition of the book.

The book is essentially "self-contained." The treatment is such that an instructor generally need not resort to another reference to answer any question that a student might normally be expected to ask. This has required more detailed discussion than would have been necessary if the presentation of certain topics had been more superficial. A list of selected references is provided at the end of the book to serve as a guide for those students who wish to probe deeper into the various fields of fluid mechanics. The Appendix contains information on dimensions and units, conversion factors, physical properties of fluids, and other useful tables.

For a brief course it is possible to omit certain chapters of the book without loss of continuity. For example, an excellent first course in fluid mechanics can be achieved by covering Chapters 1 through 8; however, Chapter 5 could be omitted if desired. One might wish to include parts of Chapter 12 (Fluid Measurements) in a first course. Schools having stringent requirements in fluid mechanics might wish to cover the entire text in their course or courses required of all engineers. At other schools only partial coverage of the text might suffice for the course required of all engineers, and other portions of the text might be covered in a second course for students in a particular branch of engineering. Thus mechanical engineers might wish to include Chapters 9 and 10 in a second course while civil engineers would emphasize Chapter 11 and perhaps Chapter 13 in a second course. The book has been used at a number of schools for courses in hydraulic machinery.

The authors wish to acknowledge the many comments and suggestions that have been received from users of the book throughout the years. These have influenced the content and mode of presentation of the material. Further comments and suggestions are always welcome.

Joseph B. Franzini
E. John Finnemore

THE METRIC (SI) SYSTEM OF UNITS

As of 1984 nearly every major country in the world, except the United States, was using, or had officially decided to use the modernized system of metric units as their official mode of measurement. Conversion to the metric system is being given serious consideration in the United States and it appears very likely that the metric system will be officially adopted in the United States within a few years. Because of the imminence of metrification in the United States and because of the need to be able to readily interact with users of the metric system, both sets of units—English and metric—are used in this edition of this book. Quantities are generally expressed in English units, followed by the corresponding metric units enclosed in parentheses.

Since users of this book may not be familiar with the metric system, some introductory remarks seem appropriate. In 1960, the Eleventh General Conference on Weights and Measures on the International System of Units, at which the United States was represented, adopted the Système Internationale d'Unités (SI). The SI is a complete system of units based on the meter-kilogram-second (MKS) system. The SI is an absolute system where *mass* (*kilogram*) is a basic unit, and force (newton) is a derived unit. The analogous units in the English system are for mass (slug) and for force (pound).

In terms of the basic dimensions, the basic units and their symbols are:

Dimension	English unit	SI unit
Length (L)	Foot (ft)	Meter (m)
Mass (M)	Slug[1]	Kilogram (kg)
Time (T)	Second (s)	Second (s)
Force (F)	Pound (lb)	Newton (N)[2]
Temperature		
Absolute	Rankine (°R)	Kelvin (K)
Ordinary	Fahrenheit (°F)	Celsius (°C)

[1] Derived unit ($lb \cdot s^2/ft$)
[2] Derived unit ($kg \cdot m/s^2$)

A partial list of derived quantities encountered in fluid mechanics and their commonly used dimensions in terms of L, M, T, and F is as follows:

Quantity	Commonly used dimensions	English unit	SI unit
Acceleration (a)	LT^{-2}	ft/s^2	m/s^2
Area (A)	L^2	ft^2	m^2
Density (ρ)	ML^{-3}	slug/ft^3	kg/m^3
Energy, work or quantity of heat	FL	ft·lb	N·m
Flowrate (Q)	L^3T^{-1}	cfs (ft^3/s)	m^3/s
Frequency	T^{-1}	cycle/s (s^{-1})	Hz (hertz) s^{-1}
Kinematic viscosity (ν)	L^2T^{-1}	ft^2/s	m^2/s
Power	FLT^{-1}	ft·lb/s	N·m/s
Pressure (p)	FL^{-2}	psi (lb/in^2)	N/m^2 = Pa
Specific weight (γ)	FL^{-3}	lb/ft^3 (pcf)	N/m^3
Velocity (V)	LT^{-1}	fps (ft/s)	m/s
Viscosity (μ)	FTL^{-2}	lb·s/ft^2	N·s/m^2
Volume (vol)	L^3	ft^3	m^3

Other derived quantities will be dealt with when they are encountered in the text.

When dealing with unusually large or very small numbers, a series of prefixes have been adopted for use with SI units. The most commonly used prefixes are:

Multiple	Prefix	Symbol
10^9	giga	G
10^6	mega	M
10^3	kilo	k
10^{-2}	centi	c
10^{-3}	milli	m
10^{-6}	micro	μ

Hence Mg (megagram) represents 10^6 grams, mm (millimeter) represents 10^{-3} meters, and kN (kilonewton) represents 10^3 newtons.

In the metric system lengths are commonly expressed in millimeters (mm), centimeters (cm), meters (m) or kilometers (km), depending on the distance being measured. A meter is about 40 inches and a kilometer is approximately five-eighths of a mile. Areas are usually expressed in square centimeters (cm^2), square meters (m^2) or hectares (100 m × 100 m = 10^4 m^2), depending on the area being measured. The hectare, used for measuring large areas, is equivalent to about 2.5 acres. A newton is the force required to accelerate one kilogram of mass one meter per second squared. That is, N = kg·m/s^2. A newton is equivalent to approximately 0.225 lb. The SI unit of stress (or pressure), newton per square meter (N/m^2), is referred to as the pascal (Pa).

In SI units energy, work or quantity of heat are ordinarily expressed in joules (J). A joule is equal to a newton-meter, i.e., J = N·m. The unit of power is the watt (W) which is equivalent to a joule per second, i.e., W = J/s = N·m/s. Important quantities and conversion factors are presented on the insides of the front and back covers of the book and in Appendix 1. Lists of symbols and abbreviations is presented on the following pages.

LIST OF SYMBOLS

The following table lists the letter symbols generally used throughout the text. Because there are so many more concepts than there are English and suitable Greek letters, certain conflicts are unavoidable. However, where the same letter has been used for different concepts, the topics are so far removed from each other that no confusion should result. Occasionally a particular letter will be used in one special case only, but this local deviation from the table will be clearly indicated, and the usage will not be employed elsewhere. The customary units of measurement for each item are given in the English system while the corresponding SI unit is given in parentheses or brackets.

With respect to symbols, the authors have for the most part attempted to adhere to generally accepted ones, but not always.

A = any area, ft^2 (m^2)
= cross-sectional area of a stream normal to the velocity, ft^2 (m^2)
= area in turbines or pumps normal to the direction of absolute velocity of the fluid, ft^2 (m^2)
A_s = area of a liquid surface as for a tank or reservoir, ft^2 or acre (m^2 or hectare)
a = area in turbines or pumps normal to the relative velocity of the fluid, ft^2 (m^2)
= linear acceleration, fps/s (m/s^2)
B = any width, ft (m)
= width of open channel at water surface, ft (m)
= width of turbine runner or pump impeller at periphery, in (cm)
b = bottom width of open channel, ft (m)
C = any coefficient [dimensionless]
= Chézy coefficient [$ft^{1/2}s^{-1}$ ($m^{1/2}s^{-1}$)]

C_c = coefficient of contraction
C_d = coefficient of discharge
C_v = coefficient of velocity
} for orifices, tubes, and nozzles [all dimensionless]

C_D = drag coefficient [dimensionless]
C_f = average friction-drag coefficient for total surface [dimensionless]
C_L = lift coefficient [dimensionless]
c = sonic (i.e., acoustic) velocity (celerity), fps (m/s)
c_f = local friction-drag coefficient [dimensionless]
c_p = specific heat at constant pressure, ft·lb/(slug·°R) [N·m/(kg·K)]
c_v = specific heat at constant volume, ft·lb/(slug·°R) [N·m/(kg·K)]
D = diameter of pipe, turbine runner, or pump impeller, ft or in (m or cm)
E = specific energy in open channels = $y + V^2/2g$, ft (m)
= linear modulus of elasticity, psi (N/m^2)
E_v = volume modulus of elasticity, psi (N/m^2)
e = height of surface roughness, ft (mm)
$\mathbf{F}$ = Froude number = $V/\sqrt{gL}$ [dimensionless]
F = any force, lb (N)
F_D = drag force, lb (N)
F_L = lift force, lb (N)
f = friction factor for pipe flow [dimensionless]
G = weight rate of flow = γQ, lb/s (N/s)
G' = weight rate of flow of a gas through a nozzle under sonic conditions, lb/s (N/s)
g = acceleration of gravity = 32.174 fps/s (standard)
= 32.2 fps/s (9.81 m/s^2) for usual computation
H = total energy head = $p/\gamma + z + V^2/2g$, ft (m)
= head on weir, ft (m)
h = any head, ft (m)
= enthalpy per unit mass = $gI + p/\rho$, ft·lb/slug (N·m/kg)
h' = minor head loss, ft (m)
$\hat{h}$ = enthalpy per unit weight = $I + p/\gamma$, ft·lb/lb (N·m/N)
h_c = depth to centroid of area, ft (m)
h_L = head lost in friction, ft (m)
h_p = head put into flow by pump, ft (m)
h_t = head taken from flow by turbine, ft (m)
I = moment of inertia of area, ft^4 or in^4 (m^4, cm^4, or mm^4)
= internal thermal energy per unit weight, ft·lb/lb (N·m/N)
i = internal thermal energy per unit mass, ft·lb/slug (N·m/kg)
K = any constant [dimensionless]
= equivalent volume modulus for fluid in an elastic pipe, psi (N/m^2)
k = any loss coefficient [dimensionless]
= c_p/c_v, specific heat ratio [dimensionless]

L = length, ft (m)
$L_r = 1/\lambda$ = scale ratio $= L_p/L_m$ [dimensionless]
l = Prandtl mixing length, ft or in (cm or mm)
$\mathbf{M}$ = Mach number $= V/c$ [dimensionless]
M = mass rate of flow, slug/s (kg/s)
MR = manometer reading, ft or in (m, cm, or mm)
m = mass $= W/g$, slugs (kg)
= molecular weight
n = an exponent or any number in general
= Manning coefficient of roughness
= revolutions per minute, min^{-1}
n_e = rotative speed of hydraulic machine at maximum efficiency, rev/min
N_s = specific speed $= n_e\sqrt{\text{gpm}}/h^{3/4}$ for pumps [dimensionless]
n_s = specific speed $= n_e\sqrt{\text{bhp}}/h^{5/4}$ for turbines [dimensionless]
$= n_e\sqrt{Q}/h^{3/4}$ for pumps and fans [dimensionless]
NPSH = net positive suction head, ft (m)
P = power, ft·lb/s (N·m/s)
= height of weir crest above channel bottom, ft (m)
= wetted perimeter, ft (m)
p = fluid pressure, lb/ft^2 or psi (N/m^2 = Pa)
p_{atm} = atmospheric pressure, psia (N/m^2, abs)
p_v = vapor pressure, psia (N/m^2, abs)
Q = volume rate of flow, cfs (m^3/s)
Q_H = heat transferred per unit weight of fluid, ft·lb/lb (N·m/N)
q = volume rate of flow per unit width of rectangular channel, cfs/ft = ft^2/s (m^2/s)
q_H = heat transferred per unit mass of fluid, ft·lb/slug (N·m/kg)
$\mathbf{R}$ = Reynolds number $= LV\rho/\mu = LV/\nu$ [dimensionless]
R = gas constant, ft·lb/(slug·°R) or N·m/(kg·K)
R_h = hydraulic radius $= A/P$, ft (m)
r = any radius, ft or in (m or cm)
r_0 = radius of pipe, ft or in (m or cm)
S = slope of energy grade line $= h_L/L$ [dimensionless]
S_0 = slope of channel bed [dimensionless]
S_w = slope of water surface [dimensionless]
s = specific gravity of a fluid = ratio of its density to that of some standard fluid [dimensionless]
T = temperature, °F or °R (°C or K)
= period of time for travel of a pressure wave, s
= torque, ft·lb (N·m)
t = time, s (s)
= thickness, ft or in (m or mm)

U, U_0 = uniform velocity of fluid, fps (m/s)
u = velocity of a solid body, fps (m/s)
= linear velocity of a point on a rotating body = $r\omega$, fps (m/s)
= local velocity of fluid, fps (m/s)
u' = turbulent velocity fluctuation in the direction of flow, fps (m/s)
V = mean velocity of fluid, fps (m/s)
= absolute velocity of fluid in hydraulic machines, fps (m/s)
V_m = meridional velocity, fps (m/s)
V_r = radial component of velocity = $V \sin \alpha = v \sin \beta$, fps (m/s)
V_u = tangential component of velocity = $V \cos \alpha = u + v \cos \beta$, fps (m/s)
V_L = volume of liquid, ft^3 (m^3)
v = relative velocity in hydraulic machines, fps (m/s)
= specific volume = $1/\rho$, ft^3/slug (m^3/kg)
v' = turbulent velocity fluctuation normal to the direction of flow, fps (m/s)
u, v, w = components of velocity in x, y, z, directions, fps (m/s)
$\mathbf{W}$ = Weber number = $V/\sqrt{\sigma/\rho L}$ [dimensionless]
W = total weight, lb (N)
= work, ft·lb (N·m)
x = a distance, usually parallel to flow, ft (m)
y = a distance along a plane in hydrostatics, ft (m)
= total depth of open channel flow, ft (m)
y_c = critical depth of open channel flow, ft (m)
y_0 = depth for uniform flow in open channel (normal depth), ft (m)
z = elevation above any arbitrary datum plane, ft (m)
α (alpha) = angle between V and u in rotating machinery, measured between their positive directions
= kinetic energy correction factor [dimensionless]
β (beta) = angle between v and u in rotating machinery, measured between $+v$ and $+u$
= momentum correction factor [dimensionless]
Γ (gamma) = circulation, ft^2/s (m^2/s)
γ (gamma) = specific weight, lb/ft^3 (N/m^3)
δ (delta) = thickness of boundary layer, in (mm)
δ_l = thickness of viscous sublayer in turbulent flow, in (mm)
ε (epsilon) = kinematic eddy viscosity, ft^2/s (m^2/s)
η (eta) = eddy viscosity, lb·s/ft^2 (N·s/m^2)
= efficiency ($= \eta_h \times \eta_m \times \eta_v$)
η_h = hydraulic efficiency
η_m = mechanical efficiency
η_v = volumetric efficiency

θ (theta) = any angle
λ (lambda) = model ratio = L_m/L_p [dimensionless]
μ (mu) = absolute or dynamic viscosity, lb·s/ft^2 (N·s/m^2)
ν (nu) = kinematic viscosity, $=\mu/\rho$, ft^2/s (m^2/s)
ξ (xi) = vorticity, s^{-1}
Π (pi) = dimensionless parameter
ρ (rho) = density, mass per unit volume = γ/g, slug/ft^3 (kg/m^3)
Σ (sigma) = summation
σ (sigma) = surface tension, lb/ft (N/m)
= cavitation factor in turbomachines [dimensionless]
τ (tau) = shear stress, lb/ft^2 (N/m^2)
ϕ (phi) = ratio $u_1/\sqrt{2gh}$ for turbines and $u_2/\sqrt{2gh}$ for centrifugal pumps
ϕ_e = value of the above ϕ at point of maximum efficiency
ϕ = velocity potential, ft^2/s (m^2/s) for two-dimensional flow
ψ (psi) = stream function, ft^2/s (m^2/s) for two-dimensional flow
ω (omega) = angular velocity = u/r = $2\pi n/60$, rad/s

Values at specific points will be indicated by suitable subscripts. In the use of subscripts 1 and 2, the fluid is always assumed to flow from 1 to 2.

LIST OF ABBREVIATIONS

abs = absolute
atm = atmospheric
avg = average
bhp = brake (or shaft) horsepower
Btu = British thermal units
cfm = cubic feet per minute
cfs = cubic feet per second
d = day or days
fpm = feet per minute
fps = feet per second
ft = foot or feet
gpm = gallons per minute
h = hour or hours (SI)
ha = hectare
hp = horsepower
hr = hour or hours (English system)
Hz = hertz (cycles per second)
in = inch or inches
J = joules = N·m
kg = kilograms = 10^3 grams
L = liter
lb = pounds of force
ln = $\log_e$
log = $\log_{10}$
m = meter or meters
mb = millibars = 10^{-3} bar
mb, abs = millibars, absolute
mgd = million gallons per day
mm = millimeters = 10^{-3} meter
mph = miles per hour
N = newton or newtons = $kg \cdot m/s^2$
N/m^2, abs = newtons per square meter, absolute
P = poise = 0.10 $N \cdot s/cm^2$
Pa = pascal = N/m^2
pcf = pounds per cubic foot
psi = pounds per square inch
psia = pounds per square inch, absolute
psig = pounds per square inch, gage
rpm = revolutions per minute
rps = revolutions per second
s = second or seconds
St = stoke = cm^2/s
W = watt or watts = J/s

FLUID MECHANICS WITH ENGINEERING APPLICATIONS

CHAPTER

ONE

PROPERTIES OF FLUIDS

Fluid mechanics is the science of the mechanics of liquids and gases and is based on the same fundamental principles that are employed in the mechanics of solids. Fluid mechanics is a more difficult subject, however, because with solids one deals with separate and tangible elements, while with fluids there are not separate elements to be distinguished.

1.1 DEVELOPMENT OF FLUID MECHANICS

Fluid mechanics may be divided into three branches: *fluid statics* is the study of the mechanics of fluids at rest; *kinematics* deals with velocities and streamlines without considering forces or energy; and *fluid dynamics* is concerned with the relations between velocities and accelerations and the forces exerted by or upon fluids in motion.

Classical *hydrodynamics* is largely a subject in mathematics, since it deals with an imaginary ideal fluid that is completely frictionless. The results of such studies, without consideration of all the properties of real fluids, are of limited practical value. Consequently, in the past, engineers turned to experiments, and from these developed empirical formulas which supplied answers to practical problems. This subject was called *hydraulics*.

Empirical hydraulics was confined largely to water and was limited in scope. With developments in aeronautics, chemical engineering, and the petroleum industry, the need arose for a broader treatment. This has led to the combining of classical hydrodynamics with the study of real fluids, and this new science is called *fluid mechanics*. In modern fluid mechanics the basic principles of hydrodynamics are combined with the experimental techniques of hydraulics. The experimental data can be used to verify theory or to provide information supplementary to mathematical analysis. The end product is a unified body of basic principles of fluid mechanics that can be applied to the solution of fluid-flow problems of engineering significance.

1.2 DISTINCTION BETWEEN A SOLID AND A FLUID

The molecules of a solid are usually closer together than those of a fluid. The attractive forces between the molecules of a solid are so large that a solid tends to retain its shape. This is not the case for a fluid, where the attractive forces between the molecules are smaller. There are plastic solids which flow under the proper circumstances, and even metals may flow under high pressures. On the other hand, there are certain very viscous liquids which do not flow readily, and it is easy to confuse them with the plastic solids. The distinction is that any fluid, no matter how viscous, will yield in time to the slightest stress. But a solid, no matter how plastic, requires a certain magnitude of stress to be exerted before it will flow.

Also, when the shape of a solid is altered by external forces, the tangential stresses between adjacent particles tend to restore the body to its original configuration. With a fluid, these tangential stresses depend on the velocity of deformation and vanish as the velocity approaches zero. When motion ceases, the tangential stresses disappear and the fluid does not tend to regain its original shape.

1.3 DISTINCTION BETWEEN A GAS AND A LIQUID

A fluid may be either a gas or a liquid. The molecules of a gas are much farther apart than those of a liquid. Hence a gas is very compressible, and when all external pressure is removed, it tends to expand indefinitely. A gas is therefore in equilibrium only when it is completely enclosed. A liquid is relatively incompressible, and if all pressure, except that of its own vapor pressure, is removed, the cohesion between molecules holds them together, so that the liquid does not expand indefinitely. Therefore a liquid may have a free surface, i.e., a surface from which all pressure is removed, except that of its own vapor.

A *vapor* is a gas whose temperature and pressure are such that it is very near the liquid phase. Thus steam is considered a vapor because its state is normally not far from that of water. A gas may be defined as a highly superheated vapor; that is, its state is far removed from the liquid phase. Thus air is considered a gas because its state is normally very far from that of liquid air.

The volume of a gas or vapor is greatly affected by changes in pressure or temperature or both. It is usually necessary, therefore, to take account of changes in volume and temperature in dealing with gases or vapors. Whenever significant temperature or phase changes are involved in dealing with vapors and gases, the subject is largely dependent on heat phenomena (*thermodynamics*). Thus fluid mechanics and thermodynamics are interrelated.

1.4 DENSITY, SPECIFIC WEIGHT, SPECIFIC VOLUME, AND SPECIFIC GRAVITY

The *density* ρ of a fluid is its *mass* per unit volume, while the *specific weight* γ is its *weight* per unit volume. In the English engineers', or gravitational, system density ρ will be in slugs per cubic foot (kg/m^3 in SI units), which may also be expressed as units of $lb \cdot s^2/ft^4$ ($N \cdot s^2/m^4$ in SI units) (Appendix 1).

Specific weight γ represents the force exerted by gravity on a unit volume of fluid and therefore must have the units of force per unit volume, such as pounds per cubic foot (N/m^3 in SI units).

Density and specific weight of a fluid are related as follows:

$$\rho = \frac{\gamma}{g} \qquad \text{or} \qquad \gamma = \rho g \tag{1.1}$$

Since the physical equations are dimensionally homogeneous, the dimensions of density are

$$\text{Dimensions of } \rho = \frac{\text{dimensions of } \gamma}{\text{dimensions of } g} = \frac{lb/ft^3}{ft/s^2} = \frac{lb \cdot s^2}{ft^4} = \frac{\text{mass}}{\text{volume}} = \frac{\text{slugs}}{ft^3}$$

In SI units,

$$\text{Dimensions of } \rho = \frac{\text{dimensions of } \gamma}{\text{dimensions of } g} = \frac{N/m^3}{m/s^2} = \frac{N \cdot s^2}{m^4} = \frac{\text{mass}}{\text{volume}} = \frac{kg}{m^3}$$

It should be noted that density ρ is absolute since it depends on mass which is independent of location. Specific weight γ, on the other hand, is not absolute for it depends on the value of the gravitational acceleration g which varies with location, primarily latitude and elevation above mean sea level.

Specific volume v is the volume occupied by a unit mass of fluid. It is commonly applied to gases and is usually expressed in cubic feet per slug (m^3/kg in SI units). Specific volume is the reciprocal of density. Thus

$$v = \frac{1}{\rho} \tag{1.2}$$

Specific gravity s of a liquid is the ratio of its density to that of pure water at a standard temperature. Physicists use 39.2°F (4°C) as the standard, but engineers often use 60°F. In the metric system the density of water at 4°C is 1.00 g/cm^3 (or 1.00 g/mL),[1] equivalent to 1000 kg/m^3, and hence the specific gravity (which is dimensionless) has the same numerical value for a liquid in that system as its density expressed in g/mL or in Mg/m^3.

The specific gravity of a gas is the ratio of its density to that of either hydrogen or air at some specified temperature and pressure, but there is no

[1] One cubic centimeter (cm^3) is equivalent to one milliliter (mL).

general agreement on these standards, and so they must be stated in any given case.

Since the density of a fluid varies with temperature, specific gravities must be determined and specified at particular temperatures.

Illustrative Example 1.1 The specific weight of water at ordinary pressure and temperature is 62.4 lb/ft^3 (9.81 kN/m^3). The specific gravity of mercury is 13.55. Compute the density of water and the specific weight and density of mercury.

$$\rho_{\text{water}} = \frac{\gamma_{\text{water}}}{g} = \frac{62.4 \text{ lb/ft}^3}{32.2 \text{ ft/s}^2} = 1.94 \text{ slugs/ft}^3 = \frac{9.81 \text{ kN/m}^3}{9.81 \text{ m/s}^2} = 1.00 \text{ Mg/m}^3 = 1.00 \text{ g/mL}$$

$$\gamma_{\text{mercury}} = s_{\text{mercury}}\gamma_{\text{water}} = 13.55(62.4) = 846 \text{ lb/ft}^3 = 13.55(9.81) = 133 \text{ kN/m}^3$$

$$\rho_{\text{mercury}} = s_{\text{mercury}}\rho_{\text{water}} = 13.55(1.94) = 26.3 \text{ slugs/ft}^3 = 13.55(1.00) = 13.55 \text{ Mg/m}^3$$

1.5 COMPRESSIBLE AND INCOMPRESSIBLE FLUIDS

Fluid mechanics deals with both incompressible and compressible fluids, that is, with fluids of either constant or variable density. Although there is no such thing in reality as an incompressible fluid, this term is applied where the change in density with pressure is so small as to be negligible. This is usually the case with liquids. Gases, too, may be considered incompressible when the pressure variation is small compared with the absolute pressure.

Liquids are ordinarily considered incompressible fluids, yet sound waves, which are really pressure waves, travel through them. This is evidence of the elasticity of liquids. In problems involving *water hammer* (Sec. 13.6), it is necessary to consider the compressibility of the liquid.

The flow of air in a ventilating system is a case where a gas may be treated as incompressible, for the pressure variation is so small that the change in density is of no importance. But for a gas or steam flowing at high velocity through a long pipeline, the drop in pressure may be so great that change in density cannot be ignored. For an airplane flying at speeds below 250 mph (100 m/s), the air may be considered to be of constant density. But as an object moving through the air approaches the velocity of sound, which is of the order of 700 mph (300 m/s), the pressure and density of the air adjacent to the body become materially different from those of the air at some distance away, and the air must then be treated as a compressible fluid (Chap. 9).

1.6 COMPRESSIBILITY OF LIQUIDS

The compressibility (change in volume due to change in pressure) of a liquid is inversely proportional to its volume modulus of elasticity, also known as the *bulk modulus*. This modulus is defined as $E_v = -v\,dp/dv = -(v/dv)\,dp$, where $v =$

specific volume and p = pressure. As v/dv is a dimensionless ratio, the units of E_v and p are identical. The bulk modulus is analogous to the modulus of elasticity for solids; however, for fluids it is defined on a volume basis rather than in terms of the familiar one-dimensional stress-strain relation for solid bodies.

In most engineering problems the bulk modulus at or near atmospheric pressure is the one of interest. The bulk modulus is a property of the fluid and for liquids is a function of temperature and pressure. In Table 1.1 are shown a few values of the bulk modulus for water. At any temperature it can be noted that the value of E_v increases continuously with pressure, but at any one pressure the value of E_v is a maximum at about 120°F (50°C). Thus water has a minimum compressibility at about 120°F (50°C).

The volume modulus of mild steel is about 26,000,000 psi (170,000 MN/m^2). Taking a typical value for the volume modulus of cold water to be 320,000 psi (2,200 MN/m^2), it is seen that water is about 80 times as compressible as steel. The compressibility of liquids covers a wide range. Mercury,[1] for example, is approximately 8 percent as compressible as water, while the compressibility of nitric acid is nearly six times greater than that of water.

Table 1.1 shows that at any one temperature the bulk modulus of water does not vary a great deal for a moderate range in pressure. Rearranging the definition of E_v, as an approximation, one may use for the case of a fixed mass of liquid at constant temperature,

$$\frac{\Delta v}{v} \approx -\frac{\Delta p}{E_v}$$

or

$$\frac{v_2 - v_1}{v_1} \approx -\frac{p_2 - p_1}{E_v} \tag{1.3}$$

Table 1.1 Bulk modulus of water, psi*

Pressure, psia	Temperature, °F				
	32°	68°	120°	200°	300°
15	292,000	320,000	332,000	308,000	
1,500	300,000	330,000	342,000	319,000	248,000
4,500	317,000	348,000	362,000	338,000	271,000
15,000	380,000	410,000	426,000	405,000	350,000

* R. L. Daugherty, Some Physical Properties of Water and Other Fluids, *Trans. ASME*, vol. 57, no. 5, July, 1935. These values can be transformed to meganewtons per square meter by multiplying them by 0.0069.

[1] "Handbook of Chemistry and Physics," Chemical Rubber Publishing Company, Cleveland, Ohio.

where E_v is the mean value of the modulus for the pressure range and the subscripts 1 and 2 refer to the before and after conditions.

Assuming E_v to have a value of 320,000 psi, it may be seen that increasing the pressure of water by 1,000 psi will compress it only $\frac{1}{320}$, or 0.3 percent, of its original volume. Therefore it is seen that the usual assumption regarding water as incompressible is justified.

1.7 SPECIFIC WEIGHT OF LIQUIDS

The specific weight of some common liquids at 68°F (20°C) and standard sea-level atmospheric pressure[1] with $g = 32.2$ ft/s^2 (9.81 m/s^2) is given in Table 1.2. The specific weight of a *liquid* varies only slightly with pressure, depending on the bulk modulus of the liquid (Sec. 1.6); it also depends on temperature, and the variation may be considerable. Since specific weight γ is equal to ρg, the specific weight of a *fluid* depends on the local value of the acceleration of gravity in addition to the variations with temperature and pressure. The variation of the specific weight of water with temperature and pressure, where $g =$ 32.2 ft/s^2 (9.81 m/s^2), is shown in Fig. 1.1. The presence of dissolved air, salts in solution, and suspended matter will increase these values a very slight amount. Ocean water may ordinarily be assumed to weigh 64.0 lb/ft^3 (10.1 kN/m^3). Unless otherwise specified or implied by some specific temperature being given, the value to use for water in the problems in the text is $\gamma = 62.4$ lb/ft^3 (9.81 kN/m^3). Under extreme conditions the specific weight of water is quite different. For example, at 500°F (260°C) and 6,000 psi (42 MN/m^2) the specific weight of water is 51 lb/ft^3 (8.0 kN/m^3).

Table 1.2 Specific weights of common liquids at 68°F (20°C), 14.7 psia (1,013 mbar, abs) with $g = 32.2$ ft/s^2 (9.81 m/s^3)

	lb/ft^3	kN/m^3
Carbon tetrachloride	99.4	15.6
Ethyl alcohol	49.3	7.76
Gasoline	42	6.6
Glycerin	78.7	12.3
Kerosene	50	7.9
Motor oil	54	8.5
Water	62.4	9.81

[1] The standard sea-level atmospheric pressure is 14.7 psia (1,013 mbar, abs). A millibar is equivalent to 100 N/m^2. The psia and mbar, abs represent absolute pressure which is the actual pressure on the fluid in contrast to gage pressure which refers to the difference in pressure between one region and another (Sec. 2.4).

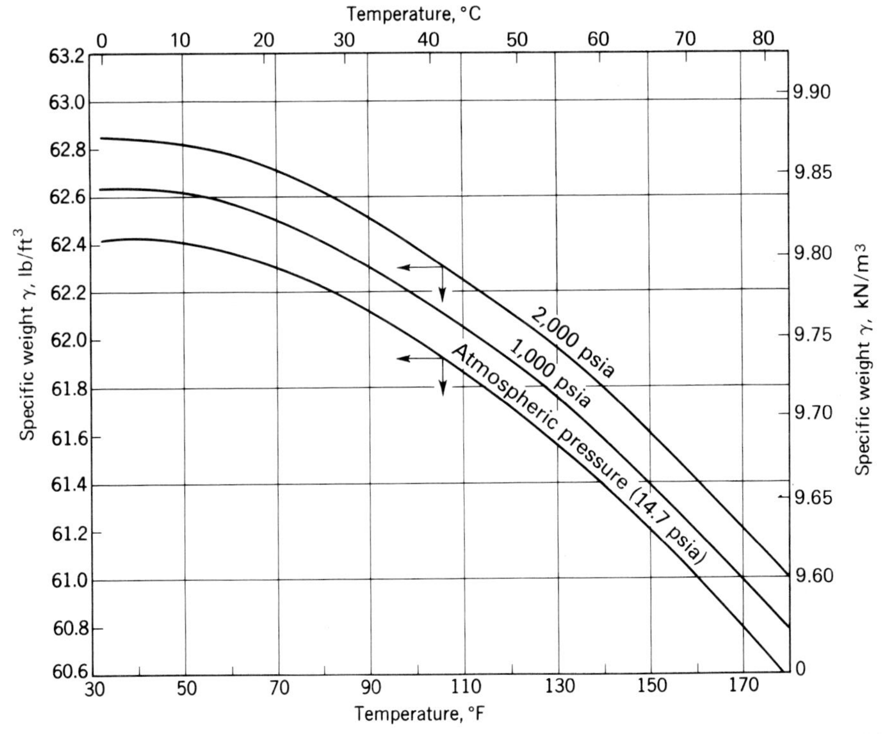

Figure 1.1 Specific weight γ of pure water as a function of temperature and pressure for condition where $g = 32.2\ \text{ft/s}^2$ ($9.81\ \text{m/s}^2$).

1.8 EQUATIONS OF STATE FOR GASES

There is no such thing as a perfect gas, but air and other real gases that are far removed from the liquid phase may be so considered. For a perfect gas the equation of state is

$$\frac{p}{\rho} = pv = RT \tag{1.4}$$

where p = absolute pressure
ρ = density (mass per unit volume)
v = specific volume ($1/\rho$)
R = a gas constant, the value of which depends upon the particular gas
T = absolute temperature in degrees Rankine or Kelvin[1]

For air the value of R is 1,715 ft·lb/(slug·°R) [287 N·m/(kg·K)]. Since $\gamma = \rho g$, Eq. (1.4) may also be written

$$\gamma = \frac{gp}{RT} \tag{1.5}$$

[1] Absolute temperature is measured above absolute zero. It may be recalled that absolute zero on the Fahrenheit scale occurs at approximately −460°F (0° Rankine). On the Celsius scale absolute zero is at −273°C (0 Kelvin).

from which the specific weight of any gas at any temperature and pressure can be computed if R and g are known.

Avogadro's law states that all gases at the same temperature and pressure under the action of a given value of g have the same number of molecules per unit of volume, from which it follows that the specific weight of a gas[1] is proportional to its molecular weight. Thus, if m denotes molecular weight, $\gamma_2/\gamma_1 = m_2/m_1$ and from Eq. (1.5) $\gamma_2/\gamma_1 = R_1/R_2$ for the same temperature, pressure, and value of g. Hence

$$m_1 R_1 = m_2 R_2 = \text{constant}$$

But this is strictly true for perfect gases only. The exact equation for any real gas is more complicated than Eq. (1.4), and hence mR is not strictly constant. For the perfect gas $mR = 49{,}710$ ft·lb/(slug·°R) [8,312 N·m/(kg·K)], while for actual gases the values of mR range between 48,700 and 49,800 ft·lb/(slug·°R) which is a variation of less than 3 percent.

Values of mR and R may be found in texts on thermodynamics and in handbooks, but a value of R may always be estimated by dividing an assumed value of mR by molecular weight. Thus water vapor in the atmosphere, because of its low partial pressure, may be treated as a perfect gas with $R = 49{,}710/18 = 2{,}760$ ft·lb/(slug·°R) [462 N·m/(kg·K)]. For steam at higher pressures this value is not applicable. In dealing with water vapor in the atmosphere Dalton's law of partial pressures states that the air and the water vapor each exert their own pressure as if the other were not present. Hence it is the partial pressure of each that is used in Eq. (1.4).

As the pressure is increased and the temperature simultaneously lowered, a gas becomes a vapor, and as gases depart more and more from the gas phase and approach the liquid phase, the equation of state becomes much more complicated than Eq. (1.4) and specific weight and other properties must then be obtained from vapor tables or charts. Such tables and charts exist for steam, ammonia, sulfur dioxide, freon, and other vapors in common engineering use.

Another fundamental equation for a perfect gas is

$$pv^n = p_1 v_1^n = \text{constant} \tag{1.6}$$

where p is absolute pressure, v $(=1/\rho)$ is specific volume, and n may have any nonnegative value from zero to infinity, depending upon the process to which the gas is subjected. If the process is at constant temperature (*isothermal*), $n = 1$. If there is no heat transfer to or from the gas, the process is known as *adiabatic*. A frictionless adiabatic process is called an *isentropic* process and n is denoted by k, where $k = c_p/c_v$, the ratio of specific heat at constant pressure to that at constant volume.[2] For expansion with friction n is less than k and for compression with friction it is greater than k. Values for k may be found in Appendix 3, Table A.5, and in thermodynamics texts and in handbooks. For air and diatomic gases at usual temperatures, k may be taken as 1.4.

[1] The specific weight of air (molecular weight $\approx$ 29.0) at 68°F (20°C) and 14.7 psia (1,013 mbar, abs) with $g = 32.2$ ft/s^2 (9.81 m/s^2) is 0.076 lb/ft^3 (0.012 kN/m^3).

[2] Specific heat and other thermodynamic properties of gases are discussed in Sec. 9.1.

By combining Eqs. (1.4) and (1.6), it is possible to obtain other useful relations such as

$$\frac{T_2}{T_1} = \left(\frac{v_1}{v_2}\right)^{n-1} = \left(\frac{p_2}{p_1}\right)^{(n-1)/n} \tag{1.7}$$

1.9 COMPRESSIBILITY OF GASES

Differentiating Eq. (1.6) gives $npv^{n-1}\,dv + v^n\,dp = 0$. Inserting the value of dp from this in $E_v = -(v/dv)\,dp$ from Sec. 1.6 yields

$$E_v = np$$

so that for an isothermal process of a gas $E_v = p$ and for an isentropic process $E_v = kp$.

Thus, at a pressure of 15 psia, the isothermal modulus of elasticity for a gas is 15 psi, and for air in an isentropic process it is 1.4×15 psi. Assuming from Table 1.1 a typical value of the modulus of elasticity of cold water to be 320,000 psi, it is seen that air at 15 psia is $320{,}000/15 = 21{,}000$ times as compressible as cold water isothermally, or 15,000 times as compressible isentropically. This emphasizes the great difference between the compressibility of normal atmospheric air and that of water.

Illustrative Example 1.2 (*a*) Calculate the density, specific weight, and specific volume of oxygen at 100°F and 15 psia (pounds per square inch absolute). (*b*) What would be the temperature and pressure of this gas if it were compressed isentropically to 40 percent of its original volume? (*c*) If the process described in (*b*) has been isothermal, what would the temperature and pressure have been?

(*a*) Molecular weight of oxygen (O_2) is 32,

$$R \approx \frac{49{,}710}{32} = 1{,}553\ \text{ft}\cdot\text{lb/(slug}\cdot{}^\circ\text{R)} \text{ as in Table A.5}a$$

$$\rho = \frac{p}{RT} = \frac{15 \times 144\ \text{lb/ft}^2}{[1{,}553\ \text{ft}\cdot\text{lb/(slug}\cdot{}^\circ\text{R)}][(460 + 100)^\circ\text{R}]}$$

$$\rho = 0.00248\ \text{slug/ft}^3$$

With $g = 32.2\ \text{ft/s}^2$, $\gamma = \rho g = 0.08\ \text{lb/ft}^3$,

$$v = \frac{1}{\rho} = \frac{1.0}{0.00248} = 403\ \text{ft}^3\text{/slug}$$

(*b*) $pv^k = (15 \times 144)(403)^{1.4} = (p_2 \times 144)(0.4 \times 403)^{1.4}$

$$p_2 = 54.1\ \text{psia}$$

$$p_2 = 54.1 \times 144 = \rho RT = \frac{0.00248}{0.40}(1553)(460 + T_2)$$

$$T_2 = 349^\circ\text{F}$$

(*c*) If isothermal, $T_2 = T_1 = 100^\circ$F and $pv = $ constant,

$$(15 \times 144)(403) = (p_2 \times 144)(0.4 \times 403)$$

$$p_2 = 37.5\ \text{psia}$$

Illustrative Example 1.3 Calculate the density, specific weight, and specific volume of chlorine gas at 25°C and pressure of 600 kN/m², abs (kilonewtons per square meter absolute).

Molecular weight of chlorine (Cl_2) = 71

$$R = \frac{8{,}312}{71} = 117\ \text{N}\cdot\text{m/(kg}\cdot\text{K)}$$

$$\rho = \frac{p}{RT} = \frac{600\ \text{kN/m}^2}{[117\ \text{N}\cdot\text{m/(kg}\cdot\text{K)}][(273 + 25)\text{K}]} = 17.2\ \text{kg/m}^3$$

With $g = 9.81\ \text{m/s}^2$, $\gamma = \rho g = 169\ \text{N/m}^3$,

$$v = \frac{1}{\rho} = \frac{1}{17.2} = 0.058\ \text{m}^3/\text{kg}$$

1.10 IDEAL FLUID

An *ideal* fluid may be defined as one in which there is *no friction*; that is, its viscosity is zero. Thus the internal forces at any internal section are always normal to the section, even during motion. Hence the forces are purely pressure forces. Such a fluid does not exist in reality.

In a *real* fluid, either liquid or gas, tangential or shearing forces always come into being whenever motion takes place, thus giving rise to fluid friction, because these forces oppose the movement of one particle past another. These friction forces are due to a property of the fluid called *viscosity*.

1.11 VISCOSITY

The viscosity of a fluid is a measure of its resistance to shear or angular deformation. The friction forces in fluid flow result from the cohesion and momentum interchange between molecules in the fluid. The viscosities of typical fluids are shown in Figs. 1.2 and 1.3. As the temperature increases, the viscosities of all liquids decrease, while the viscosities of all gases increase. This is because the force of cohesion, which diminishes with temperature, predominates with liquids, while with gases the predominating factor is the interchange of molecules between the layers of different velocities. Thus a rapidly moving molecule shifting into a slower-moving layer tends to speed up the latter. And a slow-moving molecule entering a faster-moving layer tends to slow down the faster-moving layer. This molecular interchange sets up a shear, or produces a friction force between adjacent layers. Increased molecular activity at higher temperatures causes the viscosity of gases to increase with temperature.

Consider two parallel plates (Fig. 1.4), sufficiently large so that edge conditions may be neglected, placed a small distance Y apart, the space between being filled with the fluid. The lower surface is assumed to be stationary, while the upper one is moved parallel to it with a velocity U by the application of a force F corresponding to some area A of the moving plate. Such a condition is

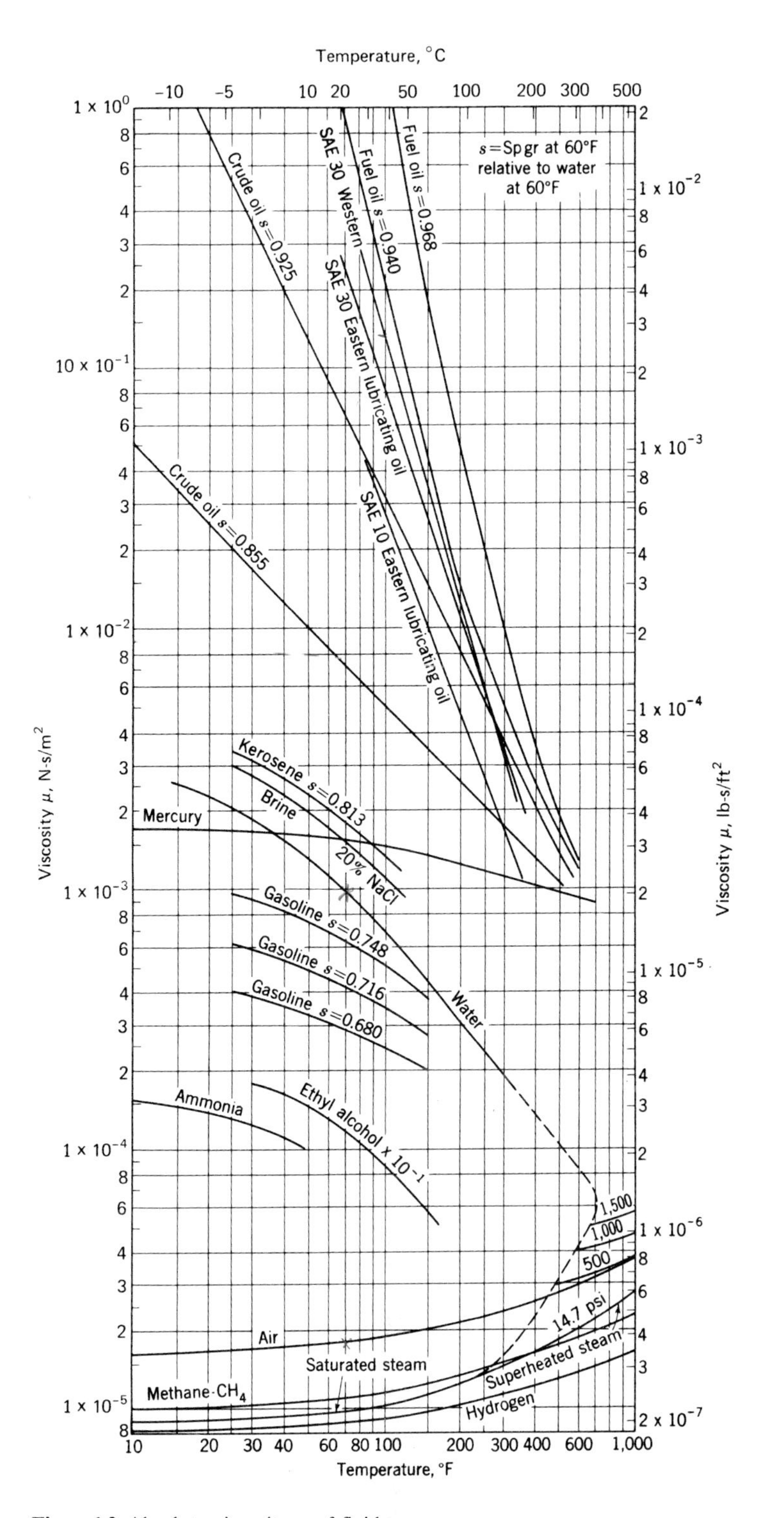

Figure 1.2 Absolute viscosity μ of fluids.

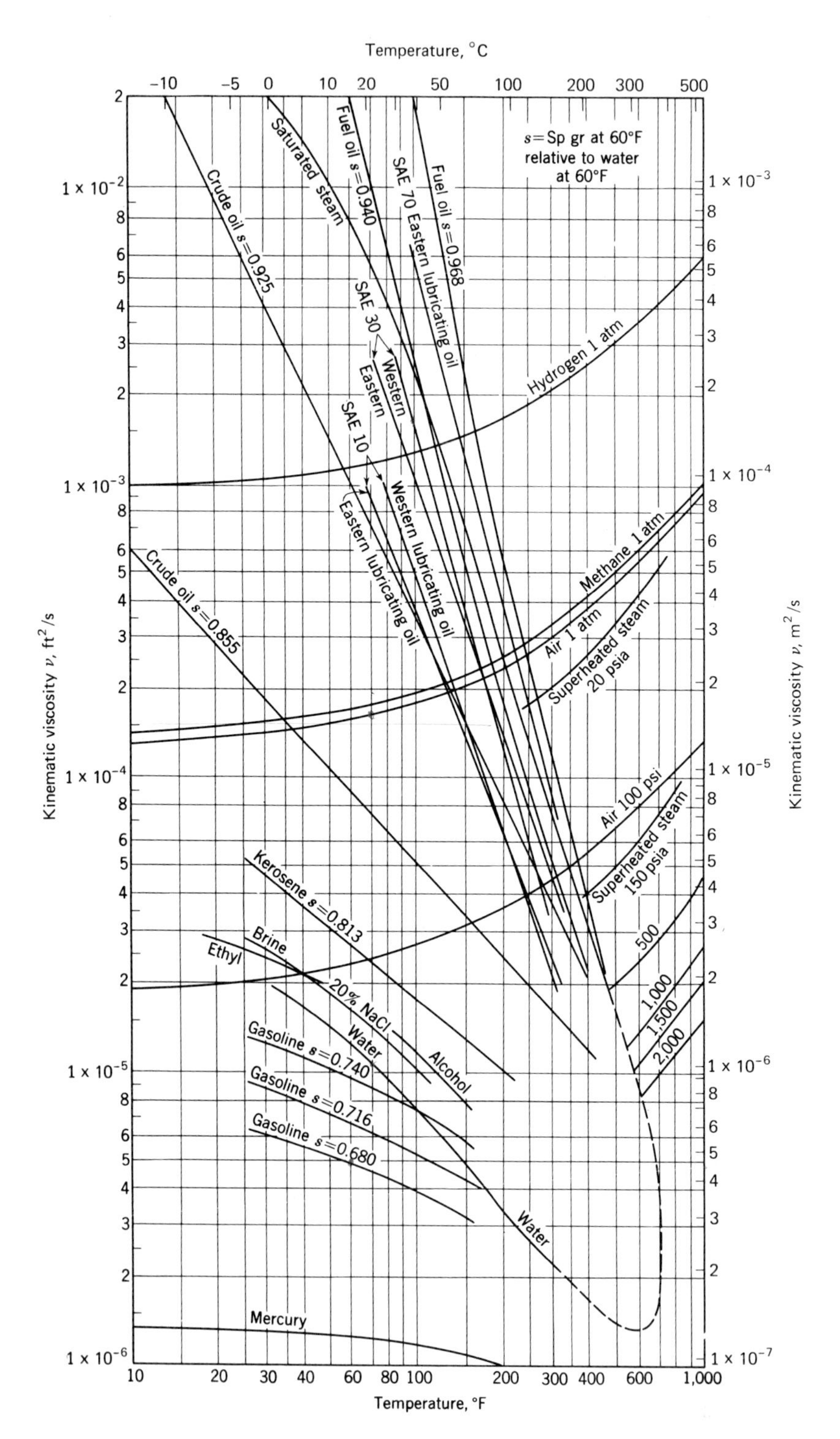

Figure 1.3 Kinematic viscosity ν of fluids.

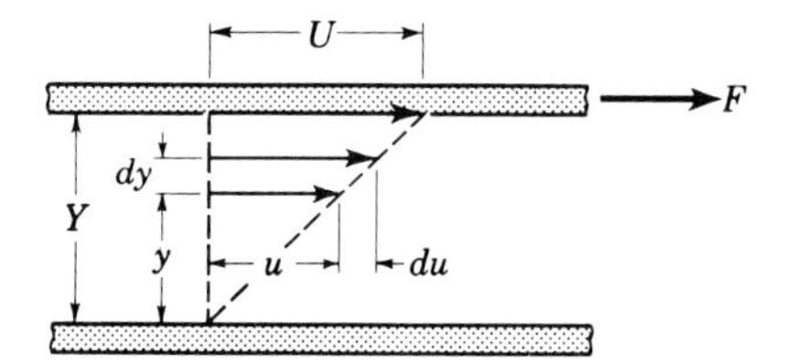

Figure 1.4

approximated, for instance, in the clearance space of a flooded journal bearing (any radial load being neglected).

Particles of the fluid in contact with each plate will adhere to it, and if the distance Y is not too great or the velocity U too high, the velocity gradient will be a straight line. The action is much as if the fluid were made up of a series of thin sheets, each of which would slip a little relative to the next. Experiment has shown that for a large class of fluids

$$F \propto \frac{AU}{Y}$$

It may be seen from similar triangles in Fig. 1.4 that U/Y can be replaced by the velocity gradient du/dy. If a constant of proportionality μ is now introduced, the shearing stress τ between any two thin sheets of fluid may be expressed by

$$\tau = \frac{F}{A} = \mu \frac{U}{Y} = \mu \frac{du}{dy} \tag{1.8}$$

Equation (1.8) is called *Newton's equation of viscosity*, and in transposed form it serves to define the proportionality constant

$$\mu = \frac{\tau}{du/dy} \tag{1.9}$$

which is called the *coefficient of viscosity*, the *absolute viscosity*, the *dynamic viscosity* (since it involves force), or simply the *viscosity* of the fluid.

It has been explained in Sec. 1.2 that the distinction between a solid and a fluid lies in the manner in which each can resist shearing stresses. A further distinction among various kinds of fluids and solids will be clarified by reference to Fig. 1.5. In the case of a solid, shear stress depends on the *magnitude* of the deformation; but Eq. (1.8) shows that in many fluids the shear stress is proportional to the *time rate* of (angular) deformation.

A fluid for which the constant of proportionality (i.e., the viscosity) does not change with rate of deformation is said to be a *Newtonian fluid* and can be represented by a straight line in Fig. 1.5. The slope of this line is determined by the viscosity. The ideal fluid, with no viscosity, is represented by the horizontal axis, while the true elastic solid is represented by the vertical axis. A plastic which sustains a certain amount of stress before suffering a plastic flow can be shown by a straight line intersecting the vertical axis at the yield stress. There are certain

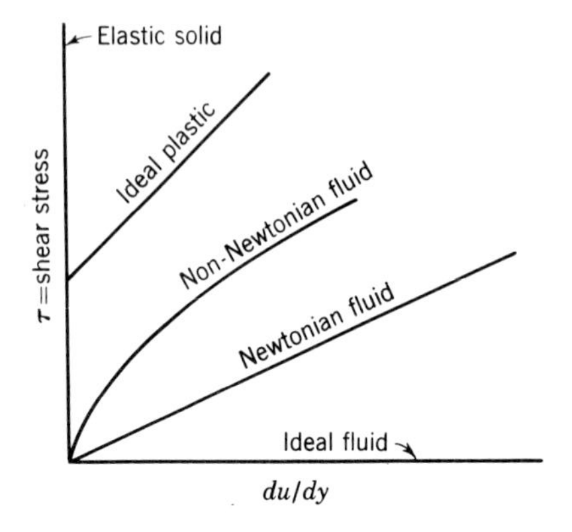

Figure 1.5

non-Newtonian fluids[1] in which μ varies with the rate of deformation. These are relatively uncommon, hence the remainder of this text will be restricted to the common fluids which obey Newton's law.

In the case of two parallel plates (Fig. 1.4), if U is constant, the shear stress on both plates is the same if end conditions are neglected. And since $\tau \propto (du/dy)$, the velocity gradient is constant throughout the fluid. However, for coaxial cylinders (Fig. 1.6) with rotative speed ω constant, the shear stress on the inner cylinder will be larger than that on the outer because of the different radii, and

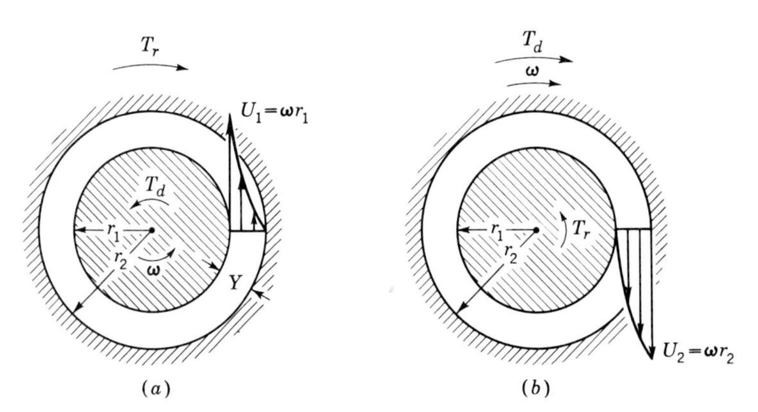

Figure 1.6 Velocity profile, rotating coaxial cylinders with gap completely filled with fluid. (*a*) Inner cylinder rotating. (*b*) Outer cylinder rotating. Z is the dimension at right angles to the plane of the sketch. Resisting torque = driving torque and $\tau \propto (du/dy)$.

$$\tau_1(2\pi r_1 Z)r_1 = \tau_2(2\pi r_2 Z)r_2$$
$$(du/dy)_1 = (du/dy)_2(r_2^2/r_1^2)$$

[1] Typical non-Newtonian fluids include paints, sludges, and certain plastics. An excellent treatment of the subject may be found in W. L. Wilkinson, "NonNewtonian Fluids," Pergamon Press, New York, 1960.

thus the velocity gradient will not be constant across the gap. By equating the resisting torque T_r to the driving torque T_d, the relationship between du/dy at r_1 and r_2 can be determined. As gap distance $Y \to 0$, $du/dy \to U/Y =$ constant, so when the gap distance is very small, the velocity profile can be assumed to be a straight line.

The dimensions of absolute viscosity are force per unit area divided by velocity gradient. In the English gravitational, or engineers', system the dimensions of absolute viscosity are as follows:

$$\text{Dimensions of } \mu = \frac{\text{dimensions of } \tau}{\text{dimensions of } du/dy} = \frac{\text{lb/ft}^2}{\text{fps/ft}} = \text{lb}\cdot\text{s/ft}^2$$

In SI units,

$$\text{Dimensions of } \mu = \frac{\text{N/m}^2}{\text{s}^{-1}} = \frac{\text{N}\cdot\text{s}}{\text{m}^2}$$

A widely used unit for viscosity in the metric system is the *poise* (P), after Poiseuille, who was one of the first investigators of viscosity. The poise = 0.10 $\text{N}\cdot\text{s/m}^2$. The *centipoise* (cP) (=0.01 P = $\text{mN}\cdot\text{s/m}^2$) is frequently a more convenient unit. It has a further advantage in that the viscosity of water at 68.4°F is 1 cP. Thus the value of the viscosity in centipoises is an indication of the viscosity of the fluid relative to that of water at 68.4°F.

In many problems involving viscosity there frequently appears the value of viscosity divided by density. This is defined as *kinematic viscosity* ν, so called because force is not involved, the only dimensions being length and time, as in kinematics. Thus

$$\nu = \frac{\mu}{\rho} \tag{1.10}$$

In the English system, kinematic viscosity is usually measured in ft^2/s while in the metric system the common units are cm^2/s, also called the *stoke* (St), after G. G. Stokes. The *centistoke* (cSt) (0.01 St) is often a more convenient unit.

The absolute viscosity of all fluids is practically independent of pressure for the range that is ordinarily encountered in engineering work. For extremely high pressures the values are somewhat higher than those shown in Fig. 1.2. The kinematic viscosity of gases varies with pressure because of changes in density.

1.12 SURFACE TENSION

Capillarity

Liquids have cohesion and adhesion, both of which are forms of molecular attraction. *Cohesion* enables a liquid to resist tensile stress, while *adhesion* enables

Table 1.3 Surface tension of water

English units		SI units		
°F	Surface tension, lb/ft	°C	Surface tension, σ: mN/m = dyn/cm	Surface tension, σ: N/m
32	0.00518	0	75.6	0.0756
40	0.00514	10	74.2	0.0742
60	0.00504	20	72.8	0.0728
80	0.00492	30	71.2	0.0712
100	0.00480	40	69.6	0.0696
140	0.00454	60	66.2	0.0662
180	0.00427	80	62.6	0.0626
212	0.00404	100	58.9	0.0589

it to adhere to another body.[1] The attraction between molecules forms an imaginary film capable of resisting tension at the interface between two immiscible liquids or at the interface between a liquid and a gas. The liquid property that creates this capability is known as *surface tension.* The surface tension of liquids covers a wide range. Typical values of the surface tension of water are presented in Table 1.3. *Capillarity* is due to both cohesion and adhesion. When the former is of less effect than the latter, the liquid will wet a solid surface with which it is in contact and rise at the point of contact; if cohesion predominates, the liquid surface will be depressed at the point of contact. For example, capillarity makes water rise in a glass tube, while mercury is depressed below the true level, as is shown by the insert in Fig. 1.7, which is drawn to scale and reproduced actual size.

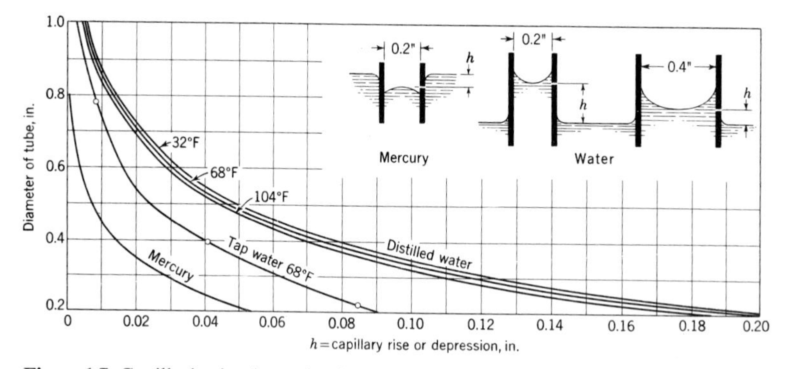

Figure 1.7 Capillarity in clean circular glass tubes.

[1] In 1877 Osborne Reynolds demonstrated that a column of mercury $\frac{1}{4}$ in in diameter could withstand a tensile stress of 3 atm for a time but that it would separate upon external jarring of the tube. Liquid tension (said to be as high as 400 atm) accounts for the rise of water in the very small channels of xylem tissue in tall trees. For practical engineering purposes, however, liquids are assumed to be incapable of resisting any direct tensile stress.

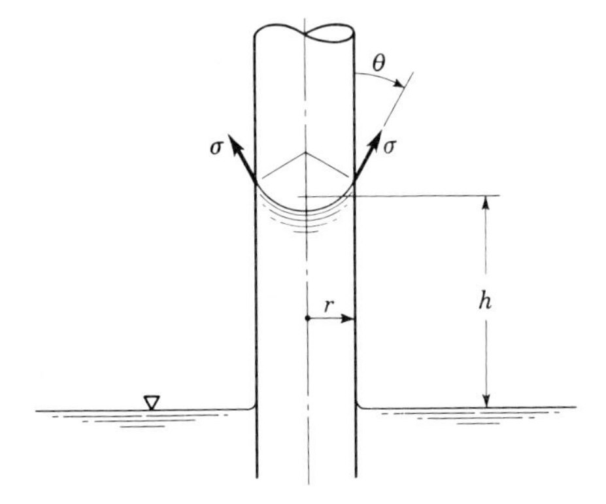

Figure 1.8 Capillary rise.

Capillary rise (or depression) in a tube is depicted in Fig. 1.8. From free-body considerations, equating the lifting force created by surface tension to the gravity force,

$$2\pi r\sigma \cos\theta = \pi r^2 h\gamma$$

$$h = \frac{2\sigma \cos\theta}{\gamma r} \tag{1.11}$$

where σ = surface tension in units of force per unit length
θ = wetting angle
γ = specific weight of liquid
r = radius of tube
h = capillary rise[1]

This expression can be used to compute the approximate capillary rise or depression in a tube. If the tube is clean, $\theta = 0°$ for water and about 140° for mercury. For tube diameters larger than $\frac{1}{2}$ in (12 mm), capillary effects are negligible. The curves of Fig. 1.7 are for water or mercury in contact with air. If mercury is in contact with water, the surface-tension effect is slightly less than when in contact with air.

Surface tension decreases slightly with increasing temperature. Surface-tension effects are generally negligible in most engineering situations; however, they may be important in problems involving capillary rise, the formation of drops and bubbles, the breakup of liquid jets, and in hydraulic model studies where the model is small.

1.13 VAPOR PRESSURE OF LIQUIDS

All liquids tend to evaporate or vaporize, which they do by projecting molecules into the space above their surfaces. If this is a confined space, the partial pressure

[1] Capillary rise is usually measured to the bottom of the meniscus.

Table 1.4 Vapor pressure of selected liquids at 68°F (20°C)

	psia	N/m^2, abs	mbar, abs
Mercury	0.000025	0.17	0.0017
Water	0.339	2,340	23.4
Kerosene	0.46	3,200	32
Carbon tetrachloride	1.76	12,100	121
Gasoline	8.0	55,000	550

exerted by the molecules increases until the rate at which molecules reenter the liquid is equal to the rate at which they leave. For this equilibrium condition the vapor pressure is known as the *saturation pressure.*

Molecular activity increases with temperature, and hence saturation pressure increases with temperature. At any one temperature the pressure on the liquid surface may be higher than this value, but it cannot be any lower, as any slight reduction induces a rapid rate of evaporation known as *boiling.* Thus the saturation pressure may be known as the *boiling pressure* for a given temperature.

The value of saturation vapor pressure is of practical interest in the case of liquids, for if the confining pressure on the liquid becomes less than this value, the liquid will vaporize.[1] The wide variation in vapor pressure of various liquids is shown in Table 1.4 and in Appendix 3, Table A.4. The very low vapor pressure of mercury makes it particularly suitable for use in barometers. Values for the vapor pressure of water at different temperatures are presented in Appendix 3, Table A.1.

Illustrative Example 1.4 At approximately what temperature will water boil if the elevation is 10,000 ft?

From Appendix 3, Table A.3, the pressure of the standard atmosphere at 10,000-ft elevation is 10.11 psia. From Appendix 3, Table A.1, the saturation pressure of water is 10.11 psia at about 193°F. Hence the water will boil at 193°F; this explains why it takes longer to cook at high elevations.

PROBLEMS[2]

1.1 If a certain gasoline weighs 46 lb/ft^3 (7,000 N/m^3), what are the values of its density, specific volume, and specific gravity relative to water at 60°F (15°C)? Use Appendix 3.

1.2 A certain gas weighs 0.10 lb/ft^3 (16 N/m^3) at a certain temperature and pressure. What are the values of its density, specific volume, and specific gravity relative to air weighing 0.075 lb/ft^3 (12 N/m^3)?

[1] Values of the saturation pressure for water for temperatures from 32 to 705.4°F may be found in J. H. Keenan, "Thermodynamic Properties of Water including Vapor, Liquid and Solid States," John Wiley & Sons, Inc., New York, 1969, and in other steam tables. There are similar vapor tables published for ammonia, carbon dioxide, sulfur dioxide, and other vapors of engineering interest.

[2] In some of the English unit problems the data are also given in parentheses in rounded SI units. Such problems may be solved using SI units rather than English units. Due to the rounding, the answers for the two systems will not be equivalent, i.e., they cannot be obtained from each other through the conversion of units.

1.3 The specific weight of glycerin in 78.7 lb/ft^3. Compute its density and specific gravity. What is its specific weight in kN/m^3?

1.4 If the specific weight of a liquid is 50 lb/ft^3 (8.0 kN/m^3), what is its density?

1.5 If the specific volume of a gas is 350 ft^3/slug (0.72 m^3/kg), what is its specific weight in lb/ft^3 (N/m^3)?

1.6 Initially when 1000.00 mL of water at 10°C are poured into a glass cylinder the depth of the water column is 100.00 cm. The water and its container are heated to 80°C. Assuming no evaporation, what then will be the depth of the water column if the coefficient of thermal expansion for the glass is 3.6 $\times 10^{-6}$ mm/mm per °C?

1.7 Water in a hydraulic press is subjected to a pressure of 15,000 psia at 68°F. If the initial pressure is 15 psia, what will be the percentage decrease in specific volume? Use Table 1.1.

1.8 At a depth of 5 miles (8 km) in the ocean the pressure is 11,930 psi (81.8 MPa). Assume specific weight at the surface is 64 lb/ft^3 (10.05 kN/m^3) and that the average volume modulus is 340,000 psi (2.34×10^9 N/m^2) for that pressure range. (*a*) What will be the change in specific volume between that at the surface and at that depth? (*b*) What will be the specific volume at that depth? (*c*) What will be the specific weight at that depth?

1.9 (*a*) What is the percentage change in the specific volume in Prob. 1.8? (*b*) What is the percentage change in the specific weight in Prob. 1.8?

1.10 To two significant figures what is the bulk modulus of water in MN/m^2 at 50°C under a pressure of 30 MN/m^2? Use Table 1.1.

1.11 Approximately what pressure in psi must be applied to water to reduce its volume 2 percent?

1.12 A vessel contains 3 ft^3 (85 L) of water at 50°F (10°C) and atmospheric pressure. If it is heated to 160°F (70°C) what will be the percentage change in its volume? What weight of water must be removed to maintain the volume at the original value? Use Fig. 1.1 or Appendix 3.

1.13 A cylindrical tank (diameter = 10 m and depth = 5.00 m) contains water at 20°C and is brimful. If the water is heated to 50°C, how much water will spill over the edge of the tank? Use Appendix 3.

1.14 A closed heavy steel chamber is filled with water at 50°F and at atmospheric pressure. If the temperature of the water and the chamber is raised to 90°F, what will be the new pressure of the water? The coefficient of thermal expansion of the steel is 6.5×10^{-6} in/in per °F; assume the chamber is unaffected by the water pressure. Use Fig. 1.1.

1.15 A hydrogen-filled cellophane balloon of the type used in cosmic-ray studies is to be expanded to its full size, which is a 100-ft-diameter sphere, without stress in the wall at an altitude of 150,000 ft. If the pressure and temperature at this altitude are 0.14 psia and −67°F respectively, find the volume of hydrogen at 14.7 psia and 68°F which should be added on the ground.

1.16 If natural gas has a specific gravity of 0.6 relative to air at 14.7 psia and 60°F, what are its specific weight and specific volume at that same pressure and temperature. What is the value of R for the gas?

1.17 A gas at 40°C under a pressure of 20,000 mbar, abs has a unit weight of 340 N/m^3. What is the value of R for the gas? What gas might this be? Refer to Appendix 3, Table A.5*b*.

1.18 If water vapor in the atmosphere has a partial pressure of 0.50 psia (3,500 Pa) and the temperature is 90°F (30°C), what is its specific weight?

1.19 If the barometer reads 14.50 psia in Prob. 1.18, what is the partial pressure of the air, and what is its specific weight? What is the specific weight of the atmosphere? (*Note:* Atmosphere here means the air plus the water vapor present.)

1.20 (*a*) Calculate the density, specific weight and specific volume of oxygen at 10°C and 30 kN/m^2, abs. (*b*) If the oxygen is enclosed in a rigid container of constant volume, what will be the pressure if the temperature is reduced to −120°C?

1.21 Calculate the density, specific weight and specific volume of air at 100°F (38°C) and 70 psia (4800 mbar, abs).

1.22 Prove that Eq. (1.7) follows from Eqs. (1.4) and (1.6).

1.23 Helium at 20 psia (140 kPa, abs) is compressed isothermally, and hydrogen at 15 psia (100 kPa, abs) is compressed isentropically. What is the modulus of elasticity of each gas? Which is the more compressible?

1.24 Helium at 20 psia (140 kN/m^2, abs) and 52°F (11°C) is isentropically compressed to one-fifth of its original volume. What is its final pressure?

1.25 (*a*) If 10 ft^3 (300 L) of carbon dioxide at 80°F (25°C) and 20 psia (140 kN/m^2, abs) is compressed isothermally to 2 ft^3 (60 L), what is the resulting pressure? (*b*) What would the pressure and temperature have been if the process had been isentropic? The isentropic exponent k for carbon dioxide is 1.28.

1.26 (*a*) If 10 m^3 of nitrogen at 30°C and 150 kPa is permitted to expand isothermally to 25 m^3, what is the resulting pressure? (*b*) What would the pressure and temperature have been if the process had been isentropic? The isentropic exponent k for nitrogen is 1.40.

1.27 A liquid has an absolute viscosity of 4.8×10^{-4} lb·s/ft^2. It weighs 54 lb/ft^3. What are its absolute and kinematic viscosities in SI units?

1.28 (*a*) What is the ratio of the viscosity of water at a temperature of 70°F to that of water at 200°F? (*b*) What is the ratio of the viscosity of the crude oil in Fig. 1.2 ($s = 0.925$) to that of the gasoline ($s = 0.680$), both being at a temperature of 60°F? (*c*) In cooling from 300 to 80°F, what is the ratio of the change of the viscosity of the SAE 30 western oil to that of the SAE 30 eastern oil?

1.29 At 60°F what is the kinematic viscosity of the gasoline in Fig. 1.3, the specific gravity of which is 0.680? Give the answer in both English and SI units.

1.30 To what temperature must the fuel oil with the higher specific gravity in Fig. 1.3 be heated in order that its kinematic viscosity may be reduced to three times that of water at 40°F?

1.31 The absolute viscosity of a certain gas is 0.0107 cP while its kinematic viscosity is 164 cSt, both measured at 1,013 mbar, abs and 95°C. Calculate its approximate molecular weight, and suggest what gas it may be.

1.32 Compare the ratio of the absolute viscosities of air and water at 70°F with that of their kinematic viscosities at the same temperature and at 14.7 psia.

1.33 A flat plate 30 cm × 50 cm slides on oil ($\mu = 0.8$ N·s/m^2) over a large plane surface. What force is required to drag the plate at 2 m/s, if the separating oil film is 0.4 mm thick?

1.34 A space of 1-in width between two large plane surfaces is filled with SAE 30 western lubricating oil at 80°F. (*a*) What force is required to drag a very thin plate of 4-ft^2 area between the surfaces at a speed of 20 fpm if this plate is equally spaced between the two surfaces? (*b*) If it is at a distance of 0.33 in from one surface?

1.35 A hydraulic lift of the type commonly used for greasing automobiles consists of a 10.000-in-diam ram which slides in a 10.006-in-diam cylinder, the annular space being filled with oil having a kinematic viscosity of 0.004 ft^2/s and specific gravity of 0.85. If the rate of travel of the ram is 0.5 fps, find the frictional resistance when 10 ft of the ram is engaged in the cylinder.

1.36 A journal bearing consists of a 6.00-in shaft in a 6.01-in sleeve 8 in long, the clearance space (assumed to be uniform) being filled with SAE 30 eastern lubricating oil at 100°F. Calculate the rate at which heat is generated at the bearing when the shaft turns at 100 rpm. Express answer in Btu/hr.

1.37 Repeat Prob. 1.36 for the case where the sleeve has a diameter of 6.48 in. Compute as accurately as possible the velocity gradient in the fluid at the shaft and sleeve.

1.38 A journal bearing consists of an 8.00-cm shaft in an 8.03-cm sleeve 10 cm long, the clearance space (assumed to be uniform) being filled with SAE 30 western lubricating oil at 40°C. Calculate the rate at which heat is generated at the bearing when the shaft turns at 120 rpm. Express answer in kN·m/s, J/s, Btu/hr, ft·lb/s, and hp.

1.39 In using a rotating-cylinder viscometer, a bottom correction must be applied to account for the drag on the flat bottom of the inner cylinder. Calculate the theoretical amount of this torque correction, neglecting centrifugal effects, for a cylinder of diameter d, rotated at a constant angular velocity ω, in a liquid of viscosity μ, with a clearance Δh between the bottom of the inner cylinder and the floor of the outer one.

1.40 Assuming a velocity distribution as shown in the diagram, which is a parabola having its vertex 4 in from the boundary, calculate the velocity gradients for $y = 0$, 1, 2, 3, and 4 in. Also calculate the shear stresses in lb/ft^2 at these points if the fluid viscosity is 400 cP.

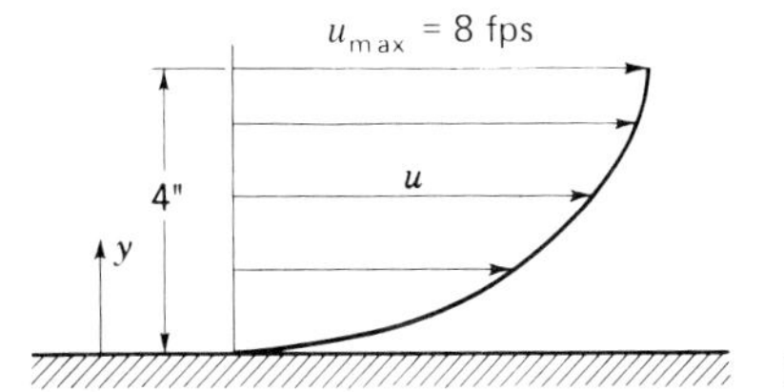

Problem 1.40

1.41 In the figure, oil of viscosity μ fills the small gap of thickness Y. Determine an expression for the torque T required to rotate the truncated cone at constant speed ω. Neglect fluid stress exerted on the circular bottom.

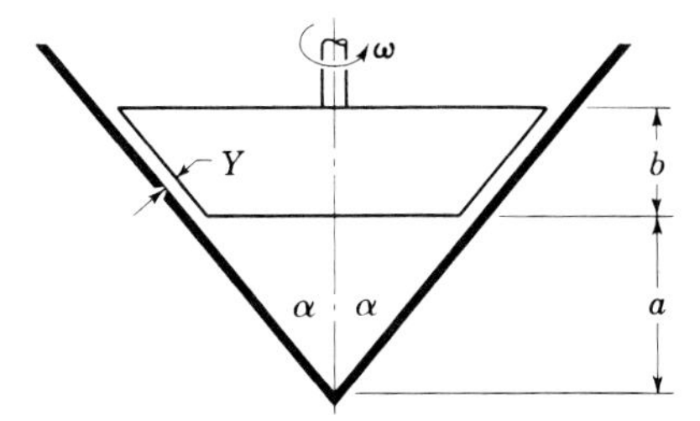

Problem 1.41

1.42 Distilled water at 10°C stands in a glass tube of 8.0-mm diameter at a height of 25.0 mm. What is the true static height?

1.43 Tap water at 68°F stands in a glass tube of 0.32-in diameter at a height of 4.50 in. What is the true static height?

1.44 Use Eq. (1.11) to compute the capillary rise of water to be expected in a 0.20-in-diameter tube. Assume pure water at 68°F. Compare the result with Fig. 1.7.

1.45 Use Eq. (1.11) to compute the capillary depression of mercury ($\theta = 140°$) to be expected in a 0.10-in-diameter tube. At 68°F the surface tension of mercury is 0.0318 lb/ft.

1.46 Derive an expression for capillary rise (or depression) between two vertical parallel plates.

1.47 At approximately what temperature will water boil in Mexico City (elevation 7,400 ft)? Refer to Appendix 3.

1.48 Water at 100°F is placed in a beaker within an airtight container. Air is gradually pumped out of the container. What reduction below standard atmospheric pressure of 14.7 psia must be achieved before the water boils?

1.49 At what pressure in millibars absolute will 40°C water boil?

CHAPTER

TWO

FLUID STATICS

There are no shear stresses in fluids at rest; hence only normal pressure forces are present. The *average pressure intensity* is defined as the force exerted on a unit area. If F represents the total force on some finite area A, while dF represents the force on an infinitesimal area dA, the pressure is

$$p = \frac{dF}{dA} \tag{2.1}$$

If the pressure is uniform over the total area, then $p = F/A$. In the English system, pressure is generally expressed in pounds per square inch (psi) or pounds per square foot (lb/ft^2) while in SI units the N/m^2 (pascal) or kN/m^2 (kPa) is commonly used.

2.1 PRESSURE AT A POINT THE SAME IN ALL DIRECTIONS

In a solid, because of the possibility of tangential stresses between adjacent particles, the stresses at a given point may be different in different directions. But in a fluid at rest, no tangential stresses can exist, and the only forces between adjacent surfaces are pressure forces normal to the surfaces. Therefore the pressure at any point in a fluid at rest is the same in every direction.

This can be proved by reference to Fig. 2.1, a very small wedge-shaped element of fluid at rest whose thickness perpendicular to the plane of the paper is constant and equal to dy. Let p be the average pressure in any direction in the plane of the paper, let α be defined as shown, and let p_x and p_z be the average pressures in the horizontal and vertical directions. The forces acting on the

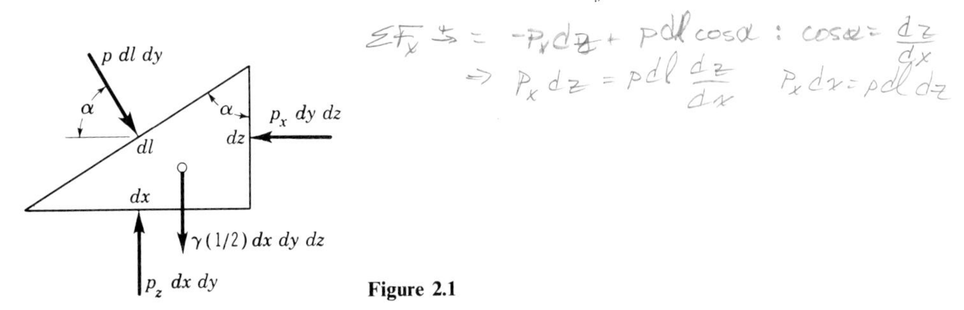

Figure 2.1

element of fluid, with the exception of those in the y direction on the two faces parallel to the plane of the paper, are shown in the diagram. For our purpose, forces in the y direction need not be considered. Since the fluid is at rest, no tangential forces are involved. As this is a condition of equilibrium, the sum of the force components on the element in any direction must be equal to zero. Writing such an equation for the components in the x direction, $p\,dl\,dy \cos \alpha - p_x\,dy\,dz = 0$. Since $dz = dl \cos \alpha$, it follows that $p = p_x$. Similarly, summing up forces in the z direction gives $p_z\,dx\,dy - p\,dl\,dy \sin \alpha - \frac{1}{2}\gamma\,dx\,dy\,dz = 0$. The third term is of higher order than the other two terms and may be neglected. From this it follows that $p = p_z$. It can also be proved that $p = p_y$ by considering a three-dimensional case. The results are independent of α; hence the pressure at any point in a fluid at rest is the same in all directions.

2.2 VARIATION OF PRESSURE IN A STATIC FLUID

Consider the differential element of static fluid shown in Fig. 2.2. Since the element is very small, we can assume that the density of the fluid within the element is constant. Assume the pressure at the center of the element is p and that the dimensions of the element are δx, δy and δz.[1] The forces acting on the fluid element in the vertical direction are the *body force*, the action of gravity on the mass within the element, and the *surface forces* transmitted from the surrounding fluid and acting at right angles against the top, bottom, and sides of the element. Since the fluid is at rest, the element is in equilibrium and the summation of forces acting on the element in any direction must be zero. If forces are summed up in the horizontal direction, that is, x or y, the only forces acting are the pressure forces on the vertical faces of the element. To satisfy $\sum F_x = 0$ and $\sum F_y = 0$, the pressure on the opposite vertical faces must be equal. Thus $\partial p/\partial x = \partial p/\partial y = 0$ for the case of the fluid at rest.

Summing up forces in the vertical direction and setting equal to zero,

$$\sum F_z = \left(p - \frac{\partial p}{\partial z}\frac{\delta z}{2}\right)\delta x\,\delta y - \left(p + \frac{\partial p}{\partial z}\frac{\delta z}{2}\right)\delta x\,\delta y - \gamma\,\delta x\,\delta y\,\delta z = 0$$

[1] In this instance a left-handed coordinate system is employed. Hence $+x$ is horizontally to the right and $+z$ is vertically upward.

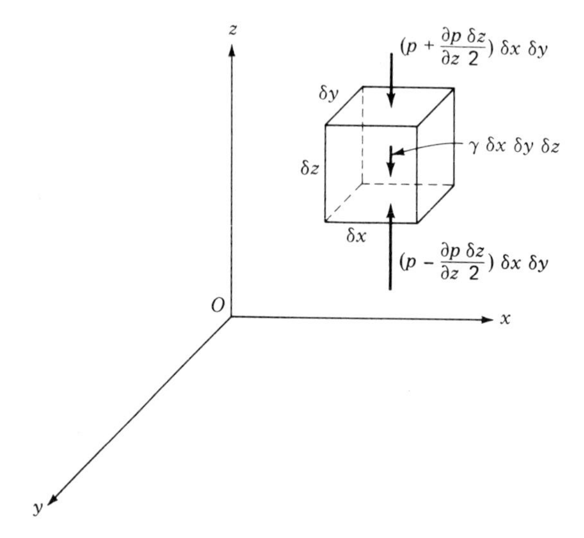

Figure 2.2

This results in $\partial p/\partial z = -\gamma$, which, since p is independent of x and y, can be written as

$$\frac{dp}{dz} = -\gamma \tag{2.2}$$

This is the general expression that relates variation of pressure in a static fluid to vertical position. The minus sign indicates that as z gets larger (increasing elevation), the pressure gets smaller.

To evaluate pressure variation in a fluid at rest one must integrate Eq. (2.2) between appropriately chosen limits. For incompressible fluids (γ = constant), Eq. (2.2) can be integrated directly. For compressible fluids, however, γ must be expressed algebraically as a function of z or p if one is to determine pressure accurately as a function of elevation. The variation of pressure in the earth's atmosphere is an important problem, and several approaches are illustrated in the following example.

Illustrative Example 2.1 Compute the atmospheric pressure at elevation 20,000 ft, considering the atmosphere as a static fluid. Assume standard atmosphere at sea level. Use four methods: (*a*) air of constant density; (*b*) constant temperature between sea level and 20,000 ft; (*c*) isentropic conditions; and (*d*) air temperature decreasing linearly with elevation at the standard lapse rate of 0.00356°F/ft.

From Appendix 3, Table A.3*a*, the conditions of the standard atmosphere at sea level are $T = 59°\text{F}$, $p = 14.7$ psia, $\gamma = 0.076\ \text{lb/ft}^3$.

(*a*) Constant density:

$$\frac{dp}{dz} = -\gamma \qquad dp = -\gamma\, dz \qquad \int_{p_1}^{p} dp = -\gamma \int_{z_1}^{z} dz$$

$$p - p_1 = \gamma(z - z_1)$$

$$p = 14.7 \times 144 - 0.076(20{,}000) = 597\ \text{lb/ft}^2\text{, abs} = 4.15\ \text{psia}$$

(*b*) Isothermal:

$$pv = \text{constant, hence } \frac{p}{\gamma} = \frac{p_1}{\gamma_1} \text{ if } g \text{ is constant}$$

$$\frac{dp}{dz} = -\gamma \qquad \text{where} \qquad \gamma = \frac{p\gamma_1}{p_1}$$

$$\frac{dp}{p} = -\frac{\gamma_1}{p_1}\,dz$$

$$\int_{p_1}^{p} = \ln\frac{p}{p_1} = -\frac{\gamma_1}{p_1}\int_{z_1}^{z} dz = -\frac{\gamma_1}{p_1}(z - z_1)$$

$$\frac{p}{p_1} = \exp\left[-\left(\frac{\gamma_1}{p_1}\right)(z - z_1)\right]$$

$$p = 14.7\exp\left[-\frac{0.076}{14.7 \times 144}(20{,}000)\right] = 7.17 \text{ psia}$$

(*c*) Isentropic:

$$pv^{1.4} = \frac{p}{\rho^{1.4}} = \text{constant} \qquad \text{hence } \frac{p}{\gamma^{1.4}} = \text{constant} = \frac{p_1}{\gamma_1^{1.4}}$$

$$\frac{dp}{dz} = -\gamma \qquad \text{where } \gamma = \gamma_1\left(\frac{p}{p_1}\right)^{0.715}$$

$$\int_{p_1}^{p} p^{-0.715}\,dp = -\gamma_1 p_1^{-0.715}\int_{z_1}^{z} dz$$

$$p^{0.285} - p_1^{0.285} = -0.285\gamma_1 p_1^{-0.715}(z - z_1)$$

$$p^{0.285} = (14.7 \times 144)^{0.285} - 0.285(0.076)(14.7 \times 144)^{-0.715}(20{,}000)$$

$$p = 950 \text{ lb/ft}^2\text{, abs} = 6.60 \text{ psia}$$

(*d*) Temperature decreasing linearly with elevation:

$$T = (460 + 59) + Kz \qquad \text{where} \qquad K = -0.00356°\text{F/ft}$$

$$dT = K\,dz \qquad \text{hence} \qquad dz = \frac{dT}{K}$$

$$\frac{pv}{T} = R = \frac{p}{\rho T} = \frac{p_1}{\rho_1 T_1}$$

$$\frac{dp}{dz} = -\gamma \qquad \text{where} \qquad \gamma = \frac{\gamma_1 T_1 p}{T p_1} \text{ if } g \text{ is constant}$$

$$\frac{dp}{p} = -\frac{\gamma_1 T_1}{p_1 T}\,dz = -\frac{\gamma_1 T_1}{p_1 K}\frac{dT}{T}$$

$$\int_{p_1}^{p}\frac{dp}{p} = -\frac{\gamma_1 T_1}{p_1 K}\int_{T_1}^{T}\frac{dT}{T}$$

$$\ln\frac{p}{p_1} = -\frac{\gamma_1 T_1}{p_1 K}\ln\frac{T}{T_1} = \ln\left(\frac{T_1}{T}\right)^{\gamma_1 T_1/p_1 K}$$

$$p = p_1\left(\frac{T_1}{T_1 + Kz}\right)^{\gamma_1 T_1/p_1 K}$$

$$p = 14.7\left(\frac{519}{447.8}\right)^{(0.076)(519)/14.7(144)(-0.00356)} = 6.8 \text{ psia (470 mbar, abs)}$$

The latter approach corresponds to the standard atmosphere, Appendix 3, Table A.3, where it is assumed the temperature varies linearly from 59°F (288 K) at sea level to −69.7°F (216.5 K) at elevation 36,150 ft (11,000 m). This region of the atmosphere is known as the troposphere. Beyond elevation 36,150 ft (11,000 m) and up to 80,000 ft (24,000 m) (stratosphere) the temperature has been observed to be approximately constant at −69.7°F (216.5 K). This standard atmosphere is generally used in design calculations where the performance of high-altitude aircraft is of interest; it serves as a good approximation of conditions to be expected in the atmosphere.

In Illustrative Example 2.1*a* it was shown that, for the case of an incompressible fluid,

$$p - p_1 = -\gamma(z - z_1) \tag{2.3}$$

where p is the pressure at an elevation z. This expression is generally applicable to liquids since they are only slightly compressible. Only where there are large changes in elevation, as in the ocean, need the compressibility of the liquid be considered, to arrive at an accurate determination of pressure variation. For small changes in elevation, Eq. (2.3) will give accurate results when applied to gases.

For the case of a liquid at rest, it is convenient to measure distances vertically downward from the free liquid surface. If h is the distance below the free liquid surface and if the pressure of air and vapor on the surface is arbitrarily taken to be zero, Eq. (2.3) can be written as

$$p = \gamma h \tag{2.4}$$

As there must always be some pressure on the surface of any liquid, the total pressure at any depth h is given by Eq. (2.4) plus the pressure on the surface. In many situations this surface pressure may be disregarded, as pointed out in Sec. 2.4.

From Eq. (2.4) it may be seen that all points in a connected body of constant density fluid at rest are under the same pressure if they are at the same depth below the liquid surface (Pascal's law). This indicates that a surface of equal pressure for a liquid at rest is a horizontal plane. Strictly speaking, it is a surface everywhere normal to the direction of gravity and is approximately a spherical surface concentric with the earth. For practical purposes a limited portion of this surface may be considered a plane area.

2.3 PRESSURE EXPRESSED IN HEIGHT OF FLUID

In Fig. 2.3 imagine an open tank of liquid upon whose surface there is no pressure, though in reality the minimum pressure upon any liquid surface is the pressure of its own vapor. Disregarding this for the moment, by Eq. (2.4) the pressure at any depth h is $p = \gamma h$. If γ is assumed constant, there is a definite relation between p and h. That is, pressure (force per unit area) is equivalent to a

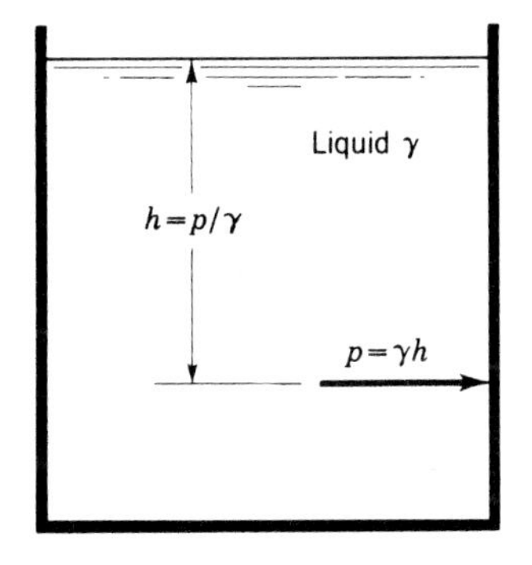

Figure 2.3

height h of some fluid of constant specific weight γ. It is often more convenient to express pressure in terms of a height of a column of fluid rather than in pressure per unit area.

Even if the surface of the liquid is under some pressure, it is necessary only to convert this pressure into an equivalent height of the fluid in question and add this to the value of h shown in Fig. 2.3, to obtain the total pressure.

The preceding discussion has been applied to a liquid, but it is equally possible to use it for a gas or vapor by specifying some *constant* specific weight γ for the gas or vapor in question. Thus pressure p may be expressed in the height of a column of *any* fluid by the relation

$$h = \frac{p}{\gamma} \tag{2.5}$$

This relationship is true for any consistent system of units. If p is in pounds per square foot, γ must be in pounds per cubic foot, and then h will be in feet. In SI units, p may be expressed in kilonewtons per square meter, in which case if γ is expressed in kilonewtons/cubic meter, h will be in meters. When pressure is expressed in this fashion, it is commonly referred to as *pressure head.* Because pressure is commonly expressed in pounds per square inch (or kN/m^2 in SI units), and as the value of γ for water is usually assumed to be 62.4 lb/ft^3 (9.81 kN/m^3), a convenient relationship is

$$h \text{ (ft of } H_2O) = \frac{144 \times \text{psi}}{62.4} = 2.308 \times \text{psi}$$

or

$$h \text{ (m of } H_2O) = \frac{kN/m^2}{9.81} = 0.102 \times kN/m^2$$

It is convenient to express pressures occurring in one fluid in terms of height of another fluid, e.g., barometric pressure in millimeters of mercury.

Equation (2.3) may be expressed as follows:

$$z + \frac{p}{\gamma} = z_1 + \frac{p_1}{\gamma} = \text{constant} \tag{2.6}$$

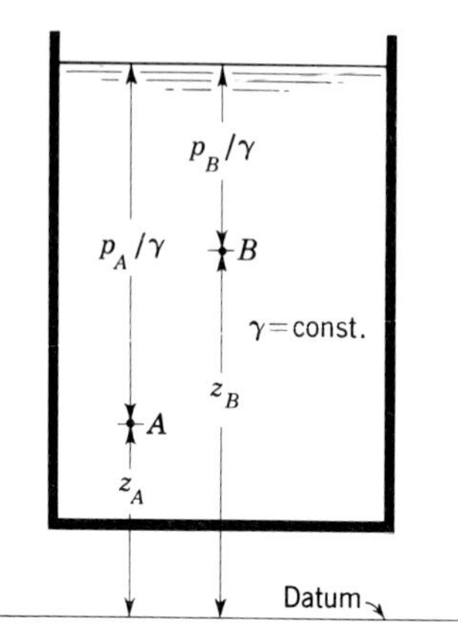

Figure 2.4

This shows that for an incompressible fluid at rest the summation of the elevation z at any point in the fluid plus the pressure head p/γ at that point is equal to the sum of these two quantities at any other point. The significance of this statement is that, in a fluid at rest with an increase in elevation, there is a decrease in pressure head, and vice versa. This concept is depicted in Fig. 2.4.

2.4 ABSOLUTE AND GAGE PRESSURES

If pressure is measured relative to absolute zero, it is called *absolute* pressure; when measured relative to atmospheric pressure as a base, it is called *gage* pressure. This is because practically all pressure gages register zero when open to the atmosphere and hence measure the difference between the pressure of the fluid to which they are connected and that of the surrounding air.

If the pressure is below that of the atmosphere, it is designated as a *vacuum* and its gage value is the amount by which it is *below* that of the atmosphere. What is called a "high vacuum" is really a low absolute pressure. A perfect vacuum would correspond to absolute zero pressure.

All values of absolute pressure are positive, since a negative value would indicate tension, which is normally considered impossible in any fluid.[1] Gage pressures are positive if they are above that of the atmosphere and negative if they are vacuum (Fig. 2.5).

From the foregoing discussion it can be seen that the following relation holds:

$$p_{\text{abs}} = p_{\text{atm}} + p_{\text{gage}} \tag{2.7}$$

where p_{gage} may be positive or negative (vacuum).

[1] For an exception to this statement, see the footnote on p. 16.

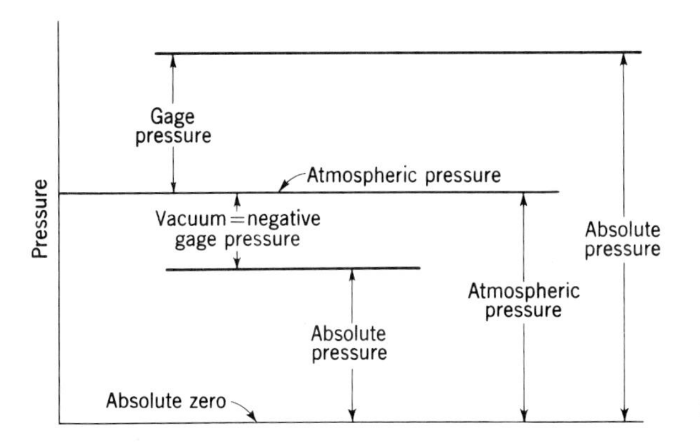

Figure 2.5

The atmospheric pressure is also called the *barometric* pressure and varies with elevation above sea level. Also, at a given place it varies slightly from time to time because of changes in meteorological conditions.

In thermodynamics it is essential to use absolute pressure, because most thermal properties are functions of the actual pressure of the fluid, regardless of the atmospheric pressure. For example, the equation of state of gas [Eq. (1.4)] is one in which absolute pressure must be used. In fact, absolute pressures must be employed in most problems involving gases and vapors.

The properties of liquids are usually not much affected by pressure, and hence gage pressures are commonly employed in problems dealing with liquids. Also it will usually be found that the atmospheric pressure appears on both sides of an equation and hence cancels. Thus the value of atmospheric pressure is usually of no significance when dealing with liquids, and for this reason as well, gage pressures are almost universally used with liquids. About the only case where the absolute pressure of a liquid needs to be considered is where conditions are such that the pressure may approach or equal the saturated vapor pressure. Throughout this text all numerical pressures will be understood to be gage pressures unless specifically given as absolute values.

2.5 BAROMETER

The absolute pressure of the atmosphere is measured with a barometer. If a tube such as that in Fig. 2.6*a* has its open end immersed in a liquid which is exposed to atmospheric pressure, and if air is exhausted from the tube, the liquid will rise in it. If the tube is long enough and if the air is completely removed, the only pressure on the surface of the liquid in the tube will be that of its own vapor pressure and the liquid will have reached its maximum height.

From concepts developed in Sec. 2.2 the pressure at O within the tube and at a at the surface of the liquid outside the tube must be the same; that is, $p_0 = p_a$.

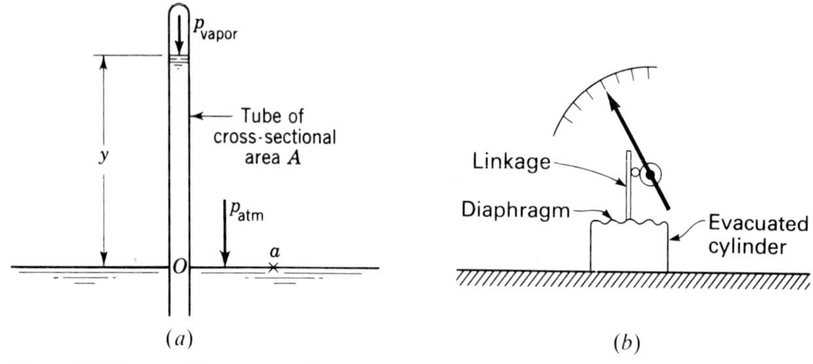

Figure 2.6 Types of barometers. (*a*) Mercury barometer. (*b*) Aneroid barometer.

From the conditions of statical equilibrium of the liquid above O in the tube of cross-sectional area A (Fig. 2.6*a*)

$$p_{\text{atm}} A - p_{\text{vapor}} A - \gamma A y = 0$$

$$p_{\text{atm}} = \gamma y + p_{\text{vapor}} \tag{2.8}$$

If the vapor pressure on the surface of the liquid in the tube were negligible, then we have

$$p_{\text{atm}} = \gamma y$$

The liquid employed for barometers of this type is usually mercury, because its density is sufficiently great to enable a reasonably short tube to be used, and also because its vapor pressure is negligibly small at ordinary temperatures. If some other liquid were used, the tube necessarily would be so high as to be inconvenient and its vapor pressure at ordinary temperatures would be appreciable; hence a nearly perfect vacuum at the top of the column would not be attainable. The height attained by the liquid would consequently be less than the true barometric height and would necessitate applying a correction to the reading. When using a mercury barometer, to get as accurate a measurement of atmospheric pressure as possible, corrections for capillarity and vapor pressure should be applied to the reading. An *aneroid* barometer measures the difference in pressure between the atmosphere and an evacuated cylinder by means of a sensitive elastic diaphragm and linkage system as indicated in Fig. 2.6*b*.

Since atmospheric pressure at sea level is so widely used, it is well to have in mind equivalent forms of expression. Application of Eq. (2.5) shows that sea-level atmospheric pressure may be expressed as follows:

14.7 psia or 101.3 kN/m^2, abs (1,013 mbars, abs)

29.9 in of Hg or 760 mm of Hg (0.76 m of Hg)

33.9 ft of H_2O or 10.3 m of H_2O

These are approximate equivalents and accurate enough for most engineering work.

2.6 MEASUREMENT OF PRESSURE

There are many ways by which pressure in a fluid may be measured. Some are discussed below.

Bourdon Gage

Pressures or vacuums are commonly measured by the bourdon gage of Fig. 2.7. In this gage a curved tube of elliptical cross section will change its curvature with changes in pressure within the tube. The moving end of the tube rotates a hand on a dial through a linkage system. A pressure and vacuum gage combined into one is known as a *compound gage* and is shown in Fig. 2.8. The pressure indicated by the gage is assumed to be that at its center. If the connecting piping is filled completely with fluid of the same density as that in A of Fig. 2.7 and if the pressure gage is graduated to read in pounds per square inch, as is customary, then

$$p_A\,(\text{psi}) = \text{gage reading (psi)} + \frac{\gamma z}{144}$$

where γ is expressed in pounds per cubic foot and z in feet.

A vacuum gage, or the negative-pressure portion of a compound gage, is traditionally graduated to read in inches or mm of mercury. For vacuums,

$$\text{in of Hg vacuum at } A = \text{gage reading (in of Hg vacuum)} - \frac{\gamma z}{144}\left(\frac{29.9}{14.7}\right)$$

Here, once again, it is assumed that this fluid completely fills the connecting tube of Fig. 2.7. The elevation-correction terms, i.e., those containing z, may be positive or negative, depending on whether the gage is above or below the point at which the pressure determination is desired. The expressions given are for the situation depicted in Fig. 2.7. When measuring liquid pressures, the gage is usually set to measure the pressure at the centerline of the pipe. When measuring gas pressures, the elevation correction terms are generally negligible.

The above expressions, when written in SI units, require no conversion factors; however, care must be taken in dealing with decimal points when adding terms.

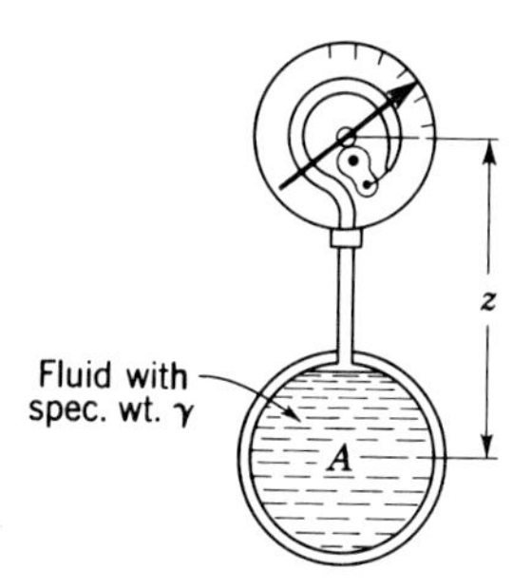

Figure 2.7 Bourdon gage.

Figure 2.8 Compound pressure and vacuum gage. Pressures in pounds per square inch, vacuums in inches of mercury.

Pressure Transducer

A *transducer* is a device which transfers energy (in any form) from one system to another. A bourdon gage, for example, is a mechanical transducer in that it has an elastic element that converts energy from the pressure system to a displacement in the mechanical measuring system. An *electrical pressure transducer* converts the displacement of a mechanical system (usually a metal diaphragm) to an electric signal either actively if it generates it own electrical output or passively if it requires an electrical input which it modifies as a function of the mechanical displacement. In one type of pressure transducer (Fig. 2.9) an electrical strain gage is attached to a diaphragm. As the pressure changes, the deflection of the diaphragm changes. This, in turn, changes the electrical output which through proper calibration can be related to pressure. Such a device when connected to a strip-chart recorder can be used to give a continuous record of pressure. In lieu of a strip-chart recorder the data may be recorded at fixed time intervals on a tape or disk using a computer data acquisition system and/or it may be displayed on a panel in digital form.

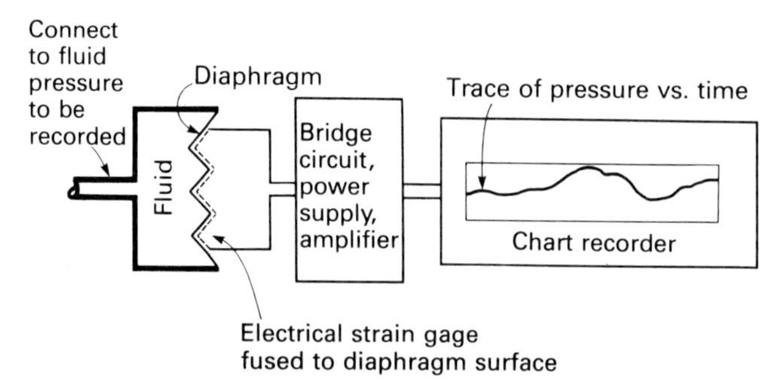

Figure 2.9 Schematic of an electrical strain-gage pressure transducer with a strip-chart recorder.

Piezometer Column

A piezometer column is a simple device for measuring moderate pressures of liquids. It consists of a tube (Fig. 2.10) in which the liquid can freely rise without overflowing. The height of the liquid in the tube will give the value of the pressure head directly. To reduce capillary error the tube diameter should be at least 0.5 in (12 mm).

If the pressure of a *flowing* fluid is to be measured, special precautions should be taken in making the connection. The hole must be drilled absolutely normal to the interior surface of the wall, and the piezometer tube or the connection for any other pressure-measuring device must not project beyond the surface. All burrs and surface roughness near the hole must be removed, and it is well to round the edge of the hole slightly. Also, the hole should be small, preferably not larger than $\frac{1}{8}$ in (3 mm).

Simple Manometer

Since the open piezometer tube is cumbersome for use with liquids under high pressure and cannot be used with gases, the simple manometer or mercury U-tube of Fig. 2.11 is a convenient device for measuring pressures. To determine the *gage pressure head* at A, in terms of the liquid at A, one may write a *gage equation* based on the fundamental relations of hydrostatic pressures [Eq. (2.3)]. Although any units of pressure or pressure head may be used in the gage equation, it is generally advantageous to express all terms in *feet* (*or meters*) *of the fluid whose pressure is to be measured.* Thus, if s is defined as s_M/s_F, the ratio of the specific gravity (or density)[1] of the manometer or gage fluid to that of the fluid whose pressure is being measured, the gage pressure head at point C is sy. This is also the head at B, while the head at A is greater than this by z, assuming the fluid in the connecting tube $A'B$ is of the same specific weight as that of the

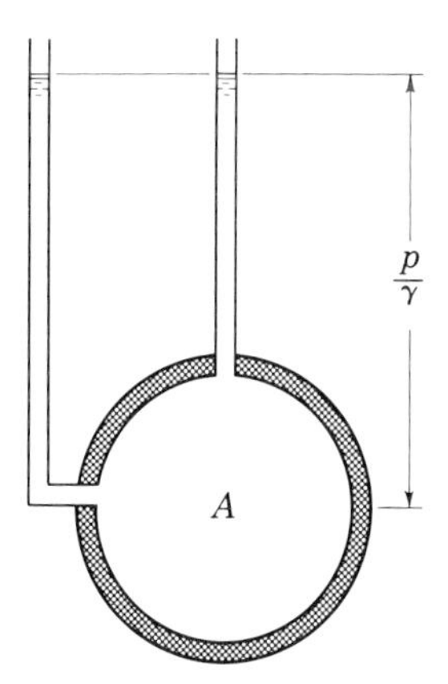

Figure 2.10 Piezometer (for measuring p/γ in liquids only).

[1] In SI units the density of a liquid when expressed in grams per milliliter has the same numerical value as the specific gravity.

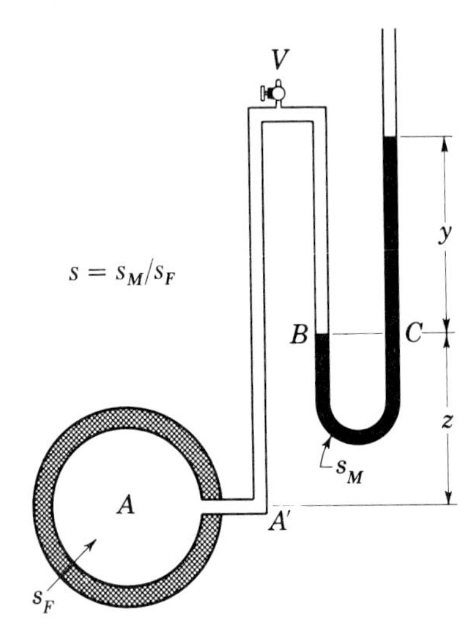

Figure 2.11 Open-end manometer (for measuring p/γ in liquids or gases).

fluid at A. For this simple case the head at A can be written down directly, but for more complicated gages it is helpful to commence the equation at the open end of the manometer with the pressure head there, then proceed through the tube to A, adding terms when descending and subtracting when ascending, equating the result to the head at A. Thus, for this example (Fig. 2.11),

$$0 + sy + z = \frac{p_A}{\gamma} \tag{2.9}$$[1]

where γ is the specific weight of the liquid at A.

If the *absolute pressure head* at A is desired, then the zero of the first term will be replaced by the atmospheric pressure head expressed in feet (or meters) of the fluid whose pressure is to be measured. For measuring the pressure in liquids, an air-relief valve V (Fig. 2.11) will provide a means for the escape of gas should any become trapped in tube $A'B$. If the fluid in A is a gas, the pressure head contribution from the distance z is generally negligible and can be neglected because of the relatively small specific gravity (or density) of the gas. If desired, the analysis of Eq. (2.9) could have been accomplished by expressing the terms in units of pressure rather than head.

In measuring a vacuum, for which the arrangement in Fig. 2.12 might be used, the resulting gage equation, subject to the same conditions as in the preceding case, is

$$0 - sy + z = \frac{p_A}{\gamma} \tag{2.10}$$

Again, it would simplify the equation if one were measuring pressure in a gas because the z term could be neglected. In measuring vacuums in liquids the

[1] In terms of gage pressure this equation can be expressed as: $0 + (s\gamma)y + \gamma z = p_A$.

arrangement in Fig. 2.13 is advantageous since gas and vapors cannot become trapped in the tube. For this case

$$0 - sy - z = \frac{p_A}{\gamma}$$

or

$$\frac{p_A}{\gamma} = -(z + sy) \tag{2.11}$$

Although mercury is generally used as the measuring fluid in the simple manometer, other liquids (for example, carbon tetrachloride) can be used. As the specific gravity of the measuring fluid approaches that of the fluid whose pressure is being measured, the reading becomes larger for a given pressure, thus increasing the accuracy of the instrument, provided the specific gravities are accurately known.

Differential Manometers

In many cases only the difference between two pressures is desired, and for this purpose differential manometers, such as shown in Fig. 2.14, may be used. In Fig. 2.14*a* the measuring fluid is of greater density than that of the fluid whose

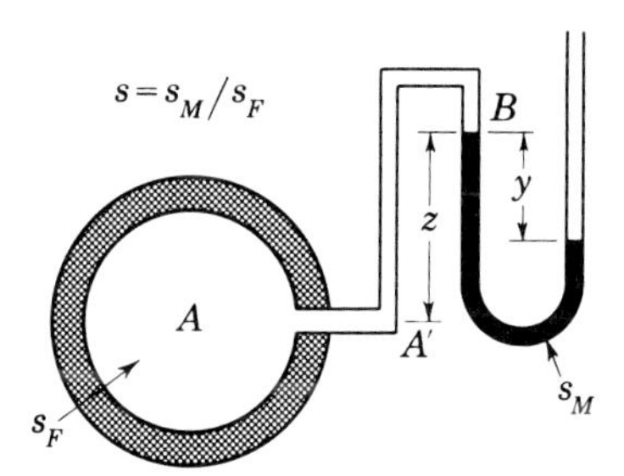

Figure 2.12

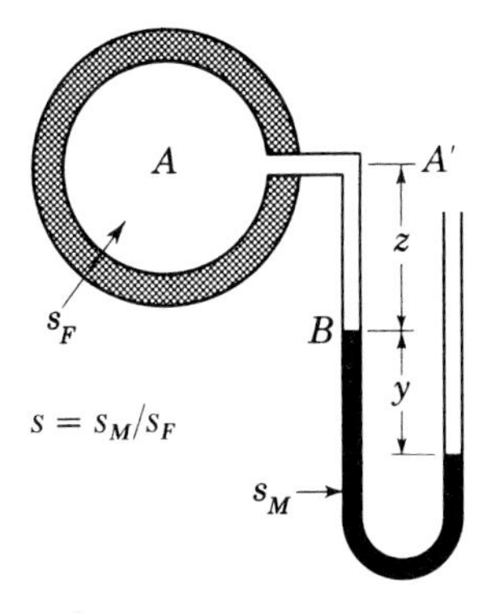

Figure 2.13 Negative-pressure manometer.

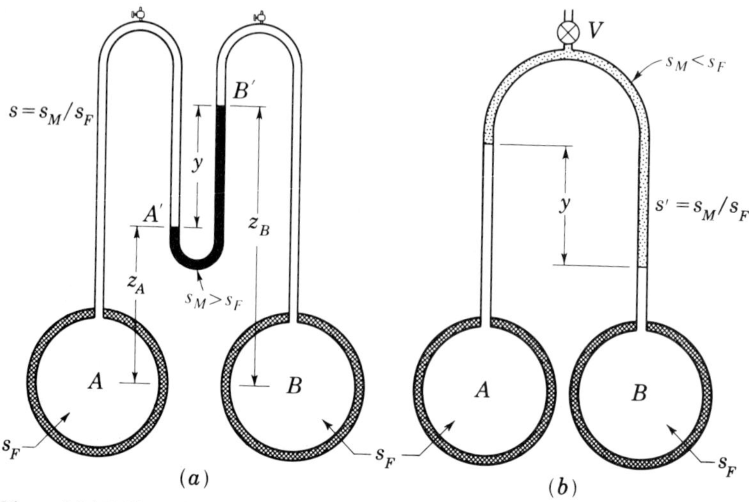

Figure 2.14 Differential manometers. (*a*) For measuring Δp in liquids or gases. (*b*) For measuring Δp in liquids only.

pressure difference is involved. If the fluids in A and B (Fig. 2.14*a*) are of the same density,

$$\frac{p_A}{\gamma} - z_A - sy + z_b = \frac{p_B}{\gamma}$$

and

$$\frac{p_A}{\gamma} - \frac{p_B}{\gamma} = z_A - z_B + sy$$

Hence

$$\frac{p_A}{\gamma} - \frac{p_B}{\gamma} = -y + sy = (s - 1)y \tag{2.12}$$

where $s = s_M/s_F$ as before.

Equation (2.12) is applicable only if A and B are at the same elevation with fluids in A and B having the same density. If the densities of the fluids in A and B are different, the pressure head differential can be found by expressing all heads in terms of one or the other of the fluids as done in Illustrative Example 2.2. If their elevations are different, an elevation-difference term must be added to the equation. It should be emphasized that by far the most common mistake made in working differential-manometer problems is to omit the $s - 1$ multiplier for the gage difference y. The reason for this term is that the effect of the column of fluid of specific gravity s_M between the levels of A' and B' (distance y) is offset by the effect of a column of fluid of specific gravity s_F of the same length in the other leg of the tube.

The differential manometer, when used with a heavy liquid such as mercury, is suitable for measuring large pressure differences. For a small pressure difference a light fluid, such as oil, or even air, may be used, and in this case the manometer is arranged as in Fig. 2.14*b*. Naturally, the manometer fluid must be one that will not mix with the fluid in A or B. By the same method of analysis as the preceding, it may be shown that for the simple case with identical liquids in A and B, and with both A and B at the same elevation,

$$\frac{p_A}{\gamma} - \frac{p_B}{\gamma} = (1 - s')y \tag{2.13}$$

where $s' = s_M/s_F$, the ratio of the specific gravities, and has a value less than 1. As the density of the measuring fluid approaches that of the fluid being measured, $1 - s'$ approaches zero and larger values of y are obtained for small pressure differences, thus increasing the sensitivity of the gage. Once again, the equation must be modified if A and B are not at the same elevation or if their densities are different.

To determine pressure difference between liquids, it is often satisfactory to use air or some other gas as the measuring fluid (Fig. 2.14*b*). Air can then be pumped through valve V until the pressure is such as to bring the two liquid columns to a suitable level. Any change in pressure raises or lowers both liquid columns by the same amount so that the difference between them is constant. In this case the value of s' may be considered to be zero, since the density of gas is so much less than that of a liquid, and the difference in pressure head between A and B is given by y directly. But for high pressures in A and B, with correspondingly high gas pressure and small pressures differences, the value of s' may not be negligible.

Another scheme for obtaining increased sensitivity is simply to incline the gage tube so that a vertical gage difference y is transposed into a reading which is magnified by $1/\sin\alpha$, where α is the angle of inclination with the horizontal.

Illustrative Example 2.2 Liquid A weighs 53.5 lb/ft^3 (8.4 kN/m^3). Liquid B weighs 78.8 lb/ft^3 (12.3 kN/m^3). Manometer liquid M is mercury. If the pressure at B is 30 psi (200 kN/m^2), find the pressure at A. Express all pressure heads in terms of the liquid in bulb B.

$$\frac{p_A}{\gamma_B} - \Delta z_{c-a}\frac{\gamma_A}{\gamma_B} + \Delta z_{a-b}\frac{\gamma_M}{\gamma_B} + \Delta z_{b-d}\frac{\gamma_B}{\gamma_B} = \frac{p_B}{\gamma_B}$$

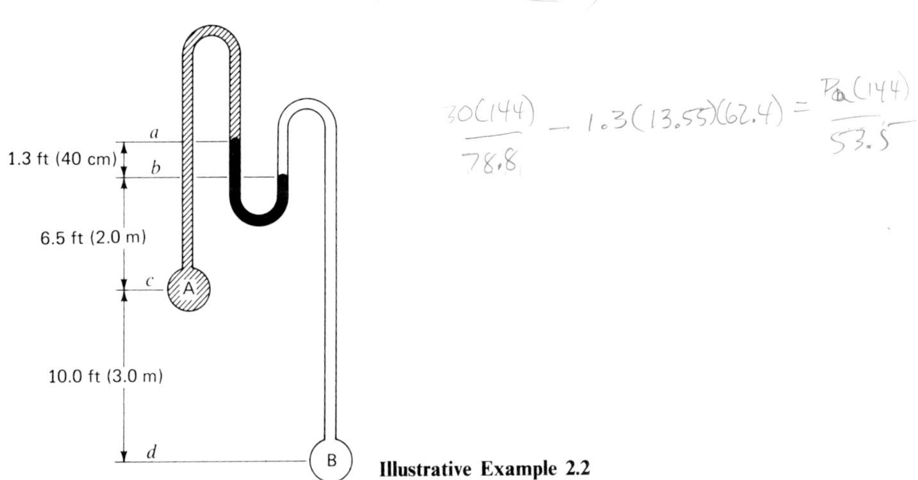

Illustrative Example 2.2

English units:

$$\frac{p_A}{\gamma_B} - 7.8\frac{53.5}{78.8} + 1.3\frac{13.55 \times 62.4}{78.8} + 16.5 = \frac{p_B}{\gamma_B}$$

$$\frac{p_A}{\gamma_B} - 5.29 + 13.95 + 16.5 = \frac{30 \times 144}{78.8} = 54.8 \text{ ft}$$

$$\frac{p_A}{\gamma_B} = 29.6 \text{ ft} \qquad p_A = 29.6\frac{78.8}{144} = 16.2 \text{ psi}$$

SI units:

$$\frac{p_A}{\gamma_B} - 2.0\frac{8.4}{12.3} - 0.4\frac{8.4}{12.3} + 0.4\frac{13.55 \times 9.81}{12.3} + 5.0 = \frac{p_B}{\gamma_B}$$

$$\frac{p_A}{\gamma_B} - 1.37 - 0.27 + 4.32 + 5.00 = \frac{200 \text{ kN/m}^2}{12.3 \text{ kN/m}^3} = 16.3 \text{ m}$$

$$\frac{p_A}{\gamma_B} = 8.6 \text{ m} \qquad p_A = 8.6 \times 12.3 = 106 \text{ kN/m}^2 = 106 \text{ kPa}$$

2.7 FORCE ON PLANE AREA

When a fluid is at rest, no tangential force can exist within the fluid. All forces are then normal to the surfaces in question. If the pressure is uniformly distributed over an area, the force is equal to the pressure times the area, and the point of application of the force is at the centroid of the area. In the case of compressible fluids (gases), the pressure variation with vertical distance is very small because of the low specific weight; hence, when computing the static fluid force exerted by a gas, p may usually be considered constant. Thus, for this case,

$$F = \int p\, dA = p \int dA = pA \tag{2.14}$$

In the case of liquids the distribution of pressure is not uniform; hence further analysis is necessary. In Fig. 2.15 consider a vertical plane whose upper edge lies in the free surface of a liquid. Let this plane be perpendicular to the plane of the paper, so that MN is merely its trace. The pressure will vary from zero at M to NK at N. Thus the total force on one side is the summation of the products of the elementary areas and the pressure upon them. It is apparent that the resultant of this system of parallel forces must be applied at a point *below* the centroid of the area, because the centroid of an area is the point of application of the resultant of a system of *uniform* parallel forces.

If the plane is lowered to $M'N'$, the proportionate change of pressure from M' to N' is less than that from M to N. Hence the center of pressure will be nearer to the centroid of the plane surface, and the deeper the plane is submerged, the more uniform the pressure becomes and the closer these two points will be together.

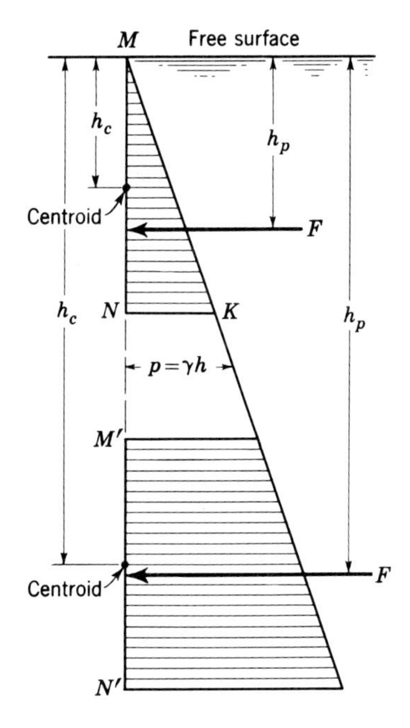

Figure 2.15

In Fig. 2.16 let MN be the trace of a plane area making an angle θ with the horizontal. To the right is the projection of this area upon a vertical plane. Let h be the variable depth to any point and y be the corresponding distance from OX, the intersection of the plane produced and the free surface.

Consider an element of area so chosen that the pressure is uniform over it. Such an element is a horizontal strip. If x denotes the width of the area at any depth, then $dA = x\,dy$. As $p = \gamma h$ and $h = y \sin\theta$, the force dF on a horizontal strip is

$$dF = p\,dA = \gamma h\,dA = \gamma y \sin\theta\,dA$$

The pressure distribution over the area forms a *pressure prism*, the volume of which is equal to the total force acting on the area.

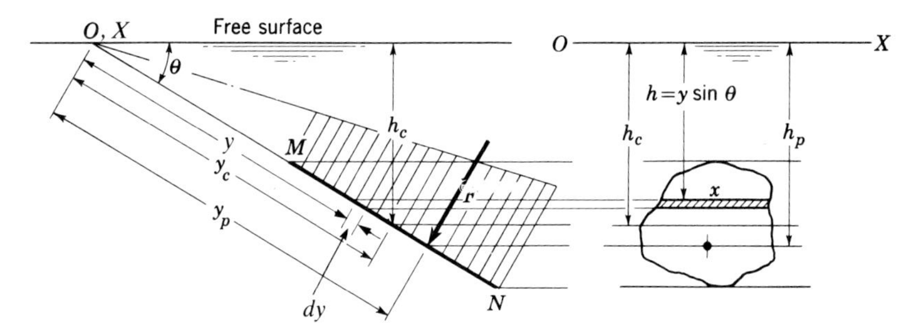

Figure 2.16

Integrating the preceding expression,

$$F = \gamma \sin \theta \int y \, dA = \gamma \sin \theta \, y_c A \tag{2.15}$$

where y_c is the distance from OX along the slope plane to the centroid of the area A. If the vertical depth of the centroid is denoted by h_c, then $h_c = y_c \sin \theta$ and

$$F = \gamma h_c A \tag{2.16}$$

Thus the total force on any plane area submerged in a liquid is found by multiplying the specific weight of the liquid by the product of the area and the depth of its centroid. The value of F is independent of the angle of inclination of the plane so long as the depth of its centroid is unchanged.[1]

Since γh_c is the pressure at the centroid, another statement is that the total force on any plane area submerged in a liquid is the product of the area and the pressure at its centroid.

2.8 CENTER OF PRESSURE

The point of application of the resultant force on an area is called the *center of pressure*. Taking OX in Fig. 2.16 as an axis of moments, the moment of an elementary force $\gamma y \sin \theta \, dA$ is

$$y \, dF = \gamma \sin \theta \, y^2 \, dA$$

and if y_p denotes the distance to the center of pressure, using the concept that the moment of the resultant force equals the summation of the moments of the components forces,

$$y_p F = \gamma \sin \theta \int y^2 \, dA = \gamma \sin \theta \, I_O$$

where I_O is the moment of inertia of the plane area about an axis through O.

If the preceding expression is divided by the value of F as given in Eq. (2.16), the result is

$$y_p = \frac{\gamma \sin \theta \, I_O}{\gamma \sin \theta \, y_c A} = \frac{I_O}{y_c A} \tag{2.17}$$

That is, the distance of the center of pressure from the axis where the plane or plane produced intersects the liquid surface is obtained by dividing the moment of inertia of the area A about the surface axis by its static moment about the same axis.

[1] For a plane submerged as in Fig. 2.16 it is obvious that Eq. (2.16) applies to one side only. As the pressure forces are identical, but opposite in direction, on the two sides, the net force on the plane is zero. In most practical cases where the thickness of the plane is not negligible, the pressures on the two sides are not the same.

This may also be expressed in another form, by noting that

$$I_O = y_c^2 A + I_c$$

where I_c is the moment of inertia of an area about its centroidal axis. Thus

$$y_p = \frac{Ay_c^2 + I_c}{y_c A} = y_c + \frac{I_c}{y_c A} \tag{2.18}$$

From this equation it may be seen that the location of the center of pressure is independent of the angle θ; that is, the plane area may be rotated about axis OX without affecting the location of the center of pressure. Also, it may be seen that the center of pressure is always below the centroid and that, as the depth of immersion is increased, the center of pressure approaches the centroid.

The lateral location of the center of pressure may be determined by considering the area to be made up of a series of elemental horizontal strips. The center of pressure for each strip would be at the midpoint of the strip. Since the moment of the resultant force, F, must be equal to the moment of the distributed force system about any axis, say, the y axis,

$$X_p F = \int x_p p \, dA$$

where X_p is the lateral distance from the selected y axis to the center of pressure of the resultant force F, and x_p is the lateral distance to the center of any elemental horizontal strip of area dA on which the pressure is p.

Another way of looking at the problem of forces on a plane area is through use of the *pressure prism concept*. The pressure acting on a plane area forms a pressure prism, the volume of which is equivalent to the magnitude of the force on the area. The center of gravity of the pressure prism is a point through which the line of action of the center of pressure must pass. Application of the pressure prism concept is convenient for determining the magnitude and location of forces on simple areas such as rectangles.

Illustrative Example 2.3 This gate is 2 feet wide perpendicular to the sketch. It is pivoted at O. The gate weighs 500 lb. Its center of gravity is 1.2 ft to the right of and 0.9 ft above O. For what values of water depth x above O will the gate remain closed? Neglect friction at the pivot and neglect thickness of the gate.

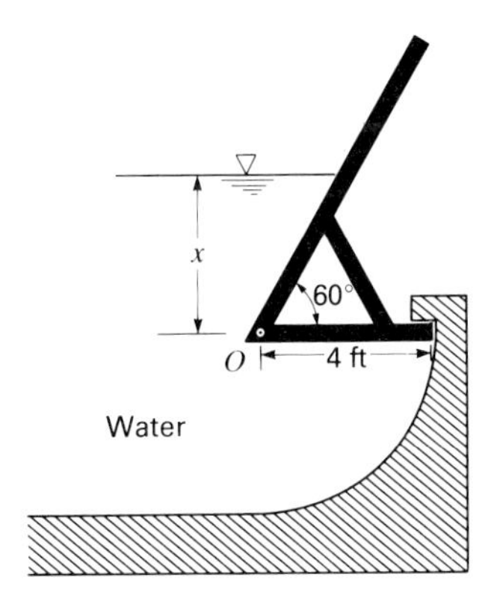

Illustrative Example 2.3

In addition to the reactive force at the hinge O, there are three forces acting on the gate: its weight W, the vertical hydrostatic force F_v upward on the rectangular bottom of the gate, and the slanting hydrostatic force F acting at right angles to the sloping rectangular portion of the gate. The magnitudes of these forces are as follows:

$$W = 500 \text{ lb}$$

$$F_v = \gamma h_c A = \gamma(x)(4 \times 2) = 8\gamma x$$

$$F = \gamma h_c A = \gamma(x/2)\left(\frac{x}{\cos 30^\circ} \times 2\right) = 1.55\gamma x^2$$

A diagram showing these three forces is as follows:

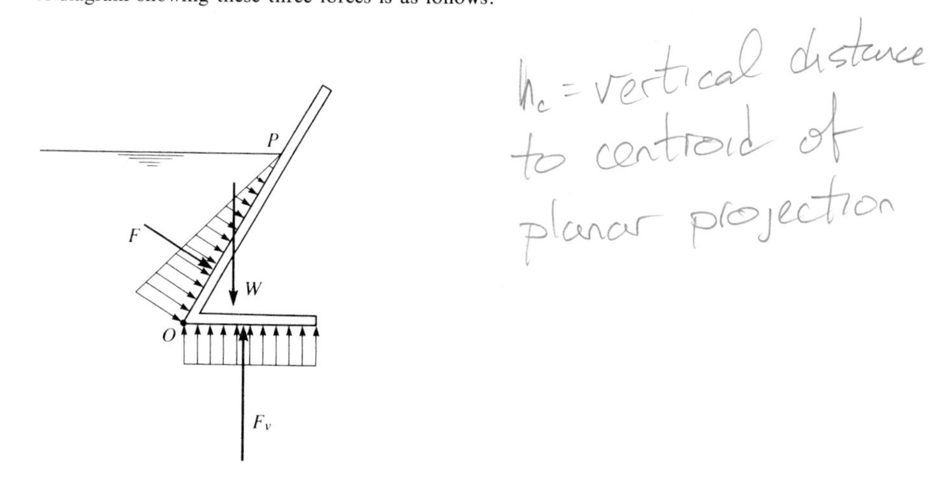

The lever arms of W and F_v with respect to O are 1.2 ft and 2.0 ft respectively. The lever arm of F gets larger as the water depth increases because the location of the center of pressure changes. The location of the center of pressure of the slanting force F may be found from Eq. 2.18:

$$y_p = y_c + \frac{I_c}{y_c A} \qquad I_c = \frac{bh^3}{12}$$

$$y_p = \frac{x/2}{\cos 30^\circ} + \frac{1/12(2)(x/\cos 30^\circ)^3}{(x/2 \cos 30^\circ)(x/\cos 30^\circ \times 2)}$$

$$y_p = 0.577x + \frac{2x}{12 \cos 30^\circ} = 0.770x$$

Hence the lever arm of F with respect to O is $x/\cos 30^\circ - 0.770x = 0.385x$. [*Note:* In this case Eq. (2.18) need not be used to find the lever arm of F because the line of action of F for the triangular distributed load on the rectangular area is known to be at the third point between O and P. (See sketch.)] For incipient rotation we take moments about O and equate them to zero:

$$\sum M = 1.155\gamma x^2(0.385x) + 500(1.2) - 8\gamma x(2) = 0$$

Substituting $\gamma = 62.4 \text{ lb/ft}^3$ gives

$$27.75x^3 + 600 - 998.4x = 0$$

By trial $x = 0.61$ ft and 5.67 ft and a negative (impossible) root. From inspection of the moment equation, the gate will remain closed when $0.61 \text{ ft} < x < 5.67 \text{ ft}$.

Illustrative Example 2.4 Water, oil, and air are present in the cylindrical tank as shown. (*a*) Find the pressure at the bottom of the tank. (*b*) Find the force exerted by the fluids on the end of the tank and determine the location of the center of pressure.

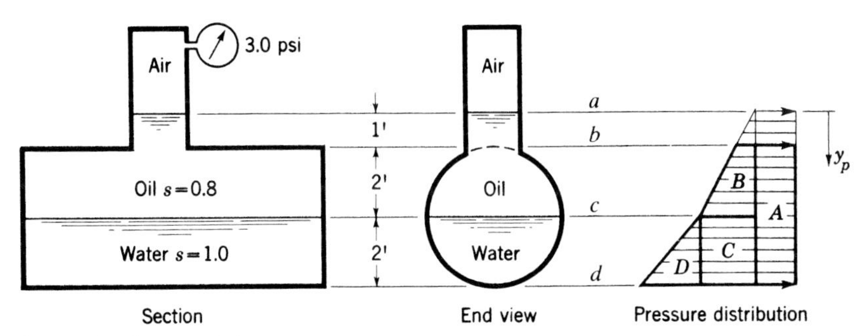

Illustrative Example 2.4

$$(a)\ p_d = p_a + \gamma_{\text{oil}} h_{a-c} + \gamma_{\text{H}_2\text{O}} h_{c-d} = 3.0 + \frac{0.80 \times 62.4}{144} 3 + \frac{62.4}{144} 2$$

$$p_d = 3.0 + 1.04 + 0.87 = 4.9 \text{ psi}$$

(*b*) The confined gas exerts a pressure of 3.0 psi that is transmitted through the other fluids to the end of the tank.

$$F_A = 3.0 \times 144(\pi 2^2) = 5{,}430 \text{ lb}$$

The force F_B on the upper half of the end of the tank due to the presence of oil is

$$F_B = \gamma h_c A = (0.8 \times 62.4)\left(3 - \frac{4 \times 2}{3\pi}\right)\left(\frac{1}{2}\pi 2^2\right) = 675 \text{ lb}$$

The force F_C on the lower half of the end of the tank due to the presence of a 3-ft depth of oil above midheight is

$$F_C = pA = (0.8 \times 62.4)3(\tfrac{1}{2}\pi 2^2) = 940 \text{ lb}$$

The force F_D on the lower half of the end of the tank due to water is

$$F_D = \gamma h_c A = 62.4\left(\frac{4 \times 2}{3\pi}\right)\left(\frac{1}{2}\pi 2^2\right) = 333 \text{ lb}$$

The total force F on the end of the tank is therefore

$$F = F_A + F_B + F_C + F_D = 7{,}380 \text{ lb}$$

The locations below the free oil surface of the centers of pressure of the component forces are

$$(y_p)_A = 3.0 \text{ ft} \qquad \text{(at center of circular area)}$$

$$(y_p)_B = y_c + \frac{I_c}{y_c A}$$

where

$$(y_c)_B = \left(3 - \frac{4 \times 2}{3\pi}\right) = 3 - 0.85 = 2.15 \text{ ft}$$

and

$$(I_c)_B = I - A(0.85)^2 = \frac{\pi 4^4}{128} - \frac{1}{2}\pi 2^2(0.85)^2 = 1.74 \text{ ft}^4$$

(Expression for I from Appendix 3, Table A.7.)

Thus $$(y_p)_B = 2.15 + \frac{1.74}{2.15(\frac{1}{2}\pi 2^2)} = 2.28 \text{ ft}$$

$$(y_p)_C = 3 + \frac{4 \times 2}{3\pi} = 3.85 \text{ ft} \qquad \text{(at centroid of lower half of circle)}$$

Note that $(I_c)_D = (I_c)_B = 1.74 \text{ ft}^4$

$$(y_p)_D = 3 + y_c + \frac{I_c}{y_c A} = 3 + 0.85 + \frac{1.74}{0.85(\frac{1}{2}\pi 2^2)} = 4.18 \text{ ft}$$

Finally $$f(y_p) = F_A(y_p)_A + F_B(y_p)_B + \cdots$$

$$y_p = 3.08 \text{ ft}$$

2.9 FORCE ON CURVED SURFACE

On any curved or warped surface such as MN in Fig. 2.17a, the force on the various elementary areas that make up the curved surface are different in direction and magnitude, and an algebraic summation is impossible. Hence Eq. (2.16) can be applied only to a plane area. But for any nonplanar area the resultant forces in certain directions can be found.

Horizontal Force on Curved Surface

Any irregular curved area MN (Fig. 2.17a) may be projected upon a vertical plane whose trace is $M'N'$ (Fig. 2.17b). The projecting elements, which are all horizontal, enclose a volume whose ends are the vertical plane $M'N'$ and the irregular area MN. This volume of liquid is in static equilibrium. Acting on the vertical projection $M'N'$ is a force F'. The horizontal force F'_x acts on the irregular end area MN and is equal and opposite to the F_x of Fig. 2.17a. Gravity W' is vertical, and the lateral forces on all the horizontal projecting elements are

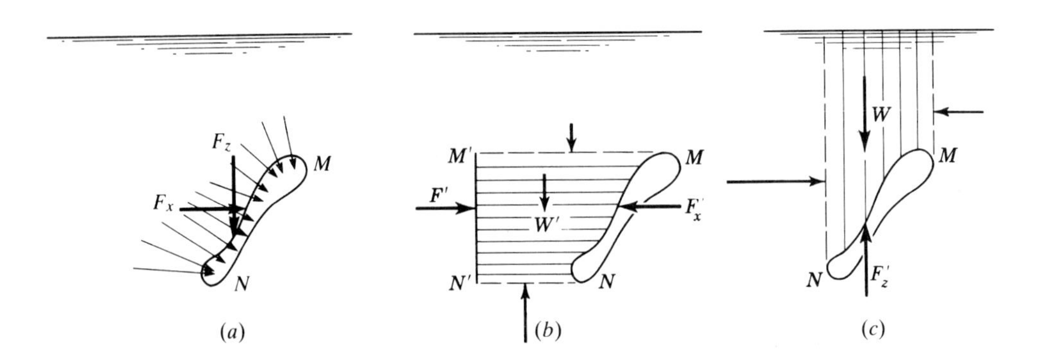

Figure 2.17 Hydrostatic forces on curved surfaces.

normal to these elements and hence normal to F'. Thus the only horizontal forces on $MNN'M'$ are F' and F'_x, and therefore

$$F' - F'_x = 0$$

and
$$F_x = F'_x = F' \tag{2.19}$$

Hence the horizontal force in any given direction on any area is equal to the force on the projection of that area upon a vertical plane normal to the given direction. The line of action of F_x must be the same as that of F'. Equation (2.19) is applicable to gases as well as liquids. In the case of a gas, the horizontal force on a curved surface is given by the pressure multiplied by the area of the vertical projection of the curved surface.

Vertical Force on Curved Surface

The vertical force F_z on a curved or warped area, such as MN in Fig. 2.17*a*, can be found by considering the volume of liquid enclosed by the area and vertical elements extending to the level of the free surface (Fig. 2.17*c*). This volume of liquid is in static equilibrium. Disregarding the pressure on the free surface, the only vertical forces on this volume of liquid are the gravity force W downward and F'_z, the upward vertical force on the irregular area MN. The force F'_z (Fig. 2.17*c*) is equal and opposite to the force F_z (Fig. 2.17*a*). The forces on the vertical elements are normal to the elements and hence are horizontal. Therefore

$$F'_z - W = 0$$

and
$$F_z = F'_z = W \tag{2.20}$$

Hence the vertical force upon any area is equal to the weight of the volume of liquid extending above that area to the level of the free surface. The line of action of F_z must be the same as that of W; that is, it must pass through the center of gravity of the volume. The vertical force exerted by a gas on a curved surface or from pressure acting on the free liquid surface of Fig. 2.17*c* is equal to the pressure multiplied by the area of the horizontal projection of the curved surface.

In case the lower side of the surface is subjected to a force while the upper side is not, the vertical force is equal to the weight of the imaginary volume of liquid above the area up to the level of the free surface. That is, the result is the same numerically as that given by Eq. (2.20). Once again, if a gas is involved, the vertical force is computed by multiplying the pressure by the area of the horizontal projection of the curved surface.

Resultant Force on Curved Surface

In general, there is no single resultant force on an irregular area, for the horizontal and vertical forces, as found in the foregoing discussion, may not be in the same plane. But in certain cases these two forces will lie in the same plane and then can be combined into a single force.

Illustrative Example 2.5 Find the horizontal and vertical components of the force exerted by the fluids on the horizontal cylinder in the accompanying figure if (*a*) the fluid to the left of the cylinder is a gas confined in a closed tank at a pressure of 35.0 kN/m^2; (*b*) the fluid to the left of the cylinder is water with a free surface at an elevation coincident with the uppermost part of the cylinder. Assume in both instances that atmospheric pressure occurs to the right of the cylinder.

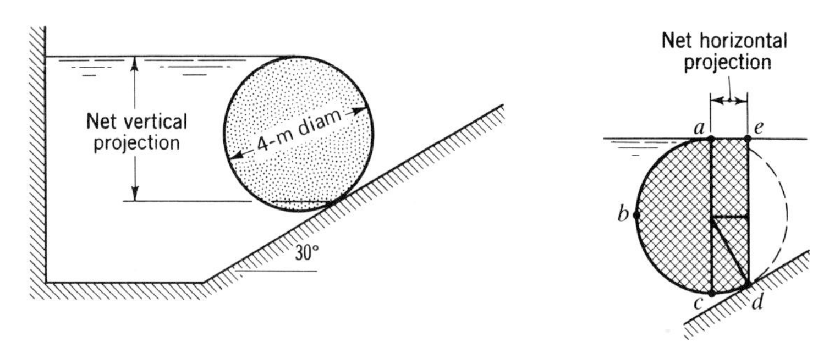

Illustrative Example 2.5

(*a*) The net vertical projection of the portion of the cylindrical surface under consideration is, as indicated in the left-hand figure, $4 - 2(1 - \cos 30°) = 3.73$ m. Hence

$$F_x = pA_z = 35.0 \text{ kN/m}^2[4 - 2(1 - \cos 30°)] \text{ m} = 35.0 \times 3.73 = 130.5 \text{ kN/m to the right}$$

Note that the vertical force of the gas on surface *ab* is equal and opposite to that on surface *bc*. Hence the net horizontal projection with regard to the gas is $ae = 2 \sin 30°$. Thus

$$F_z = pA_x = 35.0 \text{ kN/m}^2 \times 2 \sin 30° \text{ m} = 35.0 \text{ kN/m upward}$$

(*b*) $F_x = \gamma h_c A = 9.81 \text{ kN/m}^3(\frac{1}{2} \times 3.73 \text{ m})(3.73 \text{ m}) = 68.2 \text{ kN/m to the right}$

F_z = weight of cross-hatched volume of liquid

$$F_z = 9.81 \text{ kN/m}^3(\tfrac{210}{360}\pi 2^2 + \tfrac{1}{2}1 \times 1.732 + 1 \times 2) \text{ m} = 100.0 \text{ kN/m upward}$$

2.10 BUOYANCY AND STABILITY OF SUBMERGED AND FLOATING BODIES

Submerged Body

The body *DHCK* immersed in the fluid in Fig. 2.18 is acted upon by gravity and the pressures of the surrounding fluid. On its upper surface the vertical component of the force is F_z and is equal to the weight of the volume of fluid *ABCHD*. In similar manner the vertical component of force on the undersurface is F'_z and is equal to the weight of the volume of fluid *ABCKD*. The difference between these two volumes is the volume of the body *DHCK*.

The buoyant force of a fluid is denoted by F_B, and it is vertically upward and equal to $F'_z - F_z$, which is equal to the weight of the volume of fluid *DHCK*. That is, *the buoyant force on any body is equal to the weight of fluid displaced.* If

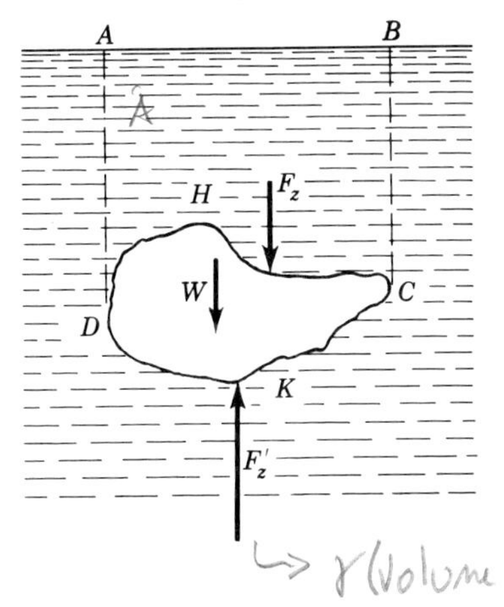

Figure 2.18

the body in Fig. 2.18 is in equilibrium, $W = F_B$, which means that the densities of body and fluid are equal. If W is greater than F_B, the body will sink. If W is less than F_B, the body will rise until its density and that of the fluid are equal, as in the case of a balloon in the air or, in the case of a free liquid surface, the body will rise until the weight of the displaced liquid equals the weight of the body. If the body is less compressible than the fluid, there is a definite level at which it will reach equilibrium. If it is more compressible than the fluid, it will rise indefinitely, provided the fluid has no definite limit of height.

When a body in equilibrium is given a slight displacement, if the forces thereby created tend to restore the body to its original position, the body is said to be in stable equilibrium. The stability of submerged or floating bodies depends on the relative position of the buoyant force and the weight of the body. The buoyant force acts through the *center of buoyancy*, which is coincident with the center of mass of the displaced fluid. *The criterion for stability of a submerged body* (balloon or submarine) *is that the center of buoyancy must be above the center of mass of the body*. The validity of this statement may be confirmed by inspecting Fig. 2.19.

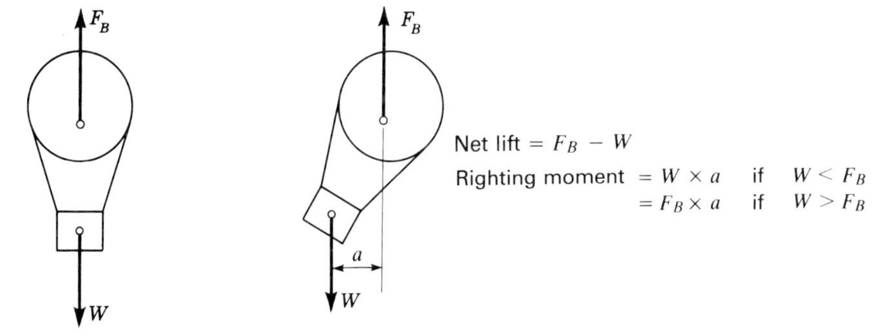

Figure 2.19 Submerged body (balloon).

Floating Body

For a body in a liquid with a free surface, if its weight W is less than that of the same volume of liquid, it will rise and float on the surface so that $W = F_B$. Hence a *floating body displaces a volume of liquid equivalent to its weight.* If a righting moment is developed when a floating body *lists*, the body will be stable regardless of whether the center of buoyancy is above or below the center of mass. An example of stable and unstable floating bodies is shown in Fig. 2.20. In this example the stable body is the one where the center of buoyancy B is above the center of gravity C. Figure 2.21 shows the section of a hull of a ship that is stable. Its center of buoyancy B is below its center of gravity C. For equilibrium the two forces W and F_B must be equal and must lie in the same vertical line. Suppose the body is rolled through the angle θ. The center of gravity C of the body is usually fixed in position, but the center of buoyancy B will generally change as shown. Thus W and F_B constitute a couple of magnitude $W \times a$. For the case under consideration this is a righting couple since it tends to restore the body to the upright position. If liquid in the hull of a ship were unconstrained, the center of mass of the floating body would move toward the center of buoyancy when the ship rolled, thus decreasing the righting couple and the stability. For this reason liquid ballast or fuel oil in floating vessels should be stored in tanks or bulkheaded compartments.

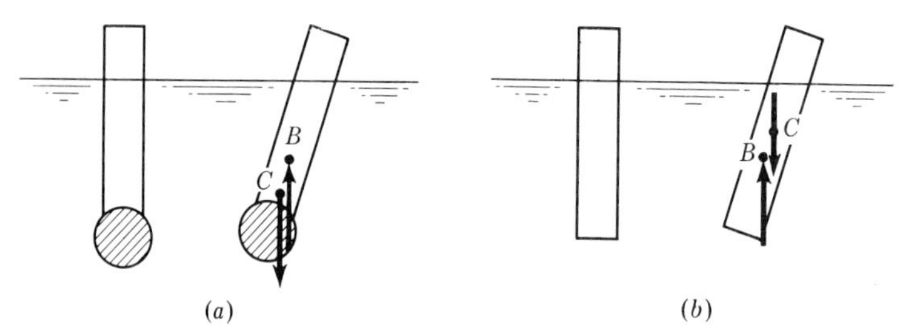

Figure 2.20 (*a*) Stable. (*b*) Unstable.

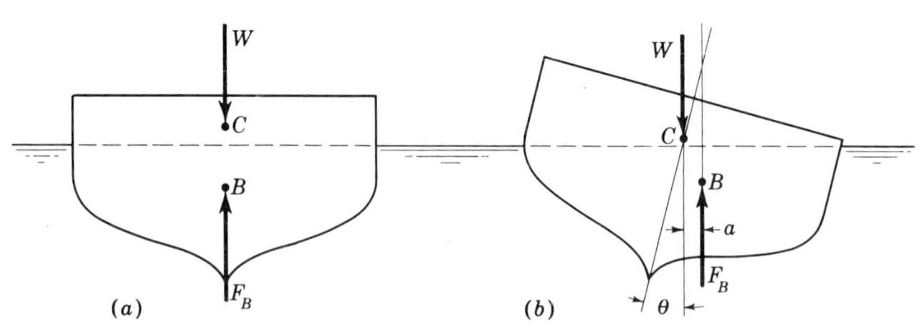

Figure 2.21

Illustrative Example 2.6 A 4-in diameter solid cylinder of height 3.75 in weighing 0.85 lb is immersed in liquid ($\gamma = 52\ \mathrm{lb/ft^3}$) contained in a tall, upright metal cylinder having a diameter of 5 in. Before immersion the liquid was 3 in deep. At what level will the solid cylinder float?

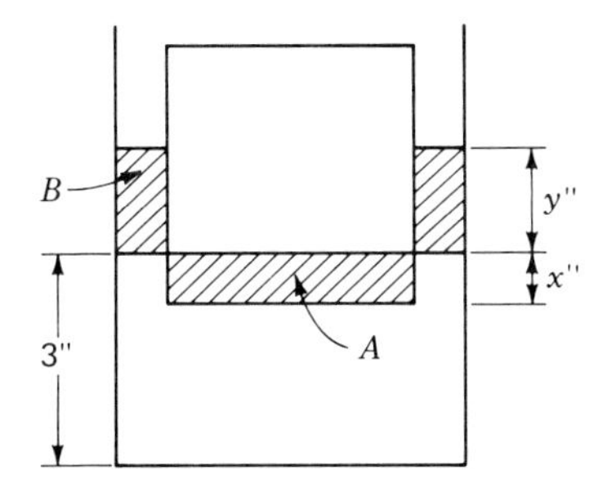

Illustrative Example 2.6

x = distance solid cylinder falls below original liquid surface
y = distance liquid rises above original liquid surface
$x + y$ = depth of submergence

$$\text{Volume } A = \text{volume } B \qquad \pi 2^2 x = \pi(2.5^2 - 2.0^2)y$$

$$4x = 2.25y \qquad x = 0.56y$$

$$F_B = \text{weight} = 0.85 = 52\pi\left(\frac{2}{12}\right)^2 \frac{x+y}{12}$$

$$x + y = 2.24 \text{ in} \qquad x = 0.81 \text{ in} \qquad y = 1.43 \text{ in}$$

Bottom of solid cylinder will be 3.0 − 0.81 = 2.19 in above bottom of hollow cylinder

2.11 FLUID MASSES SUBJECTED TO ACCELERATION

Under certain conditions there may be no relative motion between the particles of a fluid mass yet the mass itself may be in motion. If a body of fluid in a tank is transported at a uniform velocity, the conditions are those of ordinary fluid statics. But if it is subjected to acceleration, special treatment is required. Consider the case of a liquid mass in an open tank moving horizontally with a linear acceleration a_x, as shown in Fig. 2.22*a*. A free-body diagram (Fig. 2.22*b*) of a small particle (mass m) of liquid on the surface indicates that the forces exerted by the surrounding fluid on the particle are such that $F_z = W$ and $F_x = ma_x$; the latter is required to produce acceleration a_x of the particle.[1] Equal and opposite to these forces are F'_x and F'_z of Fig. 2.22*a*, the forces exerted by the particle on the surrounding fluid. The resultant of these forces is F'. The liquid surface must be at right angles to F', for if it were not, the particle would not maintain its relative position in the liquid. Hence $\tan\theta = a_x/g$. The liquid surface and all planes of equal hydrostatic pressure must be inclined at angle θ with the horizontal as shown in Fig. 2.22*a*.

[1] Note that W and F_z counterbalance one another; hence there is no acceleration in the z direction.

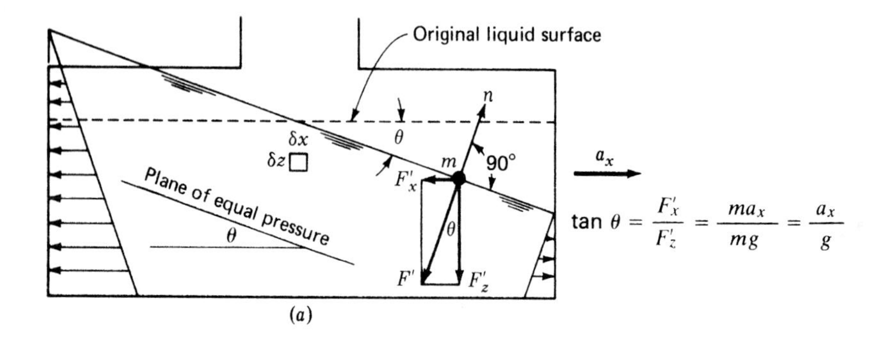

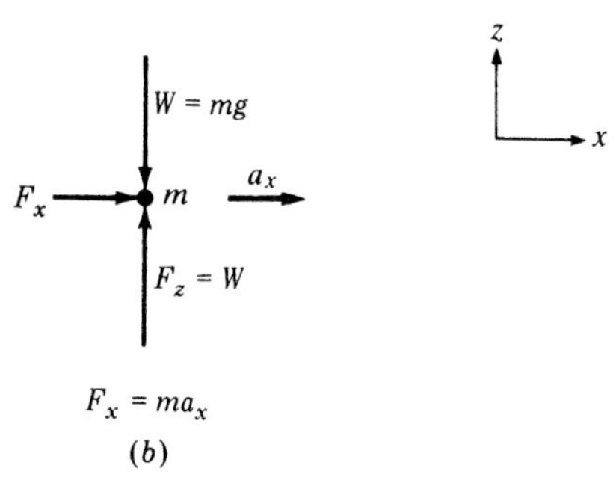

Figure 2.22 Liquid mass subjected to horizontal acceleration.

Next consider the more general case where a fluid mass is subject to acceleration in both the x and z directions. In Fig. 2.23 is shown a free-body diagram of an elemental cube of fluid at the center of which the pressure is p. Applying the equation of motion in the x-direction

$$\sum F_x = ma_x$$

$$\left(p - \frac{\partial p}{\partial x}\frac{\partial x}{2}\right)\delta y\,\delta z - \left(p + \frac{\partial p}{\partial x}\frac{\delta x}{2}\right)\delta y\,\delta z = \frac{\gamma}{g}\,\delta x\,\delta y\,\delta z\,a_x$$

which reduces to

$$\frac{\partial p}{\partial x} = -\frac{\gamma}{g}a_x = -\rho a_x \tag{2.21}$$

In the vertical direction

$$\sum F_z = ma_z$$

$$\left(p - \frac{\partial p}{\partial z}\frac{\delta z}{2}\right)\delta x\,\delta y - \left(p + \frac{\partial p}{\partial z}\frac{\delta z}{2}\right)\delta x\,\delta y - \gamma\,\delta x\,\delta y\,\delta z = \frac{\gamma}{g}\,\delta x\,\delta y\,\delta z\,a_z$$

which yields

$$\frac{\partial p}{\partial z} = \frac{\gamma}{g}(a_z + g) = -\rho(a_z + g) \tag{2.22}$$

Equations (2.21) and (2.22) can be employed to develop a generalization applicable to a fluid mass that is subject to acceleration in the x and z directions. The chain rule for the total differential of dp in terms of its partial derivatives is

$$dp = \frac{\partial p}{\partial x}\,dx + \frac{\partial p}{\partial z}\,dz$$

Substituting the expressions for $\partial p/\partial x$ and $\partial p/\partial z$ from Eqs. (2.21) and (2.22) gives

$$dp = -\frac{\gamma}{g}(a_x)\,dx - \frac{\gamma}{g}(a_z + g)\,dz \tag{2.23}$$

Along a line of constant pressure, $dp = 0$. From Eq. (2.23) if $dp = 0$,

$$\frac{dz}{dx} = -\frac{a_x}{a_z + g} \tag{2.24}$$

This then defines the slope dz/dx of a line of constant pressure within the accelerated fluid mass.

Another expression that can be developed from Eqs. (2.21) and (2.22) is

$$\frac{\partial p}{\partial n} = -\rho[a_x^2 + (a_z + g)^2]^{1/2} \tag{2.25}$$

where n is at right angles to the lines of equal pressure and in the direction of decreasing pressure. When $a_x = a_z = 0$, this equation reduces to $\partial p/\partial n = -\gamma$, which is essentially the same as the basic hydrostatic equation (Eq. 2.2). Application of Eq. (2.25) indicates that, if fluid in a container is subjected to an upward acceleration, there will be an increase of pressure within the fluid. A downward acceleration results in a decrease in pressure.

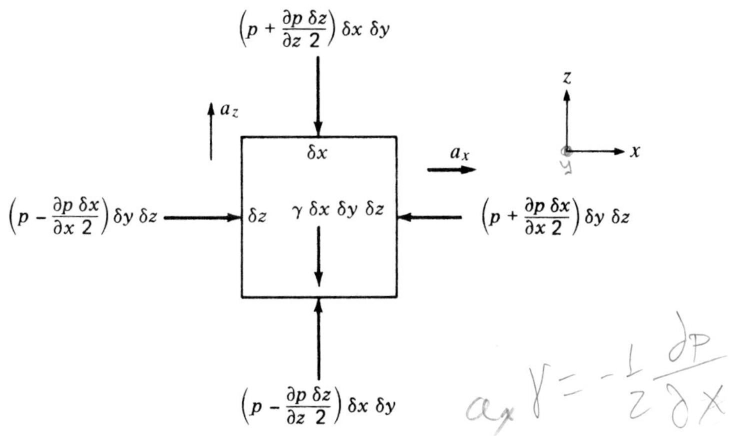

Figure 2.23

Illustrative Example 2.7 At a particular instant an airplane is traveling upward at a velocity of 180 m/s in a direction that makes an angle of 40° with the horizontal. At this instant the airplane is losing speed at the rate of 4 m/s². Also it is moving on a concave-upward circular path having a radius of 2,600 m. Determine for the given conditions the position of the free liquid surface in the fuel tank of this vehicle.

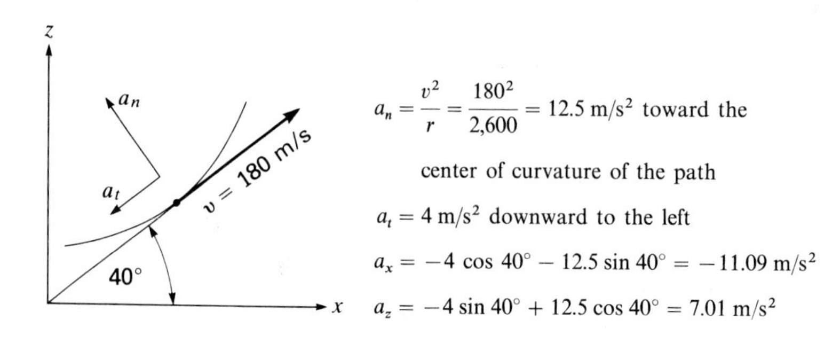

$$a_n = \frac{v^2}{r} = \frac{180^2}{2{,}600} = 12.5 \text{ m/s}^2 \text{ toward the}$$

center of curvature of the path

$$a_t = 4 \text{ m/s}^2 \text{ downward to the left}$$

$$a_x = -4 \cos 40° - 12.5 \sin 40° = -11.09 \text{ m/s}^2$$

$$a_z = -4 \sin 40° + 12.5 \cos 40° = 7.01 \text{ m/s}^2$$

Consider a liquid particle of mass m at the free surface. To achieve a_x and a_z the forces exerted by the surrounding liquid on the liquid particle are in the directions shown:

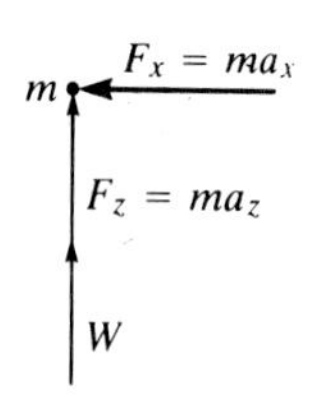

Forces exerted by the surrounding liquid on the liquid particle at the free surface

Equal and opposite to these forces are the forces F'_x, F'_z and W exerted by the liquid particle on the surrounding liquid. Thus

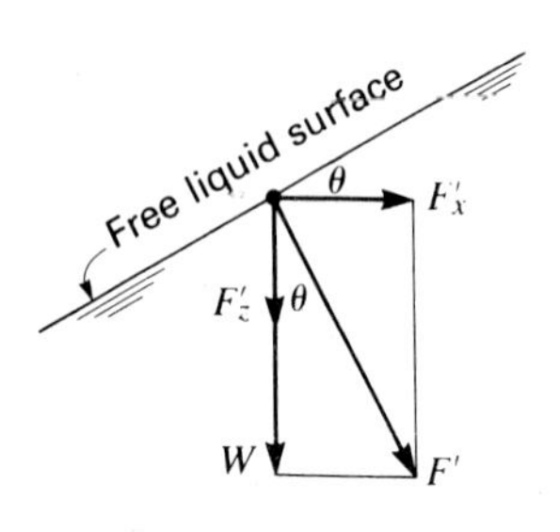

Forces exerted by the liquid particle at the free surface on the surrounding liquid

$$\tan\theta = \frac{F'_x}{F'_z + W} = \frac{ma_x}{ma_z + mg} = \frac{a_x}{a_z + g} = \frac{11.09}{7.01 + 9.81}$$

$$\theta = \tan^{-1} 0.659 = 33.4°$$

Or alternatively, employing Eq. (2.24)

$$\frac{dz}{dx} = -\frac{-11.09}{7.01 + 9.81} = 0.659, \text{ slope of the free surface}$$

PROBLEMS

2.1 Neglecting the pressure upon the surface and the compressibility of water, what is the pressure in pounds per square inch (kN/m^2) at a depth of 15,000 ft (5 km) below the surface of the ocean? The specific weight of ocean water under ordinary conditions is 64.0 lb/ft^3 (10.05 kN/m^3).

2.2 A pressure gage at elevation 8.0 m on the side of a tank containing a liquid reads 57.4 kN/m^2. Another gage at elevation 5.0 m reads 80.0 kN/m^2. Compute the specific weight and density of the liquid.

2.3 A pressure gage at elevation 20.0 ft on the side of a tank containing a liquid reads 12.8 psi. Another gage at elevation 13.0 ft reads 15.5 psi. Compute the specific weight, density, and specific gravity of the liquid.

2.4 Repeat Prob. 2.1, but consider the effects of compressibility ($E_v = 300{,}000$ psi). Neglect changes in density caused by temperature variations. (*Hint:* As a starting point, express Eq. 1.3 in terms of γ and integrate to determine γ as a function of z.)

2.5 An open tank contains 5.0 m of water covered with 2.0 m of oil ($\gamma = 8.0$ kN/m^3). Find the pressure at the interface and at the bottom of the tank.

2.6 An open tank contains 10 ft of water covered with 2 ft of oil ($s = 0.86$). Find the pressure at the interface between the liquids and at the bottom of the tank.

2.7 If air had a constant specific weight of 0.076 lb/ft^3 and were incompressible, what would be the height of air surrounding the earth to produce a pressure at the surface of 14.7 psia?

2.8 If the specific weight of a sludge can be expressed as $\gamma = 65.0 + 0.2h$, determine the pressure in psi at a depth of 15 ft below the surface. γ is in lb/ft^3, and h in ft below surface.

2.9 On a certain day the barometric pressure at sea level is 30.1 in Hg and the temperature is 70°F. The pressure gage on an airplane flying overhead indicates that the atmospheric pressure at that point is 10.6 psia and that the air temperature is 46°F. Calculate as accurately as you can the height of the airplane above sea level. Assume linear decrease of temperature with elevation.

2.10 The absolute pressure on a gas is 42.5 psia and the atmospheric pressure is 840 mbars, abs. Find the gage pressure in psi, kN/m^2 and mbars.

2.11 If the atmospheric pressure is 920 mbars, abs and a gage attached to a tank reads 400 mm Hg vacuum, what is the absolute pressure within the tank?

2.12 If the atmospheric pressure is 13.70 psia (945 mbars, abs) and a gage attached to a tank reads 8.0 in (20 cm) Hg vacuum, what is the absolute pressure within the tank?

2.13 A gage is connected to a tank in which the pressure of the fluid is 40 psi (276 kN/m^2) above atmospheric. If the absolute pressure of the fluid remains unchanged but the gage is in a chamber where the air pressure is reduced to a vacuum of 25 in (63.5 cm) Hg, what reading in psi (kN/m^2) will then be observed?

2.14 If the atmospheric pressure is 29.92 in Hg, what will be the height of water in a water barometer if the temperature of the water is 90°F; 140°F? Be as precise as possible.

2.15 The tire of an airplane is inflated at sea level to 60 psi. Assuming the tire does not expand, what is the pressure within the tire at elevation 30,000 ft? Assume standard atmosphere. Express the answer in psi and psia.

2.16 Same as Prob. 2.15 except replace 60 psi with 350 kN/m^2, replace 30,000 ft with (*a*) 12,000 m and (*b*) 18,000 m and express answers in both kN/m^2 and kN/m^2, abs.

2.17 If the atmospheric pressure were equivalent to 32.8 ft of water, what would be the reading on a barometer containing an alcohol ($s = 0.84$) if the vapor pressure of the alcohol at the temperature of observation were 2.4 psia?

2.18 If the atmospheric pressure is 940 mbars, abs what would be the reading in meters of a barometer containing water at 60°C?

2.19 In Illustrative Example 2.2 suppose the atmospheric pressure at B is 1,035 mbars, abs. What would be the absolute pressure at A? Express in mbars, abs and in m of Hg.

2.20 In the figure, originally there is a 4-in manometer reading. Atmospheric pressure is 14.7 psia. If the absolute pressure at A is doubled, what then would be the manometer reading?

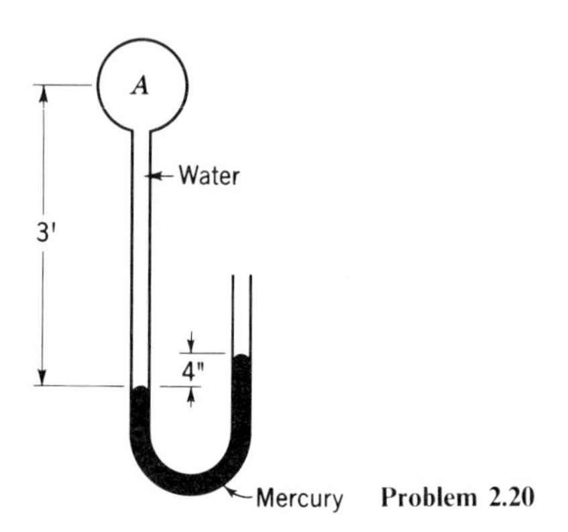

Problem 2.20

2.21 A mercury manometer (Fig. 2.11) is connected to a pipeline carrying water at 150°F and located in a room where the temperature is also 150°F. If the elevation of point B is 10 ft above A and the mercury reading is 48 in, what is the pressure in the pipe in psi? Repeat, assuming all temperatures are 68°F. Be as precise as possible, and note the effect of temperature.

2.22 Two vessels are connected to a differential manometer using mercury ($s = 13.55$), the connecting tubing being filled with water. The higher-pressure vessel is 5 ft (1.5 m) lower in elevation than the other. Room temperature prevails. If the mercury reading is 4.0 in (10 cm), what is the pressure difference in feet (m) of water, in psi (kN/m^2)? If carbon tetrachloride ($s = 1.59$) were used instead of mercury, what would be the manometer reading for the same pressure difference?

2.23 What would be the manometer reading in Illustrative Example 2.2 if $p_B - p_A = 150$ kN/m^2?

2.24 Refer to the manometer of Fig. 2.14*b*. A and B are at the same elevation. Water is contained in A and rises in the tube to a level 76 in above A. Glycerin is contained in B. The inverted U tube is filled with air at 20 psi and 70°F. Atmospheric pressure is 14.7 psia. Determine the difference in pressure between A and B if y is 14 in. Express the answer in psi. What is the absolute pressure in B in inches of mercury, feet of glycerin?

2.25 Gas confined in a rigid container exerts a pressure of 20 psi (140 kN/m^2) when its temperature is 40°F (5°C). What pressure would the gas exert if the temperature were raised to 140°F (60°C)? Barometric pressure remains constant at 28.0 in Hg.

2.26 The diameter of tube C in Fig. 2.11 is d_1, and that of tube B is d_2. Let z_0 be the elevation of the mercury when both mercury columns are at the same level. R is the distance the right-hand column of mercury rises above z_0 when the fluid in A is under pressure. Let γ' be the specific weight of the mercury (or any other measuring fluid), while γ is the specific weight of the fluid in A and the connecting tubing. Prove that

$$p_a = \gamma z_0 + \left[\gamma' + (\gamma' - \gamma)\left(\frac{d_1}{d_2}\right)^2\right]R = M + NR$$

where M and N are constants. It is seen that this equation involves only one variable, which is a reading on the scale for column C. It also shows the significance of having d_2 large compared with d_1.

2.27 In the figure, atmospheric pressure is 14.7 psia; the gage reading is 5.0 psi; the vapor pressure of the alcohol is 1.7 psia. Compute x and y.

2.28 In the sketch for Prob. 2.27 assume the following: atmospheric pressure = 850 mbars, abs; vapor pressure of the alcohol = 160 mbars, abs; $x = 2.80$ m and $y = 2.00$ m. Compute the reading on the pressure gage and on the manometer.

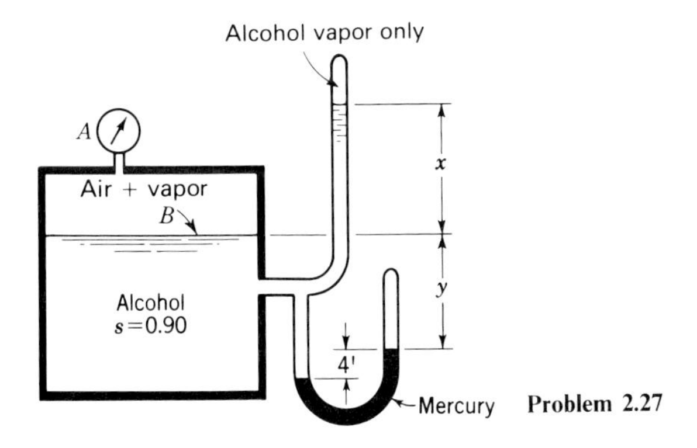

Problem 2.27

2.29 At a certain point the pressure in a pipeline containing gas ($\gamma = 0.05$ lb/ft^3) is 4.5 in of water. The gas is not flowing. What is the pressure in inches of water at another point in the line whose elevation is 500 ft greater than the first point? Make and state clearly any necessary assumptions.

2.30 If a triangle of height d and base b is vertical and submerged in liquid with its vertex at the liquid surface, derive an expression for the depth to its center of pressure.

2.31 Repeat Prob. 2.30 for the same triangle but with its vertex a distance a below the liquid surface.

2.32 If a triangle of height d and base b is vertical and submerged in a liquid with its base at the liquid surface, derive an expression for the depth to its center of pressure.

2.33 A circular area of diameter d is vertical and submerged in a liquid. Its upper edge is coincident with the liquid surface. Derive an expression for the depth to its center of pressure.

2.34 A vertical semicircular area has its diameter in a liquid surface. Derive an expression for the depth to its center of pressure.

2.35 Refer to Illustrative Example 2.4. If the air pressure were 5 psi rather than 3 psi, compute the total force and determine the location of the center of pressure.

2.36 A rectangular plate submerged in water is 5 by 6 ft, the 5-ft side being horizontal and the 6-ft side being vertical. Determine the magnitude of the force on one side of the plate and the depth to its center of pressure if the top edge is (*a*) at the water surface; (*b*) 1 ft below the water surface; (*c*) 100 ft below the water surface.

2.37 Repeat Prob. 2.36 changing all dimensions from feet to meters, i.e., plate is 5 by 6 m.

2.38 A rectangular plate 5 by 6 ft is at an angle of 30° with the horizontal, and the 5-ft side is horizontal. Find the magnitude of the force on one side of the plate and the location of its center of pressure when the top edge is (*a*) at the water surface; (*b*) 1 ft below the water surface.

2.39 A rectangular area is 5 by 6 m, with the 5-m side horizontal. It is placed with its centroid 4 m below a water surface and rotated about a horizontal axis in the plane area and through its centroid. Find the magnitude of the force on one side and the distance between the center of pressure and the centroid of the plane when the angle $\theta = 90, 60, 30, 0°$.

2.40 Repeat Prob. 2.36 for the case where the liquid consists of a 2-ft layer of oil ($s = 0.8$) resting above water.

2.41 A plane surface is circular and 4 ft (1.2 m) in diameter. If it is vertical and the top edge is 1 ft (0.3 m) below the water surface, find the magnitude of the force on one side and the depth to the center of pressure.

2.42 This Utah-shaped plate is submerged in oil ($s = 0.82$) and lies in a vertical plane. Find the magnitude and location of the hydrostatic force acting on one side of the plate.

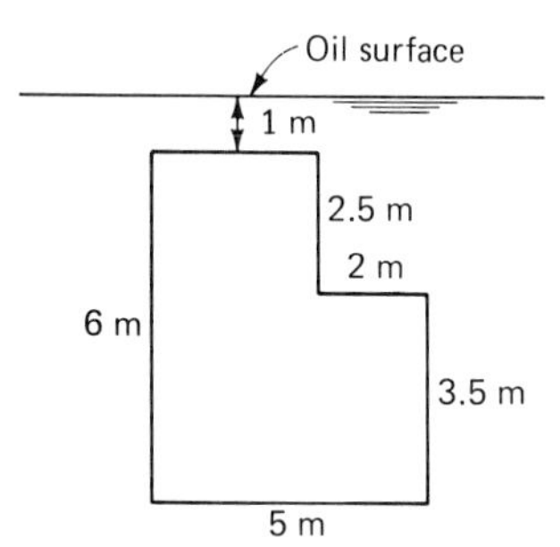

Problem 2.42

2.43 A triangle with a height of 6 ft (1.8 m) and a base of 4 ft (1.2 m) is placed vertically with its base horizontal and 1 ft (0.3 m) below a liquid surface. Determine the depth and horizontal position of the center of pressure.

2.44 Prove that for a plane area such that a straight line can be drawn through the midpoints of all horizontal elements, the center of pressure must lie on this line.

2.45 A vertical right triangle of height d and base b submerged in liquid has its vertex at the liquid surface. Find the distance from the vertical side to the center of pressure by (a) inspection; (b) calculus.

2.46 This common type of irrigation head gate is a plate which slides over the opening to a culvert. The coefficient of friction between the gate and its sliding ways is 0.6. Find the force required to slide open this 900-lb gate if it is set (a) vertically; (b) on a 2:1 slope ($n = 2$), as is common.

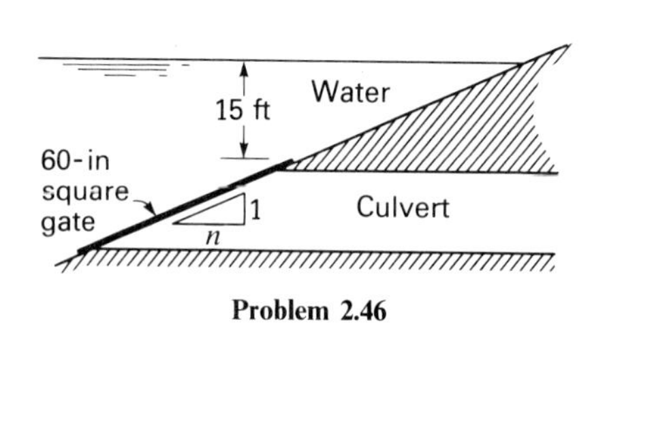

Problem 2.46

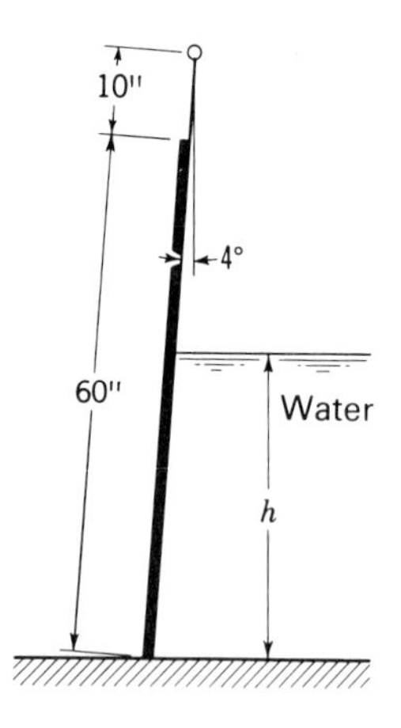

Problem 2.47

2.47 In the drainage of irrigated lands it is frequently desirable so install automatic flap gates to prevent a flood from backing up into the lateral drains from a river. Suppose a 60-in-square flap gate, weighing 2,555 lb, is hinged 40 in above the center, as shown in the figure, and the face is sloped 4° from the vertical. Find the depth to which water will rise behind the gate before it will open.

2.48 The figure shows a cylindrical tank with 0.25-in-thick walls, containing water. What is the force on the bottom? What is the force on the annular surface MM? Find the longitudinal tensile stress in the sidewalls BB if (*a*) the tank is suspended from the top; (*b*) it is supported on the bottom. Neglect the weight of the tank.

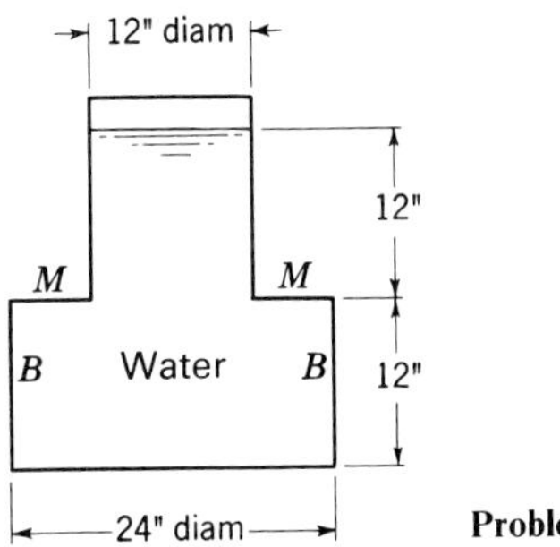

Problem 2.48

2.49 Find the magnitude and point of application of the force on the circular gate shown in the figure.

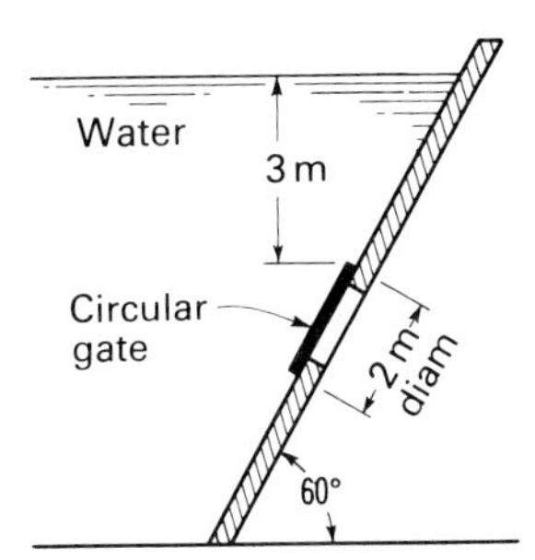

Problem 2.49

2.50 The gate MN in the figure rotates about an axis through N. If the width perpendicular to the plane of the figure is 4 ft, what torque applied to the shaft through N is required to hold the gate closed?

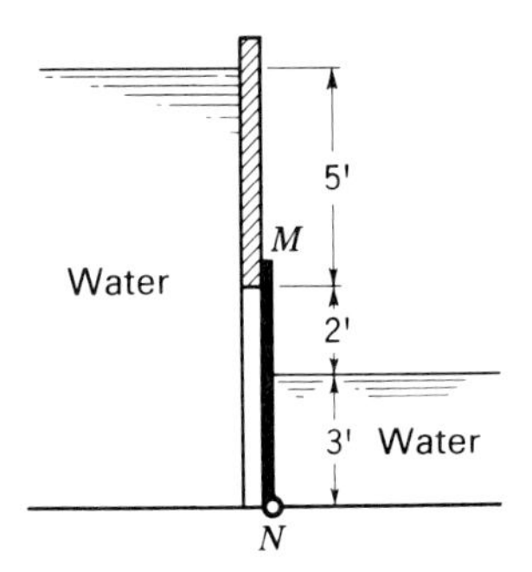

Problem 2.50

2.51 In the figure, the rectangular flashboard MN, 6 m high, is pivoted at B. (a) What must be the maximum height of B above N if the flashboard is on the verge of tipping when the water surface rises to M? (b) If the flashboard is pivoted at the location determined in (a) and the water surface is 1 m below M, what are the reactions at B and N per m of length of crest?

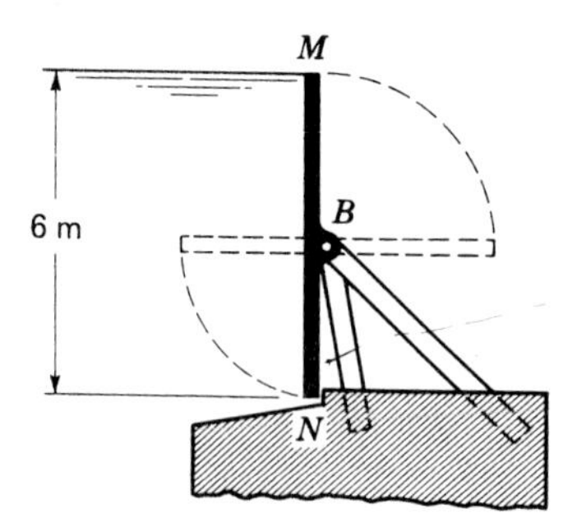

Problem 2.51

2.52 Find the minimum value of z for which the gate in the figure will rotate counterclockwise if the gate is (a) rectangular, 4 by 4 ft; (b) triangular, 4-ft base as axis, height 4 ft. Neglect friction in bearings.

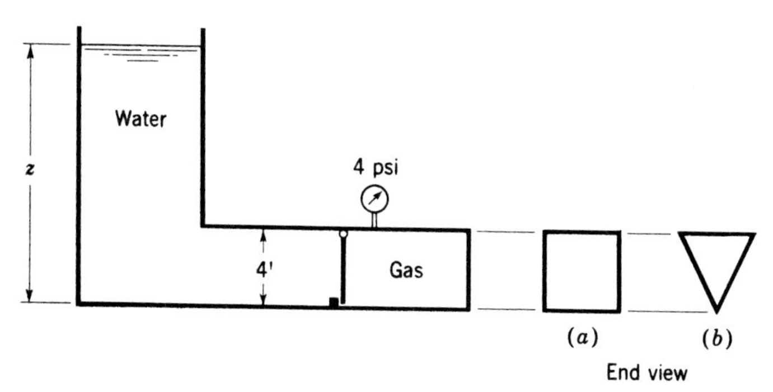

Problem 2.52

2.53 Referring to the figure, what value of b is necessary to keep the rectangular masonry wall from sliding if it weighs 150 lb/ft³ and the coefficient of friction is 0.4? Will it also be safe against overturning? Assume that water does not get underneath the block.

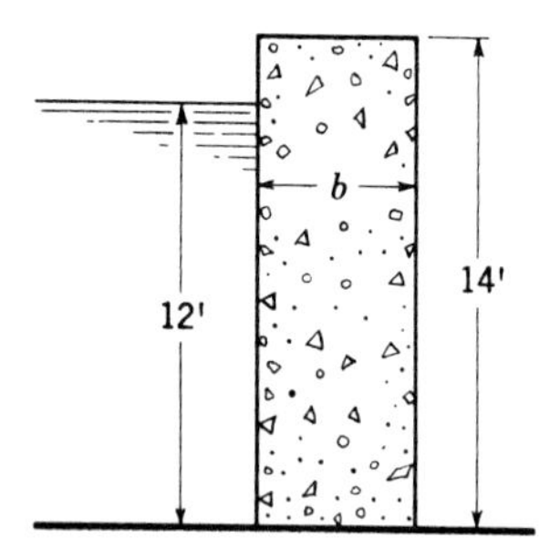

Problem 2.53

2.54 A vertical-thrust bearing for a large hydraulic gate is composed of an 11-in-radius bronze hemisphere mating into a steel hemispherical shell in the gate bottom. At what pressure must lubricant be supplied to the bearing so that a complete oil film is present if the vertical thrust on the bearing is 800,000 lb?

2.55 Find horizontal and vertical forces per foot of width on the Tainter gate shown in the figure. Locate the horizontal force and indicate the line of action of the vertical force without actually computing its location.

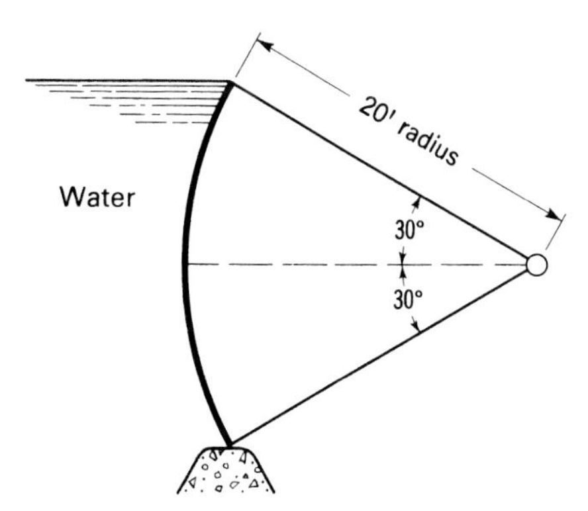

Problem 2.55

2.56 A tank with vertical ends contains water and is 8 m long normal to the plane of the figure shown. The sketch shows a portion of its cross section where MN is one-quarter of an ellipse with semiaxes b and d. If $b = 4$ m, $d = 6$ m, and $a = 1.5$ m, find for the surface represented by MN the magnitude and position of the line of action of (*a*) the horizontal component of force; (*b*) the vertical component of force; (*c*) the resultant force and its direction with the horizontal.

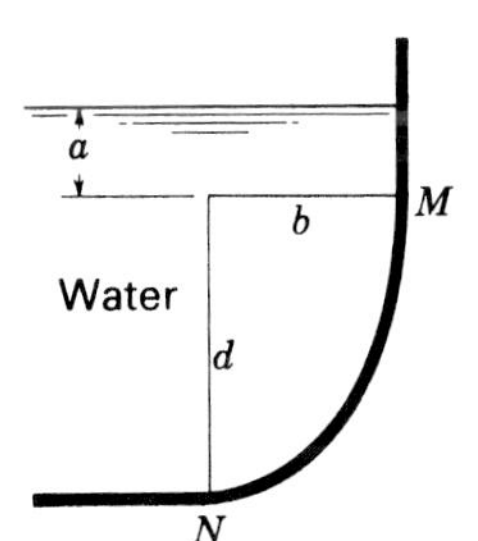

Problem 2.56

2.57 Find the answers called for in Prob. 2.56 if $a = 1.5$ ft, $b = 4$ ft, $d = 6$ ft, and MN represents a parabola with vertex at N.

2.58 The cross section of a tank is as shown in the figure. BC is a cylindrical surface. If the tank contains water to a depth of 7 ft (2 m), determine the magnitude and location of the horizontal- and vertical-force components on the wall ABC.

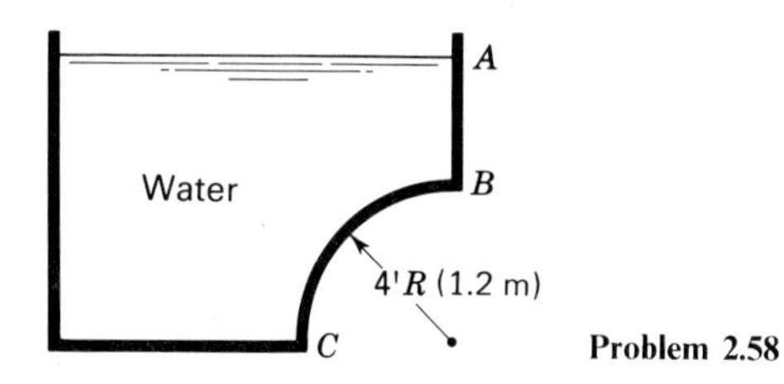

Problem 2.58

2.59 A tank has an irregular cross section as shown in the figure. Determine as accurately as possible the magnitude and location of the horizontal- and vertical-force components on a one-foot length of the wall *ABCD* when the tank contains water to a depth of 8 ft. To determine areas, use a planimeter or count squares (one-ft grid); make a cardboard cutout to locate the centroid.

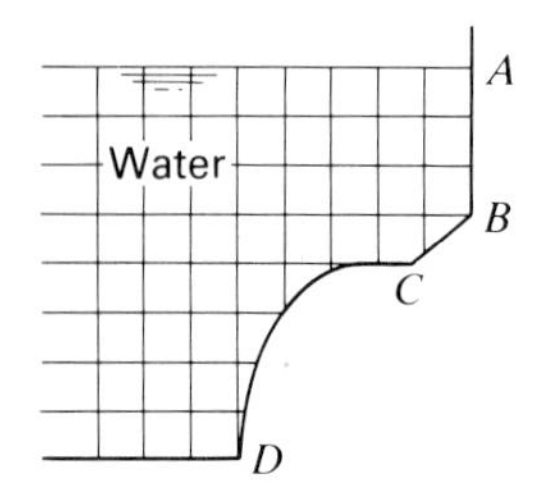

Problem 2.59

2.60 Repeat Prob. 2.58 where the tank is closed and contains gas at a pressure of 8 psi (55 kN/m^2).

2.61 Repeat Prob. 2.58 where the tank is closed and contains 3 ft of water overlain with a gas that is under a pressure of 2.0 psi.

2.62 A spherical steel tank of 20-m diameter contains gas under a pressure of 350 kN/m^2. The tank consists of two half-spheres joined together with a weld. What will be the tensile force across the weld in kN/m? If the steel is 20.0 mm thick, what is the tensile stress in the steel? Express in kN/m^2 and in psi. Neglect the effects of cross-bracing and stiffeners.

2.63 The hemispherical body shown in the figure projects into a tank. Find the horizontal and vertical forces acting on the hemispherical projection for the following cases: (*a*) tank is full of water with free surface 5 ft above *A*; (*b*) tank contains CCl_4 ($s = 1.59$) to the level of *A* overlain with water having free surface 5 ft above *A*; (*c*) tank is closed and contains gas at a pressure of 6 psi; (*d*) tank is closed and contains water to level of *A* overlain with gas at a pressure of 2 psi.

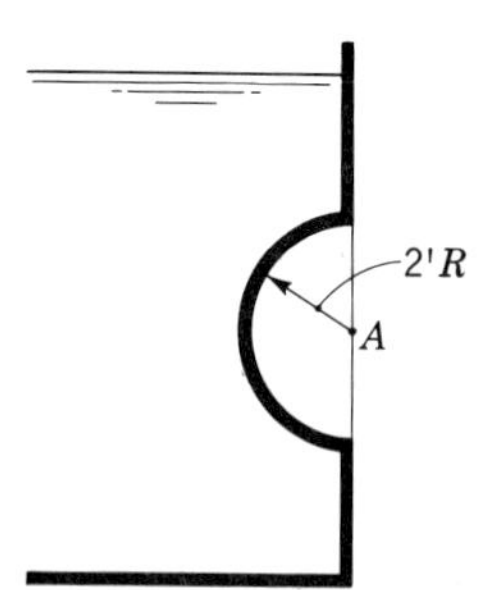

Problem 2.63

2.64 Determine the force required to hold the cone shown in the figure in position.

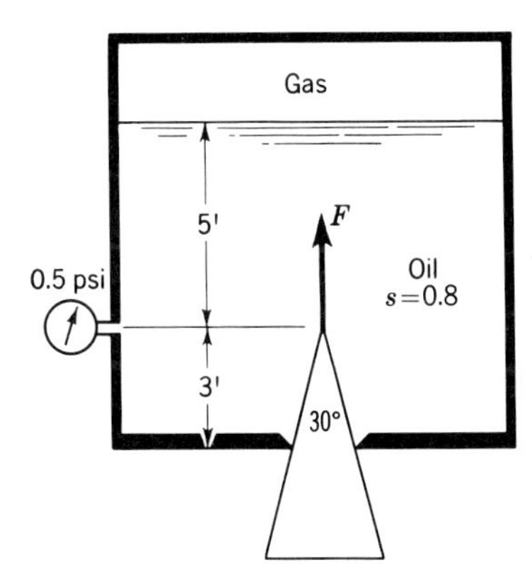

Problem 2.64

2.65 The cross section of a gate is shown in the figure. Its dimension normal to the plane of the paper is 10 m, and its shape is such that $x = 0.2y^2$. The gate is pivoted about O. Develop analytic expressions in terms of the water depth y upstream of the gate for the following: (a) horizontal force; (b) vertical force; (c) clockwise moment acting on the gate. Compute (a), (b), and (c) for the case where the water depth is 2 m.

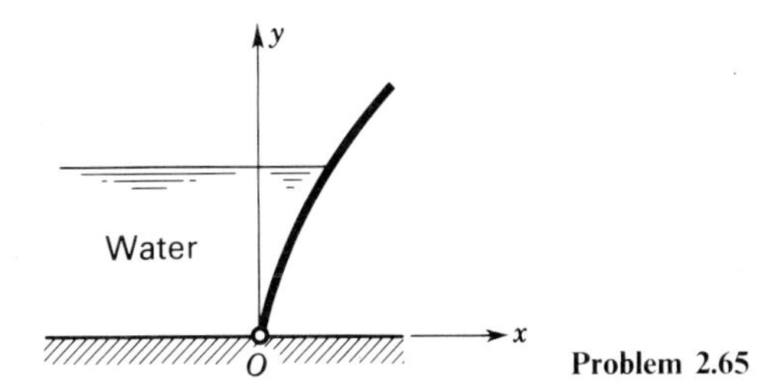

Problem 2.65

2.66 An iceberg in the ocean floats with one-seventh of its volume above the surface. What is its specific gravity relative to ocean water? What portion of its volume would be above the surface if ice were floating in pure water?

2.67 An hydrometer, Fig. 2.20a, consists of an 8-mm-diameter cylinder of length 20 cm attached to a 25-mm-diameter weighted sphere. The cylinder has a mass of 1.2 g and the mass of the sphere is 12.8 g. At what level will this device float in liquids having specific gravities 0.8, 1.0, and 1.2? Is the scale spacing on the cylindrical stem uniform? Why or why not?

2.68 Find the approximate value of the maximum specific gravity of liquid for which the device of Prob. 2.67 will be stable.

2.69 Determine the volume of an object that weighs 5 lb in water and 7 lb in oil ($s = 0.82$). What is the specific weight of the object?

2.70 A balloon weighs 250 lb and has a volume of 14,000 ft^3. It is filled with helium, which weighs 0.0112 lb/ft^3 at the temperature and pressure of the air, which weighs 0.0807 lb/ft^3. What load will the balloon support, or what force in a cable would be required to keep it from rising?

2.71 For the conditions shown in the figure, find the force F required to lift the concrete-block gate if the concrete weighs 150 lb/ft^3.

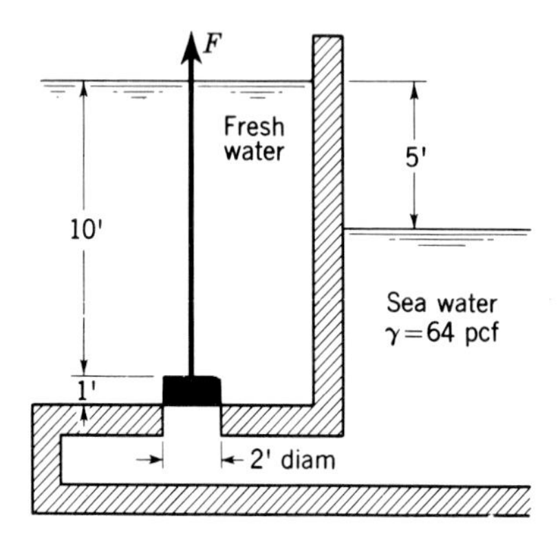

Problem 2.71

2.72 A cylindrical bucket 30 cm in diameter and 50 cm high weighing 25.0 N contains oil ($s = 0.80$) to a depth of 20 cm. (*a*) When placed in water, what will be the depth to the bottom of the bucket? (*b*) What is the maximum volume of oil the bucket can hold and still float?

2.73 A metal block 1 ft square and 10 in deep is floated on a body of liquid which consists of an 8-in layer of water above a layer of mercury. The block weighs 120 lb/ft^3. What is the position of the bottom of the block? If a downward vertical force of 250 lb is applied to the center of this block, what is the new position of the bottom of the block? Assume that the tank containing the fluid is of infinite dimensions.

2.74 Two spheres, each 1.2 m in diameter, weigh 4 and 12 kN respectively. They are connected with a short rope and placed in water. What is the tension in the rope and what portion of the lighter sphere protrudes from the water?

2.75 In Prob. 2.74 what should be the weight of the heavier sphere in order for the lighter sphere to float halfway out of the water? Assume sphere volumes remain constant.

2.76 A 4.0-ft^3 object weighing 500 lb is attached to a balloon of negligible weight and released in the ocean. The balloon was originally inflated with 2.0 lb of air to a pressure of 20 psi. To what depth will the balloon sink? Assume that air temperature within the balloon stays constant at 50°F.

2.77 Work Prob. 2.76 with all data the same except assume the balloon was originally inflated with 2.0 lb of air to a pressure of 10 psi. In this latter case the balloon is more elastic because a lower pressure is obtained with the same amount of air.

2.78 A wooden pole weighing 2 lb/ft has a cross-sectional area of 7 in^2 and is supported as shown in the figure. The hinge is frictionless. Find θ.

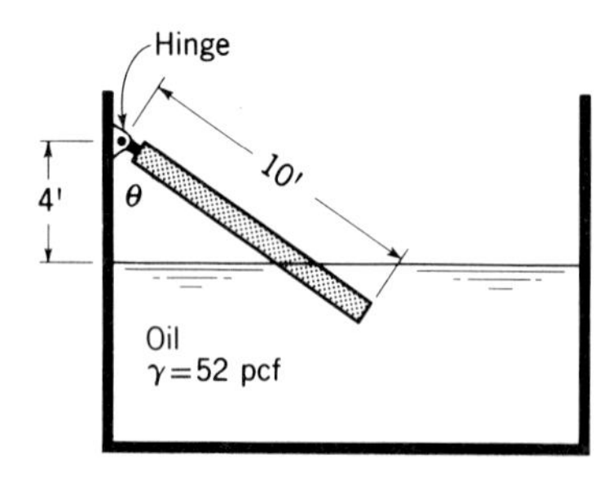

Problem 2.78

2.79 A rectangular block of uniform material and length 100 cm, width 20 cm, and depth 4 cm is floating in a liquid. It assumes the configuration shown when the uniform vertical load of 15 N/m is applied. (*a*) Find the weight of the block and the specific gravity of the liquid. (*b*) If the load is suddenly removed, what is the righting moment before the block starts to move?

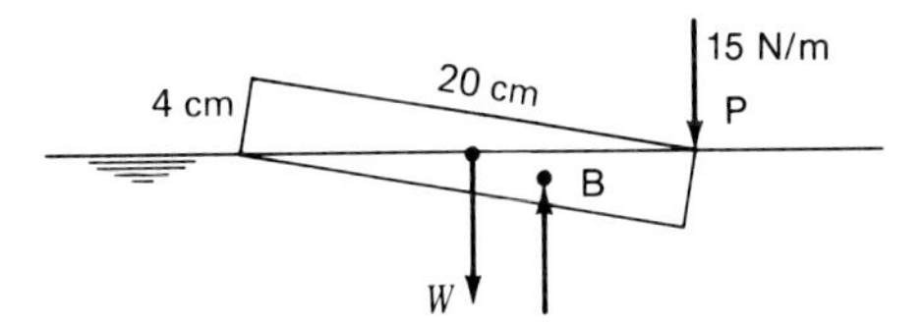

Problem 2.79

2.80 What would be the hydrostatic pressure at a depth of 9 in (25 cm) in a bucket of oil ($s = 0.82$) that is in an elevator being accelerated upward at 10 ft/s^2 (3.0 m/s^2)?

2.81 A tank containing water to a depth of 6 ft (2 m) is accelerated upward at 10 ft/s^2 (3 m/s^2). Calculate the pressure on the bottom of the tank.

2.82 Suppose the tank shown in Fig. 2.22 is rectangular and completely open at the top. It is 20 ft long, 4 ft wide, and 8 ft deep. If it is initially filled to the top, how much liquid will be spilled if it is given a horizontal acceleration $a_x = 0.3g$ in the direction of its length?

2.83 Suppose the tank of Fig. 2.22 is rectangular and completely open at the top. It is 20 m long, 4 m wide, and 3 m deep. If it is initially filled to the top, how much liquid will be spilled if it is given a horizontal acceleration $a_x = 0.3g$ in the direction of its length?

2.84 If the tank of Prob. 2.82 (2.83) is closed at the top and is completely filled, what will be the pressure difference between the left-hand end at the top and the right-hand end at the top if the liquid has a specific weight of 50 lb/ft^3 (8.0 kN/m^3) and the horizontal acceleration is $a_x = 0.3g$? Sketch planes of equal pressure, indicating their magnitude; assume zero pressure in upper right-hand corner.

2.85 At a particular instant an airplane is traveling upward at a velocity of 150 mph in a direction that makes an angle of 30° with the horizontal. At this instant the airplane is losing speed at the rate of 3 mph/s. Also, it is moving on a concave-upward circular path having a radius of 4,000 ft. Determine for the given conditions the position of the free liquid surface in the fuel tank of this vehicle.

2.86 Refer to Illustrative Example 2.7. Suppose the velocity of the airplane is 260 m/s with all other data unchanged. What then would be the position of the liquid surface in the tank?

2.87 A bubble 4 in below the water surface contains 2×10^{-7} lb of air. If the temperature is 60°F and the barometric pressure is 14.7 psia, calculate the diameter of the bubble. Refer to Secs. 1.8 and 1.12, and ignore the partial pressure of water vapor inside the bubble.

CHAPTER

THREE

KINEMATICS OF FLUID FLOW

When speaking of fluid flow, one often refers to the flow of an *ideal fluid.* Such a fluid is presumed to have no viscosity. This is an idealized situation which does not exist; however, there are instances in engineering problems where the assumption of an ideal fluid is helpful. When referring to the flow of a *real fluid*, the effects of viscosity are introduced into the problem. This results in the development of shear stresses between neighboring fluid particles when they are moving at different velocities. In the case of an ideal fluid flowing in a straight conduit, all particles move in parallel lines with equal velocity (Fig. 3.1*a*). In the flow of a real fluid the velocity adjacent to the wall will be zero; it will increase rapidly within a short distance from the wall and produce a velocity profile such as shown in Fig. (3.1*b*).

Flow may also be classified as that of an *incompressible* or *compressible* fluid. Since liquids are relatively incompressible, they are generally treated as wholly incompressible fluids. Under particular conditions where there is little pressure variation, the flow of gases may also be considered incompressible, though generally the effects of the compressibility of the gas should be considered. Some of the basic concepts governing the flow of compressible fluids are discussed in Chap. 9.

In addition to the flow of different types of fluids, i.e., real, ideal, incompressible, and compressible, there are various classifications of flow. Flow may be *steady* or *unsteady* with respect to time. It may be *laminar* or *turbulent*, as discussed in the following section. Other classifications of flow include *rotational* or *irrotational* (Chap. 5), *supercritical* or *subcritical* (Chap. 11), etc. In Table 3.1 are listed the most common ways in which flow can be classified.

Table 3.1 Classification of types of flow*

One-dimensional, two-dimensional or three-dimensional flow
See Sec. 3.6 for discussion.

Real fluid flow or ideal fluid flow (also referred to as viscid and inviscid flow)
Real fluid flow implies frictional (viscous) effects. Ideal fluid flow is hypothetical; it assumes no friction (i.e., viscosity of fluid = 0).

Incompressible fluid flow or compressible fluid flow
Incompressible fluid flow assumes the fluid has constant density (ρ = constant). Though liquids are slightly compressible they are usually assumed to be incompressible. Gases are compressible; their density is a function of absolute pressure and absolute temperature [$\rho = f(p, T)$].

Steady or unsteady flow
Steady flow means steady with respect to time. Thus the flow at every point remains constant with respect to time. In unsteady flow the flow changes with time.

Pressure flow or gravity flow
Pressure flow implies that flow occurs under pressure. Gases always flow in this manner. When a liquid flows with a free surface (for example, a partly full pipe) the flow is referred to as gravity flow because gravity is the primary moving force. Liquids also flow under pressure (for example, a pipe flowing full).

Spatially constant or spatially variable flow
Spatially constant flow occurs when the fluid density and the local average flow velocity are identical at all points in a flow field. If these quantities change along or across the flow lines, the flow is spatially variable. Examples of different types of spatially varied flow include: the local flow field around an object; flow through a gradual contraction in a pipeline; and the flow of water in a uniform gutter of constant slope receiving inflow over the length of the gutter.

Laminar or turbulent flow
See Sec. 3.1 for a discussion of the difference between these two types of flow.

Established or unestablished flow
This is discussed in Sec. 8.7.

Uniform or varied flow
This classification is ordinarily used when dealing with open-channel (gravity) flow (Chap. 11). In uniform flow the cross section (shape and area) through which the flow occurs remains constant.

Subcritical or supercritical flow
This classification is used with open-channel flow (Chap. 11).

Subsonic or supersonic flow
This classification is used with compressible flow (Chap. 9).

Rotational or irrotational flow
This is used in mathematical hydrodynamics (Chap. 5).

Other classifications of flow include converging or diverging, isothermal (constant temperature), adiabatic (no heat transfer) and isentropic (frictionless adiabatic).

* *Note:* In a given situation these different types of flow may occur in combination. For example, flow of a liquid in a pipe is usually considered to be one-dimensional, incompressible, real fluid flow that may be steady or unsteady, laminar or turbulent. Such flow is commonly spatially constant and established.

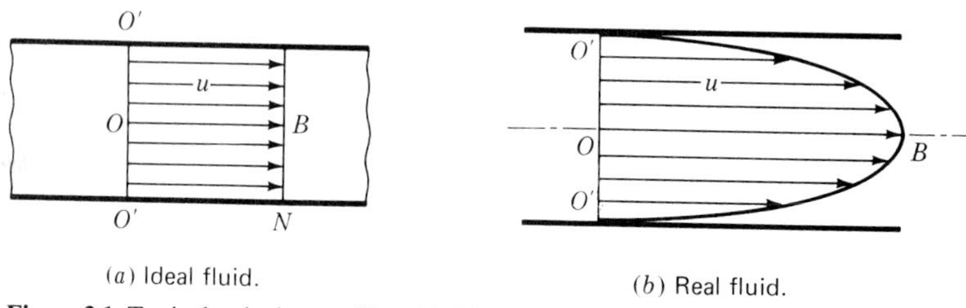

Figure 3.1 Typical velocity profiles. (*a*) Ideal fluid. (*b*) Real fluid.

3.1 LAMINAR AND TURBULENT FLOW

In this chapter we deal only with velocities and accelerations and their distribution in space without consideration of any forces involved. That there are two distinctly different types of fluid flow was demonstrated by Osborne Reynolds in 1883. He injected a fine, threadlike stream of colored liquid having the same density as water at the entrance to a large glass tube through which water was flowing from a tank. A valve at the discharge end permitted him to vary the flow. When the velocity in the tube was small, this colored liquid was visible as a straight line throughout the length of the tube, thus showing that the particles of water moved in parallel straight lines. As the velocity of the water was gradually increased by opening the valve further, there was a point at which the flow changed. The line would first become wavy, and then at a short distance from the entrance it would break into numerous vortices beyond which the color would be uniformly diffused so that no streamlines could be distinguished. Later observations have shown that in this latter type of flow the velocities are continuously subject to irregular fluctuations.

The first type is known as *laminar, streamline*, or *viscous* flow. The significance of these terms is that the fluid appears to move by the sliding of laminations of infinitesimal thickness relative to adjacent layers; that the particles move in definite and observable paths or streamlines, as in Fig. 3.2; and also that the flow is characteristic of a viscous fluid or is one in which viscosity plays a significant part (Fig. 1.4 and Sec. 1.11).

The second type is known as *turbulent* flow and is illustrated in Fig. 3.3, where (*a*) represents the irregular motion of a large number of particles during a

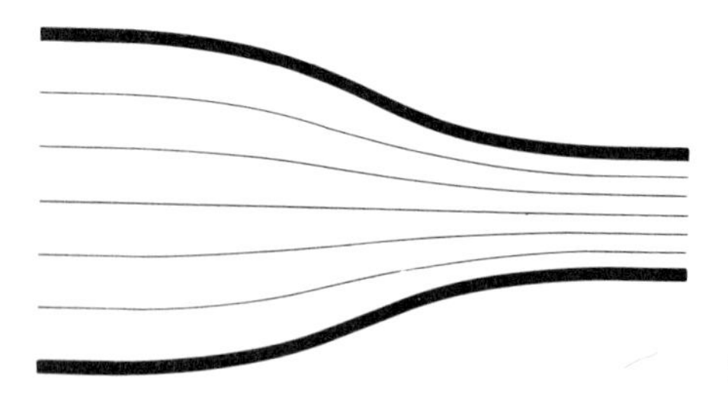

Figure 3.2

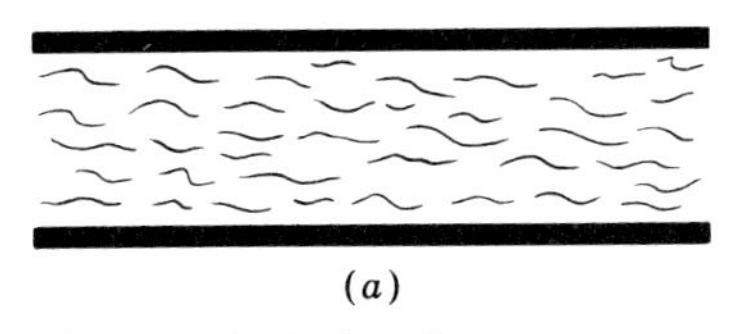
(a)

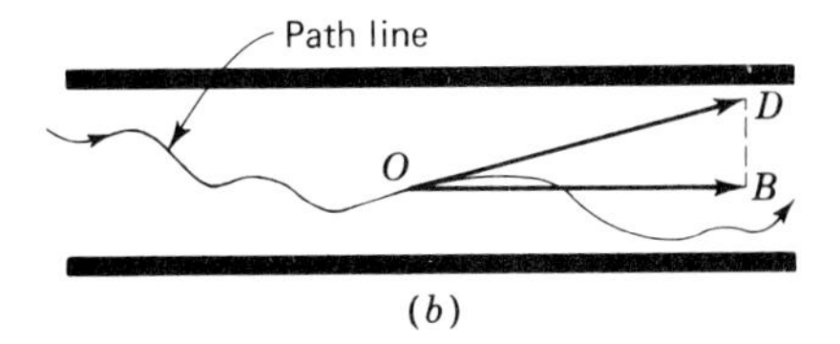

(b)

Figure 3.3 Turbulent flow.

very brief time interval, while (*b*) shows the erratic path followed by a single particle during a longer time interval. A distinguishing characteristic of turbulence is its irregularity, there being no definite frequency, as in wave action, and no observable pattern, as in the case of large eddies.

Large eddies and swirls and irregular movements of large bodies of fluid, which can be traced to obvious sources of disturbance, do not constitute turbulence, but may be described as *disturbed flow*. By contrast, turbulence may be found in what appears to be a very smoothly flowing stream and one in which there is no apparent source of disturbance. Turbulent flow is characterized by fluctuations in velocity at all points of the flow field (Figs. 3.6 and 8.6*b*). These fluctuations arise because of the presence of eddies in turbulent flow superimposed on the general flow pattern. These eddies interact with one another and with the general flow. They change shape and size with time as they move along with the flow. Each eddy dissipates its energy through viscous shear with its surroundings and eventually disappears. New eddies are continuously being formed. Large eddies (large-scale turbulence) have smaller eddies within them giving rise to small-scale turbulence. The fluctuations in velocity are rapid and irregular and often can be detected only by a fast-acting probe such as a hot-wire or hot-film anemometer (Sec. 12.4).

At a certain instant the flow passing point O in Fig. 3.3*b* may be moving with the velocity OD. In turbulent flow OD will vary continuously both in direction and in magnitude. Fluctuations of velocity are accompanied by fluctuations in pressure, which is the reason why manometers or pressure gages attached to a pipe in which fluid is flowing usually show pulsations. In this type of flow an individual particle will follow a very irregular and erratic path, and no two particles may have identical or even similar motions. Thus a rigid mathematical treatment of turbulent flow is impossible, and instead statistical means of evaluation must be employed.

Criteria governing the conditions under which the flow will be laminar and those under which it will be turbulent are discussed in Sec. 8.2.

3.2 STEADY FLOW AND UNIFORM FLOW

A *steady flow* is one in which all conditions at any point in a stream remain constant with respect to *time*, but the conditions may be different at different points. A truly *uniform flow* is one in which the velocity is the same in both

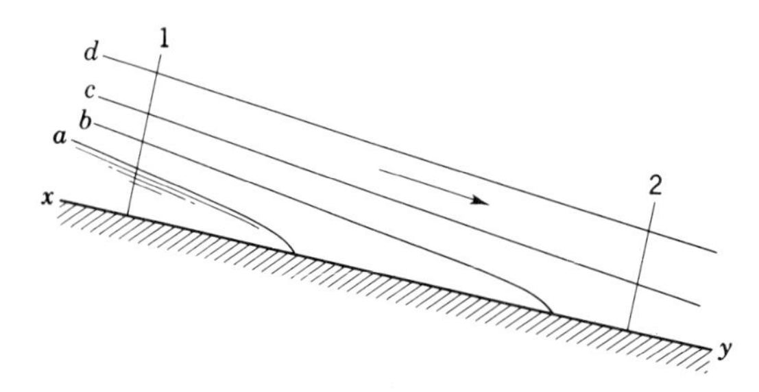

Figure 3.4 Unsteady flow in a canal.

magnitude and direction at a given instant at every point in the fluid. Both of these definitions must be modified somewhat, for true steady flow is found only in laminar flow. In turbulent flow there are continual fluctuations in velocity and pressure at every point, as has been explained. But if the values fluctuate equally on both sides of a constant average value, the flow is called steady flow. However, a more exact definition for this case would be *mean steady flow*.

Likewise, this strict definition of uniform flow can have little meaning for the flow of a real fluid where the velocity varies across a section, as in Fig. 3.1*b*. But when the size and shape of cross section are constant along the length of channel under consideration, the flow is said to be *uniform*.

Steady (or unsteady) and uniform (or nonuniform) flow can exist independently of each other, so that any of four combinations is possible. Thus the flow of liquid at a constant rate in a long straight pipe of constant diameter is *steady uniform* flow, the flow of liquid at a constant rate through a conical pipe is *steady nonuniform* flow, while at a changing rate of flow these cases become *unsteady uniform* and *unsteady nonuniform flow*, respectively.

Unsteady flow is a transient phenomenon which may in time become either steady flow or zero flow. An example may be seen in Fig. 3.4, where (*a*) denotes the surface of a stream that has just been admitted to the bed of a canal by the sudden opening of a gate. After a time the water surface will be at (*b*), later at (*c*), and finally reaches equilibrium at (*d*). The unstcady flow has then become mean steady flow. Another example of transient phenomenon is when a valve is closed at the discharge end of a pipeline (Sec. 13.6), thus causing the velocity in the pipe to decrease to zero. In the meantime there will be fluctuations in both velocity and pressure within the pipe.

Unsteady flow may also include periodic motion such as that of waves on beaches, tidal motion in estuaries, and other oscillations. The difference between such cases and that of mean steady flow is that the deviations from the mean are very much greater and the time scale is also much longer.

3.3 PATH LINES, STREAMLINES, AND STREAK LINES

A *path line* (Fig. 3.3*b*) is the trace made by a *single* particle over a *period* of time. If a camera were to take a time exposure of a flow in which a fluid particle was colored so it would register on the negative, the picture would show the course followed by the particle. This would be its path line. The path line shows the direction of the velocity of the particle at successive instants of time.

Streamlines show the mean direction of a *number* of particles at the *same* instant of time. If a camera were to take a very short time exposure of a flow in which there were a large number of particles, each particle would trace a short path, which would indicate its velocity during that brief interval. A series of curves drawn tangent to the means of the velocity vectors are streamlines.

Path lines and streamlines are identical in the steady flow of a fluid in which there are no fluctuating velocity components, in other words, for truly steady flow. Such flow may be either that of an ideal frictionless fluid or that of one so viscous and moving so slowly that no eddies are formed. This latter is the *laminar* type of flow, wherein the layers of fluid slide smoothly, one upon another. In turbulent flow, however, path lines and streamlines are not coincident, the path lines being very irregular while the streamlines are everywhere tangent to the local mean temporal velocity. The lines in Fig. 3.2 represent both path lines and streamlines if the flow is laminar; they represent only streamlines if the flow is turbulent.

In experimental fluid mechanics, a dye or other tracer is frequently injected into the flow to trace the motion of the fluid particles. If the flow is laminar, a ribbon of color results. This is called a *streak line*, or *filament line*. It is an instantaneous picture of the positions of all particles in the flow which have passed through a given point (namely, the point of injection). In utilizing fluid-tracer techniques it is important to choose a tracer with physical characteristics (especially density) the same as those of the fluid being observed. Thus the smoke rising from a cigarette, while giving the appearance of a streak line, does not properly represent the movement of the ambient air in the room because it is less dense than the air and therefore rises more rapidly.

3.4 FLOW RATE AND MEAN VELOCITY

The quantity of fluid flowing per unit time across any section is called the *flow rate*. It may be expressed in terms of volume flow rate using English units such as cubic feet per second (cfs), gallons per minute (gpm), million gallons per day (mgd), or in terms of weight flow rate (pounds per second), or mass flow rate (slugs per second). In SI units cubic meters per second (m^3/s), kilonewtons per second (kN/s), and kilograms per second (kg/s) are fairly standard for expressing volume, weight, and mass flow rate respectively. In dealing with incompressible fluids, volume flow rate is commonly used, whereas weight flow rate or mass flow rate is more convenient with compressible fluids.

Figure 3.5 presents a streamline in steady flow lying in the xz plane. Element of area dA lies in the yz plane. The mean velocity at point P is u. The volume flow rate passing through the element of area dA is

$$dQ = \mathbf{u} \cdot d\mathbf{A} = (u \cos \theta)\, dA = u(\cos \theta\, dA) = u\, dA' \tag{3.1}$$

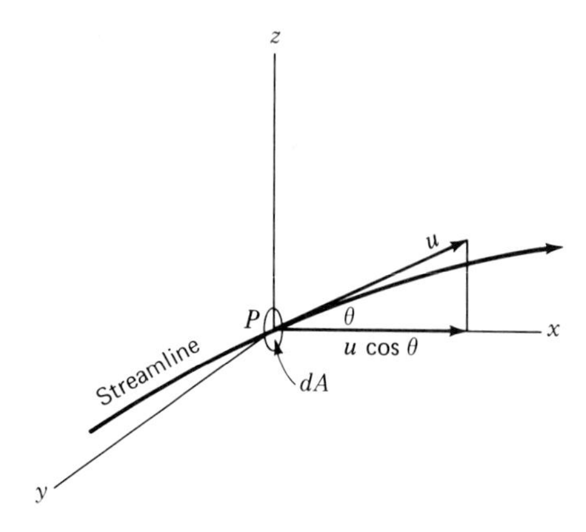

Figure 3.5

where dA' is the projection of dA on the plane normal to the direction of u. This indicates that the *volume flow rate is equal to the magnitude of the mean velocity multiplied by the flow area at right angles to the direction of the mean velocity*. The mass flow rate and the weight flow rate may be computed by multiplying the volume flow rate by the density and specific weight of the fluid respectively.

If the flow is turbulent, the *instantaneous velocity component* u'' along the streamline will fluctuate with time, even though the flow is nominally steady. A plot of u'' as a function of time is shown in Fig. 3.6. The average value of u'' over a period of time determines the temporal mean value of velocity u at point P.

The difference between u'' and u, which may be designated as u', is called the *turbulent fluctuation* of this component; it may be either positive or negative. The temporal mean value of u' must be zero, as must the temporal means of all components transverse to the channel, such as BD in Fig. 3.3. Thus at any instant

$$u'' = u + u' \tag{3.2}$$

and u may be evaluated for any finite time t as $u = (1/t) \int_0^t u'' \, dt$.

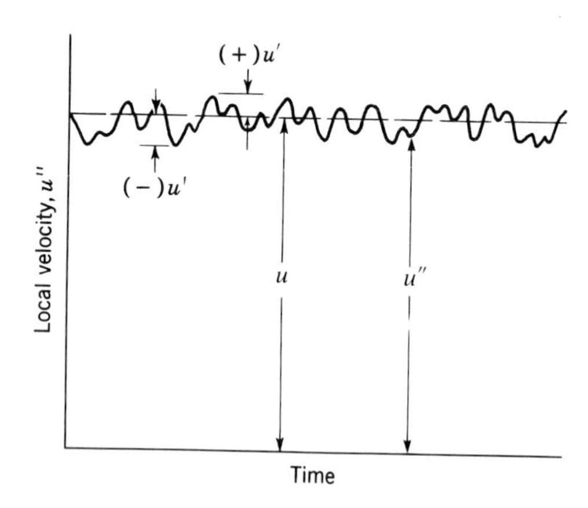

Figure 3.6 Fluctuating velocity at a point.

In a real fluid the velocity u will vary across the section in some manner, such as that shown in Fig. 3.1b, and, hence, the flow rate may be expressed as

$$Q = \int_A u\,dA = AV \tag{3.3}$$

or

$$G = \gamma \int_A u\,dA = \gamma AV \tag{3.4}$$

or

$$M = \rho \int_A u\,dA = \rho AV \tag{3.5}$$

where u is the temporal mean velocity through an infinitesimal area dA, while V is the *mean*, or average, velocity over the entire sectional area A;[1] Q is the volume flow rate (cfs or m^3/s), G is the weight flow rate (lb/s or kN/s), and M is the mass flow rate (slugs/s or kg/s).[2] If u is known as a function of A, the foregoing may be integrated. If only average values of V are known for different finite areas into which the total area may be divided, then

$$Q = A_a V_a + A_b V_b + \cdots + A_n V_n = AV \tag{3.6}$$

Similar expressions may be written for G and M. If the flow rate has been determined directly by some method, the mean velocity may be found by

$$V = \frac{Q}{A} = \frac{G}{\gamma A} = \frac{M}{\rho A} \tag{3.7}$$

3.5 EQUATION OF CONTINUITY

Figure 3.7 presents a short length of a *stream tube*, which may be assumed, for practical purposes, as a bundle of streamlines. Since the stream tube is bounded on all sides by streamlines and since there can be no net velocity normal to a streamline, no fluid can leave or enter the stream tube except at the ends. The

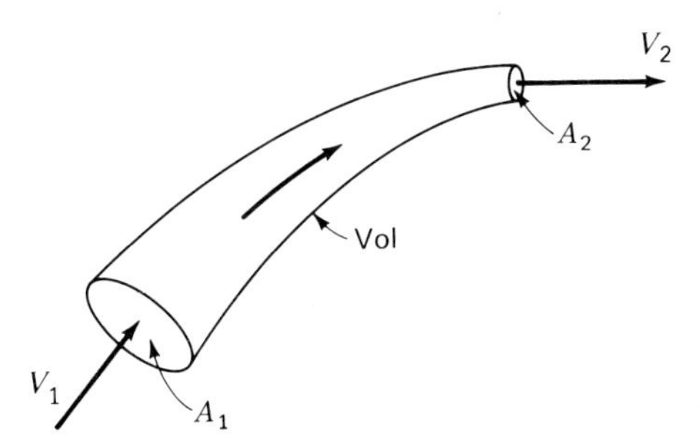

Figure 3.7 Length of stream tube as control volume.

[1] Note that area A is defined by the surface at right angles to the velocity vectors.

[2] In Eqs. (3.4) and (3.5) the γ and ρ should be to the right of the integral sign if the density of the fluid varies across the flow.

fixed volume between the two fixed sections of the stream tube is known as the *control volume* and its magnitude will be designated by vol. According to newtonian physics (i.e., disregarding the possibility of converting mass to energy), mass must be conserved. If the mass of the fluid contained in the control volume of volume (vol) at time t is mass_t, then the mass of fluid contained in vol at time $t + dt$ would be:

$$\text{mass}_{t+dt} = \text{mass}_t + \underset{\text{(inflow)}}{\rho_1 V_1 A_1\, dt} - \underset{\text{(outflow)}}{\rho_2 V_2 A_2\, dt}$$

But, the mass contained in vol at $t + dt$ can also be expressed as

$$\text{mass}_{t+dt} = \text{mass}_t + (\text{vol}) \frac{\partial \rho}{\partial t}\, dt$$

where $\partial \rho / \partial t$ is the time rate of change of the mean density of the fluid in vol. Equating these two expressions for mass_{t+dt} yields

$$\rho_1 A_1 V_1 - \rho_2 A_2 V_2 = (\text{vol}) \frac{\partial \rho}{\partial t} \tag{3.8}$$

This is the general equation of continuity for flow through regions with fixed boundaries. It states that the net rate of mass inflow to the control volume is equal to the rate of increase of mass within the control volume. This equation can be reduced to more useful forms.

For steady flow, $\partial \rho / \partial t = 0$ and

$$\rho_1 A_1 V_1 = \rho_2 A_2 V_2 = M \tag{3.9a}$$

or

$$\gamma_1 A_1 V_1 = \gamma_2 A_2 V_2 = G \tag{3.9b}$$

These are the continuity equations that apply to steady, compressible or incompressible flow within fixed boundaries.

If the fluid is incompressible, ρ = constant; hence $\rho_1 = \rho_2$ and $\partial \rho / \partial t = 0$ and thus

$$A_1 V_1 = A_2 V_2 = Q \tag{3.10}$$

This is the continuity equation that applies to incompressible fluids for both steady and unsteady flow within fixed boundaries.[1]

Equations (3.9) and (3.10) are generally adequate for the analysis of flows in conduits with solid boundaries, but for the consideration of flow in space, as that of air around an airplane, for example, it is desirable to express the continuity equation in another form, as indicated in Sec. 5.1. Or, for the case of unsteady flow of a liquid in a canal (Fig. 3.4), the principle of conservation of mass

[1] The continuity equations [Eqs. (3.9) and (3.10)] are applicable to any stream tube in a flow system. Most commonly the continuity equation is applied to the stream tube that is coincident with the boundaries of the flow.

indicates that the rate of flow past section 1 minus the rate of flow past section 2 is equal to the time rate of change of storage volume between the two sections. Thus,

$$Q_1 - Q_2 = dS/dt \tag{3.11}$$

where S is the volume of liquid contained in the canal between the two sections.

3.6 ONE-, TWO-, AND THREE-DIMENSIONAL FLOW

In true one-dimensional flow the velocity has at all points the same direction and (for an incompressible fluid) the same magnitude. Such a case is rarely of practical interest. However, the term *one-dimensional method of analysis* is applied to the flow between boundaries which are really three-dimensional, with the understanding that the "one dimension" is taken *along the central streamline* of the flow. Average values of velocity, pressure, and elevation across a section normal to this streamline are considered typical of the flow as a whole. Thus the equation of continuity in Sec. 3.5 is called the one-dimensional equation of continuity, even though it may be applied to flow in conduits which curve in space and in which the velocity varies across any section normal to the flow. It will be of increasing importance in the following chapters to recognize that, when high accuracy is desired, the equations derived by the one-dimensional method of analysis require refinement to account for the variation in conditions across the flow section.

If the flow is such that all streamlines are plane curves and are identical in a series of parallel planes, it is said to be two-dimensional. In Fig. 3.8*a* the channel has a constant dimension perpendicular to the plane of the figure. Thus every cross section normal to the flow must be a rectangle of this constant width. Three-dimensional flow is illustrated in Fig. 3.8*b*, although in this particular case the flow is axially symmetric, which simplifies the analysis. A generalized three-dimensional flow, such as the flow of cool air from an air conditioning outlet into a room, is quite difficult to analyze. Such flows are often approximated as two-dimensional or as axially symmetric flow. This offers an advantage in that it is easier to draw diagrams describing the flow and the mathematical treatment is much simpler.

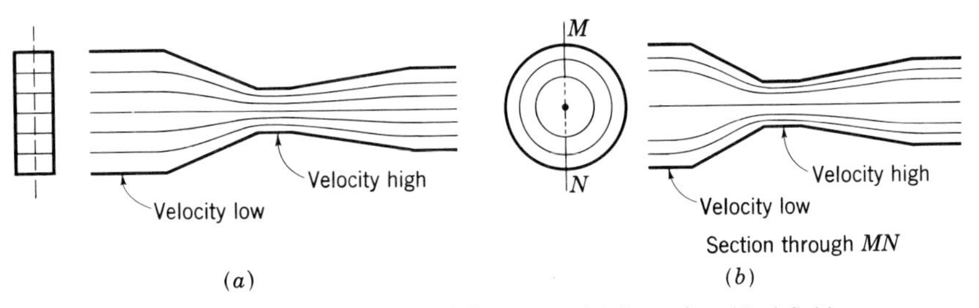

Figure 3.8 Two- and three-dimensional (axially symmetric) flow of an ideal fluid.

3.7 THE FLOW NET

The streamlines and velocity distribution in the case of steady two-dimensional flow of an *ideal* fluid within any boundary configuration may be determined by a *flow net*, such as is shown in Fig. 3.9. This is a network of streamlines and lines normal to them so spaced that the distances between both sets of lines are inversely proportional to the local velocities. The streamlines show the mean direction of flow at any point. A fundamental property of the flow net is that it provides the one and only representation of the ideal flow within the given boundaries. It is also independent of the actual magnitude of the flow and, for the *ideal* fluid, is the same whether the flow is in one direction or the reverse.

In a number of simple cases it is possible to obtain mathematical expressions, known as *stream functions* (Sec. 5.4), from which one can plot streamlines. But even the most complex cases can be solved by plotting a flow net by a trial-and-error method. Although it is possible to construct nets for three-dimensional flow, treatment here will be restricted to the simpler two-dimensional net, which will more clearly illustrate the method. Consider the two-dimensional stream tube of Fig. 3.10. Assuming a constant unit thickness perpendicular to the paper, the continuity equation gives $V_1 \, \Delta n_1 = V_2 \, \Delta n_2$.

Consider next a region of uniform flow divided into a number of strips of equal width, separated by streamlines, as in Fig. 3.8*a*. Each strip represents a stream tube, and the flow is equally divided among the tubes. As the flow approaches a bend or obstruction, the streamlines must curve so as to conform to the boundaries, but each stream tube still carries the same flow. Thus the spacing

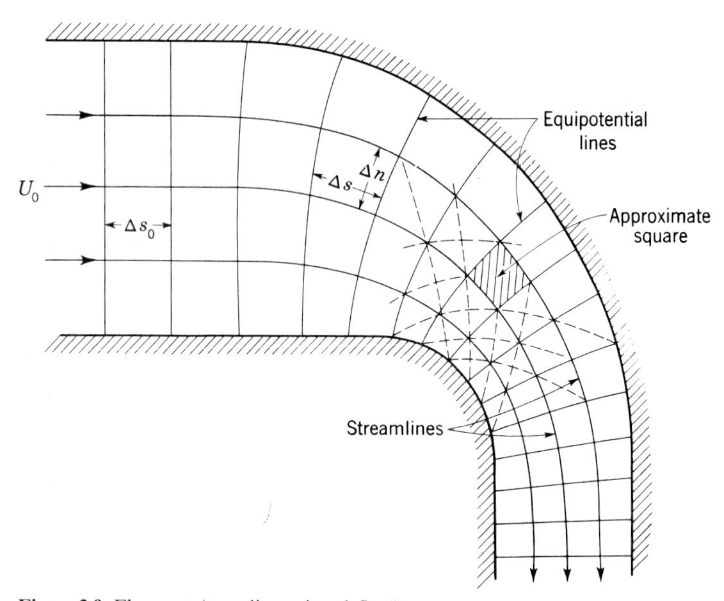

Figure 3.9 Flow net (two-dimensional flow).

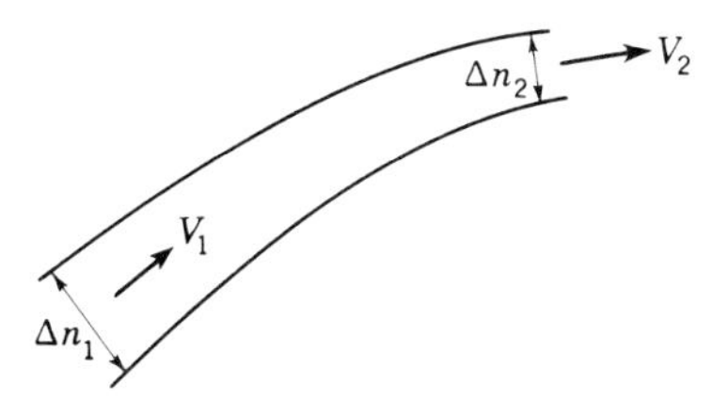

Figure 3.10

between all streamlines in the entire field is everywhere inversely proportional to the local velocities, so that, for any section normal to the velocity,

$$V\,\Delta n = \text{constant} \tag{3.12}$$

To draw the streamlines, it is necessary to start by estimating not only the spacing between them but also their directions at all points. As an aid in the latter we make use of normal, or *equipotential*, lines. As an analogy consider the flow of heat through a homogeneous material enclosed between perfectly insulated boundaries. The heat might be considered to flow along the equivalent of streamlines. As there can be no flow of heat along a line of constant temperature, it follows that the heat flow must be everywhere perpendicular to isothermal lines. In like manner streamlines must be everywhere perpendicular to equipotential lines. Because solid boundaries, across which there can be no flow, also represent streamlines, it follows that *equipotential lines must meet the boundaries everywhere at right angles.*

If the equipotential lines are spaced the same distance apart as the streamlines in the region of uniform two-dimensional flow, the net for that region is composed of perfect squares. In a region of deformed flow the quadrilaterals cannot remain square, but they will approach squares as the number of streamlines and equipotential lines are increased indefinitely. It is frequently helpful, in regions where the deformation is marked, to introduce extra streamlines and equipotential lines spaced midway between the original ones.

In drawing a flow net the beginner will make considerable use of the eraser, but with some practice a net can be sketched with fair facility to represent any boundary configuration. It is even possible to construct an approximate flow net for cases where one solid boundary does not exist and the fluid extends laterally indefinitely, as in the flow around an immersed object. Such a case reveals an advantage of the flow net that is not evident from Fig. 3.9. In the flow between confining solid boundaries it is always possible to determine the mean velocity across any section by dividing the total flow by the section area. For flow around an immersed object, as in Fig. 3.11, there is no fixed area by which to divide a definite flow, but the flow net affords a means of determining the velocity in the region of such an object.

Where a channel is curved, the equipotential lines must diverge inasmuch as they radiate from centers of curvature. The distance between the associated streamlines must vary in the same way as that between the equipotential lines. Therefore, as in Fig. 3.9, the areas are smallest along the inner radius of the bend and increase toward the outside.

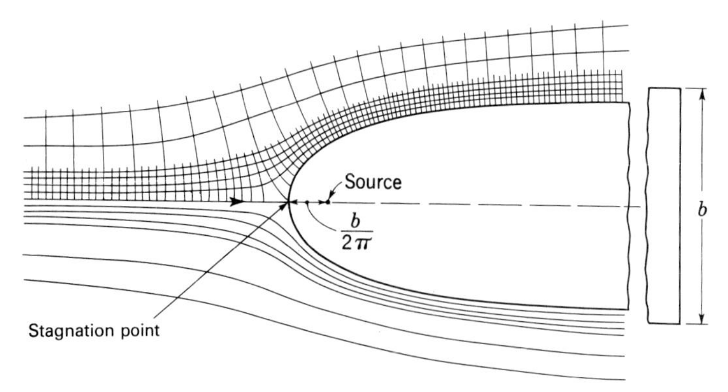

Figure 3.11 Two-dimensional flow of a frictionlesss fluid past a solid whose surface is perpendicular to the plane of the paper. Streamlines or path lines for steady flow.

The accuracy of the final flow net can be checked by drawing diagonals, as indicated by a few dashed lines in Fig. 3.9. If the net is correct, these dashed lines also form a network of lines that cross each other at right angles and produce areas that approach squares in shape.

3.8 USE AND LIMITATIONS OF FLOW NET

Although the flow net is based on an ideal frictionless fluid, it may be applied to the flow of a real fluid within certain limits. Such limits are dictated by the extent to which the real fluid is affected by factors which the ideal-fluid theory neglects, the principal one of which is fluid friction.

The viscosity effects of a real fluid are most pronounced at or near a solid boundary and diminish rapidly with distance from the boundary. Hence, for an airplane or a submerged submarine, the fluid may be considered as frictionless, except when very close to the object. The flow net always indicates a velocity next to a solid boundary, whereas a real fluid must have zero velocity adjacent to a wall. The region in which the velocity is so distorted, however, is confined to a relatively thin layer called the *boundary layer*, outside of which the real fluid behaves very much like the ideal fluid.

The effect of the boundary friction is minimized when the streamlines are converging, but in a diverging flow there is a tendency for the streamlines not to follow the boundaries if the rate of divergence is too great. In a sharply diverging flow, such as is shown schematically in Fig. 3.12, there may be a *separation* of the boundary layer from the wall,[1] resulting in eddies and even reverse flow in that region. The flow is badly disturbed in such a case, and the flow net may be of limited value.

[1] If the flow is laminar (Sec. 3.1), there will be no separation.

A practical application of the flow net may be seen in the flow around a body, as shown in Fig. 3.11, which may represent, for example, the upstream portion of a bridge pier at a distance below the surface where surface wave action is not a factor. Except for a thin layer adjacent to the body, this diagram represents the flow in front of and around the sides of the body. The central streamline is seen to branch at the forward tip of the body to form two streamlines along the walls. At the forward tip the velocity must be zero, hence this point is called a *stagnation point*.

Considering the limitations of the flow net in diverging flow, it may be seen that, while the flow net gives a fairly accurate picture of the velocity distribution in the region near the upstream part of any solid body, it may give little information concerning the flow conditions near the rear because of the possibility of separation and eddies. The disturbed flow to the rear of a body is known as a *turbulent wake* (Fig. 3.12*b*). The space occupied by the wake may be greatly diminished by streamlining the body, i.e., giving the body a long slender tail, which tapers to a sharp edge for two-dimensional flow or to a point for three-dimensional flow.

3.9 FRAME OF REFERENCE IN FLOW PROBLEMS

In flow problems we are really concerned only with the *relative* velocity between the fluid and the body. It makes no difference whether the body is at rest and the fluid flows past it or whether the fluid is a rest and the body moves through the fluid. There are thus two frames of reference. In one the observer (or the camera) is at rest with respect to the solid body. If the observer at rest with respect to a bridge pier views a steady flow past it or is on a ship moving at constant velocity

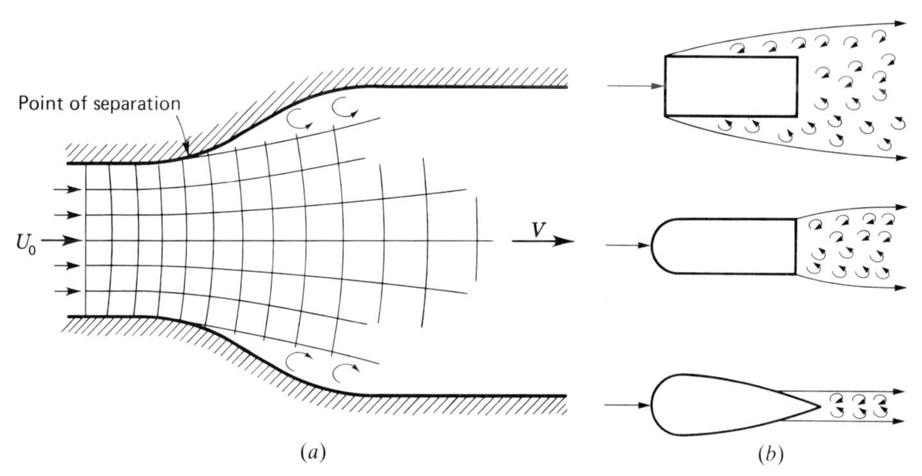

Figure 3.12 Separation in diverging flow. (*a*) Eddy formation in a diverging channel. (*b*) Turbulent wakes.

through still water, the streamlines appear to him to be unchanging and therefore the flow is steady. But if he floats with the current past the pier or views a ship going by while he stands on the bank, the flow pattern which he observes is changing with time. Thus the flow is unsteady.

The same flow may then be either steady or unsteady according to the frame of reference. The case that is usually of more practical importance is steady, ideal flow, for which the streamlines and path lines are identical. In unsteady flow streamlines and path lines are entirely different from each other and also bear no resemblance to those of steady flow.

Illustrative Example 3.1 An incompressible ideal fluid flows at 0.5 cfs through a circular pipe into a conically converging nozzle. Determine the average velocity of flow at sections A and B of the accompanying figure.

As a first step an approximate flow net is sketched to provide a general picture of the flow. *Since this is an axially symmetric flow, the net is not a true two-dimensional flow net.*

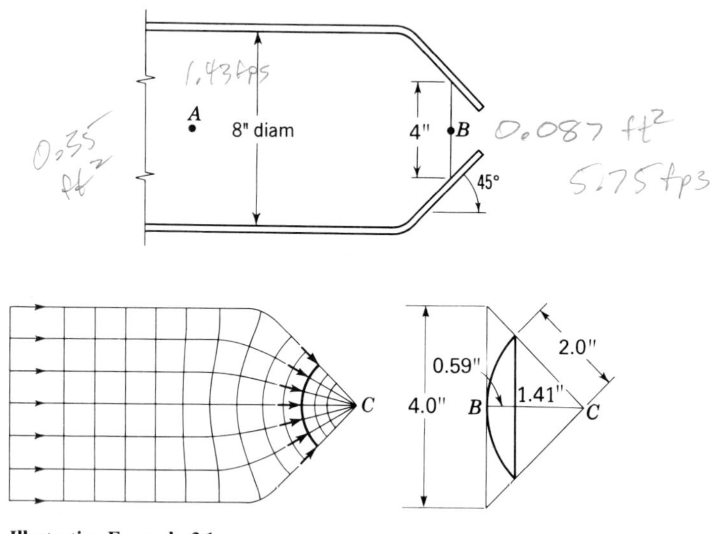

Illustrative Example 3.1

At section A the streamlines are parallel and hence the area at right angles to the velocity vectors is a plane circle. Thus

$$V_A = \frac{Q}{A_A} = \frac{0.5}{(\pi/4)(\frac{8}{12})^2} = 1.43 \text{ fps}$$

At section B, however, the area at right angles to the streamlines is not clearly defined; it is a curved, dish-shaped section. As a rough approximation it might be assumed to be the portion of the surface of a sphere of radius 2.0 in that is intersected by a circle of diameter 2.82 in. Thus

$$V_B = \frac{Q}{A_B} = \frac{Q}{2\pi Rh} = \frac{0.5}{[2\pi(2)(0.59)]/144} = 9.75 \text{ fps}$$

If the data for this example had been given in SI units, the solution would have been similar to the foregoing except the conversion from inches to feet would not have been necessary.

It should be noted that for the flow net shown in the sketch the velocity at C is infinite because the flow area at that point is zero. This, of course, cannot occur; in the real case a jet somewhat similar to that of Fig. 12.12 will form downstream of the nozzle opening.

3.10 VELOCITY AND ACCELERATION IN STEADY FLOW

In a typical three-dimensional flow field the velocities may be everywhere different in magnitude and direction. Also, the velocity at any point in the field may change with time. Let us first consider the case where the flow is steady and thus independent of time. If the velocity of a fluid particle has components u, v, and w parallel to the x, y, and z axes, then for steady flow,

$$u_{st} = u(x, y, z) \tag{3.13a}$$

$$v_{st} = v(x, y, z) \tag{3.13b}$$

$$w_{st} = w(x, y, z) \tag{3.13c}$$

Applying the chain rule of partial differentiation, the acceleration of the fluid particle for steady flow can be expressed as

$$\mathbf{a}_{st} = \frac{d}{dt}\mathbf{V}(x, y, z) = \frac{\partial \mathbf{V}}{\partial x}\frac{dx}{dt} + \frac{\partial \mathbf{V}}{\partial y}\frac{dy}{dt} + \frac{\partial \mathbf{V}}{\partial z}\frac{dz}{dt} \tag{3.14}$$

where $$|\mathbf{V}| = (u^2 + v^2 + w^2)^{1/2}$$

Noting that $dx/dt = u$, $dy/dt = v$, and $dz/dt = w$,

$$\mathbf{a}_{st} = u\frac{\partial \mathbf{V}}{\partial x} + v\frac{\partial \mathbf{V}}{\partial y} + w\frac{\partial \mathbf{V}}{\partial z} \tag{3.15}$$

This equation can be written as three scalar equations:

$$(a_x)_{st} = u\frac{\partial u}{\partial x} + v\frac{\partial u}{\partial y} + w\frac{\partial u}{\partial z} \tag{3.16a}$$

$$(a_y)_{st} = u\frac{\partial v}{\partial x} + v\frac{\partial v}{\partial y} + w\frac{\partial v}{\partial z} \tag{3.16b}$$

$$(a_z)_{st} = u\frac{\partial w}{\partial x} + v\frac{\partial w}{\partial y} + w\frac{\partial w}{\partial z} \tag{3.16c}$$

These equations show that even though the flow is steady, the fluid may possess an acceleration by virtue of a change in velocity with change in position. This type of acceleration is commonly referred to as *convective acceleration*. With incompressible fluid flow there is a convective acceleration wherever the effective flow area changes along the flow path. This is also true for compressible fluid flow, but, in addition, convective acceleration of a compressible fluid occurs wherever the density varies along the flow path irrespective of any changes in the effective flow area.

3.11 VELOCITY AND ACCELERATION IN UNSTEADY FLOW

If the flow is unsteady, then Eqs. (3.13*a*) to (3.13*c*) take the form

$$u = u(x, y, z, t)\cdots \tag{3.17}$$

Following a similar procedure to that of the preceding section results in the following set of scalar equations:

$$a_x = \left(u\frac{\partial u}{\partial x} + v\frac{\partial u}{\partial y} + w\frac{\partial u}{\partial z}\right) + \frac{\partial u}{\partial t} \tag{3.18a}$$

$$a_y = \left(u\frac{\partial v}{\partial x} + v\frac{\partial v}{\partial y} + w\frac{\partial v}{\partial z}\right) + \frac{\partial v}{\partial t} \tag{3.18b}$$

$$a_z = \left(u\frac{\partial w}{\partial x} + v\frac{\partial w}{\partial y} + w\frac{\partial w}{\partial z}\right) + \frac{\partial w}{\partial t} \tag{3.18c}$$

In the above set of equations the three terms in parentheses are recognized as the convective accelerations, while the $\partial u/\partial t$, $\partial v/\partial t$, and $\partial w/\partial t$ terms represent the acceleration caused by the unsteadiness of the flow. This type of acceleration is commonly referred to as the *local acceleration.*

In the case of uniform flow (streamlines parallel to one another) and with ρ = constant, the convective acceleration is zero and

$$\mathbf{a} = \frac{\partial \mathbf{V}}{\partial t} \tag{3.19}$$

At times it is helpful to superimpose the coordinate system on the instantaneous streamline pattern in such a fashion that the x axis is coincident with the streamline at a particular point of concern. In such a case the position s will indicate location along the instantaneous streamline. Thus, generally, $\mathbf{V} = \mathbf{V}(s, t)$ and the acceleration of the particle can be conveniently expressed as

$$\mathbf{a} = V\frac{\partial \mathbf{V}}{\partial s} + \frac{\partial \mathbf{V}}{\partial t} \tag{3.20}$$

In uniform flow with ρ = constant the first term of the above expression is zero, while in steady flow the second term becomes zero. In Eq. (3.20) **a** represents the acceleration of the fluid particle along the streamline. In the terminology of curvilinear motion this is referred to as the *tangential acceleration.* At this point in our discussion we should recall that a particle moving steadily along a curved path has a *normal acceleration* a_n toward the center of curvature of the path. From mechanics,

$$a_n = \frac{V^2}{r} \tag{3.21}$$

where r is the radius of the path. A particle moving on a curved path will always have a normal acceleration, though its tangential acceleration may be zero.

Illustrative Example 3.2 Refer to the figure of Illustrative Example 3.1. The flow is steady at 0.5 cfs. Find the acceleration in the flow at sections A and B.

$$\mathbf{a}_A = 0 \qquad \text{(because the flow is uniform at section } A \text{ and also steady)}$$

For any point in section B,

$$\mathbf{a}_B = u\frac{\partial \mathbf{V}}{\partial x} + v\frac{\partial \mathbf{V}}{\partial y}$$

For the point B on the axis of the pipe at section B,

$$\mathbf{a}_B = u\frac{\partial \mathbf{V}}{\partial x}$$

The effective area through which the flow is occurring in the converging section of the nozzle may be expressed approximately as $A = 2\pi hr$, where $h = r(1 - \cos 45°) = 0.293r$ and r is the distance from point C.

Thus $A = 2\pi(0.293r^2) = 1.84r^2$, and the velocity in the converging nozzle (assuming the streamlines flow radially toward C) may be expressed approximately as

$$V = \frac{Q}{A} = \frac{0.5}{1.84r^2}$$

At section B ($r = 2$ in $= 0.167$ ft)

$$V = \frac{0.5}{1.84r^2} = 9.75 \text{ fps}$$

and

$$\frac{\partial V}{\partial x} = -\frac{\partial V}{\partial r} = \frac{0.54}{r^3} = 118 \text{ fps/ft}$$

Thus

$$a_B = u\frac{\partial V}{\partial x} = 9.75(118) = 1{,}150 \text{ ft/s}^2 \quad \text{(convective acceleration)}$$

Illustrative Example 3.3 A two-dimensional flow field is given by $u = 2y$, $v = x$. Sketch the flow field. Derive a general expression for the velocity and acceleration (x and y are in units of length L; u and v are in units of L/T). Find the acceleration in the flow field at point A ($x = 3.5$, $y = 1.2$).

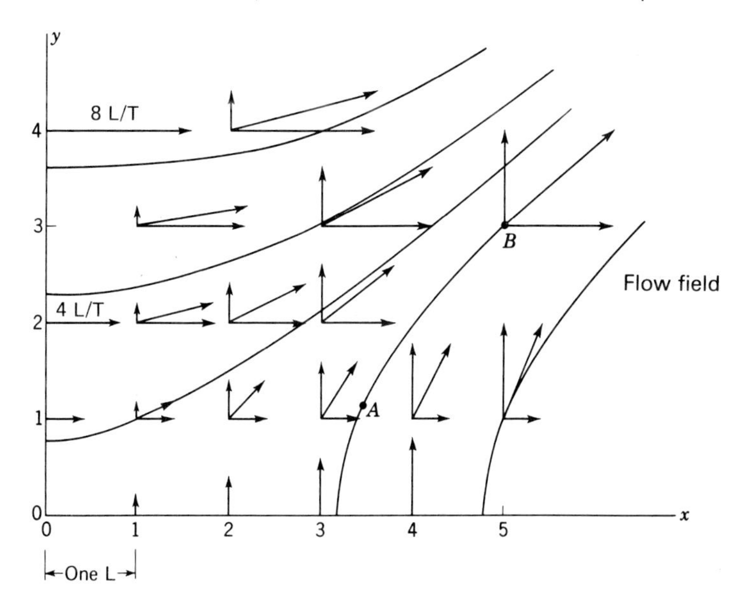

Velocity components u and v are plotted to scale, and streamlines are sketched tangentially to the resultant velocity vectors. This gives a general picture of the flow field.

$$V = (u^2 + v^2)^{1/2} = (4y^2 + x^2)^{1/2}$$

$$a_x = u\frac{\partial u}{\partial x} + v\frac{\partial u}{\partial y} = 2y(0) + x(2) = 2x$$

$$a_y = u\frac{\partial v}{\partial x} + v\frac{\partial v}{\partial y} = 2y(1) + x(0) = 2y$$

$$a = (a_x^2 + a_y^2)^{1/2} = (4x^2 + 4y^2)^{1/2}$$

$$(a_A)_x = 2x = 7.0L/T^2; (a_A)_y = 2y = 2.4L/T^2$$

$$a_A = [(a_A)_x^2 + (a_A)_y^2]^{1/2} = [(7.0)^2 + (2.4)^2]^{1/2} = 7.4L/T^2$$

To get a rough check on the acceleration imagine a velocity vector at point A. This vector would have a magnitude approximately midway between that of the adjoining vectors, or $V_A \approx 4L/T$. The radius of curvature of the sketched streamline at A is roughly $3L$. Thus $(a_A)_n \approx 4^2/3 \approx 5.3L/T^2$. The tangential acceleration of the particle at A may be approximated by noting that the velocity along the streamline increases from about $3.2L/T$, where it crosses the x axis, to about $8L/T$ at B. The distance along the streamline between these two points is roughly $4L$. Hence a very approximate value of the tangential acceleration at A is

$$(a_A)_t = V\frac{\partial V}{\partial s} \approx 4\left(\frac{8 - 3.2}{4}\right) \approx 4.8L/T^2$$

Vector diagrams of these roughly computed normal and tangential acceleration components are plotted for comparison with the true acceleration as given by the analytic expressions. It will be proved later, in Chap. 5, that the flow in this example must be that of a compressible fluid.

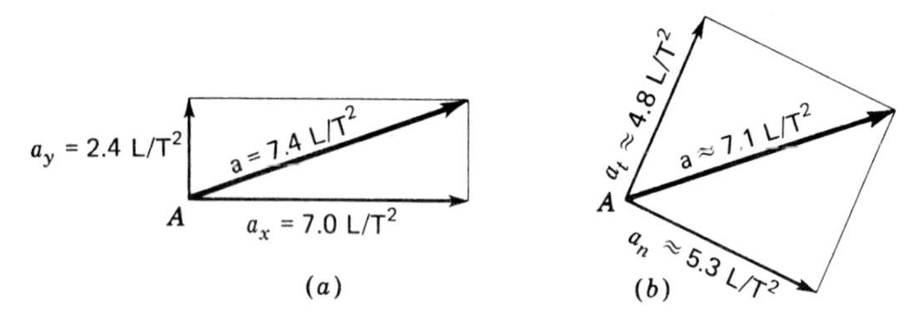

(*a*) True vector diagram of acceleration at A.
(*b*) Approximate vector diagram of acceleration at A.

PROBLEMS

3.1 Classify the following cases of flow as to whether they are steady or unsteady, uniform or nonuniform: (*a*) water flowing from a tilted pail; (*b*) flow from a rotating lawn sprinkler; (*c*) flow through the hose leading to the sprinkler; (*d*) natural stream during dry-weather flow; (*e*) natural stream during flood; (*f*) flow in a city water-distribution main in a straight section of constant diameter and no side connections. (*Note:* There is room for legitimate argument in some of the above cases, which should stimulate independent thought.)

3.2 In the laminar flow of a fluid in a circular pipe the velocity profile is exactly a true parabola. The rate of discharge is then represented by the volume of a paraboloid. Prove that for this case the ratio of the mean velocity to the maximum velocity is 0.5.

3.3 The velocities in a circular conduit 8 in (200 mm) in diameter were measured at radii 0, 1.44, 2.60, and 3.48 in (0, 36, 65, and 87 mm) and were found to be 20.3, 19.7, 17.7, and 14.5 fps (7.0, 6.8, 6.1, and 5.0 m/s) respectively. Find approximate values (graphically) of the volume flow rate and the mean velocity. Also determine the ratio of the mean velocity to the maximum velocity.

3.4 A gas ($\gamma = 0.04$ lb/ft^3) flows at the rate of 1.0 lb/s past section A through a long rectangular duct of uniform cross section 2 by 2 ft. At section B some distance along the duct the gas weighs 0.065 lb/ft^3. What is the average velocity of flow at sections A and B?

3.5 The velocity of a liquid ($s = 1.26$) in a 4-in (10-cm) pipe line is 1.5 fps (0.5 m/s). Calculate the rate of flow in cfs, gal/min, lb/s, and slugs/s. (For the metric data calculate the flow rate in L/s, m^3/s, kN/s, and kg/s.)

3.6 Oxygen flows in a 2-in by 2-in duct at a pressure of 40 psi and a temperature of 100°F. If the atmospheric pressure is 13.4 psia and the velocity of flow is 18 fps, calculate the weight flow rate.

3.7 Air at 100°F (40°C) and under a pressure of 40 psia (3,000 mbar, abs) flows in a 10-in (250-mm) diameter conduit at a mean velocity of 30 fps (10 m/s). Find the mass flow rate.

3.8 Water flows at 5 gal/min (300 cm^3/s) through a small circular hole in the bottom of a large tank. Assuming the water in the tank approaches the hole radially, what is the velocity in the tank at 2, 4, and 8 in (5, 10, and 20 cm) from the hole?

3.9 Gas flows at a steady rate in a 10-cm-diameter pipe that enlarges to a 15-cm-diameter pipe. At a certain section of the 10-cm pipe the density of the gas is 200 kg/m^3 and the velocity is 20 m/s. At a certain section of the 15-cm pipe the velocity is 14 m/s. What must be the density of the gas at that section? If these same data were given for the case of unsteady flow at a certain instant, could the problem have been solved? Discuss.

3.10 Gas is flowing in a long 6-in-diameter pipe from A to B. At section A the flow is 0.35 lb/s while at the same instant at section B the flow is 0.38 lb/s. The distance between A and B is 800 ft. Find the mean value of the time rate of change of the specific weight of the gas between sections A and B at that instant.

3.11 A compressible fluid flows in a 15-in-diameter leaky pipe. Measurements are made simultaneously at two points A and B along the pipe that are 40,000 ft apart. Two sets of measurements are taken with a span of exactly 30 min between them. The data are as follows:

Time	ρ_1(slug/ft^3)	u_1(ft/s)	ρ_2(slug/ft^3)	u_2(ft/s)
0	0.60	66	0.70	52
30 min	0.68	52	0.80	42

Assuming the flow rate is decreasing linearly with respect to time, compute the approximate average mass rate of leakage between A and B.

3.12 Water flows in a river. At 9 A.M. the flow past bridge 1 is 2,000 cfs (55 m^3/s). At the same instant the flow past bridge 2 is 1,600 cfs (45 m^3/s). At what rate is water being stored in the river between the two bridges at this instant? Assume zero seepage and negligible evaporation.

3.13 Make an approximate plot of the frictionless velocity along both the inner and the outer boundaries of Fig. 3.9. By what percent is the ideal maximum inner velocity greater than the ideal minimum outer velocity?

3.14 Consider the two-dimensional flow about a 2-in-diameter cylinder. Sketch the flow net for the ideal flow around one-quarter of the cylinder. Start with a uniform net of $\frac{1}{2}$-in squares, and fill in with $\frac{1}{4}$-in squares where desirable. (*Note:* The velocity at the stagnation point is zero, and it can be proved by classical hydrodynamics that the velocity tangent to the cylinder at a point 90° from the stagnation point is twice the uniform velocity.) Make a plot of the velocity along the center streamline from a point upstream where the velocity is uniform to the stagnation point, and then along the boundary of the cylinder from the stagnation point to the 90° point. Compare the result thus obtained with the value given by the equation $V = 2U_0 \sin\theta$, where U_0 is the undisturbed stream velocity and θ is the angle subtended by the arc from the stagnation point to any point on the cylinder where V is desired.

3.15 The figure shows the flow net for two-dimensional flow from a rounded, long-slotted exit from a tank. If a is 3 in and U_0 is 10 fps, approximately how long will it take a particle to move from point A to point B on the same streamline? (*Note:* Between each pair of equipotential lines, measure Δs, and then compute the average velocity and time increment.)

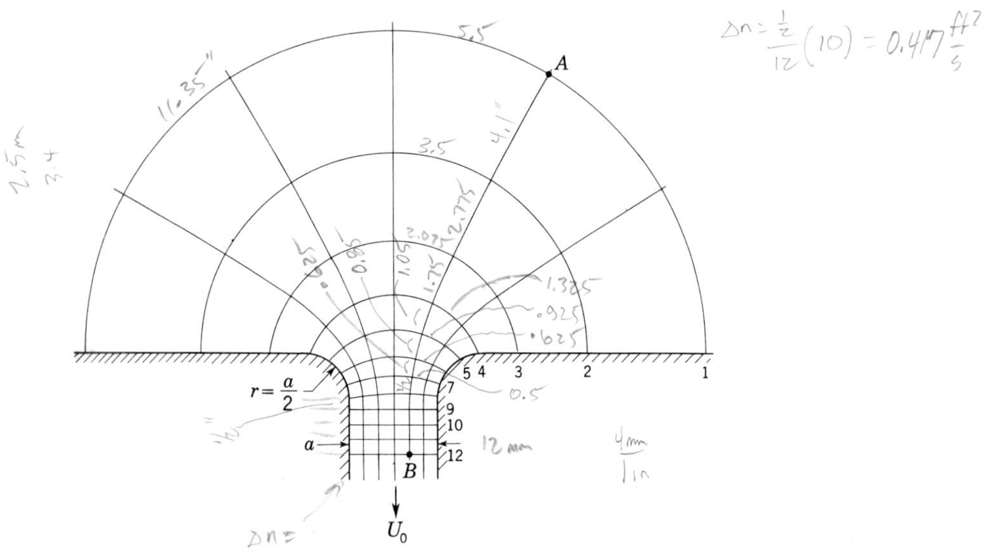

Problem 3.15

3.16 Repeat Prob. 3.15 using the following data: $a = 15$ cm and $U_0 = 0.5$ m/s. Find also the approximate velocity where the flow crosses equipotential line 3.

3.17 In Fig. 3.11 assume that $b = 5$ ft and the uniform velocity is 10 fps. Make a plot of the velocity along the boundary of the solid. By what percent does the maximum velocity along the boundary exceed the uniform velocity?

3.18 The velocity along a streamline lying on the x axis is given by $u = 10 + x^{1/2}$. What is the convective acceleration at $x = 3$? Specify units in terms of L and T. Assuming the fluid is incompressible, is the flow converging or diverging? Sketch to approximate scale the adjoining streamlines. If the fluid were compressible, what could one say about its density?

3.19 A flow field is defined by $u = 2$, $v = 3$, $w = 4$. What is the velocity of flow? Specify units in terms of L and T.

3.20 A flow is defined by $u = 2(1 + t)$, $v = 3(1 + t)$, $w = 4(1 + t)$. What is the velocity of flow at the point (3, 1, 4) at $t = 2$? What is the acceleration at that point at $t = 2$? Specify units in terms of L and T.

3.21 A two-dimensional flow field is given by $u = 3 + 2xy + 4t^2$, $v = xy^2 + 3t$. Find the velocity and acceleration of a particle of fluid at point (2, 1) at $t = 5$. Specify units in terms of L and T.

3.22 Sketch the flow field defined by $u = 3y$, $v = 2$, and derive expressions for the x and y components of acceleration. Find the magnitude of the velocity and acceleration for the point having the coordinates (3, 4). Specify units in terms of L and T.

3.23 Sketch the flow field defined by $u = 0$, $v = 3xy$, and derive expressions for the x and y components of acceleration. Find the acceleration at the point (2, 2). Specify units in terms of L and T.

3.24 Sketch the flow field defined by $u = -2y$, $v = 3x$, and derive expressions for the x and y components of acceleration. As in Illustrative Example 3.3, find approximate values of the normal and tangential accelerations of the particle at the point (2, 3). Specify units in terms of L and T. Compare the value of $(a_n^2 + a_t^2)^{1/2}$ with the computed value $(a_x^2 + a_y^2)^{1/2}$.

3.25 The velocity along a circular streamline of radius 5 ft is 3 fps. Find the normal and tangential components of the acceleration if the flow is steady.

3.26 A large tank contains an ideal liquid which flows out of the bottom of the tank through a 6-in-diameter hole. The rate of steady outflow is 10 cfs. Assume that the liquid approaches the center of the hole radially. Find the velocities and convective accelerations at points that are 2 and 3 ft from the center of the hole.

3.27 Refer to Prob. 3.26. Suppose the flow is unsteady and $Q = 10 - 0.5t$, where Q is in cfs and t is in s. Find the local acceleration at a point 2 ft from the center of the hole at time $t = 10$ s. What is the local acceleration at this point at $t = 15$ s? Find the total acceleration at a point 3 ft from the center of the hole at $t = 15$ s.

3.28 An ideal liquid flows out the bottom of a large tank through a 10-cm-diameter hole at a steady rate of 0.40 m^3/s. Assume the liquid approaches the center of the hole radially. Find the velocities and convective accelerations at points 0.5 and 1.0 m from the center of the hole.

3.29 Refer to Prob. 3.28. Suppose the flow is unsteady and $Q = 0.40 - 0.02t^{0.5}$, where Q is in m^3/s and t in s. Find the local and convective accelerations at a point 0.5 m from the center of the hole at time $t = 20$ s. What is the total acceleration?

3.30 The figure shows to scale a two-dimensional stream tube. (*a*) If the flow rate is 40 m^3/s per meter perpendicular to the plane of the sketch, determine approximate values of the normal and tangential accelerations of a fluid particle at A. What is the resultant acceleration of a particle at A? (*b*) If the flow rate is $(20 - 3t)$ m^3/s per meter with t in s, find the approximate values of normal and tangential accelerations of a fluid particle at A when $t = 4$ s. What is the new resultant acceleration of a fluid particle at A?

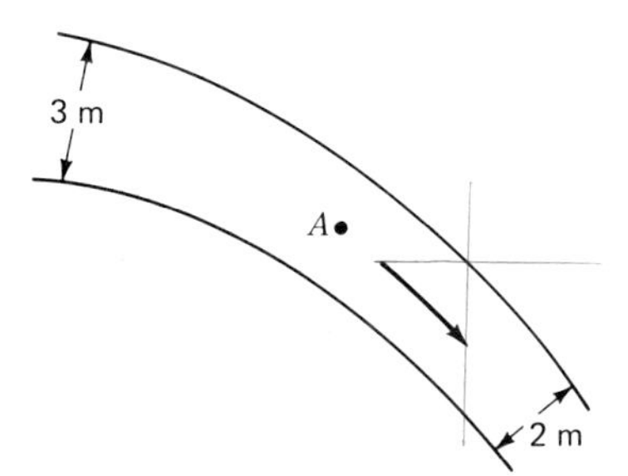

Problem 3.30

3.31 Assume that the streamlines for a two-dimensional flow of a frictionless incompressible fluid against a flat plate normal to the initial velocity may be represented by the equation $xy =$ constant and that the flow is symmetrical about the plane through $x = 0$. A different streamline may be plotted for each value of the constant. Using a scale of 1 in $=$ 6 units of distance, plot streamlines for values of the constant of 16, 64, and 128.

3.32 For the case in Prob. 3.31, it can be shown that the velocity components at any point are $u = ax$ and $v = -ay$, where a is a constant. Thus the actual velocity is $V = a\sqrt{x^2 + y^2} = ar$, where r is the radius to the origin. Let $a = \frac{1}{3}$; then if 1 in = 6 ft for the streamlines, 1 in = 2 fps for the velocity scale. Draw curves of equal velocity for values of 2, 4, 6, 8, and 10 fps. How does the velocity vary along the surface of the plate?

3.33 For three-dimensional flow with the y axis as the centerline, assume that the equation for the bounding streamline of a jet impinging vertically downward on a flat plate is $x^2y = 64$. Plot the flow showing the centerline and bounding streamlines of the jet. (a) What is the approximate average velocity in the vertical jet at $y = 10$ if the average velocity in the vertical jet is 5.0 m/s at $y = 16$? (b) For the above conditions find the approximate velocity along the plate at $r = 12$, 24, and 36.

CHAPTER

FOUR

ENERGY CONSIDERATIONS IN STEADY FLOW

In this chapter fluid flow is approached from the viewpoint of energy considerations. The first law of thermodynamics tells us that energy can be neither created nor destroyed. Moreover, all forms of energy are equivalent. In the following sections the various forms of energy present in fluid flow are briefly discussed.

4.1 KINETIC ENERGY OF A FLOWING FLUID

A body of mass m when moving at a velocity V possesses a kinetic energy, KE = $\frac{1}{2}mV^2$. Thus if a fluid were flowing with all particles moving at the same velocity, its kinetic energy would also be $\frac{1}{2}mV^2$; this can be written as:

$$\frac{\text{KE}}{\text{Weight}} = \frac{\frac{1}{2}mV^2}{(\gamma)(\text{vol})} = \frac{\frac{1}{2}[\rho(\text{vol})]V^2}{(\gamma)(\text{vol})} = \frac{V^2}{2g} \tag{4.1}$$

In English units $V^2/2g$ is expressed in ft·lb/lb = ft and in SI units as N·m/N = m.

In most situations the velocities of the different fluid particles are not the same, so it is necessary to integrate all portions of the stream to obtain the true value of the kinetic energy. It is convenient to express the true value in terms of the mean velocity V and a factor α. Hence,

$$\frac{\text{True KE}}{\text{Weight}} = \alpha\frac{V^2}{2g} \tag{4.2}$$

In order to obtain an expression for α, consider the case where the axial components of the velocity vary across a section (Fig. 3.1b). If u is the local axial velocity component at a point, the mass flow per unit of time through an elementary area dA is $\rho\, dQ = \rho u\, dA$. Thus the true flow of kinetic energy per unit of time across area dA is $(\rho u\, dA)(u^2/2) = (\gamma/2g)u^3\, dA$. The weight rate of flow through dA is $\gamma Q = \gamma u\, dA$. Thus for the entire section

$$\frac{\text{True KE/time}}{\text{Weight/time}} = \frac{\text{true KE}}{\text{weight}} = \frac{\gamma/2g \int u^3\, dA}{\gamma \int u\, dA} = \frac{\int u^3\, dA}{2g \int u\, dA} \tag{4.3}$$

Comparing Eq. (4.3) with Eq. (4.2) we get

$$\alpha = \frac{1}{V^2}\frac{\int u^3\, dA}{\int u\, dA} = \frac{1}{AV^3}\int u^3\, dA \tag{4.4}$$

As the average of cubes is greater than the cube of the average, the value of α will always be more than 1. The greater the variation in velocity across the section, the larger will be the value of α. For laminar flow in a circular pipe, $\alpha = 2$; for turbulent flow in pipes, α ranges from 1.01 to 1.15, but it is usually between 1.03 and 1.06.

In some instances it is very desirable to use the proper value of α, but in most cases the error neglecting its divergence from 1.0 is negligible. As precise values of α are seldom known, it is customary for turbulent flow to assume that the kinetic energy is $V^2/2g$ per unit weight of fluid, i.e., ft·lb/lb = ft (N·m/N = m in SI units). In laminar flow the velocity is usually so small that the kinetic energy per unit weight of fluid is negligible.

4.2 POTENTIAL ENERGY

The potential energy of a particle of fluid depends on its elevation above any arbitrary datum plane. We are usually interested only in differences of elevation, and therefore the location of the datum plane is determined solely by considerations of convenience. A fluid particle of weight W situated a distance z above datum possesses a potential energy of Wz. Thus its potential energy per unit weight is z, i.e., ft·lb/lb = ft (N·m/N = m in SI units).

4.3 INTERNAL ENERGY

Internal energy is more fully presented in texts on thermodynamics since it is thermal energy, but in brief, it is energy due to the motion of molecules and forces of attraction between them. Internal energy[1] is a function of temperature; it can be expressed in terms of energy i per unit of mass or in terms of energy I per unit of weight. Note that $i = gI$.

[1] In the technical literature internal energy per unit mass is commonly represented by the symbol u. In this text, however, we use i for internal energy per unit mass since u is used in several situations for velocity.

The zero of internal energy may be taken at any arbitrary temperature, since we are usually concerned only with differences. For a unit mass, $\Delta i = c_v \Delta T$, where c_v is the specific heat at constant volume whose units are ft·lb/(slug·°R) [N·m/(kg·K) in SI units]. Thus Δi is expressed in ft·lb/slug (N·m/kg in SI units). Internal energy I per unit of weight is usually expressed in ft·lb/lb = ft (N·m/N = m in SI units).[1]

4.4 GENERAL EQUATION FOR STEADY FLOW OF ANY FLUID

The first law of thermodynamics states that for steady flow the external work done on any system plus the thermal energy transferred into or out of the system is equal to the change of energy of the system.

Thus, for steady flow,

$$\text{Work} + \text{heat} = \Delta \text{ energy}$$

It should be noted that work, heat, and energy all have the same units and thus are interchangeable under certain conditions.

Let us now apply the first law of thermodynamics to the fluid system defined by the fluid mass contained at time t in the *control volume* between sections 1 and 2 of the stream tube in Fig. 4.1. *The control volume is fixed in position and does not move or change shape* (Fig. 4.1*b*). The fluid system we are dealing with consists of the fluid that was contained between sections 1 and 2 at time t. This fluid

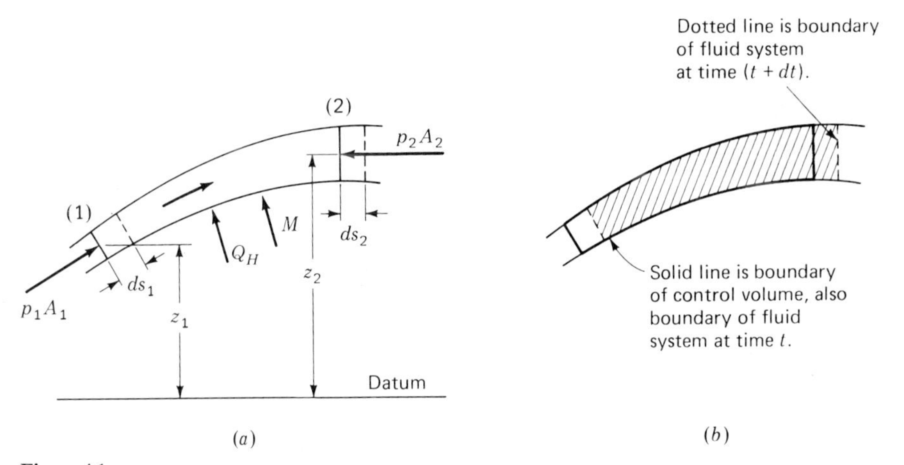

Figure 4.1

[1] In the English system of measurement, internal energy I is sometimes expressed in Btu/lb; however those units are rarely used today. Nevertheless it is important to be familiar with such units when reading technical papers that were written a number of years ago. One Btu = 778 ft·lb.

system moves to a new position during time interval dt, as indicated in Fig. 4.1. During this short time interval we shall assume that the fluid moves a short distance ds_1 at section 1 and ds_2 at section 2. In this discussion we restrict ourselves to steady flow so that $\gamma_1 A_1\, ds_1 = \gamma_2 A_2\, ds_2$ (Eq. 3.9*b*). In moving these short distances, work is done on the fluid system by the pressure forces $p_1 A_1$ and $p_2 A_2$. This work is referred to as *flow work*. It may be expressed as

$$\text{Flow work} = p_1 A_1\, ds_1 - p_2 A_2\, ds_2$$

The minus sign in the second term indicates that the force and displacement are in opposite directions.

In addition to *flow work*, if there is a machine between sections 1 and 2 there will be *shaft work*. During the short time interval dt we can write

$$\text{Shaft work} = \frac{\text{weight}}{\text{time}} \times \frac{\text{energy}}{\text{weight}} \times \text{time}$$

$$= \left(\gamma_1 A_1 \frac{ds_1}{dt}\right) h_M\, dt = (\gamma_1 A_1\, ds_1) h_M$$

where h_M is the energy put into the flow by the machine per unit weight of flowing fluid. If the machine is a pump, h_M is positive; if the machine is a turbine, h_M is negative. It should be noted that frictional shear stresses at the boundary of the fluid system do work on the fluid within the system. The shear stresses are not external to the system and the work they do is converted to heat which tends to increase the temperature of the fluid within the system.

The heat transferred from an external source into the fluid system over time interval dt is

$$\text{Heat} = \left(\gamma_1 A_1 \frac{ds_1}{dt}\right) Q_H\, dt = (\gamma_1 A_1\, ds_1) Q_H$$

where Q_H is the energy put into the flow by the external heat source per unit weight of flowing fluid. If the heat flow is out of the fluid, the value of Q_H is negative.

In using the concept of the control volume, we consider a fluid system defined by the mass of fluid contained in the control volume at t. At time $t + dt$ this same mass of fluid has moved to a new position as indicated in Fig. 4.1*b*. At that instant the energy E_2 of the fluid system (cross-hatched area of Fig. 4.1*b*) equals the energy E_1 that was possessed by the fluid mass when it was coincident with the control volume at time t plus the energy ΔE_{out} that flowed out of the control volume during time interval dt minus the energy ΔE_{in} that flowed into the control volume during time interval dt. Thus,

$$E_2 = E_1 + \Delta E_{out} - \Delta E_{in}$$

Hence the change in energy ΔE of the fluid system under consideration during time interval dt is

$$\Delta E = E_2 - E_1 = \Delta E_{out} - \Delta E_{in}$$

During time interval dt the weight of fluid entering at section 1 is $\gamma_1 A_1\,ds_1$, and for steady flow an equal weight must leave section 2 during the same time interval. Hence the energy ΔE_{in} which enters at section 1 during time dt is

$$\gamma_1 A_1\,ds_1(z_1 + \alpha V_1^2/2g + I_1),$$

while that which leaves (ΔE_{out}) at section 2 is represented by a similar expression. Thus

$$\Delta\text{ energy} = \Delta E = \gamma_2 A_2\,ds_2\left(z_2 + \alpha_2\frac{V_2^2}{2g} + I_2\right) - \gamma_1 A_1\,ds_1\left(z_1 + \alpha_1\frac{V_1^2}{2g} + I_1\right)$$

Applying the first law of thermodynamics (work + heat = Δ energy), at the same time factoring out $\gamma_1 A_1\,ds_1 = \gamma_2 A_2\,ds_2$ for steady flow and rearranging, we get

$$\frac{p_1}{\gamma_1} - \frac{p_2}{\gamma_2} + h_M + Q_H = \left(z_2 + \alpha_2\frac{V_2^2}{2g} + I_2\right) - \left(z_1 + \alpha_1\frac{V_1^2}{2g} + I_1\right)$$

or
$$\left(z_1 + \frac{p_1}{\gamma_1} + \alpha_1\frac{V_1^2}{2g} + I_1\right) + h_M + Q_H = \left(z_2 + \frac{p_2}{\gamma_2} + \alpha_2\frac{V_2^2}{2g} + I_2\right) \tag{4.5}$$

This equation applies to liquids, gases, and vapors, and to ideal fluids as well as to real fluids with friction. The only restriction is that it is for steady flow. The p/γ terms represent energy possessed by the fluid per unit weight of fluid by virtue of the pressure under which the fluid exists. Under proper circumstances this pressure will be released and transformed to other forms of energy, i.e., kinetic, potential, or internal energy. Likewise it is possible for these other forms of energy to be transformed into pressure energy.

In turbulent flow there are other forms of kinetic energy besides that of translation described in Sec. 4.1. These are the rotational kinetic energy of eddies initiated by fluid friction and the kinetic energy of the turbulent fluctuations of velocity. They are not represented by any specific terms in Eq. (4.5) because their effect appears indirectly. The kinetic energy of translation can be converted into increases in p/γ or z, but these other forms of kinetic energy can never be transformed into anything but thermal energy. Thus they appear as an increase in the numerical value of I_2 over the value it would have if there were no friction.

The general energy equation (4.5) and the continuity equation are two important keys to the solution of many problems in fluid mechanics. For compressible fluids it is necessary to have a third equation, which is the equation of state which provides a relationship between density (or specific volume) and the absolute values of the pressure and temperature.

In many cases Eq. (4.5) is greatly shortened because certain quantities are equal and thus cancel each other, or are zero. Thus, if two points are at the same elevation, $z_1 - z_2 = 0$. If the conduit is well insulated or if the temperature of the fluid and that of its surroundings are practically the same, Q_H may be taken as zero. On the other hand, Q_H may be very large, as in the case of flow of water through a boiler tube. If there is no machine between sections 1 and 2, then the term h_M drops out. If there is a machine present, the work done by or upon it may be determined by solving Eq. (4.5) for h_M.

4.5 ENERGY EQUATION FOR STEADY FLOW OF INCOMPRESSIBLE FLUIDS

For liquids, and even for gases and vapors when the change in pressure is very small, the fluid may be considered as incompressible for all practical purposes, and thus we may take $\gamma_1 = \gamma_2 = \gamma =$ constant. In turbulent flow the value of α is only a little more than unity, and as a simplifying assumption, it will be assumed equal to unity. If the flow is laminar, $V^2/2g$ is usually very small compared to the other terms in Eq. (4.5), hence little error is introduced if α is set equal to 1.0 rather than 2.0, its true value. Thus, for an incompressible fluid, Eq. (4.5) becomes

$$\left(\frac{p_1}{\gamma} + z_1 + \frac{V_1^2}{2g}\right) + h_M + Q_H = \left(\frac{p_2}{\gamma} + z_2 + \frac{V_2^2}{2g}\right) + (I_2 - I_1) \tag{4.6}$$

Fluid friction produces eddies and turbulence, and these forms of kinetic energy are eventually transformed into thermal energy. If there is no heat transfer, the effect of friction is to produce an increase in temperature so that I_2 becomes greater than I_1.

Suppose there is a loss of heat Q_H at such a rate as to maintain the temperature constant so that $I_2 = I_1$. In this event there is an actual loss of energy from the system equal to the mechanical energy which has been converted into thermal energy by friction.

A change in the internal energy of a fluid is accompanied by a change in temperature and is equal to the external heat added to or taken away from the fluid plus the heat generated by fluid friction. Thus

$$\frac{\Delta \text{ internal energy}}{\text{Unit of mass}} = \Delta i = i_2 - i_1 = c(T_2 - T_1)$$

$$\frac{\Delta \text{ internal energy}}{\text{Unit of weight}} = \Delta I = \frac{\Delta i}{g} = I_2 - I_1 = \frac{c}{g}(T_2 - T_1)$$

$$= Q_H + h_L \tag{4.7}$$

where c is the specific heat[1] of the incompressible fluid and h_L is the fluid-friction energy loss per unit weight of fluid. The foregoing can be expressed as

$$h_L = (I_2 - I_1) - Q_H = \frac{c}{g}(T_2 - T_1) - Q_H \tag{4.8}$$

[1] For water, $c = 1$ Btu/(mass of standard lb·°R) = 1 Btu(32.2 ft/s²)/(lb·°R) = 32.2 Btu/(slug·°R). In SI units, c for water = 1 cal/(g·K). These can also be expressed as 25,000 ft·lb/(slug·°R) and 4,187 N·m/(kg·K), equivalent to 25,000 ft²/(s²·°R) and 4,187 m²/(s²·K) respectively.

If the loss of heat (Q_H negative) is greater than h_L, then T_2 will be less than T_1. If there is any absorption of heat (Q_H positive), T_2 will be greater than the value which would have resulted from friction alone. A large value of h_L produces only a very small rise in temperature if there is no heat transfer or, stated another way, only a very small transfer of heat is required to maintain isothermal flow.

If there is no machine between sections 1 and 2 and if no heat is gained or lost, by substituting Eq. (4.8) into Eq. (4.6) the energy equation for an incompressible fluid becomes

$$\frac{p_1}{\gamma} + z_1 + \frac{V_1^2}{2g} = \frac{p_2}{\gamma} + z_2 + \frac{V_2^2}{2g} + h_L \tag{4.9}$$

where h_L (commonly referred to as head loss) represents the energy loss per unit weight of fluid. In some cases the value of h_L may be very large, and although for any real fluid it can never be zero, there are cases when it is so small that it may be neglected with small error.[1] In such special cases

$$\frac{p_1}{\gamma} + z_1 + \frac{V_1^2}{2g} = \frac{p_2}{\gamma} + z_2 + \frac{V_2^2}{2g} \tag{4.10}$$

and from this it follows that

$$\frac{p}{\gamma} + z + \frac{V^2}{2g} = \text{constant} \tag{4.11}$$

The equation in either of these last two forms is known as Bernoulli's theorem, in honor of Daniel Bernoulli, who presented it in 1738. Note that Bernoulli's theorem is for a *frictionless incompressible* fluid. However, it can be applied to real incompressible fluids with good results in situations where frictional effects are very small.

Illustrative Example 4.1 A liquid with a specific gravity of 1.26 flows in a pipe at a rate of 25 cfs (700 L/s). At a point where the pipe diameter is 24 in (60 cm), the pressure is 45 psi (300 kN/m^2). Find the pressure at a second point where the pipe diameter is 12 in (30 cm) if the second point is 3 ft (1.0 m) lower than the first point. Neglect head loss.

English units:

$$V_1 = \frac{25}{\pi} = 7.96 \text{ fps} \qquad V_2 = \frac{25}{\pi/4} = 31.8 \text{ fps}$$

From Eq. (4.10)

$$0 + \frac{45(144)}{1.26(62.4)} + \frac{(7.96)^2}{64.4} = -3 + \frac{p_2(144)}{1.26(62.4)} + \frac{(31.8)^2}{64.4}$$

$$p_2 = 38.6 \text{ psi}$$

[1] Recognizing when frictional effects are so small that frictionless flow may be assumed is important. For example, the pressure around the nose of a streamlined body (Fig. 3.11) may be determined quite accurately by assuming frictionless flow; however frictional effects must be considered if the shear stresses at the boundary are to be determined.

SI units:

$$\gamma_{\text{water}} = 9{,}810 \text{ N/m}^3 = 9.81 \text{ kN/m}^3$$

$$V_1 = \frac{0.70 \text{ m}^3/\text{s}}{\pi(0.3)^2 \text{ m}^2} = 2.48 \text{ m/s} \qquad V_2 = 4V_1 = 9.92 \text{ m/s}$$

$$0 + \frac{300}{1.26(9.81)} + \frac{(2.48)^2}{2(9.81)} = -1.0 + \frac{p_2}{1.26(9.81)} + \frac{(9.92)^2}{2(9.81)}$$

$$p_2 = 254 \text{ kN/m}^2$$

Illustrative Example 4.2 Water flows at 10 m^3/s in a 150-cm-diameter pipe; the head loss in a 1,000-m length of this pipe is 20 m. Find the increase in water temperature assuming no heat enters or leaves the pipe.

Eq. (4.8)

$$h_L = 20 \text{ m} = \frac{c}{g}(T_2 - T_1)$$

$$c \text{ for water} = 4{,}187 \text{ N·m/(kg·K)}$$

$$\Delta T = T_2 - T_1 = \frac{gh_L}{c} = \frac{(9.81 \text{ m/s}^2)(20 \text{ m})}{4{,}187[(\text{kg·m/s}^2)\text{·m}]/(\text{kg·K})}$$

$$= 0.047 \text{ K}$$

4.6 HEAD

In Eq. (4.9) each term has the dimensions of *length.* Thus p/γ, called *pressure head,* represents the energy per unit weight stored in the fluid by virtue of the pressure under which the fluid exists; z, called *elevation head,* represents the potential energy per pound of fluid; and $V^2/2g$, called *velocity head,* represents the kinetic energy per pound of fluid. The sum of these three terms is called the *total head* and is denoted by H, thus

$$H = \frac{p}{\gamma} + z + \frac{V^2}{2g} \tag{4.12}$$

Each term in this equation, although ordinarily expressed in feet (or meters), represents *foot pounds of energy per pound of fluid flowing* (newton meters of energy per newton of fluid flowing in SI units).

For a frictionless incompressible fluid with no machine between 1 and 2, $H_1 = H_2$, but for a real fluid,

$$H_1 = H_2 + h_L \tag{4.13}$$

which is merely a brief way of writing Eq. (4.9). For a real fluid it is obvious that if there is no input of energy head h_M by a machine between sections 1 and 2, the total head must decrease in the direction of flow.

If there is a machine between sections 1 and 2, then

$$H_1 + h_M = H_2 + h_L \tag{4.14}$$

If the machine is a pump, $h_M = h_p$, where h_p is the energy head put into the flow by the pump. If the machine is a turbine, $h_M = -h_t$, where h_t is the energy head extracted from the flow by the turbine.

4.7 POWER CONSIDERATIONS IN FLUID FLOW

In deriving Eq. (4.5), the term $\gamma A\,ds$ representing weight of fluid was factored out; thus every term of the equation represents energy per unit weight (i.e., energy head). If the energy head is multiplied by the weight rate of flow, the resulting product represents power, for

$$\text{Power} = \frac{\text{energy}}{\text{time}} = \frac{\text{energy}}{\text{weight}} \times \frac{\text{weight}}{\text{time}} = H \times G = H\gamma Q \tag{4.15}$$

In English units,

$$\text{Horsepower} = \frac{\gamma Q H}{550} \tag{4.16}$$

while in metric units

$$\text{Kilowatts} = \frac{\gamma Q H}{1{,}000} \tag{4.17}$$

where γ = the unit weight of fluid, lb/ft^3 (N/m^3 in SI units)
Q = the rate of flow, ft^3/s (m^3/s in SI units)
H = the energy head, ft (m in SI units)
Note: 1 hp = 550 ft·lb/s = 0.746 kW.

In these equations H may be any head for which the corresponding power is desired. For example, to find the power extracted from the flow by a turbine substitute h_t for H, to find the power of a jet substitute $V_j^2/2g$ for H where V_j is the jet velocity, to find the power lost because of fluid friction substitute h_L for H.

With respect to power, it may be recalled from mechanics that the power developed when a force F acts on a translating body, or when a torque T acts on a rotating body, is given by

$$\text{Power} = Fu = T\omega \tag{4.18}$$

where u is linear velocity in feet per second (or meters per second) and ω is angular velocity in radians per second. The force F represents the component force in the direction of the velocity u. These equations will be referred to in

Chap. 6, where the dynamic forces exerted by moving fluids are discussed, and again in Chaps. 14 through 17, in the discussion of turbomachinery.

Illustrative Example 4.3 A liquid with a specific gravity of 1.26 is being pumped in a pipeline from A to B. At A the pipe diameter is 24 in (60 cm) and the pressure is 45 psi (300 kN/m^2). At B the pipe diameter is 12 in (30 cm) and the pressure is 50 psi (330 kN/m^2). Point B is 3 ft (1.0 m) lower than A. Find the flow rate if the pump puts 22 hp (16 kW) into the flow. Neglect head loss.

English units:

$$HP = 22 = \frac{(1.26 \times 62.4)Qh_p}{550}$$

$$h_p = \frac{154}{Q}$$

$$0 + \frac{45(144)}{1.26(62.4)} + \frac{(Q/\pi)^2}{64.4} + \frac{154}{Q} = -3 + \frac{50(144)}{1.26(62.4)} + \frac{[Q/(0.25\pi)]^2}{64.4}$$

By trial $Q = 14.2$ cfs.

SI units:

$$kW = 16 = \frac{(1.26 \times 9.810)Qh_p}{1{,}000}$$

$$h_p = \frac{1.29}{Q}$$

$$0 + \frac{300}{1.26(9.81)} + \frac{[Q/\pi(0.3)^2]^2}{2(9.81)} + \frac{1.29}{Q} = -1.0 + \frac{330}{1.26(9.81)} + \frac{[Q/\pi(0.15)^2]^2}{2(9.81)}$$

By trial $Q = 0.42$ m^3/s.

Illustrative Example 4.4 Find the rate of energy loss due to pipe friction for the pipe of Illustrative Example 4.2.

$$\text{Rate of energy loss} = \frac{\gamma QH}{1{,}000} \qquad \text{where} \qquad H = h_L$$

$$= \frac{(9{,}810\ \text{N/m}^3)(10\ \text{m}^3\text{/s})(20\ \text{m})}{1{,}000}$$

$$= 1{,}960\ \text{kW}$$

4.8 CAVITATION

In liquid flow problems the possibility of cavitation must be investigated.[1] According to the Bernoulli theorem [Eq. (4.11)], if at any pressure the velocity head increases, there must be a corresponding decrease in the pressure head. For any

[1] The phenomenon of cavitation is not possible in gas flow because a gas does not change state at low pressure, whereas a liquid will change to a gas (vapor) if the pressure is low enough.

liquid there is a minimum absolute pressure possible, namely, the vapor pressure of the liquid. The vapor pressure depends upon the liquid and its temperature. If the conditions are such that a calculation results in a lower absolute pressure than the vapor pressure, this simply means that the assumptions upon which the calculations are based no longer apply. The criterion with respect to cavitation is as follows:

$$\left(\frac{p_{\text{crit}}}{\gamma}\right)_{\text{abs}} = \frac{p_v}{\gamma}$$

But

$$\left(\frac{p_{\text{crit}}}{\gamma}\right)_{\text{abs}} = \frac{p_{\text{atm}}}{\gamma} + \left(\frac{p_{\text{crit}}}{\gamma}\right)_{\text{gage}}$$

Thus

$$\left(\frac{p_{\text{crit}}}{\gamma}\right)_{\text{gage}} = -\left(\frac{p_{\text{atm}}}{\gamma} - \frac{p_v}{\gamma}\right) \tag{4.19}$$

where p_{atm}, p_v, and p_{crit} represent the atmospheric pressure, the vapor pressure, and the critical (or minimum) possible pressure, respectively, in liquid flow. Equation (4.19) shows that the gage pressure head in a flowing liquid can be negative, but no more negative than $p_{\text{atm}}/\gamma - p_v/\gamma$.

If at any point the local velocity is so high that the pressure in a liquid is reduced to its vapor pressure, the liquid will then vaporize (or boil) at that point and bubbles of vapor will form. As the fluid flows into a region of higher pressure, the bubbles of vapor will suddenly condense; in other words, they may be said to *collapse*. This action may produce very high dynamic pressure upon the adjacent solid walls, and since this action is continuous and has a high frequency, the material in that zone may be damaged. Turbine runners, pump impellers, and ship screw propellers are often severely and quickly damaged by such action, because holes are rapidly produced in the metal. Spillways and other types of hydraulic structures are also subject to damage from cavitation. The damaging action is commonly referred to as *pitting*. Not only is cavitation destructive, but it may produce a drop in efficiency of the machine or propeller or other device.

In order to avoid cavitation, it is necessary that the absolute pressure at every point be above the vapor pressure. To ensure this, it is necessary to raise the general pressure level, either by placing the device below the intake level so that the liquid flows to it by gravity rather than being drawn up by suction or by designing the machine so that there are no local velocities high enough to produce such a low pressure.

Figure 4.2 shows photographs of blades for an axial-flow pump set up in a transparent-lucite working section where the pressure level can be varied. For *a*, *b*, and *c* there was the same water velocity on the same vane but with decreasing absolute pressures. This resulted in the formation of a vapor pocket of increasing size. The stream flow and the pressure for *d* were the same as for *b*, but the nose of the blade was slightly different in shape, which gave a different type of bubble formation. This shows the effect of a slight change in design.

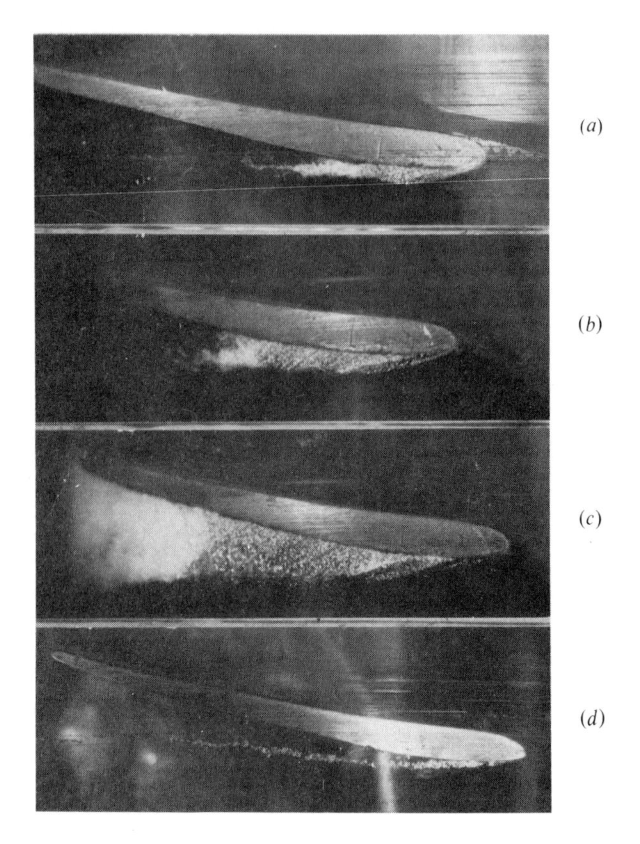

Figure 4.2 Cavitation phenomena: flow around an axial-flow pump blade illustrating the effect of reduced absolute pressure (*a*), (*b*), and (*c*) and the effect of a slight change of shape (*d*). *(Photographs by Hydrodynamics Laboratory, California Institute of Technology.)*

Illustrative Example 4.5 A liquid ($s = 0.86$) with a vapor pressure of 3.8 psia flows through the horizontal constriction in the accompanying figure. Atmospheric pressure is 26.8 in Hg. Find the maximum theoretical flow rate (i.e., at what Q does cavitation occur?). Neglect head loss.

Since the standard atmosphere is equivalent to 29.9 inches of mercury and 14.7 psia (Sec. 2.5),

$$p_{\text{atm}} = \frac{26.8}{29.9}(14.7) = 13.2 \text{ psia}$$

$$\left(\frac{p_{\text{crit}}}{\gamma}\right)_{\text{gage}} = -\left[\frac{13.2 - 3.8}{0.86(62.4)}\right]144 = -25.2 \text{ ft}$$

$$0 + \frac{10(144)}{0.86(62.4)} + \frac{(Q/2.25\pi)^2}{64.4} = 0 - 25.2 + \frac{(Q/0.25\pi)^2}{64.4}$$

$$Q = 45.5 \text{ cfs}$$

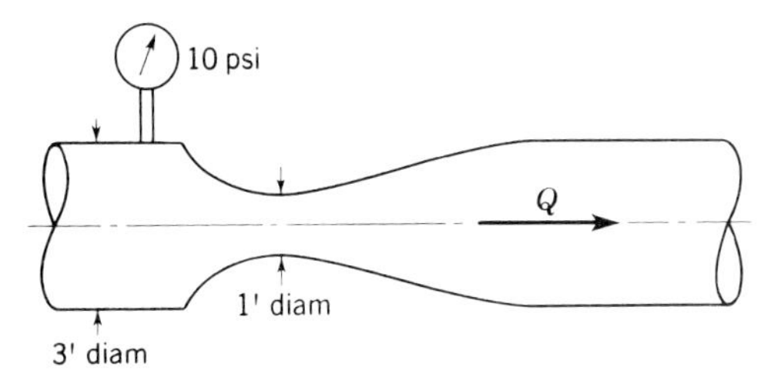

Illustrative Example 4.5

4.9 ENERGY EQUATION FOR STEADY FLOW OF COMPRESSIBLE FLUIDS

If sections 1 and 2 are so chosen that there is no machine between them, and if α is assumed as unity, Eq. (4.5) becomes

$$\left(\frac{p_1}{\gamma_1} + I_1 + z_1 + \frac{V_1^2}{2g}\right) + Q_H = \left(\frac{p_2}{\gamma_2} + I_2 + z_2 + \frac{V_2^2}{2g}\right) \tag{4.20}$$

For most compressible fluids, i.e., gases or vapors, the quantity p/γ is usually very large compared with $z_1 - z_2$ because of the small value of γ, and therefore the z terms are usually omitted. But $z_1 - z_2$ should not be ignored unless it is known to be negligible compared with the other quantities.

The p/γ and the I terms are usually combined for gases and vapors into a single term called *enthalpy*, indicated by a single symbol such as $\hat{h} = I + p/\gamma$, where $\hat{h}$ is energy per unit weight.[1] With these changes Eq. (4.20) becomes

$$\hat{h}_1 + \frac{V_1^2}{2g} + Q_H = \hat{h}_2 + \frac{V_2^2}{2g} \tag{4.21}$$

This equation may be used for any gas or vapor and for any process. Some knowledge of thermodynamics is required to evaluate the $\hat{h}$ terms, and in the case of vapors it is necessary to use vapor tables or charts, because their properties cannot be expressed by any simple equations. Various aspects of the flow of compressible fluids are discussed in Chap. 9

[1] Enthalpy represents the energy possessed by a given mass (or weight) of gas or vapor by virtue of the absolute temperature under which it exists. In thermodynamics enthalpy is commonly expressed in terms of energy per unit mass (h) rather than energy per unit weight ($\hat{h}$). Thus $h = g\hat{h} = gI + p/\rho = i + pv$. Values of h for vapors commonly used in engineering such as steam, ammonia, freon, and others, may be obtained from vapor tables or charts. For a perfect gas and practically for real gases $\Delta h = g(\Delta\hat{h}) = c_p \Delta T$, where c_p is specific heat at constant pressure. For air at usual pressures c_p has a value of 6,000 ft·lb/(slug·°R) [or 1,003 N·m/(kg·K)]. These are equivalent to 6,000 ft^2/(s^2·°R) [or 1,003 m^2/(s^2·K)].

4.10 EQUATION OF STEADY MOTION ALONG A STREAMLINE FOR AN IDEAL FLUID

Referring to Fig. 4.3, let us consider frictionless steady flow of a fluid along the streamline. We shall consider the forces acting on a small cylindrical element of the fluid in the direction of the streamline and apply Newton's second law, that is, $F = ma$. The forces tending to accelerate the fluid mass are pressure forces on the two ends of the element

$$p\,dA - (p + dp)\,dA = -dp\,dA$$

where dA is the cross section of the element at right angles to the streamline, and the weight component in the direction of motion

$$-\rho g\,ds\,dA(dz/ds) = -\rho g\,dA\,dz$$

The mass of the element is $\rho\,dA\,ds$, while its acceleration for steady flow can be expressed by Eq. (3.15) as $V(dV/ds)$. Thus, applying $F = ma$, we get

$$-dp\,dA - \rho g\,dA\,dz = \rho\,ds\,dA\,V\frac{dV}{ds}$$

Dividing by $-\rho\,dA$,

$$\frac{dp}{\rho} + V\,dV + g\,dz = 0 \tag{4.22}$$

This equation is commonly referred to as the one-dimensional Euler equation, because it was first derived by Leonhard Euler in about 1750. It applies to both compressible and incompressible flow since the variation of ρ over the elemental length ds is small. Equation (4.22) can also be expressed as

$$\frac{dp}{\gamma} + d\frac{V^2}{2g} + dz = 0 \tag{4.23}$$

For the case of a compressible ideal fluid, since $\gamma \neq$ constant, an equation of state relating γ to p and T must be introduced before integrating Eq. (4.23).

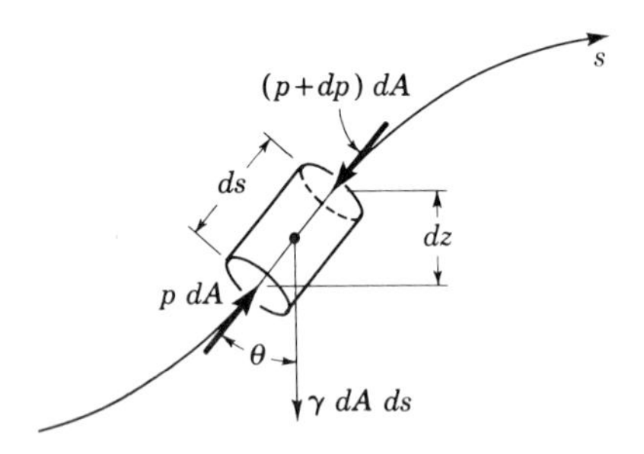

Figure 4.3 Element on streamtube (ideal fluid).

For the case of an incompressible fluid (γ = constant), Eq. (4.23) can be integrated to give,

$$\int \frac{dp}{\gamma} + \int d\frac{V^2}{2g} + \int dz = \text{constant} \tag{4.24}$$

Thus
$$\frac{p}{\gamma} + \frac{V^2}{2g} + z = \text{constant} = \text{total head} = H \tag{4.25}$$

This is Bernoulli's equation [Eq. (4.11)] for steady flow of a frictionless incompressible fluid along a streamline. Thus we have developed the Bernoulli equation from two viewpoints, first from energy considerations and now from Newton's second law.

If there is no flow,

$$z + \frac{p}{\gamma} = \text{constant} \tag{4.26}$$

This equation is identical to Eq. (2.6); it shows that for an incompressible fluid at rest, the summation of the elevation z at any point in the fluid plus the pressure head p/γ at that point is equal to the sum of these two quantities at any other point.

4.11 EQUATION OF STEADY MOTION ALONG A STREAMLINE FOR A REAL FLUID

Let us now follow the same procedure as in the last section, except that now we shall consider a real fluid. The fluid element (Fig. 4.4) is similar to the one of Fig. 4.3, except that with a real fluid there is an additional force acting because of fluid friction, namely, $\tau(2\pi r)\,ds$, where τ is the shear stress at the boundary of the element and $2\pi r\,ds$ is the area over which the shear stress acts, r being the radius of the cylindrical element under consideration. Writing $F = ma$, for steady flow we get

$$-dp\,dA - \rho g\,dA\,dz - \tau(2\pi r)\,ds = \rho\,ds\,dA\,V\frac{dV}{ds}$$

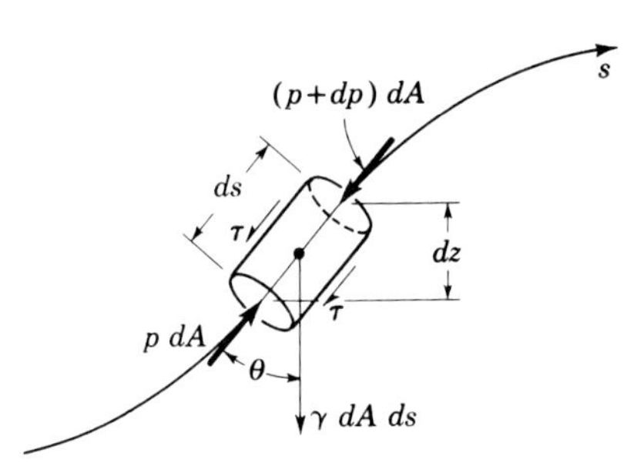

Figure 4.4 Element on streamtube (real fluid).

In this case $dA = \pi r^2$. Making this substitution for dA and dividing through by $-\rho\pi r^2$ gives

$$\frac{dp}{\rho} + V\,dV + g\,dz = -\frac{2\tau\,ds}{\rho r} \tag{4.27}$$

This equation is similar to Eq. (4.22), except that it has an extra term. The extra term $-(2\tau\,ds)/\rho r$ accounts for fluid friction.

Equation (4.27) may also be expressed as

$$\frac{dp}{\gamma} + d\frac{V^2}{2g} + dz = -\frac{2\tau\,ds}{\gamma r} \tag{4.28}$$

This equation applies to steady flow of both compressible and incompressible real fluids. However, once again an equation of state relating γ to p and T must be introduced before integration if we are dealing with a compressible fluid. Energy equations for flow of compressible fluids are developed in Secs. 9.3 and 9.5.

For an incompressible fluid (γ = constant), we can integrate directly. Integrating from some section 1 to another section 2, where the distance between them is L, we get for an incompressible real fluid

$$\frac{p_2}{\gamma} - \frac{p_1}{\gamma} + \frac{V_2^2}{2g} - \frac{V_1^2}{2g} + z_2 - z_1 = -\frac{2\tau L}{\gamma r}$$

or

$$\left(\frac{p_1}{\gamma} + \frac{V_1^2}{2g} + z_1\right) - \frac{2\tau L}{\gamma r} = \left(\frac{p_2}{\gamma} + \frac{V_2^2}{2g} + z_2\right) \tag{4.29}$$

Comparing Eq. (4.29) with Eq. (4.9), we see that

$$h_L = \frac{2\tau L}{\gamma r} \tag{4.30}$$

This expression will be referred to in Chap. 8.

4.12 DEFINITION OF HYDRAULIC GRADE LINE AND ENERGY LINE

When dealing with flow problems involving liquids the concept of energy line and hydraulic grade line is usually advantageous. Even with gas flow the concept can be useful. The term $z + p/\gamma$ is referred to as the *static head*, or *piezometric head*, because it represents the level to which liquid will rise in a piezometer tube (Sec. 2.6). The *piezometric head line*, or *hydraulic grade line* (HGL), is a line drawn through the tops of the piezometer columns. A pitot tube (Sec. 12.3), a small open tube with its open end pointing upstream, will intercept the kinetic energy of the flow and hence indicate the *total energy head*, $z + p/\gamma + u^2/2g$. Referring to Fig. 4.5, which depicts the flow of an ideal fluid, the vertical distance from point A on the stream tube to the level of the piezometric head at that point represents the pressure head in the flow at point A. The vertical distance from the liquid level

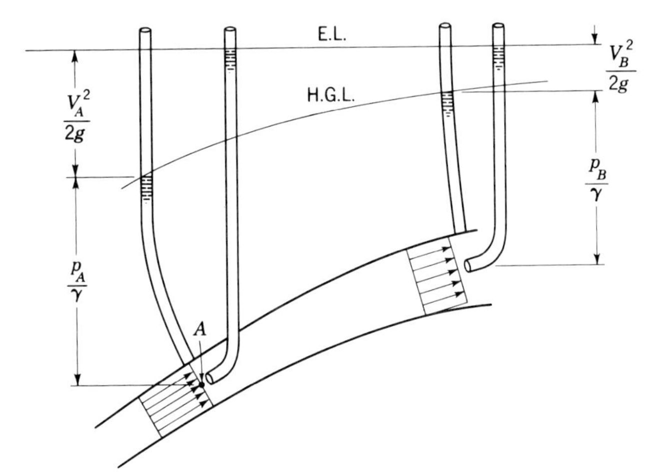

Figure 4.5 Ideal fluid.

in the piezometer tube to that in the pitot tube is $V^2/2g$. In Fig. 4.5, the line sketched through the pitot-tube liquid levels is known as the *energy line* (EL). For flow of an ideal fluid, the energy line is horizontal since there is no head loss.

A pitot tube intercepts the total energy in the flow field at the point at which it is located (Fig. 4.6). Hence the level above datum to which liquid will rise in a pitot tube is $z + p/\gamma + u^2/2g$, where u is the local velocity. For a pitot tube to indicate the true level of the energy line, it must be placed in the flow at a point where $u^2/2g = \alpha(V^2/2g)$, or where $u = \sqrt{\alpha}V$. If α (Sec. 4.1) is assumed to have a value of 1.0, then to indicate the true energy line, the tube must be placed in the flow at a point where $u = V$. One rarely knows ahead of time where in the flow $u = V$; hence the correct positioning of a pitot tube, in order that it indicate the true position of the energy line, is generally unknown.

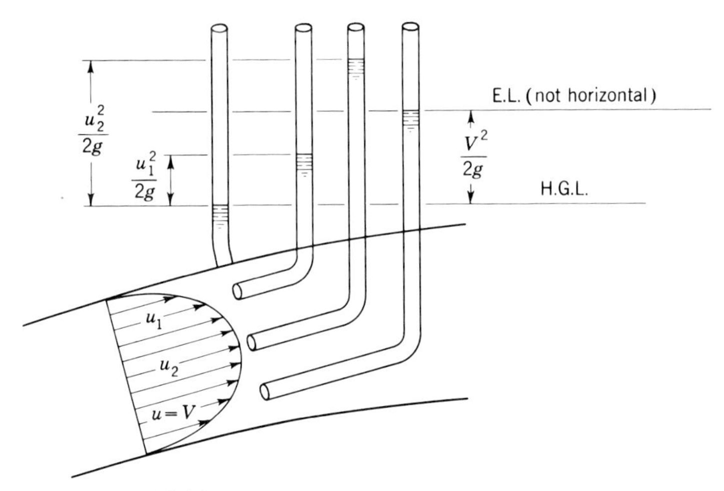

Figure 4.6 Real fluid.

Illustrative Example 4.6 Water flows in a wide open channel as shown in the accompanying figure. Two pitot tubes are connected to a differential manometer containing a liquid ($s = 0.82$). Find u_A and u_B.

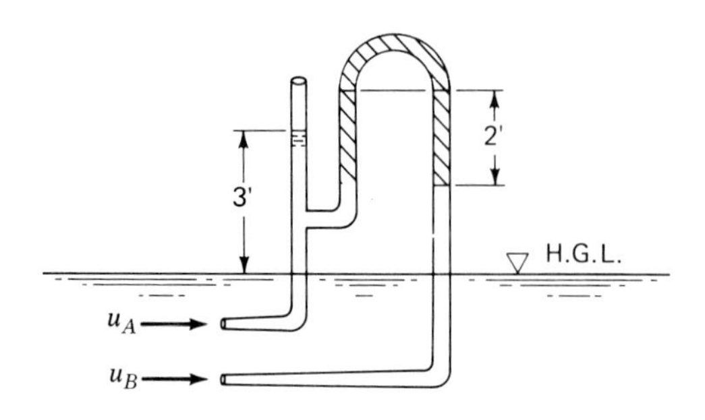

Illustrative Example 4.6

The water surface is coincident with the hydraulic grade line. Hence,

$$\frac{u_A^2}{2g} = 3 \text{ ft}$$

$$u_A = 13.9 \text{ fps}$$

Applying Eq. (2.13), noting that a pitot tube intercepts $z + p/\gamma + u^2/2g$,

$$\frac{u_A^2}{2g} - \frac{u_B^2}{2g} = \frac{24}{12}(1 - 0.82)$$

from which

$$u_B = 13.0 \text{ fps}$$

4.13 LOSS OF HEAD AT SUBMERGED DISCHARGE

When a fluid with a velocity V is discharged from the end of a pipe into a tank or reservoir which is so large that the velocity within it is negligible, the entire kinetic energy of the flow is dissipated. That this is so can be seen by examining Fig. 4.7. In the pipe up to point (*a*) the kinetic energy of the flowing fluid per unit weight of fluid is $V^2/2g$, but at point (*b*) in the tank the velocity is zero and hence the kinetic energy per unit weight of fluid is also zero. Thus the loss of head in this case, *submerged discharge*, is $V^2/2g$. The loss occurs after the fluid leaves the end of the pipe. This is a situation where fast moving fluid impinges on stationary fluid. It is an impact situation not unlike the case where a fast moving mudball collides with an immovable wall. The loss of head at submerged discharge is $V^2/2g$ irrespective of whether the fluid is ideal or real, compressible or incompressible.[1]

[1] A more detailed discussion of the loss of head at submerged discharge is presented in Sec. 8.17.

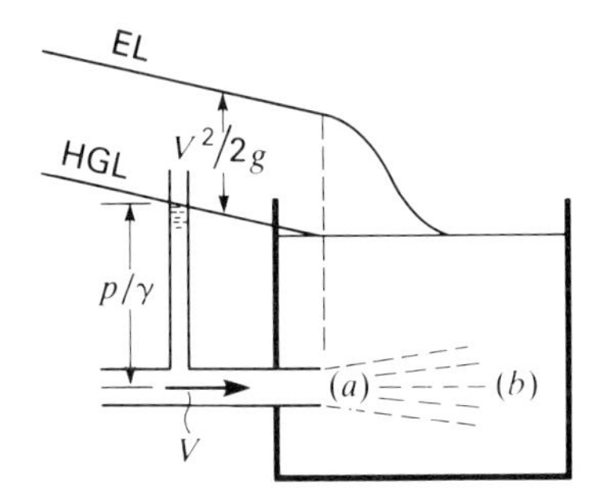

Figure 4.7 Submerged discharge ($h_L = V^2/2g$).

4.14 APPLICATION OF HYDRAULIC GRADE LINE AND ENERGY LINE

Familiarity with the concept of the energy line and hydraulic grade line is useful in the solution of flow problems involving incompressible fluids. If a piezometer tube is erected at B in Fig. 4.8, the liquid will rise in it to a height BB' equal to the pressure head existing at that point. If the end of the pipe at E were closed so that no flow would take place, the height of this column would then be BM. The drop from M to B' when flow occurs is due to two factors, one of these being that a portion of the pressure head has been converted into the velocity head which the liquid has at B, and the other than there has been a loss of head due to fluid friction between A and B.

If a series of piezometers were erected along the pipe, the liquid would rise in them to various levels. The line drawn through the summits of such an imaginary series of liquid columns is called the *hydraulic grade line*. It may be observed that the hydraulic grade line represents what would be the free surface if one could exist and maintain the same conditions of flow.

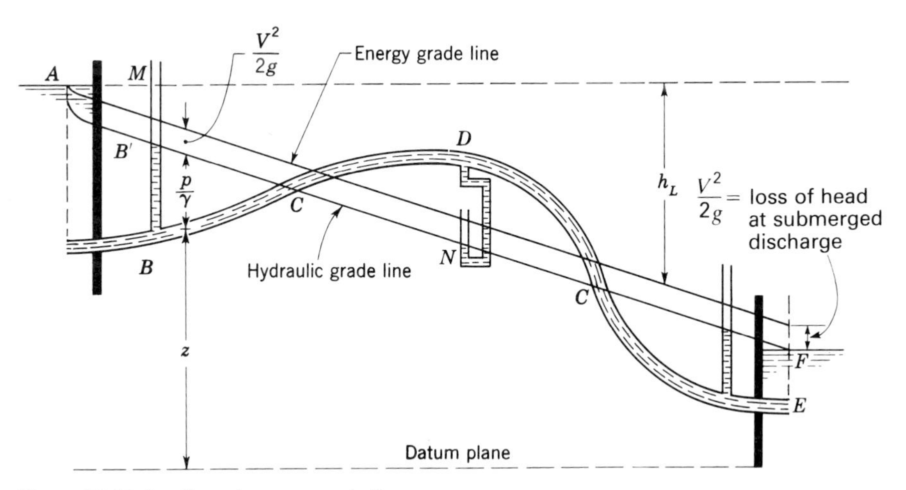

Figure 4.8 Hydraulic and energy grade lines.

The hydraulic grade line indicates the pressure along the pipe, as at any point the vertical distance from the pipe to the hydraulic grade line is the pressure head at that point, assuming the profile to be drawn to scale. At C this distance is zero, thus indicating that the absolute pressure within the pipe at that point is atmospheric. At D the pipe is above the hydraulic grade line, indicating that there the pressure head is $-DN$, or a vacuum of DN ft (or m) of liquid.

If the profile of a pipeline is drawn to scale, not only does the hydraulic grade line enable the pressure head to be determined at any point by measurement on the diagram, but it shows by mere inspection the variation of the pressure in the entire length of the pipe. The hydraulic grade line is a straight line only if the pipe is straight and of uniform diameter. But for the gradual curvatures that are often found in long pipelines, the deviation from a straight line will be small. Of course, if there are local losses of head, aside from those due to normal pipe friction, there may be abrupt drops in the hydraulic grade line. Changes in diameter with resultant changes in velocity will also cause abrupt changes in the hydraulic grade line.

If the velocity head is constant, as in Fig. 4.8, the drop in the hydraulic grade line between any two points is the value of the loss of head between those two points, and the slope of the hydraulic grade line is then a measure of the rate of loss. Thus in Fig. 4.9 the rate of loss in the larger pipe is much less than in the smaller pipe. If the velocity changes, the hydraulic grade line might actually rise in the direction of flow, as shown in Figs. 4.9 and 4.10.

The vertical distance from the level of the surface at A in Fig. 4.8 down to the hydraulic grade line represents $V^2/2g + h_L$ from A to any point in question. Hence the position of the grade line is independent of the position of the pipe. Thus it is not necessary to compute pressure heads at various points in the pipe to plot the hydraulic grade line. Instead, values of $V^2/2g + h_L$ from A to various points can be laid off below the horizontal line through A, and this procedure is often more convenient. If the pipe is of uniform diameter, it is necessary to locate only a few points, and often only two are required.

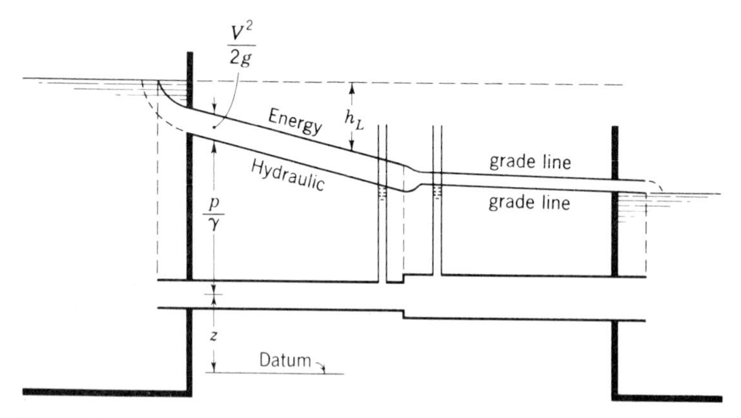

Figure 4.9 *(Plotted to scale from measurements made by Daugherty.)*

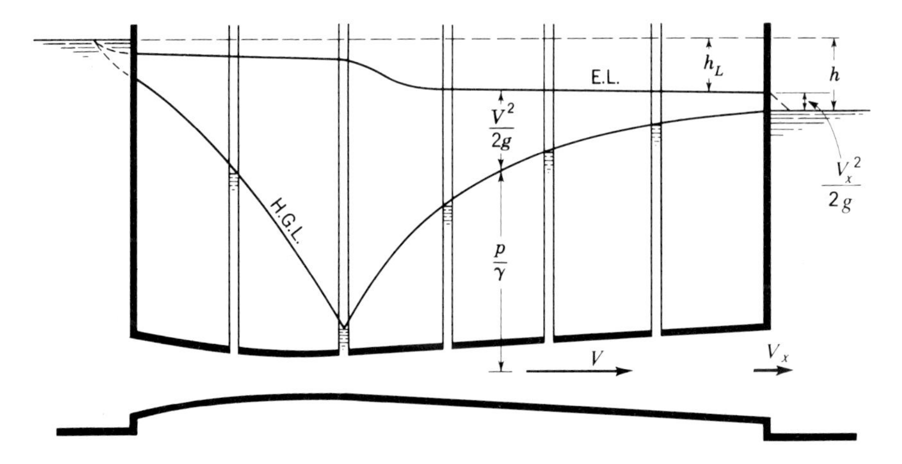

Figure 4.10 *(Plotted to scale from measurements made by Daugherty.)*

If Fig. 4.8 represents to scale the profile of a pipe of uniform diameter, the hydraulic grade line can be drawn as follows. At the intake to the pipe there will be a drop below the surface at A, which should be laid off equal to $V^2/2g$ plus a local entrance loss. (This latter is explained in Chap. 8.) At E the pressure is EF, and hence the grade line must end at F. If the pipe discharged freely into the air at E, the line would pass through E. The location of other points, such as B' and N, may be computed if desired. In the case of a long pipe of uniform diameter the error is very small if the hydraulic grade line is drawn as a straight line from the liquid surface directly above the intake to the liquid surface directly above the discharge end of the pipe if the latter is submerged or to the end of the pipe if there is a free discharge into the atmosphere.

If values of h_L are laid off below the horizontal line through A, the resulting line represents values of the total energy head H measured above any arbitrary datum plane inasmuch as the line is above the hydraulic grade line a distance equal to $V^2/2g$. This line is the *energy grade line*[1]. It shows the rate at which the energy decreases, and it must always drop downward in the direction of flow unless there is an energy input from a pump. The energy grade line is also independent of the position of the pipeline.

Energy grade lines are shown in Figs. 4.8 to 4.10. The last one, plotted to scale from measurements made by Daugherty, shows that the chief loss of head is in the diverging portion and just beyond the section of minimum diameter. In all three of these cases there is a submerged discharge and hence the velocity head is lost at discharge. In Fig. 4.10 the loss is greatly reduced by the conical diffuser which results in an enlarged discharge area and hence a reduced velocity at discharge.

[1] The energy grade line is usually referred to as simply the energy line.

4.15 METHOD OF SOLUTION OF FLOW PROBLEMS

For the solutions of problems of liquid flow there are two fundamental equations, the equation of continuity (3.10) and the energy equation in one of the forms from Eqs. (4.5) to (4.10). The following procedure may be employed:

1. Choose a datum plane through any convenient point.
2. Note at what sections the velocity is known or is to be assumed. If at any point the section area is great compared with its value elsewhere, the velocity head is so small that it may be disregarded.
3. Note at what points the pressure is known or is to be assumed. In a body of liquid at rest with a free surface the pressure is known at every point within the body. The pressure in a jet is the same as that of the medium surrounding the jet.
4. Note whether or not there is any point where all three terms, pressure, elevation, and velocity, are known.
5. Note whether or not there is any point where there is only one unknown quantity.

It is generally possible to write an energy equation that will fulfill conditions 4 and 5. If there are two unknowns in the equation, then the continuity equation must be used also. The application of these principles is shown in the following illustrative examples.

Illustrative Example 4.7 A pipeline with a pump leads to a nozzle as shown in the accompanying figure. Find the flow rate when the pump develops a head of 80 ft. Assume that the head loss in the 6-in-diameter pipe may be expressed by $h_L = 5V_6^2/2g$, while the head loss in the 4-in-diameter pipe is $h_L = 12V_4^2/2g$. Sketch the energy line and hydraulic grade line, and find the pressure head at the suction side of the pump.

Select the datum as the elevation of the water surface in the reservoir. Note from continuity that

$$V_6 = (\tfrac{3}{6})^2 V_3 = 0.25V_3 \qquad \text{and} \qquad V_4 = (\tfrac{3}{4})^2 V_3 = 0.563V_3$$

where V_3 is the jet velocity. Writing an energy equation from the surface of the reservoir to the jet,

$$\left(z_1 + \frac{p_1}{\gamma} + \frac{V_1^2}{2g}\right) - h_{L_6} + h_p - h_{L_4} = z_3 + \frac{p_3}{\gamma} + \frac{V_3^2}{2g}$$

$$0 + 0 + 0 - 5\frac{V_6^2}{2g} + 80 - 12\frac{V_4^2}{2g} = 10 + 0 + \frac{V_3^2}{2g}$$

Express all velocities in terms of V_3:

$$-\frac{5(0.25V_3)^2}{2g} + 80 - 12\frac{(0.563V_3)^2}{2g} = 10 + \frac{V_3^2}{2g}$$

$$V_3 = 29.7 \text{ fps}$$

$$Q = A_3 V_3 = \frac{\pi}{4}\left(\frac{3}{12}\right)^2 29.7 = 1.45 \text{ cfs}$$

Head loss in suction pipe:

$$h_L = 5\frac{V_6^2}{2g} = \frac{5(0.25V_3)^2}{2g} = \frac{0.312V_3^2}{2g}$$

$$= 4.3 \text{ ft}$$

Head loss in discharge pipe:

$$h_L = 12\frac{V_4^2}{2g} = \frac{12(0.563V_3)^2}{2g} = 52.1 \text{ ft}$$

$$\frac{V_3^2}{2g} = 13.7 \text{ ft} \qquad \frac{V_4^2}{2g} = 4.3 \text{ ft} \qquad \frac{V_6^2}{2g} = 0.86 \text{ ft} \approx 0.9 \text{ ft}$$

The energy line and hydraulic grade line are drawn on the figure to scale. Inspection of the figure shows that the pressure head on the suction side of the pump is $p_B/\gamma = 14.8$ ft. Likewise, the pressure head at any point in the pipe may be found if the figure is to scale.

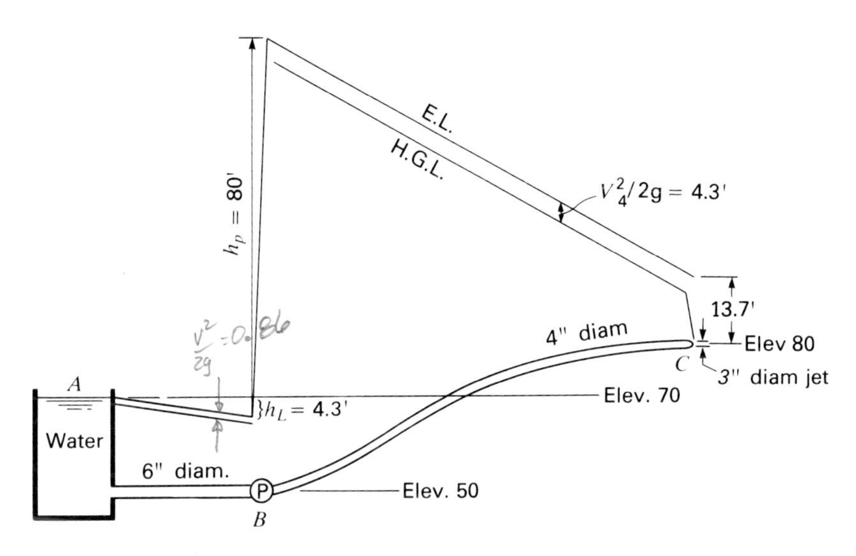

Illustrative Example 4.7

Illustrative Example 4.8 Given the two-dimensional flow as shown in the accompanying figure. Determine the flow rate. Assume no head loss.

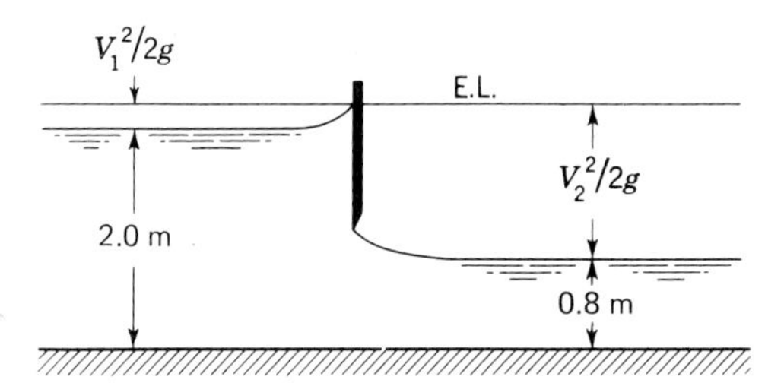

Illustrative Example 4.8

The hydraulic grade line is represented by the water surface in the region where the streamlines are parallel. The energy line is a distance $V^2/2g$ above the water surface, assuming $\alpha = 1.0$. If there is no head loss, the energy line is horizontal. Writing the energy equation from section 1 to 2, we have

$$2.0 + \frac{V_1^2}{2g} = 0.8 + \frac{V_2^2}{2g} \qquad (a)$$

But from continuity,

$$(2 \times 1)V_1 = (0.8 \times 1)V_2 \qquad (b)$$

Substituting Eq. (b) into Eq. (a), and using $g = 9.81\ \text{m/s}^2$

$$V_1 = 2.12\ \text{m/s} \qquad V_2 = 5.30\ \text{m/s}$$

$$Q = (2 \times 1)2.12 = 4.24\ \text{m}^3/\text{s} \text{ (for 1 m of width perpendicular to figure)}$$

$$\frac{V_1^2}{2g} = 0.23\ \text{m} \qquad \frac{V_2^2}{2g} = 1.43\ \text{m}$$

4.16 PRESSURE IN FLUID FLOW

Strictly speaking the equations that have been derived apply to flow along a single streamline. They may, however, be used for stream tubes of large cross-sectional area by taking average values. The case of pressure variation over a section will now be considered.

Figure 4.11 shows a small prism of a flowing fluid. The forces acting on the faces of the prism at right angles to the direction of flow and in the plane of the sketch are p_1A and p_2A as shown. Forces in the direction of motion balance out if the flow is steady and uniform. Summing forces in the direction at right angles to the flow, we get

$$p_1A + \gamma A y \cos\alpha - p_2A = 0$$

where y is the dimension of the prism as shown, and A is cross-sectional area. From this we get $p_2 - p_1 = \gamma y \cos\alpha = \gamma h = \gamma(z_1 - z_2) = -\gamma(\Delta z)$, which is similar to Eq. (2.3). That is, in any plane normal to the direction of flow, the pressure varies according to the hydrostatic law if the flow is uniform and steady. The

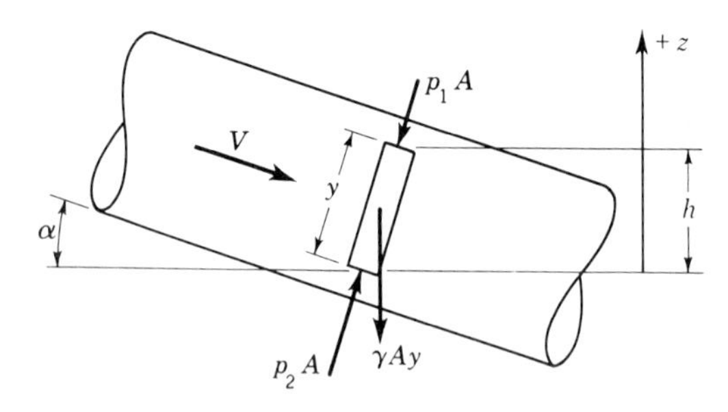

Figure 4.11

average pressure is then the pressure at the center of gravity of such an area. The pressure is lowest near the top of the pipe, and cavitation, if it were to occur, would appear there first. On a horizontal diameter through the pipe the pressure is everywhere the same. Since the velocity is higher near the center than near the walls, it follows that the local energy head is also higher near the center. This emphasizes the fact that a flow equation applies along the same streamline, but not between two streamlines, any more than between two streams in two separate channels.

Static Pressure

In a flowing fluid the pressure measured at right angles to the flow is called the *static pressure*. This is the value given by piezometer tubes and other devices explained in Sec. 12.2.

Stagnation Pressure

The center streamline in Fig. 3.11 shows that the velocity becomes zero at the stagnation point. If p_0/γ denotes the static-pressure head at some distance away where the velocity is V_0, while p_s/γ denotes the pressure head at the stagnation point, then, applying Eq. (4.10) to these two points, $p_0/\gamma + 0 + V_0^2/2g = p_s/\gamma + 0 + 0$, or the stagnation pressure is

$$p_s = p_0 + \gamma \frac{V_0^2}{2g} = p_0 + \rho \frac{V_0^2}{2} \tag{4.31}$$

The quantity $\gamma V_0^2/2g$, or $\rho V_0^2/2$, is sometimes called the *dynamic pressure*.

Equation (4.31) applies to a fluid where compressibility may be disregarded. In Sec. 9.4 it is shown that for a compressible fluid.

$$p_s = p_0 + \gamma_0 \frac{V_0^2}{2g}\left(1 + \frac{V_0^2}{4c^2} + \cdots\right) \tag{4.32}$$

where c is the sonic velocity (Appendix 2). For air at 68°F(20°C), $c \approx 1{,}130$ fps (345 m/s). If $V_0 = 226$ fps (69 m/s) the error in neglecting the compressibility factor, which is the value in the parentheses, is only 1 per cent. But for higher values of V_0, the effect becomes much more important. Equation (4.32) is, however, restricted to values of V_0/c less than 1.

4.17 JET TRAJECTORY

A free liquid jet in air will describe a *trajectory*, or path under the action of gravity, with a vertical velocity component which is continually changing. The trajectory is a streamline, and consequently, if air friction is neglected, Bernoulli's

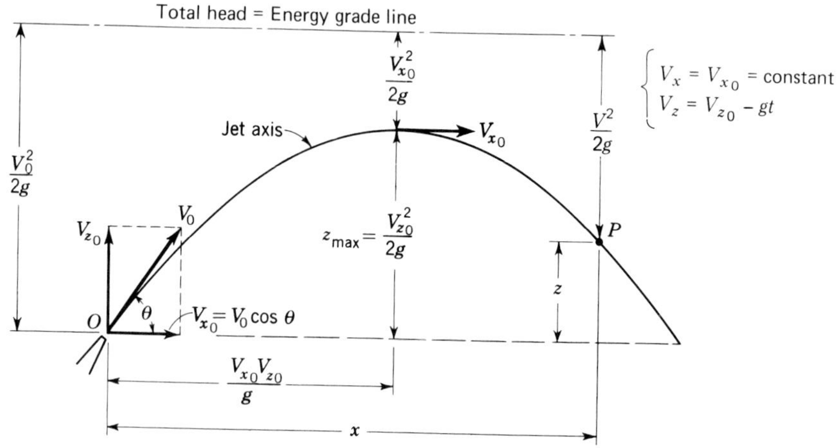

Figure 4.12 Jet trajectory

theorem may be applied to it, with all the pressure terms zero. Thus the sum of the elevation and velocity head must be the same for all points of the curve. The energy grade line is a horizontal line at distance $V_0^2/2g$ above the nozzle, where V_0 is the initial velocity of the jet as it leaves the nozzle (Fig. 4.12).

The equation for the trajectory may be obtained by applying Newton's equations of uniformly accelerated motion to a particle of the liquid passing from the nozzle to point P, whose coordinates are x, z, in time t. Then $x = V_{x_0}t$ and $z = V_{z_0}t - \frac{1}{2}gt^2$. Evaluating t from the first equation and substituting it in the second gives

$$z = \frac{V_{z_0}}{V_{x_0}}x - \frac{g}{2V_{x_0}^2}x^2 \tag{4.33}$$

By setting $dz/dx = 0$, we find that z_{max} occurs when $x = V_{x_0}V_{z_0}/g$. Substituting this value for x in Eq. (4.33) gives $z_{max} = V_{z_0}^2/2g$. Thus Eq. (4.33) is that of an inverted parabola having its vertex at $x_0 = V_{x_0}V_{z_0}/g$ and $z_0 = V_{z_0}^2/2g$. Since the velocity at the top of the trajectory is horizontal and equal to V_{x_0}, the distance from this point to the energy gradient is evidently $V_{x_0}^2/2g$. This may be obtained in another way by considering that $V_0^2 = V_{x_0}^2 + V_{z_0}^2$. Dividing each term by $2g$ gives the relations shown in Fig. 4.12.

If the jet is initially horizontal, as in the flow from a vertical orifice, $V_{x_0} = V_0$ and $V_{z_0} = 0$. Equation (4.33) is then readily reduced to an expression for the initial jet velocity in terms of the coordinates from the vena contracta (Fig. 12.13) to any point of the trajectory, z now being positive downward:

$$V_0 = x\sqrt{\frac{g}{2z}} \tag{4.34}$$

Illustrative Example 4.9 If a jet is inclined upward 30° from the horizontal, what must be its velocity to reach over a 10-ft wall at a horizontal distance of 60 ft, neglecting friction?

$$V_{x_0} = V_0 \cos 30° = 0.866\, V_0$$

$$V_{z_0} = V_0 \sin 30° = 0.5\, V_0$$

From Newton's laws,

$$x = 0.866\, V_0 t = 60$$

$$z = 0.5\, V_0 t - 16.1 t^2 = 10$$

From the first equation, $t = 69.3/V_0$. Substituting this in the second equation.

$$0.5\, V_0 \frac{69.3}{V_0} - \frac{1}{2} \times 32.2 \left(\frac{69.3}{V_0}\right)^2 = 10$$

from which $V_0^2 = 3{,}140$, or $V_0 = 56$ fps.

4.18 FLOW IN A CURVED PATH

The energy equations previously developed apply fundamentally to flow along a streamline or along a stream of large cross section if certain average values are used. Now conditions will be investigated in a direction normal to a streamline. In Fig. 4.13 is shown *an element of fluid moving in a horizontal plane*[1] with a velocity V along a curved path of radius r. The element has a linear dimension dr in the plane of the paper and an area dA normal to the plane of the paper. The mass of this fluid element is $\rho\, dA\, dr$, and the normal component of acceleration is V^2/r. Thus the centripetal force acting upon the element toward the center of curvature is $\rho\, dA\, dr V^2/r$. As the radius increases from r to $r + dr$, the pressure

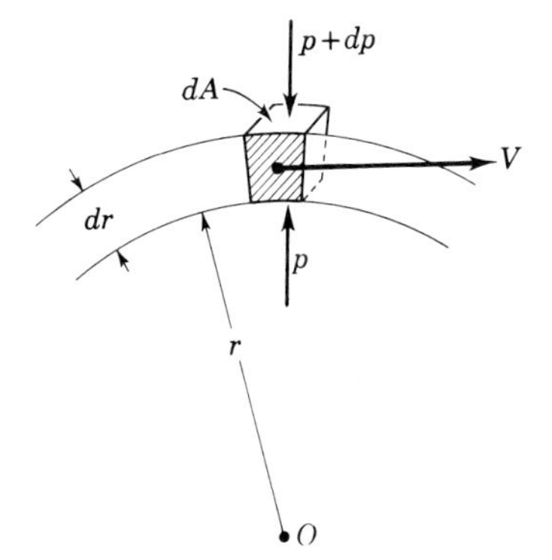

Figure 4.13

[1] A more generalized analysis of flow along a curved path in a vertical or inclined plane leads to a result that includes z terms.

will change from p to $p + dp$. Thus the resultant force in the direction of the center of curvature is $dp\,dA$. Equating these two forces,

$$dp = \rho \frac{V^2}{r} dr \tag{4.35}$$

When horizontal flow is in a straight line for which r is infinity, the value of dp is zero. That is, no difference in pressure can exist in the horizontal direction transverse to horizontal flow in a straight line.

As dp is positive if dr is positive, the equation shows that pressure increases from the concave to the convex side of the stream, but the exact way in which it increases depends upon the way in which V varies with the radius. In the next two sections two important practical cases will be presented in which V varies in two different ways.

4.19 FORCED OR ROTATIONAL VORTEX

A fluid may be made to rotate as a solid body without relative motion theoretically between particles, either by the rotation of a containing vessel or by stirring the contained fluid, so as to force it to rotate. Thus an *external torque* is applied. A common example is the rotation of liquid within a centrifugal pump or of gas in a centrifugal compressor.

Cylindrical Forced Vortex

If the entire body of fluid rotates as a solid, then V varies directly with r; that is, $V = r\omega$, where ω is the imposed angular velocity. Inserting this value in Eq. (4.35), for the case of rotation about a vertical axis in any horizontal plane we have

$$dp = \rho\omega^2 r\,dr = \frac{\gamma}{g}\omega^2 r\,dr$$

Between any two radii r_1 and r_2, this integrates as

$$\frac{p_2}{\gamma} - \frac{p_1}{\gamma} = \frac{\omega^2}{2g}(r_2^2 - r_1^2) \tag{4.36}$$

If p_0 is the value of the pressure when $r_1 = 0$, this becomes

$$\frac{p}{\gamma} = \frac{\omega^2}{2g}r^2 + \frac{p_0}{\gamma} \tag{4.37}$$

which is seen to be the equation of a parabola. In Fig. 4.14a it is seen that, if the fluid is a liquid, the pressure head p/γ at any point is equal to z, the depth of

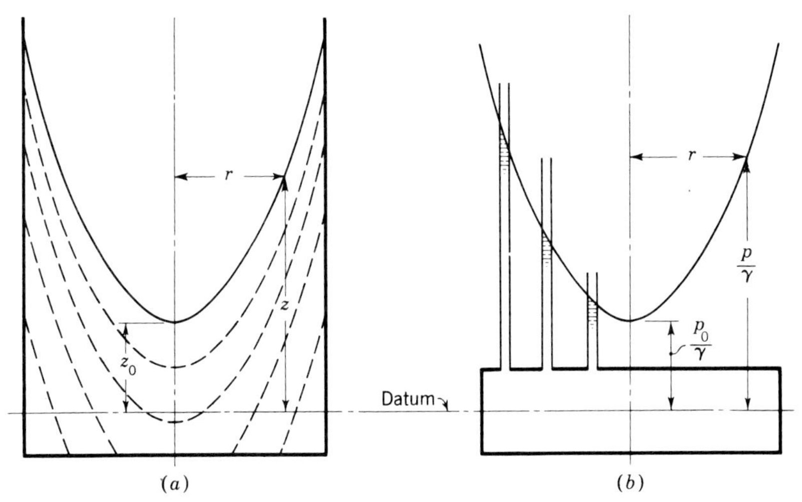

Figure 4.14 Forced vortex. (*a*) Open vessel. (*b*) Closed vessel.

the point below the free surface. Hence the preceding equations may also be written as

$$z_2 - z_1 = \frac{\omega^2}{2g}(r_2^2 - r_1^2) \tag{4.38}$$

and

$$z = \frac{\omega^2}{2g} r^2 + z_0 \tag{4.39}$$

where z_0 is the elevation when $r_1 = 0$. Equations (4.38) and (4.39) are the equations of the free surface, if one exists, or in any case are the equations for any surface of equal pressure; these are a series of paraboloids as shown by the dotted lines in Fig. 4.14*a*.

For the open vessel shown in Fig. 4.14*a*, the pressure head at any point is equal to its depth below the free surface. If the liquid is confined within a vessel, as shown in Fig. 4.14*b*, the pressure along any radius will vary in just the same way as if there were a free surface. Hence the two are equivalent.

In the preceding discussion the axis of the vessel was assumed to be vertical; however the axis might be inclined. Since pressure varies with elevation as well as radius, a more general equation applicable to fluid in a closed tank with an inclined axis is

$$\frac{p_2}{\gamma} - \frac{p_1}{\gamma} + z_2 - z_1 = \frac{\omega^2}{2g}(r_2^2 - r_1^2) \tag{4.40}$$

Equation (4.36) is the special case where $z_1 = z_2$ (closed tank with vertical axis), and Eq. (4.38) is the special case where $p_1 = p_2$ (open tank with vertical axis). It should be noted that Eqs. (4.36) through (4.40) are not energy equations since they represent conditions across streamlines rather than along a streamline.

Spiral Forced Vortex

So far the discussion has been confined to the rotation of all particles in concentric circles. Suppose that there is now superimposed a flow with a velocity having radial components, either outward or inward. If the height of the walls of the open vessel in Fig. 4.14*a* were less than that of the liquid surface, and if liquid were supplied to the center at the proper rate by some means, then it is obvious that liquid would flow outward. If, on the other hand, liquid flowed into the tank over the rim from some source at a higher elevation and were drawn out at the center, the flow would be inward. The combination of this approximately radial flow with the circular flow would result in path lines that were some form of spirals.

If the closed vessel in Fig. 4.14*b* is arranged with suitable openings near the center and also around the periphery, and if it is provided with vanes, as shown in Fig. 4.15, it becomes either a centrifugal pump impeller or a turbine runner, as the case may be. These vanes constrain the flow of the liquid and determine both its relative magnitude and its direction. If the area of the passages normal to the direction of flow is a, the equation of continuity fixes the relative velocities, since

$$Q = a_1 v_1 = a_2 v_2 = \text{constant}$$

This relative flow is the flow as it would appear to an observer or a camera, revolving with the rotor. The pressure difference due to this superimposed flow alone is found by the energy equation, neglecting friction losses and assuming vertical axis of rotation, to be $p_2/\gamma - p_1/\gamma = (v_1^2 - v_2^2)/2g$.

Hence, for the case of rotation with flow (i.e., spiral forced vortex), the total pressure difference between two points is found by adding together the pressure differences due to the two flows considered separately. That is, for the case of a vertical axis,

$$\frac{p_2}{\gamma} - \frac{p_1}{\gamma} = \frac{\omega^2}{2g}(r_2^2 - r_1^2) + \frac{v_1^2 - v_2^2}{2g} \tag{4.41}$$

Of course, friction losses will modify this result to some extent. If the axis is inclined, z terms must be added to the equation. It is seen that Eq. (4.36) is a special case of Eq. (4.41) when $v_1 = v_2$ either when both are finite or when $v_1 = v_2 = 0$.

For a forced vortex with spiral flow, energy is put into the fluid in the case of a pump and extracted from it in the case of a turbine. In the limiting case of zero

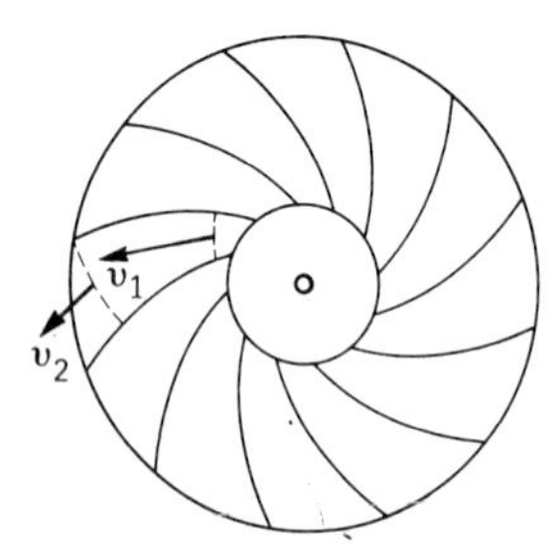

Figure 4.15 Flow through rotor.

flow, when all path lines become concentric circles (i.e., cylindrical forced vortex), energy input from some external source is still necessary for any real fluid in order to maintain the rotation.

4.20 FREE OR IRROTATIONAL VORTEX

In the free vortex there is no expenditure of energy whatever from an outside source, and the fluid rotates by virtue of some rotation previously imparted to it or because of some internal action. Some examples are a whirlpool in a river, the rotary flow that often arises in a shallow vessel when liquid flows out through a hole in the bottom (as is often seen when water empties from a bathtub), and the flow in a centrifugal-pump case just outside the impeller or that in a turbine case as the water approaches the guide vanes.

As no energy is imparted to the fluid, it follows that neglecting friction, H is constant throughout; that is, $p/\gamma + z + V^2/2g = \text{constant}$.

Cylindrical Free Vortex

The angular momentum with respect to the center of rotation of a particle of mass m moving along a circular path of radius r at a velocity V_u is mV_ur where V_u is the velocity along the circular path (i.e., tangential velocity).[1] Newton's second law states that, for the case of rotation, the torque is equal to the time rate of change of angular momentum. Hence, torque $= d(mV_ur)/dt$. In the case of a free vortex (frictionless fluid) there is no torque applied; therefore, $mVr = \text{constant}$ and thus $V_ur = C$, where the value of C is determined by knowing the value of V at some radius r. Assuming a vertical axis of rotation and inserting $V_u = C/r$ in Eq. (4.35), we obtain

$$dp = \rho \frac{C^2}{r^2} \frac{dr}{r} = \frac{\gamma}{g} \frac{C^2}{r^3} dr$$

Between any two radii r_1 and r_2 this integrates as

$$\frac{p_2}{\gamma} - \frac{p_1}{\gamma} = \frac{C^2}{2g}\left(\frac{1}{r_1^2} - \frac{1}{r_2^2}\right) = \frac{V_{u1}^2}{2g}\left[1 - \left(\frac{r_1}{r_2}\right)^2\right] \tag{4.42}$$

If there is a free surface, the pressure head p/γ at any point is equal to the depth below the surface. Also, at any radius the pressure varies in a vertical direction according to the hydrostatic law. Hence this equation is merely a special case where $z_1 = z_2$.

As $H = p/\gamma + z + V_u^2/2g = \text{constant}$, it follows that at any radius r

$$\frac{p}{\gamma} + z = H - \frac{V_u^2}{2g} = H - \frac{C^2}{2gr^2} = H - \frac{V_{u1}^2}{2g}\left(\frac{r_1}{r}\right)^2 \tag{4.43}$$

[1] In this chapter V_u is used to represent the tangential component of velocity. Other symbols commonly used to represent tangential velocity are V_t and V_θ.

Assuming the axis to be vertical, the pressure along the radius can be found from this equation by taking z constant; and for any constant pressure p, values of z determining a surface of equal pressure, can be found. If p is zero, the values of z determine the free surface (Fig. 4.16*a*), if one exists.

Equation (4.43) shows that H is the asymptote approached by $p/\gamma + z$ as r approaches infinity and V_u approaches zero. On the other hand, as r approaches zero, V_u approaches infinity, and $p/\gamma + z$ approaches minus infinity. Since this is physically impossible, the free vortex cannot extend to the axis of rotation. In reality, since high velocities are attained as the axis is approached, the friction losses, which vary as the square of the velocity, become of increasing importance and are no longer negligible. Hence the assumption that H is constant no longer holds. The core of the vortex tends to rotate as a solid body as in the central part of Fig. 4.16*b*.

Spiral Free Vortex

If a radial flow is superimposed upon the concentric flow previously described, the path lines will then be spirals. If the flow is out through a circular hole in the bottom of a shallow vessel, the surface of a liquid takes the form shown in Fig. 4.16*a*, with an air *core* sucked down the hole. If an outlet symmetrical with the axis is provided in the arrangement shown in Fig. 4.16*b*, we might have a flow component either radially inward or radially outward. If the two plates shown are a constant distance B apart, the radial flow component with a velocity V_r is then across a series of concentric cylindrical surfaces whose area is $2\pi rB$. Thus

$$Q = 2\pi rBV_r = \text{constant}$$

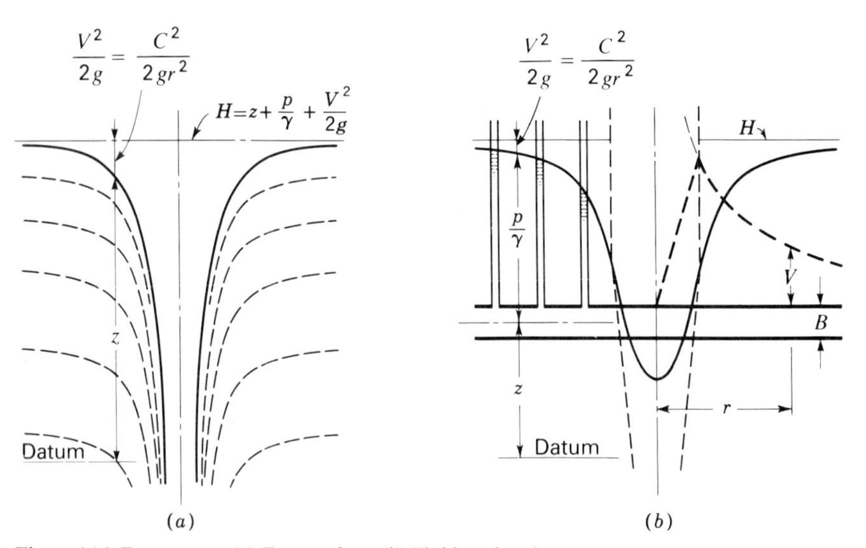

Figure 4.16 Free vortex. (*a*) Free surface. (*b*) Fluid enclosed.

from which it is seen that rV_r = constant. Thus the radial velocity varies in the same way with r that the circumferential velocity did in the preceding discussion of the free cylindrical vortex. The pressure variation in a spiral free vortex (Fig. 4.16*b*) is given by

$$\frac{p_2}{\gamma} - \frac{p_1}{\gamma} = \frac{V_1^2}{2g} - \frac{V_2^2}{2g} \tag{4.44}$$

where $V = \sqrt{V_r^2 + V_u^2}$, the velocity of flow.

Illustrative Example 4.10 This centrifugal pump with a 12-in-diameter impeller is surrounded by a casing which has a constant height of 1.5 in between sections a and b and then enlarges into a volute at c. Water leaves the impeller with a velocity of 60 fps at an angle of 15° with the tangent. (*a*) At what rate is water flowing through the pump? (*b*) Neglecting friction, what will be the magnitude and direction of the velocity at b and what will be the gain in pressure head from a to b?

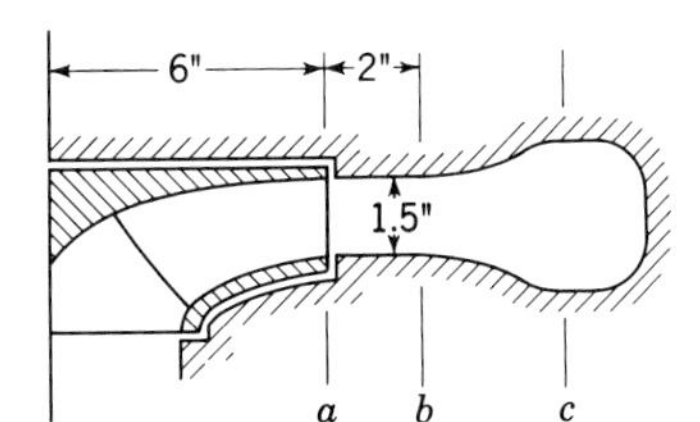

Illustrative Example 4.10

(*a*) Flow through the pump $Q = 2\pi r_a B(V_r)_a$ where $(V_r)_a = 60 \sin 15° = 15.53$ fps

$$Q = 2\pi(6/12)(1.5/12) \times 15.53 = 6.10 \text{ cfs}$$

(*b*) From continuity: $Q = 2\pi r_a B(V_r)_a = 2\pi r_b B(V_r)_b$,

hence
$$(V_r)_b/(V_r)_a = r_a/r_b = 6/8$$

Since torque = 0 in the space between a and b, angular momentum must be conserved.

Thus
$$m(V_u)_a r_a = m(V_u)_b r_b$$

and
$$(V_u)_b/(V_u)_a = r_a/r_b = 6/8$$

The region between a and b is a spiral free vortex.

Since V_u and V_r are both seen to decrease in the same proportion as flow occurs from a to b, the angle α does not change and $V_b/V_a = 6/8$; $V_b = (6/8) \times 60 = 45$ fps. Finally, writing the energy equation along the flow lines gives,

$$p_b/\gamma - p_a/\gamma = V_a^2/2g - V_b^2/2g = \frac{60^2 - 45^2}{2g} = 24.5 \text{ ft}$$

PROBLEMS

4.1 In laminar flow through a circular pipe the velocity profile is a parabola, the equation of which is $u = u_m[1 - (r/r_0)^2]$, where u is the velocity at any radius r, u_m is the maximum velocity in the center of the pipe where $r = 0$, and r_0 is the radius to the wall of the pipe. From Prob. 3.2, $V = 0.5u_m$. Prove that $\alpha = 2$. (*Note:* Let $dA = 2\pi r\, dr$.)

4.2 Assume the velocity profile for turbulent flow in a circular pipe to be approximated by a parabola from the axis to a point very close to the wall where the local velocity is $u = 0.7u_m$, where u_m is the maximum velocity at the axis. The equation for this parabola is $u = u_m[1 - 0.3(r/r_0)^2]$. Prove that $\alpha = 1.03$.

4.3 Assume an open rectangular channel with the velocity at the surface twice that at the bottom and with the velocity varying as a straight line from top to bottom. Prove that $\alpha = \frac{10}{9}$.

4.4 Find α for the case of a two-dimensional laminar flow.

4.5 Assume frictionless flow in a long, horizontal, conical pipe, the diameter of which is 2 ft at one end and 4 ft at the other. The pressure head at the smaller end is 16 ft of water. If water flows through this cone at the rate of 125.6 cfs, find the velocities at the two ends and the pressure head at the larger end.

4.6 Water flows through a long, horizontal, conical diffuser at the rate of 4.0 m^3/s. The diameter of the diffuser changes from 1.0 m to 1.5 m. The pressure at the smaller end is 7.5 kN/m^2. Find the pressure at the downstream end of the diffuser, assuming frictionless flow. Assume also, that the angle of the cone is so small that separation of the flow from the walls of the diffuser does not occur.

4.7 A vertical pipe 6 ft (2.0 m) in diameter and 60 ft (20 m) long has a pressure head at the upper end of 18 ft (5 m) of water. When the flow of water through it is such that the mean velocity is 15 fps (5 m/s), the friction loss is $h_L = 4$ ft (1.25 m). Find the pressure head at the lower end of the pipe when the flow is (*a*) downward; (*b*) upward.

4.8 A conical pipe has diameters at the two ends of 1.5 and 4.5 ft and is 50 ft long. It is vertical, and the friction loss is $h_L = 8$ ft for flow of water in either direction when the velocity at the smaller section is 30 fps. If the smaller section is at the top and the pressure head there is 6.5 ft of water, find the pressure head at the lower end when the flow is (*a*) downward; (*b*) upward.

4.9 A pipeline supplies water to a hydroelectric power plant, the elevation of which is 2,000 ft (600 m) below the level of the water at intake to the pipe. If 10 per cent of this total, or 200 ft (60 m), is lost in friction in the pipe, what will be the value of ΔI in Btu/lb (J/N), and what will be the rise in temperature if there is no heat transfer?

4.10 Water is flowing at 12 m^3/s through a long pipe. The temperature of the water drops 0.25°C even though heat is transferred to the water at the rate of 6000 kN·m/s. Find the head loss in the pipe.

4.11 A pipeline supplies water to a hydroelectric plant from a reservoir in which the water temperature is 60°F. (*a*) Suppose that in the length of the pipe there is a total loss of heat to the surrounding air of 0.30 Btu/lb of water and the temperature of the water at the power house is 59.9°F. What is the friction loss per pound of water? (*b*) With the same flow rate as in (*a*) what will be the temperature of the water at the power house if there is absorption of heat from hot sunshine at the rate of 3.0 Btu/lb of water?

4.12 In the figure, the pipe *AB* is of uniform diameter. The pressure at *A* is 20 psi and at *B* is 30 psi. In which direction is the flow, and what is the friction loss in feet of the fluid if the liquid has a specific weight of (*a*) 30 lb/ft^3; (*b*) 100 lb/ft^3?

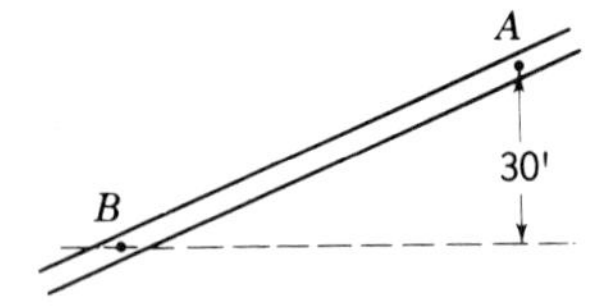

Problem 4.12

4.13 Refer to the figure for Prob. 4.12. If the difference in elevation between *A* and *B* is 8 m and the pressures at *A* and *B* are 150 and 250 kN/m^2 respectively, find the direction of flow and the head loss in meters of liquid. Assume the liquid has a specific gravity of 0.85.

4.14 A pipeline conducts water from a reservoir to a powerhouse, the elevation of which is 800 ft (250 m) lower than that of the surface of the reservoir. The water is discharged through a nozzle with a jet velocity of 220 fps (68 m/s) and the diameter of the jet is 8 in (20 cm). Find the horsepower (kW) of the jet and the horsepower (kW) lost in friction between reservoir and jet.

4.15 A pump lifts water at the rate of 200 cfs (6.0 m^3/s) to a height of 400 ft (120 m) and the friction loss in the pipe is 30 ft (10 m). What is the horsepower (kW) required if the pump efficiency is 90 per cent? Sketch the energy line and hydraulic grade line.

4.16 A turbine is located at an elevation 600 ft below that of the surface of the water at intake. The friction in the pipeline leading to it is 25 ft, and the turbine efficiency is 90 per cent. What will be the horsepower delivered by it if the flow is 100 cfs? Sketch energy line and hydraulic grade line.

4.17 A pump circulates water at the rate of 2,000 gpm (125 L/s) in a closed circuit holding 10,000 gal (40 m^3). The net head developed by the pump is 300 ft (100 m) and the pump efficiency is 90 per cent. Assuming the bearing friction to be negligible and that there is no loss of heat from the system, find the temperature rise in the water in 1 hr.

4.18 The diameters of the suction and discharge pipes of a pump are 6 and 4 in, respectively. The discharge pressure is read by a gage at a point 5 ft above the center line of the pump, and the suction pressure is read by a gage 2 ft below the center line. If the pressure gage reads 20 psi and the suction gage reads a vacuum of 10 in Hg when gasoline ($s = 0.75$) is pumped at the rate of 1.2 cfs, find the power delivered to the fluid. Sketch the energy line and hydraulic grade line.

4.19 Water enters a pump through a 10-in-diameter pipe at 5 psi. It leaves the pump at 20 psi through a 6-in-diameter pipe. If the flow rate is 5 cfs, find the horsepower delivered to the water by the pump. Assume suction and discharge sides of pump are at the same elevation.

4.20 Oil ($s = 0.82$) enters a pump through a 20-cm-diameter pipe at 40 kN/m^2. It leaves the pump at 125 kN/m^2 through a 15-cm-diameter pipe. If the flow rate is 75 L/s, find the rate at which energy is delivered to the oil by the pump. Assume suction and discharge sides of the pump are at the same elevation.

4.21 In Fig. 4.10, neglect all head losses except at discharge and assume water is flowing. If $h = 16$ ft (5 m) and the water surface in the lower reservoir is 15 ft (4.5 m) higher than the constriction, find the highest permissible temperature of the water in order that there be no cavitation. The diameter of the constriction is three-fourths the diameter of the pipe where it joins the downstream tank. Atmospheric pressure is 14.2 psia (98 kN/m^2, abs).

4.22 Repeat Prob. 4.21 but in this case let the water temperature be 60°F. Find the minimum permissible diameter of the constriction (expressed as a fraction of the outlet diameter) in order that there be no cavitation.

4.23 In Fig. 4.10 neglect all head losses except at discharge and assume water is flowing. If $h = 10$ ft (3 m), find the flow rate. What is the gage pressure and what is the absolute pressure in the constriction if the atmospheric pressure is equal to standard atmospheric pressure at 5,000 ft (2,000 m) elevation? Assume that the water surface in the lower reservoir is 15 ft (5 m) above the constriction and that the diameter of the constriction is three-fourths the diameter of the pipe where it joins the downstream tank. Diameter of constriction = 12 in (30 cm).

4.24 Repeat Prob. 4.23 assuming head losses as follows: head loss in converging section of pipe equals to 8 in (0.2 m), head loss in diverging section of pipe equals 27 in (0.7 m).

4.25 Refer to the figure for Illustrative Example 4.5. Find the maximum theoretical flow rate for water at 80°C if the diameters are 60 cm and 20 cm respectively, the upstream pressure is 50 kN/m^2, and the atmosphere pressure is 695 mm of mercury. Neglect head loss.

4.26 Air flows isothermally through a long, horizontal pipe of uniform diameter. At a section where the pressure is 150 psia (1,000 kPa) the velocity is 80 fps (25 m/s). Because of fluid friction, the pressure at a distant section is 30 psia (200 kPa). (*a*) What is the increase in kinetic energy of the air? (*b*) What is the amount of thermal energy in Btu/lb (J/N) of air that must be transferred in order to maintain the temperature constant?

4.27 If the temperature of the air in Prob. 4.26 is 70°F and the diameter of the pipe is 3 in, find the total heat transferred in Btu/hr.

4.28 Air ($\gamma = 0.075\ \text{lb/ft}^3$) is flowing. If $u = 13.0$ fps, and $V_c = 15$ fps, determine the readings on manometers a and b in the figure.

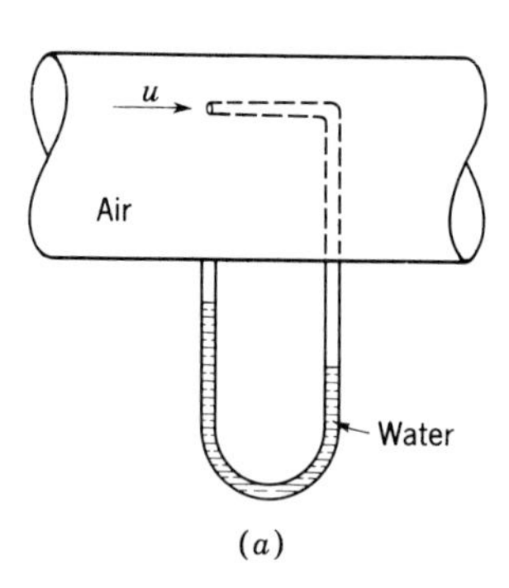

(a)

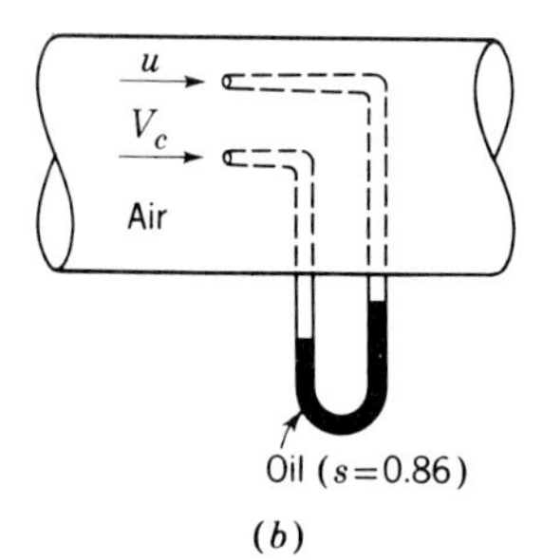

(b) **Problem 4.28**

4.29 Refer to the figure for Prob. 4.28. A gas ($\rho = 1.30\ \text{kg/m}^3$) is flowing. If $u = 50$ m/s and $V_c = 70$ m/s, determine the readings on the manometers in (a) and (b).

4.30 Referring to the figure, assume the flow to be frictionless in the siphon. Find the rate of discharge in cfs (m^3/s) and the pressure head at B if the pipe has a uniform diameter of 6 in (15 cm).

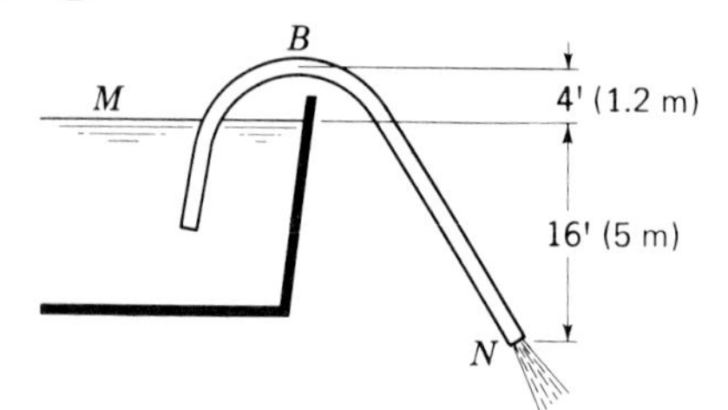

Problem 4.30

4.31 In Prob. 4.30 assume that the friction loss between intake and B is 2 ft (0.6 m) and between B and N is 3 ft (0.9 m). What is the rate of discharge and the pressure head at B?

4.32 Neglect friction and assume that the minimum pressure allowable in the siphon of Prob. 4.30 is a vacuum of -32 ft (-10 m) of water. What would then be the maximum difference in elevation between M and N, instead of the 16 ft (5 m) shown in the sketch?

4.33 Referring to the figure, assume that liquid flows from A to C at the rate of 200 L/s and that the friction loss between A and B is negligible but that between B and C it is $0.1\ V_B^2/2g$. Find the pressure heads at A and C.

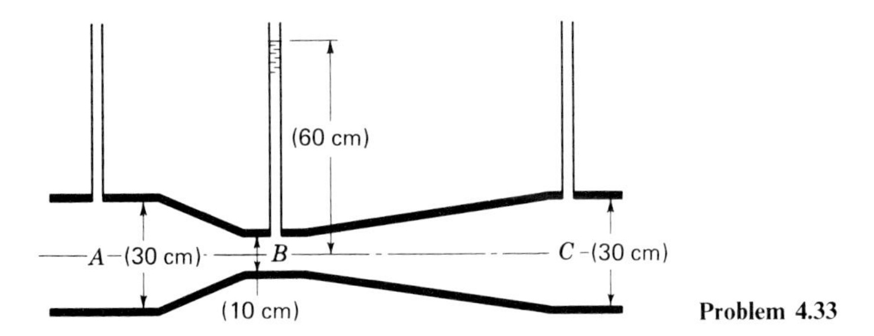

Problem 4.33

4.34 In the figure, the diameter of the vertical pipe is 4 in (10 cm), and that of the stream discharging into the air at E is 3 in (7 cm). Neglecting all losses of energy, what are the pressure heads at B, C, and D?

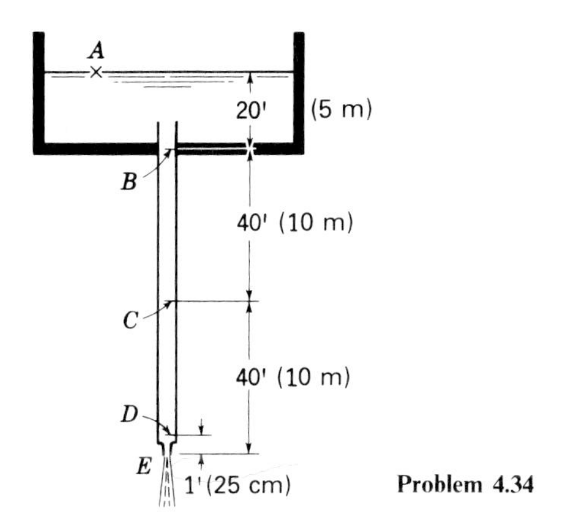

Problem 4.34

4.35 Referring to the figure, at B the diameter of the tube is 1 in, and the diameter of the water jet discharging into the air at C is 1.5 in. (*a*) If $h = 6$ ft and all friction losses are neglected, what are the values of the velocity and the pressure head at B? Assume the tube flows full. (*b*) What is the rate of discharge in cfs? What would it be if the tube were cut off at B?

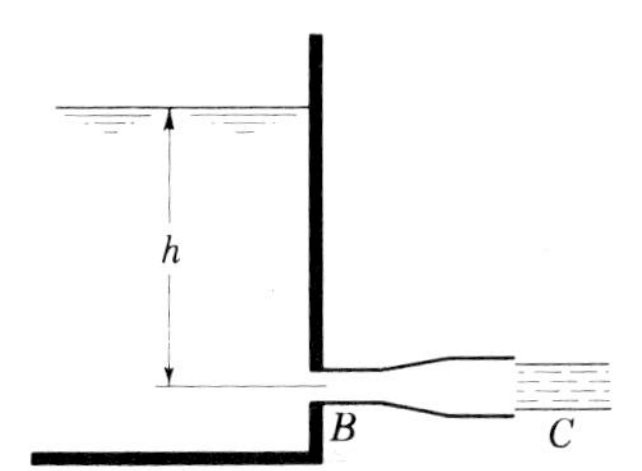

Problem 4.35

4.36 Refer to the figure for Prob. 4.35. Assume the diameter at B is 5 cm, and the diameter of the jet discharging into the air is 6.5 cm. If $h = 3.5$ m and all friction losses are neglected, what is the flow rate assuming the tube flows full? What is the pressure head at B? What would be the flow rate if the tube were cut off at B?

4.37 Pump P in the figure draws up liquid in a suction pipe at a velocity of 10 fps (3 m/s). Assume that the friction losses in the pipe are $2V^2/2g$ and that the barometric pressure is 14.50 psia (100 kN/m², abs). What would be the maximum allowable value of z if the liquid were (*a*) water at a temperature of 60°F (15°C); (*b*) gasoline with a vapor pressure of 8 psia (50 kN/m², abs) and a specific weight of 45 lb/ft³ (7 kN/m³)?

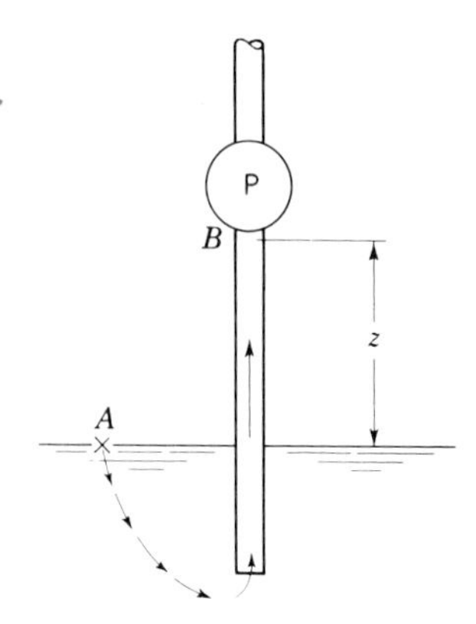

Problem 4.37

4.38 Take the situation depicted in Illustrative Example 4.7. Suppose all data are the same except for the pump, which, instead of developing 80 ft of head, delivers 100 hp to the water. Determine the flow rate. Plot the energy line and hydraulic grade line, and determine the pressure on the suction side of the pump.

4.39 The upper circular plate in the figure is horizontal and is fixed in position, while the lower annular plate is free to move vertically and is not supported by the pipe in the center. Water is admitted at the center at the rate of 2 cfs and discharges into the air around the periphery. The annular plate weights 5 lb, and the weight of water on it should be considered. (*a*) If the distance between the two plates is maintained at 1 in, what total weight W can be supported? (*b*) What is the pressure head where the radius is 3 in, and what is it at a radius of 6 in?

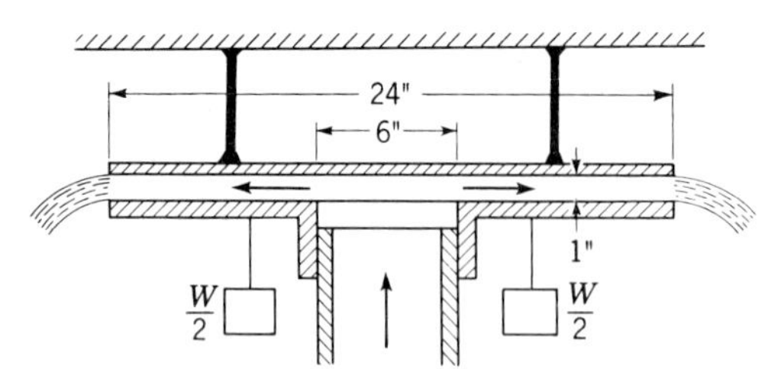

Problem 4.39

4.40 In Fig. 3.11 the velocity of the undisturbed field is 20 fps (5 m/s) and the velocities very near the surface at radii from the "source" making angles with the axis of 0, 60, 120, 150° are 0, 18.6, 25.2,

23.3 fps (0, 4.65, 6.3, 5.8 m/s) respectively. What will be the elevation of the surface of a liquid relative to that of the free surface of the undisturbed field? (This problem illustrates the way in which the water surface drops alongside a bridge pier or past the side of a moving ship.)

4.41 If the body shown in Fig. 3.11 is not two-dimensional but is a solid of revolution about a horizontal axis, the flow will be three-dimensional and the streamlines will be differently spaced. Also, the distance between the stagnation point and the "source" will be $d/4$, where d is the diameter at a great distance from the stagnation point. At points very near the surface at radii from the source making angles with the axis of 0, 60, 120, and 150°, the velocities are 0, 15.0, 22.8, and 21.2 fps respectively when the velocity of the undisturbed field is 20 fps. If the body is an airship and the atmospheric pressure in the undisturbed field is 12 psia, what will be the pressures at these points, for air temperature of 39.3°F?

4.42 In Prob. 4.41 assume the body is a submarine with diameters at the four points of 0, 8.0, 13.86, and 15.44 ft respectively. If the submarine is submerged in the ocean ($\gamma = 64.1$ lb/ft^3) with its axis 50 ft below the surface, find the pressures in pounds per square inch at these points along the top and along the bottom.

4.43 Refer to Illustrative Example 4.8. If the depths upstream and downstream of the gate were 5.0 ft and 2.0 ft respectively, find the flow rate per foot of channel width.

4.44 Refer to Illustrative Example 4.8. Suppose the gate opening is reduced so the depth downstream is 0.6 m. Find the upstream depth under these conditions if the flow rate remains constant at 4.24 m^3/s per m of width.

4.45 Find the stagnation pressure on the nose of a submarine moving at 15 knots in seawater ($\gamma = 64$ lb/ft^3) when it is 80 ft below the surface.

4.46 Plot the stagnation pressure on an object as it passes through air at sea level (standard atmosphere) as a function of velocity. Repeat for movement through air at 5,000-ft elevation. Let V vary from zero to c using 0, 25, 50, 75, and 100% of c.

4.47 Wind blows at a velocity of 15 m/s against the trunk of a tree at an elevation of 2,000 m above sea level. What is the stagnation pressure assuming standard atmospheric conditions? Express answer as a gage pressure and as an absolute pressure in kN/m^2, Pa, and mm of Hg.

4.48 Find the stagnation pressure on a chimney at elevation 500 m if the wind speed is 12 m/s.

4.49 By manipulation of Eq. (4.33), demonstrate that it represents a standard parabola of the form $z - z_0 = a(x - x_0)^2$, where a is a constant and x_0 and z_0 are the coordinates of the vertex.

4.50 Find the maximum ideal horizontal range of a jet having an initial velocity of 100 fps. At what angle of inclination is this obtained?

4.51 It is required to throw a fire stream so as to reach the window in the wall shown in the figure. Assuming a jet velocity of 80 fps and neglecting air friction, find the angle (or angles) of inclination which will achieve this result.

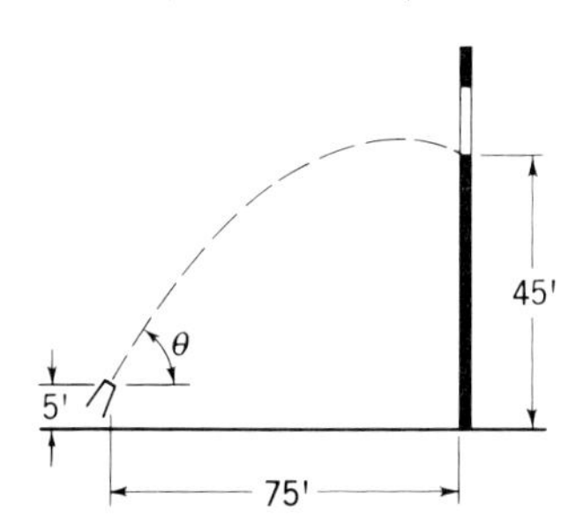

Problem 4.51

4.52 A jet issues horizontally from an orifice in the vertical wall of a large tank. Neglecting air resistance, determine the velocity of the jet for the following variety of trajectories: (*a*) $x = 4.0$ m, $y = 1.0$ m; (*b*) $x = 4.0$ m, $y = 2.0$ m; (*c*) $x = 4.0$ m; $y = 4.0$ m; (*d*) $x = 4.0$ m, $y = 8.0$ m. Express answers in m/s.

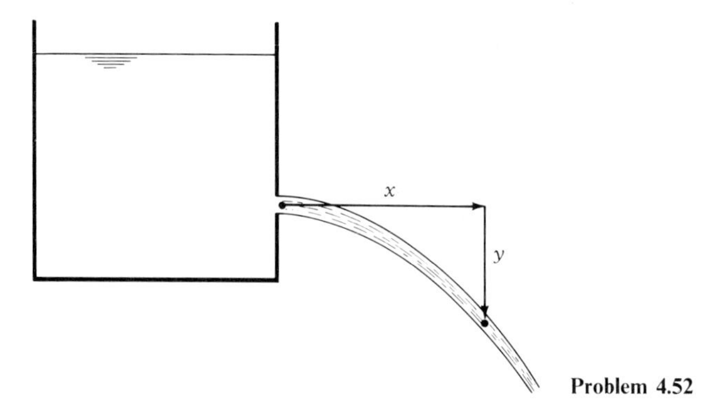

Problem 4.52

4.53 Freshwater sewage effluent is discharged from a horizontal outfall pipe on the floor of the ocean at a point where the depth is 100 ft. The jet is observed to rise to the surface at a point 85 ft horizontally from the end of the pipe. Assuming the ocean water to have a specific gravity of 1.03 and neglecting fluid friction and mixing of the jet with the ocean water, find the velocity at the end of the outfall. [*Note:* In this case the jet is submerged, and it is no longer possible to neglect the density of the surrounding medium; hence the value of g in Eqs. (4.33) and (4.34) must be adjusted accordingly.]

4.54 In the figure is shown a two-dimensional ideal flow in vertical plane. Data are as follows: $r = 10$ ft, $b = 4$ ft, $\gamma = 62.4$ lb/ft^3, $V = 20$ fps. If the pressure at A is 5 psi, find the pressure at B.

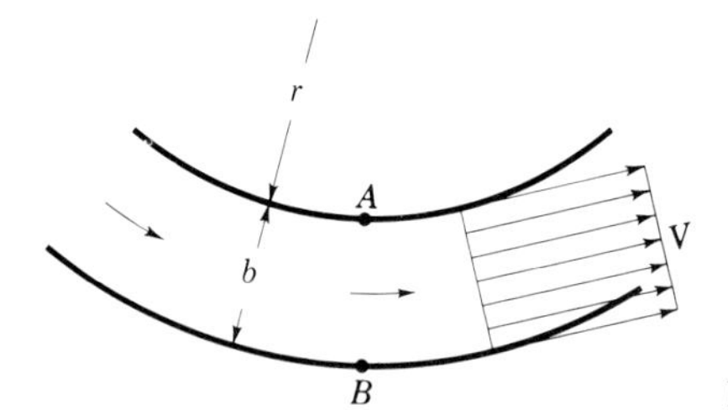

Problem 4.54

4.55 Repeat Prob. 4.54. Let $V = Q/A = 20$ fps, but assume a parabolic velocity profile.

4.56 Refer to the figure for Prob. 4.54. Flow occurs in a vertical plane. Data are as follows: $r = 5$ m, $b = 2$ m, $\gamma = 9.81$ kN/m^3, $V = 6$ m/s. Find the pressure at A if the pressure at B is 125 kN/m^2.

4.57 A closed vessel 18 in in diameter completely filled with fluid is rotated at 1,700 rpm. What will be the pressure difference between the circumference and the axis of rotation in feet of the fluid and in pounds per square inch if the fluid is (*a*) air with a specific weight of 0.075 lb/ft^3; (*b*) water at 68°F; (*c*) oil with a specific weight of 50 lb/ft^3?

4.58 A closed vessel 100-cm in diameter is completely filled with oil ($\gamma = 8.3$ kN/m^3) and is rotated at 600 rpm. What will be the pressure difference between the circumference and axis of rotation? Express answer in Pa.

4.59 An open cylindrical vessel partially filled with water is 3 ft in diameter and rotates about its axis, which is vertical. How many revolutions per minute would cause the water surface at the periphery to be 4 ft higher than the water surface at the axis? What would be the necessary speed for the same conditions if the fluid were mercury?

4.60 In Fig. 4.15 suppose that the vanes are all straight and radial, that $r_1 = 0.25$ ft (8 cm), $r_2 = 0.75$ ft (24 cm), and that the height perpendicular to the plane of the figure is constant at $b = 0.2$ ft (6 cm). Then $a = 2\pi rb$. If the speed is 1,200 rpm and the flow of liquid is 7.54 cfs (0.2 m^2/s), find the difference in the pressure head between the outer and the inner circumferences, neglecting friction losses. Does it make any difference whether the flow is outward or inward?

4.61 Refer to Illustrative Example 4.10. If the impeller diameter is 25 cm, the casing height is 3 cm between *a* and *b*, and water leaves the impeller with a velocity of 15 m/s at an angle of 18° with the tangent, find the flow rate, the magnitude and direction of the velocity at *b* (where $r = 18$ cm), and the pressure increase from *a* to *b*. Neglect friction.

4.62 An air duct of 2 ft by 2 ft (1 m by 1 m) square cross section turns a bend of radius 4 ft (2 m) as measured to the center line of the duct. If the measured pressure difference between the inside and outside walls of the bend is 1 in (4 cm) of water, estimate the rate of air flow in the duct. Assume standard sea-level conditions in the duct and assume ideal flow around the bend.

4.63 Assume ideal fluid. The pressure at section 1 in the figure is 10 psi, $V_1 = 12$ fps, $V_2 = 45$ fps, and $\gamma = 54$ lb/ft^3. (*a*) Determine the reading on the manometer. (*b*) If the downstream piezometer were replaced with a pitot tube, what would be the manometer reading? Comment on the practicality of these arrangements.

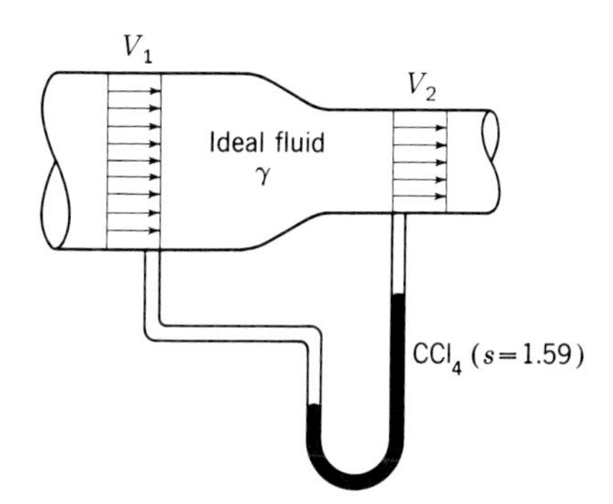

Problem 4.63

4.64 Refer to the figure for Prob. 4.63. Assume an ideal fluid with $\rho = 850$ kg/m^3. The pressure at section 1 is 100 kN/m^2; $V_1 = 5$ m/s, $V_2 = 18$ m/s. (*a*) Determine the reading on the manometer. (*b*) If the downstream piezometer were replaced with a pitot tube, what would be the manometer reading?

CHAPTER

FIVE

BASIC HYDRODYNAMICS

In this chapter we discuss various mathematical methods of describing the flow of fluids. The presentation here provides only an introduction to the vast subject of hydrodynamics, but gives some notion of the possibilities of a rigorous mathematical approach to flow problems.

5.1 DIFFERENTIAL EQUATION OF CONTINUITY

In Chap. 3 a very practical, but special, form of the equation of continuity was presented. For some purposes a more general three-dimensional form is desired. Also, in that chapter the concept of the flow net was explained largely on an intuitive basis. To reach a more fundamental understanding of the mechanics of the flow net, it is necessary to consider the differential equations of continuity and irrotationality (Sec. 5.2) which give rise to the orthogonal network of streamlines and equipotential lines.

Aside from application to the flow net, the differential form of the continuity equation has an important advantage over the one-dimensional form which was derived in Sec. 3.5 in that it is perfectly general for two- or three-dimensional fluid space and for either steady or unsteady flow.

Figure 5.1 shows three coordinate axes x, y, z mutually perpendicular and fixed in space. Let the velocity components in these three directions be u, v, w, respectively. Consider now a small parallelepiped, having sides Δx, Δy, Δz. In the x direction the rate of mass flow into this box through the left-hand face is approximately $\rho u \, \Delta y \, \Delta z$, this expression becoming exact in the limit as the box is shrunk to a point. The corresponding rate of mass flow out of the box through

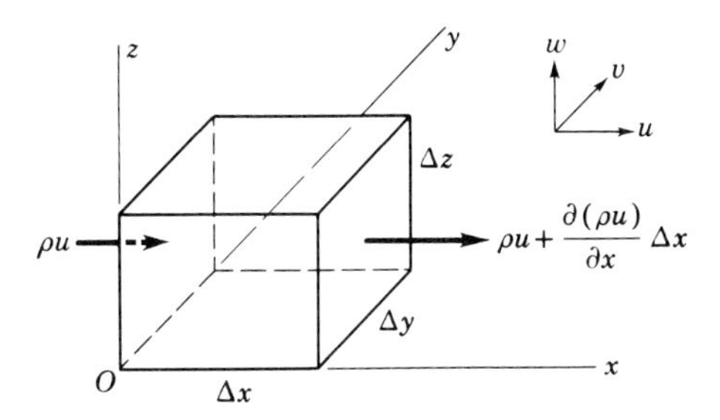

Figure 5.1

the right-hand face is $\{\rho u + [\partial(\rho u)/\partial x]\,\Delta x\}\,\Delta y\,\Delta z$. Thus the net rate of mass flow into the box in the x direction is $-[\partial(\rho u)/\partial x]\,\Delta x\,\Delta y\,\Delta z$. Similar expressions may be obtained for the y and z directions. The sum of the rates of mass inflow in the three directions must equal the time rate of change of the mass in the box, or $(\partial\rho/\partial t)\,\Delta x\,\Delta y\,\Delta z$. Summing up, applying the limiting process, and dividing both sides of the equation by the volume of the parallelepiped, which is common to all terms, we get

Unsteady compressible flow:
$$-\frac{\partial(\rho u)}{\partial x}-\frac{\partial(\rho v)}{\partial y}-\frac{\partial(p w)}{\partial z}=\frac{\partial\rho}{\partial t} \tag{5.1}$$

which is the equation of continuity in its most general form. This equation as well as the other equations in this section are, of course, valid regardless of whether the fluid is a real one or an ideal one. If the flow is steady, ρ does not vary with time, but it may vary in space. Since $\partial(\rho u)/\partial x = \rho(\partial u/\partial x) + u(\partial\rho/\partial x)$, it follows that for steady flow the equation may be written

Steady compressible flow:
$$u\frac{\partial\rho}{\partial x}+v\frac{\partial\rho}{\partial y}+w\frac{\partial\rho}{\partial z}+\rho\left(\frac{\partial u}{\partial x}+\frac{\partial v}{\partial y}+\frac{\partial w}{\partial z}\right)=0 \tag{5.2}$$

In the case of an incompressible fluid (ρ = constant), whether the flow is steady or not, the equation of continuity becomes

Steady incompressible flow:
$$\frac{\partial u}{\partial x}+\frac{\partial v}{\partial y}+\frac{\partial w}{\partial z}=0 \tag{5.3}$$

For two-dimensional flow, application of the same procedure to an elemental volume in polar coordinates yields for steady flow the following equations:

For compressible fluid:

$$\frac{1}{r}(\rho v_r)+\frac{\partial}{\partial r}(\rho v_r)+\frac{\partial}{r\,\partial\theta}(\rho v_t)=0 \tag{5.4}$$

For incompressible fluid:

$$\frac{v_r}{r}+\frac{\partial v_r}{\partial r}+\frac{\partial v_t}{r\,\partial\theta}=0 \tag{5.5}$$

where v_r and v_t represent the velocities in the radial and tangential[1] directions, respectively.

Illustrative Example 5.1 Assuming ρ to be constant, do these flows satisfy continuity? (*a*) $u = -2y$, $v = 3x$; (*b*) $u = 0$, $v = 3xy$; (*c*) $u = 2x$, $v = -2y$. Continuity for incompressible fluids is satisfied if $\partial u/\partial y + \partial v/\partial y = 0$.

(*a*) $\dfrac{\partial(-2y)}{\partial x} + \dfrac{\partial(3x)}{\partial y} = 0 + 0 = 0$ (continuity is satisfied)

(*b*) $\dfrac{\partial(0)}{\partial x} + \dfrac{\partial(3xy)}{\partial y} = 0 + 3x \neq 0$ (continuity is not satisfied)

(*c*) $\dfrac{\partial(2x)}{\partial x} + \dfrac{\partial(-2y)}{\partial y} = 2 - 2 = 0$ (continuity is satisfied)

Note: If (*b*) did indeed describe a flow field, the fluid must be compressible.

5.2 ROTATIONAL AND IRROTATIONAL FLOW

The discussion in the remainder of this chapter is restricted to incompressible fluids. Irrotational flow may be briefly described as flow in which each element of the moving fluid suffers no *net* rotation from one instant to the next, with respect to a given frame of reference. The classic example of irrotational motion (although not a fluid) is that of the carriages on a Ferris wheel. Each carriage describes a circular path as the wheel revolves, but does not rotate with respect to the earth. In irrotational flow, however, a fluid element may deform as shown in Fig. 5.2*a*, where the axes of the element rotate equally toward or away from each other. As long as the algebraic average rotation is zero, the motion is irrotational.

In Fig. 5.2*b* is depicted an example of rotational flow. In this case there is a net rotation of the fluid element. Actually, the deformation of the element in Fig. 5.2*b* is less than that of Fig. 5.2*a*.

Let us now express the condition of irrotationality in mathematical terms. It will help to restrict the discussion at first to two-dimensional motion in the *xy*

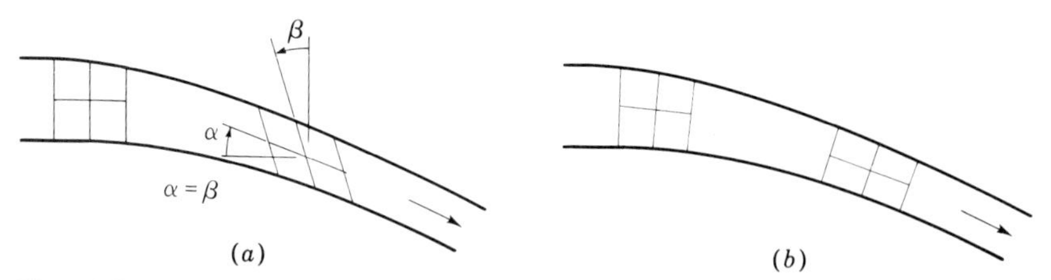

Figure 5.2 Two-dimensional flow along a curved path. (*a*) Irrotational flow. (*b*) Rotational flow.

[1] In this chapter v_t is used to represent the tangential component of velocity.

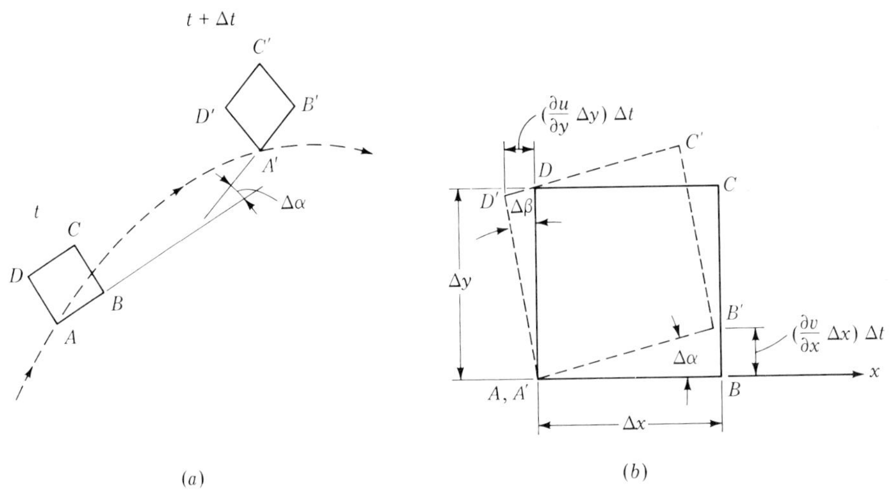

Figure 5.3

plane. Consider a small fluid element moving as depicted in Fig. 5.3*a*. During a short time interval Δt the element moves from one position to another and in the process it deforms as indicated. Superimposing A' on A, defining an x axis along AB, and enlarging the diagram, we get Fig. 5.3*b*. The angle $\Delta\alpha$ between AB and $A'B'$ can be expressed as

$$\Delta\alpha = \frac{BB'}{\Delta x} = \frac{[(\partial v/\partial x)\,\Delta x]\,\Delta t}{\Delta x} = \frac{\partial v}{\partial x}\,\Delta t$$

Hence the rate of rotation of the edge of the element that was originally aligned with AB is

$$\omega_\alpha = \frac{\Delta\alpha}{\Delta t} = \frac{\partial v}{\partial x}$$

Likewise

$$\Delta\beta = \frac{DD'}{\Delta y} = \frac{[-(\partial u/\partial y)\,\Delta y]\,\Delta t}{\Delta y} = -\frac{\partial u}{\partial y}\,\Delta t$$

and the rate of rotation of the edge of the element that was originally aligned with AD is

$$\omega_\beta = \frac{\Delta\beta}{\Delta t} = -\frac{\partial u}{\partial y}$$

with the negative sign because $+u$ is directed to the right. The rate of rotation of the element about the z axis is now defined to be ω_z, the average of ω_α and ω_β, thus:

$$\omega_z = \frac{1}{2}\left(\frac{\partial v}{\partial x} - \frac{\partial u}{\partial y}\right) \tag{5.6}$$

But the criterion we originally stipulated for irrotational flow was that the rate of rotation be zero. Therefore the flow is irrotational in the xy plane if $\partial v/\partial x$

$-\partial u/\partial y = 0$. In three-dimensional flow there are corresponding expressions for the components of angular-deformation rates about the x and y axes. Finally, for the general case, *irrotational flow* is defined to be that for which

$$\omega_x = \omega_y = \omega_z = 0 \tag{5.7}$$

The primary significance of irrotational flow is that it is defined by a velocity potential. This is discussed in Sec. 5.6.

Illustrative Example 5.2 Determine whether these flows are rotational or irrotational. (*a*) $u = -2y$, $v = 3x$; (*b*) $u = 0$, $v = 3xy$; (*c*) $u = 2x$, $v = -2y$.

If $\qquad \dfrac{\partial v}{\partial x} - \dfrac{\partial u}{\partial y} = 0 \qquad$ (flow is irrotational)

(*a*) $\dfrac{\partial(3x)}{\partial x} - \dfrac{\partial(-2y)}{\partial y} = 3 + 2 \neq 0 \qquad$ (flow is rotational)

(*b*) $\dfrac{\partial(3xy)}{\partial x} - \dfrac{\partial(0)}{\partial y} = 3y - 0 \neq 0 \qquad$ (flow is rotational)

(*c*) $\dfrac{\partial(-2y)}{\partial x} - \dfrac{\partial(2x)}{\partial y} = 0 - 0 = 0 \qquad$ (flow is irrotational)

5.3 CIRCULATION AND VORTICITY

To get a better understanding of the character of a flow field, we should acquaint ourselves with the concept of *circulation*.[1] Let the streamlines of Fig. 5.4 represent a two-dimensional *flow field*, while L represents any closed path in this field. The circulation Γ (gamma) is defined mathematically as a *line integral* of the velocity about a closed path. Thus

$$\Gamma = \oint_L \mathbf{V} \cdot d\mathbf{L} = \oint_L V \cos \beta \, dL \tag{5.8}$$

where $\mathbf{V}$ is the velocity in the flow field at the element $d\mathbf{L}$ of the path, and β is the angle between $\mathbf{V}$ and the tangent to the path (in the positive direction along the path) at that point. Equation (5.8) is analogous to the common equation in mechanics for work done as a body moves along a curved path while the force makes some angle with the path. The only difference here is the substitution of a velocity for a force.

Evaluation of Eq. (5.8) about any closed curve generally involves a tedious step-by-step integration. Some valuable information is acquired, however, by evaluating the circulation of the two-dimensional flow field of Fig. 5.5 by taking the line integral around the boundary of the indicated element. Since the element

[1] In Secs. 10.10 and 10.11 the concept of circulation is utilized to develop an expression for lift force on an air foil.

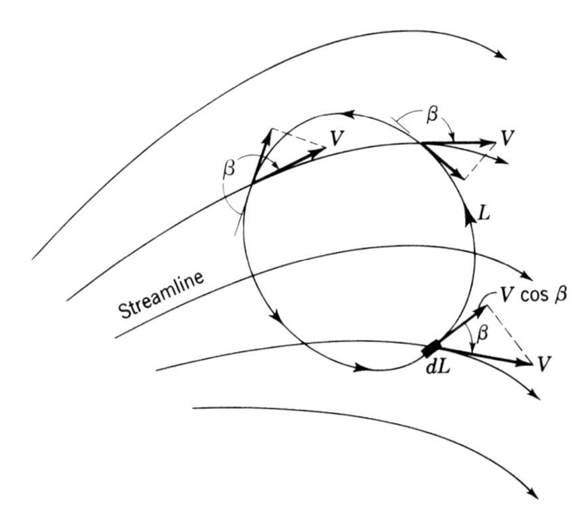

Figure 5.4 Circulation around a closed path in a two-dimensional field.

is of differential size, the resulting circulation is also differential. Thus, starting at A and proceeding counterclockwise,

$$d\Gamma = \frac{u_A + u_B}{2} dx + \frac{v_B + v_C}{2} dy - \frac{u_C + u_D}{2} dx - \frac{v_D + v_A}{2} dy \tag{5.9}$$

where the values of u_A, u_B, u_C, u_D and v_A, v_B, v_C, v_D are as indicated in Fig. 5.5. Substituting these values into Eq. (5.9), expanding, combining terms, and disregarding those of higher order yields

$$d\Gamma = \left(\frac{\partial v}{\partial x} - \frac{\partial u}{\partial y}\right) dx\, dy \tag{5.10}$$

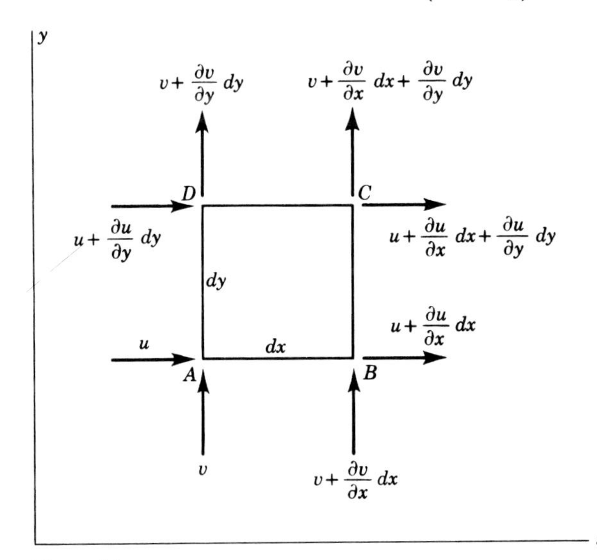

Figure 5.5

The *vorticity* ξ (*xi*) is defined as the circulation per unit of enclosed area. Thus

$$\xi = \frac{d\Gamma}{dx\,dy} = \frac{\partial v}{\partial x} - \frac{\partial u}{\partial y} \tag{5.11}$$

Comparing Eq. (5.11) with Eq. (5.7), we see that an irrotational flow is one for which the vorticity $\xi = 0$. Similarly, if the flow is rotational, $\xi \neq 0$.

Using a similar procedure for polar coordinates, we find

$$\xi = \frac{\partial v_t}{\partial r} + \frac{v_t}{r} - \frac{\partial v_r}{r\,\partial \theta} \tag{5.12}$$

Illustrative Example 5.3 Check these flows for continuity and determine the vorticity of each: (*a*) $v_t = 6r$, $v_r = 0$; (*b*) $v_t = 0$, $v_r = -5/r$.

Applying Eqs. (5.5) and (5.12),

(*a*) $\frac{0}{r} + \frac{\partial(0)}{\partial r} + \frac{\partial(6r)}{r\,\partial\theta} = 0$ (continuity is satisfied)

$\xi = \frac{\partial(6r)}{\partial r} + \frac{6r}{r} - \frac{\partial(0)}{r\,\partial\theta} = 6 + 6 - 0 = 12$ (flow is rotational)

(*b*) $-\frac{5/r}{r} + \frac{\partial(-5r^{-1})}{\partial r} + \frac{\partial(0)}{r\,\partial\theta} = -\frac{5}{r^2} + \frac{5}{r^2} + 0 = 0$ (continuity is satisfied)

$\xi = \frac{\partial(0)}{\partial r} + \frac{0}{r} - \frac{\partial(-5/r)}{r\,\partial\theta} = 0$ (flow is irrotational)

5.4 THE STREAM FUNCTION

The stream function ψ, based on the continuity principle, is a mathematical expression that describes a flow field. In Fig. 5.6 are shown two adjacent streamlines of a two-dimensional flow field. Let $\psi(x, y)$ represent the streamline nearest the origin. Then $\psi + d\psi$ is representative of the second streamline. Since there is no flow across a streamline, we can let ψ be indicative of the flow carried through

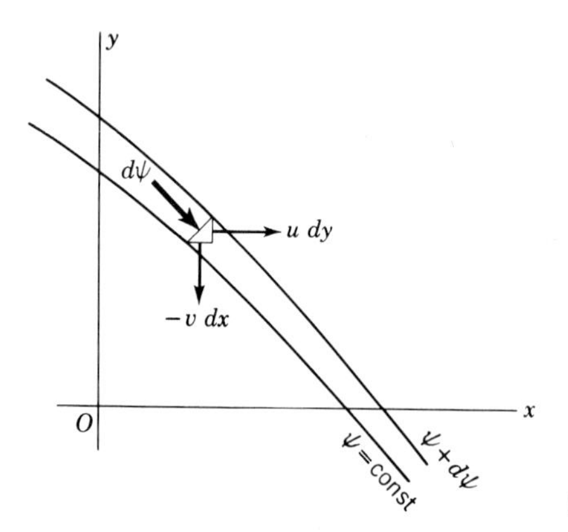

Figure 5.6 Stream function.

the area from the origin O to the first streamline. And thus $d\psi$ represents the flow carried between the two streamlines of Fig. 5.6. For continuity referring to the triangular fluid element of Fig. 5.6, we see that for an incompressible fluid

$$d\psi = -v\,dx + u\,dy \tag{5.13}$$

The total derivative $d\psi$ may also be expressed as

$$d\psi = \frac{\partial\psi}{\partial x}\,dx + \frac{\partial\psi}{\partial y}\,dy \tag{5.14}$$

Comparing these last two equations, we note that

$$u = \frac{\partial\psi}{\partial y} \qquad \text{and} \qquad v = -\frac{\partial\psi}{\partial x} \tag{5.15}$$

Thus, if ψ can be expressed as a function of x and y, we can find the velocity components (u and v) at any point of a two-dimensional flow field by application of Eq. (5.15). Conversely, if u and v are expressed as functions of x and y, we can find ψ by integrating Eq. (5.13). However, it should be noted that since the derivation of ψ is based on the principle of continuity, *it is necessary that continuity be satisfied for the stream function to exist.* Also, since vorticity was not considered in the derivation of ψ, *the flow need not be irrotational for the stream function to exist.*

The equation of continuity

$$\frac{\partial u}{\partial x} + \frac{\partial v}{\partial y} = 0$$

may be expressed in terms of ψ by substituting the expressions for u and v from Eq. (5.15); doing so we get

$$\frac{\partial}{\partial x}\left(\frac{\partial\psi}{\partial y}\right) - \frac{\partial}{\partial y}\left(\frac{\partial\psi}{\partial x}\right) = 0 \qquad \text{or} \qquad \frac{\partial^2\psi}{\partial x\,\partial y} = \frac{\partial^2\psi}{\partial y\,\partial x}$$

which shows that, if $\psi = \psi(x, y)$, the derivatives taken in either order give the same result and that a flow described by a stream function automatically satisfies the continuity equation.

5.5 BASIC FLOW FIELDS

In this section several basic flow fields that are commonly encountered will be discussed. Though these flow fields imply an ideal fluid, they closely depict the flow of a real fluid if outside the zone of viscous influence provided there is no separation of the flow from the boundaries with which the fluid is in contact (see Sec. 3.8). The simplest of all flows is that where the streamlines are straight, parallel, and evenly spaced as indicated in Fig. 5.7. In this case $v = 0$ and $u =$ constant. Thus, from Eq. (5.13), $d\psi = u\,dy$, and hence $\psi = Uy$, where U is the

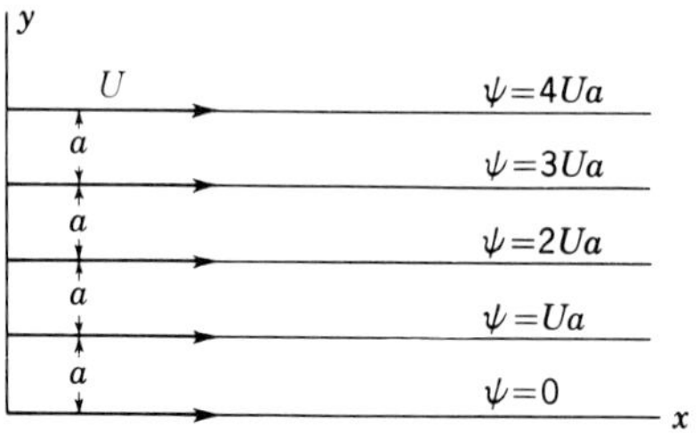

Figure 5.7 Rectilinear flow field.

velocity of flow. If the distance between streamlines is a, the values of ψ for the streamlines are as indicated in Fig. 5.7.

Another flow field of general interest is that of a *source* or a *sink*. In the case of a source, the flow field consists of radial streamlines symmetrically spaced as shown in Fig. 5.8. If q is the *source strength*, or rate of flow from the source, it is at once apparent that $\psi = q\theta/2\pi$. Customarily, for this case, the $\psi = 0$ streamline is defined as that coincident with the direction of the x axis. From inspection of the flow field it is obvious that $v_t = 0$ and $v_r = q/2\pi r$. Thus $v_r \to 0$ as $r \to \infty$. For a sink (inward flow), the stream function is expressible as $\psi = -q\theta/2\pi$.

Flow fields may be combined by superposition to give other fields of importance. For example, let us combine a source and sink of equal strength with a rectilinear flow. Let $2a$ be the distance between the source and sink. Referring to Fig. 5.9 and defining θ_1 and θ_2 as shown, we can write for the combined field

$$\psi = Uy + \frac{q\theta_1}{2\pi} - \frac{q\theta_2}{2\pi} \tag{5.16}$$

Transforming the last two terms of this equation to cartesian coordinates by replacing the θ's with appropriate trigonometric function, we get

$$\psi = Uy + \frac{q}{2\pi}\left(\arctan\frac{y}{x+a} - \arctan\frac{y}{x-a}\right) \tag{5.17}$$

This equation will permit one to plot streamlines by determining values of ψ at various points in the flow field having coordinates (x, y). Lines of constant ψ are

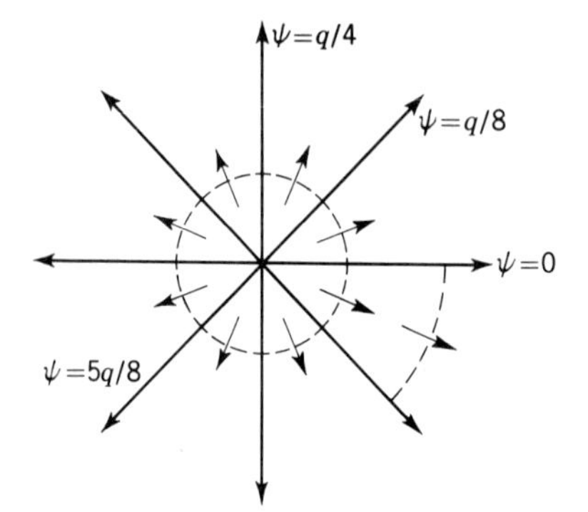

Figure 5.8 Source flow field.

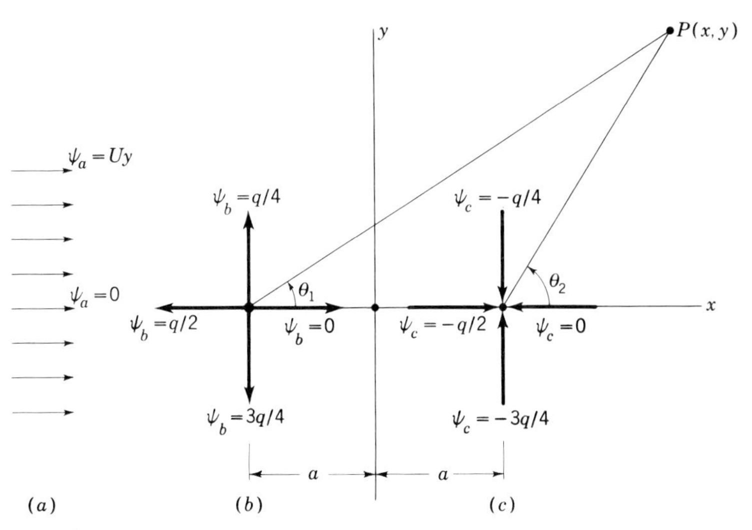

Figure 5.9 Superposition of flow fields. (*a*) Uniform rectilinear flow. (*b*) Source. (*c*) Sink. Source and sink are of equal strength and a distance $2a$ apart along the x *axis*.

streamlines. The resulting flow field for this case is shown in Illustrative Example 5.4. The $\psi = 0$ line produces a closed curve (oval), and thus the flow field represents ideal flow past a body of that shape. By using different values of a and different relationships between U and q, it is possible to describe a whole array of two-dimensional flow fields about ovals of various shapes. As $2a$, the distance between the source and sink gets smaller, the oval approaches a circle. However, when $a = 0$, the flow field reduces mathematically to uniform rectilinear flow since the source and sink will cancel each other out. The location of stagnation points S may be found by differentiating Eq. (5.17) to obtain an expression for $u = \partial\psi/\partial y$ and then to determine the values of x for which $u = 0$.

The flow field of Illustrative Example 5.4 is for an ideal fluid and, of course, does not represent the flow picture for a real fluid, where there may be separation[1] with the formation of a wake on the downstream side of the body (Fig. 10.12). However, on the upstream side of the body where the boundary layer is thin, the flow of a real fluid is well represented by this example.

Illustrative Example 5.4 A flow field for a source and sink of equal strength is combined with a uniform rectilinear flow. As an example let $U = 0.80$, $q = 2\pi$, $a = 2$. Thus,

$$\psi = Uy + \frac{q}{2\pi}\left(\tan^{-1}\frac{y}{x+a} - \tan^{-1}\frac{y}{x-a}\right)$$

Thus
$$\psi = 0.80y + \tan^{-1}\frac{y}{x+2} - \tan^{-1}\frac{y}{x-2}$$

[1] Refer to Sec 10.6 for a discussion of the conditions under which separation will take place on the back side of a solid body.

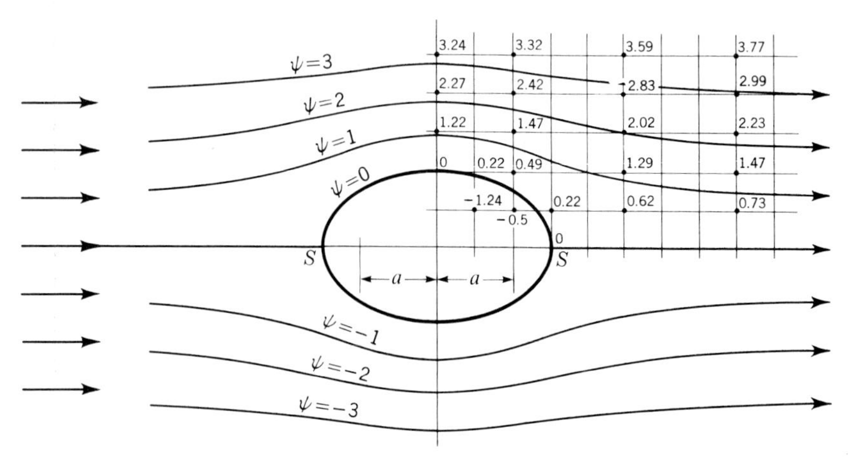

Illustrative Example 5.4 Flow field for source and sink of equal strength in a uniform rectilinear flow field. Note point S represents a stagnation point.

Let

$$A = \frac{y}{x+2} \quad \text{and} \quad B = \frac{y}{x-2}$$

x	y	$\frac{y}{x+2}$	$\frac{y}{x-2}$	Degrees			Radians		
				$\tan^{-1} A$	$\tan^{-1} B$	$0.8y$	$\tan^{-1} A$	$\tan^{-1} B$	ψ
0	2	$\frac{2}{2}$	$-\frac{2}{2}$	45°00′	135°00′	1.60	0.78	2.36	0.00
0	3	$\frac{3}{2}$	$-\frac{3}{2}$	56°19′	123°41′	2.40	0.98	216	1.22
0	4	$\frac{4}{2}$	$-\frac{4}{2}$	63°26′	116°34′	3.20	1.11	2.04	2.27
2	2	$\frac{2}{4}$	∞	26°34′	90°00′	1.60	0.46	1.57	0.49
2	3	$\frac{3}{4}$	∞	36°54′	90°00′	2.40	0.64	1.57	1.47
5	1	$\frac{1}{7}$	$\frac{1}{3}$	8°08′	18°26′	0.80	0.14	0.32	0.62
5	2	$\frac{2}{7}$	$\frac{2}{3}$	15°55′	33°42′	1.60	0.28	0.59	1.29
8	1	$\frac{1}{10}$	$\frac{1}{6}$	5°43′	9°28′	0.80	0.10	0.17	0.73
8	2	$\frac{2}{10}$	$\frac{2}{6}$	11°19′	18°26′	1.60	0.19	0.32	1.47

Suppose we define a *doublet* as a source-sink combination for which $2qa = m$, a constant. Permitting a to approach zero, the stream function of the doublet imposed on the uniform field is then

$$\psi = Uy - \frac{m \sin \theta}{2\pi r} \tag{5.18}$$

Taking $\psi = 0$ to determine the form of the closed-body contour and noting that $y = r \sin \theta$, we get

$$0 = Ur \sin \theta - \frac{m \sin \theta}{2\pi r}$$

or
$$r_{\psi=0} = R = \sqrt{\frac{m}{U2\pi}} = \text{constant}$$

Therefore the closed-body contour for this case is a circle with radius R and Eq. (5.18) is the stream function for two-dimensional flow about a circular cylinder. Further mention of this flow is made in Sec. 10.10, where lift and circulation are discussed. Other basic flow fields of interest include the forced vortex ($\psi = \omega r^2/2$), the free vortex ($\psi = \Gamma \ln r/2\pi$), and various combinations such as the source in rectilinear flow (see Prob. 5.31), the source and sink of equal strength, and the doublet in rectilinear flow with circulation (see Fig. 10.18).

5.6 VELOCITY POTENTIAL

For two-dimensional flow the velocity potential $\phi(x, y)$ may be defined[1] in cartesian coordinates as

$$u = -\frac{\partial \phi}{\partial x} \quad \text{and} \quad v = -\frac{\partial \phi}{\partial y} \tag{5.19}$$

The corresponding expressions in polar coordinates are

$$v_r = -\frac{\partial \phi}{\partial r} \quad \text{and} \quad v_t = -\frac{\partial \phi}{r\,\partial \theta} \tag{5.20}$$

If we substitute the expressions of Eq. (5.19) into the continuity equation [Eq. (5.3)], we get

$$\frac{\partial^2 \phi}{\partial x^2} + \frac{\partial^2 \phi}{\partial y^2} = 0 \tag{5.21}$$

This is known as the Laplace equation; it is of importance in both solid mechanics and fluid mechanics.

If the expressions of Eq. (5.19) are substituted into the equation for vorticity [Eq. (5.11)], we get

$$\xi = \frac{\partial v}{\partial x} - \frac{\partial u}{\partial y} = \frac{\partial}{\partial x}\left(-\frac{\partial \phi}{\partial y}\right) - \frac{\partial}{\partial y}\left(-\frac{\partial \phi}{\partial x}\right) = -\frac{\partial^2 \phi}{\partial x\,\partial y} + \frac{\partial^2 \phi}{\partial y\,\partial x} = 0$$

Since $\xi = 0$, the flow is irrotational, and thus, *if a velocity potential exists, the flow must be irrotational.* Because of the existence of a velocity potential, such flow is often referred to as *potential flow.*

The rotation of fluid particles requires the application of torque, which in turn depends on shearing forces. Such forces are possible only in a viscous fluid. In inviscid (or ideal) fluids there can be no shears and hence no torques.

[1] Some authors prefer for the potential to increase in the direction of flow by defining $u = \partial\phi/\partial x$ and $v = \partial\phi/\partial y$.

5.7 ORTHOGONALITY OF STREAMLINES AND EQUIPOTENTIAL LINES

It was noted in Eq. (5.14) that

$$d\psi = \frac{\partial \psi}{\partial x} dx + \frac{\partial \psi}{\partial y} dy$$

Similarly,

$$d\phi = \frac{\partial \phi}{\partial x} dx + \frac{\partial \phi}{\partial y} dy$$

From Eqs. (5.15) and (5.19) we can express these two equations as

$$d\psi = -v\,dx + u\,dy$$

and

$$d\phi = -u\,dx - v\,dy$$

Along a streamline, $\psi =$ constant, so $d\psi = 0$, and from the first equation we get $dy/dx = v/u$. Along an equipotential line, $\phi =$ constant, so $d\phi = 0$, and from the second equation we get $dy/dx = -u/v$. Geometrically, this tells us that the streamlines and equipotential lines are *orthogonal*, or everywhere perpendicular to each other.

The equipotential lines $\phi = C_i$ and the streamlines $\psi = K_i$, where the C_i and the K_i have equal increments between adjacent lines, form a network of intersecting perpendicular lines which is called a flow net (Fig. 5.10). The small quadrilaterals must evidently be squares as their size approaches zero, for $|\partial\phi/\partial x| = |\partial\psi/\partial y|$, or for finite increments, $|\Delta\phi/\Delta x| = |\Delta\psi/\Delta y|$. The difference in value of the stream function between adjacent streamlines is called the *strength* of the stream tube bounded by the two streamlines, and it represents the two-dimensional flow through the tube.[1]

Referring to Fig. 5.10, the maximum velocity at any point O is seen to be tangential to the streamline. This velocity is given by $V = -\partial\phi/\partial s$, where s is measured along the streamline. The expression $\partial\phi/\partial s$ is also known as the *gradient* of the velocity potential. Thus the velocity is often written in convenient vector notation as $\mathbf{V} = -\text{grad}\,\phi$, which holds for either two- or three-dimensional flow. The absolute velocity may always be written as the vector sum of its components; thus, in three dimensions,

$$V = |\mathbf{V}| = \sqrt{u^2 + v^2 + w^2} \tag{5.22}$$

Stream functions can exist in the absence of irrotationality and potential functions are possible even though continuity is not satisfied. But, since lines of ϕ

[1] Consider flow in the direction of the x axis in a stream tube (of unit thickness perpendicular to the plane of the paper) bounded by $\psi = K_1$ and $\psi = K_2$. Let y_1 and y_2 represent the graphical locations of the two streamlines. The flow through the tube is then given by

$$Q = \int_{y_1}^{y_2} u\,dy = -\int_{y_1}^{y_2} \frac{\partial \psi}{\partial y} dy = -\int_{K_1}^{K_2} d\psi = K_1 - K_2$$

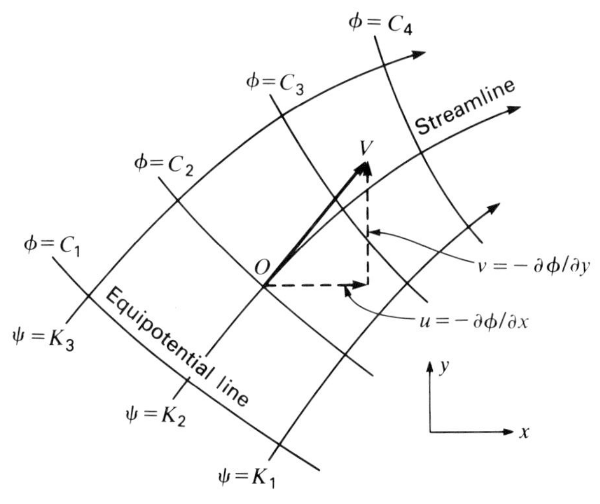

Figure 5.10 Flow net

and ψ are required to form an orthogonal network, a flow net can only exist if irrotationality (the condition for the existence of ϕ) and continuity (the condition for the existence of ψ) are satisfied. The Laplace equation [Eq. (5.21)] was derived assuming the existence of velocity potentials and the satisfaction of continuity. Thus, if a given flow satisfies the Laplace equation, a flow net can be constructed for that flow. Such flows which satisfy continuity and irrotationality are referred to as *potential flow*. Because of the irrotationality requirement such flows are usually those of ideal fluids. An exception where a real fluid satisfies the conditions for potential flow is that of *laminar flow through porous media*. In such a case the velocity head is negligible and the energy equation may be written as

$$\left(\frac{p_1}{\gamma} + z_1\right) - \left(\frac{p_2}{\gamma} + z_2\right) = h_L$$

where the head loss h_L is directly proportional to velocity for laminar flow.

Taking the differential, we get

$$-d\left(\frac{p}{\gamma} + z\right) = dh_L = \frac{V}{K}\,ds$$

where ds is the distance along the streamline, and $1/K$ is a constant of proportionality. Hence,

$$V = -K\frac{d(p/\gamma + z)}{ds} = -\frac{\partial \phi}{ds} \tag{5.23}$$

Thus we see that $\phi = K(p/\gamma + z)$. The constant K is commonly referred to as the coefficient of permeability. Equation (5.23) is known as Darcy's law. The equation was first proposed by Darcy in the form $V = K\,dh_L/ds$; he observed experimentally that, in the case of flow through a porous medium, V is proportional to

dh_L/ds. Later it was observed that this proportionality holds only if the flow is laminar. A distinction should be made between V, the *apparent velocity* defined as $V = Q/A$ where A is the total cross-sectional flow area, and V_p, the *average velocity through the pores*, defined as $V_p = V/n$ where n is the porosity of the medium. Excellent discussions of the application of flow nets to flow through porous media are available in the literature.[1]

Illustrative Example 5.5 A flow is defined by $u = 2x$ and $v = -2y$. Find the stream function and potential function for this flow and plot the flow net.

Check continuity:

$$\frac{\partial u}{\partial x} + \frac{\partial v}{\partial y} = 2 - 2 = 0$$

Hence continuity is satisfied and it is possible for a stream function to exist:

$$d\psi = -v\,dx + u\,dy = 2y\,dx + 2x\,dy$$

$$\psi = 2xy + C_1$$

Check to see if the flow is irrotational:

$$\frac{\partial v}{\partial x} - \frac{\partial u}{\partial y} = 0 - 0 = 0$$

Hence the flow is irrotational and a potential function exists:

$$d\phi = -u\,dx - v\,dy = -2x\,dx + 2y\,dy$$

$$\phi = -(x^2 - y^2) + C_2$$

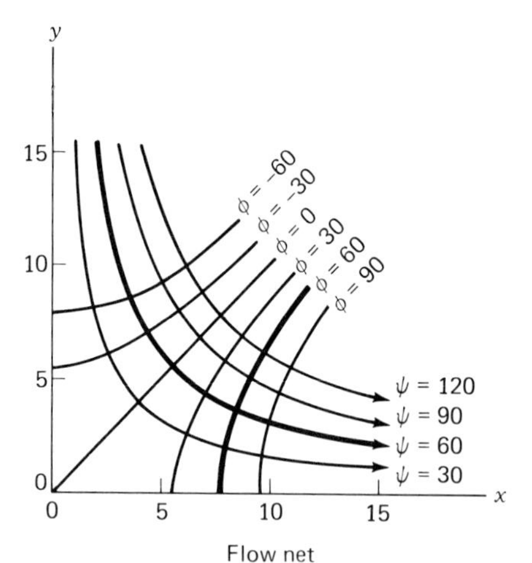

Illustrative Example 5.5

[1] H. R. Vallentine, "Applied Hydrodynamics," 2d ed., chap 3, Butterworth & Co. (Publishers), Ltd., London, 1967. A. Casagrande, "Seepage through Dams," *J. New Engl. Water Works Assoc.*, vol. 51, pp. 131–172, 1937.

The location of lines of equal ψ can be found by substituting values of ψ into the expression $\psi = 2xy$. Thus for $\psi = 60$, $x = 30/y$. This line is plotted (in the upper right-hand quadrant) on the adjoining figure. In a similar fashion lines of equal potential can be plotted. For example, for $\phi = 60$ we have $-(x^2 - y^2) = 60$ and $x = \pm\sqrt{y^2 - 60}$. This line is also plotted on the figure. The flow net depicts *flow in a corner*. Mathematically the net will plot symmetrically in all four quadrants.

PROBLEMS

5.1 Why are Eqs. (5.2) and (5.3) applicable to real fluids as well as ideal fluids?

5.2 Derive Eqs. (5.4) and (5.5).

5.3 Given a flow defined by $u = 3 + 2x$. If this flow satisfies continuity, what can be said about the density of the fluid?

5.4 The flow of an incompressible fluid is defined by $u = 2$, $v = 8x$. Does a stream function exist for this flow? If so, determine the expression for the stream function.

5.5 Sketch streamlines ($\psi = 0$, 1, 2, 3) for the following flow fields, note the values of u and v, and verify that continuity is satisfied in all cases. (*a*) $\psi = 10y$, (*b*) $\psi = -20x$, (*c*) $\psi = 10y - 20x$.

5.6 Write an expression for the stream function of each of the following flows. Assume ρ = constant. (*Note:* they may not all have stream functions.)

(*a*) $u = 2$
(*b*) $u = 2 \quad v = 3$
(*c*) $u = 2 + 3x \quad v = 4$
(*d*) $u = 2y \quad v = 3x$
(*e*) $u = 2y \quad v = -3x$
(*f*) $u = 3xy \quad v = 1.5x^2$
(*g*) $u = 3y \quad v = 0$
(*h*) $u = 3x \quad v = 3y$
(*i*) $u = 3x \quad v = -3y$
(*j*) $u = 4 + 2x \quad v = -6 - 2y$

5.7 A flow field is described by the equation $\psi = 1.2xy$. Sketch the streamlines in one quadrant for $\psi = 0$, 1, 2, 3, 4.

5.8 Plot the streamlines in the upper right-hand quadrant for the flow defined by $\psi = 1.5x^2 + y^2$ and determine the value of the velocity at $x = 4$, $y = 2$.

5.9 The components of the velocities of a certain flow system are

$$u = -\frac{Q}{2\pi}\left(\frac{x}{x^2 + y^2}\right) + By + C$$

$$v = -A\left(\frac{y}{x^2 + y^2}\right) + Dx + E$$

(*a*) Calculate a value of A consistent with continuous flow. (*b*) Sketch the streamlines for this flow system, assuming $B = C = D = E = 0$.

5.10 Which of the flows in Prob. 5.6 are irrotational?

5.11 A flow field is described by $\psi = x^2 - y$. Sketch the streamlines for $\psi = 0$, 1, and 2. Derive an expression for the velocity at any point in the flow field and determine the vorticity of the flow.

5.12. A source discharging 10 cfs/ft is at $(-1, 0)$ and a sink taking in 10 cfs/ft is at $(+1, 0)$. If a uniform flow with velocity 5 fps from left to right is superimposed on the source-sink combination, what is the length of the resulting closed body contour?

5.13 A source of strength 8π is located at (2, 0). Another source of strength 16π is located at $(-3, 0)$. For the combined flow field produced by these two sources: (*a*) find the location of the stagnation point; (*b*) plot the $\psi = 0$, $\psi = 4\pi$, $\psi = 8\pi$ lines; (*c*) find the values of ψ at (0, 2) and at $(3, -1)$; (*d*) find the velocity at $(-2, 5)$.

5.14 A source discharging 20 m^3/s per m is located at the origin and a uniform flow with a velocity of 3 m/s from left to right is superimposed on the source flow. Determine the stream function of the flow in polar and rectangular coordinates.

5.15 For the flow of the preceding problem find the location of the stagnation points and find the velocity at $x = 3$ m, $y = 4$ m.

5.16 Refer to Prob. 5.14. Find the difference in pressure head between point A (-10 m, 0) and point B (0, 1.67 m).

5.17 Using the method described in Illustrative Example 5.4, plot the boundary of the body and a set of streamlines for a steady two-dimensional flow past a body such as that of Fig. 3.11, for $b = 15$ m using a scale of 1 cm = 2 m.

5.18 Combine the uniform flow defined by $u = 16$ fps with the doublet $2qa = m$, where $q = 10$ cfs/ft and $a = 2$ in. Sketch the streamlines for $\psi = -3, -2, -1, -\frac{1}{2}, 0, \frac{1}{2}, 1, 2,$ and 3 cfs/ft. Use a scale of 1 in = 1 in.

5.19 A flow is defined by the stream function $\psi = 15r \sin\theta - 30 \ln r - (20/r)\sin\theta$. Sketch this flow field. Calculate the velocities at $r = 3$ for $\theta = 0$, 45, 90, 150, 210, and 315°.

5.20 Given is the two-dimensional flow described by $u = x^2 + 2x - 4y$, $v = -2xy - 2y$. (*a*) Does this satisfy continuity? (*b*) Compute the vorticity. (*c*) Plot the velocity vectors for $0 < x < 5$ and $0 < y < 4$ and sketch the general flow pattern. (*d*) Find the location of all stagnation points in the entire flow field. (*e*) Find the expression for the stream function.

5.21 Given the stream function $\psi = 3x - 2y$. Is this a potential flow? Does it satisfy the Laplace equation?

5.22 Which of the flows in Prob. 5.6 can be described by a flow net? Write expressions for the stream functions and the potential functions.

5.23 An ideal fluid flows in a two-dimensional 90° bend. The inner and outer radii of the bend are 0.4 and 1.4 ft (0.1 and 0.4 m). Sketch the flow net and estimate the velocity at the inner and outer walls of the bend if the velocity in the 1.0-ft (0.3-m)-wide straight section is 10 fps (3 m/s). Develop an analytic expression for the stream function, in this case noting that $v_t = -\partial\psi/\partial r$ and $v_r = \partial\psi/r\,\partial\theta$. Determine the inner and outer velocities accurately.

5.24 A cylindrical drum with a 2-ft radius is securely held in position in an open channel of rectangular section. The channel is 10 ft wide, and the flow rate is 240 cfs. Water flows beneath the drum as shown in the figure. Sketch the flow net, and determine from flow net measurements the pressure at the points indicated along the wetted drum surface. Neglect fluid friction. Sketch the pressure distribution, and by numerical integration determine an approximate value of the horizontal thrust on the cylinder.

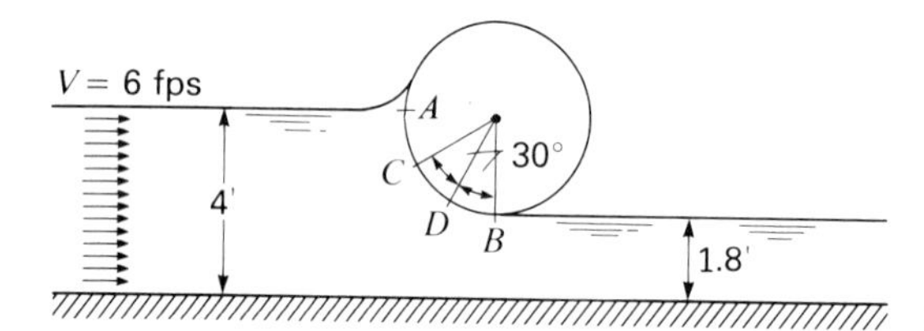

Problem 5.24

5.25 Refer to Illustrative Example 6.7. Sketch a flow net. Using the given dimensions in English units through application of Bernoulli's principle, determine the approximate pressure distribution along the channel bottom and around the curved structure. By numerical integration estimate the magnitude of the horizontal and vertical components of the force of the water on the structure.

5.26 Work Prob. 5.25 using the dimensions as given in SI units.

5.27 For the two-dimensional flow of a frictionless incompressible fluid against a flat plate normal to the initial velocity, the stream function is given by $\psi = -2axy$, while its *conjugate function*, the velocity potential, is

$$\phi = a(x^2 - y^2)$$

where a is a constant and the flow is symmetrical about the yz plane (Fig. 5.1). By direct differentiation demonstrate that these functions satisfy Eq. (5.21). Using a scale of 1 in = 1 unit of distance, plot the streamlines given by $\psi = \pm 2a, \pm 4a, \pm 6a, \pm 8a$, and the equipotential lines given by $\phi = 0$, $\pm 2a, \pm 4a, \pm 6a, \pm 8a$. Observe that this flow net also gives the ideal flow around an inside square corner. Compare your results with Illustrative Example 5.5 and note the effect of changing the sign of ψ and ϕ.

5.28 In Prob. 5.27 determine the velocity components u and v, and demonstrate that they satisfy the differential equations for continuity and irrotational flow. In which direction is the flow? Prove that the absolute velocity is given by $V = 2ar$, where r is the radius to the point from the origin. Now assume that the linear scale is 1 in = 1 ft. Determine the constant a such that the flow net of Prob. 5.27 will represent a flow of 3 ft^2/s between any two adjacent streamlines. What are the dimensions of a? Draw curves of equal velocity for values of 3, 6, 9, 12 fps. How does the velocity vary along the surface of the plate?

5.29 The three-dimensional counterpart of the flow in Probs. 5.27 and 5.28 is that of flow along the y axis approaching the plate in the xz plane. As the flow must be symmetrical about the y axis, the traces of stream and equipotential surfaces in the xy plane will be representative of those in all planes containing the y axis. The velocity potential is now given by $\phi = -a(0.5x^2 - y^2)$, and the stream function by $\psi = -ax^2y$. Notice that these functions no longer satisfy Eq. (5.21). Why not? Again plot streamlines and equipotential lines for the values given in Prob. 5.27. The velocities u and v may still be determined by Eq. (5.19). Prove that the absolute velocity for this case is given by $V = a\sqrt{x^2 + 4y^2}$. With the value of $a = 1.5\ \text{s}^{-1}$ found in Prob. 5.28, draw curves of equal velocity for values of 3, 6, 9, 12 fps. How does the velocity vary along the surface? What is the total flow between any two adjacent stream surfaces?

5.30 For the two-dimensional flow around any angle α, the velocity potential and stream function are given in polar coordinates as $\phi = -ar^{\pi/\alpha} \cos(\pi\theta/\alpha)$ and $\psi = -ar^{\pi/\alpha} \sin(\pi\theta/\alpha)$, respectively. Prove that the functions given in Prob. 5.27 are a specialization of these expressions for $\alpha = \pi/2$. Take the case of $\alpha = 3\pi/2$, and plot streamlines and equipotential lines for the values given in Prob. 5.27. Compare the velocity at the corner with that at the corner in Prob. 5.27.

5.31 The flow around the body of Fig. 3.11 may be considered as that due to the sum of two velocity potentials, $\phi_1 = -Ux$, representing an undisturbed flow of velocity U in the x direction, and $\phi_2 = -S \ln r$, representing the radial flow from a source located inside the body behind the stagnation point. To relate U and S, it is observed that the total flow $2\pi S$ from the source (which is hydrodynamically equivalent to the body itself) must be equal to the flow of the main stream which is not passing through the body of width b, or $2\pi S = Ub$. This gives

$$\phi_2 = -\frac{Ub}{2\pi} \ln r$$

(*a*) The distance from the stagnation point to the source is determined by setting the radial velocity from the source, $v_r = -\partial\phi/\partial r$, equal and opposite to the undisturbed velocity U. Prove that this establishes the source at a distance $b/2\pi$ behind the stagnation point. The absolute velocity at any point of the field may be determined by the vector sum of the components U and v_r.

There follows an ingenious method of plotting the boundary of such a streamlined body, as shown in the figure. Suppose that the streamlines in the undisturbed flow are spaced a distance a

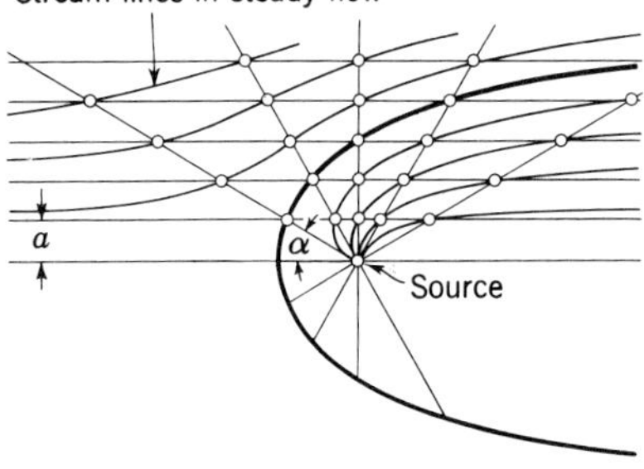

Problem 5.31

apart, where $b/2a = n$, an integer. Next divide the upper half of the source into n radial sectors, each of angle α, that is, $n\alpha = \pi$. Then the undisturbed flow between the x axis and the first streamline is associated with the source flow in the first sector from the stagnation point. Thus the intersection of the first streamline with the first line must be a point on the boundary of the body, through which there can be no flow. Similarly, the intersection of the horizontal line at $2a$ with the radial line at 2α forms another point, and so on. Further streamlines can be plotted by connecting successive intersections of the original horizontal lines with the radial lines, recognizing that the same flow must exist between any adjacent pair of streamlines. Thus the intersection of a horizontal line ea above the axis with a radial line at $f\alpha$ from the stagnation point must lie on a streamline which is $(e - f)a$ distant from the axis in the undisturbed region, where e and f are integers.

(*b*) Assume a value of $U = 20$ fps (5 m/s) and a two-dimensional flow past a streamlined body for which $b = 36$ ft (12 m). Compute the distance from the source to the stagnation point and to the surface of the body at a radius 90° to the axis. What is the value of the source velocity at the latter point?

(*c*) What is the magnitude of the velocity of the fluid along the surface at the 90° point? (Compare with the results of Prob. 3.17.) What is its direction relative to the axis?

5.32 Find the distance to the surface of the body and the two velocities called for in the preceding problem for an angle of 30°. (Compare with Prob. 3.17.)

5.33 Superimpose a point source ($Q = 100$ cfs) on a rectilinear flow field ($U = 20$ fps). Plot the body contour at $\theta = 30, 60, 90, 120, 150, 180°$ using a scale of 1 in = 1 ft. Compute the velocities along the body contour at these points. Determine the pressures at these points assuming $\rho = 1.94$ slug/ft^3 with zero pressure in the undisturbed rectilinear flow field. What is the velocity and pressure in the combined flow field at the following points? *Hint:* Refer to Prob. 5.31.

(*a*) $\theta = 45°$ $r = 4.0$ ft
(*b*) $\theta = 90°$ $r = 2.0$ ft
(*c*) $\theta = 90°$ $r = 4.0$ ft
(*d*) $\theta = 135°$ $r = 2.0$ ft

CHAPTER

SIX

MOMENTUM AND FORCES IN FLUID FLOW

Previously, two important fundamental concepts of fluid mechanics were presented: the continuity equations and the energy equation. In this chapter a third basic concept, the *impulse-momentum principle*, will be developed. This concept is of particular importance in flow problems where the determination of forces is involved. Such forces occur whenever the velocity of a stream of fluid is changed in either direction or magnitude. By the law of action and reaction, an equal and opposite force is exerted by the fluid upon the body producing the change. After developing the impulse-momentum principle its application to a number of important problems is discussed.

6.1 DEVELOPMENT OF THE IMPULSE-MOMENTUM PRINCIPLE

The impulse-momentum principle will be derived from Newton's second law. The flow may be compressible or incompressible, real (with friction) or ideal (frictionless), steady or unsteady, and the equation need not be applied along a streamline. In Chap. 4 when applying the energy equation to real fluids we found that the energy loss must be computed. This difficulty is not encountered in momentum analysis.

Newton's second law may be expressed as

$$\sum \mathbf{F} = \frac{d(m\mathbf{V})}{dt} \tag{6.1}$$

Thus, the sum of the external forces on a body is equal to the rate of change of momentum of that body. The bold face symbols **F** and **V** represent vectors and hence the change in momentum is in the same direction as the force. Equation (6.1) can also be expressed as $\sum(\mathbf{F})\,dt = d(m\mathbf{V})$, i.e., impulse equals change of momentum, hence the terminology *impulse-momentum principle* is used.

Let us apply Eq. (6.1) to a body defined by the mass of fluid contained at time t in the *control volume* of Fig. 6.1. Henceforth we shall refer to this mass of fluid as the *fluid system.* The control volume is fixed in position; it does not move, nor does it change shape or size. At time $(t + \Delta t)$ the fluid mass we are dealing with (i.e., the fluid system) has moved to a new position indicated by the shaded area of Fig. 6.1. Let us now define some terms.

$(m\mathbf{V})_t$ = momentum at time t of the fluid system (coincident with the control volume at time t)

$(m\mathbf{V})_{t+\Delta t}$ = momentum at time $(t + \Delta t)$ of the fluid system (coincident with the shaded area of Fig. 6.1 at time $t + \Delta t$)

$(m'\mathbf{V}')_t$ = momentum of the fluid mass contained within the control volume at time t

$(m'\mathbf{V}')_{t+\Delta t}$ = momentum of the fluid mass contained within the control volume at time $(t + \Delta t)$

$\Delta(m\mathbf{V})_{\text{out}}$ = momentum of the fluid mass that leaves the control volume during time interval Δt

$\Delta(m\mathbf{V})_{\text{in}}$ = momentum of the fluid mass that enters the control volume during time interval Δt

At time t the momentum of the fluid system is equal to the momentum of the fluid mass contained in the control volume at time t because the same fluid mass is involved in both cases. Thus

$$(m\mathbf{V})_t = (m'\mathbf{V}')_t$$

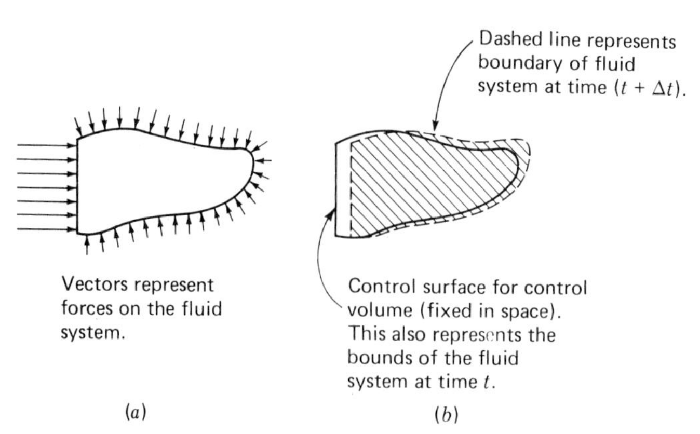

Figure 6.1 Control volume for general case. (*a*) Fluid mass acted on by certain forces. (*b*) Location of the fluid system at times t and $(t + \Delta t)$.

At time $(t + \Delta t)$ the momentum of the fluid system (shaded area of Fig. 6.1*b*) is equal to the momentum of the fluid mass in the control volume at $(t + \Delta t)$ plus the momentum of the mass that has flowed out of the control volume during time interval Δt minus that which has flowed into the control volume (unshaded area of Fig. 6.1*b*) during time interval Δt. Thus

$$(m\mathbf{V})_{t+\Delta t} = (m'\mathbf{V}')_{t+\Delta t} + \Delta(m\mathbf{V})_{\text{out}} - \Delta(m\mathbf{V})_{\text{in}}$$

The change of momentum of the fluid system is

$$\Delta(m\mathbf{V}) = (m\mathbf{V})_{t+\Delta t} - (m\mathbf{V})_t \tag{6.2}$$

Substituting the two preceding expressions into Eq. (6.2), we get

$$\Delta(m\mathbf{V}) = (m'\mathbf{V}')_{t+\Delta t} - (m'\mathbf{V}')_t + \Delta(m\mathbf{V})_{\text{out}} - \Delta(m\mathbf{V})_{\text{in}}$$

Applying Eq. (6.1), dividing through by Δt, rearranging, and noting that the limit of $\Delta(m\mathbf{V})/\Delta t = d(m\mathbf{V})/dt$ as $\Delta t \to 0$, we get

$$\begin{aligned}\sum \mathbf{F} &= \lim_{t \to 0} \frac{\Delta(m\mathbf{V})}{\Delta t} = \frac{d(m\mathbf{V})}{dt} \\ &= \frac{d(m\mathbf{V})_{\text{out}} - d(m\mathbf{V})_{\text{in}}}{dt} + \frac{(m'\mathbf{V}')_{t+\Delta t} - (m'\mathbf{V}')_t}{dt}\end{aligned} \tag{6.3}$$

The above equation states that the force acting on a fluid mass is equal to the rate of change of the momentum of the fluid mass which, in turn, is equal to the sum of the two terms on the right-hand side of the equation. The first term on the right side of the equation represents the net rate of outflow of momentum across the control surfaces while the second term represents the rate of accumulation of momentum within the control volume. Equation (6.3) is perfectly general. It applies to compressible or incompressible, real or ideal, and steady or unsteady flow. In the case of steady flow, the last term of Eq. (6.3) is equal to zero and the equation is

$$\sum \mathbf{F} = \frac{d(m\mathbf{V})_{\text{out}} - d(m\mathbf{V})_{\text{in}}}{dt} = \frac{d(m\mathbf{V})_{\text{out}}}{dt} - \frac{d(m\mathbf{V})_{\text{in}}}{dt} \tag{6.4}$$

Thus, for steady flow the force on the fluid mass is equal to the net rate of outflow of momentum across the control surface.

Since Eqs. (6.1) through (6.4) are vectorial equations they can also be expressed as scalar equations in terms of forces and velocities in the x, y, and z directions respectively.

It is advantageous to select a control volume such that the control surface is normal to the velocity where it cuts the flow. Consider such a situation in Fig. 6.2. Also, let the velocity be constant where it cuts across the control surface. In Fig. 6.2 the fluid system we are dealing with is contained between sections 1 and 2 at time t. This fluid system moves to a new position during time interval dt, as indicated in Fig. 6.2. During this short interval we will assume the fluid moves a short distance ds_1 at section 1 and ds_2 at section 2. Also, we are restricting

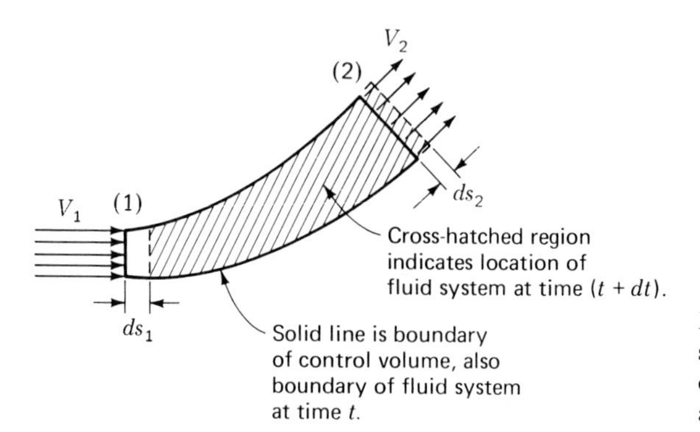

Figure 6.2 Control volume for steady flow with control surface cutting a constant velocity stream at right angles.

ourselves to steady flow so that Eq. (6.4) is applicable. The momentum crossing the control surface at section 1 during the interval dt is $(\rho_1 A_1\, ds_1)\mathbf{V}_1$ while that crossing section 2 is $(\rho_2 A_2\, ds_2)\mathbf{V}_2$. Substituting these expressions into Eq. (6.4), noting that since the control surface cuts the velocity at right angles, $V = ds/dt$ and $Q = AV$, we get for steady flow along a stream tube

$$\sum \mathbf{F} = \rho_2 Q_2 \mathbf{V}_2 - \rho_1 Q_1 \mathbf{V}_1 \tag{6.5}$$

From continuity, for steady flow, $\rho Q = \rho_1 Q_1 = \rho_2 Q_2$; thus we can write

$$\sum \mathbf{F} = \rho Q(\mathbf{V}_2 - \mathbf{V}_1) = \rho Q(\Delta \mathbf{V}) \tag{6.6}$$

The direction of $\sum \mathbf{F}$ will be the same as that of $\Delta\mathbf{V}$. The $\sum \mathbf{F}$ represents the vectorial summation of all forces acting on the fluid mass including gravity forces, shear forces, and pressure forces including those exerted by fluid surrounding the fluid mass under consideration as well as the pressure forces exerted by the solid boundaries in contact with the fluid mass. The right-hand side of Eq. (6.6) represents the change in momentum per unit time.

Since Eq. (6.6) is vectorial it can be expressed by the following scalar equations

$$\sum F_x = \rho_2 Q_2 V_{2_x} - \rho_1 Q_1 V_{1_x} = \rho Q(\Delta V_x) \tag{6.7a}$$

$$\sum F_y = \rho_2 Q_2 V_{2_y} - \rho_1 Q_1 V_{1_y} = \rho Q(\Delta V_y) \tag{6.7b}$$

$$\sum F_z = \rho_2 Q_2 V_{2_z} - \rho_1 Q_1 V_{1_z} = \rho Q(\Delta V_z) \tag{6.7c}$$

In Sec. 6.3 and succeeding sections these equations will be applied to several situations that are commonly encountered in engineering practice. If the flow in a single stream tube splits up into several streamtubes, the ρQV's of each stream tube are computed separately and then substituted into Eqs. (6.5) to (6.7). (See Illustrative Example 6.1.) The big advantage of the impulse-momentum principle is that one need not be concerned with the details of what is occurring within the flow; only the conditions at the end sections of the control volume govern the analysis.

6.2 MOMENTUM CORRECTION FACTOR

If the velocity is not uniform over a section, the momentum per unit time transferred across that section is greater than that computed by using the mean velocity. Thus the rate of momentum transfer (*momentum flux*) across an elementary area dA, where the local velocity is u, is $(\rho u\, dA)u = \rho u^2\, dA$, and the rate of momentum transfer across the entire section is $\rho \int_A u^2\, dA$, while that computed by using the mean velocity is $\rho QV = \rho AV^2$. Hence the correction factor by which ρQV should be multiplied to obtain the true momentum per unit time is

$$\beta = \frac{1}{AV^2} \int_A u^2\, dA \tag{6.8}$$

For laminar flow in a circular pipe, $\beta = \frac{4}{3}$, but for turbulent flow in circular pipes, it usually ranges from 1.005 to 1.05, as shown by Eq. (8.35*b*). For open-channel flow it may be greater. Unless otherwise specified, the value of β in the ensuing discussion will be taken as 1.0.

6.3 FORCE EXERTED ON PRESSURE CONDUITS

Consider the case of horizontal flow to the right through the reducer of Fig. 6.3*a*. A free-body diagram of the forces acting on the fluid mass contained in the reducer is shown in Fig. 6.3*b*. We shall apply Eq. (6.7*a*) to this fluid mass to examine the forces that are acting in the x direction. The forces $p_1 A_1$ and $p_2 A_2$ represent pressure forces exerted by fluid located just upstream and just downstream of the fluid mass under consideration. The force $(F_{R/F})_x$ represents the force exerted by the reducer on the fluid in the x direction. Neglecting shear forces at the boundary of the reducer, the force $(F_{R/F})_x$ is the integrated effect of the normal pressure forces that are exerted on the fluid by the wall of the reducer. The intensity of pressure at the wall will decrease as the diameter decreases because of the increase in velocity head. A typical pressure diagram is shown in Fig. 6.4.

Applying Eq. (6.7*a*) and assuming the fluid to be ideal with $(F_{R/F})_x$ directed as shown, we get

$$\sum F_x = p_1 A_1 - p_2 A_2 - (F_{R/F})_x = \rho Q(V_2 - V_1) \tag{6.9}$$

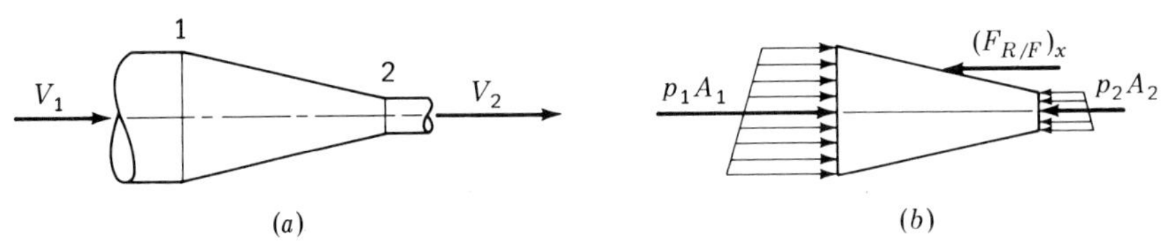

Figure 6.3

In Eq. (6.9) each term can be evaluated independently from the given flow data, except $(F_{R/F})_x$, which is the quantity we wish to find. Rewriting Eq. (6.9), the result is

$$(F_{R/F})_x = p_1 A_1 - p_2 A_2 - \rho Q(V_2 - V_1) \tag{6.10}$$

This gives the value of the total force exerted by the *reducer on the fluid* in the x direction.[1] This force acts to the left as assumed in Fig. 6.3*b* and as applied in Eq. (6.9). The force of the *fluid on the reducer* is, of course, equal and opposite to that of the reducer on the fluid. If the flow were to the left in Fig. 6.3, a similar analysis would apply, but it is necessary to be consistent in regard to plus and minus signs. The usual convention is to consider the direction in which the flow is occurring as the positive direction.

Consideration of the weight of fluid between sections 1 and 2 in Fig. 6.3 results in the conclusion that pressures are larger on the bottom half of the pipe than on the upper half. It should be noted that it is the conditions at the end sections of the control volume that govern the analysis. What occurs within the flow between sections 1 and 2 is unimportant so far as the determination of forces is concerned. Figure 6.4 gives a schematic representation of the pressure distribution on the fluid within the reducer. The integrated effect of the pressures exerted by the reducer itself is equivalent in the x direction to $(F_{R/F})_x$ and in the z direction to the weight of fluid between sections 1 and 2.

If the fluid undergoes a change in both direction and velocity, as in the reducing pipe bend in Fig. 6.5, the procedure is similar to that of the preceding case, except that it is convenient to deal with components. Assuming the flow is in a horizontal plane so that the weight can be neglected, applying Eq. (6.7*a*) by summing up forces acting on the fluid in the x direction, and equating them to the change in fluid momentum in the x direction gives

$$(F_{B/F})_x = p_1 A_1 - p_2 A_2 \cos\theta - \rho Q(V_2 \cos\theta - V_1) \tag{6.11}$$

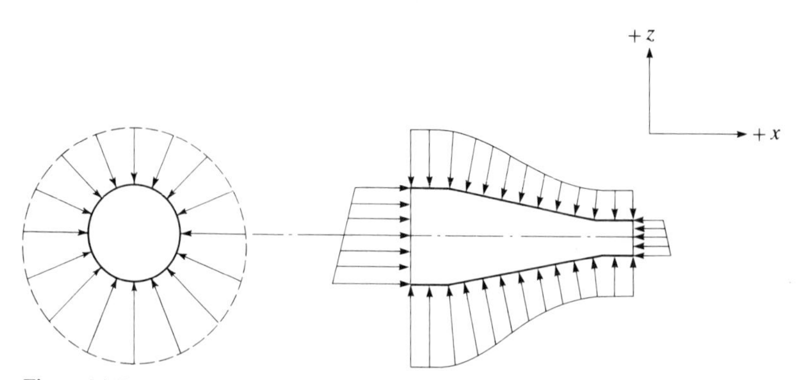

Figure 6.4 Pressure distribution on the fluid in a reducer.

[1] The gage pressure must be used in these equations. If the absolute pressure were used, the effect of atmospheric pressure on the external boundary of the conduit would have to be considered.

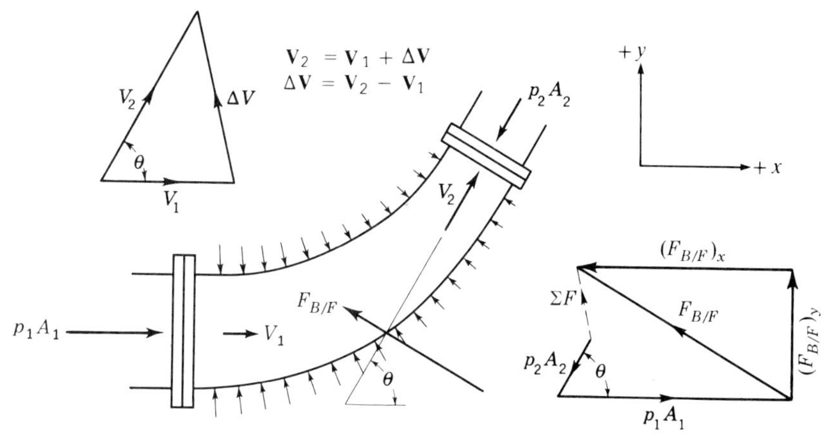

Figure 6.5 Forces on the fluid in a reducing bend.

Similarly, in the y direction,

$$(F_{B/F})_y = p_2 A_2 \sin \theta + \rho Q V_2 \sin \theta \tag{6.12}$$

In a specific case, if the numerical values determined by these equations are positive, then the assumed directions are correct. A negative value for either one merely indicates that that component is in the direction opposite from that assumed.

Note that $\sum \mathbf{F} = \rho Q \, \Delta \mathbf{V}$ is the resultant of all the forces acting on the fluid, which includes the pressure forces on the two ends, while $F_{B/F}$ is the force exerted by the bend on the fluid. The value of $F_{B/F}$ is $\sqrt{(F_{B/F})_x^2 + (F_{B/F})_y^2}$, and it is represented by the closing line in the force diagram shown in Fig. 6.5.

The total force $\mathbf{F}_{F/B}$ exerted by the fluid on the bend is equal in magnitude but opposite in direction to the force $\mathbf{F}_{B/F}$ of the bend on the fluid. The force of the fluid on the bend tends to move the portion of the pipe under consideration. Hence, where such changes in velocity or alignment occur, a large pipe will usually be "anchored" by attaching it to a concrete block of sufficient size or weight to provide the necessary resistance.

If the flow had been in the vertical plane the weight of the fluid between sections 1 and 2 would have to be estimated and included in Eq. (6.12). The effect of shear stresses due to fluid friction could be introduced into the problem; however, these effects are usually small.

Illustrative Example 6.1 Determine the magnitude and direction of the resultant force exerted on this double nozzle. Both nozzle jets have a velocity of 12 m/s. The axes of the pipe and both nozzles lie in a horizontal plane. $\gamma = 9.81$ kN/m^3. Neglect friction.

Continuity:

$$A_1 V_1 = A_2 V_2 + A_3 V_3$$

$$15^2 V_1 = 10^2(12) + 7.5^2(12) \qquad V_1 = 8.33 \text{ m/s}$$

$$Q_1 = \frac{\pi}{4}(0.15)^2 8.33 = 0.147 \text{ m}^3\text{/s} \qquad Q_2 = 0.094 \text{ m}^3\text{/s} \qquad Q_3 = 0.053 \text{ m}^3\text{/s}$$

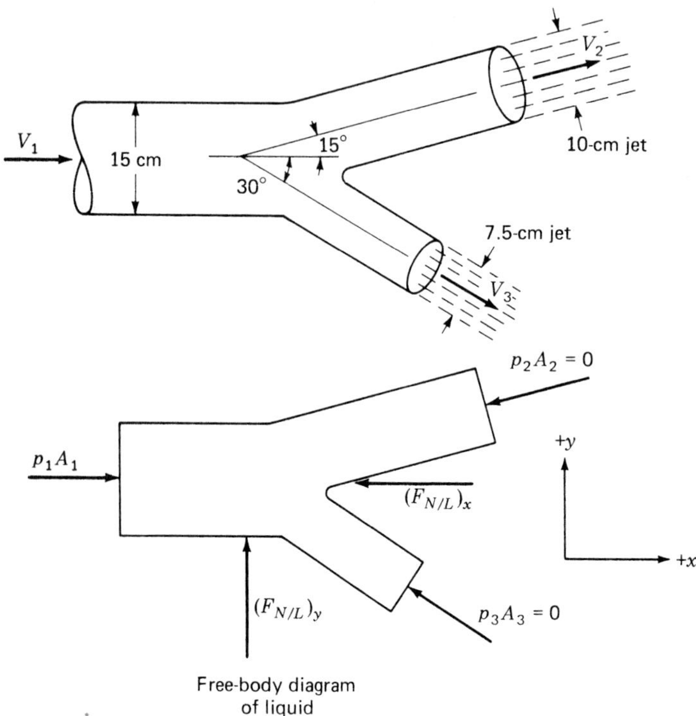

Illustrative Example 6.1

Energy equation:

$$\frac{p_1}{\gamma} + \frac{8.33^2}{2(9.81)} = 0 + \frac{12^2}{2(9.81)}$$

$$\frac{p_1}{\gamma} = 3.80 \text{ m} \qquad p_1 = 37.3 \text{ kN/m}^2 \qquad p_1A_1 = 0.659 \text{ kN}$$

$$\sum F_x = p_1A_1 - (F_{N/L})_x = (\rho Q_2 V_{2_x} + \rho Q_3 V_{3_x}) - \rho Q_1 V_{1_x}$$

$$\rho = \frac{\gamma}{g} = \frac{9.81 \text{ kN/m}^3}{9.81 \text{ m/s}^2} = 1.0 \frac{\text{kN}\cdot\text{s}^2}{\text{m}^4} = 10^3 \frac{\text{kg}}{\text{m}^3}$$

$$V_{2_x} = V_2 \cos 15^\circ = 12(0.966) = 11.6 \text{ m/s}$$

$$V_{3_x} = V_3 \cos 30^\circ = 12(0.866) = 10.4 \text{ m/s} \qquad V_{1_x} = V_1 = 8.33 \text{ m/s}$$

$$0.659 - (F_{N/F})_x = 10^3(0.094)(11.6) + 10^3(0.053)(10.4) - 10^3(0.147)8.33 = 0.417 \text{ kN}$$

$$(F_{N/L})_x = 0.659 - 0.417 = 0.242 \text{ kN}$$

$$\sum F_y = (F_{N/L})_y = (\rho Q_2 V_{2_y} + \rho Q_3 V_{3_y}) - \rho Q_1 V_{1_y}$$

$$V_{2_y} = V_2 \sin 15^\circ = 12(0.259) = 3.1 \text{ m/s}$$

$$V_{3_y} = -V_3 \sin 30^\circ = -12(0.50) = -6.0 \text{ m/s} \qquad V_{1_y} = 0$$

$$(F_{N/L})_y = 10^3(0.094)(3.1) + 10^3(0.053)(-6.0) - 10^3(1.47)(0)$$

$$= 0.291 - 0.318 = -0.027 \text{ kN}$$

The minus sign indicates that the assumed direction of $(F_{N/L})_y$ was wrong. Therefore $(F_{N/L})_y$ acts in the negative y direction. Equal and opposite to $F_{N/L}$ is $F_{L/N}$

$$(F_{L/N})_x = 0.242 \text{ kN} \qquad \text{(in positive } x \text{ direction)}$$

$$(F_{L/N})_y = 0.027 \text{ kN} \qquad \text{(in positive } y \text{ direction)}$$

6.4 FORCE EXERTED ON A STATIONARY VANE OR BLADE

A procedure similar to that of Sec. 6.3 may be employed to find the force exerted on a stationary vane or blade. The main difference is that with a vane or blade the fluid is in contact with the atmosphere; hence the pA forces disappear. Another difference is that in many types of fluid machinery where vanes or blades are used the velocities are often so high that the neglect of friction may introduce a sizeable error. In such cases, for accurate results, friction should be considered. The following example illustrates these points.

Illustrative Example 6.2 In the figure suppose that $\theta = 30°$, $V_1 = 100$ fps, and the stream is a jet of water with an initial diameter of 2 in. Assume friction losses such that $V_2 = 95$ fps. Find the resultant force on the blade. Assume that flow occurs in a horizontal plane.

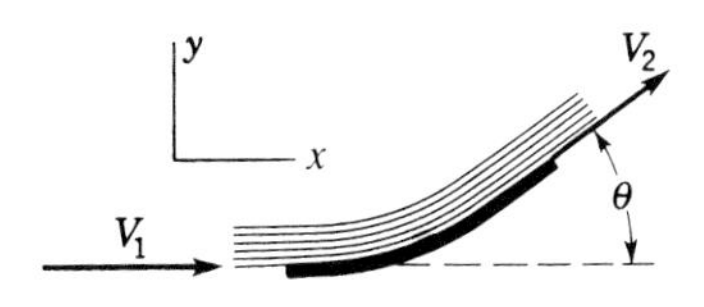

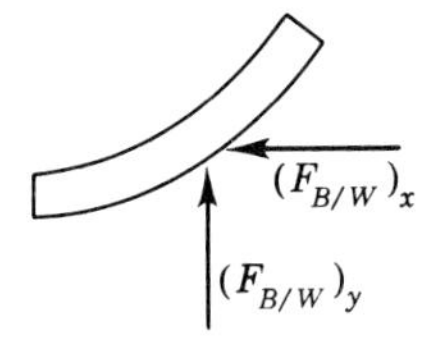

Illustrative Example 6.2

This problem is best solved by taking a free-body diagram of the element of fluid in contact with the blade. The forces acting on this element are as shown in the sketch. The forces $(F_{B/W})_x$ and $(F_{B/W})_y$ represent the components of force of blade on water in the x and y directions. These forces include shear stresses tangential to the blade as well as pressure forces normal to the blade.

Applying Eq. (6.7*a*) along the x axis and noting that $A = 0.0218 \text{ ft}^2$ (Appendix 3, Table A.6*a*)

$$-(F_{B/W})_x = \rho Q(V_{2_x} - V_{1_x}) = 1.94(0.0218 \times 100)(0.866 \times 95 - 100)$$
$$= 4.23(-17.7) = -74.9 \text{ lb}$$

Hence $\qquad (F_{B/W})_x = +74.9 \text{ lb}$

The plus sign indicates that the assumed direction of $(F_{B/W})_x$ was correct.

Applying Eq. (6.7*b*) along the y axis,

$$+(F_{B/W})_y = \rho Q(V_{2_y} - V_{1_y}) = 4.23(0.50 \times 95 - 0) = 201 \text{ lb}$$

Thus the force of the blade on the fluid is the resultant of a 74.9-lb component to the left and a 201-lb component upward in the y direction. Equal and opposite to this is the force of the fluid on the blade (downward and to the right). The resultant force is 215 lb at an angle of 69.5° below the horizontal.

If friction were neglected (i.e., $V_2 = V_1 = 100$ fps), the forces would have been calculated as $(F_{B/W})_x = 56.7$ lb and $(F_{B/W})_y = 212$ lb. When the angle of deflection θ from the initial direction of the jet is less than 90°, friction increases the value of $(F_{B/W})_x$ over the value it would have if there were no friction. When θ is greater than 90°, friction decreases the value of F_x. On the other hand, friction decreases the value of F_y for any value of angle θ.

If the flow had been in the vertical plane, the effect on V_2 of the higher elevation at exit from the blade would have to be considered and the weight of the liquid on the blade would have to be estimated and added to $\rho Q(\Delta V_z)$ to get the total value of $(F_{B/W})_z$.

6.5 RELATION BETWEEN ABSOLUTE AND RELATIVE VELOCITIES

In much of the work that follows it will be necessary to deal with both absolute and relative velocities of the fluid. The absolute velocity **V** of a body (Fig. 6.6) is its velocity relative to the earth. The relative velocity **v** of a body is its velocity relative to a second body, which may in turn be in motion relative to the earth.

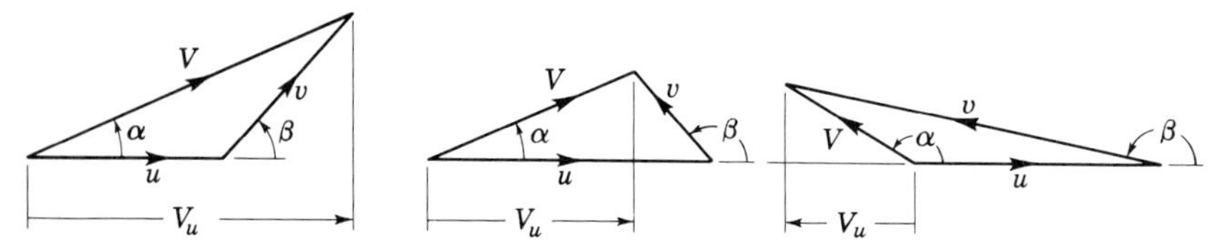

Figure 6.6 Relative and absolute velocity relations.

The absolute velocity **V** of the first body is the vector sum of its velocity **v** relative to the second body and the absolute velocity **u** of the latter. The relation of the three is thus

$$\mathbf{V} = \mathbf{u} + \mathbf{v} \tag{6.13}$$

Let us define α and β as the angles made by the absolute and relative velocities of a fluid, respectively, with the positive direction of the linear velocity u of some point on a solid body. It is seen from Fig. 6.6 that, whatever the shape of the velocity vector triangle,

$$V \sin \alpha = v \sin \beta$$

$$V_u = V \cos \alpha = u + v \cos \beta$$

where V_u is defined as the component of the absolute velocity V of the fluid in the direction of u.

6.6 FORCE UPON A MOVING VANE OR BLADE

The force exerted by a stream upon a single moving object can be determined by Eq. (6.6) provided the flow is steady and the body has a motion of translation along the line of action of the stream initially. If the latter condition is not fulfilled, the case becomes the complex one of unsteady flow.

There are two principal differences between the action upon a stationary and a moving object. One is that for the case of a moving object, it is necessary to consider both absolute and relative velocities, which makes the determination of $\Delta \mathbf{V}$ more difficult. The other is that the amount of fluid that strikes a single moving object in any time interval Δt is different from that which strikes a stationary object.

If the cross-sectional area of a stream is A_1 and its velocity is V_1, then the rate at which fluid is emitted from the nozzle is $G = \gamma Q = \gamma A_1 V_1$. But for the case of a single object, the amount of fluid which strikes the body per unit time will be less than this if the body is moving away from the nozzle and more than this if it is moving toward the nozzle. As an extreme case, suppose the object is moving in the same direction as the jet and with the same or higher velocity. It is clear that none of the fluid will act upon the body. If it is moving with a velocity less than that of the jet, the amount of fluid that strikes the body per unit time will be proportional to the difference between the two velocities, i.e., to $v_1 = V_1 - u$. If G' denotes the weight of fluid per second striking a single object moving with a velocity u in the same direction as V_1, then

$$G' = \gamma Q' = \gamma A_1(V_1 - u) = \gamma A_1 v_1 \tag{6.14}$$

An explanation for the difference between G and G' may also be seen by considering Fig. 6.7, where the fluid issues from a nozzle at the rate of $G = \gamma A_1 V_1$ per unit time. But in this unit of time the object will have moved away from the nozzle the distance u, and the volume of fluid between the two will have been increased by the amount $A_1 u$.

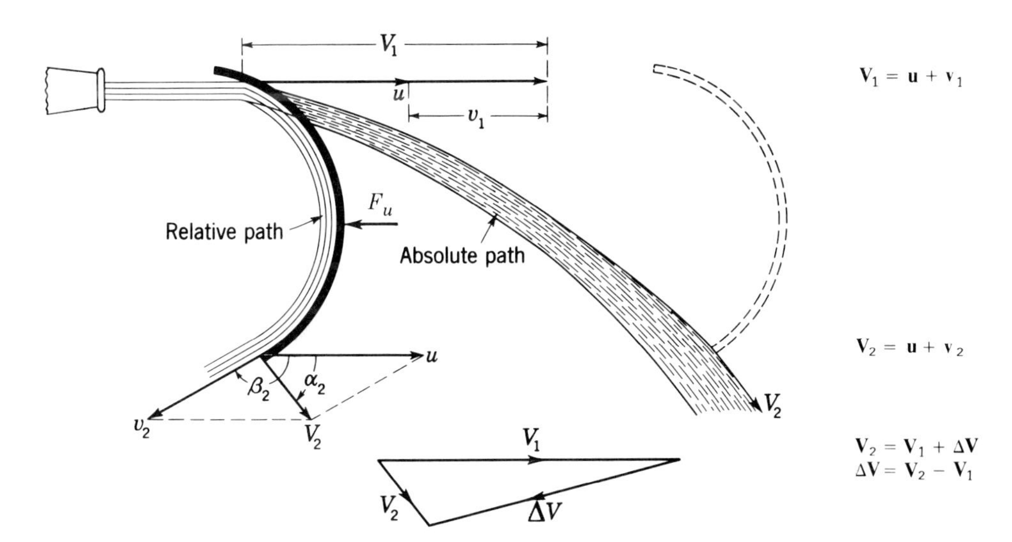

Figure 6.7 Jet acting on a vane in translation.

Thus, for this special case of a single object moving in the same direction as the stream initially, the component of the force exerted by the vane on the fluid is

$$F_u = \frac{G'}{g} \Delta V_u = \frac{\gamma A_1 v_1}{g} \Delta V_u \tag{6.15}$$

acting to the left in Fig. 6.7. Equal and opposite to this is the force of the jet on the vane acting to the right. The subscript u represents the component in the same direction as u. It will be seen later that this same subscript is used to represent the tangential direction at a point on a rotating body.

It should be noted that there is a force exerted in the direction at right angles to u but we will not concern ourselves with it here.

It can be proved that $\Delta V_u = \Delta v_u$. Referring to Fig. 6.7, it is seen that $V_{2_u} = V_2 \cos \alpha_2$ and $V_{1_u} = V_1 \cos \alpha_1 = V_1$. Therefore

$$\Delta V_u = V_{2_u} - V_{1_u} = V_2 \cos \alpha_2 - V_1$$

But $V_1 = u + v_1$, so $\Delta V_u = V_2 \cos \alpha_2 - v_1 - u$. Now, $v_{2_u} = v_2 \cos \beta_2$ and $v_{1_u} = v_1 \cos \beta_1 = v_1$. Therefore $\Delta v_u = v_{2_u} - v_{1_u} = v_2 \cos \beta_2 - v_1$. But

$$v_2 \cos \beta_2 = V_2 \cos \alpha_2 - u.$$

So

$$\Delta v_u = V_2 \cos \alpha_2 - u - v_1$$

Hence $\Delta V_u = \Delta v_u$. Thus the change of the absolute velocity in the u direction is identical to the change of the relative velocity in that direction, and hence Eq. (6.15) for the u component of the force on a single body may also be written as

$$F_u = \frac{G'}{g} \Delta v_u = \frac{\gamma A_1 v_1}{g} \Delta v_u \tag{6.16}$$

In Fig. 6.7, by the time a particle of fluid which strikes the moving vane at the instant it is in the position shown by the solid line has reached the point of outflow from the vane, the latter will have reached the position shown by the dotted line. Thus two paths may be traced for the fluid, one relative to the moving vane, which is as it would appear to an observer (or a camera) moving with the vane, and the other relative to the earth, termed the *absolute path*, as it would appear to an observer (or a camera) stationary with respect to the earth.

A study of Fig. 6.7 shows that the direction of the relative velocity at outflow from the vane is determined by the shape of the latter, but the relative velocity at entrance, just *before* the fluid strikes the vane, is determined solely by the relation between V_1 and u. Just *after* the fluid strikes the vane, its relative velocity must be tangent to the vane surface. To avoid excess energy loss, these two directions should agree; otherwise there will be an abrupt change in velocity and direction of flow at this point.

There are few instances where a stream of fluid impinges on a single body or vane. More commonly, the jet is directed on a series of vanes as with a Pelton wheel (Fig. 15.1). In such a case the effective flow impinging on each of the series of closely spaced buckets is $Q = A_1 V_1$, since whatever flow does not impinge on the first bucket will impinge on the second one, and so on around the circle. Thus the u component of the force exerted by the fluid on a series of vanes is expressible as

$$F_u = \frac{G}{g} \Delta V_u = \frac{\gamma A_1 V_1}{g} \Delta V_u = \frac{\gamma A_1 V_1}{g} \Delta v_u \tag{6.17}$$

Once again, the force of the vanes on the fluid is in the direction of ΔV_u, that is, to the left in Fig. 6.7.

Illustrative Example 6.3 A 2-in-diameter water jet with a velocity of 100 fps impinges on a single vane moving in the same direction (thus $F_x = F_u$) at a velocity of 60 fps. If $\beta_2 = 150°$ and friction losses over the vane are such that $v_2 = 0.9v_1$, compute the force exerted by the water on the vane.

The velocity vector diagrams at entrance and exit to the vane are shown in the accompanying figure. Since $v_2 = 0.9 \times 40 = 36$ fps,

$$V_2 \sin \alpha_2 = v_2 \sin \beta_2 = 36 \times 0.5 = 18 \text{ fps} \tag{a}$$

$$V_2 \cos \alpha_2 = u + v_2 \cos \beta_2 = 60 + 36(-0.866) = 28.8 \text{ fps} \tag{b}$$

Solving (*a*) and (*b*) simultaneously yields $V_2 = 34$ fps, $\alpha_2 = 32°$. Hence

$$-F_x = \rho Q'(V_2 \cos \alpha_2 - V_1) = 1.94(0.0218)(100 - 60)(28.8 - 100) = -120.4 \text{ lb}$$

So $F_x = 120.4$ lb. The force of vane on water is to the left as assumed; hence force of water on vane is 120.4 lb to the right.

$$-F_y = \rho Q'(V_2 \sin \alpha_2 - 0) = 1.94(0.0218)(40)(-18) = -30.5 \text{ lb}$$

Thus $F_y = 30.5$ lb in the direction shown. The force of water on the vane is equal and opposite and thus 30.5 lb upward. If the blade were one of a series of blades,

$$-F_x = \rho Q(V_2 \cos \alpha_2 - V_1) = 1.94(0.0218)(100)(28.8 - 100) = -301 \text{ lb}$$

For the case of a series of blades, energy considerations could have been used for the solution. The horsepower of the original jet is

$$\text{HP}_{\text{in}} = \frac{\gamma Q(V_1^2/2g)}{550} = \frac{62.4(0.0218 \times 100)[(100)^2/64.4]}{550} = 38.4$$

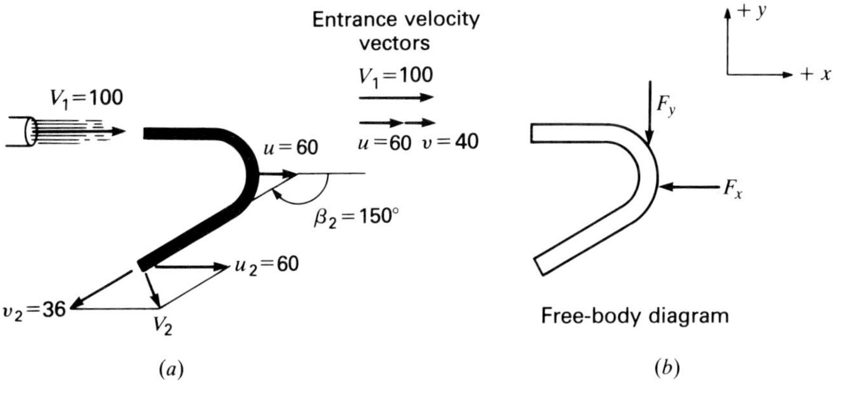

Illustrative Example 6.3

The horsepower of the water as it leaves the system is

$$\text{HP}_{\text{out}} = \frac{\gamma Q(V_2^2/2g)}{550} = 4.44$$

The horsepower transferred to the blades (i.e., out of the fluid) is

$$\text{HP}_{\text{transfer}} = \frac{Fu}{550} = \frac{(301)(60)}{550} = 32.8$$

An equation for conservation of energy expressed in terms of power is

$$\text{HP}_{\text{in}} - \text{HP}_{\text{out}} - \text{HP}_{\text{transfer}} - \text{HP}_{\text{friction loss}} = 0$$

Thus

$$38.4 - 4.4 - 32.8 = \text{HP}_{\text{friction loss}}$$

Therefore

$$\text{HP}_{\text{friction loss}} = 1.2$$

That this is so may be verified by computing

$$\frac{\gamma Q(v_1^2/2g) - \gamma Q(v_2^2/2g)}{550} = \frac{62.4(0.0218 \times 100)[(40)^2 - (36)^2]}{550(64.4)} = 1.2$$

It should be noticed that the horsepower loss due to friction is small. Commonly, in problems of this type, an assumption that $v_1 = v_2$ will give reasonably good results.

6.7 REACTION OF A JET

In Fig. 6.8 consider a jet issuing steadily from a tank which is large enough so that the velocity within it may be neglected. Let the area of the jet be A_2 and its velocity V_2, and assuming an ideal fluid, $V_2 = \sqrt{2gh}$. In this case with the jet flowing to the right, a force equal to $\rho Q_2 V_2$ is exerted to the left on the tank. That this is so may be seen by applying Eq. (6.7*a*) to the free-body diagram (Fig. 6.8*b*) of the liquid in the tank. In Fig. 6.8*b* the boldface vectors represent the

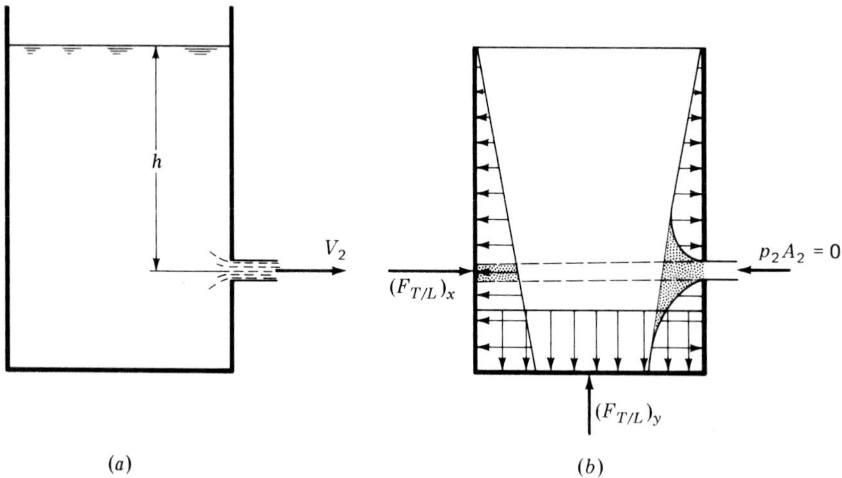

Figure 6.8

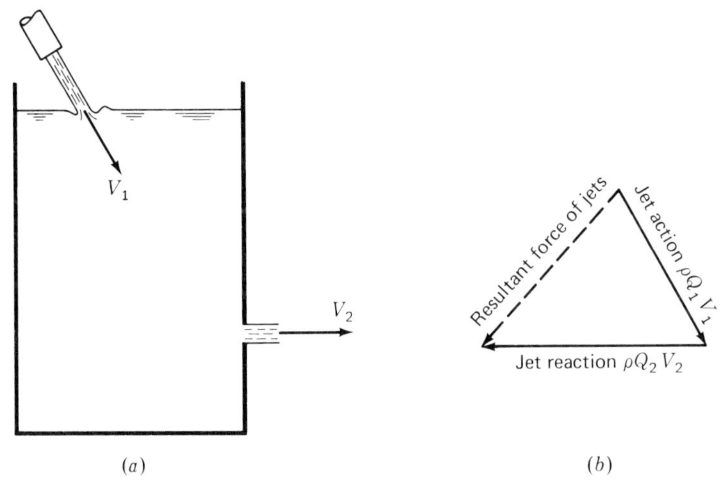

Figure 6.9

forces of the tank on the liquid while the distributed load represents the force of the liquid on the tank. In the figure it is the distributed load in the vertical plane of the centerline of the jet that is shown.

Applying Eq. (6.7*a*) to the liquid, we get

$$F_x = (F_{T/L})_x = \rho Q_2(V_2 - 0) = \rho A_2 V_2^2 = \rho A_2(2gh) = 2\gamma h A_2 \qquad (6.18)$$

This is the net force of the tank on the liquid in the x direction; it acts to the right and causes the change of velocity of the flowing liquid from zero to V_2. Equal and opposite to this force is the force of the liquid on the tank, often referred to as the *jet reaction.* If the tank were supported on frictionless rollers, it would be moved to the left by this action. The net force $\rho Q_2 V_2$ is equal to the difference in the magnitude of the pressure forces on the two ends of the tank. On the left end of the tank a normal hydrostatic pressure exists while on the right end of the tank there is a lowering of the pressure near the orifice because of the increase in velocity within the tank in that region. By observing the last term of Eq. (6.18) we see that this net force is equal to twice the hydrostatic force on A_2. Thus the net force (shaded area at right end of tank) is equal to twice the hydrostatic force on A_2 (shaded area at left end of the tank).

Refer now to Fig. 6.9 where a jet of liquid of cross-sectional area A_1 is discharged into the tank with a velocity V_1. In this case a force $\mathbf{F} = \rho Q_1 \mathbf{V}_1$ is exerted by the jet on the liquid which, in turn, transmits the force to the tank. This is referred to as *jet action.*

The resultant force on the tank caused by one jet entering the tank at section 1 and the other jet leaving the tank at section 2 is the vector sum of $\rho Q_1 \mathbf{V}_1$ and $\rho Q_2 \mathbf{V}_2$ where the first vector (jet action) acts in the direction of V_1 (downward to the right in Fig. 6.9) and the second vector (jet reaction) acts in the direction opposite to that of V_2. Thus a jet entering a system acts on the system in the direction in which the jet is traveling while a jet leaving a system acts on the system in the direction opposite to that in which the jet is traveling.

Illustrative Example 6.4 In the accompanying figure (*a*) is shown a curved pipe section of length 40 feet that is attached to the straight pipe section as shown. Determine the resultant force on the curved pipe, and find the horizontal component of the jet reaction. All significant data are given in the figure. Assume an ideal liquid with $\gamma = 55\ \text{lb/ft}^3$.

The energy equation between sections 1 and 3 gives

$$\frac{30 \times 144}{55} + 35 = 20 + \frac{V_3^2}{2g}$$

and

$$V_3 = 77.6\ \text{fps} \qquad \text{(jet velocity)}$$

$$Q = A_3 V_3 = 3.81\ \text{cfs}$$

$$V_2 = \frac{Q}{A_2} = 43.6\ \text{fps}$$

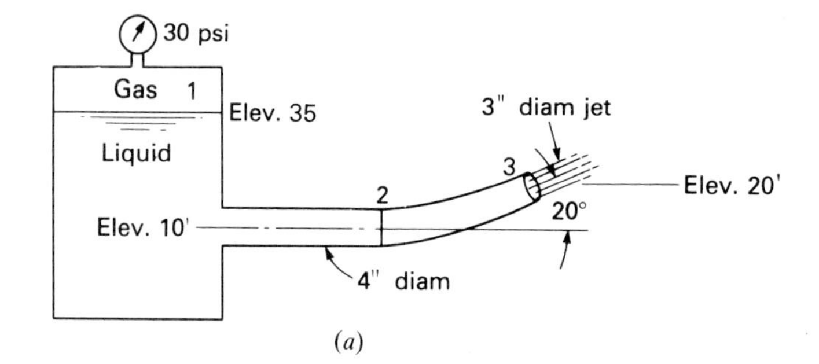

(*a*)

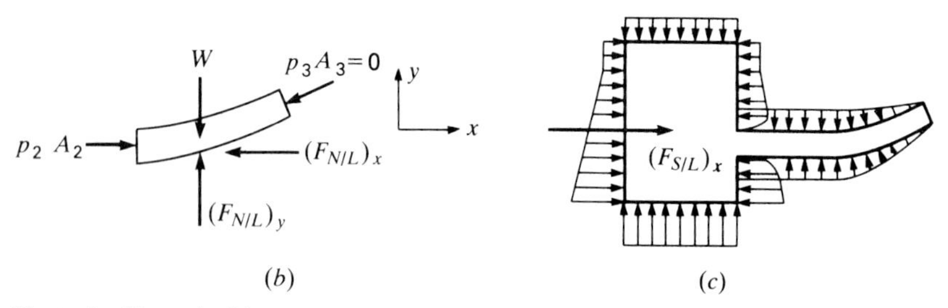

(*b*) (*c*)

Illustrative Example 6.4

Energy equation between section 2 and 3 gives

$$10 + \frac{p_2(144)}{55} + \frac{(43.6)^2}{64.4} = 20 + \frac{(77.6)^2}{2g}$$

and

$$p_2 = 28.3\ \text{psi}$$

The free-body diagram of the forces acting on the liquid contained in the curved pipe is shown in (*b*) of the figure. Applying Eq. (6.7*a*),

$$p_2 A_2 - p_3 A_3 \cos 20^\circ - (F_{N/L})_x = \rho Q(V_3 \cos 20^\circ - V_2)$$

where $(F_{N/L})_x$ represents the force of the curved pipe on the liquid in the x direction. Since section 3 is a jet in contact with the atmosphere, $p_3 = 0$. Thus

$$28.3\left(\frac{\pi}{4} \times 4^2\right) - (F_{N/L})_x = \left(1.94 \times \frac{55}{62.4}\right)(3.81)(77.6 \times 0.94 - 43.6)$$

$$356 - (F_{N/L})_x = 191$$

$$(F_{N/L})_x = +165 \text{ lb}$$

where the plus sign indicates that the assumed direction is correct. In the y direction the $p_2 A_2$ force has no component. Estimating the weight of liquid W as 150 lb,

$$(F_{N/L})_y = \rho Q(77.6 \times 0.342 - 0) + 150 = +323 \text{ lb}$$

The resultant force of liquid on the curved pipe is equal and opposite to that of the curved pipe on liquid. The resultant force of liquid on the curved pipe is $[(165)^2 + (323)^2]^{1/2} = 363$ lb downward and to the right at an angle of 62°56′ with the horizontal.

The horizontal jet reaction is best found by taking a free-body diagram of the liquid in the system as shown in (*c*) of the figure:

$$(F_{S/L})_x = \rho Q(V_3 \cos 20^\circ - 0) = 475 \text{ lb}$$

where $(F_{S/L})_x$ represents the force of the system on the liquid in the x direction. $(F_{S/L})_x$ is equivalent to the integrated effect of the x components of the pressure vectors shown in (*c*). Equal and opposite to $(F_{S/L})_x$ is the force of the liquid on the system, i.e., the jet reaction. Hence the horizontal jet reaction is a 475-lb force to the left. Thus there is a 165-lb force to the right tending to separate the curved pipe section from the straight pipe section, while at the same time there is a 475-lb force tending to move the entire system to the left.

6.8 JET PROPULSION

In Sec. 6.7 an expression was derived for the reaction of a jet from a stationary tank. Assume now that the tank in Fig. 6.8 is moving to the left with a velocity u. If the orifice is small compared with the size of the tank, the relative velocity within the latter may be disregarded, as may also any change in h for a short interval of time. Thus the absolute velocity of the fluid within the tank is $V_1 = u$ to the left. If the jet issues from the orifice with a relative velocity v_2, taking velocities to the right as positive, the absolute velocity of the jet will be $V_2 = v_2 - u$. Hence

$$\Delta V = V_2 - V_1 = (v_2 - u) - (-u) = v_2$$

The same result is obtained by referring to Fig. 6.8 for the case of a stationary tank (i.e., $u = 0$). In this instance $\Delta V = V_2 - 0 = v_2$. Thus the force of reaction is independent of the velocity of the tank, and Eq. (6.18) applies for either rest or motion.

Rocket

Both the fuel and the oxygen for combustion are contained within a rocket which is analogous to the tank of Fig. 6.8. The only difference is that the exit pressure

p_0 of the gases leaving the orifice or nozzle at section 2 may exceed the atmospheric pressure p_a. If A_2 equals the area of the jet, the rocket thrust is

$$F = \rho A_2 v_2^2 + (p_0 - p_a)A_2 \tag{6.19}$$

where v_2 is the velocity at which the jet issues from the rocket. The thrust F is independent of the speed of the rocket.

Jet Engine

By *jet engine* is meant a device which carries only its fuel and takes in the air for combustion from the atmosphere. It is analogous to the tank of Fig. 6.9, including the intake of fluid at section 1, except that the velocity of the air received is usually in the same straight line as the velocity of the exit jet at section 2. There are three forms of jet engines, but the equation is the same for all three. The *ram jet* must be brought up to a high speed by rockets or some other means, and then it scoops in the air from in front and compresses it by virtue of the stagnation pressure due to its speed. The *turbojet* can take off from the ground, for in it the air is compressed by a compressor driven by a gas turbine, the exhaust from which supplies the jet propulsion. Then there is a *pulsating* machine, which scoops in air in cycles. The inlet is then closed, the fuel-air mixture is exploded; a jet then gives the device a spurt; and the process is repeated.

The thrust of a jet engine is

$$F = \frac{(G_a + G_f)v_2 - G_a u}{g} \tag{6.20}$$

where G_a = weight of air taken in per second
G_f = weight of fuel consumed per second
v_2 = velocity of exhaust with respect to the engine
u = velocity of flight = velocity of air entry with respect to the engine

The thrust will vary with the speed of flight. Usually $p_0 = p_a$, and so the second term of Eq. (6.19) is not included in Eq. (6.20).

6.9 TORQUE IN ROTATING MACHINES

When a fluid flows through a rotor, its radius usually varies along its path. Hence it is desirable to compute torque rather than a force. The resultant torque is the summation of the torques produced by all the elementary forces, but it has been shown that the latter may be considered as equivalent to two single forces, one concentrated at the entrance to and the other at the exit from any device. For steady flow these equivalent forces have been shown to be $\rho Q V_1$ and $\rho Q V_2$. Referring to Figs. 6.10 and 6.11 and taking moments, the torque produced is

$$T = \rho Q(r_1 V_1 \cos \alpha_1 - r_2 V_2 \cos \alpha_2) \tag{6.21}$$

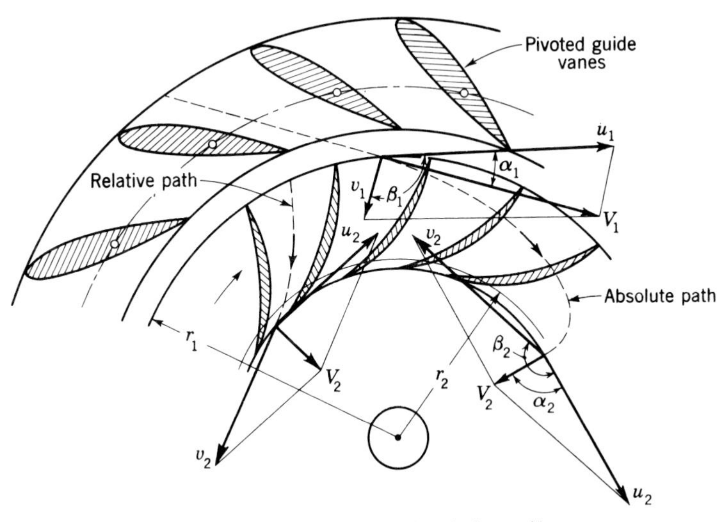

Figure 6.10 Radial-flow hydraulic turbine. (Flow is inward.)

This equation is a statement that torque equals the time rate of change of moment of momentum. The angles α and β are defined as before (Sec. 6.5) as the angles made by the absolute and relative velocities of the fluid, respectively, with the positive direction of the linear velocity u of a point on the moving body (tip or root of the blade).

If T as given by this equation is positive, it is the value of the torque exerted by the fluid on the runner of a turbine. The torque output from the shaft of the turbine is less than this by virtue of the mechanical friction. If the value of T is negative, it represents the torque exerted on the fluid by the impeller of a pump

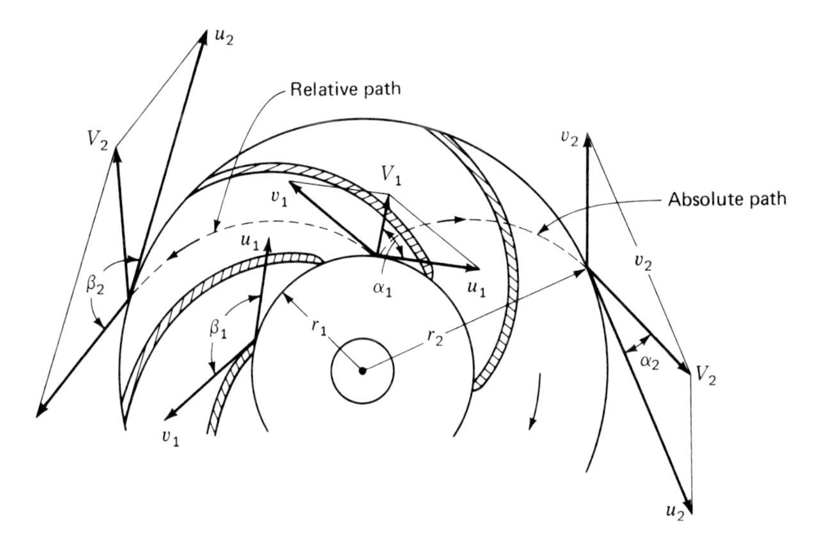

Figure 6.11 Centrifugal-pump impeller with radial flow. (Flow is outward.)

or compressor or fan. The torque input to the shalft of such a machine is greater than this because of mechanical friction.

Equation (6.21) and subsequent equations that may be derived from it are correct, but it is difficult to determine the numerical values to be used in them. Thus fluid particles in different streamlines may flow with different velocities, and it is necessary to estimate what the average velocity may be. Also, it is known that the average direction of a stream is often different from that of the vane which it is supposed to follow, but as yet there is no exact knowledge as to the amount of deviation in every case. Thus even the average velocity is not given precisely by dividing the flow by the cross-sectional area of a rotor passage. Furthermore, the entrance or exit edges of vanes are not always parallel to the axis of rotation, and thus the radii will be different for different streamlines.

Despite these defects, the theory is useful. It indicates the shape or nature of the performance curves of a given machine; it shows the influence of each separate factor; and it suggests the direction in which changes in design should be made in order to alter the characteristics which have been found by test of an existing machine.

An approximate analysis of the behavior of a radial-flow turbine can be made by assuming that all elements of the vanes are parallel to the axis of rotation and that water enters and leaves the vanes smoothly. An example of such an analysis is presented in Illustrative Example 6.5. An important aspect of such an analysis involves consideration of the principle of continuity of flow in the radial direction. Namely, $Q = (A_c V_r)_1 = (A_c V_r)_2 = (m2\pi rz V_r)_1 = (m2\pi rz V_r)_2$, where A_{c_1} and A_{c_2} represent circumferential areas; V_{r_1} and V_{r_2} represent the radial components of the velocity of the water at r_1 and r_2; m_1 and m_2 represent that portion of the circumferential area that is available for flow; and z is the width of the flow passage between the sides of the turbine. Usually m_1 and m_2 are assumed to have a value of 1.0. In actuality m is less than 1.0, perhaps about 0.8, because the vanes occupy a portion of the available flow area. Most commonly the passage widths z_1 and z_2 are equal.

It is immaterial in the use of Eq. (6.21) and subsequent equations whether the fluid flows radially inward, as in Fig. 6.10, or radially outward, as in Figs. 6.11 and 6.12, or remains at a constant distance from the axis, as in Figs. 6.13*a* and 6.13*b*. In any case, r_1 is the radius at entrance and r_2 is that at exit. In Figs. 6.10 through 6.13 the absolute and the relative flow paths are shown.

Figures 6.10 through 6.12 show two-dimensional flow in planes normal to the axis of rotation. This is known as *radial flow*. The streamlines and velocity triangles lie in the plane of the paper and are readily represented. In *axial flow*, as in Figs. 6.13*a* and 6.13*b*, a particle of fluid remains at a constant distance from the axis, and the streamlines are helices on coaxial cylinders. A streamline and its velocity triangles are shown on a developed cylinder for the corresponding radius.

Mixed flow is intermediate between these two extremes, and velocities have radial, axial, and tangential components. A streamline is a conical helix with a varying radius from the axis of rotation. Needless to say, this is a complicated three-dimensional-flow situation.

Figure 6.12 Radial-flow-pump impeller rotating at 200 rpm. (*a*) Instantaneous photo showing relative flow. (*b*) Time exposure showing absolute flow. *(Photographs from Hydrodynamics Laboratory, California Institute of Technology.)*

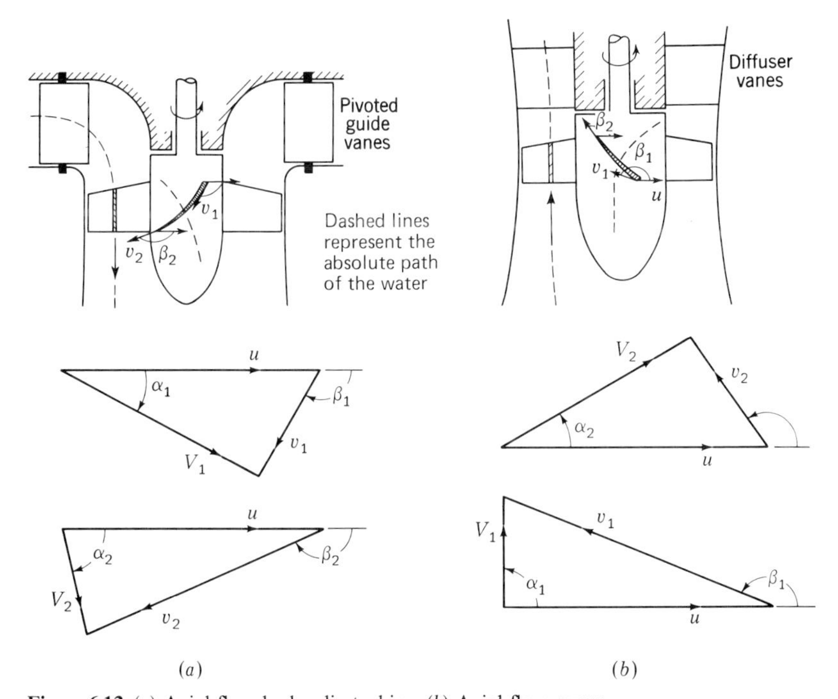

Figure 6.13 (*a*) Axial-flow hydraulic turbine. (*b*) Axial-flow pump.

6.10 HEAD EQUIVALENT OF MECHANICAL WORK

If Eq. (6.21) is multiplied by angular velocity ω, the product represents the rate at which mechanical energy is delivered *by* the fluid to a turbine or at which mechanical energy is delivered *to* the fluid by a pump. From Eqs. (4.15) and (4.18), $\gamma QH = T\omega$. Replacing H by a specific value h'' and noting that, when Eq. (6.21) is multiplied by ω, $r_1\omega = u_1$ and $r_2\omega = u_2$, we have $\gamma Qh'' = T\omega = \rho Q(u_1 V_1 \cos\alpha_1 - u_2 V_2 \cos\alpha_2)$, or

$$h'' = \frac{u_1 V_1 \cos\alpha_1 - u_2 V_2 \cos\alpha_2}{g} \tag{6.22}$$

which is the *head utilized* by a turbine or, when h'' is negative, the *head imparted* to the fluid by the impeller of a pump.

If the value of h'', as determined by Eq. (6.22), is positive, it is the mechanical work done by the fluid on the vanes of a turbine runner per unit weight of fluid. If the value is negative, it is the mechanical work done on the fluid by the impeller of a pump or similar device per unit weight of fluid. Obviously, the work done by or on the fluid is equal to the loss or gain of energy, respectively, of the fluid.

Illustrative Example 6.5 A radial-flow turbine has the following dimensions: $r_1 = 1.6$ ft, $r_2 = 1.0$ ft, and $\beta_1 = 80°$. The width of the flow passage between the two sides of the turbine is 0.8 ft. At 300 rpm the flow rate through the turbine is 120 cfs. Find: (*a*) the blade angle β_2 such that the water exits from the turbine in the radial direction, (*b*) the torque exerted by the water on the runner and the horsepower thus developed, (*c*) the head utilized by the runner and the power resulting therefrom. Assume that water enters and leaves the blades smoothly. Assume the blades are so thin that they do not occupy any of the available flow area.

(*a*) At the outer periphery ($r_1 = 1.6$ ft):

$$u_1 = \omega r_1 = (2\pi/60)300 \times 1.6 = 50.3 \text{ fps}$$

From continuity $Q = 120 = 2\pi r_1(z)V_{r_1} = 2\pi \times 1.6 \times 0.8 \times V_{r_1}$

$$V_{r_1} = 120/8.04 = 14.92 \text{ fps} = v_1 \sin\beta_1 = v_1 \sin 80°$$

$$v_1 = 14.92/\sin 80° = 15.15 \text{ fps}$$

$$v_1 \cos 80° = 15.15 \times 0.1736 = 2.63 \text{ fps}$$

$$V_1 \cos\alpha_1 = u_1 + v_2 \cos\beta_1 = 50.3 + 2.6 = 52.9 \text{ fps}$$

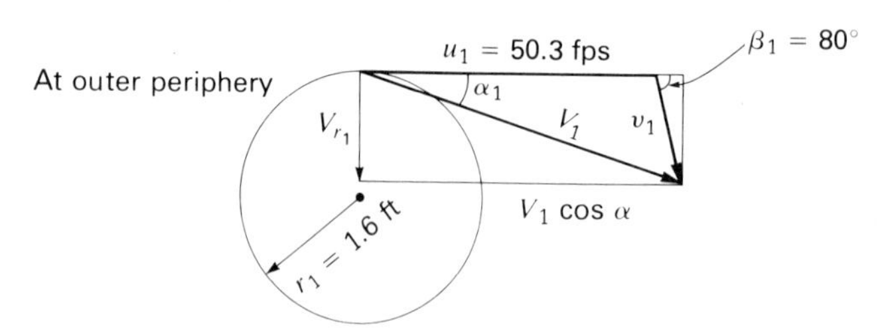

Illustrative Example 6.5

At the inner periphery ($r_2 = 1.0$ ft):

$$u_2 = \omega r_2 = (2\pi/60)300 \times 1.0 = 31.4 \text{ fps}$$

$$V_{r_2} = V_{r_1}(r_1/r_2) = 14.92(1.6/1.0) = 23.9 \text{ fps}$$

Because the water exits from the turbine in the radial direction, $\alpha_2 = 90°$ and $V_2 = V_{r_2}$.

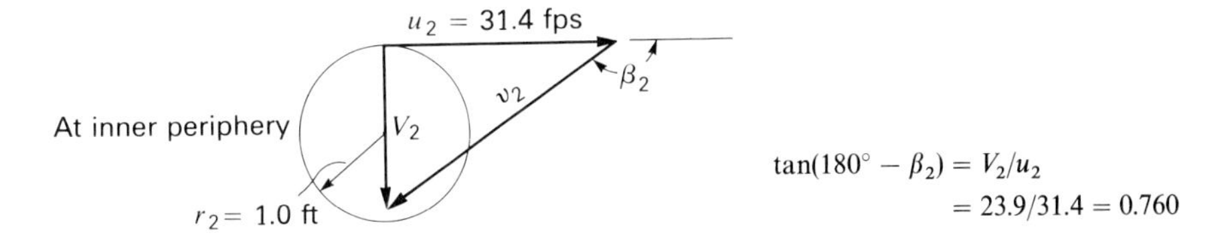

$$\tan(180° - \beta_2) = V_2/u_2 = 23.9/31.4 = 0.760$$

Thus $\beta_2 = 180° - 37.2° = 142.8°$, the required blade angle

(*b*) Equation 6.21: $T = \rho Q(r_1 V_1 \cos\alpha_1 - r_2 V_2 \cos\alpha_2)$
$= 1.94(120)(1.6 \times 52.9 - 0) = 19{,}700$ ft·lb

And power $= T\omega = 19{,}700 \times 31.4 = 619{,}000$ ft·lb/s
$= 1125$ hp

(*c*) Equation 6.22: $h'' = \dfrac{u_1 V_1 \cos\alpha_1 - u_2 V_2 \cos\alpha_2}{g}$

$$= \frac{50.3 \times 52.9 - 0}{32.2} = 82.6 \text{ ft}$$

Power $= \gamma Q h'' = 62.4 \times 120 \times 82.6 = 619{,}000$ ft·lb/s
$= 1125$ hp

Further calculations indicate that the absolute velocity of the water changed from $V_1 = 55.0$ fps at entry to $V_2 = 23.9$ fps at exit while the velocity of the water relative to the blades changed from $v_1 = 15.15$ fps at entry to $v_2 = 39.5$ fps at exit.

6.11 FLOW THROUGH ROTATING CHANNEL

The equation to be derived is sometimes called the equation of *relative velocities*, because the absolute velocities of the energy equation are replaced by relative velocities. The usual energy equation may be written between entrance to and exit from a passage which is itself rotating about some axis, but in addition to the friction loss h_L there is an additional loss h'', due to the fact that the fluid is delivering mechanical work and losing energy thereby. (If the passage is that of a pump, the numerical value of h'' will be negative.) Thus

$$\left(\frac{p_1}{\gamma} + z_1 + \frac{V_1^2}{2g}\right) - \left(\frac{p_2}{\gamma} + z_2 + \frac{V_2^2}{2g}\right) = h_L + \frac{u_1 V_1 \cos\alpha_1 - u_2 V_2 \cos\alpha_2}{g}$$

By trigonometry, $V^2 = v^2 + u^2 + 2vu \cos \beta$, and

$$uV \cos \alpha = u(u + v \cos \beta)$$

Inserting these values, the equation is reduced to

$$\left(\frac{p_1}{\gamma} + z_1 + \frac{v_1^2 - u_1^2}{2g}\right) - \left(\frac{p_2}{\gamma} + z_2 + \frac{v_2^2 - u_2^2}{2g}\right) = h_L \tag{6.23}$$

If there is no flow, both v_1 and v_2 become zero and the equation reduces to that of a forced vortex [Eq. (4.40)] since $u = \omega r$. If there is no rotation, both u_1 and u_2 become zero, the relative velocities become absolute velocities, and the equation becomes the usual energy equation. The frame of reference having been changed, the mechanical work done does not appear as a separate term in Eq. (6.23).

6.12 REACTION WITH ROTATION

The force of reaction of a jet from a stationary body is given in Sec. 6.7 and from a body in translation in Sec. 6.8. Since Sec. 6.11 develops the equation for the flow through a channel in rotation, we are now ready to consider the force of reaction of a fluid discharged from a rotating body.

A familiar object to illustrate this subject is the rotating lawn sprinkler. In Fig. 6.14 assume that the cross-sectional area of the arms is so large relative to the area of the jets that fluid-friction loss in the arms may be neglected. Water enters at the center, where $r_1 = 0$, so that in Eq. (6.23), $u_1 = 0$. With the sprinkler arms lying in a horizontal plane, $z_1 - z_2 = 0$, and for the jets discharging into the air, p_2 is atmospheric pressure and will be regarded as zero. Since friction is neglected, $h_L = 0$, and if we let $h = p_1/\gamma + v_1^2/2g$, Eq. (6.23) applied to Fig. 6.14 becomes

$$v_2 = \sqrt{2gh + u_2^2} \tag{6.24}$$

where h is the sum of the pressure head and velocity head at entry to the sprinkler.

If a_2 denotes the sum of the areas of all the jets (two in the figure shown), then $Q = a_2 v_2$. This shows that the discharge is a function of the rotative speed,

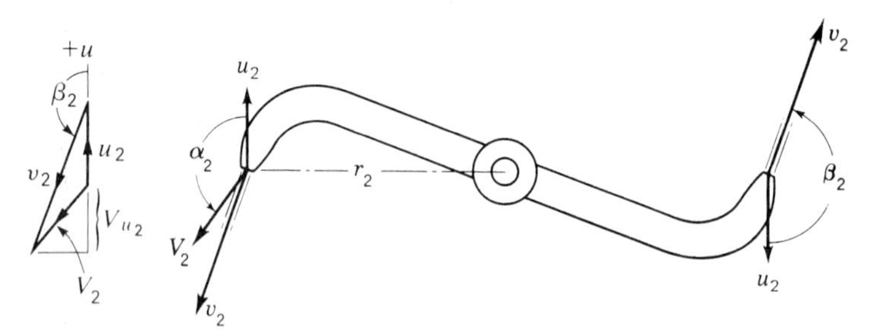

Figure 6.14

since $u_2 = r_2 \omega$. The tangential component of the absolute velocity of discharge is $V_{u_2} = u_2 + v_2 \cos \beta_2$, and hence the tangential component of the force of reaction is

$$F_u = \frac{\gamma Q}{g} \Delta V_u = \frac{\gamma a_2 v_2}{g} (0 - V_{u_2}) = -\frac{\gamma a_2 v_2}{g} (u_2 + v_2 \cos \beta_2)$$

As the radius is a factor in any rotating body, it is usually better to compute torque rather than a force. In this case the torque is

$$T = F_u r_2 = -\frac{\gamma a_2 v_2}{g} r_2 (u_2 + v_2 \cos \beta_2) \tag{6.25}$$

The ideal maximum, or runaway, speed is when $T = 0$, and this will be the case when $u_2 = -v_2 \cos \beta_2$ and when $V_2 \cos \alpha_2 = 0$ or $\alpha_2 = 90°$. Because of mechanical friction this condition will never be realized. Of the total power supplied to the sprinkler, the greater part is lost in the kinetic energy of the jets. The total power *developed* by the sprinkler is used in overcoming friction in the bearings and air resistance. If there were more arms, with larger orifices, so as to discharge more water, there could be a surplus of power which would be useful power delivered. A primitive turbine constructed in this manner was known as Barker's mill.

6.13 MOMENTUM PRINCIPLE APPLIED TO PROPELLERS AND WINDMILLS

In the case of a fan in a duct, the cross section of the fluid affected by the fan is the same upstream as it is downstream, and the principal effect of a fan is to increase the pressure in the duct. In the case of a propeller revolving in free air, however, this is not so. The pressure must necessarily be the same at a distance either upstream or downstream from the propeller. How, then, may the revolving blades be considered to do work on the air? This situation may be analyzed by consideration of the *slipstream*, or *propeller race*, which is nothing more than the body of air affected by the propeller (Fig. 6.15). The flow is undisturbed at section 1 upstream from the propeller and is accelerated as it approaches the propeller. Additional increase in velocity occurs downstream of the propeller until it reaches a value of V_4 at section 4. It is customary to replace the propeller in simple slipstream theory with a stationary *actuating disk* across which the pressure is made to rise, as shown in the pressure profile below the slipstream of Fig. 6.15 and also in Fig. 6.16. We thereby neglect the rotational effect of the propeller, together with the helical path of vortices shed from the blade tips (Sec. 10.8). The thrust force F_T will be given by the pressure change at the disk times the area of the disk,

$$F_T = \frac{\pi D^2}{4} (p_3 - p_2) = A(\Delta p) \tag{6.26}$$

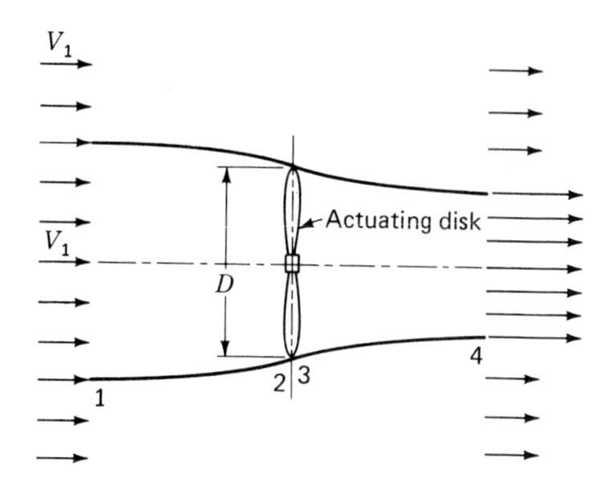

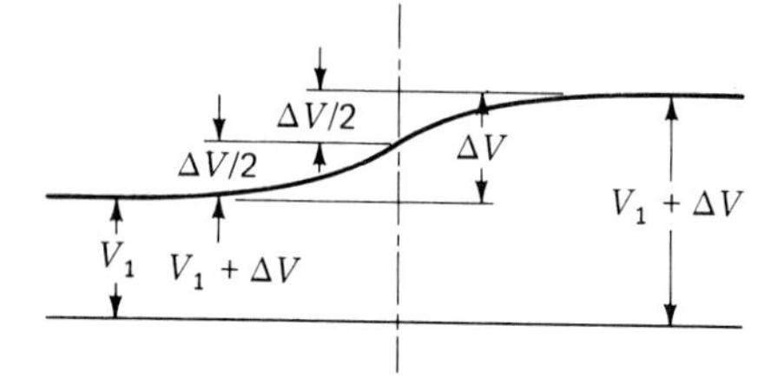

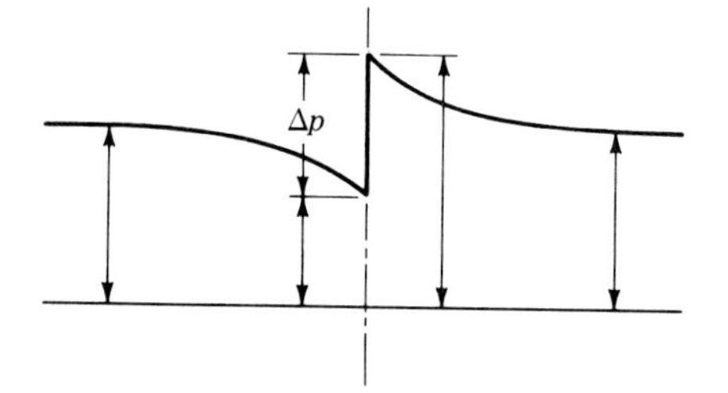

Figure 6.15 Slipstream of propeller in free fluid. V_1 represents the velocity of the undisturbed fluid relative to the propeller; p_0 represents the undisturbed pressure in the fluid.

where D and A represent the diameter and area of the actuating disk and p_2 and p_3 represent the pressures just upstream and downstream of the propeller, as indicated in Figs. 6.15 and 6.16. It should be noted that the pressures exerted on the boundary of the slipstream between sections 1 and 2 balance one another out and need not be considered.

By Newton's second law, the force F_T must equal the rate of change of momentum of the fluid upon which it acts. If we let Q be the rate of flow through the slipstream,

$$F_T = AV\rho(V_4 - V_1) = Q\rho(\Delta V) \tag{6.27}$$

where V represents the mean velocity through the actuating disk and V_1 and V_4 are the velocities in the slipstream at sections 1 and 4 of Fig. 6.15 where the pressures correspond to the normal undisturbed pressure p_0 in the flow field.

The propeller we are considering could be a stationary one like a fan or a moving one such as the propeller of a moving aircraft or ship. If it were a fan, V_1 would generally be equal to zero and the slipstream upstream of the fan would have a much larger diameter than that shown in Fig. 6.15. If we were dealing with a propeller of a moving aircraft or ship the craft would be moving to the left with a velocity V_1 through a stationary fluid, in which case Fig. 6.15 shows velocities relative to the craft.

Writing the Bernoulli equation from a point upstream where the velocity is V_1 to a point downstream where it is $V_1 + \Delta V$, recognizing that the pressure terms at these points cancel and (assuming an ideal fluid) that the disk adds $\Delta p/\gamma$ units of energy to the fluid per unit weight of fluid, we get

$$\frac{V_1^2}{2g} + \frac{\Delta p}{\gamma} = \frac{(V_1 + \Delta V)^2}{2g} \tag{6.28}$$

Equating Eqs. (6.26) and (6.27) and solving for Q in terms of Δp, and then substituting into this expression for Q, the expression for Δp that results from solving Eq. (6.28) gives

$$Q = A\left(V_1 + \frac{\Delta V}{2}\right) \tag{6.29}$$

This may be expressed as

$$Q = A\left(V_1 + \frac{V_4 - V_1}{2}\right) = A\left(\frac{V_1 + V_4}{2}\right) = AV$$

This shows that the velocity V at the disk is the average of the upstream and downstream velocities. It also shows that one-half of ΔV occurs upstream of the propeller while the other half of ΔV occurs downstream.

Solving Eq. (6.28) for ΔV and substituting F_F/A for Δp gives

$$\Delta V = -V_1 + \sqrt{V_1^2 + \frac{2F_T}{A\rho}} \tag{6.30}$$

We may use the slipstream analysis to determine the maximum possible efficiency of a propeller. The power output P_{out} is given by

$$P_{\text{out}} = F_T V_1 = (\rho Q\, \Delta V)V_1$$

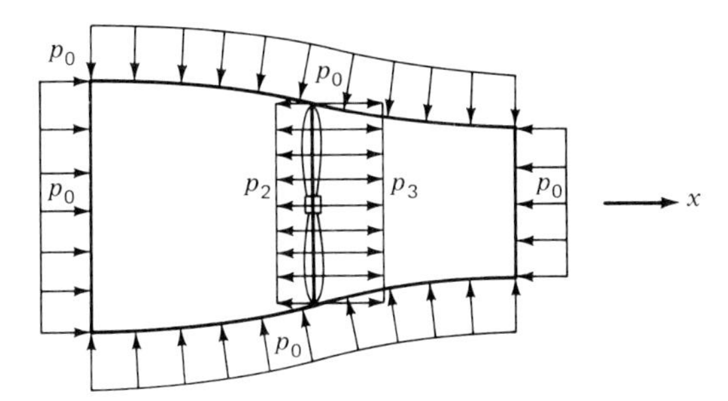

Figure 6.16 Forces acting on the fluid within the slip-stream of Fig. 6.15. Net force on the fluid is:

$$\sum F_x = p_3 A - p_2 A = (\Delta p)A$$

The power input P_{in} is that required to increase the velocity of the fluid in the slipstream from V_1 to V_4. Applying Eq. (4.15) we get

$$P_{in} = \gamma Q(V_4^2/2g - V_1^2/2g) = \frac{\rho Q}{2}(V_4^2 - V_1^2)$$

$$= \rho Q\left(\frac{V_4 + V_1}{2}\right)(V_4 - V_1) = (\rho Q V)(\Delta V)$$

The efficiency η is given by the ratio of the power output to the power input. Thus

$$\eta = \frac{P_{out}}{P_{in}} = \frac{(\rho Q\,\Delta V)V_1}{(\rho Q V)\,\Delta V} = \frac{V_1}{V} = \frac{V_1}{V_1 + \Delta V/2} = \frac{1}{1 + \frac{1}{2}(\Delta V/V)} \tag{6.31}$$

The efficiency is seen to be a function of the ratio $\Delta V/V_1$. The efficiency approaches 100 percent as ΔV approaches zero, but if $\Delta V = 0$, the propeller produces no force. The actual maximum efficiency of aircraft propellers is about 85 percent. However the efficiency of an airplane propeller drops rapidly at speeds in excess of 400 mph (600 km/hr) because of compressibility effects. For ships, propeller efficiencies of only 60 or 70 percent are attainable.

A windmill is essentially the opposite of a propeller in that the function of a windmill is to extract energy from the wind. The slipstream for a windmill expands as it passes the *actuated disk*, and the pressure drops as does the velocity. By a procedure similar to the one for a propeller, it can be shown that the maximum theoretical efficiency of a windmill is 59.3 percent. Because of friction and other losses the actual efficiency of windmills rarely exceeds 40 percent.

Illustrative Example 6.6 Find the thrust and efficiency of two 6.5-ft-diameter propellers through which flows a total of 20,000 cfs of air (0.072 lb/ft³). The propellers are attached to an airplane moving at 150 mph through still air. Find also the pressure rise across the propellers and the horsepower input to each propeller. Neglect eddy losses.

Velocity of air relative to airplane is

$$V_1 = 150 \text{ mph} = \frac{150 \times 44}{30} = 220 \text{ fps}$$

Velocity of air through the actuating disk is

$$V = V_1 + \frac{\Delta V}{2} = \frac{Q}{A} = \frac{20{,}000/2}{(\pi/4)(6.5)^2} = 301 \text{ fps}$$

Thus

$$\Delta V = 2(301 - 220) = 162 \text{ fps}$$

$$F_T = \rho Q\,\Delta V = \frac{0.072}{32.2}(20{,}000)(162)$$

$$= 7{,}240 \text{ lb} \quad \text{(total thrust of both propellers)}$$

$$\eta = \frac{1}{1 + \Delta V/2V_1} = \frac{1}{1 + 162/440} = 0.73 = 73 \text{ percent}$$

F_T on one propeller = 7240/2 = 3620 lb. But $F_T = (\Delta p)(A)$, thus $3620 = \Delta p(\pi/4)(6.5)^2$

$$\Delta p = 109 \text{ psf} = 0.756 \text{ psi}$$

$$\text{hp/propeller} = \frac{\gamma Q(\Delta p/\gamma)}{550} = \frac{Q(\Delta p)}{550} = \frac{10{,}000(109)}{550} = 1980$$

Check:

$$\text{hp/propeller} = \frac{F_T(V_1 + \Delta V/2)}{550} = \frac{3620(301)}{550} = 1980$$

6.14 OTHER APPLICATIONS OF THE MOMENTUM PRINCIPLE

In addition to the cases that have already been discussed, there are numerous other fluid-flow situations where the momentum principle is useful. It is used to develop an expression for the head loss in an expansion (Sec. 8.19) and for the conjugate depths of a hydraulic jump (Sec. 11.19). The momentum principle is also employed in the development of the relationships in a shock wave (Sec. 9.9). Another application is that of finding the forces exerted on open-flow structures. The magnitudes of such forces may generally be found by application of the momentum principle. The application to this type of problem can best be discussed with an illustrative example

Illustrative Example 6.7 This water passage is 10 ft (3 m) wide normal to the accompanying figure. Determine the horizontal force acting on the shaded structure. Assume ideal flow.

In free-surface flow such as this where the streamlines are parallel, the water surface is coincident with the hydraulic grade line. Writing an energy equation from the upstream section to the downstream section,

$$6 + \frac{V_1^2}{2g} = 3 + \frac{V_2^2}{2g} \qquad (a)$$

From continuity,

$$6(10)V_1 = 3(10)V_2 \qquad (b)$$

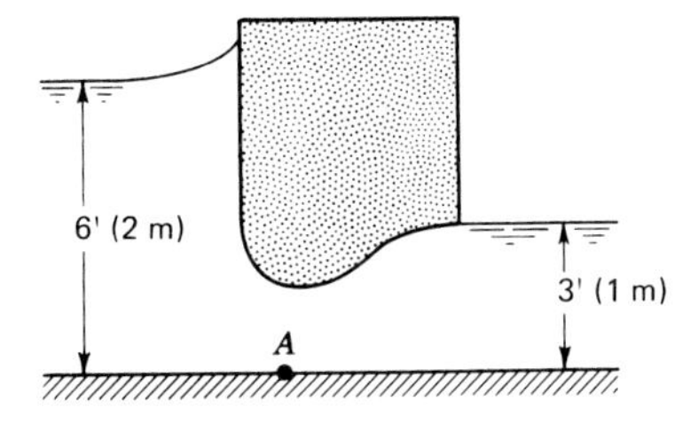

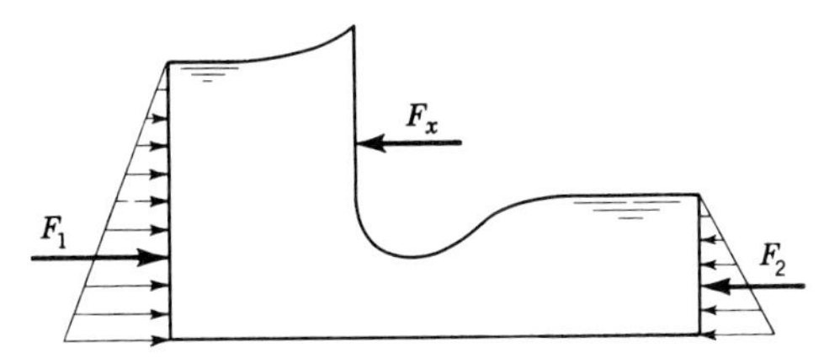

Illustrative Example 6.7

Substituting Eq. (*b*) into Eq. (*a*) yields

$$V_1 = 8.02 \text{ fps} \qquad V_2 = 16.04 \text{ fps}$$

$$Q = A_1V_1 = A_2V_2 = 481 \text{ cfs}$$

Next take a free-body diagram of the element of water shown in the figure and apply Eq. (6.7*a*),

$$F_1 - F_2 - F_x = \rho Q(V_2 - V_1)$$

where F_x represents the force of the structure on the water in the horizontal direction.

From Eq. (2.16) we have $F_1 = \gamma h_{c_1} A_1$ and $F_2 = \gamma h_{c_2} A_2$

Hence $$62.4(3)(10 \times 6) - 62.4(1.5)(10 \times 3) - F_x = 1.94(481)(16.04 - 8.02)$$

and $$F_x = 940 \text{ lb}$$

The positive sign means that the assumed direction is correct. Hence the force of the water on the structure is equal and opposite, namely, 940 lb to the right.

The momentum principle will not permit one to obtain the vertical component of the force of the water on the shaded structure because the pressure distribution along the bottom of the channel is unknown. The pressure distribution along the boundary of the structure and along the bottom of the channel can be estimated by sketching a flow net and applying Bernoulli's principle. The horizontal and vertical components of the force can be found by computing the integrated effect of the pressure-distribution diagram.

Employing the given SI units the solution to the problem is as follows:

$$2 + \frac{V_1^2}{2(9.81)} = 1 + \frac{V_2^2}{2(9.81)} \tag{a}$$

$$2(3)V_1 = 1(3)V_2 \tag{b}$$

Substituting Eq. (*b*) into Eq. (*a*) yields:

$$V_1 = 2.56 \text{ m/s} \qquad V_2 = 5.12 \text{ m/s}$$

$$Q = A_1 V_1 = A_2 V_2 = 15.4 \text{ m}^2\text{/s}$$

From the free-body diagrams,

$$F_1 - F_2 - F_x = \rho Q(V_2 - V_1)$$

$$9.81(1)(2)(3) - 9.81(0.5)(1)(3) - F_x = 1.0(15.4)(5.12 - 2.56)$$

$$F_x = 4.7 \text{ kN}$$

PROBLEMS

6.1 For laminar flow as in Prob. 4.1, prove that $\beta = \frac{4}{3}$.

6.2 For the turbulent-flow case as approximated in Prob. 4.2, prove that $\beta = 1.014$.

6.3 For laminar flow between two stationary parallel plates such as to give two-dimensional flow, find (*a*) the ratio of mean velocity to maximum velocity; (*b*) α; (*c*) β. Once again, for this case the velocity profile is parabolic as in Prob. 6.1.

6.4 On the right end of a 6-in (15-cm)-diameter pipe is a nozzle which discharges a 2-in (5-cm)-diameter jet. The pressure in the pipe is 57 psi (380 kN/m^2), and the pipe velocity is 10 fps (3 m/s). The jet discharges into the air. If the fluid is water, what is the axial force exerted upon the nozzle? Find also the head loss in the nozzle.

6.5 On the right end of a 6-in-dia horizontal pipe is a diverging nozzle which discharges a 6.5-in-dia jet. The velocity in the pipe is 10 fps and the jet discharges into the air. If the fluid is water, what is the axial force on the nozzle? In which direction does this force act? Neglect fluid friction.

6.6 Water under a pressure of 65 psia (450 kN/m^2, abs) flows with a velocity of 10 fps (3 m/s) through a right-angle bend having a uniform diameter of 12 in (30 cm). The bend lies in a horizontal plane; water enters from the west and leaves toward the north. Assuming no drop in pressure, what is the resultant force acting upon the bend? What is its direction?

6.7 Water enters a reducing right-angle bend (same orientation as in Prob. 6.6) with a velocity of 8 fps (2.5 m/s) and a pressure of 5.0 psi (35 kN/m^2). The diameter of the bend at entrance is 24 in (60 cm), and at exit it is 18 in (45 cm). Neglecting any friction loss, find the magnitude and the direction of the resultant force on the bend.

6.8 In Fig. 6.3 the diameters are 36 in (90 cm) and 24 in (60 cm). At the larger end the pressure is 100 psi (700 kN/m^2) and the velocity is 8 fps (2.5 m/s). Find the resultant force on the conical reducer, neglecting any friction, if water flows (*a*) to the right; (*b*) to the left. (*c*) If friction were not neglected, what effect would this have on the two previous answers?

6.9 Determine the magnitude and direction of the force on the double nozzle of the figure. Both nozzle jets have a velocity of 40 fps. The axes of the pipe and both nozzles all lie in a horizontal plane. $\gamma = 62.4$ lb/ft^3. Neglect friction.

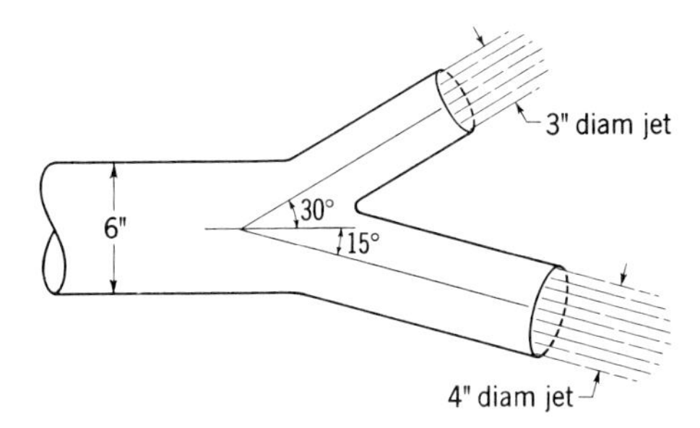

Problem 6.9

6.10 In Prob. 6.9, what angle should the 4-in jet make with the axis of the 6-in-dia pipe so that the resultant force is along the axis of the 6-in-dia pipe?

6.11 Find the pull on the bolts in the figure. Assume ideal flow.

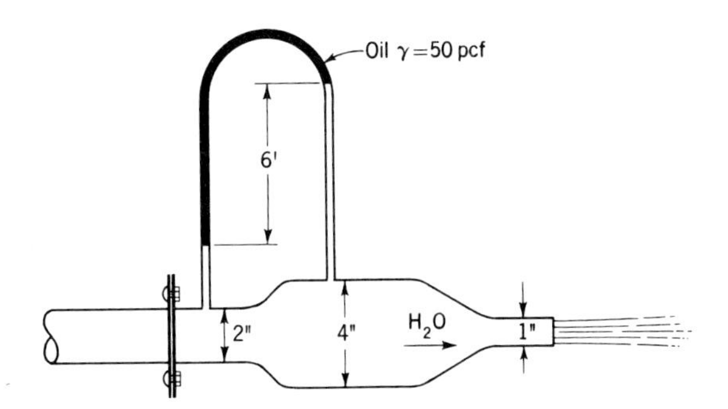

Problem 6.11

6.12 Repeat Prob. 6.11 for the case where the diameters are 5, 10, and 2.5 cm respectively. Assume a 180-cm manometer reading with a manometer liquid having a specific gravity of 0.80.

6.13 If a jet of any fluid of an area A and with a velocity V is deflected through an angle θ without any change in the magnitude of the velocity, prove that $F = (2\gamma A/g)V^2 \sin(\theta/2)$.

6.14 If a jet of any fluid is deflected through an angle θ and fluid friction reduces V_2 to $0.8V_1$, derive an equation for the dynamic force exerted.

6.15 In Illustrative Example 6.2 assume that $\theta = 120°$, that the jet is water with a velocity of 100 fps (30 m/s), and that the jet diameter is 2 in (5 cm). If friction loss is neglected, find (*a*) the component of force in the same direction as the jet; (*b*) the component of the force normal to the jet; (*c*) the magnitude and direction of the resultant force exerted on the body.

6.16 Solve Prob. 6.15 assuming that friction is such as to reduce V_2 to 80 fps (25 m/s).

6.17 Suppose the jet in Prob. 6.15 were to strike a large flat plate normally. Approximately, what would be the force on the plate?

6.18 In Prob. 6.17 what would be the stagnation pressure, and what would be the average pressure on a circular plate if the area of the plate were 20 times the area of the jet? Assume that the center of the jet is coincident with the center of the plate.

6.19 Plotted to scale are streamlines in the plane of the center of a free jet impinging vertically on a horizontal circular plate. Determine as accurately as you can by scaling off the pertinent dimensions the velocity of the water as it leaves the plate and the total force exerted by the water on the plate. Include the weight of the water. The jet diameter is 30 cm and stagnation pressure at point S is 6.0 kN/m^2.

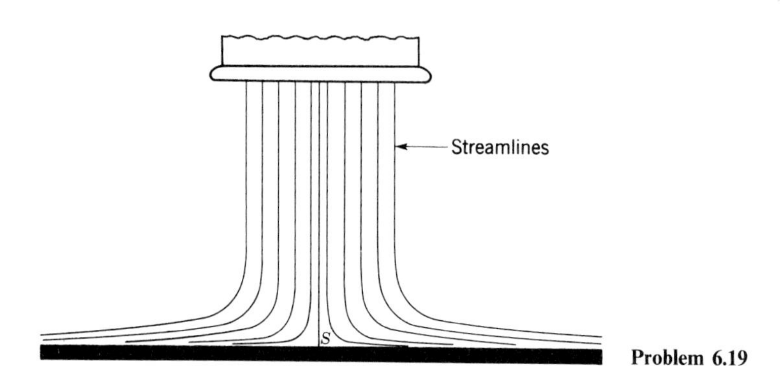

Problem 6.19

6.20 Repeat Prob. 6.19 for the case where the jet diameter is 4 in and the jet velocity is 20 fps.

6.21 A horizontal jet of water issues from an orifice in the side of a tank under a head h_1 and strikes a large plate a short distance away which covers the end of a horizontal tube in the side of a second tank. The second tank contains oil of specific weight 52 lb/ft^3 (8,000 N/m^3) at rest. The height of the oil above the tube is h_2. The jet diameter is three-fourths of the inside diameter of the tube. The jet and tube are at the same elevation. (*a*) If the impact of the water is just sufficient to hold the plate in place, find the relation between h_1 and h_2. Neglect the weight of the plate and assume ideal flow. (*b*) Consider the effect of the weight of the plate. Find h_2 if $h_1 = 10$ ft (3 m), weight of plate = 50 lb (200 N), jet diameter = 1.5 in (3.6 cm), and coefficient of friction between plate and tube = 0.6.

6.22 A 6-in-diameter water jet having a velocity of 30 fps at section P is directed vertically upward against the cone as shown in the figure. Assume streamlines are parallel at a section Q. What must be the weight of the cone if it remains in the position shown with $a = 1.5$ ft, $b = 0.5$ ft, and $c = 3$ ft? Neglect friction.

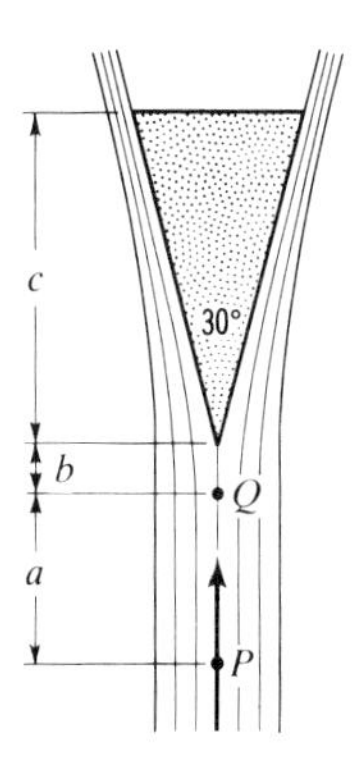

Problem 6.22

6.23 Repeat Prob. 6.22 for the case where the jet velocity is 20 m/s at P and $a = 1.5$ m, $b = 0.5$ m, and $c = 3$ m. Jet diameter = 0.5 m.

6.24 Repeat Prob. 6.22 for the case where the jet velocity is 15 fps and $a = 0$, $b = 1$ ft, $c = 3$ ft.

6.25 A jet of water 3 in in diameter has a velocity of 120 fps. It strikes a single vane, which has an angle $\beta_2 = 90°$ and which is moving in the same direction as the jet, with a velocity u. When u has values of 0, 40, 60, 80, 100, and 120 fps, find values of (*a*) G'; (*b*) $V_2 \cos \alpha_2$; (*c*) ΔV_u; (*d*) Δv_u; (*e*) F_u. Assume $v_2 = 0.9v_1$. Present answers in a neat tabular form.

6.26 If the jet in Prob. 6.25 strikes a single vane for which $\beta_2 = 180°$, all other data remaining the same, find values of (*a*) G'; (*b*) v_2; (*c*) V_2; (*d*) ΔV; (*e*) Δv; (*f*) F_u. Assume $v_2 = 0.9v_1$. Present answers in a neat tabular form.

6.27 In Prob. 6.26 assume all data the same except that friction loss in flow over the vane is such that $v_2 = 0.8v_1$. Find the results called for in Prob. 6.26. Present answers in a neat tabular form.

6.28 Assume that all data are the same as in Prob. 6.27 except that $\beta_2 = 150°$. Find (*a*) $v_2 \cos \beta_2$; (*b*) $V_2 \cos \alpha_2$; (*c*) ΔV_u; (*d*) Δv_u; (*e*) F_u. Present answers in a neat tabular form.

6.29 Suppose the single vane of Illustrative Example 6.3 is traveling to the left toward the nozzle at 20 fps. What then would be the force components on the vane?

6.30 A jet of water with an area of 3 in^2 (20 cm^2) and a velocity of 200 fps (60 m/s) strikes a single vane which reverses it through 180° without friction loss. Find the force exerted if the vane moves (*a*) in the same direction as the jet with a velocity of 80 fps (25 m/s); (*b*) in a direction opposite to that of the jet with a velocity of 80 fps (25 m/s).

6.31 A series of vanes is acted on by a 3-in water jet having a velocity of 100 fps, $\alpha_1 = \beta_1 = 0°$. Find the required blade angle β_2 in order that the force acting on the vane in the direction of the jet is 200 lb. Neglect friction. Solve using vane velocities of 0, 20, 50, and 80 fps. Also find the maximum possible vane velocity.

6.32 A 2-in (5-cm)-diameter air jet impinges on a series of blades. The absolute velocities are shown in the figure. Assume $\gamma = 0.076$ lb/ft^3 (12 N/m^3) and neglect friction. (*a*) What power is transmitted to the blades? Find also the velocity of the blades. (*b*) Determine the necessary blade angles at entrance and exit. Assume the air enters the blade smoothly.

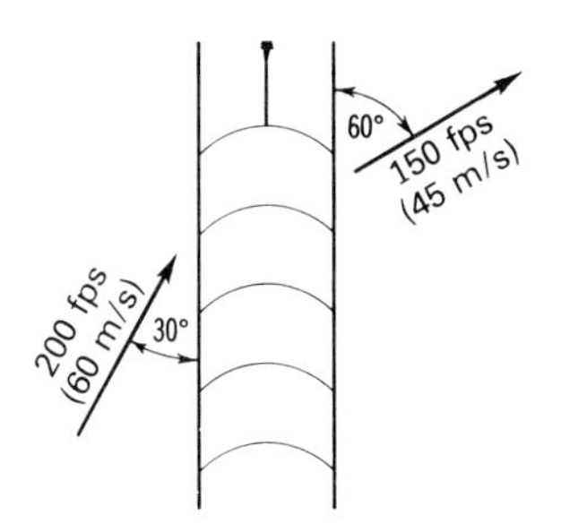

Problem 6.32

6.33 The water jet in the figure, moving at 50 fps (12 m/s), is divided by the splitter so that one-third of the water is diverted toward A. Calculate the *magnitude* and *direction* of the resultant force on this single stationary blade. Assume ideal flow in a horizontal plane.

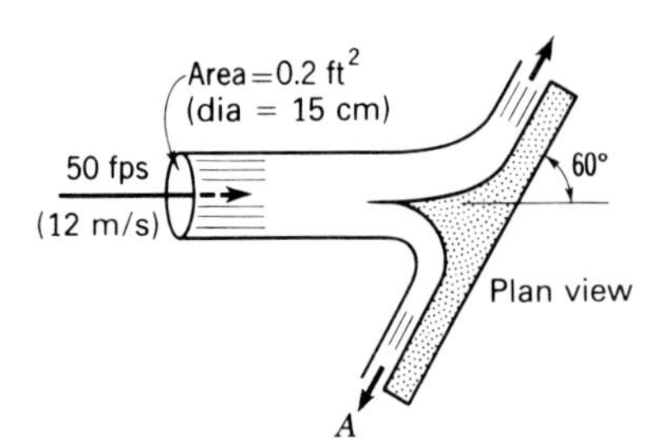

Problem 6.33

6.34 For the conditions of Prob. 6.33 compute the magnitude and direction of the resultant force on the single blade if it is moving to the right at a velocity of 10 fps (3 m/s).

6.35 Suppose the blade of Prob. 6.33 is one of a series of blades which are moving to the right at 10 fps (3 m/s). Determine the horizontal force on the blade system, and compute the power transferred to the blades. Compute the power of the jet and of the water leaving the blade system to verify an energy balance (Illustrative Example 6.3).

6.36 If a jet of fluid strikes a single body moving in the same direction with a velocity u, flows over it without friction loss, and leaves with a relative velocity in the direction of β_2, prove that $F_u = (\gamma A_1/g)(1 - \cos \beta_2)(V_1 - u)^2$.

6.37 A locomotive tender running at 20 mph (9 m/s) scoops up water from a trough between the rails, as shown in the figure. The scoop delivers water at a point 8 ft (2.5 m) above its original level and in the direction of motion. The area of the stream of water at entrance is 50 in^2 (300 cm^2). The water is everywhere under atmospheric pressure. Neglecting all losses, what is the absolute velocity of the water as it leaves the scoop? What is the force acting on the tender caused by this? What is the minimum speed at which water will be delivered to the point 8 ft (2.5 m) above the original level?

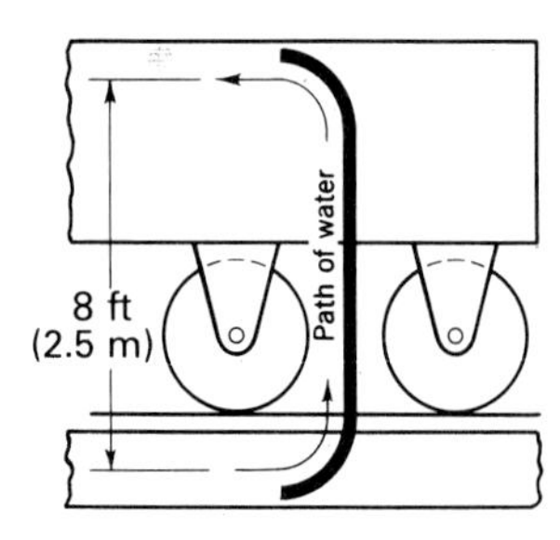

Problem 6.37

6.38 Find the thrust developed when water is pumped in through a 10-in (25-cm)-diameter pipe in the bow of a boat at $v = 5$ fps (1.5 m/s) and emitted through a 6-in (15-cm)-diameter pipe in the stern of the boat.

6.39 An ideal liquid ($\gamma = 62.4$ lb/ft^3) flows from a 2-ft-diameter tank as shown in the figure. The jet diameter is 3 in. If the static coefficient of friction between the tank and floor is 0.56, determine the minimum value of h at which the tank will commence to move to the left. The tank itself weighs 100 lb.

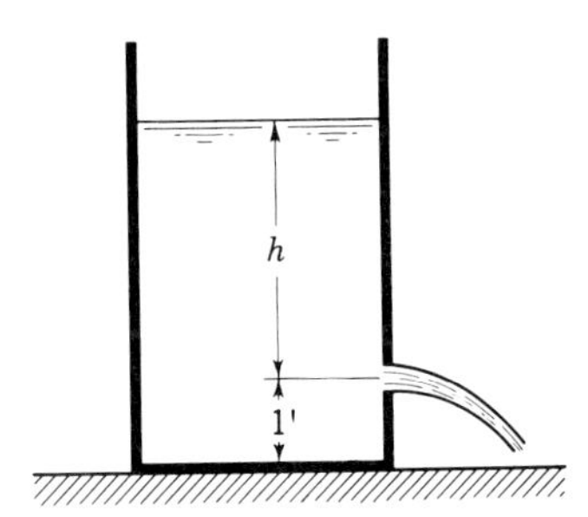

Problem 6.39

6.40 Find the magnitude and direction of the resultant force of the fluid on the compressor shown in the figure. Air ($\gamma = 0.075\ \text{lb/ft}^3$) enters at A through a 3-ft^2 area at a velocity of 10 fps. Air is discharged at B through a 2-ft^2 area at a velocity of 12 fps.

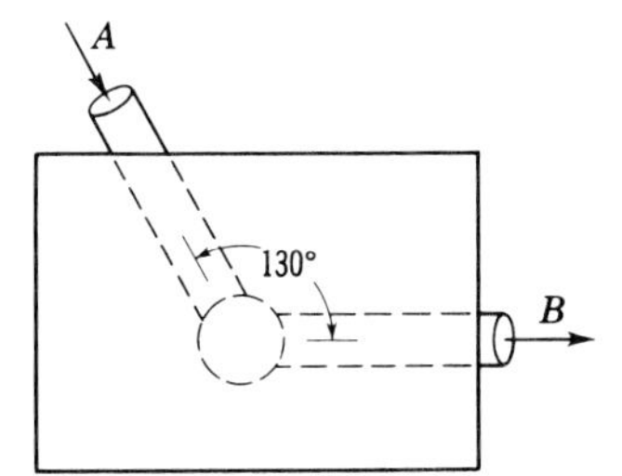

Problem 6.40

6.41 Repeat Prob. 6.40 for the case where a gas ($\gamma = 12.1\ \text{N/m}^3$) enters at A through a 60-cm-diameter pipe at 3 m/s and leaves at B through a 50-cm-diameter pipe at 4 m/s.

6.42 A typical rocket has a propellant flow rate of 24.1 lb/s through a nozzle with a throat area of 10.3 in^2. The exit area of the nozzle is 51.5 in^2, the exhaust velocity is 6,670 fps. The nozzle is designed to expand the gases down to 14.7 psia at exit. Find the rocket thrust (*a*) at sea level and (*b*) at an elevation of 20,000 ft where the barometer pressure is 6.75 psia; (*c*) find the specific weight of the exhaust gas.

6.43 A turbojet at a speed of 600 fps takes in air at the rate of 40 lb/s. The air/fuel ratio is 30:1, and the exhaust velocity is 2,000 fps. Find the thrust.

6.44 A turbojet moving at 200 m/s takes in air at a rate of 10 kg/s. The air/fuel ratio is 22:1 and the exhaust velocity is 550 m/s. Find the thrust.

6.45 The absolute velocity of a jet of steam impinging upon the blades of a steam turbine is 4,000 fps (1000 m/s), and that leaving is 2,800 fps (800 m/s). $\alpha_1 = 20°$, $\alpha_2 = 150°$, $u_1 = u_2 = 500$ fps (180 m/s), and $r_1 = r_2 = 0.5$ ft (15 cm). Find the torque exerted on the rotor and the power delivered to it if $G = 0.4$ lb/s (2 N/s).

6.46 When a turbine runner is held so that it cannot rotate, the discharge under a head of 50 ft is found to be 29.5 cfs. $\alpha_1 = 35°$, $\beta_2 = 155°$, $r_1 = 0.70$ ft, $r_2 = 0.42$ ft, $A_1 = 0.837\ \text{ft}^2$, $a_2 = 0.882\ \text{ft}^2$. What is the value of the torque at zero speed? Neglect shock loss.

6.47 Repeat Prob. 6.46 when the data are given in SI units as follows: $h = 15$ m, $Q = 0.85\ \text{m}^3/\text{s}$, $r_1 = 0.2$ m, $r_2 = 0.12$ m, $A_1 = 0.078\ \text{m}^2$, $a_2 = 0.082\ \text{m}^2$.

6.48 Refer to Illustrative Example 6.5. Calculate values for v_1 and v_2. Prove that continuity is satisfied by showing that $2\pi r_1 V_{r_1} = 2\pi r_2 V_{r_2}$ where V_r is the radial component of V.

6.49 A radial-flow turbine has the following dimensions: $r_1 = 1.6$ ft, $r_2 = 1.0$ ft, $\beta_1 = 76°$, and $\beta_2 = 130°$. The width of the flow passage between the two sides of the turbine is 0.65 ft. When operating at 150 rpm the flow rate through the turbine is 50 cfs. Find (*a*) the torque exerted by the water, (*b*) the horsepower delivered to the shaft, and (*c*) the head converted into mechanical work.

6.50 Develop Eq. (6.23) by making the substitutions indicated in the text.

6.51 A paddlewheel with vanes that are all straight and radial is to be used as a crude centrifugal pump for water. $r_1 = 3$ in, $r_2 = 9$ in, and the height perpendicular to the plane of the figure is 0.2 ft. If the speed is 1,200 rpm and the flow is 3,380 gpm, find the difference in pressure between the inner and outer circumferences, neglecting friction losses. Express the answer in pounds per square inch. Which point is at the higher pressure? Compute the torque required to drive the pump. What is the horsepower requirement? Verify that the horsepower requirement is equal to the difference between the horsepower of the outflow minus the horsepower of the inflow.

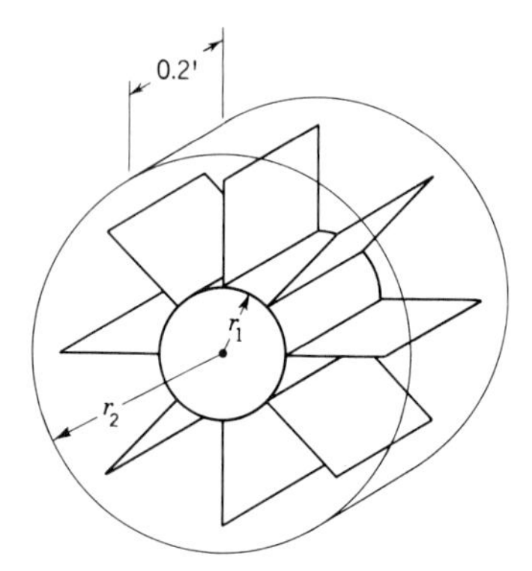

Problem 6.51

6.52 Repeat Prob. 6.51 where the data are given in SI units as follows: $r_1 = 7.5$ cm, $r_2 = 22.5$ cm, height perpendicular to plane of figure = 6.0 cm. Speed is 1,200 rpm and flow is 200 L/s. Express pressure difference in kN/m^2 and power in kW.

6.53 Given a lawn sprinkler such as that in Fig. 6.14 with $\beta_2 = 160°$, and the total area of the jets at a radius at 15 in (40 cm) is 0.0008 ft^2 (0.7 cm^2). When $h = 144$ ft (42 m), compute the rate of discharge, the torque exerted by the water, and the power developed if the rotative speed of the sprinkler is 400 rpm. Neglect fluid friction, but note that the calculated torque is that required to overcome mechanical friction and air resistance.

6.54 Repeat Prob. 6.53 for the case where some external object prevents the sprinkler from rotating.

6.55 How fast would the sprinkler of Prob. 6.53 rotate if there were no mechanical friction or air resistance (i.e., consider the case where $T = 0$)? This is known as runaway speed.

6.56 At what approximate speed will the sprinkler of Prob. 6.53 (English units) develop maximum horsepower?

6.57 The flow from a lawn sprinkler such as Fig. 6.14 is 120 L/min, $\beta_2 = 180°$, and the total area of the jets is 110 mm^2. The jets are located 25 cm from the center of rotation. Determine the speed of rotation if there is no friction.

6.58 Consider the case of a windmill (essentially the opposite of a propeller), apply the momentum and energy principles, and determine the maximum theoretical efficiency based on an input energy available from the wind velocity in a stream tube having a cross section equivalent to that of the windmill blade circle.

6.59 A 20-in-diameter household fan drives air ($\gamma = 0.076$ lb/ft^3) at a rate of 1.60 lb/s. Find the thrust exerted by the fan. What is the pressure difference on the two sides of the fan? Find the required horsepower to drive the fan. Neglect losses.

6.60 A 2.0-m-diameter fan drives air ($\gamma = 12$ N/m^3) at a rate of 45 N/s. Find the thrust exerted by the fan. What is the pressure difference on the two sides of the fan? Find the required kilowatts to drive the fan. Neglect losses.

6.61 A 14-in (35 cm) electric fan is placed on a frictionless mount and is observed to exert a thrust of 0.5 lb (2.25 N). Find the approximate velocity of the slipstream of standard air (sea level) which it produces. If 45 percent of the power supplied to the blades is lost in eddies and friction and if the driving motor has an efficiency of 60 percent, find the required electrical input in watts.

6.62 A fan sucks air from outside to inside a building through an 18-in-diameter duct. The density of the air is 0.0022 slug/ft^3. If the pressure difference across the two sides of the fan is 3.6 in of water, determine the flow rate of the air in cubic feet per second. What thrust must the fan support be designed to withstand?

6.63 Determine the horizontal thrust on the cylinder of Prob. 5.24 using impulse-momentum.

6.64 Find the horizontal thrust of the water on each meter of width of the sluice gate shown in the figure.

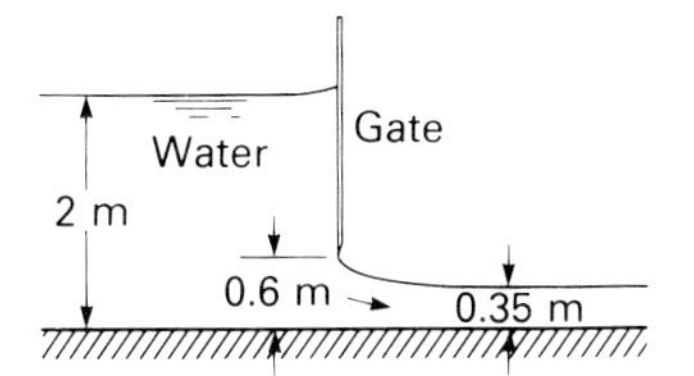

Problem 6.64

6.65 In Illustrative Example 6.7 suppose the passage narrowed down to a width of 8 ft (2.5 m) at the second section. With the same depths find the flow rate and the horizontal force on the structure.

6.66 Flow occurs over a spillway of constant section as shown in the figure. Determine the horizontal force on the spillway per foot of spillway width (perpendicular to the spillway section). Assume ideal flow.

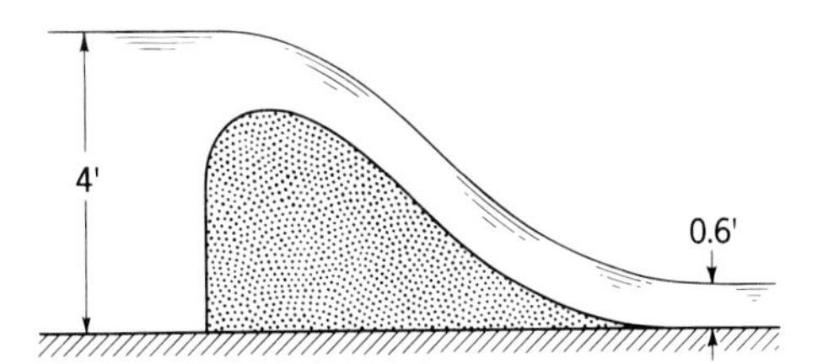

Problem 6.66

6.67 Repeat Prob. 6.66 for the case where the water depths are 4.0 and 0.6 m respectively. In this case find the force per meter of spillway width.

6.68 An hydraulic jump (Sec. 11.19) occurs in a "diamond-shaped" transparent closed conduit as shown in the figure. The conduit is horizontal, and the water depth just upstream of the jump is 2.0 ft. The conduit is completely full of water downstream of the jump. Pressure-gage readings are as shown in the figure. (*a*) Compute the flow rate. Note that, because of turbulence in the jump, there is a substantial energy loss. Hence ideal flow cannot be assumed. Shear forces along the boundary may be neglected however. (*b*) Determine the horsepower loss in the jump.

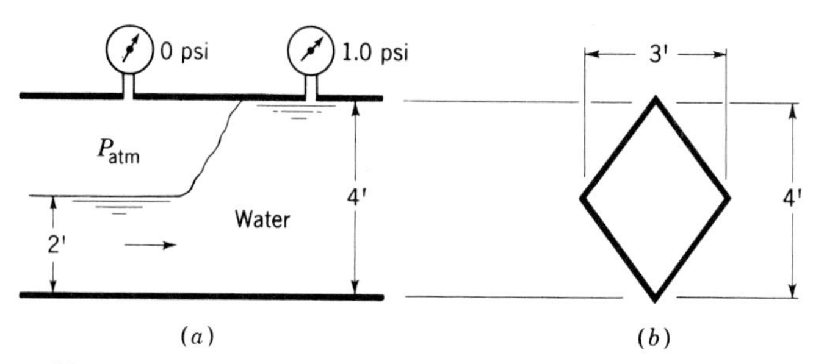

Problem 6.68

CHAPTER

SEVEN

SIMILITUDE AND DIMENSIONAL ANALYSIS

7.1 DEFINITION AND USES OF SIMILITUDE

It is usually impossible to determine all the essential facts for a given fluid flow by pure theory, and hence dependence must often be placed upon experimental investigations. The number of tests to be made can be greatly reduced by a systematic program based on dimensional analysis and specifically on the laws of similitude or similarity, which permit the application of certain relations by which test data can be applied to other cases.

Thus the similarity laws enable us to make experiments with a convenient fluid such as water or air, for example, and then apply the results to a fluid which is less convenient to work with, such as hydrogen, steam, or oil. Also, in both hydraulics and aeronautics, valuable results can be obtained at a minimum cost by tests made with small-scale models of the full-size apparatus. The laws of similitude make it possible to predict the performance of the *prototype*, which means the full-size device, from tests made with the model. It is not necessary that the same fluid be used for the model and its prototype. Neither is the model necessarily smaller than its prototype. Thus the flow in a carburetor might be studied in a very large model. And the flow of water at the entrance to a small centrifugal-pump runner might be investigated by the flow of air at the entrance to a large model of the runner.

A few examples where models may be used are ships in towing basins, airplanes in wind tunnels, hydraulic turbines, centrifugal pumps, spillways of dams, river channels, and the study of such phenomena as the action of waves and tides on beaches, soil erosion, and the transportation of sediment.

It should be emphasized that the model need not necessarily be different in size from is prototype. In fact, it may be the same device, the variables in the case being the velocity and the physical properties of the fluid.

7.2 GEOMETRIC SIMILARITY

One of the desirable features in model studies is that there be geometric similarity, which means that the model and its prototype be identical in shape but differ only in size. The important consideration is that the flow patterns be geometrically similar. If the scale ratio[1] is denoted by L_r, which means the ratio of the linear dimensions of the prototype to corresponding dimensions in the model, it follows that areas vary as L_r^2 and volumes as L_r^3. Complete geometric similarity is not always easy to attain. Thus the surface roughness of a small model may not be reduced in proportion unless it is possible to make its surface very much smoother than that of the prototype. In the study of sediment transportation, it may not be possible to scale down the bed materials without having material so fine as to be impractical. Thus fine powder, because of cohesive forces between the particles, does not simulate the behavior of sand. Again in the case of a river, the horizontal scale is usually limited by the available floor space, and this same scale used for the vertical dimensions may produce a stream so shallow that capillarity has an appreciable effect and also the slope may be such that the flow is laminar. In such cases it is necessary to use a distorted model, which means that the vertical scale is larger than the horizontal scale. If the horizontal scale ratio is denoted by L_r and the vertical scale ratio by $L_{r'}$, the cross section area ratio is $L_r L_{r'}$.

7.3 KINEMATIC SIMILARITY

Kinematic similarity implies geometric similarity and in addition it implies that the ratio of the velocities at all corresponding points in the flow is the same. If subscripts p and m denote prototype and model, respectively, the velocity ratio V_r is

$$V_r = \frac{V_p}{V_m} \tag{7.1}$$

and its value in terms of L_r will be determined by dynamic considerations, as explained in the following section.

[1] In this text we shall define $L_r = L_p/L_m$ as the *scale ratio*. The reciprocal of this, $\lambda = L_m/L_p$, will be referred to as the *model ratio*, or model scale. Thus a model ratio of 1:20 corresponds to a scale ratio of 20:1.

As time T is dimensionally L/V, the time scale is

$$T_r = \frac{L_r}{V_r} \tag{7.2}$$

and in a similar manner the acceleration scale is

$$a_r = \frac{L_r}{T_r^2} = \frac{V_r^2}{L_r} \tag{7.3}$$

7.4 DYNAMIC SIMILARITY

If two systems are dynamically similar, corresponding forces must be in the same ratio in the two. Forces that may act on a fluid element include those due to gravity F_G, pressure F_P, viscosity F_V, and elasticity F_E. Also, if the element of fluid is at a liquid-gas interface, there are forces due to surface tension F_T. If the summation of forces on a fluid element does not add up to zero, the element will accelerate in accordance with Newton's law. Such an unbalanced force system can be transformed into a balanced system by adding an inertia force F_I that is equal and opposite to the resultant of the acting forces. Thus, generally,

$$\sum \mathbf{F} = \mathbf{F}_G + \mathbf{F}_P + \mathbf{F}_V + \mathbf{F}_E + \mathbf{F}_T = \text{Resultant}$$

and $$\mathbf{F}_I = -\text{Resultant}$$

Thus $$\mathbf{F}_G + \mathbf{F}_P + \mathbf{F}_V + \mathbf{F}_E + \mathbf{F}_T + \mathbf{F}_I = 0$$

These forces may be expressed in simplest terms as follows:

Gravity: $F_G = mg = \rho L^2 g$

Pressure: $F_P = (\Delta p)A = (\Delta p)L^2$

Viscosity: $F_V = \mu\left(\frac{du}{dy}\right)A = \mu\left(\frac{V}{L}\right)L^2 = \mu V L$

Elasticity: $F_E = E_v A = E_v L^2$

Surface tension: $F_T = \sigma L$

Inertia: $F_I = ma = \rho L^3 \frac{L}{T^2} = \rho L^4 T^{-2} = \rho V^2 L^2$

In many flow problems some of these forces are either not present or insignificant. In Fig. 7.1 are depicted two geometrically similar flow systems. Let it be assumed that they also possess kinematic similarity and that the forces acting on any fluid element are F_G, F_P, F_V, and F_I. Then dynamic similarity will be achieved if

$$\frac{F_{G_p}}{F_{G_m}} = \frac{F_{P_p}}{F_{P_m}} = \frac{F_{V_p}}{F_{V_m}} = \frac{F_{I_p}}{F_{I_m}}$$

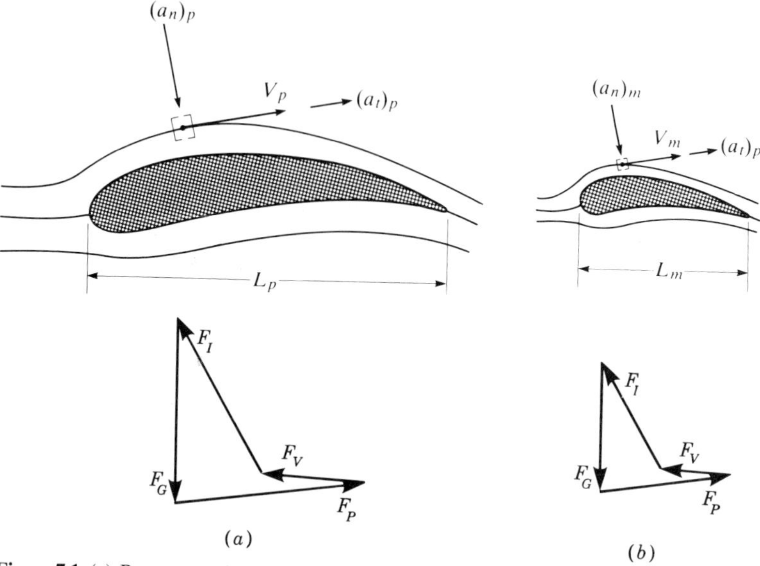

Figure 7.1 (*a*) Prototype. (*b*) Model. $L_r = L_p/L_m$; $V_r = V_p/V_m$.

where subscripts p and m refer to prototype and model as before. These relations can be expressed as

$$\left(\frac{F_I}{F_G}\right)_p = \left(\frac{F_I}{F_G}\right)_m \qquad \left(\frac{F_I}{F_P}\right)_p = \left(\frac{F_I}{F_P}\right)_m \qquad \left(\frac{F_I}{F_V}\right)_p = \left(\frac{F_I}{F_V}\right)_m$$

Each of the quantities is dimensionless. With four forces acting, there are three independent expressions that must be satisfied; for three forces there are two independent expressions; and so on. The significance of the dimensionless ratios is discussed in the following paragraphs.

Reynolds Number[1]

In the flow of a fluid through a completely filled conduit, gravity does not affect the flow pattern. It is also obvious that capillarity is of no practical importance, and hence the significant forces are inertia and fluid friction due to viscosity. The same is true of an airplane traveling at speeds below that at which compressibility of the air is appreciable. Also, for a submarine submerged far enough so as not to produce waves on the surface, the only forces involved are those of friction and inertia.

Considering the ratio of inertia forces to viscous forces, the parameter obtained is called the *Reynolds number*, or **R**, in honor of Osborne Reynolds, who

[1] It is now standard practice to represent Reynolds number, Froude number, etc. by **R**, **F**, etc. For blackboard, overhead projector, and ordinary typing it is suggested that R, F, etc be used to represent these dimensionless numbers. The symbols Re and N_R and Fr and N_F are also used to represent Reynolds number and Froude number respectively.

presented this in a publication of his experimental work in 1882, but it was Lord Rayleigh ten years later who developed the theory of dynamic similarity. The ratio of these two forces is

$$\mathbf{R} = \frac{F_I}{F_V} = \frac{L^2V^2\rho}{LV\mu} = \frac{LV\rho}{\mu} = \frac{LV}{\nu} \tag{7.4}$$

For any consistent system of units, **R** is a dimensionless number. The linear dimension L may be any length that is significant in the flow pattern. Thus, for a pipe completely filled, it might be either the diameter or the radius, and the numerical value of **R** will vary accordingly. General usage prescribes L as the pipe diameter. Thus for a pipe flowing full $\mathbf{R} = DV\rho/\mu = DV/\nu$ where D is the diameter of the pipe.

If two systems, such as a model and its prototype, or two pipelines with different fluids, are to be dynamically equivalent so far as inertia and viscous forces are concerned, they must both have the same value of **R**. Thus for such a case, dynamic similarity is achieved when

$$\left(\frac{LV}{\nu}\right)_m = \mathbf{R}_m = \mathbf{R}_p = \left(\frac{LV}{\nu}\right)_p \tag{7.5}$$

For the same fluid in model and prototype, Eq. (7.5) shows that for dynamic similarity a high velocity must be used with a model of small linear dimensions. The fluid used in the model need not be the same as that in the prototype provided L and V are so chosen as to give the same value of **R** in model and prototype.

Illustrative Example 7.1 If the Reynolds number of a model and its prototype are the same, find an expression for V_r, T_r, and a_r.

$$\mathbf{R} = \frac{L_m V_m}{\nu_m} = \frac{L_p V_p}{\nu_p}$$

$$V_r = \frac{V_p}{V_m} = \frac{L_m \nu_p}{L_p \nu_m} = \frac{\nu_r}{L_r} = \left(\frac{\nu}{L}\right)_r$$

$$T_r = \frac{L_r}{V_r} = \left(\frac{L^2}{\nu}\right)_r \qquad a_r = \frac{V_r}{T_r} = \left(\frac{\nu^2}{L^3}\right)_r$$

Froude Number[1]

Considering inertia and gravity forces alone, a ratio is obtained called a *Froude number*, or **F**, in honor of William Froude, who experimented with flat plates towed lengthwise through water in order to estimate the resistance of ships due to wave action. The ratio of inertia forces to gravity forces is

$$\frac{\rho L^2 V^2}{\rho g L^3} = \frac{V^2}{gL}$$

[1] Froude is pronounced (fro͞od).

Although this is sometimes defined as a Froude number, it is more common to use the square root so as to have V in the first power, as in the Reynolds number. Thus a Froude number is

$$\mathbf{F} = \frac{V}{\sqrt{gL}} \tag{7.6}$$

Systems involving gravity and inertia forces include the wave action set up by a ship, the flow of water in open channels, the forces of a stream on a bridge pier, the flow over a spillway, the flow of a jet from an orifice, and other cases where gravity is the dominant factor.

For the computation of **F**, the length L must be some linear dimension that is significant in the flow pattern. For a ship it is commonly taken as the length at the waterline. For an open channel it is taken as the depth of flow. For situations where inertia and gravity forces predominate, dynamic similarity is achieved when

$$\left(\frac{V}{\sqrt{gL}}\right)_m = \mathbf{F}_m = \mathbf{F}_p = \left(\frac{V}{\sqrt{gL}}\right)_p \tag{7.7}$$

In some flow situations fluid friction is a factor as well as gravity and inertia. In such cases to achieve dynamic similarity the Reynolds number and the Froude number criteria must both be satisfied simultaneously. A comparison of Eqs. (7.5) and (7.7) shows that the two cannot be satisfied at the same time with fluids of the same viscosity. The only way to satisfy Eqs. (7.5) and (7.7) for both model and prototype is to use fluids of different viscosities. Simultaneous solution of Eqs. (7.5) and (7.7) yields $(\nu_m/\nu_p) = (L_m/L_p)^{3/2}$. Sometimes a fluid with the proper viscosity can be found, but usually it is either impractical or impossible. Consequently the usual technique is to operate the model so as to satisfy one of the dimensionless numbers and then correct the results with experimental data dependent on the other dimensionless number. In the case of a ship, the towing of a model will give the total resistance, from which must be subtracted the empirically computed skin friction to determine the wave-making resistance of the model. Applying the Froude number criterion then permits one to determine the wave-making resistance of the full-size ship. A computed skin friction for the ship is then added to this value to give the total ship resistance. The details of such calculations are deferred to Chap. 10.

When water flows in an open channel, fluid friction as well as gravity and inertia may be a factor. However, for the flow of water in an open channel, there is often fully developed turbulence in which case the hydraulic friction is independent of the Reynolds number (Sec. 8.11). Thus for this case, identical Froude numbers will give dynamic similarity. When a high viscosity liquid flows in an open channel or when water flows at a relatively low Reynolds number the effect of Reynolds number cannot be neglected.

The scale ratios for Froude number similarity will now be discussed. From Eq. (7.6), V varies as $\sqrt{gL}$, and if g is considered to be the same in prototype and model, as is usually the case, then from Eq. (7.1),

$$V_r = \frac{V_p}{V_m} = \sqrt{\frac{L_r}{1}} \qquad \text{(for same } \mathbf{F} \text{ and same } g\text{)}$$

and from Eq. (7.2) the ratio of time for prototype to mode is

$$T_r = \frac{T_p}{T_m} = \frac{L_r}{V_r} = \sqrt{\frac{L_r}{1}} \qquad \text{(for same } \mathbf{F} \text{ and same } g\text{)}$$

and

$$a_r = \frac{V_r}{T_r} = 1 \qquad \text{(for same } \mathbf{F} \text{ and same } g\text{)}$$

A knowledge of the time scale is useful in the study of cyclic phenomena such as waves and tides.

Since the velocity varies as $\sqrt{L_r}$ and the cross section area as L_r^2, it follows that

$$Q_r = \frac{Q_p}{Q_m} = \frac{L_r^{5/2}}{1} \qquad \text{(for same } \mathbf{F} \text{ and same } g\text{)}$$

As mentioned in Sec. 7.2 for river models it is usually necessary to use an enlarged vertical scale to provide the model with adequate depth.[1] Application of the Froude number indicates that for this case V varies as $\sqrt{L_{r'}}$ where $L_{r'}$ is the scale ratio in the vertical direction. Thus

$$\frac{Q_p}{Q_m} = \frac{L_r L_{r'}^{3/2}}{1} \qquad \text{(for same } \mathbf{F} \text{ and same } g\text{)}$$

Mach Number

Where compressibility is important, it is necessary to consider the ratio of the inertia to the elastic forces. The *Mach number* **M** is defined as the square root of this ratio. Thus

$$\mathbf{M} = \left[\frac{\rho V^2 L^2}{E_v L^2}\right]^{1/2} = \frac{V}{\sqrt{E_v/\rho}} = \frac{V}{c} \tag{7.8}$$

where c is the sonic velocity (or celerity) in the medium in question. (See Appendix 2). Thus the Mach number is the ratio of the fluid velocity (or the velocity of the body through a stationary fluid) to that of a sound wave in the same medium. If **M** is less than 1, the flow is *subsonic*; if it is equal to 1, it is *sonic*; if it is greater than 1, the flow is called *supersonic*; and for extremely high values of **M** the flow is called *hypersonic*.

[1] Enlarged vertical scales are also commonly employed in models of large water bodies such as lakes, reservoirs, estuaries, and bays.

Weber Number

In a few cases of flow, surface tension may be important, but normally it is negligible. The ratio of inertia forces to surface tension is $\rho V^2 L^2/\sigma L$, the square root of which is known as the *Weber number*:

$$\mathbf{W} = \frac{V}{\sqrt{\sigma/\rho L}} \tag{7.9}$$

An illustration of its application is at the leading edge of a very thin sheet of liquid flowing over a surface.

Euler Number[1]

A dimensionless quantity related to the ratio of the inertia forces to the pressure forces is known as the *Euler number*. It is expressed in a variety of ways, one form being

$$\mathbf{E} = \frac{V}{\sqrt{2(\Delta p/\rho)}} = \frac{V}{\sqrt{2g(\Delta p/\gamma)}} \tag{7.10}$$

If only pressure and inertia influence the flow, the Euler number for any boundary form will remain constant. If other parameters (viscosity, gravity, etc.) cause the flow pattern to change, however, **E** will also change. The expression for **E** [Eq. (7.10)] may be recognized as being equivalent to the coefficient of velocity, discussed in Chapter 12.

Illustrative Example 7.2 A certain submerged body is to move horizontally through oil ($\gamma = 52$ lb/ft^3, $\mu = 0.0006$ lb·s/ft^2) at a velocity of 45 fps. To study the characteristics of this motion, an enlarged model of the body is tested in 60°F water. The model ratio λ is 8:1. Determine the velocity at which this enlarged model should be pulled through the water to achieve dynamic similarity. If the drag force on the model is 0.80 lb, predict the drag force on the prototype.

Body is submerged, hence there is no wave action. Reynolds criterion must be satisfied.

$$\left(\frac{DV}{\nu}\right)_p = \left(\frac{DV}{\nu}\right)_m \qquad \text{where} \qquad \frac{D_m}{D_p} = \frac{8}{1}$$

$$\nu_m = 1.22 \times 10^{-5} \text{ lb·s/ft}^2 \qquad \text{(Appendix 3, Table A.1)}$$

$$\nu_p = \frac{\mu}{\rho} = \frac{0.0006}{52/32.2} = 0.000322 \text{ lb·s/ft}^2$$

$$\frac{D_p(45)}{0.000322} = \frac{(8D_p)V_m}{1.22 \times 10^{-5}}$$

$$V_m = 0.213 \text{ fps}$$

$$F \propto \rho V^2 L^2 \qquad \text{hence} \qquad \frac{F_p}{F_m} = \frac{\rho_p V_p^2 L_p^2}{\rho_m V_m^2 L_m^2}$$

$$\frac{F_p}{F_m} = \frac{(52/32.2)(45)^2 1}{1.94(0.213)^2(8)^2} = 580$$

$$F_p = 580 F_m \qquad F_p = 580(0.8) = 464 \text{ lb}$$

[1] Euler is pronounced (oi'lər).

7.5 SCALE RATIOS

The Reynolds number, the Froude number, and the Mach number are the dimensionless parameters most commonly encountered in fluid mechanics. In the preceding section the scale ratios for velocity, time, and acceleration for the Reynolds and Froude numbers were developed. Scale ratios for other quantities can be developed in a similar fashion. Such relations are presented in Table 7.1. These enable one to quickly calculate the scale ratio (prototype divided by model) of any desired quantity for the case where the given dimensionless number is the same in both prototype and model. The computed ratio, of course, gives a realistic result only if the flow is predominantly governed by the particular dimensionless number. Thus an important aspect of physical modeling of fluid phenomena is the need to know which dimensionless number is most important.

Table 7.1 Flow characteristics and similitude scale ratios (ratio of prototype quantity to model quantity)

		Scale ratios for laws of		
Characteristic	Dimension	Reynolds	Froude	Mach
Geometric				
Length	L	L_r	L_r	L_r
Area	L^2	L_r^2	L_r^2	L_r^2
Volume	L^3	L_r^3	L_r^3	L_r^3
Kinematic				
Time	T	$\left(\frac{L^2\rho}{\mu}\right)_r$	$(L^{1/2}g^{-1/2})_r$	$\left(\frac{L\rho^{1/2}}{E_v^{1/2}}\right)_r$
Velocity	LT^{-1}	$\left(\frac{\mu}{L\rho}\right)_r$	$(L^{1/2}g^{1/2})_r$	$\left(\frac{E_v^{1/2}}{\rho^{1/2}}\right)_r$
Acceleration	LT^{-2}	$\left(\frac{\mu^2}{\rho^2L^3}\right)_r$	g_r	$\left(\frac{E_v}{L\rho}\right)_r$
Discharge	L^3T^{-1}	$\left(\frac{L\mu}{\rho}\right)_r$	$(L^{5/2}g^{1/2})_r$	$\left(\frac{L^2E_v^{1/2}}{\rho^{1/2}}\right)_r$
Dynamic				
Mass	M	$(L^3\rho)_r$	$(L^3\rho)_r$	$(L^3\rho)_r$
Force	MLT^{-2}	$\left(\frac{\mu^2}{\rho}\right)_r$	$(L^3\rho g)_r$	$(L^2E_v)_r$
Pressure	$ML^{-1}T^{-2}$	$\left(\frac{\mu^2}{L^2\rho}\right)_r$	$(L\rho g)_r$	$(E_v)_r$
Impulse and momentum	MLT^{-1}	$(L^2\mu)_r$	$(L^{7/2}\rho g^{1/2})_r$	$(L^3\rho^{1/2}E_v^{1/2})_r$
Energy and work	ML^2T^{-2}	$\left(\frac{L\mu^2}{\rho}\right)_r$	$(L^4\rho g)_r$	$(L^3E_v)_r$
Power	ML^2T^{-3}	$\left(\frac{\mu^3}{L\rho^2}\right)_r$	$(L^{7/2}\rho g^{3/2})_r$	$\left(\frac{L^2E_v^{3/2}}{\rho^{1/2}}\right)_r$

Note: Usually g is the same in model and prototype.

7.6 COMMENTS ON MODELS

In the use of models it is essential that the fluid velocity should not be so low that laminar flow exists when the flow in the prototype is turbulent. Also, conditions in the model should not be such that surface tension is important if such conditions do not exist in the prototype. For example, the depth of water flowing over the crest of a model spillway should not be too small.

While model studies are very important and valuable, it is necessary to exercise some judgment in transferring results from the model to the prototype. It is not always necessary or desirable that these various dimensionless ratios be adhered to in every case. Thus, in tests of model centrifugal pumps, geometric similarity is essential, but it is desirable to operate at such a rotative speed that the peripheral velocity and all fluid velocities are the same as in the prototype, since only in this way may cavitation be detected.

The roughness of a model should be scaled down in the same ratio as the other linear dimensions, which means that a small model should have surfaces that are much smoother than those in its prototype. But this requirement imposes a limit on the scale that can be used if true geometric similarity is to be achieved. However, in the case of a distorted model with a vertical scale larger than the horizontal scale, it may be necessary to make the model surface rough in order to simulate the flow conditions in the prototype. As any distorted model lacks the proper similitude, no simple rule can be given for this; the roughness should be adjusted by trial until the flow conditions are judged to be typical of those in the prototype. In most distorted models, metal tabs (Fig. 7.2) are used to provide proper frictional boundary effects. The size and spacing of the tabs is determined by trial so as to create flow conditions in the model identical to those observed in the prototype. Vertical distortion disturbs circulation patterns and the metal tabs create large-scale eddies. Hence mixing (the disposition of pollutants) in distorted models must be interpreted with caution.

In models of systems, such as siphons, involving liquids where large negative pressures are expected and hence the possibility of cavitation exists, the model must be placed in an air-tight chamber in which a partial vacuum is maintained so as to produce an absolute pressure in the model identical to that in the prototype.[1]

When modeling a subsonic airplane in a wind tunnel, it is commonly necessary to conduct the test under high pressure in order to satisfy the Reynolds criterion

$$\left(\frac{DV\rho}{\mu}\right)_m = \left(\frac{DV\rho}{\mu}\right)_p$$

without introducing compressibility effects. For example, suppose $L_r = D_p/D_m = 20$. If the viscosity μ and density ρ of the air were the same in the

[1] Hydraulic Models, *Manual of Engineering Practice*, no. 25, American Society of Civil Engineers. 1942.

Figure 7.2 View of the Corps of Engineers Model of San Francisco Bay ($L_r = 1000$, $L_{r'} = 100$) showing the vertical metal tabs that were installed to provide proper frictional flow resistance.

model and prototype, then to satisfy Reynolds' criterion, $V_m = 20 \times V_p$. For an airplane operating at normal speed this would make the model Mach number much greater than one, and compressibility effects would invalidate the behavior of the model. If, however, the test were conducted under a pressure of 20 atm with identical model and prototype temperatures, $\rho_m = 20 \times \rho_p$ and $\mu_m \approx \mu_p$ since the viscosity of air changes very little with pressure (or density). In this case the model should be operated at a velocity equal to that of the prototype in order for the Reynolds numbers to be the same.

Illustrative Example 7.3 A 1:50 model of a boat has a wave resistance of 0.02 N when operating in water at 1.0 m/s. Find the corresponding prototype wave resistance. Find also the horsepower requirement for the prototype. What velocity does this test represent in the prototype?

Gravity and inertia forces predominate; hence the Froude criterion is applicable.

$$\mathbf{F}_p = \mathbf{F}_m = \left(\frac{V}{\sqrt{gL}}\right)_p = \left(\frac{V}{\sqrt{gL}}\right)_m$$

Since both the model and prototype are acted upon by the earth's gravitational field, the g's can be canceled out. Thus

$$\frac{V_p^2}{L_p} = \frac{V_m^2}{L_m}$$

and
$$\frac{V_p^2}{V_m^2} = \frac{L_p}{L_m} = L_r = 50$$

$$F \propto pV^2L^2$$

Since
$$\frac{F_p}{F_m} = \frac{\rho L_p^2 V_p^2}{\rho L_m^2 V_m^2} = L_r^2 L_r = L_r^3$$

Therefore
$$F_p = L_r^3 F_m = (50)^3(0.02) = 2{,}500 \text{ N} = 562 \text{ lb}$$

$$V_p = \sqrt{L_r} \times V_m = \sqrt{50} \times 1 = 7.1 \text{ m/s} = 23.3 \text{ fps}$$

$$HP_p = \frac{F_p V_p}{550} = \frac{562 \times 23.3}{550} = 23.8$$

7.7 DIMENSIONAL ANALYSIS

Fluid-mechanics problems may be approached by *dimensional analysis*, a mathematical technique making use of the study of dimensions. Dimensional analysis is related to similitude; however, the approach is different. In dimensional analysis, from a general understanding of fluid phenomena, one first predicts the physical parameters that will influence the flow, and then, by grouping these parameters in dimensionless combinations, a better understanding of the flow phenomena is made possible. Dimensional analysis is particularly helpful in experimental work because it provides a guide to those things that significantly influence the phenomena; thus it indicates the direction in which experimental work should go.

Physical quantities may be expressed in either the force-length-time (FLT) system or in the mass-length-time (MLT) system. These two systems are interrelated through Newton's law which states that force equals mass time acceleration, $F = ma$, or

$$F = M\frac{L}{T^2}$$

Through this relation, conversion can be made from one system to the other. The dimensions used in either system may be in English units or in metric units. Details on the English and metric (SI) systems of units and conversion factors are presented in the front matter and also in Appendix 1.

To illustrate the steps in a dimensional-analysis problem, let us consider the drag force F_D exerted on a sphere as it moves through a viscous liquid. We must visualize the physical problem to consider what physical factors influence the drag force. Certainly, the size of the sphere must enter the problem; also, the velocity of the sphere must be important. The fluid properties involved are the density ρ and the viscosity μ. Thus we can write

$$F_D = f(D, V, \rho, \mu)$$

Here D, the sphere diameter, is used to represent sphere size.

We want to determine how these variables are interrelated. Our approach is to satisfy dimensional homogeneity. That is, we want the dimensions on one side of the equation to correspond to those on the other. The preceding expression may be written as a power equation

$$F_D = CD^a V^b \rho^c \mu^d$$

where C is a dimensionless constant. Using the MLT system and substituting the proper dimensions,

$$\frac{ML}{T^2} = L^a \left(\frac{L}{T}\right)^b \left(\frac{M}{L^3}\right)^c \left(\frac{M}{LT}\right)^d$$

To satisfy dimensional homogeneity the exponents of each dimension must be identical on both sides of the equation. Thus

For M: $\quad 1 = c + d$

For L: $\quad 1 = a + b - 3c - d$

For T: $\quad -2 = -b - d$

Since we have three equations with four unknowns, we must express three of the unknowns in terms of the fourth. Solving for a, b, and c in terms of d, we get

$$a = 2 - d \qquad b = 2 - d \qquad c = 1 - d$$

Thus $$F_D = CD^{2-d}V^{2-d}\rho^{1-d}\mu^d$$

and grouping variables according to their exponents,

$$F_D = C(\rho)(D^2V^2)\left(\frac{VD\rho}{\mu}\right)^{-d}$$

It may be noted that the quantity $VD\rho/\mu$ is a Reynolds number. Thus the original power equation can be expressed as

$$F_D = f'(\mathbf{R})\rho D^2 V^2$$

or $$\frac{F_D}{\rho D^2 V^2} = f'(\mathbf{R})$$

The result indicates that the drag on a sphere is equal to some coefficient times $\rho D^2 V^2$, where the coefficient is a function of the Reynolds number. This is indeed true, as indicated by the discussion of drag on a sphere in Sec. 10.7.

The foregoing approach to dimensional analysis is commonly referred to as the *Rayleigh method*, after Lord Rayleigh, who originally proposed it. Another more generalized approach is through use of the *Buckingham Π theorem*.[1] This

[1] E. Buckingham, Model Experiments and the Form of Empirical Equations, *Trans. ASME*, vol. 37, pp. 263–296, 1915.

theorem states that if there are n dimensional variables in a dimensionally homogeneous equation, described by m fundamental dimensions, they may be grouped in $n - m$ dimensionless groups. Thus, in the preceding example, $n = 5$ and $m = 3$ (M, L, and T) and $n - m = 2$; these dimensionless groups were **R** and $F_D/\rho D^2 V^2$. Buckingham referred to these dimensionless groups as Π terms. The advantage of the Π theorem is that it tells one ahead of time how many dimensionless groups are to be expected.

Applying the Π theorem to the preceding example, one would proceed as follows:

$$f'(F_D, D, V, \rho, \mu) = 0$$

where $n = 5$, $m = 3$, so $n - m = 2$. Thus we can write

$$\phi(\Pi_1, \Pi_2) = 0$$

The problem now is to find the Π's by arranging the five parameters into two dimensionless groups. Taking ρ, D, and V as the primary variables,[1] the Π terms are:

$$\Pi_1 = \rho^{a_1} D^{b_1} V^{c_1} \mu^{d_1}$$

$$\Pi_2 = \rho^{a_2} D^{b_2} V^{c_2} F_D^{d_2}$$

The values of the exponents are determined as before, noting that since the Π's are dimensionless, they can be replaced with $M^0 L^0 T^0$. Experience in fluid mechanics has shown that these dimensionless groups commonly take the form of a Reynolds number, Froude number, or Mach number. Hence one should always be on the lookout for them when using dimensional analysis. Working with Π_1,

$$M^0 L^0 T^0 = \left(\frac{M}{L^3}\right)^{a_1} L^{b_1} \left(\frac{L}{T}\right)^{c_1} \left(\frac{M}{LT}\right)^{d_1}$$

M: $\quad 0 = a_1 + d_1$

L: $\quad 0 = -3a_1 + b_1 + c_1 - d_1$

T: $\quad 0 = -c_1 - d_1$

Solving for a_1, b_1, and c_1 in terms of d_1,

$$a_1 = -d_1 \qquad b_1 = -d_1, \qquad c_1 = -d_1$$

Thus

$$\Pi_1 = \rho^{-d_1} D^{-d_1} V^{-d_1} \mu^{d_1} = \left(\frac{\mu}{\rho D V}\right)^{d_1} = \left(\frac{\rho D V}{\mu}\right)^{-d_1}$$

Hence, noting that $(\rho DV/\mu)$ is a Reynolds number,

$$\Pi_1 = \mathbf{R}$$

[1] It is generally advantageous to choose primary variables that relate to mass, geometry, and kinematics.

Working in a similar fashion with Π_2, one gets

$$\Pi_2 = \frac{F_D}{\rho D^2 V^2}$$

Finally, $\phi(\Pi_1, \Pi_2) = 0$ may be expressed as

$$\Pi_1 = \phi'(\Pi_2) \qquad \text{or} \qquad \Pi_2 = \phi''(\Pi_1)$$

So

$$\frac{F_D}{\rho D^2 V^2} = \phi''(\mathbf{R})$$

and

$$F_D = \phi''(\mathbf{R}) \rho D^2 V^2$$

It should be emphasized that dimensional analysis does not provide a complete solution to fluid problems. It provides a partial solution only. The success of dimensional analysis depends entirely on the ability of the individual using it to define the parameters that are applicable. If one omits an important variable, the results are incomplete and this may lead to incorrect conclusions. For example, with a compressible fluid at high velocities, compressibility effects may be significant in which case the volume modulus E_v of the fluid must be considered an important physical property. Introducing E_v into the previous example of dimensional analysis of the drag on a sphere will show that for the more general case the drag may depend on the Mach number as well as the Reynolds number. If one includes a variable that is totally unrelated to the problem, an additional insignificant dimensionless group will result. Thus, to use dimensional analysis successfully, one must be familiar with the fluid phenomena involved.

Illustrative Example 7.4 Derive an expression for the flow rate q over the spillway shown in the accompanying figure per foot of spillway perpendicular to the sketch. Assume that the sheet of water is relatively thick so that surface-tension effects may be neglected. Assume also that gravity effects predominate so strongly over viscosity that viscosity may be neglected.

Under the assumed conditions the variables that effect q would be the head H, the acceleration of gravity g, and possibly the spillway height P. Thus

$$q = f(H, g, P)$$

or

$$f'(q, H, g, P) = 0$$

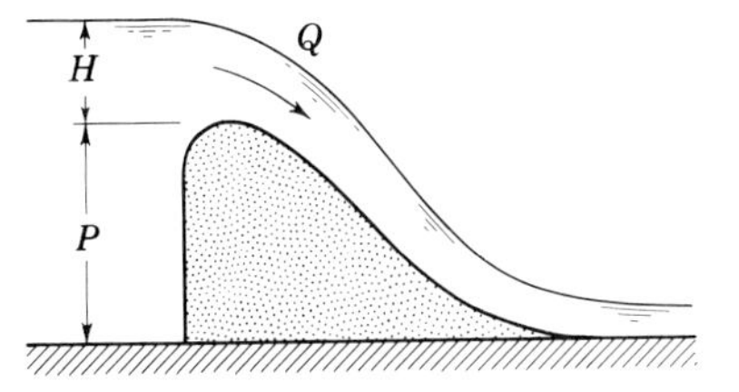

Illustrative Example 7.4

In this case $n = 4$, and $m = 2$, since only kinematic properties are involved. Hence, according to the Π theorem, there are $n - m = 2$ dimensionless groups, and

$$\phi(\Pi_1, \Pi_2) = 0$$

Using q and H as the primary variables,

$$\Pi_1 = q^{a_1} H^{b_1} g^{c_1}$$

$$\Pi_2 = q^{a_2} H^{b_2} P^{c_2}$$

Working with Π_1,

$$L^0 T^0 = \left(\frac{L^3}{TL}\right)^{a_1} L^{b_1} \left(\frac{L}{T^2}\right)^{c_1}$$

L: $\qquad 0 = 2a_1 + b_1 + c_1$

T: $\qquad 0 = -a_1 - 2c_1$

Hence $\qquad c_1 = -\tfrac{1}{2}a_1 \qquad b_1 = -\tfrac{3}{2}a_1$

$$\Pi_1 = q^{a_1} H^{-3/2a_1} g^{-1/2a_1} = \left(\frac{q}{g^{1/2}H^{3/2}}\right)^{a_1}$$

Working with Π_2,

$$L^0 T^0 = \left(\frac{L^3}{TL}\right)^{a_2} L^{b_2} L^{c_2}$$

L: $\qquad 0 = 2a_2 + b_2 + c_2$

T: $\qquad 0 = -a_2$

Hence $\qquad a_2 = 0 \qquad c_2 = -b_2$

$$\Pi_2 = q^0 H^{b_2} P^{-b_2} = \left(\frac{H}{P}\right)^{b_2}$$

Finally, $\phi(\Pi_1, \Pi_2) = 0$ can be written as

$$\Pi_1 = \phi'(\Pi_2)$$

$$\frac{q}{\sqrt{g}H^{3/2}} = \phi'\left(\frac{H}{P}\right)$$

or

$$q = \phi'\left(\frac{H}{P}\right)\sqrt{g}H^{3/2}$$

Thus dimensional analysis indicates that the flow rate per unit length of spillway is proportional to $\sqrt{g}$ and to $H^{3/2}$. The flow rate also is affected by the H/P ratio. This relationship is discussed in Sec. 12.11.

If viscosity were included as one of the variables, another dimensionless group would have resulted. This dimensionless group would have had the form of a Reynolds number. With surface tension included as a variable, the resulting dimensionless group would have been a Weber number.

PROBLEMS

7.1 What is the value of Reynolds number for water at 68°F flowing with a velocity of 5 fps in a 6-in-diameter pipe? Note that $L = D$. See Appendix 3 for the physical properties of water.

7.2 What is the Reynolds number for oil ($s = 0.85$ and $\mu = 0.24$ N·s/m^2) flowing with a velocity of 3.6 m/s in a pipe having a diameter of 10 cm?

7.3 What is the value of Reynolds number for air at a pressure of 100 psia (700 kN/m^2, abs) and a temperature of 150°F (65°C) flowing at a velocity of 80 fps (25 m/s) in a pipe having a 6-in (15 cm) diameter? See Appendix 3 for the physical properties of air.

7.4 What is the Reynolds number for air at an absolute pressure of 200 kN/m^2, abs and a temperature of 150°C flowing at a velocity of 15 m/s in a 20-cm-diameter pipe?

7.5 Models are to be built of the following prototypes. For dynamic similarity, indicate which single dimensionless ratio will govern, given reasons why. (*a*) Oil flowing through a full pipeline, (*b*) a water jet, (*c*) an airplane flying at low speed, (*d*) a supersonic aircraft, (*e*) flow over the spillway of a dam, (*f*) a deep submersible vehicle, (*g*) a missile (supersonic), (*h*) tides.

7.6 A model airplane has linear dimensions that are one-twentieth those of its prototype. If the plane is to fly at 400 mph, what must be the air velocity in the wind tunnel for the same Reynolds number if the air temperature and pressure are the same?

7.7 A model airplane has dimensions that are one-twentieth those of its prototype. It is desired to test it in a pressure wind tunnel at a speed the same as that of the prototype. If the air temperature is the same and the Reynolds number is the same, what must be the pressure in the wind tunnel relative to the atmospheric pressure?

7.8 What weight flow rate of 70°F air at 50 psi in a 1-in-diameter pipe will give dynamic similarity to a 250-gpm flow of 60°F water in a 4-in-diameter pipe?

7.9 A drag force of 10 N is exerted on a submerged sphere when it moves through 20°C water at 1.5 m/s. Another sphere having three times the diameter is placed in a wind tunnel where the air pressure and temperature are 1.5 MN/m^2, abs and 300 K respectively. What air velocity is required for dynamic similarity and what will be the drag force on the larger sphere?

7.10 A 1:30 scale model of a submarine is tested in a wind tunnel. It is desired to know the drag on the submarine when it is operating at 10 knots in 40°F ocean water. At what velocity should the object be tested in a wind tunnel containing 70°F air at atmospheric pressure? If the drag on the model is 80 lb, what would be the drag on the prototype? At what velocity should the test be conducted when testing in a water tunnel if the water temperature is 65°F? What would be the drag on this model?

7.11 What flow rate (kg/s) of 80°C air in a 5-cm-diameter pipe will give dynamic similarity to a 50 L/s flow of 60°C water in a 40-cm-diameter pipe if the pressure on the air is 400 kN/m^2, abs?

7.12 A ship 600 ft long is to operate at a speed of 25 mph. If a model is 10 ft long, what should be its speed in fps to give the same Froude number? What is the value of the Froude number?

7.13 The flow over a spillway is 5,000 cfs. For dynamic similarity, what should be the model scale if the model flow rate is to be 45 cfs? The force on a certain area of the model is measured to be 1.0 lb. What would be the force on the corresponding area of the prototype?

7.14 In a 1:40 model of the flow over the crest of a spillway the velocity at a particular point is 0.5 m/s. What velocity does this represent in the prototype? The force exerted on a certain area in the model is 0.12 N. What would be the force on the corresponding area in the prototype? Develop your own dimensionless ratios.

7.15 A ship 600 ft long is to operate at a speed of 25 mph in ocean water whose viscosity is 1.2 cP and specific weight is 64 lb/ft^3. What should be the kinematic viscosity of a fluid used with the model so that both the Reynolds number and the Froude number would be the same? Does such a liquid exist? Assume the model is 10 ft long.

7.16 A 1:500 model is constructed to study tides. What length of time in the model corresponds to a day in the prototype? Suppose this model could be transported to the moon and tested there. What then would be the time relationship between the model and prototype? g of earth equals six times g of moon.

7.17 On the earth a vertical water jet issuing upward from a nozzle at a velocity of 80 fps will rise to a height of approximately 100 ft. To get a water jet to rise to a height of 100 ft on the moon, what must be its velocity? Neglect atmospheric resistance. Gravity of moon equals $\frac{1}{6}$ × gravity of earth.

7.18 A sectional model of a spillway 3 ft high is placed in a laboratory flume of 10-in width. Under a head of 0.375 ft the flow is 0.70 cfs. What flow does this represent in the prototype if the model scale is 1:25 and the spillway is 650 ft long?

7.19 The flow over a model spillway is 0.086 m^3/s per m of width. What flow does this represent in the prototype spillway if the model scale is 1:18?

7.20 One wishes to model the flow about a missile when traveling at 1,000 mph through the atmosphere at elevation 10,000 ft. The model is to be tested in a wind tunnel at standard atmospheric conditions with 70°F air. What air speed in the wind tunnel is required for dynamic similarity?

7.21 A model of a supersonic aircraft is tested in a variable density wind tunnel at 1,200 fps (360 m/s). The air is at 100°F (38°C) with a pressure of 18 psia (125 kN/m^2, abs). At what velocity should this model be tested to maintain dynamic similarity if the air temperature is raised to 120°F (50°C) and the pressure increased to 24 psia (170 kN/m^2, abs)?

7.22 The flow about a ballistic missile which travels at 1,500 fps (450 m/s) through air at 60°F (15°C) and 14.7 psia (101.3 kN/m^2, abs) is to be modeled in a high-speed wind tunnel with a 1:8 model. If the air in the wind tunnel test section has a temperature of 5°F (−15°C) at a pressure of 11 psia (75 kN/m^2, abs), what velocity is required in the model test section? If the drag force on the model is 80 lb (360 N), approximately what is the drag force on the prototype?

7.23 A ship's model with a scale of 1:40 has a wave resistance of 0.25 lb when traveling at a velocity of 1.8 fps which is kinematically similar to the design velocity of the ship. What is the design velocity of the ship and what is its wave resistance at that velocity?

7.24 Develop the scale ratios given in Table 7.1 for the case where prototype and model Reynolds numbers are the same.

7.25 Develop the scale ratios given in Table 7.1 for the case where prototype and model Froude numbers are identical.

7.26 Develop the scale ratios given in Table 7.1 for the case where prototype and model Mach numbers are the same.

7.27 A flowmeter for gas measurement registers a pressure drop of 1.0 psi when the flow through it is 0.16 lb/s. The gas ($\gamma = 0.35$ lb/ft^3, $\mu = 2.4 \times 10^{-6}$ $lb\cdot s/ft^2$) is flowing in a $\frac{3}{4}$-in-diameter pipe. An enlarged model that is geometrically similar is to be tested in a 6-in-diameter pipe. What flow rate of 80°F water will achieve dynamic similarity? What would be the pressure drop across the water meter?

7.28 Use dimensional analysis to arrange the following groups into dimensionless parameters: (*a*) τ, V, ρ; (*b*) Δp, V, γ, g; (*c*) F, ρ, L, V; (*d*) V, L, ρ, σ. Use the MLT system.

7.29 Find the dimensions of torque, energy, power, force, and momentum in the FLT system. Repeat for the MLT system.

7.30 By dimensional analysis derive an expression for the power developed by an engine in terms of the torque and rotative speed.

7.31 Derive an expression for the velocity of rise of an air bubble in a stationary liquid. Consider the effect of surface tension as well as other variables.

7.32 Derive an expression for the drag on a submerged torpedo. The parameters involved are the size of the torpedo L, the velocity of the torpedo V, the viscosity of the water μ, and the density of the water ρ. The size of a torpedo may be represented by its diameter or its length.

7.33 Derive an expression for the drag on a surface vessel. Use the same parameters as in Prob. 7.32, and add the acceleration due to gravity g, to account for the effect of wave action.

7.34 Derive an expression for the drag on an aircraft flying at supersonic speed.

7.35 Using dimensional analysis, derive an expression for small flow rates over a spillway, in the form of a function including dimensionless quantities. The parameters involved are height of spillway P, head on the spillway H, acceleration due to gravity g, viscosity of liquid μ, density of liquid ρ, and surface tension σ.

7.36 Use dimensional analysis to derive an expression for the height of capillary rise in a glass tube.

7.37 By dimensional analysis determine the expression for the shear stress at the wall when an incompressible fluid flows in a pipe under pressure. The significant parameters are velocity of flow V, diameter of pipe D, and viscosity μ and density ρ of the fluid.

CHAPTER

EIGHT

STEADY INCOMPRESSIBLE FLOW IN PRESSURE CONDUITS

In this chapter some of the aspects of steady flow in pressure conduits are discussed. The discussion is limited to *incompressible fluids*, that is, to those for which $\rho \approx$ constant. This includes all liquids. In this chapter isothermal conditions are assumed so as to eliminate thermodynamic effects, some of which are discussed in Chap. 9. Gases flowing with very small pressure changes may be considered *incompressible*, for then $\rho \approx$ constant.

8.1 LAMINAR AND TURBULENT FLOW

If the head loss in a given length of uniform pipe is measured at different values of the velocity, it will be found that, as long as the velocity is low enough to secure laminar flow, the head loss, due to friction, will be directly proportional to the velocity, as shown in Fig. 8.1. But with increasing velocity, at some point B, where visual observation in a transparent tube would show that the flow changes from laminar to turbulent, there will be an abrupt increase in the rate at which the head loss varies. If the logarithms of these two variables are plotted on linear scales or if the values are plotted directly on log-log paper, it will be found that, after a certain transition region has been passed, lines will be obtained with slopes ranging from about 1.75 to 2.00.

It is thus seen that for laminar flow the drop in energy due to friction varies as V, while for turbulent flow the friction varies as V^n, where n ranges from about 1.75 to 2. The lower value of 1.75 for turbulent flow is found for pipes with very smooth walls; as the wall roughness increases, the value of n increases up to its maximum value of 2.

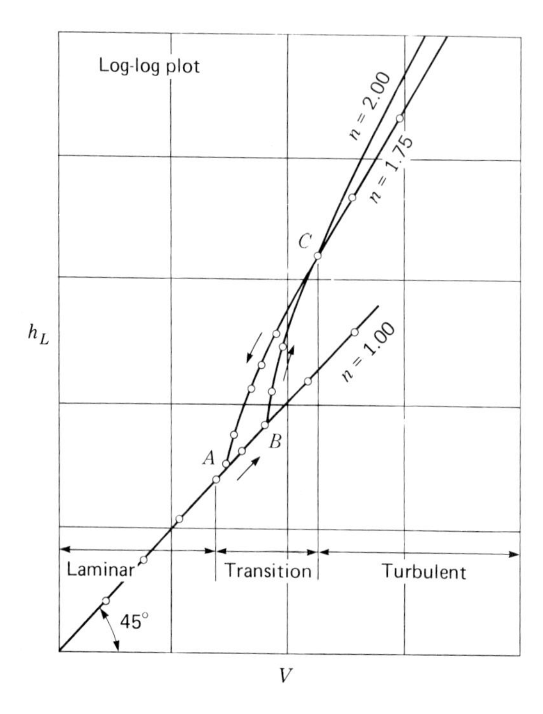

Figure 8.1 Log-log plot for flow in a uniform pipe. ($n = 2.00$, rough-wall pipe; $n = 1.75$ smooth-wall pipe.)

The points in Fig. 8.1 were plotted directly from Osborne Reynolds' measurements and show decided curves in the transition zone where values of n are even greater than 2. If the velocity is gradually reduced from a high value, the line BC will not be retraced. Instead, the points lie along curve CA. Point B is known as the *higher critical point*, and A as the *lower critical point*.

However, velocity is not the only factor that determines whether the flow is laminar or turbulent. The criterion is Reynolds number, which has been discussed in Sec. 7.4. For a circular pipe the significant linear dimension L is usually taken as the diameter D, and thus

$$\mathbf{R} = \frac{DV\rho}{\mu} = \frac{DV}{\nu} \tag{8.1}$$

where any consistent system of units may be used, since $\mathbf{R}$ is a dimensionless number.[1]

[1] It is sometimes convenient to use a "hybrid" set of units and compensate with a correction factor. Thus by substituting $V = Q/A$ and $V = G/\gamma A$ into Eq. (8.1), we get

$$\mathbf{R} = 1.27Q/\nu D = 1.27G/\mu g D,$$

where Q and G are defined in the Notation in the front of the book. The last form is especially convenient in the case of gases; it shows that in a pipe of uniform diameter the Reynolds number is constant along the pipe, even for a compressible fluid where the density and velocity vary, if there is no appreciable variation in temperature to alter the viscosity of the gas.

8.2 CRITICAL REYNOLDS NUMBER

The upper critical Reynolds number, corresponding to point B of Fig. 8.1, is really indeterminate and depends upon the care taken to prevent any initial disturbance from affecting the flow. Its value is normally about 4,000, but laminar flow in circular pipes has been maintained up to values of $\mathbf{R}$ as high as 50,000. However, in such cases this type of flow is inherently unstable, and the least disturbance will transform it instantly into turbulent flow. On the other hand, it is practically impossible for turbulent flow in a straight pipe to persist at values of $\mathbf{R}$ much below 2,000, because any turbulence that is set up will be damped out by viscous friction. This lower value is thus much more definite than the higher one and is really the dividing point between the two types of flow. Hence this lower value will be defined as the *true critical Reynolds number*. However, this lower critical value is subject to slight variations. Its value will be higher in a converging pipe and lower in a diverging pipe than in a straight pipe. Also, its value will be less for flow in a curved pipe than in a straight one, and even for a straight uniform pipe its value may be as low as 1,000, where there is an excessive degree of roughness. However, for normal cases of flow in straight pipes of uniform diameter and usual roughness, the critical value may be taken as $\mathbf{R} = 2{,}000$.

For water at 75°F the kinematic viscosity is 1.00×10^{-5} ft²/s, and for this case the critical Reynolds number is obtained when

$$DV_{\text{crit}} = \mathbf{R}\nu = 2{,}000 \times 10^{-5}\ \text{ft}^2/\text{s} = 0.020\ \text{ft}^2/\text{s}.$$

Thus, for a pipe 1 in (25 mm) in diameter,

$$V_{\text{crit}} = 0.02/(1/12) = 0.24\ \text{fps}\ (0.073\ \text{m/s})$$

Or if the velocity were 2.4 fps (0.73 m/s) the diameter would be only 0.1 in (2.5 mm). Velocities or pipe diameters as small as these are not often encountered with water flowing in practical engineering, though they may be found in certain laboratory instruments. Hence, for such fluids as water and air, practically all cases of engineering importance are in the turbulent-flow region. But if the fluid is a viscous oil, laminar flow is often encountered.

Illustrative Example 8.1 An oil ($s = 0.85$, $\nu = 1.8 \times 10^{-5}$ m²/s) flows in a 10-cm-diameter pipe at 0.50 L/s. Is the flow laminar or turbulent?

$$V = \frac{Q}{A} = \frac{500\ \text{cm}^3/\text{s}}{\pi(10)^2\ \text{cm}^2/4} = 6.37\ \text{cm/s} = 0.0637\ \text{m/s}$$

$$\mathbf{R} = \frac{DV}{\nu} = \frac{0.10\ \text{m}(0.0637\ \text{m/s})}{1.8 \times 10^{-5}\ \text{m}^2/\text{s}} = 354$$

Since $\mathbf{R} < 2{,}000$, the flow is laminar.

8.3 HYDRAULIC RADIUS

For conduits having noncircular cross sections, some value other than the diameter must be used for the linear dimension in the Reynolds number. Such a characteristic is the *hydraulic radius*, defined as

$$R_h = \frac{A}{P} \tag{8.2}$$

where A is the cross-sectional area of the flowing fluid, and P is the *wetted perimeter*, that portion of the perimeter of the cross section where there is contact between fluid and solid boundary. For a circular pipe flowing full, $R_h = \pi r^2/2\pi r = r/2$, or $D/4$. Thus R_h is not the radius of the pipe, and hence the term "radius" is misleading. If a circular pipe is exactly half full, both the area and the wetted perimeter are half the preceding values; so R_h is $r/2$, the same as if it were full. But if the depth of flow in a circular pipe is 0.8 times the diameter, for example, $A = 0.674D^2$ and $P = 2.21D$, then $R_h = 0.304D$, or $0.608r$.

The hydraulic radius is a convenient means for expressing the shape as well as the size of a conduit, since for the same cross-sectional area the value of R_h will vary with the shape.

In evaluating the Reynolds number for flow in a noncircular conduit (Sec. 8.13) it is customary to substitute $4R_h$ for D in Eq. (8.1).

8.4 GENERAL EQUATION FOR CONDUIT FRICTION

The following discussion applies to either laminar or turbulent flow and to any shape of cross section.

Consider steady flow in a conduit[1] of uniform cross section A (Fig. 8.2). The pressures at sections 1 and 2 are p_1 and p_2, respectively. The distance between

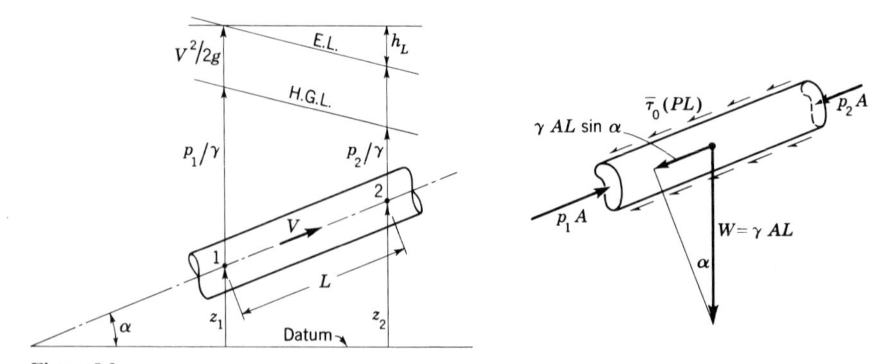

Figure 8.2

[1] This conduit can have any shape of cross section; it need not be circular.

sections is L. For equilibrium in steady flow, the summation of forces acting on any fluid element must be equal to zero (i.e., $\sum F = ma = 0$). Thus, in the direction of flow,

$$p_1 A - p_2 A - \gamma L A \sin \alpha - \bar{\tau}_0(PL) = 0 \tag{8.3}$$

where $\bar{\tau}_0$, the *average shear stress* (average shear force per unit area) at the conduit wall, is defined by

$$\bar{\tau}_0 = \frac{\int_0^P \tau_0 \, dP}{P} \tag{8.4}$$

in which τ_0 is the local shear stress[1] acting over a small incremental portion dP of the wetted perimeter.

Noting that $\sin \alpha = (z_2 - z_1)/L$ and dividing each term in Eq. (8.3) by γA gives

$$\frac{p_1}{\gamma} - \frac{p_2}{\gamma} - z_2 + z_1 = \bar{\tau}_0 \frac{PL}{\gamma A} \tag{8.5}$$

From the left-hand sketch of Fig. 8.2 it can be seen that

$$h_L = (z_1 + p_1/\gamma) - (z_2 + p_2/\gamma)$$

Substituting h_L for the right side of this expression and R_h for A/P in Eq. (8.5) we get,

$$h_L = \bar{\tau}_0 \frac{L}{R_h \gamma} \tag{8.6}$$

This equation is applicable to any shape of uniform cross section regardless of whether the flow is laminar or turbulent.

For a smooth-walled conduit, where wall roughness (discussed in Sec. 8.9) may be neglected, it might be assumed that the average fluid shear stress $\bar{\tau}_0$ at the wall is some function of ρ, μ, V and some characteristic linear dimension, which will here be taken as the hydraulic radius R_h. Thus

$$\bar{\tau}_0 = K R_h^a \rho^b \mu^c V^n \tag{8.7}$$

where K is a dimensionless number. Substituting in Eq. (8.7) dimensional values of F, L, and T for force, length, and time, we get[2]

$$FL^{-2} = KL^a(FL^{-4}T^2)^b(FL^{-2}T)^c(LT^{-1})^n$$

[1] The local shear stress varies from point to point around the perimeter of all conduits (irrespective of whether the wall is smooth or rough) except for the case of a circular pipe flowing full where the shear stress at the wall is the same at all points of the perimeter.

[2] Here we are using the *FLT* system, while in Chap. 7 the *MLT* system was used. It makes no difference which system is used since the results are the same.

As the dimensions on the two sides of the equation must be the same,

For F: $$1 = b + c$$

For L: $$-2 = a - 4b - 2c + n$$

For T: $$0 = 2b + c + n$$

The solution of these three simultaneous expressions in terms of n is $a = n - 2$, $b = n - 1$, $c = 2 - n$.

Inserting these values of the exponents in Eq. (8.7), the result is

$$\bar{\tau}_0 = KR_h^{n-2}\rho^{n-1}\mu^{2-n}V^n \tag{8.8}$$

This may be rearranged as

$$\bar{\tau}_0 = K\left(\frac{R_h V\rho}{\mu}\right)^{n-2}\rho V^2 = 2K\mathbf{R}^{n-2}\rho\frac{V^2}{2} \tag{8.9}$$

for $R_h V\rho/\mu$ is a Reynolds number with R_h as the characteristic length.

Grouping the dimensionless terms on the right side of Eq. (8.9) into a single term C_f, we get

$$C_f = 2K\mathbf{R}^{n-2} \tag{8.10}$$

Hence $$\bar{\tau}_0 = C_f\rho\frac{V^2}{2} \tag{8.11}$$

Inserting this value of $\bar{\tau}_0$ in Eq. (8.6) and noting that $\gamma = \rho g$,

$$h_L = C_f\frac{L}{R_h}\frac{V^2}{2g} \tag{8.12}$$

which may be applied to any shape of smooth-walled cross section. Later it will be shown (Sec. 8.11) that this equation also applies to rough-walled conduits.

8.5 PIPES OF CIRCULAR CROSS SECTION

In Sec. 8.3 it is shown that for a circular pipe flowing full $R_h = D/4$. Substituting this value in Eq. (8.12), the result (for both smooth-walled and rough-walled conduits) is

$$h_L = f\frac{L}{D}\frac{V^2}{2g} \tag{8.13}$$

where $$f = 4C_f = 8K\mathbf{R}^{n-2} \tag{8.14}$$

Equation (8.13) is known as the *pipe-friction equation*, and is also commonly referred to as the Darcy-Weisbach equation.[1] Like the coefficient C_f, the friction factor f is dimensionless and is also some function of Reynolds number. Much

[1] In a slightly different form where D is replaced by the hydraulic radius R_h, Eq. (8.13) is known as the Fanning equation which is widely used by chemical engineers.

research has been directed toward determining the way in which f varies with **R** and also with pipe roughness. The pipe-friction equation expresses the fact that the head lost in friction in a given pipe can be expressed in terms of the velocity head. The equation is dimensionally homogeneous and may be used with any consistent system of units.

Dimensional analysis gives us the proper form for an equation but does not yield a numerical result since it is not concerned with abstract numerical factors. Hence it shows in Eq. (8.8) that whatever the value of the exponent of V, the exponents of all the other quantities involved are then determined. It also shows that Eq. (8.13) is a rational expression for pipe friction. But the numerical values of such quantities as K, n, and f must be determined by experiment or other means.

For a circular pipe flowing full, Eq. (8.6) may be written as

$$h_L = \bar{\tau}_0 \frac{L}{R_h \gamma} = \frac{\tau_0 2L}{r_0 \gamma} \tag{8.15}$$

where the shear stress at the wall, $\tau_0 = \bar{\tau}_0$, because of symmetry, and $R_h = r_0/2$ where r_0 is the radius of the pipe.

Following a development similar to that of Eqs. (8.3) to (8.6) and noting for a circle that $A = \pi r^2$ and $P = 2\pi r$, it can be shown that for a cylindrical body of fluid concentric to the pipe, $h_L = \tau 2L/r\gamma$ where τ is the shear stress in the fluid at radius r. Relating this to Eq. (8.15) it follows that the shear stress in the flow in a circular pipe at any radius r is

$$\tau = \tau_0 \frac{r}{r_0} \tag{8.16}$$

or the shear stress is zero at the center of the pipe and increases linearly with the radius to a maximum value τ_0 at the wall as in Fig. 8.3. This is true regardless of whether the flow is laminar or turbulent.

From Eqs. (8.6) and (8.13) and substituting $R_h = D/4$ for a circular pipe, we obtain

$$\tau_0 = \frac{f}{4} \rho \frac{V^2}{2} = \frac{f}{4} \gamma \frac{V^2}{2g} \tag{8.17}$$

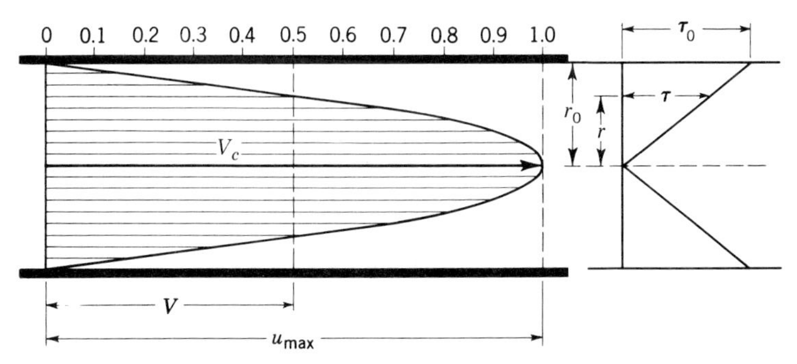

Figure 8.3 Velocity profile in laminar flow and distribution of shear stress.

With this equation, τ_0 for flow in a circular pipe may be computed for any experimentally determined value of f.

8.6 LAMINAR FLOW IN CIRCULAR PIPES

In Sec. 1.11 it was noted that for laminar flow $\tau = \mu\, du/dy$, where u is the value of the velocity at a distance y from the boundary. As $y = r_0 - r$, it is also seen that $\tau = -\mu\, du/dr$; in other words, the minus sign indicates that u decreases as r increases. The coefficient of viscosity μ is a constant for any particular fluid at a constant temperature, and therefore if the shear varies from zero at the center of the pipe to a maximum at the wall, it follows that the velocity profile must have a zero slope at the center and have a continuously steeper velocity gradient as the wall is approached.

To determine the velocity profile for laminar flow in a circular pipe the expression $\tau = \mu\, du/dy$ is substituted into the expression $h_L = \tau 2L/r\gamma$.

Thus
$$h_L = \frac{\tau 2L}{r\gamma} = \mu \frac{du}{dy}\frac{2L}{r\gamma} = -\mu \frac{du}{dr}\frac{2L}{r\gamma}$$

From this
$$du = -\frac{h_L \gamma}{2\mu L} r\, dr$$

Integrating and determining the constant of integration from the fact that $u = u_{max}$ when $r = 0$, we obtain

$$u = u_{max} - \frac{h_L \gamma}{4\mu L} r^2 = u_{max} - kr^2 \tag{8.18}$$

From this equation it is seen that the velocity profile is a parabola, as shown in Fig. 8.3. Note that $k = h_r\gamma/4\mu L$.

Substituting the boundary condition that $u = 0$ when $r = r_0$ into the second expression of Eq. (8.18) and noting that $u_{max} = V_c$, the centerline velocity, we find $k = V_c/r_0^2$. Thus Eq. (8.18) can be expressed as

$$u = V_c - (V_c/r_0^2)r^2 = V_c(1 - r^2/r_0^2) \tag{8.19}$$

Combining Eqs. (8.18) and (8.19) we get an expression for V_c as follows

$$V_c = u_{max} = \frac{h_L \gamma}{4\mu L} r_0^2 = \frac{h_L \gamma}{16\mu L} D^2 \tag{8.20}$$

Equation (8.18) may be multiplied by a differential area $dA = 2\pi r\, dr$ and the product integrated from $r = 0$ to $r = r_0$ to find the rate of discharge. As in previous cases, the rate of discharge is equivalent to the volume of a solid bounded by the velocity profile. In this case the solid is a paraboloid with a maximum height of u_{max}. The mean height of a paraboloid is one-half the maximum height, and hence the mean velocity V is $0.5u_{max}$. Thus

$$V = \frac{h_L \gamma}{32\mu L} D^2 \tag{8.21}$$

From this last equation, noting that $\gamma = g\rho$ and $\mu/\rho = \nu$, the loss of head in friction is given by

$$h_L = 32\frac{\mu}{\gamma}\frac{L}{D^2}V = 32\nu\frac{L}{gD^2}V \tag{8.22}$$

which is the Hagen-Poiseuille law for laminar flow in tubes. Hagen, a German engineer, experimented with water flowing through small brass tubes and published his results in 1839. Poiseuille, a French scientist, experimented with water flowing through capillary tubes in order to determine the laws of flow of blood through the veins of the body and published his studies in 1840.

From Eq. (8.22) it is seen that in laminar flow the loss of head is proportional to the first power of the velocity. This is verified by experiment, as shown in Fig. 8.1. The striking feature of this equation is that it involves no empirical coefficients or experimental factors of any kind, except for the physical properties of the fluid such as viscosity and density (or specific weight). From this it would appear that in laminar flow the friction is independent of the roughness of the pipe wall. That this is true is also borne out by experiment.

Dimensional analysis shows that the friction loss may also be expressed by Eq. (8.13). Equating (8.13) and (8.22) and solving for the friction factor f, we obtain for laminar flow under pressure in a circular pipe,

$$f = \frac{64\nu}{DV} = \frac{64}{\mathbf{R}} \tag{8.23}$$

Hence, if $\mathbf{R}$ is less than 2,000, we may use Eq. (8.22) to find pipe friction head loss or we may use Eq. (8.13) with the value of f as given by Eq. (8.23).

8.7 ENTRANCE CONDITIONS IN LAMINAR FLOW

In the case of a pipe leading from a reservoir, if the entrance is rounded so as to avoid any initial disturbance of the entering stream, all particles will start to flow with the same velocity, except for a very thin film in contact with the wall. Particles next to the wall have a zero velocity, but the velocity gradient is here extremely steep, and with this slight exception, the velocity is uniform across the diameter, as shown in Fig. 8.4. As the fluid progresses along the pipe, the streamlines in the vicinity of the wall are slowed down by friction emanating from the wall, but as Q is constant for successive sections, the velocity in the center must be accelerated, until the final velocity profile is a parabola, as shown in Fig. 8.3. Theoretically, an infinite distance is required for this, but it has been established both by theory and by observation that the maximum velocity in the center of the pipe will reach 99 per cent of its ultimate value in the distance $L' = 0.058\mathbf{R}D$.[1] Thus, for the critical value $\mathbf{R} = 2{,}000$, the distance L' of Fig. 8.4 equals 116 pipe

[1] H. L. Langhaar, Steady Flow in the Transition Length of a Straight Tube, *J. Appl. Mech.*, vol. 10, p. 55, 1942.

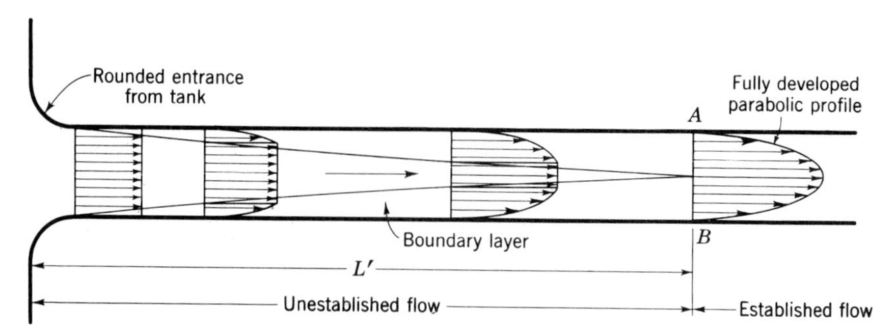

Figure 8.4 Velocity profiles along a pipe in laminar flow.

diameters. In other cases of laminar flow with Reynolds numbers less than 2,000, the distance L' will be correspondingly less in accordance with the expression $L' = 0.058\mathbf{R}D$.

In the entry region of length L' the flow is *unestablished*; that is, the velocity profile is changing. In this region the flow can be visualized as consisting of a central core in which there are no frictional effects and an annular zone extending from the core outward to the pipe wall. This outer zone increases in thickness as it moves along the wall and is known as the *boundary layer*. Viscosity in the boundary layer acts to transmit the effect of boundary shear inwardly into the flow. At section AB the boundary layer has grown until it occupies the entire section of the pipe. At this point, for laminar flow, the velocity profile is a perfect parabola. Beyond section AB the velocity profile does not change, and the flow is known as *established flow*.

As shown in Prob. 4.1 for a circular pipe, the kinetic energy of a stream with a parabolic velocity profile is $2V^2/2g$, where V is the mean velocity. At the entrance to the pipe the velocity is uniformly V across the diameter, except for an extremely thin layer next to the wall. Thus, at the entrance to the pipe, the kinetic energy per unit weight is practically $V^2/2g$. Hence, in the distance L', there is a continuous increase in kinetic energy accompanied by a corresponding decrease in pressure head. Therefore, at a distance L' from the entrance, the piezometric head is less than the static value in the reservoir by $2V^2/2g$ plus the friction loss in this distance.

Laminar flow has been dealt with rather fully, not merely because it is of importance in problems involving fluids of very high viscosity, but especially because it permits a simple and accurate rational analysis. The general approach used here is of some assistance in the study of turbulent flow, where conditions are so complex that rigid mathematical treatment is impossible.

Illustrative Example 8.2 For the case of Illustrative Example 8.1 find the centerline velocity, the velocity at $r = 2$ cm, the friction factor, the shear stress at the pipe wall, and the head loss per meter of pipe length.

Since the flow is laminar,

$$V_c = 2V = 12.7 \text{ cm/s}$$

$$u = u_{\max} - kr^2 \qquad u_{\max} = V_c = 12.7 \text{ cm/s}$$

When $r = r_0 = 5$ cm, $u = 0$, hence $0 = 12.7 - k(5)^2$

$$k = 0.51/(\text{cm}\cdot\text{s}) \qquad u_{2\,\text{cm}} = 12.7 - 0.51(2)^2 = 10.7\ \text{cm/s}$$

$$f = \frac{64}{\mathbf{R}} = \frac{64}{354} = 0.18$$

$$\tau_0 = \frac{f}{4}\rho\frac{V^2}{2} = \frac{0.18}{4}(0.85\ \text{g/cm}^3)\frac{(6.37\ \text{cm/s})^2}{2}$$

$$\tau_0 = 0.77\frac{g}{(\text{cm}\cdot\text{s}^2)}\frac{\text{N}\cdot\text{s}^2}{\text{kg}\cdot\text{m}}\frac{100\ \text{cm}}{\text{m}} = 0.077\ \text{N/m}^2$$

$$h_L/L = f\frac{1}{D}\frac{V^2}{2g} = 0.18\frac{1}{0.10\ \text{m}}\frac{(0.0637\ \text{m/s})^2}{2(9.81\ \text{m/s}^2)} = 0.00037\ \text{m/m}$$

8.8 TURBULENT FLOW

In Sec. 3.1 it was explained that in laminar flow the fluid particles move in straight lines while in turbulent flow they follow random paths. Consider the case of laminar flow as shown in Figs. 8.5*a* and 8.5*b* where the velocity u increases with y. Even though the fluid particles are moving horizontally to the right, because of molecular motion, molecules will cross line *ab* and will thereby transport momentum. On the average, the velocities of the molecules in the slower moving fluid below the line will be less than those of the faster moving fluid above; the result is that the molecules which cross from below tend to slow down the faster moving fluid. Likewise, the molecules which cross the line *ab* from above tend to speed up the slower moving fluid below. The result is the production of a shear stress along the surface whose trace is *ab*, the value of which is given in Sec. 1.11 as $\tau = \mu\, du/dy$. This equation is applicable to laminar flow only.

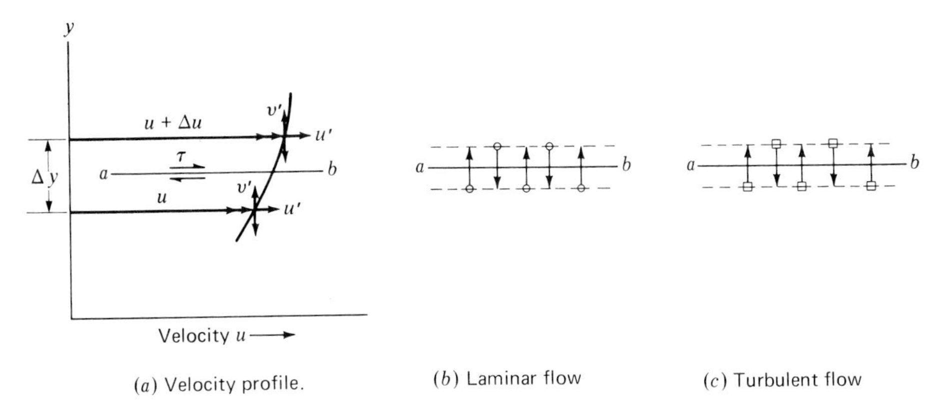

Figure 8.5 (*a*) Velocity profile. (*b*) Laminar flow (transfer of molecules across *ab*). (c) Turbulent flow (transfer of finite fluid masses across *ab*).

Let us examine some of the characteristics of turbulent flow to see how it differs from laminar flow. In turbulent flow the velocity at a point in the flow field fluctuates in both magnitude and direction.[1] As a consequence a multitude of small eddies are created by the viscous shear between adjacent particles. These eddies grow in size and then disappear as their particles merge into adjacent eddies. Thus there is a continuous mixing of particles, with a consequent transfer of momentum.

First Expression

In the modern conception of turbulent flow, a mechanism similar to that described in the foregoing for laminar flow is assumed. However, the molecules are replaced by minute but finite masses (Fig. 8.5*c*). Hence, by analogy, the shear stress along the plane through *ab* in Fig. 8.5 may be defined in the case of turbulent flow as

$$\text{Turbulent shear stress} = \eta \frac{du}{dy} \tag{8.24}$$

But unlike μ, the *eddy-viscosity* η is not a constant for a given fluid at a given temperature, but depends upon the turbulence of the flow. It may be viewed as a coefficient of momentum transfer, expressing the transfer of momentum from points where the velocity is low to points where it is higher, and vice versa. Its magnitude may range from zero to many thousand times the value of μ. However, its numerical value is of less interest than its physical concept. In dealing with turbulent flow it is sometimes convenient to use *kinematic eddy viscosity* $\varepsilon = \eta/\rho$ which is a property of the flow alone, analogous to kinematic viscosity.

In general, the total shear stress in turbulent flow is the sum of the laminar shear stress plus the turbulent shear stress, i.e.,

$$\tau = \mu \frac{du}{dy} + \eta \frac{du}{dy} = \rho(\nu + \varepsilon) \frac{du}{dy} \tag{8.25}$$

In turbulent flow the second term of this equation is usually many times larger than the first term.

In turbulent flow the local axial velocity has been shown, in Sec. 3.4 (see Fig. 3.6), to have fluctuations of plus and minus u', and there are also fluctuations of plus and minus v' and w' normal to u as shown in Fig. 8.6*b*. As it is obvious that there can be no values of v' next to and perpendicular to a smooth wall, turbulent flow cannot exist there. Hence, near a smooth wall, the shear is due to laminar flow alone and $\tau = \mu\, du/dy$. It should be noted that the shear stress always acts to cause the velocity distribution to become more uniform.

[1] The velocity at a point in a so-called "steady" turbulent flow can be best visualized as a vector that fluctuates in both direction and magnitude. The mean temporal velocity at that point corresponds to the "average" of those vectors.

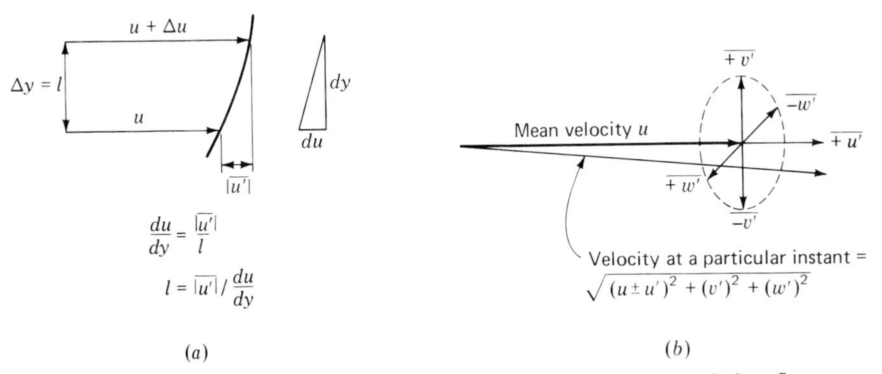

Figure 8.6 (*a*) Prandtl's mixing length *l*. (*b*) Instantaneous local velocity in turbulent flow.

At some distance from the wall, such as $0.2r$, the value of du/dy becomes small in turbulent flow, and hence the viscous shear becomes negligible in comparison with the turbulent shear. The latter can be large, even though du/dy is small, because of the possibility of η being very large. This is due to the great turbulence that may exist at an appreciable distance from the wall. But at the center of the pipe, where du/dy is zero, there can be no shear at all. Hence, in turbulent flow as well as in laminar flow, the shear stress is a maximum at the wall and decreases linearly to zero at the axis, as shown in Fig. 8.3 and proved in Sec. 8.5.

Second Expression

Another expression for turbulent shear stress may be obtained which is different from that in Eq. (8.24). Thus in Fig. 8.5*a*, if a mass m of fluid below *ab*, where the temporal mean axial velocity is u, moves upward into a zone where the temporal mean axial velocity is $u + \Delta u$, its initial momentum in the axial direction must be increased by $m\,\Delta u$. Conversely, a mass m which moves from the upper zone to the lower will have its axial momentum decreased by $m\,\Delta u$. Hence this transfer of momentum back and forth across *ab* will produce a shear in the plane through *ab* proportional to Δu. This shear is possible only because of the velocity profile shown. If the latter were vertical, Δu would be zero and there could be no shear.

If the distance Δy in Fig. 8.5*a* is so chosen that the average value of $+u'$ in the upper zone over a time period long enough to include many velocity fluctuations is equal to Δu, i.e., $\Delta u = \overline{|u'|}$, the two streams will be separated by what is known as the *Prandtl mixing length* l, which will be referred to later. Consider, over a short time interval, a mass moving upward from below *ab* with a velocity v'; it will transport into the upper zone, where the velocity is $u + u'$, a momentum per unit time which is on the average equal to $\rho(v'\,dA)(u)$. The slower moving mass from below *ab* will tend to retard the flow above *ab*; this creates a shear force along the plane of *ab*. This force can be found by applying the momentum principle [Eq. (6.6)], $F = \tau\, dA = \rho Q(\Delta V) = \rho(v'\, dA)(u + u' - u) = \rho u'v'\, dA$. Thus, over a time period of sufficient length to permit a large number of velocity fluctuations, the shear stress given by

$$\tau = F/dA = -\rho\overline{u'v'} \tag{8.26}$$

where $\overline{u'v'}$ is the temporal average of the product of u' and v'. This is an alternate form for Eq. (8.24), and in modern turbulence theory $-\rho\overline{u'v'}$ is referred to as the *Reynolds stress.*

The minus sign appears in Eq. (8.26) because the product $\overline{u'v'}$ on the average is negative. By inspecting Fig. 8.5*a* it can be seen that $+v'$ is associated with $-u'$ values more than with $+u'$ values. The opposite is true for $-v'$. Even though the temporal mean values of u' and v' are individually equal to zero, the temporal mean value of their product is not zero. This is so because combinations of $+v'$ and $-u'$ and of $-v'$ and $+u'$ predominate over combinations of $+v'$ and $+u'$ and $-v'$ and $-u'$ respectively.

Prandtl reasoned that in any turbulent flow $\overline{|u'|}$ and $\overline{|v'|}$ must be proportional to each other and of the same order of magnitude. He also introduced the concept of mixing length l, which is defined as the distance one must move transversely to the direction of flow such that $\Delta u = \overline{|u'|}$. From Fig. 8.6*a* it can be seen that $\Delta u = l\, du/dy$ and hence $\overline{|u'|} = l\, du/dy$. If $\overline{|u'|} \propto \overline{|v'|}$ and if one permits l to account for the constant of proportionality, Prandtl[1] has shown that $-\overline{u'v'}$ varies as $l^2(du/dy)^2$. Thus

$$\tau = -\rho\overline{u'v'} = \rho l^2\left(\frac{du}{dy}\right)^2 \tag{8.27}$$

This equation expresses terms that can be measured. Thus in any experiment where the pipe friction is determined, τ_0 can be computed by Eq. (8.6), and τ at any radius is then found by Eq. (8.16). A traverse of the velocity across a pipe diameter will give u at any radius, and the velocity profile will give du/dy at any radius. Thus Eq. (8.27) enables the Prandtl mixing length l to be found as a function of the pipe radius. The purpose of all of this is to enable us to develop theoretical equations for the velocity profile in turbulent flow, and from this in turn to develop theoretical equations for f, the friction coefficient.

8.9 VISCOUS SUBLAYER IN TURBULENT FLOW

In Fig. 8.4 it is shown that, for laminar flow, if the fluid enters with no initial disturbance, the velocity is uniform across the diameter except for an exceedingly thin film at the wall, inasmuch as the velocity next to any wall is zero. But as flow proceeds down the pipe, the velocity profile changes because of the growth of a laminar boundary layer which continues until the boundary layers from

[1] Hermann Schlichting: "Boundary Layer Theory", 4th ed., McGraw-Hill Book Co., 1960, p. 479.

opposite sides meet at the pipe axis and then there is fully developed laminar flow.

If the Reynolds number is above the critical value, so that the flow is turbulent, the initial condition is much like that in Fig. 8.4. But as the laminar boundary layer increases in thickness, a point is soon reached where a transition occurs and the boundary layer becomes turbulent. This turbulent boundary layer generally increases in thickness much more rapidly, and soon the two from opposite sides meet at the pipe axis, and there is then fully developed turbulent flow.

This initial laminar boundary layer may be given a Reynolds number such as $\mathbf{R}_x = Ux/\nu$, where U is the uniform velocity and x is the distance measured from the initial point. When x has such a value that this $\mathbf{R}_x$ is about 500,000, the transition occurs to the turbulent boundary layer. Fully developed turbulent flow will be found at about 50 pipe diameters from the pipe entrance for a smooth pipe with no special disturbance at entrance; otherwise the turbulent boundary layers from the two sides will meet within a shorter distance. It is this fully developed turbulent flow that we shall consider in all that follows.

There can be no turbulence next to a smooth wall since it is impossible for v' to have any value at a solid boundary. Therefore immediately adjacent to a smooth wall there will be a laminar or *viscous* sublayer, as shown in Fig. 8.7, within which the shear is due to viscosity alone. This viscous sublayer is extremely thin, usually only a few hundredths of a millimeter, but its effect is great because of the very steep velocity gradient within it and because $\tau = \mu\, du/dy$ in that region. At a distance from the wall the viscous effect becomes negligible, but the turbulent shear is then large. Between the two there must be a transition zone where both types of shear are significant. It is evident that there can be no sharp lines of demarcation separating these three zones, inasmuch as one must merge gradually into the other.

By plotting a velocity profile from the wall on the assumption that the flow is entirely laminar (Sec. 8.6) and plotting another velocity profile on the assumption that the flow is entirely turbulent (Sec. 8.10), the two will intersect, as shown in Fig. 8.8. It is obvious that there can be no abrupt change in profile at this point

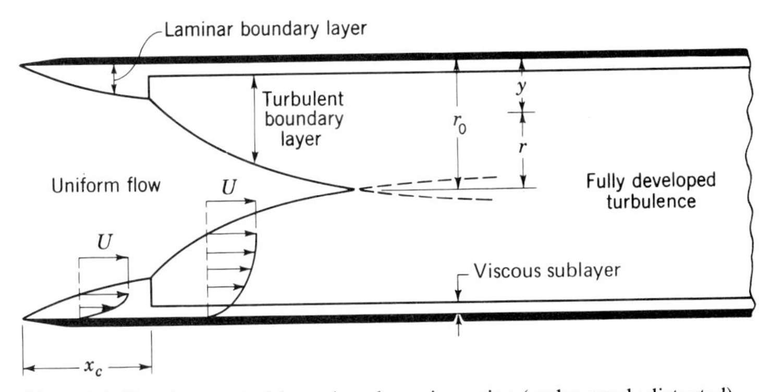

Figure 8.7 Development of boundary layer in a pipe (scales much distorted).

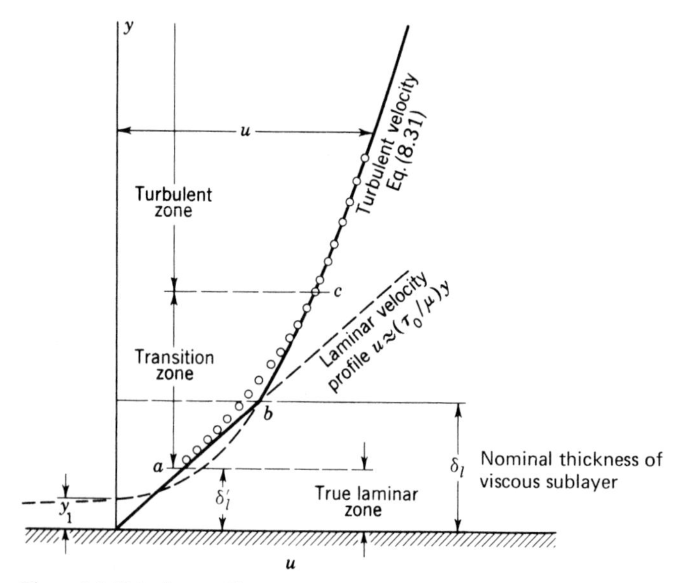

Figure 8.8 Velocity profile near a solid wall (vertical scale greatly exaggerated).

of intersection, but that one curve must merge gradually into the other, as shown by the experimental points.

Any value taken for the thickness of this viscous sublayer must be purely arbitrary. The simultaneous solution of the equations for the two curves, together with some experimental factors, will give the value of y for point b as follows

$$\delta_l = 11.6 \frac{\nu}{\sqrt{\tau_0/\rho}} \tag{8.28}$$

where δ_l is referred to as the *nominal thickness of the viscous sublayer*. The transition curve ac determined by measurements indicates that a is a better limit of the viscous-sublayer thickness. Present information is that the thickness of the viscous sublayer out to point a is approximately

$$\delta_l' = 3.5 \frac{\nu}{\sqrt{\tau_0/\rho}} \tag{8.29}$$

In a circular pipe the laminar velocity profile has been shown to be a parabola, but in this extremely thin region near the wall it can scarcely be distinguished from a straight line.

The transition zone may be said to extend from a to c in Fig. 8.8. For the latter point the value of y has been estimated to be about $70\nu/\sqrt{\tau_0/\rho}$. Beyond this the flow is so turbulent that viscous shear is negligible.

From Eq. (8.15)

$$\sqrt{\frac{\tau_0}{\rho}} = \sqrt{\frac{fV^2}{8}}$$

and making this substitution in Eq. (8.28), we obtain

$$\delta_l = \frac{32.8\nu}{V\sqrt{f}} \tag{8.30}$$

from which it is seen that the higher the velocity or the lower the kinematic viscosity, the thinner the viscous sublayer. Thus, for a given constant pipe diameter, the thickness of the viscous sublayer decreases as the Reynolds number increases.

It is now in order to discuss what is meant by a smooth wall. There is no such thing in reality as a mathematically smooth surface. But if the irregularities on any actual surface are such that the effects of the projections do not pierce through the viscous sublayer (Fig. 8.8), the surface is *hydraulically smooth* from the fluid-mechanics viewpoint. If the effects of the projections extend beyond the sublayer, the laminar layer is broken up and the surface is no longer hydraulically smooth. To be more specific, in Fig. 8.9 if $\delta_l > 6e$, the pipe will behave as though it is hydraulically smooth, while if $\delta_l < 0.3e$, the pipe will behave as *wholly rough*, the significance of which is discussed in Sec. 8.10. In between these values, i.e., with $6e > \delta_l > 0.3e$ the pipe will behave in a transitional mode; that is, neither hydraulically smooth nor wholly rough.

Inasmuch as the thickness of the viscous sublayer in a given pipe will decrease with an increase in Reynolds number, it is seen that the same pipe may be hydraulically smooth at low Reynolds numbers and rough at high Reynolds numbers. Thus, even a relatively smooth pipe may behave as a rough pipe if the Reynolds number is high enough. It is also apparent that, with increasing Reynolds number, there is a gradual transition from smooth to rough pipe flow. These concepts are depicted schematically in Fig. 8.9, where e is the equivalent height of the roughness projection.

8.10 VELOCITY PROFILE IN TURBULENT FLOW

Prandtl reasoned that turbulent flow in a pipe is strongly influenced by the flow phenomena near the wall. In the vicinity of the wall, $\tau \approx \tau_0$. He assumed that the mixing length l near the wall was proportional to the distance from the wall, that

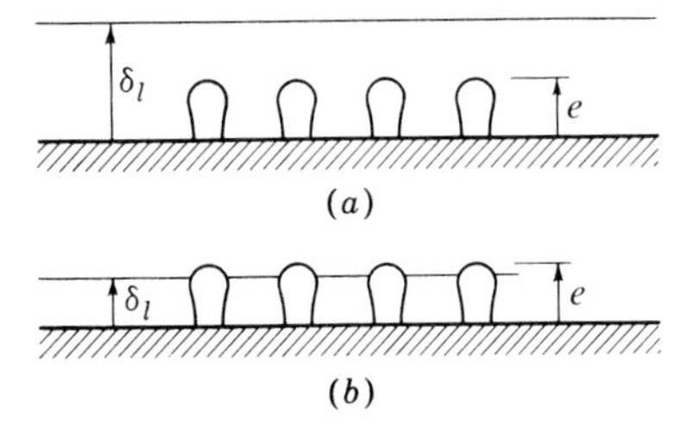

Figure 8.9 Turbulent flow near boundary. (*a*) Relatively low **R**, $\delta_l > e$. If $\delta_l > 6e$ pipe behaves as a smooth pipe. (*b*) Relatively high **R**, $\delta_l < e$. If $\delta_l < 0.3e$ pipe behaves as a wholly rough pipe.

is, $l = Ky$. By experiment it has been determined that K has a value of 0.40.[1] Using these assumptions and applying Eq. (8.27), we get

$$\tau \approx \tau_0 = \rho l^2\left(\frac{du}{dy}\right)^2 = \rho K^2 y^2\left(\frac{du}{dy}\right)^2$$

or

$$du = \frac{1}{K}\sqrt{\frac{\tau_0}{\rho}}\frac{dy}{y}$$

from which

$$u = 2.5\sqrt{\frac{\tau_0}{\rho}}\ln y + C$$

The constant C may be evaluated by noting that $u = u_{max}$ when $y = r_0$. Substituting the expression for C, replacing y by $r_0 - r$, and transforming to log, the equation becomes

$$u = u_{max} - 2.5\sqrt{\frac{\tau_0}{\rho}}\ln\frac{r_0}{r_0 - r} = u_{max} - 5.75\sqrt{\frac{\tau_0}{\rho}}\log\frac{r_0}{r_0 - r} \tag{8.31}$$

Although this equation is derived by assuming certain relations very near to the wall, it has been found to hold practically to the axis of the pipe.

Starting with the derivation of Eq. (8.27), this entire development is open to argument at nearly every step. But the fact remains that Eq. (8.31) agrees very closely with actual measurements of velocity profiles for both smooth and rough pipes. However, there are two zones in which the equation is defective. At the axis of the pipe du/dy must be zero. But Eq. (8.31) is logarithmic and does not have a zero slope at $r = 0$, and hence the equation gives a velocity profile with a sharp point (or cusp) at the axis, whereas in reality it is rounded at the axis. This discrepancy affects only a very small area and involves very slight error in computing the rate of discharge when using Eq. (8.31).

Equation (8.31) is also not applicable very close to the wall. In fact it indicates that when $r = r_0$, the value of u is minus infinity. The equation indicates that $u = 0$, not at the wall, but at a small distance from it, shown as y_1 in Fig. 8.8. However, this discrepancy is well within the confines of the viscous sublayer, where the equation is not supposed to apply. Moreover as the viscous sublayer is very thin, the flow within it has very little effect upon the total rate of discharge.

Hence, although Eq. (8.31) is not perfect, it is reliable except for these two small areas, and thus the rate of discharge may be determined with a high degree of accuracy by using the value of u given by it and integrating over the area of the pipe. Thus

$$Q = \int u\,dA = 2\pi\int_0^{r_0} ur\,dr$$

[1] If the fluid is not clear, i.e., if it is carrying particles in suspension, K will have a value less than 0.40.

Substituting the first expression of Eq. (8.31) for u, integrating and dividing by the pipe area πr_0^2, the mean velocity is[1]

$$V = u_{\max} - 2.5\sqrt{\frac{\tau_0}{\rho}}\left[\ln r_0 - \frac{2}{r_0^2}\int_0^{r_0} r\ln(r_0 - r)\,dr\right]$$

This equation reduces to

$$V = u_{\max} - \frac{3}{2} \times 2.5\sqrt{\frac{\tau_0}{\rho}} = u_{\max} - 1.33V\sqrt{f} \tag{8.32}$$

From this last equation the *pipe factor*, which is the ratio of the mean to the maximum velocity, may be obtained. It is

$$\frac{V}{u_{\max}} = \frac{1}{1 + 1.33\sqrt{f}} \tag{8.33}$$

Using the relation of Eq. (8.33) in Eq. (8.31) and replacing $\sqrt{\tau_0/\rho}$ by $\sqrt{fV^2/8}$, the result is

$$u = (1 + 1.33\sqrt{f})V - 2.04\sqrt{f}\,V\log\frac{r_0}{r_0 - r} \tag{8.34}$$

which enables a velocity profile to be plotted for any mean velocity and any value of f in turbulent flow. In Fig. 8.10 may be seen profiles for both a smooth and a rough pipe plotted from this equation. The only noticeable difference

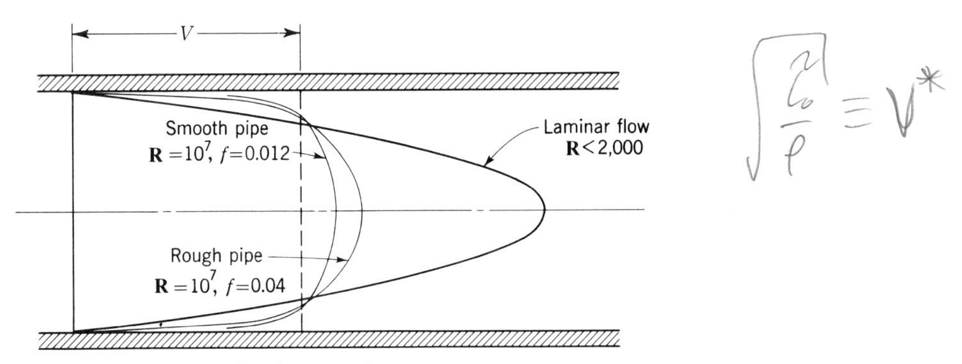

Figure 8.10 Velocity profiles for equal flow rates.

[1] The integral results in indeterminate values at $r = r_0$, as we should expect, inasmuch as the equation for u does not really apply close to the wall. However, these have been shown to reduce to negligible quantities. See B. A. Bakhmeteff, "The Mechanics of Turbulent Flow," p. 70, Princeton University Press, Princeton, N.J., 1941.

between these and measured profiles is that the latter are more rounded at the axis of the pipe.[1]

Comparing the turbulent-flow-velocity profiles with the laminar-flow-velocity profile (Fig. 8.10) shows the turbulent-flow profiles to be much flatter near the central portion of the pipe and steeper near the wall. It is also seen that the turbulent profile for the smooth pipe is flatter near the central section (i.e., blunter) than for the rough pipe. In contrast, the velocity profile in laminar flow is independent of pipe roughness.

As a theoretical equation has now been derived for the velocity profile for turbulent flow in circular pipes, it is also possible to derive equations for the kinetic-energy and momentum-correction factors when mean velocities are used. These equations are[2]

$$\alpha = 1 + 2.7f \tag{8.35a}$$

$$\beta = 1 + 0.98f \tag{8.35b}$$

Illustrative Example 8.3 The head loss in 200 ft of 6-in-diameter pipe is known to be 25 ft·lb/lb when oil ($s = 0.90$) of viscosity 0.0008 lb · s/ft² flows at 2.0 cfs. Determine the centerline velocity, the shear stress at the wall of the pipe, and the velocity at 2 in from the centerline.

The first step is to determine whether the flow is laminar or turbulent.

$$V = \frac{Q}{A} = \frac{2}{0.196} = 10.2 \text{ fps}$$

$$\mathbf{R} = \frac{DV\rho}{\mu} = \frac{0.5(10.2)(0.9 \times 1.94)}{0.0008} = 11{,}200$$

Since $\mathbf{R} > 2{,}000$, the flow is turbulent. Using Eq. (8.12), the friction factor can be found:

$$f = \frac{h_L D(2g)}{LV^2} = \frac{25(0.5)64.4}{200(10.2)^2} = 0.039$$

From Eq. (8.33),

$$u_{\max} = V(1 + 1.33\sqrt{f}) = 12.9 \text{ fps}$$

Equation (8.15) yields

$$\tau_0 = \frac{f\rho V^2}{8} = \frac{0.039(0.9 \times 1.94)(10.2)^2}{8} = 0.89 \text{ lb/ft}^2$$

[1] Although the preceding theory agrees very well with experimental data, it is not absolutely correct throughout the entire range from the axis to the pipe wall, and present indications are that some slight shift in the numerical constants will agree somewhat more closely with test data. Thus, in Eqs. (8.33) and (8.34) the 1.33 may be replaced by 1.44, and in Eq. (8.34), although many writers use 2 instead of 2.04, a better practical value seems to be 2.15.

[2] L. F. Moody, Some Pipe Characteristics of Engineering Interest, *Houille Blanche*, May–June, 1950.

Finally, from Eq. (8.31),

$$u_{2\,\text{in}} = 12.9 - 5.75\sqrt{\frac{\tau_0}{\rho}}\log\frac{3}{1}$$

$$u_{2\,\text{in}} = 12.9 - 1.96 = 10.94 \text{ fps}$$

Note that if the flow had been laminar, the velocity profile would have been parabolic and the centerline velocity would have been twice the average velocity.

8.11 PIPE ROUGHNESS

Unfortunately, there is as yet no scientific way of measuring or specifying the roughness of commercial pipes. Several experimenters have worked with pipes with artificial roughness produced by various means so that the roughness could be measured and described by geometrical factors, and it has been proved that the friction is dependent not only upon the size and shape of the projections, but also upon their distribution or spacing. Much remains to be done before this problem is completely solved.

The most noteworthy efforts in this direction were made by a German engineer Nikuradse, a student of Prandtl's. He coated several different sizes of pipe with sand grains which had been segregated by sieving so as to obtain different sizes of grain of reasonably uniform diameters. The diameters of the sand grains may be represented by e, which is known as the *absolute roughness*. In Sec. 8.4 dimensional analysis of pipe flow showed that for a smooth-walled pipe the friction factor f is a function of Reynolds number. A general approach, including e as a parameter, reveals that $f = \phi(\mathbf{R}, e/D)$. The term e/D is known as the *relative roughness*. In his experimental work Nikuradse had values of e/D ranging from 0.000985 to 0.0333.

In the case of artificial roughness such as this, the roughness is uniform, whereas in commercial pipes it is irregular both in size and in distribution. However, the roughness of commercial pipe may be described by e, which means that the pipe has the same value of f at a high Reynolds number that would be obtained if a smooth pipe were coated with sand grains of uniform size e.

For pipes it has been found that if $\delta_l > 6e$, the viscous sublayer completely submerges the effect of e. Von Kármán, using information from Eq. (8.31) and data from Nikuradse's experiments, developed an equation for friction factor for such a case:

"Smooth-pipe" flow $\delta_l > 6e$:

$$\frac{1}{\sqrt{f}} = 2\log\mathbf{R}\sqrt{f} - 0.8 \tag{8.36}$$

This equation applies to any pipe as long as $\delta_l > 6e$; when this condition prevails, the flow is known as *smooth flow*. The equation has been found to be reliable for smooth pipes for all values of $\mathbf{R}$ over 4,000. For such pipes, i.e., drawn tubing, brass, glass, etc., it can be extrapolated with confidence for values of $\mathbf{R}$ far beyond any present experimental values because it is functionally correct, assuming wall surface so smooth that the effects of the projections do not pierce the viscous sublayer, which becomes increasingly thinner with increasing **R**. That this is so is evident from the fact that the formula yields a value of $f = 0$ for $\mathbf{R} = \infty$. This is in accord with the facts because **R** is infinite for a fluid of zero viscosity, and for such a case f must be zero.

Blasius[1] has shown that for Reynolds numbers between 3,000 and 100,000 the friction factor for a *very smooth pipe* may be expressed approximately as

$$f = \frac{0.316}{\mathbf{R}^{0.25}} \tag{8.37}$$

He also found that over this range of Reynolds numbers the velocity profile in a smooth pipe is closely approximated by the expression

$$\frac{u}{u_{\max}} = \left(\frac{y}{r_0}\right)^{1/7} \tag{8.38}$$

where $y = r_0 - r$, the distance from the pipe wall. This equation is commonly referred to as the *seventh-root law* for turbulent-velocity distribution. Though it is not absolutely accurate, it is useful because it is easy to work with mathematically. At Reynolds numbers above 100,000 a somewhat smaller exponent must be used to give good results.

At high Reynolds numbers δ_l becomes smaller. If $\delta_l < 0.3e$, it has been found that the pipe behaves as a *wholly rough* pipe; i.e., its friction factor is independent of the Reynolds number. For such a case von Kármán found that the friction factor could be expressed as

"Rough-pipe" flow
$\delta_l < 0.3e$:

$$\frac{1}{\sqrt{f}} = 2 \log \frac{D}{e} + 1.14 \tag{8.39}$$

The values of f from this equation correspond to the values from the chart (Fig. 8.11), where the lines become horizontal.

In the gap where $6e > \delta_l > 0.3e$ neither smooth flow [Eq. (8.36)] nor wholly rough flow [Eq. (8.39)] applies. Colebrook[2] found that in this intermediate range an approximate relationship was

Transitional flow
$6e > \delta_l > 0.3e$:

$$\frac{1}{\sqrt{f}} = -2 \log\left(\frac{e/D}{3.7} + \frac{2.51}{\mathbf{R}\sqrt{f}}\right) \tag{8.40}$$

[1] H. Blasius, Das Ähnlichkeitsgesetz bei Reibungsvorgängen in Flüssigkeiten, *Forsch. Gebiete Ingenieurw.*, vol. 131, 1913.

[2] C. F. Colebrook, Turbulent Flow in Pipes, with Particular Reference to the Transition Region between the Smooth and Rough Pipe Laws, *J. Inst. Civil Engrs. (London)*, February, 1939.

8.12 CHART FOR FRICTION FACTOR

As the preceding equations for f are very inconvenient to use, it is preferable to obtain numerical values from a chart,[1] such as Fig. 8.11, prepared by Moody. This chart is based on the best information available and has been plotted with the aid of the equations of the preceding section. As a matter of convenience, values for air and water at 60°F have been placed at the top of the chart to save the necessity of computing Reynolds number for those two typical cases.

The chart shows that there are four zones: laminar flow; a critical range where values are uncertain because the flow might be either laminar or turbulent; a transition zone, where f is a function of both Reynolds number and relative pipe roughness; and a zone of complete turbulence (*rough pipe flow*) where the value of f is independent of Reynolds number and depends solely upon the relative roughness.

There is no sharp line of demarcation between the transition zone and the zone of complete turbulence. The dashed line of Fig. 8.11 that separates the two zones was suggested by R. J. S. Pigott; the equation of this line is $\mathbf{R} = 3500/(e/D)$.

For use with this chart, values of e may be obtained from Table 8.1. As the ratio e/D is dimensionless, any units may be used provided they are the same for

Table 8.1 Values of absolute roughness e for new pipes

	Feet	Millimeters
Drawn tubing, brass, lead, glass, centrifugally spun cement, bituminous lining, transite	0.000005	0.0015
Commercial steel or wrought iron	0.00015	0.046
Welded-steel pipe	0.00015	0.046
Asphalt-dipped cast iron	0.0004	0.12
Galvanized iron	0.0005	0.15
Cast iron, average	0.00085	0.25
Wood stave	0.0006 to 0.003	0.18 to 0.9
Concrete	0.001 to 0.01	0.3 to 3
Riveted steel	0.003 to 0.03	0.9 to 9

Note: $\frac{e}{D} = \frac{e \text{ in feet}}{D \text{ in feet}} = \frac{e \text{ in mm}}{D \text{ in mm}} = 10^{-1} \times \frac{e \text{ in mm}}{D \text{ in cm}}$

[1] Fig. 8.11 is often referred to as a *Stanton diagram* as Stanton was the first person to propose such a plot.

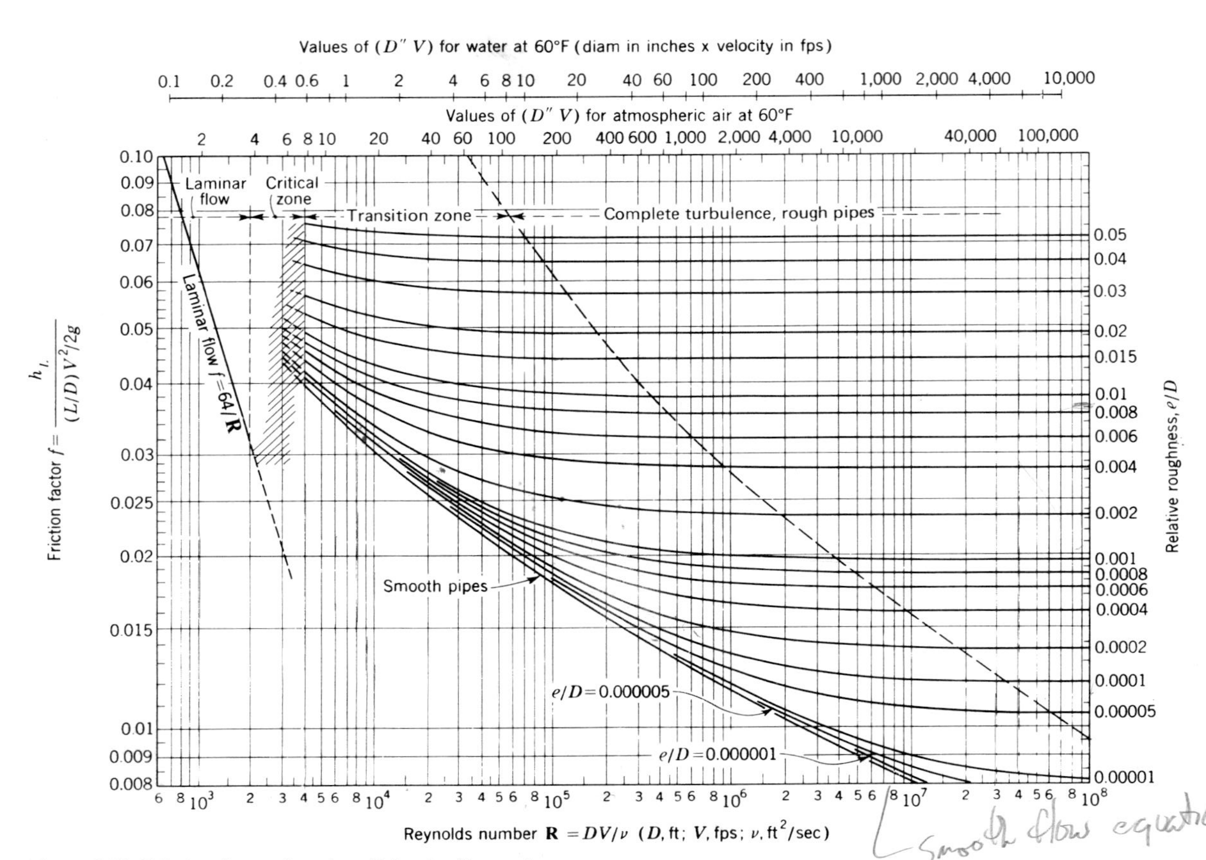

Figure 8.11 Friction factor for pipes (Moody diagram).

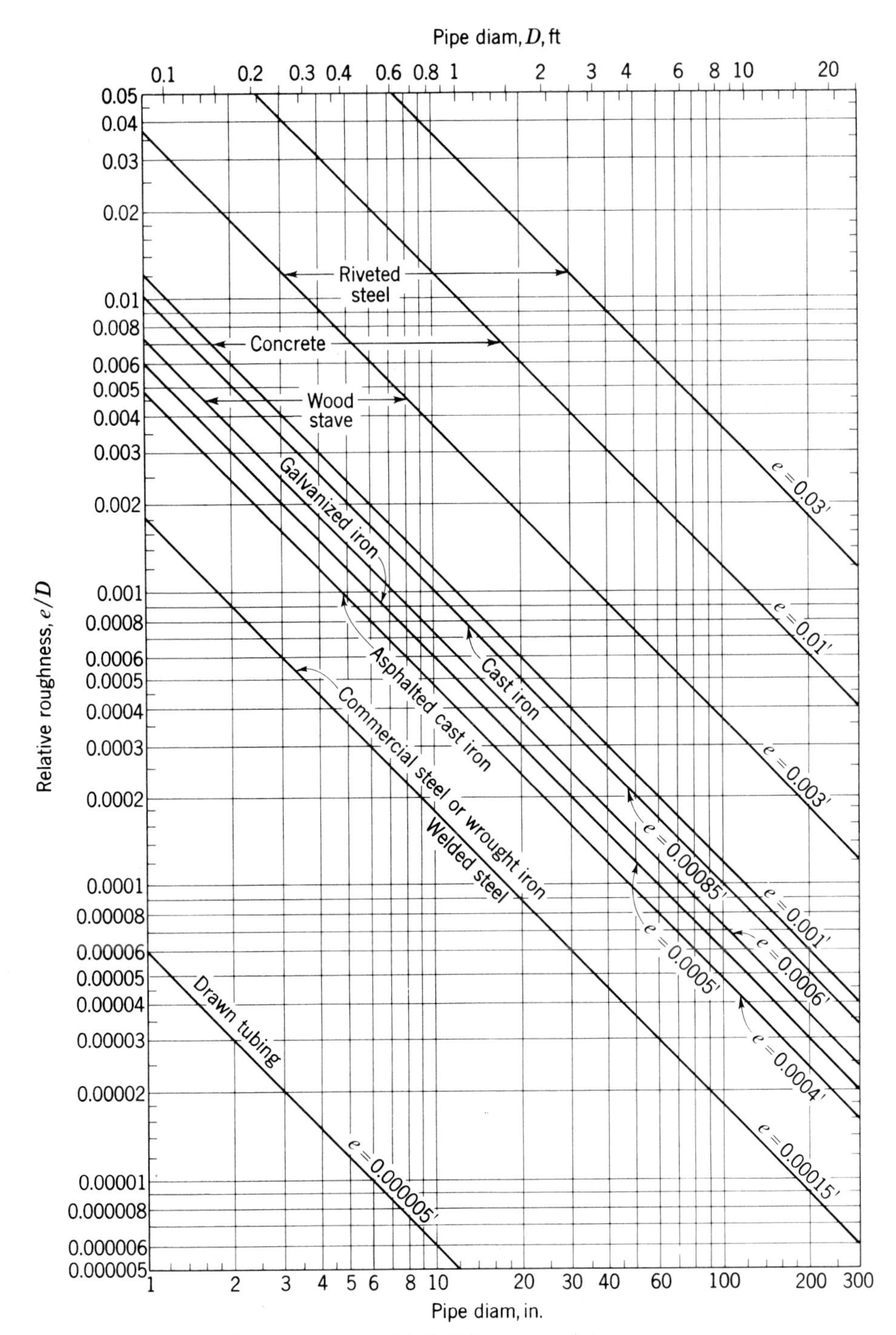

Figure 8.12 Roughness factors (e expressed in feet) for commercial pipes.

both. Values of e/D for commercial pipe may conveniently be found from Fig. 8.12, which has also been prepared by Moody. In the use of these charts, as well as in Eq. (8.13), the exact value of the internal diameter of the pipe should be used. Except in large sizes, these values differ somewhat from the nominal sizes, and especially so in the case of very small pipes.

With reference to the values of e, it must be observed that these are given here for new, clean pipes, and even in such cases there may be considerable variation in the values. Consequently, in practical cases, the value of f may be in error by ± 5 per cent for smooth pipes and by ± 10 per cent for rough ones. For old pipes values of e may be much higher, but there is much variation in the degree with which pipe roughness increases with age, since so much depends upon the nature of the fluid being transported. In small pipes there is the added factor that deposits materially reduce the internal diameter. In addition, the effect of the roughness of pipe joints may increase the value of f substantially. Hence judgment must be used in estimating a value of e, and consequently of f.

For complete turbulence, where the friction is directly proportional to V^2 and independent of Reynolds number, values of f may be determined for any assumed relative roughness. Most practical problems come within the transition zone, and there it is necessary to have also a definite value of Reynolds number. Hence, if the problem is to determine the friction loss for a given size of pipe with a given velocity, the solution is a direct one. But if the unknown quantities are either the diameter or the velocity or both, the Reynolds number is unknown. However, the value of f changes very slowly with large changes in Reynolds number; so the problem may readily be solved by assuming either a Reynolds number or a value of f to start with and then obtaining the final solution by trial. Since f will generally have a value between 0.01 and 0.07, it is best to assume f initially and work from there (Illustrative Example 8.4). Only one or two trials will usually suffice. This procedure is practically the only one that can be employed where other losses in addition to pipe friction enter into the problem.[1]

Illustrative Example 8.4 Water at 20°C flows in a 50-cm-diameter welded-steel pipe. If the energy gradient is 0.006, determine the flow rate. Find also the nominal thickness of the viscous sublayer. (*Note:* $e/D = 0.046/500 = 0.00009$.)

From Eq. (8.12)

$$\frac{h_L}{L} = 0.006 = f\frac{1}{D}\frac{V^2}{2g} = f\frac{1}{0.5}\frac{V^2}{2(9.81)}$$

from which $V = 0.243/f^{1/2}$.

Try $f = 0.030$, then $$V = 1.4 \text{ m/s}$$

and $$\mathbf{R} = \frac{DV}{\nu} = \frac{0.5(1.4)}{1 \times 10^{-6}} = 7 \times 10^5$$

[1] Charts involving these same functional relations may be plotted with different coordinates from those in Fig. 8.11 and may be more convenient for certain specific purposes, but it is believed that the form shown is best both for instruction purposes and for general use.

For $\mathbf{R} = 7 \times 10^5$ and $e/D = 0.00009$ the pipe friction chart (Fig. 8.11) indicates $f = 0.0136$. Since the f versus $\mathbf{R}$ curve is relatively flat, we will assume $f = 0.0136$ for the next trial.

For this case, $V = 0.243/f^{1/2} = 2.08$ m/s and $\mathbf{R} \approx 10^6$

For $R = 10^6$ the chart indicates $f = 0.0131$.

For the next trial, let $f = 0.0131$. This gives $V = 2.12$ m/s and $\mathbf{R}$ is still $\approx 10^6$, hence $V = 2.12$ m/s is the answer.

$$Q = AV = \frac{\pi(0.5)^2}{4}(2.12) = 0.416 \text{ m}^3/\text{s}$$

Eq. (8.30)

$$\delta_l = \frac{32.8\nu}{V\sqrt{f}} = \frac{32.8(10^{-6} \text{ m}^2/\text{s})}{2.12 \text{ m/s}\sqrt{0.0131}}$$

$$\delta_l = 135 \times 10^{-6} \text{ m} = 0.135 \text{ mm}$$

Note $\delta_l = 2.9e$, therefore the flow is in the transition zone which is typical.

8.13 FLUID FRICTION IN NONCIRCULAR CONDUITS

Most closed conduits used in engineering practice are of circular cross section; however, rectangular ducts and cross sections of other geometry are occasionally used. Some of the foregoing equations may be modified for application to noncircular sections by use of the hydraulic-radius concept.

The hydraulic radius was defined (Sec. 8.3) as $R_h = A/P$, where A is the cross-sectional area and P is the wetted perimeter. For a circular pipe flowing full,

$$R_h = \frac{A}{P} = \frac{\pi D^2/4}{\pi D} = \frac{D}{4} \tag{8.41}$$

or

$$D = 4R_h \tag{8.42}$$

These values may be substituted into Eq. (8.13) and into the expression for Reynolds number. Thus

$$h_L = \frac{f}{4}\frac{L}{R_h}\frac{V^2}{2g} \tag{8.43}$$

$$\mathbf{R} = \frac{(4R_h)V\rho}{\mu} \tag{8.44}$$

From these two expressions the head loss in noncircular conduits can be estimated by use of Fig. 8.11, where e/D is replaced by $e/4R_h$. This approach gives good results for turbulent flow, but for laminar flow the results are poor, because in such flow frictional phenomena are caused by viscous action throughout the body of the fluid, while in turbulent flow the frictional effect is accounted for largely by the region close to the wall; i.e., it depends on the wetted perimeter.

8.14 EMPIRICAL EQUATIONS FOR PIPE FLOW

The presentation of friction loss in pipes given in Secs. 8.1 to 8.12 incorporates the best knowledge available on this subject, as far as application to Newtonian fluids (Sec. 1.11) is concerned. Admittedly, however, the trial-and-error type of solution, especially when encumbered with computations for relative roughness and Reynolds number, becomes tedious when repeated often for similar conditions, as with a single fluid such as water. It is natural, therefore, that empirical design formulas have been developed, applicable only to specific fluids and conditions but very convenient in a certain range. Perhaps the best example of such a formula is that of Hazen and Williams, applicable only to the flow of water in pipes larger than 2 in (5 cm) and at velocities less than 10 fps (3 m/s), but widely used in the waterworks industry. This formula is given in the form

English units:

$$V = 1.32 C_{HW} R_h^{0.63} S^{0.54} \tag{8.45a}$$

SI units:

$$V = 0.85 C_{HW} R_h^{0.63} S^{0.54} \tag{8.45b}$$

where R_h (ft or m) is the hydraulic radius (Sec. 8.3), and $S = h_L/L$, the energy gradient. The advantage of this formula over the standard pipe-friction formula is that the roughness coefficient C_{HW} is not a function of the Reynolds number and trial solutions are therefore eliminated. Values of C_{HW} range from 140 for very smooth, straight pipe down to 110 for new riveted-steel and vitrified pipe and to 90 or 80 for old and tuberculated pipe.

Another empirical formula, which is discussed in detail in Sec. 11.3, is the Manning formula, which in English units is

$$V = \frac{1.49}{n} R_h^{2/3} S^{1/2} \tag{8.46}$$

where n is a roughness coefficient, varying from 0.009 for the smoothest brass or glass pipe, to 0.014 for average drainage tile or vitrified sewer pipe, to 0.021 for corrugated metal, and up to 0.035 for tuberculated cast-iron pipe (Table 11.1). The Manning formula applies to about the same flow range as does the Hazen-Williams formula.

Nomographic charts and diagrams have been developed for the application of Eqs. (8.45) and (8.46). The attendant lack of accuracy in using these formulas is not important in the design of water-distribution systems, since it is seldom possible to predict the capacity requirements with high precision.

8.15 MINOR LOSSES IN TURBULENT FLOW

Losses due to the *local* disturbances of the flow in conduits such as changes in cross section, projecting gaskets, elbows, valves, and similar items are called *minor losses*. In the case of a very long pipe or channel, these losses are usually insignificant in comparison with the fluid friction in the length considered. But if the length of pipe or channel is very short, these so-called minor losses may actually be major losses. Thus, in the case of the suction pipe of a pump, the loss of head at entrance, especially if a strainer and a foot valve are installed, may be very much greater than the friction loss in the short inlet pipe.

Whenever the velocity of a flowing stream is altered either in direction or in magnitude in turbulent flow, eddy currents are set up and a loss of energy in excess of the pipe friction in that same length is created.[1] Head loss in decelerating (i.e., diverging) flow is much larger than that in accelerating (i.e., converging) flow (Sec. 8.19). In addition, head loss generally increases with an increase in the geometric distortion of the flow. Though minor losses are usually confined to a very short length of path, the effects may not disappear for a considerable distance downstream. Thus an elbow in a pipe may occupy only a small length but the disturbance in the flow will extend for a long distance downstream.

The most common sources of minor loss are described in the remainder of this chapter. Such losses may be represented in one of two ways. They may be expressed as $kV^2/2g$, where k must be determined for each case, or they may be represented as being equivalent to a certain length of straight pipe, usually expressed in terms of the number of pipe diameters.

8.16 LOSS OF HEAD AT ENTRANCE

Referring to Fig. 8.13, it may be seen that, as fluid from the reservoir enters the pipe, the streamlines tend to converge, much as though this were a jet issuing from a sharp-edged orifice, so that at B a maximum velocity and a minimum pressure are found.[2] At B the central stream is surrounded by fluid which is in a state of turbulence but has very little forward motion. Between B and C the fluid is in a very disturbed condition because the stream expands and the velocity decreases while the pressure rises. From C to D the flow is normal.

It is seen that the loss of energy at entrance is distributed along the length AC, a distance of several diameters. The increased turbulence and vortex motion in this portion of the pipe cause the friction loss to be much greater than in a corresponding length where the flow is normal, as is shown by the drop of the total-energy line. Of this total loss, a portion h' would be due to the normal pipe friction. Hence the difference between this and the total, or h'_e, is the true value of the extra loss caused at entrance.

The loss of head at entrance may be expressed as

$$h'_e = k_e \frac{V^2}{2g} \tag{8.47}$$

[1] In laminar flow these losses are insignificant because irregularities in the flow boundary create a minimal disturbance to the flow and separation is essentially nonexistent.

[2] Section B, the point of maximum contraction of the flow, is referred to as the *vena contracta*.

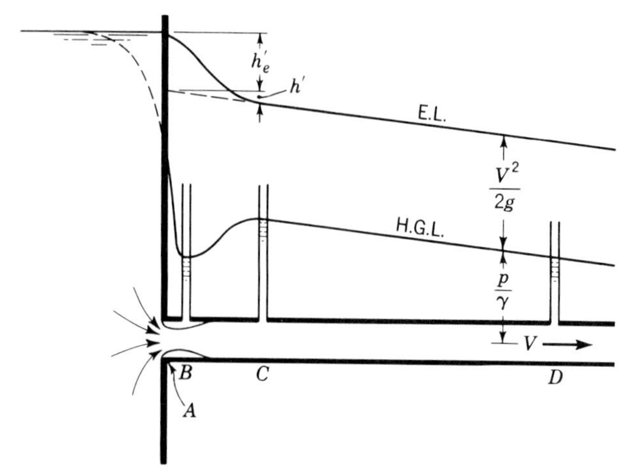

Figure 8.13 Conditions at entrance.

where V is the mean velocity in the pipe, and k_e is the loss coefficient whose general values are shown in Fig. 8.14.

The entrance loss is caused primarily by the turbulence created by the enlargement of the stream after it passes section B, and this enlargement in turn depends upon how much the stream contracts as it enters the pipe. Thus it is very much affected by the conditions at the entrance to the pipe. Values of the entrance-loss coefficients have been determined experimentally. If the entrance to the pipe is well rounded or *bell-mouthed* (Fig. 8.14*a*), there is no contraction of the stream entering and the coefficient of loss is correspondingly small. For a *flush* or *square-edged entrance*, such as shown in Fig. 8.14*b*, k_e has a value of about 0.5. A *reentrant tube*, such as shown in Fig. 8.14*c*, produces a maximum contraction of the entering stream because the streamlines come from around the outside wall of the pipe, as well as more directly from the fluid in front of the entrance. The degree of the contraction depends upon how far the pipe may project within the reservoir and also upon how thick the pipe walls are, compared with its diameter. With very thick walls, the conditions approach that of a square-edged entrance. For these reasons the loss coefficients for reentrant tubes vary; for very thin tubes $k_e \approx 0.8$.

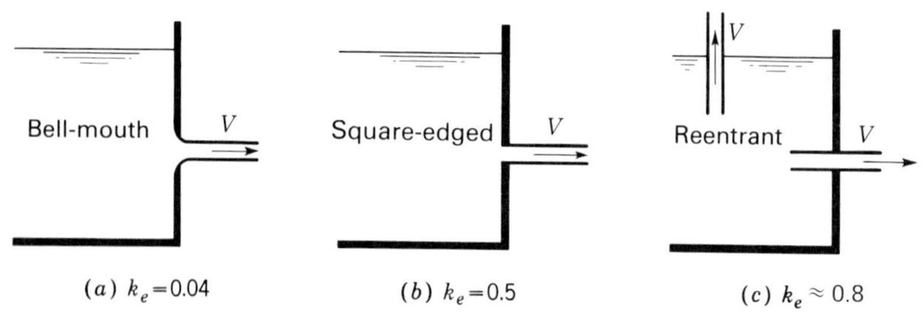

Figure 8.14 Entrance loss coefficients.

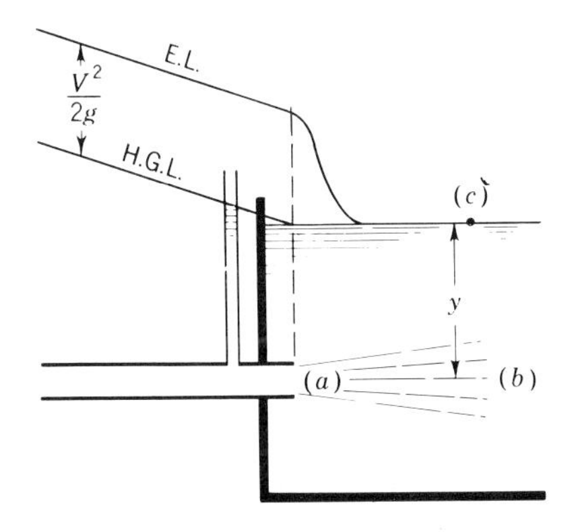

Figure 8.15 Submerged discharge loss.

8.17 LOSS OF HEAD AT SUBMERGED DISCHARGE[1]

When a fluid with a velocity V is discharged from the end of a pipe into a closed tank or reservoir which is so large that the velocity within it is negligible, the entire kinetic energy of the flow is dissipated. Hence the discharge loss is

$$h_d' = \frac{V^2}{2g} \tag{8.48}$$

That this is true may be shown by writing an energy equation between (a) and (c) in Fig. 8.15. Taking the datum plane through (a) and recognizing that the pressure head of the fluid at (a) is y, its depth below the surface, $H_a = y + 0 + V^2/2g$ and $H_c = 0 + y + 0$. Therefore

$$h_d' = H_a - H_c = \frac{V^2}{2g}$$

The discharge loss coefficient is 1.0 under all conditions; hence the only way to reduce the discharge loss is to reduce the value of V by means of a diverging tube. This is the reason for a diverging draft tube from a reaction turbine (Sec. 16.6).

As contrasted with entrance loss, it must here be emphasized that discharge loss occurs *after* the fluid *leaves* the pipe,[2] while entrance loss occurs *after* the fluid *enters* the pipe.

8.18 LOSS DUE TO CONTRACTION

Sudden Contraction

The phenomena attending the sudden contraction of a flow are shown in Fig. 8.16. There is a marked drop in pressure due to the increase in velocity and to the loss of energy in turbulence. It is noted that in the corner upstream at section C

[1] This topic was first discussed in Sec. 4.13.

[2] In a short pipe where the discharge loss may be a major factor, greater accuracy is obtained by using the correction factor α, as explained in Sec. 4.1 [see also Eq. (8.35a)].

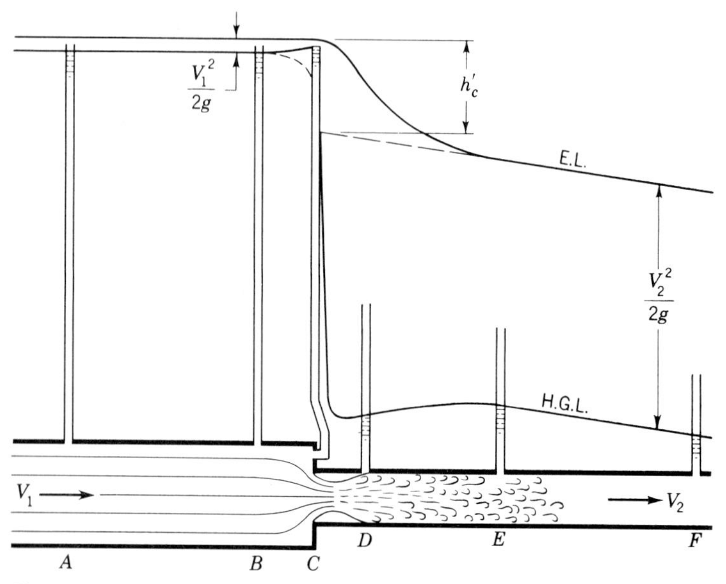

Figure 8.16 Loss due to sudden contraction. *(Plotted to scale from observations made by Daugherty.)*

there is a rise in pressure because the streamlines here are curving, so that the centrifugal action causes the pressure at the pipe wall to be greater than in the center of the stream. The dashed line indicates the pressure variation along the centerline streamline from sections B to C.

From C to E the conditions are similar to those described for entrance. The loss of head for a sudden contraction may be represented by

$$h_c' = k_c \frac{V_2^2}{2g} \tag{8.49}$$

where k_c has the values given in Table 8.2.

The entrance loss of Sec. 8.16 is a special case where $D_2/D_1 = 0$.

Gradual Contraction

In order to reduce the foregoing losses, abrupt changes of cross section should be avoided. This may be accomplished by changing from one diameter to the other by means of a smoothly curved transition or by employing the frustum of a cone. With a smoothly curved transition a loss coefficient k_c as small as 0.05 is possible. For conical reducers a minimum k_c of about 0.10 is obtained with a total cone angle of 20 to 40°. Smaller or larger total cone angles result in higher values of k_c.

Table 8.2 Loss coefficients for sudden contraction

D_2/D_1	0.0	0.1	0.2	0.3	0.4	0.5	0.6	0.7	0.8	0.9	1.0
k_c	0.50	0.45	0.42	0.39	0.36	0.33	0.28	0.22	0.15	0.06	0.00

The nozzle at the end of a pipeline (Fig. 8.23*b*) is a special case of gradual contraction. The head loss through a nozzle at the end of a pipeline is given by Eq. (8.49) where k_c is the nozzle loss coefficient whose value commonly ranges from 0.04 to 0.20 and V_j is the jet velocity.[1] The head loss through a nozzle cannot be regarded as a minor loss because the jet velocity head is usually quite large. More details on the flow through nozzles is presented in Sec. 12.6.

8.19 LOSS DUE TO EXPANSION

Sudden Expansion

The conditions at a sudden expansion are shown in Fig. 8.17. There is a rise in pressure because of the decrease in velocity, but this rise is not so great as it would be if it were not for the loss in energy. There is a state of excessive turbulence from C to F beyond which the flow is normal. The drop in pressure just beyond section C, which was measured by a piezometer not shown in the illustration, is due to the fact that the pressures at the wall of the pipe are in this case less than those in the center of the pipe because of centrifugal effects.

Figures 8.16 and 8.17 are both drawn to scale from test measurements for the same diameter ratios and the same velocities and show that the loss due to sudden expansion is greater than the loss due to a corresponding contraction. This is so because of the inherent instability of flow in an expansion where the diverging paths of the flow tend to encourage the formation of eddies within the flow. Moreover, separation of the flow from the wall of the conduit induces

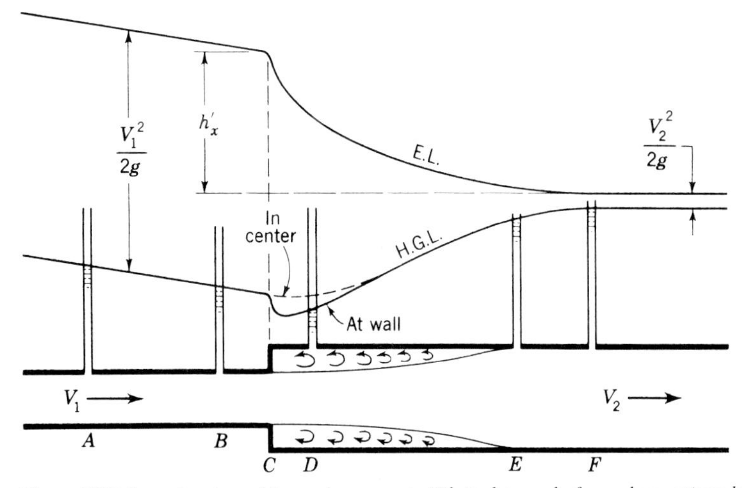

Figure 8.17 Loss due to sudden enlargement. *(Plotted to scale from observations by Daugherty. Velocity the same as in Fig. 8.16.)*

[1] See also Eq. (12.13).

pockets of eddying turbulence outside the flow region. In converging flow there is a dampening effect on eddy formation and the conversion from pressure energy to kinetic energy is quite efficient.

An expression for the loss of head in a sudden enlargement can be derived as follows. In Fig. 8.18, section 2 corresponds to section F in Fig. 8.17, which is a section where the velocity profile has become normal again and marks the end of the region of excess energy loss due to the turbulence created by the sudden enlargement. In Fig. 8.18 assume that the pressure at section 2 in the ideal case without friction is p_0. Then in this ideal case

$$\frac{p_0}{\gamma} = \frac{p_1}{\gamma} + \frac{V_1^2}{2g} - \frac{V_2^2}{2g}$$

If in the actual case the pressure at section 2 is p_2 while the average pressure on the annular ring is p', then, equating the resultant force on the body of fluid between sections 1 and 2 to the time rate of change of momentum between sections 1 and 2, we obtain

$$p_1 A_1 + p'(A_2 - A_1) - p_2 A_2 = \frac{\gamma}{g}(A_2 V_2^2 - A_1 V_1^2)$$

From this

$$\frac{p_2}{\gamma} = \frac{A_1}{A_2}\frac{p_1}{\gamma} + \frac{A_2 - A_1}{A_2}\frac{p'}{\gamma} + \frac{A_1}{A_2}\frac{V_1^2}{g} - \frac{V_2^2}{g}$$

The loss of head is given by the difference between the ideal and actual pressure heads at section 2. Thus $h_x' = (p_0 - p_2)/\gamma$, and noting that

$$A_1 V_1 = A_2 V_2$$

and that $A_1 V_1^2 = A_1 V_1 V_1 = A_2 V_2 V_1$, we obtain, from substituting the above expressions for p_0/γ and p_2/γ into $(p_0 - p_2)/\gamma$,

$$h_x' = \frac{(V_1 - V_2)^2}{2g} + \left(1 - \frac{A_1}{A_2}\right)\left(\frac{p_1}{\gamma} - \frac{p'}{\gamma}\right)$$

It is usually assumed that $p' = p_1$, in which case the loss of head due to sudden enlargement is

$$h_x' = \frac{(V_1 - V_2)^2}{2g} \tag{8.50}$$

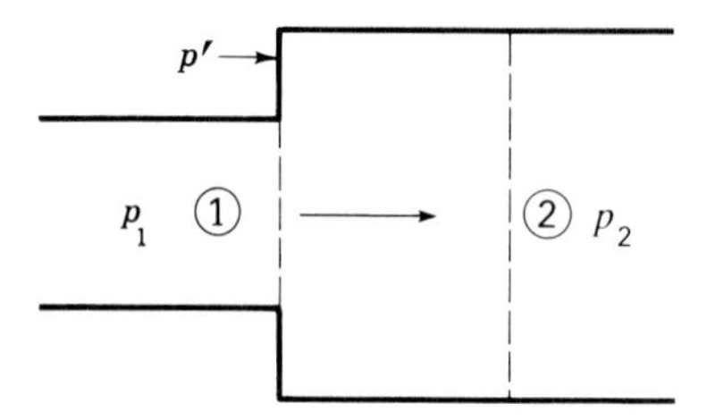

Figure 8.18

Although it is possible that under some conditions p' will equal p_1, it is also possible for it to be either more or less than that value, in which case the loss of head will be either less or more than that given by Eq. (8.50). The exact value of p' will depend upon the manner in which the fluid eddies around in the corner adjacent to this annular ring. However, the deviation from Eq. (8.50) is quite small and of negligible importance.

The discharge loss of Sec. 8.17 is seen to be a special case where A_2 is infinite compared with A_1, or $V_2 = 0$, so that Eq. (8.50) will reduce to Eq. (8.48).

Gradual Expansion

To minimize the loss accompanying a reduction in velocity, a diffuser such as shown in Fig. 8.19 may be used. The diffuser may be given a curved outline, or it may be a frustum of a cone. In Fig. 8.19 the loss of head will be some function of the angle of divergence and also of the ratio of the two areas, the length of the diffuser being determined by these two variables.

In flow through a diffuser the total loss may be considered as made up of two factors. One is the ordinary pipe friction loss, which may be represented by

$$h_L = \int \frac{f}{D} \frac{V^2}{2g} \, dL$$

In order to integrate the foregoing, it is necessary to express the variables f, D, and V as functions of L. For our present purpose it is sufficient, however, merely to note that the friction loss increases with the length of the cone. Hence, for given values of D_1 and D_2, the larger the angle of the cone, the less its length and the less the pipe friction, which is indicated by the curve marked F in Fig. 8.20*a*. However, in flow through a diffuser, there is an additional turbulence loss set up by induced currents which produce a vortex motion over and above that which normally exists. This additional turbulence loss will naturally increase with the degree of divergence, as is indicated by the curve marked T in Fig. 8.20*a*, and if the rate of divergence is great enough, there may be a separation at the walls and eddies flowing backward along the walls. The total loss in the diverging cone is then represented by the sum of these two losses, marked k'. This is seen to have a minimum value at 6° for the particular case chosen, which is for a very smooth

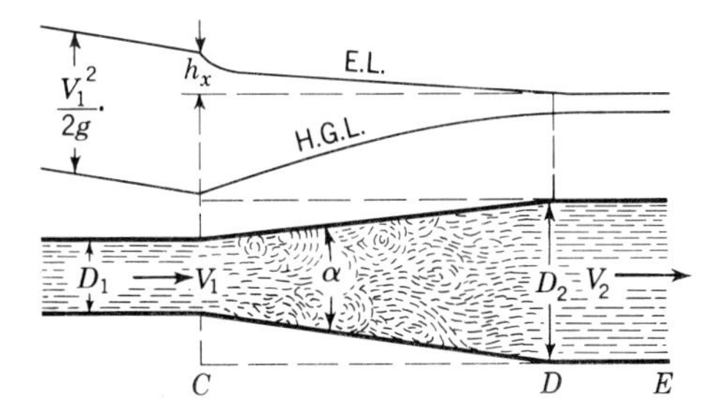

Figure 8.19 Loss due to gradual enlargement.

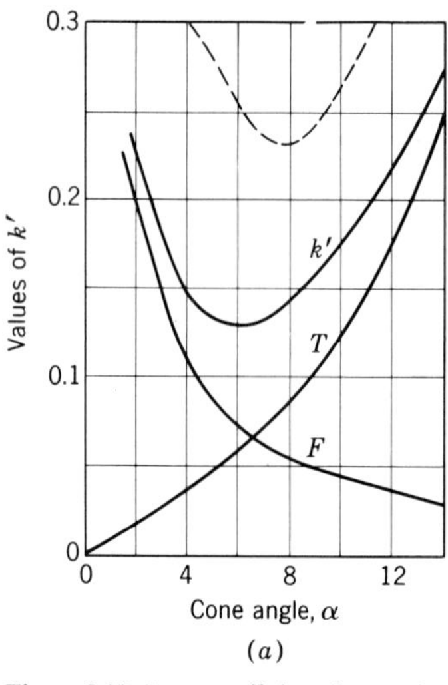

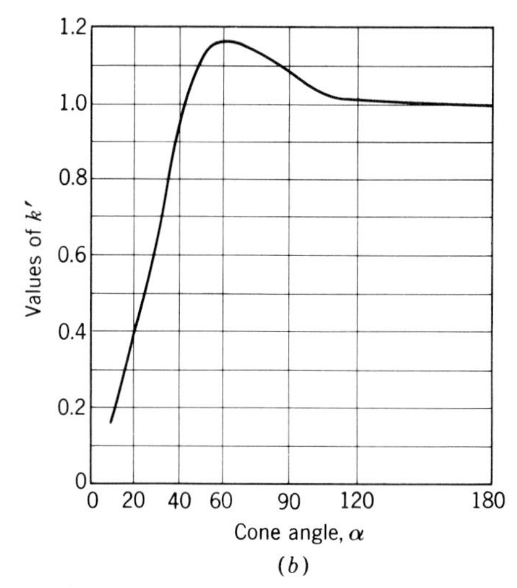

Figure 8.20 Loss coefficient for conical diffusers.

surface. If the surface were rougher, the value of the friction F would be increased. This increases the value of k', which is indicated by the dotted curve, and also shifts the angle for minimum loss to 8°. Thus the best angle of divergence increases with the roughness of the surface.

It has been seen that the loss due to a sudden enlargement is very nearly represented by $(V_1 - V_2)^2/2g$. The loss due to a gradual enlargement is expressed as

$$h' = k' \frac{(V_1 - V_2)^2}{2g} \tag{8.51}$$

Values of k' as a function of the cone angle α are shown in Fig. 8.20*b*,[1] for a wider range than appears in Fig. 8.20*a*. It is of interest to note that at an angle slightly above 40° the loss is the same as that for a sudden enlargement, which is 180°, and that between these two the loss is greater than for a sudden enlargement, being a maximum at about 60°. This is because the induced currents set up are greater within this range.

8.20 LOSS IN PIPE FITTINGS

The loss of head in pipe fittings may be expressed as $kV^2/2g$, where V is the velocity in a pipe of the nominal size of the fitting. Typical values of k are given in Table 8.3. As an alternative, the head loss due to a fitting may be found by

[1] A. H. Gibson, *Engineering (London)*, Feb. 16, 1912. These values were based on area ratios of 1:9, 1:4, 1:2.25 and gave one curve up to an angle of about 30°. Beyond that the three ratios gave three curves which differed by as much as about 18 per cent at 60°, where the turbulence was a predominating factor, and then drew together again as 180° was approached. The curve here shown is a composite of these three.

Table 8.3 Values of loss factors for pipe fittings*

Fitting	k	L/D
Globe valve, wide open	10	350
Angle valve, wide open	5	175
Close-return bend	2.2	75
T, through side outlet	1.8	67
Short-radius elbow	0.9	32
Medium-radius elbow	0.75	27
Long-radius elbow	0.60	20
45° elbow	0.42	15
Gate valve, wide open	0.19	7
half open	2.06	72

* Flow of Fluids through Valves, Fittings, and Pipe, *Crane Co., Tech. Paper* 409, p. 20, May, 1942. Values based on tests by Crane Co. and at the University of Wisconsin, the University of Texas, and Texas College.

increasing the pipe length by using values of L/D given in the table. However, it must be recognized that these fittings create so much turbulence that the loss caused by them is proportional to V^2, and hence this latter method should be restricted to the case where the pipe friction itself is in the zone of complete turbulence. For very smooth pipes, it is better to use the k values when determining the loss through fittings.

8.21 LOSS IN BENDS AND ELBOWS

In flow around a bend or elbow, because of centrifugal effects [Eq. (4.35)], there is an increase in pressure along the outer wall and a decrease in pressure along the inner wall. The centrifugal force on a number of fluid particles, each of mass m, along the diameter CD of the pipe that is normal to the plane of curvature of the pipe is shown in Fig. 8.21. The centrifugal force on the particles near the center of the pipe, where the velocities are high, is larger than the centrifugal force on the particles near the walls of the pipe, where the velocities are low. Because of this unbalanced condition a secondary flow[1] develops as shown in Fig. 8.21. This combines with the axial velocity to form a double spiral flow which persists for some distance. Thus not only is there some loss of energy within the bend itself, but this distorted flow condition persists for some distance downstream until dissipated by viscous friction. The velocity in the pipe may not become normal again within as much as 100 pipe diameters downstream from the bend. In fact, more than half the friction loss produced by a bend or elbow takes place in the straight pipe following it.

[1] Secondary flow in the bends of open channels is discussed in Sec. 11.21.

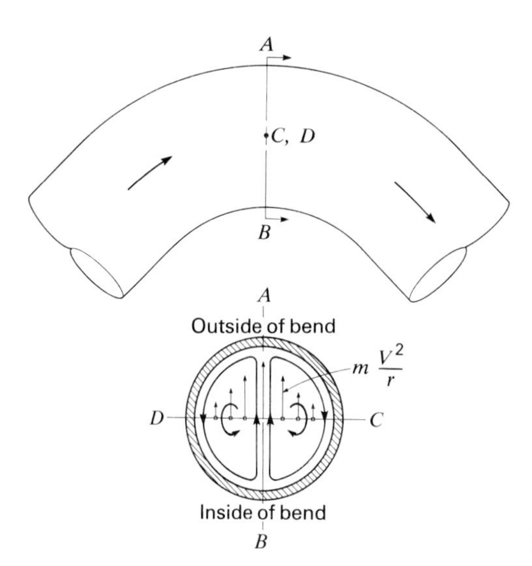

Figure 8.21 Secondary flow in bend.

Most of the loss of head in a sharp bend may be eliminated by the use of a vaned elbow, such as is shown in Fig. 8.22. The vanes tend to impede the formation of the secondary flows that would otherwise occur.

The head loss produced by a bend ($h_b = k_b V^2/2g$) in excess of the loss for an equal length of straight pipe is greatly dependent upon the ratio of the radius of curvature r to the diameter of the pipe D, and combinations of different pipe bends placed close together cannot be treated by adding up the losses of each one considered separately. The total loss depends not only upon the spacing between the bends, but also upon the relations of the directions of the bends and the planes in which they are located. Bend loss is not proportional to the angle of the bend; for 22.5 and 45° bends the losses are respectively about 40 and 80 per cent of the loss in a 90° bend. Typically for a 90° bend, k_b varies for a smooth pipe from 0.15 for $r/D = 2$ to 0.10 for $r/D = 10$, and for a pipe with $e/D = 0.0020$, k_b varies from about 0.30 to 0.20 for r/D values of 2 and 10 respectively. Information on values of k_b is available in the literature.[1]

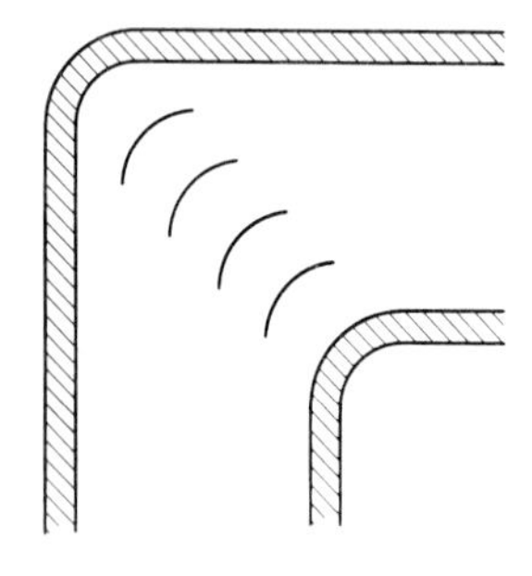

Figure 8.22 Vaned elbow.

[1] R. J. S. Pigott, Pressure Losses in Tubing, Pipe, and Fittings, *Trans. ASME*, vol. 72, p. 679, July, 1950. See also: E. F. Brater and H. W. King, "Handbook of Applied Hydraulics," 6th ed., McGraw-Hill Book Co., New York, N.Y., 1976.

8.22 SOLUTION OF PIPE-FLOW PROBLEMS

We have examined the fundamental fluid mechanics associated with the frictional loss of energy in pipe flow. While the interest of the scientist extends very little beyond this, it is the task of the engineer to apply these fundamentals to various types of practical problems. Pipe flow problems may be solved using the Hazen–Williams equation (8.45), the Manning equation (8.46) or the Darcy–Weisbach equation (8.13). The latter is to be preferred as it will provide greater accuracy since its application utilizes the basic parameters that influence pipe friction, namely, Reynolds number **R** and relative roughness e/D. To get good results with the Hazen–Williams and Manning equations the user must select proper values for C_{HW} and n respectively. This is more difficult than estimating the e/D ratio for a pipe as required by the Darcy–Weisbach equation. An advantage of the Manning equation is that all types of pipe flow problems can be solved directly by using it, while certain types of problems must be solved by trial and error when using the Darcy–Weisbach equation as discussed in Sec. 8.12. The Hazen–Williams equation is not well suited for the solution of all problems where minor losses must be considered because, upon rearranging Eq. (8.45), we find that in the Hazen–Williams equation the head loss due to pipe friction is proportional to $V^{1.85}$, while minor losses are expressed as being proportional to V^2.

In the typical direct-solution problem using the Darcy–Weisbach equation the head loss is determined for the given flow rate and pipeline and minor loss characteristics. The indirect-solution problems (trial and error) are of two principal types: (1) given the pipeline and minor loss characteristics and the head loss, find the flow rate and (2) given the flow rate and the energy gradient, find the required pipe diameter. The feature of these problems is the variation of f with Reynolds number. The usual procedure (see Illustrative Example 8.4) is to assume a reasonable value of f by referring to Fig. 8.11. This will then lead, through the pipe-friction and energy equations, to a *computed* velocity and Reynolds number. This determines a more accurate value of f, and it will generally be necessary to repeat the solution for new values of V and Q. As f varies little within a small range of **R**, a third trial will rarely be necessary.

The following example illustrates the method of solution for flow through a pipeline of uniform diameter.

Illustrative Example 8.5 Referring to Fig. 8.23, find the flow rate through a new 10-in-diameter cast-iron pipe of length 5,000 ft, with $\Delta z = 260$ ft. Consider the entrance to be sharp-cornered, nonprojecting.

From Fig. 8.12, $e/D \approx 0.001$. Referring to Fig. 8.11, assume $f = 0.020$. From Sec. 8.16 we choose a value of $k_e = 0.5$ for the loss at entrance. Then, writing the energy equation between the water surface and the free jet,

$$260 + 0 + 0 = 0 + 0 + \frac{V_2^2}{2g} + \left(0.5 + 0.02 \times \frac{5{,}000}{\frac{10}{12}}\right)\frac{V_2^2}{2g}$$

This gives $V_2^2/2g = 2.14$ ft and $V_2 = 11.74$ fps. We may now confirm the trial value of f by returning to Fig. 8.11, with $D''V = 10 \times 11.74 = 117.4$ and $e/D = 0.001$. Again, the chart shows $f = 0.020$, so no repeat solution is required. The flow is $Q = A_2 V_2 = 0.545 \times 11.74 = 6.40$ cfs.

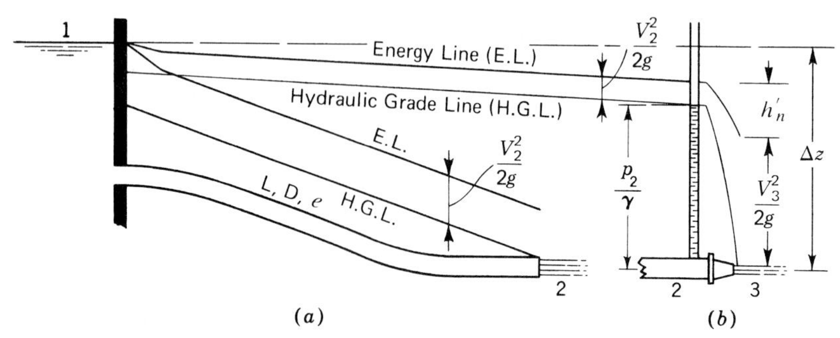

Figure 8.23 Discharge from a reservoir. (*a*) Free discharge. (*b*) With nozzle. As L/D gets larger the E.L. and H.G.L. approach one another.

In the foregoing example it may be seen that with this length of pipe it would have made very little difference if the entrance loss and also the velocity head at discharge had been neglected altogether. It is generally conceded that, for pipes of length greater than 1,000 diameters, the error incurred by neglecting minor losses is less than that inherent in selecting a value of f. In applying this rule one must of course use common sense and recall that a valve, for example, is a minor loss only when it is wide open. Partially closed, it may be the most important loss in the system.

If the pipe discharged into a fluid that was at a pressure other than atmospheric, the proper value of p_2/γ would have to be used in the energy equation.

Another example of flow from a reservoir is that of a penstock leading to an impulse turbine. In this case the pipe does not discharge freely but ends in a nozzle (Fig. 8.23b), which has a known or assumed loss coefficient. The head loss in the nozzle, h_n', is associated with the high issuing velocity head and is therefore not a minor loss. The procedure is to employ the equation of continuity to place all losses in terms of the velocity head in the pipe. This is the logical choice for the "common unknown" because the trial-and-error solution will again be built around the pipe friction loss rather than the nozzle loss.

Illustrative Example 8.6 In Fig. 8.23 suppose that the pipeline of the preceding example is now fitted with a nozzle at the end which discharges a jet 2.5 in in diameter and which has a loss coefficient of 0.11. Find the flow rate.

Let point 2 now refer to the pipe at the base of the nozzle and point 3 be in the jet. The head loss in the nozzle is $0.11V_3^2/2g$. Writing the energy equation between 1 and 3, neglecting entrance loss,

$$260 + 0 + 0 = 0 + 0 + \frac{V_3^2}{2g} + 6{,}000f\frac{V_2^2}{2g} + 0.11\frac{V_3^2}{2g}$$

By the continuity equation, $V_3^2/2g = (10/2.5)^4 V_2^2/2g = 256V_2^2/2g$. Thus

$$260 = (1.11 \times 256 + 6{,}000f)\frac{V_2^2}{2g}$$

A trial value of f is selected. Let $f = 0.02$ for the first assumption. Then $260 = (284 + 120)V_2^2/2g$, from which

$$\frac{V_2^2}{2g} = \frac{260}{404} = 0.644 \text{ ft}$$

and $V_2 = 8.02\sqrt{0.644} = 6.45$ fps. With $D''V = 10 \times 6.45 = 64.5$ and $e/D = 0.001$, Fig. 8.11 shows $f = 0.02$. In this case the first solution may be considered sufficiently accurate, but in general the value of f determined from the chart may be materially different from that assumed, and a second trial may be necessary.

The rate of discharge is $Q = A_2 V_2 = 0.545 \times 6.45 = 3.52$ cfs, and

$$V_3 = 16V_2 = 16 \times 6.45 = 103.2 \text{ fps}$$

As additional information, $H_2 = p_2/\gamma + V_2^2/2g = 260 - 0.02 \times 6{,}000 \times 0.644 = 182.72$ ft, and the pressure head $p_2/\gamma = 182.72 - 0.644 = 182.08$ ft.

This example shows that the addition of the nozzle has reduced the discharge from 6.40 to 3.52 cfs but has increased the jet velocity from 11.74 to 103.2 fps. The head loss due to pipe friction is 77.3 ft and the head loss through the nozzle is 18.1 ft. (The head loss at entrance which was neglected in the calculations is approximately 0.3 ft.)

We may change Illustrative Example 8.5 into a type-2 indirect-solution problem by specifying the rate of discharge and finding the required diameter. Although this type of problem can be attacked in exactly the same way as the foregoing, the solution is facilitated by a slightly different procedure if the length is so great that the minor losses are negligible. From the continuity equation, $V = Q/A = 4Q/\pi D^2$. Substituting this expression for V in the pipe-friction equation,

$$h_L = f\frac{L}{D}\frac{V^2}{2g}$$

and rearranging, we obtain

$$\frac{D^5}{f} = \frac{8LQ^2}{\pi^2 g h_L} = \text{constant} \tag{8.52}$$

A value of f may be assumed more or less arbitrarily and an approximate value of the pipe diameter computed by this equation. This determines the velocity, Reynolds number, and relative roughness. A new value of f is determined with the aid of Fig. 8.11, and the computation may be repeated if necessary. In general, the diameter so obtained will not be a standard pipe size, and the size selected will usually be the next largest commercially available size. In planning for the future it must be recalled that scale deposits will increase the roughness and reduce the cross-sectional area. For pipes in water service, the absolute roughness e of old pipes (twenty years and more) may increase over that of new pipes by threefold for concrete or cement-lined steel, up to twentyfold for cast iron, and even to fortyfold for tuberculated wrought-iron and steel pipe. Equation (8.52) shows that for a constant value of f, Q varies as $D^{5/2}$. Hence for the case where minor losses are negligible and f is constant, to achieve a 100 percent increase in flow, the diameter need be increased only 32 percent. This amounts to a 74 percent increase in cross-sectional area.

If the minor losses and the velocity head in the pipe are not negligible in comparison with the pipe friction, the problem may be handled by expressing such losses in equivalent lengths of pipe, and the solution reduces to the case just described. This approach can be used if the pipe behaves as wholly rough (Sec. 8.11) in which case f depends only on e/D and is independent of **R**. The length equivalence of a minor loss is obtained by equating $(fL/D)V^2/2g$ to $(k)V^2/2g$. From this one obtains the equivalent length of pipe as $L_e = kD/f$ where k is the minor loss coefficient.

8.23 PIPELINE WITH PUMP OR TURBINE

If a pump lifts a fluid from one reservoir to another, as in Fig. 8.24, not only does it do work in lifting the fluid the height Δz, but also it has to overcome the friction loss in the suction and discharge piping. This friction head is equivalent to some added lift, so that the effect is the same as if the pump lifted the fluid a height $\Delta z + \sum h_L$. Hence the power delivered to the liquid by the pump is $\gamma Q(\Delta z + \sum h_L)$. The power required to run the pump is greater than this, depending on the efficiency of the pump. The total pumping head h_p for this case is

$$h_p = \Delta z + \sum h_L \tag{8.53}$$

If the pump discharges a stream through a nozzle, as shown in Fig. 8.25, not only has the liquid been lifted a height Δz, but also it has received a kinetic energy head of $V_2^2/2g$, where V_2 is the velocity of the jet. Thus the total pumping head is now

$$h_p = \Delta z + \frac{V_2^2}{2g} + \sum h_L \tag{8.54}$$

In any case the total pumping head may be determined by writing the energy equation between any point upstream from the pump and any other point downstream, as in Eq. (4.14). For example, if the upstream reservoir were at a higher elevation than the downstream one, then the Δz's in the two foregoing equations would have negative signs.

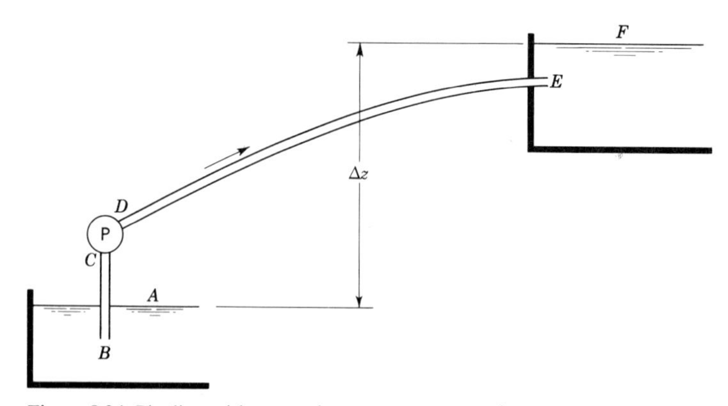

Figure 8.24 Pipeline with pump between two reservoirs.

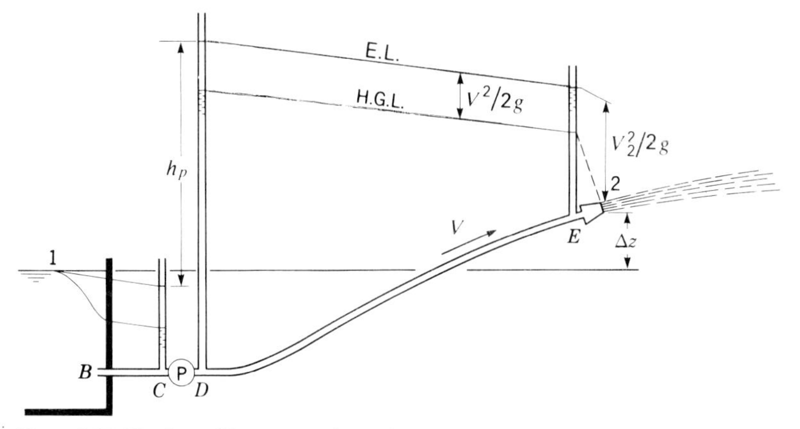

Figure 8.25 Pipeline with pump and nozzle.

The machine that is employed for converting the energy of flow into mechanical work is called a *turbine*. In flowing from the upper tank in Fig. 8.26 to the lower, the fluid loses potential energy head equivalent to Δz. This energy is expended in two ways, part of it in hydraulic friction in the pipe and the remainder in the turbine. Of that which is delivered to the turbine, a portion is lost in hydraulic friction and the rest is converted into mechanical work.

The power delivered to the turbine is decreased by the friction loss in the pipeline, and its value is given by $\gamma Q(\Delta z - \sum h_L)$. The power delivered by the machine is less than this, depending upon both the hydraulic and mechanical losses of the turbine. The head under which the turbine operates is

$$h_t = \Delta z - \sum h_L \tag{8.55}$$

where $\sum h_L$ is the loss of head in the supply pipe and does not include the head loss in the draft tube (DE in Fig. 8.26), since the draft tube is considered an integral part of the turbine. The draft tube has a gradually increasing cross-sectional area which results in a reduced velocity at discharge. This enhances the

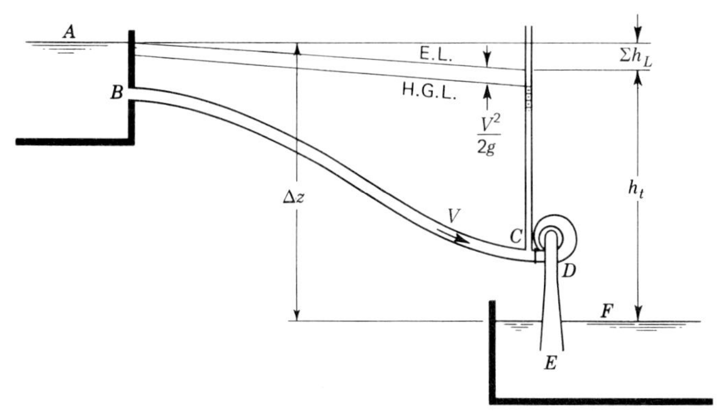

Figure 8.26 Pipeline with turbine.

efficiency of the turbine because of the reduced head loss at discharge (Sec. 16.6). It should be noted that the h_t of Eq. (8.55) represents the energy head removed from the fluid by the turbine; this, of course, is identical to the energy head transferred to the turbine from the fluid.

Illustrative Example 8.7 In this problem we will assume that the Reynolds number is high enough to assure turbulent flow. A pump is located 15 ft above the surface of a liquid ($\gamma = 52$ lb/ft^3) in a closed tank. The pressure in the space above the liquid surface is 5 psi. The suction line to the pump is 50 ft of 6-in-diameter pipe ($f = 0.025$). The discharge from the pump is 200 ft of 8-in-diameter pipe ($f = 0.030$). This pipe discharges in a submerged fashion to an open tank whose free liquid surface is 10-ft lower than the liquid surface in the pressure tank. If the pump puts 2.0 hp into the liquid, determine the flow rate and find the pressure in the pipe on the suction side of the pump.

From Eq. (4.16),

$$\text{HP} = \frac{\gamma Q h_p}{550} = 2 = \frac{52 Q h_p}{550}$$

Thus

$$h_p = \frac{21.2}{Q}$$

Writing the energy equation from one liquid surface to the other,

$$\frac{5(144)}{52} - 0.5\frac{V_6^2}{2g} - 0.025\left(\frac{50}{6/12}\right)\frac{V_6^2}{2g} + h_p - 0.030\left(\frac{200}{8/12}\right)\frac{V_8^2}{2g} = -10 + \frac{V_8^2}{2g}$$

Substituting $V_6 = Q/0.196$ and $V_8 = Q/0.349$ this reduces to

$$23.9 + h_p - 2.49Q^2 = 0$$

or

$$23.9 + \frac{21.2}{Q} - 2.49Q^2 = 0$$

By trial,

$$Q = 3.48 \text{ cfs}$$

To obtain the pressure at the suction side of the pump,

$$\frac{5(144)}{52} - 0.5\frac{V_6^2}{2g} - 0.025\left(\frac{50}{6/12}\right)\frac{V_6^2}{2g} = 15 + \frac{p}{\gamma} + \frac{V_6^2}{2g}$$

where

$$V_6 = \frac{3.48}{0.196} = 17.8 \text{ fps}$$

from which

$$\frac{p}{\gamma} = -20.9 \text{ ft}$$

or $p_2 = -20.9(52/144) = -7.55$ psi which is equivalent to $(7.55/14.70)29.9 = 15.3$ inches of mercury vacuum.

In this type of problem one should check the absolute pressure against the vapor pressure of the liquid to see that vaporization does not occur.

8.24 BRANCHING PIPES

Suppose that three reservoirs A, B, and C of Fig. 8.27 are connected to a common junction J by pipes 1, 2, and 3, in which the friction losses are h_1, h_2, and h_3, respectively. It is supposed that all pipes are sufficiently long, so that minor

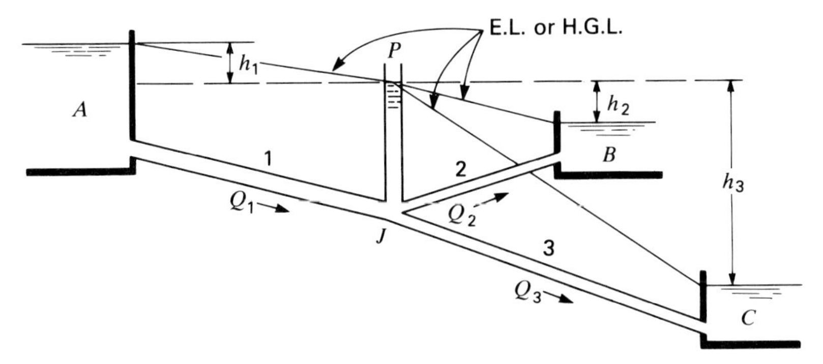

Figure 8.27 Branching pipes.

losses and velocity heads may be neglected. Actually, any one of the pipes may be considered leading to or from some destination other than the reservoir shown by simply replacing the reservoir with a piezometer tube in which the water level is the same as that of the reservoir surface. The continuity and energy equations require that the flow entering the junction equal the flow leaving it and that the pressure head at J (which may be represented schematically by the open piezometer tube shown, with water at elevation P) be common to all pipes. That is, for the condition shown:

1. $Q_1 = Q_2 + Q_3$.
2. Elevation P is common to all.

If P is below the surface of B, then the flow will be out of B and $Q_1 + Q_2 = Q_3$. The diagram suggests several problems, three of which will be discussed below:

1. Given all pipe lengths and diameters, the surface elevations of two reservoirs, and the flow to or from one of these two, find the surface elevation of the third reservoir. This is a direct-solution problem. Suppose that Q_1 and the elevations of A and B are given. The head loss h_1 is determined directly by the pipe-friction equation, using Fig. 8.11 to determine the proper value of f. This fixes P and h_2 was given. The flow in pipe 2 may then be determined, assuming a reasonable value of f and adjusting it if necessary. Condition 1 (continuity at the junction) then determines Q_3, which in turn determines h_3 and the surface elevation of C.
2. Given all pipe lengths and diameters, the elevations of water surfaces of two reservoirs, and the flow to or from the third, find the elevation of the surface in the third reservoir. Suppose Q_2 and the surface elevations of A and C are given. Then the quantities $Q_1 - Q_3$ and $h_1 + h_3$ are known. These relations are solved simultaneously for their component parts in one of two ways: (a) by assuming successive distributions of the flows Q_1 and Q_2 satisfying the first relation, until a distribution is found which also satisfies the head-loss relation; (b) by assuming successive elevations of the piezometer level P, which is

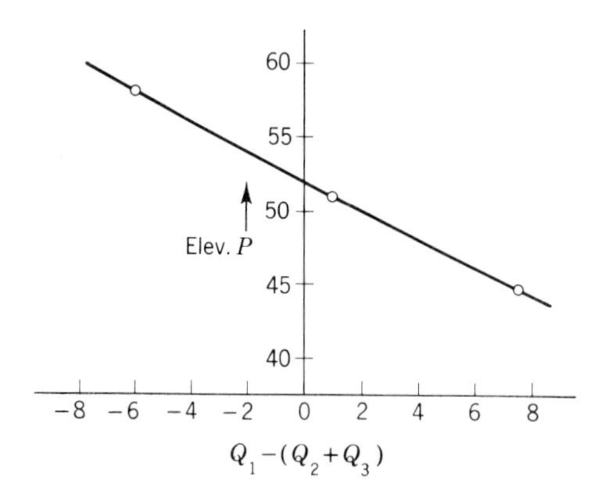

Figure 8.28

to say, distributions of h_1 and h_3 satisfying the second relation above, until a level is found which also satisfies the discharge relation. With P known and h_2 determined by the given discharge Q_2, the elevation of B is easily obtained.

3. Given all pipe lengths and diameters and the elevations of all three reservoirs, find the flow in each pipe. This is the classic *three-reservoir problem*, and it differs from the foregoing cases in that it is not immediately evident whether the flow is *into* or *out of* reservoir B. This direction is readily determined by first assuming no flow in pipe 2; that is, the piezometer level P is assumed at the elevation of the surface of B. The head losses h_1 and h_3 then determine the flows Q_1 and Q_3, and depending on whether $Q_1 > Q_3$ or $Q_1 < Q_3$, the condition of continuity is determined as $Q_1 = Q_2 + Q_3$ or $Q_1 + Q_2 = Q_3$, respectively. From this point the solution proceeds as in (*b*) of case 2 above. The piezometer level is moved up or down by trial until the resulting flow distribution satisfies the continuity relation. In reaching the final adjusted level it is helpful to make a small plot such as is shown in Fig. 8.28 for the case where $Q_1 = Q_2 + Q_3$. Two or three points, with one fairly close to the axis, determine a curve which intersects the vertical axis at the equilibrium level of P, that is, for the condition $Q_1 - (Q_2 + Q_3) = 0$.

8.25 PIPES IN SERIES[1]

The discussion in Sec. 8.22 was restricted to the case of a single pipe. If the pipe is made up of sections of different diameters, as shown in Fig. 8.29, the continuity and energy equations establish the following two simple relations which must be satisfied:

$$Q = Q_1 = Q_2 = Q_3 = \cdots \tag{8.56}$$

$$h_L = h_{L1} + h_{L2} + h_{L3} + \cdots \tag{8.57}$$

[1] Once again it should be mentioned that either the Hazen–Williams or Manning equation can be used to solve pipe-flow problems, though the Darcy–Weisbach approach is best.

If the rate of discharge is given, the head loss may be found directly by adding the contributions from the various sections, as in Eq. (8.57).

If the total head loss is given and the flow is required, either of two methods may be used, the equivalent-velocity-head method or the equivalent-length method. In the former, the procedure is to write Eq. (8.57) for the head loss in each length in terms of the dimensions applying to it; i.e.,

$$h_L = f_1 \frac{L_1}{D_1} \frac{V_1^2}{2g} + f_2 \frac{L_2}{D_2} \frac{V_2^2}{2g} + \cdots$$

where the values of the friction factor f are chosen from Fig. 8.11 to be in the range of reasonable values for the given pipes. By the equation of continuity, the individual section losses may be expressed in terms of one of the velocity heads. When minor losses are to be included, these may also be placed as additive terms in Eq. (8.57) and expressed in terms of the same velocity head. Thus, for any pipeline, no matter how complex, the total loss of head may be written as

$$h_L = K \frac{V^2}{2g} \tag{8.58}$$

This equation may now be solved for V, and then Q can be computed. For better accuracy the assumed values of f may be adjusted and a second solution obtained.

In the equivalent-length method all pipes are expressed in terms of equivalent lengths of one given pipe size, usually the one which figures most prominently in the system. By equivalent length is meant a length L_e of pipe of a certain diameter D_e and friction factor f_e which for the same flow will give the same head loss as the pipe under consideration of length L, diameter D, and friction factor f. As usual, the f values are initially assumed, but they may be improved by computing **R** from the initial results and entering the Moody diagram (Fig. 8.11). Thus from the pipe-friction and continuity equations,

$$L_e = L \frac{f}{f_e} \frac{D_e}{D} \frac{V^2/2g}{V_e^2/2g} = L \frac{f}{f_e} \left(\frac{D_e}{D}\right)^5 \tag{8.59}$$

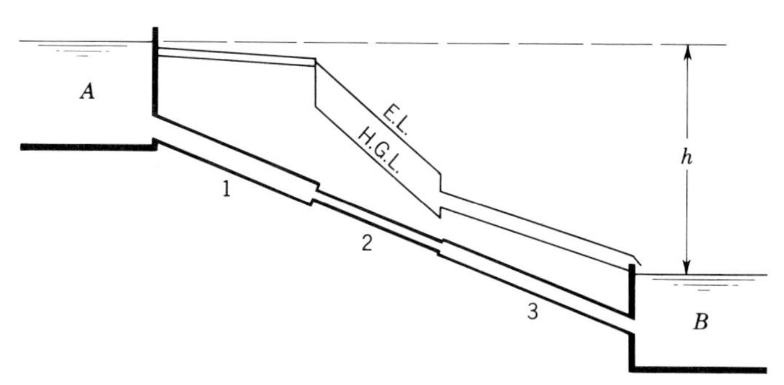

Figure 8.29 Pipes in series.

In case Eqs. (8.45) or (8.46) are being used, the equivalent length is established by the relation $S = h_L/L$, or $L_e = L(S/S_e)$, where the values of the energy gradient are obtained for any assumed rate of discharge. Minor losses can also be handled using the equivalent-length method.

Illustrative Example 8.8 Suppose in Fig. 8.29 the pipes 1, 2, and 3 are 300 m of 30-cm-diameter, 150 m of 20-cm-diameter, and 250 m of 25-cm-diameter, respectively, of new cast iron and are conveying 15°C water. If $h = 10$ m, find the rate of flow from *A to B*.

(*a*) BY THE EQUIVALENT-VELOCITY-HEAD METHOD. For cast-iron pipe $e = 0.25$ mm (Table 8.1); hence the corresponding values for e/D are: 0.00083, 0.00125, and 0.0010, and from Fig. 8.11 we will assume $f_1 = 0.019$, $f_2 = 0.021$, and $f_3 = 0.020$. Then,

$$h = 10 = 0.019\left(\frac{300}{0.3}\right)\frac{V_1^2}{2g} + 0.021\left(\frac{150}{0.2}\right)\frac{V_2^2}{2g} + 0.020\left(\frac{250}{0.25}\right)\frac{V_3^2}{2g}$$

From continuity

$$\frac{V_2^2}{2g} = \frac{V_1^2}{2g}\left(\frac{D_1}{D_2}\right)^4 = \frac{V_1^2}{2g}\left(\frac{30}{20}\right)^4 = 5.06\frac{V_1^2}{2g}$$

Similarly

$$\frac{V_3^2}{2g} = 2.07\frac{V_1^2}{2g}$$

and thus

$$10 = \frac{V_1^2}{2g}\left(0.019\frac{1{,}000}{1} + 0.021\frac{750}{1}5.06 + 0.020\frac{1{,}000}{1}2.07\right)$$

from which

$$\frac{V_1^2}{2g} = 0.071 \text{ m}$$

Hence

$$V_1 = \sqrt{2(9.81 \text{ m/s}^2)(0.072 \text{ m})} = 1.18 \text{ m/s}$$

The corresponding values of **R** are 0.31×10^6, 0.47×10^6, and 0.37×10^6; the corresponding friction factors are only slightly different from those originally assumed since the flow is at Reynolds numbers very close to those at which the pipes behave as rough pipes.

Hence

$$Q = A_1V_1 = \frac{\pi}{4}(0.30)^2 1.18 = 0.083 \text{ m}^3\text{/s}$$

Greater accuracy would have been obtained if the friction factors had been adjusted to match the pipe-friction chart more closely and if minor losses had been included. In that case, $Q = 0.081$ m^3/s.

(*b*) BY THE EQUIVALENT-LENGTH METHOD. Choose the 30-cm pipe as the standard. Using the above values of f in Eq. (8.59)

Pipe 2:

$$L_e = 150\left(\frac{0.021}{0.019}\right)\left(\frac{30}{20}\right)^5 = 1{,}260 \text{ m of 30-cm pipe}$$

Pipe 3:

$$L_e = 250\left(\frac{0.020}{0.019}\right)\left(\frac{30}{25}\right)^5 = 650 \text{ m of 30-cm pipe}$$

$$\text{Add pipe 1} = 300 \text{ m of 30-cm pipe}$$

$$\text{Total } L_e = 2{,}210 \text{ m of 30-cm pipe}$$

Thus

$$h = 10 = 0.019 \frac{2{,}210}{0.30} \frac{V_1^2}{2g}$$

$$\frac{V_1^2}{2g} = 0.071 \text{ m} \qquad V_1 = 1.18 \text{ m/s} \qquad \text{and} \qquad Q = 0.083 \text{ m}^3\text{/s}$$

as above.

8.26 PIPES IN PARALLEL

In the case of flow through two or more parallel pipes, as in Fig. 8.30, the continuity and energy equations establish the following relations which must be satisfied:

$$Q = Q_1 + Q_2 + Q_3 \tag{8.60}$$

and

$$h_L = h_{L1} = h_{L2} = h_{L3} \tag{8.61}$$

as the pressures at A and B are common to all pipes. If the head loss is given, the total discharge may be computed directly by adding the contributions from the various pipes, as in Eq. (8.60).

If the total flow is given and the head loss and distribution of flow among the pipes are required, an approximate solution may be obtained by assuming a reasonable value of h_L and computing the resulting individual flows and the *percentage* distribution of flow. This percentage distribution will not change greatly with the magnitude of the flow and may then be applied to find the actual distribution of the total discharge. The accuracy of the solution may be checked by comparing the computed head losses in the separate pipes. They should be the same. If they are not the same, the assumed values of f can be corrected to match the Moody diagram (Fig. 8.11). A more accurate procedure is to write Eq. (8.61) for the flow in each pipe in terms of the dimensions applying to it. This may be accomplished by observing that the loss of head in any pipe is

$$h_L = \left(f \frac{L}{D} + \sum k \right) \frac{V^2}{2g}$$

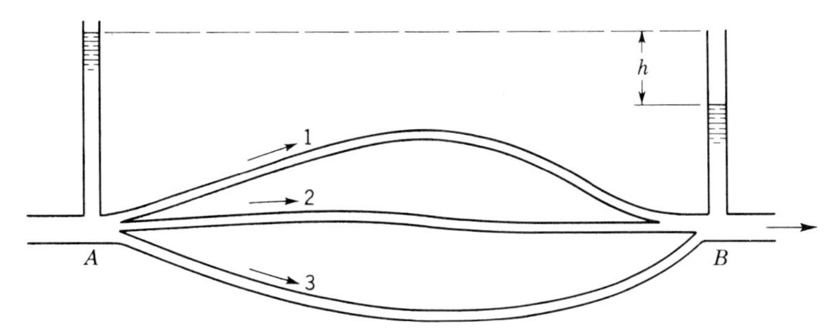

Figure 8.30 Pipes in parallel.

where $\sum k$ is the sum of the minor-loss coefficients, which may usually be neglected if the pipe is longer than 1,000 diameters. Solving for V and then Q, the following is obtained for pipe 1:

$$Q_1 = A_1 V_1 = A_1 \sqrt{\frac{2gh_L}{f_1(L_1/D_1) + \sum k}} = C_1 \sqrt{h_L} \tag{8.62}$$

where C_1 is constant for the given pipe, except for the change in f with Reynolds number. The flows in the other pipes may be similarly expressed, using reasonable values of f from Fig. 8.11. Finally, Eq. (8.57) becomes

$$Q = C_1\sqrt{h_L} + C_2\sqrt{h_L} + C_3\sqrt{h_L} = \sqrt{h_L}(C_1 + C_2 + C_3)$$

This enables a first determination of h_L and the distribution of flows and velocities in the pipes. Adjustments in the values of f may be made next, if indicated, and finally a corrected determination of h_L and the distribution of flows.

It is instructive to compare the solution methods for pipes in parallel with those for pipes in series. The role of the head loss in one case becomes that of the discharge rate in the other, and vice versa. The student is already familiar with this situation from the elementary theory of dc circuits. The flow corresponds to the electrical current, the head loss to the voltage drop, and the frictional resistance to the ohmic resistance. The outstanding deficiency in this analogy occurs in the variation of potential drop with flow, which is with the first power in the electrical case ($E = IR$) and with the second power in the hydraulic case ($h_L \propto V^2 \propto Q^2$) for fully developed turbulent flow.

Illustrative Example 8.9 Three pipes *A*, *B*, and *C* are interconnected as shown. The pipe characteristics are as follows:

Pipe	D (in)	L (ft)	f
A	6	2000	0.020
B	4	1600	0.032
C	8	4000	0.024

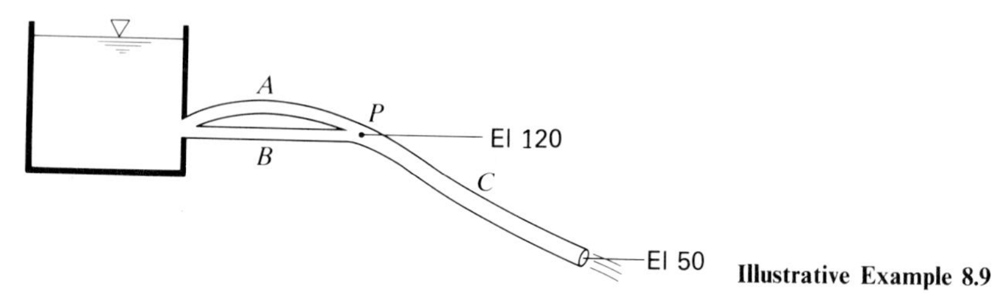

Illustrative Example 8.9

Find the rate at which water will flow in each pipe. Find also the pressure at point *P*. All pipe lengths are much greater than 1,000 diameters, therefore minor losses may be neglected.

Energy Eq:

$$200 - 0.020\frac{2000}{6/12}\frac{V_A^2}{2g} - 0.024\frac{4000}{8/12}\frac{V_C^2}{2g} = 50$$

i.e.

$$150 = 80\frac{V_A^2}{2g} + 144\frac{V_C^2}{2g}$$

Continuity: $$6^2V_A + 4^2V_B = 8^2V_C$$

i.e. $$36V_A + 16V_B = 64V_C$$

Also, $$h_{L_A} = h_{L_B} = 0.020\frac{2000}{6/12}\frac{V_A^2}{2g} = 0.032\frac{1600}{4/12}\frac{V_B^2}{2g}$$

i.e. $$80V_A^2 = 153.6V_B^2, \qquad V_B = 0.0722V_A$$

Substituting into continuity,

$$36V_A + 16(0.722V_A) = 64V_C$$

$$47.5V_A = 64V_C \qquad V_A = 1.346V_C$$

Substituting into the energy equation,

$$150 = 80\frac{(1.346V_C)^2}{2g} + 144\frac{V_C^2}{2g} = 288.9\frac{V_C^2}{2g}$$

$$V_C^2 = 64.4(150/288.9) = 33.4$$

$$V_C = 5.88 \text{ fps} \qquad Q_C = A_C V_C = (0.349)5.78 = 2.02 \text{ cfs}$$

$$V_A = 1.346V_C = 7.78 \text{ fps} \qquad Q_A = (0.196)7.78 = 1.53 \text{ cfs}$$

Continuity: $$36(7.78) + 16V_B = 64(5.78)$$

$$280.2 + 16V_B = 370.1$$

$$V_B = 89.9/16 = 5.62 \text{ fps}$$

$$Q_B = A_B V_B = (0.0873)5.62 = 0.490 \text{ cfs}$$

As a check, note that $Q_A + Q_B = Q_C$

To find the pressure at P

$$200 - 80\frac{V_A^2}{2g} = 120 + p_P/\gamma$$

$$p_P/\gamma = 80 - 80\frac{(7.78)^2}{2g} = 4.75 \text{ ft}$$

Check: $$120 + p_P/\gamma - 144\frac{V_C^2}{2g} = 50$$

$$p_P/\gamma = 144\frac{(5.78)^2}{2g} - 70 = 4.75 \text{ ft}$$

So $p_P/\gamma = 4.75$ ft and $p_P = (62.4/144)4.75 = 2.06$ psi

In this example it was assumed that the values of f for each pipe were known. Actually f depends on **R** (Fig. 8.11). Usually the absolute roughness e of each pipe is known or assumed and an accurate solution is achieved through trial and error until the f's and **R**'s for each pipe agree with the Moody diagram (Fig. 8.11).

8.27 PIPE NETWORKS

An extension of pipes in parallel is a case frequently encountered in municipal distribution systems, in which the pipes are interconnected so that the flow to a

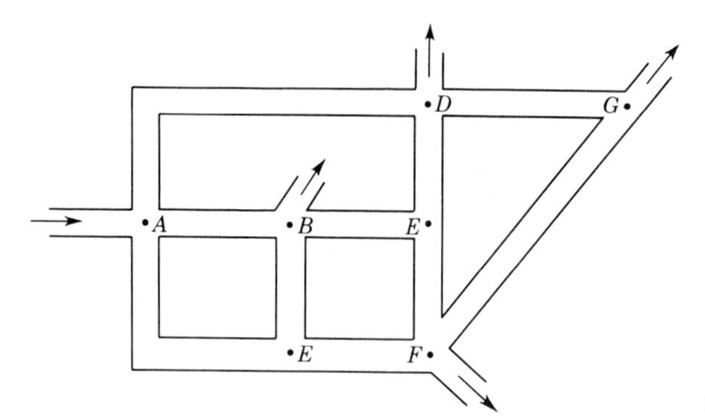

Figure 8.31 Pipe network.

given outlet may come by several different paths, as shown in Fig. 8.31. Indeed, it is frequently impossible to tell by inspection which way the flow travels, as in pipe *BE*. Nevertheless, the flow in any network, however complicated, must satisfy the basic relations of continuity and energy as follows:

1. The flow into any junction must equal the flow out of it.
2. The flow in each pipe must satisfy the pipe-friction laws for flow in a single pipe.
3. The algebraic sum of the head losses around any closed loop must be zero.

Pipe networks are generally too complicated to solve analytically, as was possible in the simpler cases of parallel pipes (Sec. 8.26). A practical procedure is the method of successive approximations, introduced by Cross.[1] It consists of the following elements, in order:

1. By careful inspection assume the most reasonable distribution of flows that satisfies conditions 1.
2. Write condition 2 for each pipe in the form

$$h_L = KQ^n \tag{8.63}$$

   where K is a constant for each pipe. For example, the standard pipe-friction equation in the form of Eq. (8.62) would yield $K = 1/C^2$ and $n = 2$ for constant f. The empirical formulas (8.45) and (8.46) are seen to be readily reducible to the desired form. Minor losses within any loop may be included, but minor losses at the junction points are neglected.
3. To investigate condition 3, compute the algebraic sum of the head losses around each elementary loop, $\sum h_L = \sum KQ^n$. Consider losses from clockwise flows as positive, counterclockwise negative. Only by good luck will these add to zero on the first trial.

[1] Hardy Cross, Analysis of Flow in Networks of Conduits or Conductors, *Univ. Ill. Eng. Expt. Sta. Bull.* 286, 1936.

4. Adjust the flow in each loop by a correction, ΔQ, to balance the head in that loop and give $\sum KQ^n = 0$. The heart of this method lies in the determination of ΔQ. For any pipe we may write

$$Q = Q_0 + \Delta Q$$

where Q is the correct discharge and Q_0 is the assumed discharge. Then, for each pipe,

$$h_L = KQ^n = K(Q_0 + \Delta Q)^n = K(Q_0^n + nQ_0^{n-1}\,\Delta Q + \cdots)$$

If ΔQ is small compared with Q_0, the terms of the series after the second one may be neglected. Now, for a circuit, with ΔQ the same for all pipes,

$$\sum h_L = \sum KQ^n = \sum KQ_0^n + \Delta Q \sum KnQ_0^{n-1} = 0$$

As the corrections of head loss in all pipes must be summed *arithmetically*, we may solve this equation for ΔQ,

$$\Delta Q = \frac{-\sum KQ_0^n}{\sum |KnQ_0^{n-1}|} = \frac{-\sum h_L}{n\sum |h_L/Q_0|} \tag{8.64}$$

as, from Eq. (8.63), $h_L/Q = KQ^{n-1}$. It must be emphasized again that the numerator of Eq. (8.64) is to be summed algebraically, with due account of sign, while the denominator is summed arithmetically. The negative sign in Eq. (8.64) indicates that when there is an excess of head loss around a loop in the clockwise direction, the ΔQ must be subtracted from clockwise Q_0's and added to counterclockwise ones. The reverse is true if there is a deficiency of head loss around a loop in the clockwise direction.

5. After each circuit is given a first correction, the losses will still not balance because of the interaction of one circuit upon another (pipes which are common to two circuits receive two independent corrections, one for each circuit). The procedure is repeated, arriving at a second correction, and so on, until the corrections become negligible.

Either form of Eq. (8.64) may be used to find ΔQ. As values of K appear in both numerator and denominator of the first form, values proportional to the actual K may be used to find the distribution. The second form will be found most convenient for use with pipe-friction diagrams for water pipes.

An attractive feature of the approximation method is that errors in computation have the same effect as errors in judgment and will eventually be corrected by the process.

The pipe-network problem lends itself well to solution by use of a digital computer.[1] Programming takes time and care, but once set up, there is great flexibility and many man-hours of labor can be saved.

[1] Lyle N. Hoag and Gerald Weinberg, Pipeline Network Analysis by Electronic Digital Computer, *J. Am. Water Works Assoc.*, vol. 49. pp. 517–529, 1957.

Illustrative Example 8.10 If the flows into and out of a two-loop pipe system are as shown, determine the flow in each pipe. The K values for each pipe were calculated from the pipe and minor loss characteristics and from an assumed value of f.

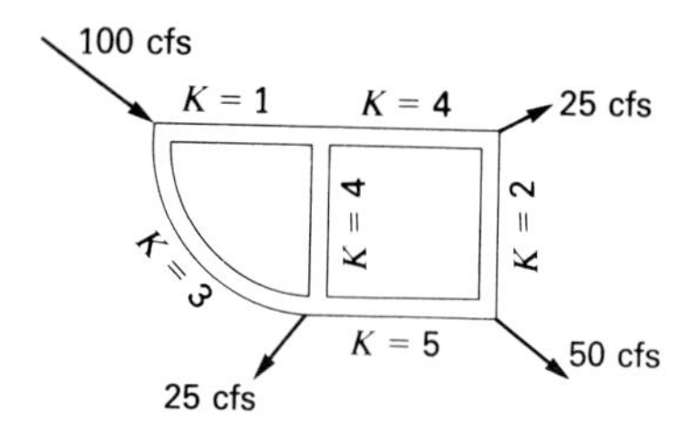

As a first step, assume flow in each pipe such that continuity is satisfied at all junctions. Calculate ΔQ for each loop, make corrections to the assumed Q's and repeat several times until the ΔQ's are quite small. As a final step the values of f for each pipe should be checked against the Moody diagram and modified, if necessary.

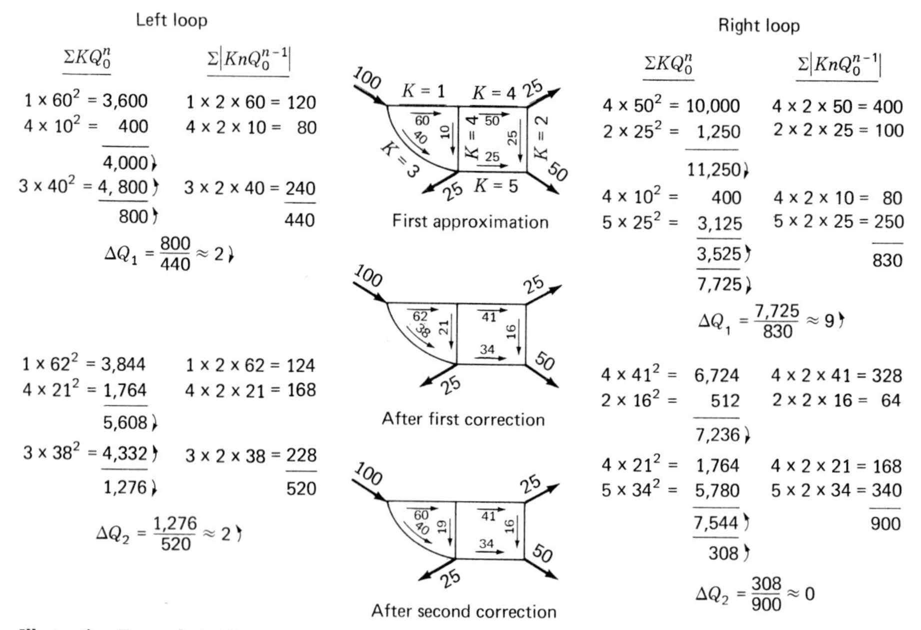

Illustrative Example 8.10

Further corrections can be made if greater accuracy is desired.

PROBLEMS

8.1 An oil with a kinematic viscosity of 0.00015 ft^2/s (0.135 St) flows through a pipe of diameter 6 in (15 cm). Below what velocity will the flow be laminar?

8.2 An oil with a kinematic viscosity of 0.005 ft^2/s flows through a 3-in-diameter pipe with a velocity of 10 fps. Is the flow laminar or turbulent?

8.3 Hydrogen at atmospheric pressure and a temperature of 50°F has a kinematic viscosity of 0.0011 ft^2/s. Determine the maximum laminar flow rate in pounds per second in a 2-in-diameter pipe. At this flow rate what is the average velocity?

8.4 Air at a pressure of approximately 1,500 kN/m^2, abs and a temperature of 100°C flows in a 1.5-cm-diameter tube. What is the maximum laminar flow rate? Express answer in liters per second, newtons per second, and kilograms per second. At this flow rate what is the average velocity?

8.5 What is the hydraulic radius of a rectangular air duct 6 by 14 in?

8.6 What is the percentage difference between the hydraulic radii of a 20-cm-diameter and a 20-cm-square duct?

8.7 Two pipes, one circular and one square, have the same cross-sectional area. Which has the larger hydraulic radius, and by what percentage?

8.8 Steam with a specific weight of 0.25 lb/ft^3 (40 N/m^3) flows with a velocity of 100 fps (30 m/s) through a circular pipe. The friction factor f was found to have a value of 0.016. What is the shearing stress at the wall?

8.9 Find the head loss per unit length when oil ($s = 0.9$) of viscosity 0.007 ft^2/s (0.00065 m^2/s) flows in a 3-in (7.5-cm)-diameter pipe at a rate of 5 gpm (0.30 L/s).

8.10 Tests made on a certain 12-in-diameter pipe showed that, when $V = 10$ fps, $f = 0.015$. The fluid used was water at 60°F. Find the unit shear at the wall and at radii of 0, 0.2, 0.3, 0.5, 0.75 times the pipe radius.

8.11 If the oil of Prob. 8.2 weighs 58 lb/ft^3, what will be the flow rate and head loss in a 3,000-ft length of 4-in-diameter pipe when the Reynolds number is 800?

8.12 With laminar flow in a circular pipe, at what distance from the centerline does the average velocity occur?

8.13 With laminar flow in a circular pipe, find the velocities at $0.1r$, $0.3r$, $0.5r$, $0.7r$, and $0.9r$. Plot the velocity profile.

8.14 Prove that the centerline velocity is twice the average velocity when laminar flow occurs in a circular pipe.

8.15 When laminar flow occurs in a two-dimensional passage, find the relation between the average and maximum velocities.

8.16 With laminar flow between two parallel, flat plates a small distance d apart, at what distance from the centerline will the velocity be equal to the mean velocity?

8.17 How much power is lost per meter of pipe length when oil with a viscosity of 0.20 N·s/m^2 flows in a 20-cm-diameter pipe at 0.50 L/s? The oil has a density of 840 kg/m^3.

8.18 In Prob. 8.2 what will be the approximate distance from the pipe entrance to the first point at which the flow is established?

8.19 The absolute viscosity of water at 60°F is 0.0000236 lb·s/ft^2. (*a*) If at a distance of 3 in from the center of the pipe of Prob. 8.10 the velocity profile gives a value for du/dy of 4.34 per second, find the viscous shear and the turbulent shear at that radius. (*b*) What is the value of the mixing length l, and what is the value of the ratio l/r_0?

8.20 Water at 60°F enters a pipe with a uniform velocity of 10 fps. (*a*) What is the distance at which the transition occurs from a laminar to a turbulent boundary layer? (*b*) If the thickness of this initial laminar boundary layer is given by $4.91\sqrt{\nu x/U}$, what is the thickness reached by it at the point of transition?

8.21 Water in a pipe is at a temperature of 60°F (15°C). (*a*) If the mean velocity is 12 fps (3.5 m/s), and the value of f is 0.015, what is the nominal thickness δ_l of the viscous sublayer? (*b*) What will be the thickness of the viscous sublayer if the velocity is increased to 20 fps (5.8 m/s) and f does not change?

8.22 For the data in Prob. 8.21(*a*), what is the distance from the wall to the assumed limit of the transition region where true turbulent flow begins?

8.23 Water at 40°C flows in a 20-cm-diameter pipe with $V = 5$ m/s and $e = 0.12$ mm. Head loss measurements indicate that $f = 0.022$. What is the thickness of the viscous sublayer? Is the pipe behaving as a wholly rough pipe?

8.24 Express Eq. (8.28) in terms of pipe diameter, friction factor, and Reynolds number.

8.25 Compute δ_l for the data of Illustrative Example 8.3.

8.26 Find the head loss in a 6-in-diameter pipe having $e = 0.042$ in when oil ($s = 0.90$) having a viscosity of 0.0008 lb·s/ft^2 flows at a rate of 15 cfs. Determine the shear stress at the wall of the pipe and the velocity at 2.0 in from centerline.

8.27 When water at 150°F (70°C) flows in a 0.75-in (2-cm)-diameter copper tube at 1.0 gpm (0.05 L/s), find the head loss per 100 ft (10 m). What is the centerline velocity, and what is the value of δ_l?

8.28 Repeat Prob. 8.27 for flow rates of 0.05 and 20 gpm.

8.29 Find the head loss in a 10-cm-diameter pipe having $e = 0.85$ mm when oil ($s = 0.82$) having a viscosity of 0.0052 N·s/m^2 flows at a rate of 40 L/s. Determine the shear stress at the wall of the pipe. Find the velocity 2 cm from the centerline. Under these conditions is this pipe behaving as a wholly rough, transitional, or smooth pipe?

8.30 The velocities in a 90-cm-diameter pipe are measured as 5.00 m/s on the centerline and 4.82 m/s at $r = 10$ cm. Approximately what is the flow rate?

8.31 The velocities in a 36-in-diameter pipe are measured as 15.0 fps at $r = 0$ in and 14.5 fps at $r = 4$ in. Approximately what is the flow rate?

8.32 With turbulent flow in a circular pipe prove that the mean velocity occurs at a distance of approximately $0.78r_0$ from the center line of the pipe.

8.33 The flow rate in a 12-in (30-cm)-diameter pipe is 8 cfs (0.225 m^3/s). The flow is known to be turbulent, and the centerline velocity is 12.0 fps (3.70 m/s). Plot the velocity profile, and determine the head loss per foot (m) of pipe.

8.34 Refer to the data of Prob. 8.29. Above what flow rate will this pipe behave as a wholly rough pipe? Below what flow rate will it behave as a smooth pipe?

8.35 Kerosene ($s = 0.81$) flows at a temperature of 80°F (26.5°C) in a 2-in (5-cm)-diameter smooth brass pipeline at a rate of 10 gpm (0.60 L/s). (*a*) Find the head loss per foot (m). (*b*) For the same head loss what would be the rate of flow if the temperature of the kerosene were raised to 120°F (50°C)?

8.36 Substitute into Eq. (8.40) the given and computed data of Illustrative Example 8.4 to verify the validity of the equation.

8.37 Water at 40°C flows in a 20-cm-diameter pipe with $V = 5$ m/s. Head loss measurements indicate that $f = 0.022$. Determine the value of e. Find the shear stress at the wall of the pipe and at $r = 4$ cm. What is the value of du/dy at $r = 4$ cm?

8.38 Water at 15°C flows through a 20-cm-diameter pipe with $e = 0.01$ mm. (*a*) If the mean velocity is 3.5 m/s, what is the nominal thickness δ_l of the viscous sublayer? (*b*) What will be the thickness of the viscous sublayer if the velocity is increased to 5.8 m/s? (Compare with Prob. 8.21.)

8.39 When water at 50°F flows at 3.0 cfs in a 24-in pipeline, the head loss is 0.0003 ft/ft. What will be the head loss when glycerin at 68°F flows through this same pipe at the same rate?

8.40 Air flows at 50 lb/min in a 4-in-diameter welded-steel pipe at 100 psia and 60°F. Determine the head loss and pressure drop in 150 ft of this pipe. Assume the air to be of constant density.

8.41 Air flows at an average velocity of 0.5 m/s through a long 3.2-m-diameter circular tunnel. Find the head-loss gradient at a point where the air temperature and pressure are 15°C and 108 kN/m^2, abs respectively. Assume $e = 2$ mm. Find also the shear stress at the wall and the thickness δ_l of the viscous sublayer.

8.42 Repeat Prob. 8.41 for the case where the average velocity is 5.0 m/s.

8.43 Air at 60°F (20°C) and atmospheric pressure flows with a velocity of 20 fps (5 m/s) through a 2-in (50-mm)-diameter pipe. Find the head loss in 50 ft (20 m) of pipe having $e = 0.0001$ in (0.0025 mm).

8.44 What is the head loss per foot of pipe when oil ($s = 0.90$), having a viscosity of 2×10^{-4} lb·s/ft^2 (9.6×10^{-3} N·s/m^2) flows in a 2-in (5-cm)-diameter welded-steel pipe at 0.15 cfs (4.2 L/s)?

8.45 Consider water at 50°F flowing in a 36-in-diameter concrete pipe ($e = 0.02$ in). Determine **R**, τ_0, u_{max}/V, δ_l, δ_l/e, and the flow regime (hydraulic smoothness) for flow rates of 200, 20, 2, 0.2, and 0.02 cfs.

8.46 Find the flow rate when 60°F water causes a head loss of 0.25 ft in 100 ft of average cast-iron pipe. Diameter of pipe is 6 in.

8.47 Gasoline with a kinematic viscosity of 6×10^{-6} ft²/s (5×10^{-7} m²/s) flows in a 12-in (30-cm)-diameter smooth pipe. Find the flow rate when the head loss is 0.4 ft per 100 ft (0.4 m per 100 m).

8.48 What size pipe is required to carry oil having a kinematic viscosity of 0.0002 ft²/s at a rate of 8.0 cfs if the head loss is to be 0.4 ft·lb/lb per 100 ft of pipe length? Assume $e = 0.00015$ ft.

8.49 A straight, new, asphalted cast-iron pipe is 42 in in diameter and 1,000 ft long. Using the value of f as determined from Fig. 8.11, find the shear force on the pipe if the fluid is water at 72°F and the average velocity is 10 fps. What will be the shear force if the average velocity is reduced to 5 fps?

8.50 A steel pipe ($e = 0.0002$ ft) of length 15,000 feet is to convey oil ($\nu = 0.0006$ ft²/s) at a rate of 10 cfs from a reservoir of surface elevation 625 ft to one of surface elevation 400 ft. What pipe size would you select?

8.51 Water at 140°F flows in 0.824-in-diameter iron pipe ($e = 0.00015$ ft) of length 400 ft between points A and B. At point A the elevation of the pipe is 104.0 ft and the pressure 8.20 psi. At point B the elevation of the pipe is 99.5 ft and the pressure is 9.05 psi. Compute the flow rate as accurately as you can.

8.52 Make a plot of the values of α and β versus **R** for smooth brass pipes. Take a **R** range from 10^2 to 10^6. On the same plot show values of u_{max}/V.

8.53 Air at 50 psia and temperature of 150°F flows in a 12- by 18-in rectangular air duct at the rate of 1 lb/min. Find the head loss per 100 ft of duct. Express answer in feet of air flowing and in pounds per square inch. Assume $e = 0.0005$ in.

8.54 Find the approximate rate at which 60°F (15°C) water will flow in a conduit shaped in the form of an equilateral triangle if the head loss is 2 ft per 100 ft (2 m per 100 m). The cross-sectional area of the duct is 120 in² (775 cm²) and $e = 0.0018$ in (0.045 mm).

8.55 When fluid of specific weight 50 lb/ft³ flows in a 6-in-diameter pipe, the frictional stress between the fluid and the pipe wall is 0.5 lb/ft². Calculate the head loss per foot of pipe. If the flow rate is 2.0 cfs, how much power is lost per foot of pipe?

8.56 Find the value of the Hazen-Williams coefficient for the water flow in Prob. 8.39.

8.57 Find the value of the Hazen-Williams coefficient for the case where water flows at 0.18 m³/s in a 60-cm-diameter pipeline with a head loss of 0.0012 m/m.

8.58 In a field test of the 16-ft-diameter Colorado River aqueduct Manning's n was found to have a value of 0.0132 when 50°F water was flowing at a Reynolds number of 10.5×10^6. Determine the average value of e for this conduit.

8.59 Prove that for a constant rate of discharge and a constant value of f the friction head loss in a pipe varies inversely as the fifth power of the diameter.

8.60 Two long pipes are used to convey water between two reservoirs whose water surfaces are at different elevations. One pipe has a diameter twice that of the other. If both pipes have the same value of f and if minor losses are neglected, what is the ratio of the flow rates through the two pipes?

8.61 A 12-in (30-cm)-diameter pipe with a friction factor $f = 0.02$ conducts fluid between two tanks at 10 fps (3 m/s). The entrance and exit conditions to and from the pipe are flush with the wall of the tank. Find the ratio of the minor losses to the pipe friction loss if the length of the pipe is (*a*) 5 ft (1.5 m); (*b*) 100 ft (30 m); (*c*) 2,000 ft (600 m).

8.62 A smooth pipe 12 in (30 cm) in diameter and 300 ft (90 m) long has a flush entrance and a submerged discharge. The velocity is 10 fps (3 m/s). If the fluid is water at 60°F (15°C), what is the total loss of head?

8.63 Suppose that the fluid in Prob. 8.62 were oil with a kinematic viscosity of 0.001 ft²/s (9.3×10^{-5} m²/s) and a specific gravity of 0.925. What would be the head loss in feet (m) of oil and in pounds per square inch (kN/m²)?

8.64 A smooth pipe consists of 50 ft (15 m) of 8-in (20-cm) pipe followed by 300 ft (90 m) of 24-in (60-cm) pipe with an abrupt change of cross section at the junction. It has a flush entrance and a submerged discharge. If it carries water at 60°F (15°C) in the smaller pipe with a velocity of 18 fps (5.5 m/s), what is the total head loss?

8.65 A 6-in-diameter pipe ($f = 0.032$) of length 100 ft connects two reservoirs whose water-surface elevations differ by 10 ft. The pipe entrance is flush, and the discharge is submerged. (*a*) Compute the flow rate. (*b*) If the last 10 ft of pipe were replaced with a conical diffuser with a cone angle of 10°, compute the flow rate.

8.66 For two pipes in series with a diameter ratio of 1:2 and a velocity of 20 fps in the smaller pipe, find the loss of head due to (*a*) sudden contraction; (*b*) sudden enlargement; (*c*) expansion in a conical diffuser with a total angle of 20° and 6°.

8.67 In a 100-ft length of 4-in-diameter wrought-iron pipe there are one open globe valve, one medium-radius elbow, and one pipe bend ($k_b = 0.10$) with a radius of curvature of 40 in. The bend is 90°, and its length is not included in the 100 ft. No entrance or discharge losses are involved. If the fluid is water at 72°F and the velocity is 6 fps, what is the total head loss?

8.68 It has been found that with great care laminar flow can be maintained up to $\mathbf{R} = 50{,}000$. Compute the friction head per 100 ft of pipe for a Reynolds number of 50,000 if (*a*) the flow is laminar; (*b*) the flow is turbulent in a smooth pipe; (*c*) the flow is turbulent in a rough pipe with $e/D = 0.05$. Consider two situations, one where the fluid is 60°F water, the other where the fluid is SAE 10 oil at 150°F. Pipe diameter is 2.0 in.

8.69 Water at 60°F flows through 10,000 ft of 12-in-diameter pipe between two reservoirs whose water-surface elevation difference is 200 ft. (*a*) Find the flow rate if $e = 0.0018$ in. (*b*) Find the flow rate if e were twenty times larger.

8.70 How large a wrought-iron pipe is required to convey oil ($s = 0.9$, $\mu = 0.0008$ lb·s/ft^2) from one tank to another at a rate of 1.0 cfs if the pipe is 3,000 ft long and the difference in elevation of the free liquid surfaces is 40 ft?

8.71 If the diameter of a pipe is doubled, what effect does this have on the flow rate for a given head loss? Consider (*a*) laminar flow; (*b*) turbulent flow.

8.72 A 6-in (15-cm)-diameter pipeline 400 ft (120 m) long discharges a 2-in (5-cm)-diameter jet of water into the atmosphere at a point which is 200 ft (60 m) below the water surface at intake. The entrance to the pipe is a projecting one, with $k_e = 0.9$, and the nozzle loss coefficient is 0.05. Find the flow rate and the pressure head at the base of the nozzle, assuming $f = 0.03$.

8.73 A 2.0-m-diameter concrete pipe of length 1560 m for which $e = 1.5$ mm conveys 12°C water between two reservoirs at a rate of 8.0 m^3/s. What must be the difference in water surface elevation between the two reservoirs?

8.74 A pipe with an average diameter of 62 in is 6,272 ft long and delivers water to a powerhouse at a point 1,375 ft lower in elevation than the water surface at intake. Assume $f = 0.025$. When the pipe delivers 300 cfs, what is its efficiency? What is the horsepower delivered to the plant?

8.75 Find the kilowatt loss in 1,000 m of 50-cm-diameter pipe for which $e = 0.05$ mm when 45°C crude oil ($s = 0.855$) flows at 0.22 m^3/s.

8.76 California crude oil, warmed until its kinematic viscosity is 0.0004 ft^2/s, is pumped through a 2-in pipe ($e = 0.001$ in). Its specific weight is 59.8 lb/ft^3. (*a*) At what maximum velocity would the flow still be laminar? (*b*) What would then be the loss in energy head in pounds per square inch per 1,000 ft of pipe? (*c*) What would be the loss in energy head per 1,000 ft if the velocity were three times the value in (*a*)?

8.77 Water flows at 10 fps (3 m/s) through a vertical 6-in (15-cm)-diameter pipe standing in a body of water with its lower end 3 ft (0.9 m) below the surface. Considering all losses and with $f = 0.025$, find the pressure at a point 10 ft (3 m) above the surface of the water when the flow is (*a*) downward; (*b*) upward.

8.78 A horizontal pipe 6 in (15 cm) in diameter and for which $f = 0.025$ projects into a body of water 3 ft (1 m) below the surface. Considering all losses, find the pressure at a point 13 ft (4 m) from the end of the pipe if the velocity is 10 fps (3 m/s) and the flow is (*a*) from the pipe into the body of water; (*b*) from the body of water into the pipe.

8.79 A pipe runs from one reservoir to another, both ends of the pipe being under water. The intake is nonprojecting. The length of pipe is 500 ft (150 m), its diameter is 10.25 in (26 cm), and the difference in water levels in the two reservoirs is 110 ft (33.5 m). If $f = 0.02$, what will be the pressure at a point 300 ft (90 m) from the intake, the elevation of which is 120 ft (36 m) lower than the surface of the water in the upper reservoir?

8.80 A pipeline runs from one reservoir to another, both ends being under water, and the intake end is nonprojecting. The difference in water levels in the two reservoirs is 110 ft, and the length of pipe is 500 ft. (*a*) What is the discharge if the pipe is 10.25 in in diameter and $f = 0.022$? (*b*) When this same pipe is old, assume that the growth of tubercles has reduced the diameter to 9.5 in and that $f = 0.06$. What then will be the rate of discharge?

8.81 A pump delivers water through 300 ft (100 m) of 4-in (10 cm) fire hose to a nozzle which throws a 1-in (2.5-cm)-diameter jet. The loss coefficient of the nozzle is 0.04, and the value of f is 0.025. The nozzle is 20 ft (6 m) higher than the pump, and it is required that the jet velocity be 70 fps (20 m/s). What must be the pressure in the hose at the pump?

8.82 A jet of water is discharged through a nozzle at a point 200 ft below the water level at intake. The jet is 4 in in diameter, and the loss coefficient of the nozzle is 0.15. If the pipeline is 12 in in diameter, 500 ft long, with a nonprojecting entrance, what is the pressure at the base of the nozzle? Assume $f = 0.015$.

8.83 Refer to Fig. 8.23*b*. Suppose $\Delta z = 50$ ft (15 m) and the pipe is 600 ft (180 m) of 8-in (20-cm)-diameter pipe ($f = 0.025$). Find the jet diameter that will result in the greatest jet horsepower. Assume the nozzle loss coefficient is 0.05.

8.84 A riveted-steel pipeline 2,000 ft (600 m) long is 5 ft (1.5 m) is diameter. The lower end is 140 ft (42.4 m) below the level of the surface at intake and is joined to a turbine at this lower end. If the efficiency of the pipeline is 95 per cent, find the power delivered to the turbine. Assume the water temperature to be 60°F (15°C).

8.85 A 10-in (25-cm) pipeline is 3 miles (5 km) long. Let $f = 0.022$. If 4 cfs (0.1 m^3/s) of water is to be pumped through it, the total actual lift being 20 ft (6 m), what will be the power required if the pump efficiency is 70 percent?

8.86 In Fig. 8.24 assume pipe diameter = 10 in, $BC = 20$ ft, $DE = 3{,}000$ ft, and $\Delta z = 135$ ft. Assume $f = 0.022$. (*a*) If $Q = 7$ cfs of water and the pump efficiency is 80 per cent, what is the horsepower required? (*b*) If the elevation of C above the water surface is 13 ft, that of D is 15 ft, and that of E is 110 ft, compute the pressures at C, D, and E. (*c*) Sketch the energy line and the hydraulic grade line.

8.87 In Fig. 8.24 assume a pipe diameter of 40 cm, $BC = 10$ m, $DE = 850$ m, and $\Delta z = 45$ m. Assume $f = 0.018$. Find the maximum theoretical flow rate if 15°C water is being pumped at an altitude of 2,000 m above sea level. Point C is 5.0 m above the water surface of the lower reservoir.

8.88 In Fig. 8.24 assume that the pipe diameter is 3 in, $BC = 20$ ft, $DE = 200$ ft, and $\Delta z = 70$ ft. The elevation of C above the water surface is 15 ft. Assume $f = 0.04$. (*a*) If the pressure head at C is to be no less than -25 ft, what is the maximum rate at which the water is pumped? (*b*) If the efficiency of the pump is 60 percent, what is the horsepower required?

8.89 Refer to Fig. 8.25. Suppose that water-surface elevation, elevation of pump, and elevation of nozzle tip are 100, 90, and 120 ft, respectively. Pipe BC is 40 ft long, 8 in in diameter, $f = 0.025$; pipe DE is 200 ft long, 10 in in diameter, $f = 0.030$; jet diameter is 6 in, and nozzle loss coefficient is 0.04. Assume the pump is 80 percent efficient under all conditions of operation. Make a plot of flow rate and p_C/γ versus pump horsepower input. At what flow rate will cavitation occur in the pipe at C if the water is at 50°F and atmospheric pressure is 13.9 psia?

8.90 A 12-in (30-cm) pipe 10,000 ft (3,000 m) long for which $f = 0.02$ discharges freely into the air at an elevation 15 ft (4.5 m) lower than the surface of the water at intake. It is necessary that the flow be doubled by inserting a pump. If the efficiency of the latter is 70 per cent, what will be the power required?

8.91 When a certain pump is delivering 1.2 cfs (35 L/s) of water, a pressure gage at D (Fig. 8.25) reads 25 psi (175 kN/m^2), while a vacuum gage at C reads 10 in (25 cm) Hg. The pressure gage at D is 2 ft (60 cm) higher than the vacuum gage at C. The pipe diameters are 4 in (10 cm) for the suction pipe and 3 in (7.5 cm) for the discharge pipe. Find the power delivered to the water.

8.92 A certain turbine in a testing laboratory has been found to discharge 12 cfs (340 L/s) under a head h_t of 64 ft (19.5 m). It is to be installed near the end of a pipe 500 ft (150 m) long. The supply line (flush entrance) and discharge line (submerged exit) are both 12 in (30 cm) in diameter. The total fall from the surface of the water at intake to the surface of the tailwater is 48 ft (14.5 m). Assume $f = 0.02$ for the pipe. What will be the net head on the turbine, the rate of discharge, and the power delivered to it? Note that for turbines, $Q \propto \sqrt{h_t}$ (Eq. 14.11).

8.93 If in Illustrative Example 8.7 the vapor pressure of the liquid is 2.0 psia and the atmospheric pressure is 14.4 psia, what is the maximum theoretical flow rate?

8.94 In Fig. 8.26 assume pipe diameter = 12 in (30-cm), BC = 200 ft (60 m); Δz = 120 ft (36.5 m), and $f = 0.021$. The entrance to the pipe at the intake is flush with the wall. (*a*) If Q = 8 cfs (225 L/s) of water, what is the head supplied to the turbine? (*b*) What is the power delivered by the turbine if its efficiency is 75 per cent? (*c*) What is the efficiency of the penstock?

8.95 Assume the total fall from one body of water to another is 120 ft. The water is conducted through 200 ft of 12-in pipe with the entrance flush with the wall. Let $f = 0.021$. At the end of the pipe is a turbine and draft tube which discharged 5 cfs of water when tested under a head of 43.8 ft in another location. What would be the rate of discharge through the turbine and the net head on it under the present conditions? Note that for turbines, $Q \propto \sqrt{h_t}$ (Eq. 14.11).

8.96 A pump is installed to deliver water from a reservoir of surface elevation zero to another of elevation 200 ft. The 12-in-diameter suction pipe ($f = 0.020$) is 40 ft long, and the 10-in-diameter discharge pipe ($f = 0.032$) is 4,500 ft long. The pump head may be defined as $h_p = 300 - 20Q^2$, where h_p, the pump head, is in feet and Q in cubic feet per second. Compute the rate at which this pump will deliver the water. Also, what is the horsepower input to the water?

8.97 Suppose that, in Fig. 8.27, pipe 1 is 36-in (90 cm) smooth concrete, 5,000 ft (1500 m) long; pipe 2 is 24-in (60-cm) cast iron, 1,500 ft (450 m) long; and pipe 3 is 18-in (45-cm) cast iron, 4,000 ft (1200 m) long. The elevations of water surfaces in reservoirs A and B are 200 (60) and 150 ft (45 m), respectively, and the discharge Q_1 is 50 cfs (1.4 m^3/s). Find the elevation of the surface of reservoir C. Neglect minor losses and assume the energy line and hydraulic grade line are coincident. Use the $D''V$ scale on Fig. 8.11.

8.98 With the sizes and lengths and materials of pipes given in Prob. 8.97, suppose that the surface elevations of reservoirs A and C are 200 and 125 ft, respectively, and the discharge Q_2 is 20 cfs into reservoir B. Find the surface elevation of reservoir B.

8.99 Suppose, in Prob. 8.98, that the discharge Q_2 is known to be 20 cfs from reservoir B. Find the elevation of the surface of reservoir B.

8.100 Suppose, in Fig. 8.27, that pipes 1, 2, and 3 are 900 m of 60 cm, 300 m of 45 cm, and 1,200 m of 40 cm, respectively, of new welded-steel pipe. The surface elevations of A, B, and C are 30, 18, and 0 m, respectively. Find the approximate water flow in all pipes. Assume a normal temperature.

8.101 Suppose that, in Fig. 8.27, pipe 1 is 1,500 ft of 12-in new cast-iron pipe, pipe 2 is 800 ft of 6-in wrought-iron pipe, and pipe 3 is 1,200 ft of 8-in wrought-iron pipe. The water surface of reservoir B is 20 ft below that of A, while the junction J is 35 ft below the surface of A. In place of reservoir C, pipe 3 leads away to some other destination but its elevation at C is 60 ft below A. When the pressure head at J is 25 ft, find the flow in each pipe and the pressure head at C.

8.102 Suppose that in Fig. 8.29 pipes 1, 2, and 3 are 500 ft of 3.068-in (150 m of 78-mm), 200 ft of 2.067-in (60 m of 52-mm), and 400 ft of 2.469-in (120 m of 63-mm) wrought-iron pipe. With a total head loss of 20 ft (6 m) from A to B, find the flow of 60°F (15°C) water by the equivalent-velocity-head method.

8.103 Two pipes connected in series are respectively 100 ft of 1-in (e = 0.000005 ft) and 500 ft of 6-in (e = 0.0008 ft). With a total head loss of 25 ft, find the flow of 60°F water by the equivalent-velocity-head method.

8.104 Repeat Prob. 8.102 for the case where the fluid is an oil with $s = 0.9$, $\mu = 0.0008$ lb·s/ft^2 (0.04 N·s/m^2).

8.105 A pipeline 800 ft (240 m) long discharges freely at a point 150 ft (45 m) below the water level at intake. The pipe projects into the reservoir. The first 500 ft (150 m) is 12 in (30 cm) in diameter, and the remaining 300 ft (90 m) is 8 in (20 cm) in diameter. Find the rate of discharge, assuming $f = 0.04$. If the junction of the two sizes of pipe is 120 ft (35 m) below the intake water surface level, find the pressure just above C and just below C, where C denotes the point of junction. Assume a sudden contraction at C.

8.106 Repeat Prob. 8.105 neglecting minor losses.

8.107 Three new cast-iron pipes, having diameters of 30, 24, and 18 in, respectively, each 500 ft long, are connected in series. The 30-in pipe leads from a reservoir (flush entrance), and the 18-in discharges into the air at a point 12 ft below the water surface in the reservoir. Assuming all changes in section to be abrupt, find the rate of discharge of water at 60°F.

8.108 Suppose that in Fig. 8.30 pipes 1, 2, and 3 are 500 ft of 2-in, 300 ft of 3-in, and 700 ft of 4-in, respectively. The pipes are all smooth brass. When the total flow of 80°F crude oil ($s = 0.855$) is 0.6 cfs, find the head loss from A to B and the flow in each pipe.

8.109 Repeat Prob. 8.108 for the case where the flow rate is 0.06 cfs.

8.110 Suppose that in Fig. 8.30 pipes 1, 2, and 3 are copper tubing as follows: 80 m of 3 cm, 100 m of 4 cm and 80 m of 5 cm, respectively. When the total flow of 40°C crude oil ($s = 0.855$) is 5 L/s, find the head loss from A to B and the flow in each pipe.

8.111 Repeat Prob. 8.110 for the case where the total flow rate is 0.40 L/s.

8.112 The pipes in the system shown in the figure are all cast iron. (a) With a flow of 20 cfs, find the head loss from A to D. (b) What should be the diameter of a single pipe from B to C such that it replaces pipes 2, 3, and 4 without altering the capacity for the same head loss from A to D?

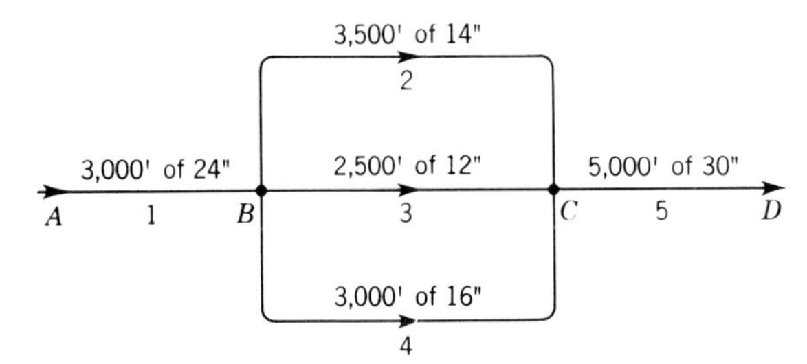

Problem 8.112

8.113 (a) With the same pipe lengths, sizes, and connections as in Prob. 8.112, find the flow in each pipe if the head loss from A to D is 100 ft and if all pipes have $f = 0.018$. Also find the head losses from A to B, B to C, and C to D. (b) Find the new head loss distributions and the percentage increase in the capacity of the system achieved by adding another 12-in pipe 2500 ft long between B and C.

8.114 Compute the flow in each pipe of the system shown in the figure, and determine the pressures at B and C. Pipe AB is 1,000 ft long, 6 in in diameter, and $f = 0.03$; pipe BC (upper) is 600 ft long, 4 in in diameter, and $f = 0.02$; pipe BC (lower) is 800 ft long, 2 in in diameter, and $f = 0.04$; pipe CD is 400 ft long, 4 in in diameter, and $f = 0.02$. The elevations are reservoir water surface, 100 ft; A, 80 ft; B, 50 ft; C, 40 ft; and D, 25 ft. There is free discharge to the atmosphere at D. Neglect velocity heads.

Problem 8.114

8.115 Refer to the figure for Prob. 8.114. Compute the flow in each pipe and determine the pressures at B and C. Pipe AB is 500 m long, 20 cm in diameter, and $f = 0.03$; pipe BC (upper) is 400 m long, 10 cm in diameter, and $f = 0.02$; pipe BC (lower) is 300 m long, 15 cm in diameter, and $f = 0.025$; pipe CD is 800 m long, 30 cm in diameter, and $f = 0.018$. The elevations are reservoir water surface, 100 m; $A = 80$ m, $B = 50$ m, $C = 40$ m, and $D = 15$ m. There is free discharge to the atmosphere at D. Neglect velocity heads..

8.116 Referring to the figure, when the pump develops 25 ft of head, the velocity of flow in pipe C is 4 fps. Neglecting minor losses, find (a) the flow rates in cubic feet per second in pipes A and B under these conditions and (b) the elevation of pipe B at discharge. The pipe characteristics are as follows: pipe A, 4,000 ft long, 2 ft in diameter, and $f = 0.03$; pipe B, 4,000 ft long, 1 ft in diameter, and $f = 0.03$; pipe C, 4,000 ft long, 2 ft in diameter, and $f = 0.02$.

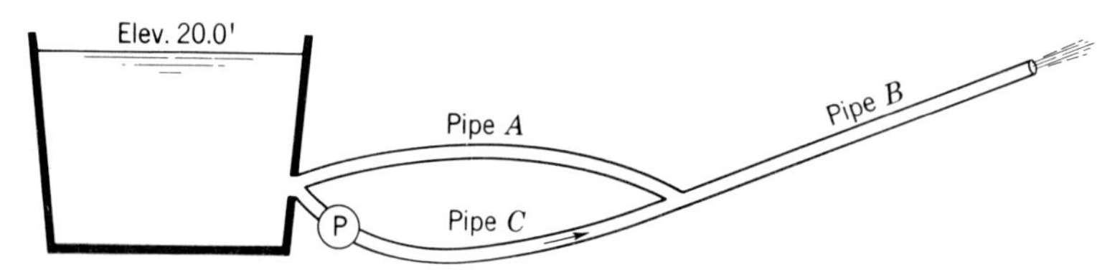

Problem 8.116

8.117 Repeat Prob. 8.116 for the case where the velocity in pipe C is 5 fps with all other data remaining the same.

8.118 Refer to the figure for Prob. 8.116. Assume the water surface in the reservoir is at elevation 100 m. Pipes A, B, and C are all 800 m long, and they all have a diameter of 60 cm with $f = 0.025$. Neglecting minor losses, find (a) the flow rate in all pipes and (b) the elevation of pipe B at discharge under conditions where the pump develops 10 m of head when the velocity in pipe C is 3.0 m/s.

8.119 An 8-in (20-cm) cast-iron pipe 1,000 ft (300 m) long forms one link of a pipe network. If the velocities to be encountered are assumed to fall within the range of 2 to 8 fps (0.6 to 2.4 m/s), derive an equation for the flow of water at 60°F (15°C) in this pipe in the form $h_L = KQ^n$. (*Hint:* Using the information in Figs. 8.11 and 8.12, set up two simultaneous equations corresponding to the ends of the desired velocity range, then solve for the unknowns K and n.)

8.120 Solve the pipe network shown in the figure using four approximations, to find the flow in each pipe. For simplicity, take $n = 2.0$ and use the value of f for complete turbulence, as given in Fig. 8.11. All pipe is cast iron. If the pressure head at a is 100 ft, find the pressure head at d (which might represent a fire demand, for example) neglecting velocity heads.

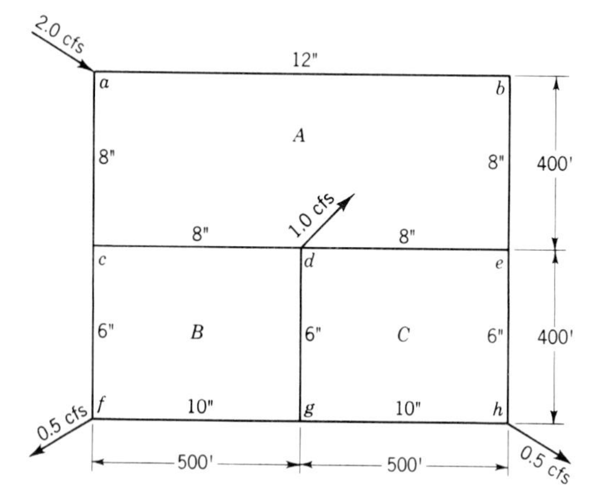

Problem 8.120

8.121 Solve the pipe network shown in the figure using five trials. The 12- and 16-in pipes are of average cast iron, while the 18- and 24-in sizes are of average concrete ($e = 0.003$ ft). Assume $n = 2.0$, and use the values of f from Fig. 8.11 for complete turbulence. If the pressure at h is 80 psi, find the pressure at f.

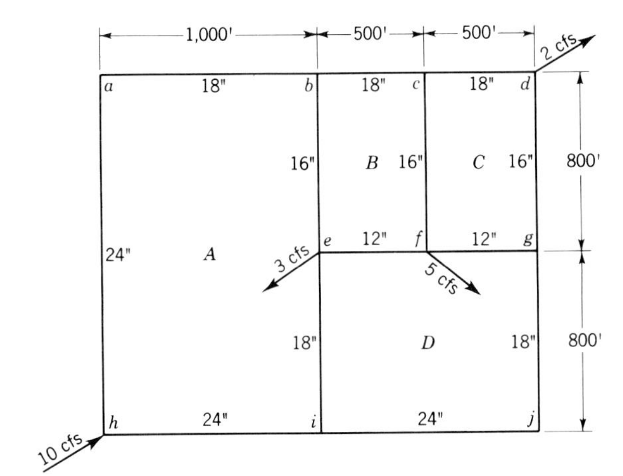

Problem 8.121

CHAPTER

NINE

STEADY FLOW OF COMPRESSIBLE FLUIDS

When dealing with a compressible fluid, if the density change is gradual and not more than a few percent, the flow may be treated as incompressible by using an average density for best results. However, if $\Delta\rho/\rho > 0.05$, the effects of compressibility must be considered. The purpose of this chapter is to investigate compressible-fluid problems that require such considerations. The discussion will be limited to steady one-dimensional flow of compressible fluids. It will be seen that such problems are more difficult than incompressible-fluid problems because of thermodynamic considerations.

9.1 THERMODYNAMIC CONSIDERATIONS

To further our understanding of the flow of compressible fluids, it will be advantageous in our discussion to review briefly some thermodynamic principles. The thermodynamic properties of a gas (Appendix 3, Table A.5) include the gas constant R, specific heat c_p at constant pressure, specific heat c_v at constant volume, and the isentropic exponent $k = c_p/c_v$. The density (or specific volume) of a gas depends on the absolute pressure and absolute temperature under which it exists; for real gases the relationship between these is closely defined by the perfect-gas law $p/\rho = pv = RT$, which was discussed in Sec. 1.8. Then we have the equation of state of the gas, which is:

For isothermal conditions, $pv = \text{constant}$
For isentropic conditions, $pv^k = \text{constant}$
For adiabatic conditions, $pv^n = \text{constant}$

An *isothermal* process is one in which there is no change in temperature, while an *adiabatic* process is one in which no heat is added to or taken away from the flow system. An *isentropic* process is a frictionless adiabatic process. For adiabatic processes with friction, $n < k$ for expansion and $n > k$ for compression. The solution of a compressible-flow problem is similar to that of an incompressible one except that the equation of state of the compressible fluid must be introduced into the problem.

The enthalpy h per unit mass of a gas is defined by $h = i + p/\rho = i + RT$, where i, the internal energy per unit mass due to the kinetic energy of molecular motion and forces between molecules, is a function of temperature. The enthalpy per unit weight is $\hat{h} = h/g = I + p/\gamma = I + RT/g$. Hence enthalpy represents the energy possessed by a gas by virtue of the absolute temperature under which it exists.

The specific heat c_p at constant pressure is defined as the increase in enthalpy per unit of mass when the temperature of a gas is increased one degree with its pressure held constant. Thus,

$$c_p = \left(\frac{\partial h}{\partial T}\right)_{p=\text{constant}} \tag{9.1}$$

where h is the enthalpy per unit of mass.

The specific heat c_v at constant volume is defined as the increase in internal energy per unit of mass when the temperature is increased one degree with its volume held constant. Thus,

$$c_v = \left(\frac{\partial i}{\partial T}\right)_{v=\text{constant}} \tag{9.2}$$

where i is the internal energy per unit of mass.

For perfect gases these equations can be written as $dh = c_p\,dT$ and $di = c_v\,dT$. Now since $h = i + p/\rho = i + RT$, $dh = di + R\,dT$. Combining these relationships leads to

$$c_p = c_v + R \tag{9.3}$$

Introducing the specific heat ratio $k = c_p/c_v$ and combining with Eq. (9.3) gives

$$c_p = \frac{k}{k-1}R \qquad \text{and} \qquad c_v = \frac{R}{k-1} \tag{9.4}$$

The first law of thermodynamics was discussed in Sec. 4.4. The second law of thermodynamics deals with reversibility of processes. A reversible process is one in which after the process the gas returns precisely to its initial state. A frictionless adiabatic (isentropic) process is reversible. The flow through a converging nozzle where there is little friction and little or no heat transfer can be approximated as a reversible process. Flow in a pipeline, however, is an irreversible process because of the pipe friction.

Illustrative Example 9.1 Compute the change in internal energy and the change in enthalpy of 15 kg of air if its temperature is raised from 20 to 30°C. The initial pressure is 95 kN/m², abs.

Properties of gases are given in Appendix 3, Table A.5.

$$\Delta i = c_v(T_2 - T_1) = 716 \frac{\text{N·m}}{(\text{kg·K})}(30 - 20)\ \text{K} = 7160\ \text{N·m/kg}$$

Change in internal energy $= \Delta i \times (15\ \text{kg}) = 107{,}400\ \text{N·m}$

$$\Delta h = c_p(T_2 - T_1) = 1{,}003(10) = 10{,}030\ \text{N·m/kg}$$

Change in enthalpy $= \Delta h \times (15\ \text{kg}) = 150{,}000\ \text{N·m}$

Illustrative Example 9.2 Suppose that 15 kg of air ($T_1 = 20°\text{C}$) of Illustrative Example 9.1 were compressed isentropically to 40 per cent of its original volume. Find the final temperature and pressure, the work required, and the changes in internal energy and enthalpy.

The following relations apply: $pv = RT$ and $pv^k = \text{constant}$, where $k = 1.40$ for air.

$$pv^k = pv\frac{v^k}{v} = \frac{RT}{v}v^k = RTv^{k-1} = \text{constant}$$

Since $R = \text{constant}$, $Tv^{k-1} = \text{constant}$ and

$$T_2 = T_1\left(\frac{v_1}{v_2}\right)^{k-1} = (273 + 20)\left(\frac{1.0}{0.4}\right)^{0.40} = 422\ \text{K} = 149°\text{C}$$

$$\frac{pv}{T} = R = \text{constant} \qquad p_1 = 95\ \text{kN/m}^2\text{, abs (from Illus. Ex. 9.1)}$$

$$\frac{p_1v_1}{T_1} = \frac{p_2v_2}{T_2} \qquad \text{and} \qquad v_2 = 0.4v_1$$

$$\frac{95v_1}{293} = \frac{p_2(0.4v_1)}{422} \qquad p_2 = 342\ \text{kN/m}^2\text{, abs}$$

Since this is an isentropic process, the work required is equal to the change in internal energy. This can be confirmed by computing the values of the pressure and corresponding volumes occupied by the gas during the isentropic process, plotting a pressure-vs.-volume curve, and finding the area under the curve and thereby determining the work done on the fluid. Thus the work required is

$$\int_{s_1}^{s_2} F\,ds = \int_{s_1}^{s_2} (F/A)A\,ds = \int_{\text{vol 1}}^{\text{vol 2}} p\,d(\text{vol})$$

or

$$\Delta i = c_v(T_2 - T_1) = 716(422 - 293) = 92{,}400\ \text{N·m/kg}$$

$$\Delta i \times 15\ \text{kg} = 1{,}385{,}000\ \text{N·m} = \text{work required} = \text{change in internal energy}$$

$$\Delta h = c_p(T_2 - T_1) = 1{,}003(129) = 129{,}400\ \text{N·m/kg}$$

$$\Delta h \times 15\ \text{kg} = 1{,}941{,}000\ \text{N·m} = \text{change in enthalpy}$$

9.2 FUNDAMENTAL EQUATIONS APPLICABLE TO THE FLOW OF COMPRESSIBLE FLUIDS

The fundamental equations for the flow of compressible fluids have already been stated in Chaps. 3, 4, and 6. For convenience we restate them here.

Continuity

The expression for continuity for one-dimensional flow of a compressible fluid is

$$G = \gamma A V = \text{constant} \tag{9.5a}$$

or

$$M = \rho A V = \text{constant} \tag{9.5b}$$

where G and M are the weight flow rate and mass flow rate respectively.

Energy Equation

For one-dimensional flow of a compressible fluid if there is no machine between sections 1 and 2 the energy equation[1] is expressible as:

$$\hat{h}_1 + \frac{V_1^2}{2g} + Q_H = \hat{h}_2 + \frac{V_2^2}{2g} \tag{9.6}$$

where the *enthalpy* per unit of weight is $\hat{h} = I + p/\gamma$.

Impulse-Momentum Equation

The impulse-momentum equation (Sec. 6.1) for steady one-dimensional flow of a compressible fluid is:

$$\sum \mathbf{F} = \rho_2 Q_2 \mathbf{V}_2 - \rho_1 Q_1 \mathbf{V}_1 \tag{9.7}$$

Euler Equation

For one-dimensional compressible flow of an ideal fluid in a circular pipe the Euler equation (Sec. 4.10) may be expressed as

$$\frac{dp}{\rho} + V\,dV = 0 \tag{9.8}$$

In this equation and in Eq. (9.6) the z terms were dropped, for in the flow of compressible fluids the z terms are almost always negligible compared with the other terms in the energy equation.

Sonic (Acoustic) Velocity

In Appendix 2 it is shown that the sonic velocity $c = \sqrt{E_v/\rho} = \sqrt{kRT}$. This represents the celerity at which a pressure wave will travel through a compressible fluid. In Chap. 7, an important dimensionless parameter, the Mach number $\mathbf{M}$, was mentioned. $\mathbf{M} = V/c$, where V is the velocity of flow. If $\mathbf{M} < 1$, the flow is subsonic; if $\mathbf{M} = 1$, the flow is sonic; if $\mathbf{M} > 1$, the flow is supersonic.

[1] The reader should review Secs. 4.4 and 4.9 for a discussion of the energy equation as it applies to compressible fluids.

9.3 ADIABATIC FLOW (WITH OR WITHOUT FRICTION)

If heat transfer Q_H is zero, the flow is adiabatic. Hence Eq. (9.6) may be written as

$$\hat{h}_1 + \frac{V_1^2}{2g} = \hat{h}_2 + \frac{V_2^2}{2g} \tag{9.9}$$

Since $\Delta\hat{h} = (c_p/g)\,\Delta T$, we get for adiabatic flow,

$$V_2^2 - V_1^2 = 2g(\hat{h}_1 - \hat{h}_2) = 2c_p(T_1 - T_2) \tag{9.10}$$

From Eq. (9.4), $c_p = kR/(k-1)$ and for a perfect gas $pv = RT$. Substituting these into Eq. (9.10) gives for adiabatic flow

$$V_2^2 - V_1^2 = \frac{2k}{k-1}(p_1v_1 - p_2v_2) = \frac{2k}{k-1}RT_1\left(1 - \frac{T_2}{T_1}\right) \tag{9.11}$$

The preceding equations are valid for flow either with or without friction. It should be noted that in these equations the temperatures and pressures are absolute.

Equation (9.9) can be written as

$$\frac{c_p}{g}T_1 + \frac{V_1^2}{2g} = \frac{c_p}{g}T_2 + \frac{V_2^2}{2g} = \frac{c_p}{g}T_s \tag{9.12}$$

where T_s is the stagnation temperature (where V is zero). Thus, in adiabatic flow, the stagnation temperature is constant along a streamline regardless of whether or not the flow is frictionless.

9.4 STAGNATION PRESSURE

An expression for stagnation pressure in compressible flow may be developed by assuming isentropic conditions ($Q_H = 0$ and $pv^k =$ constant). From Appendix 2, the sonic (acoustic) velocity $c = \sqrt{kp/\rho} = \sqrt{kRT}$. Substituting this relation in Eq. (9.11), noting that $V_1 = c_1\mathbf{M}_1$, where $\mathbf{M}_1$ is the Mach number [Eq. (7.8)], and applying Eq. (1.7) under isentropic conditions, we get

$$\frac{V_2^2}{c_1^2} = \mathbf{M}_1^2 + \frac{2}{k-1}\left[1 - \left(\frac{p_2}{p_1}\right)^{(k-1)/k}\right] \tag{9.13}$$

Refer now to Fig. 9.1, which shows a stagnation point in compressible flow. At the stagnation point s the increased pressure causes a rise in density and temperature. At this point, $V_2 = V_s = 0$. Hence, for the situation depicted in Fig. 9.1, Eq. (9.13) may be expressed as

$$0 = \left(\frac{V_0}{c_0}\right)^2 + \frac{2}{k-1}\left[1 - \left(\frac{p_s}{p_0}\right)^{(k-1)/k}\right]$$

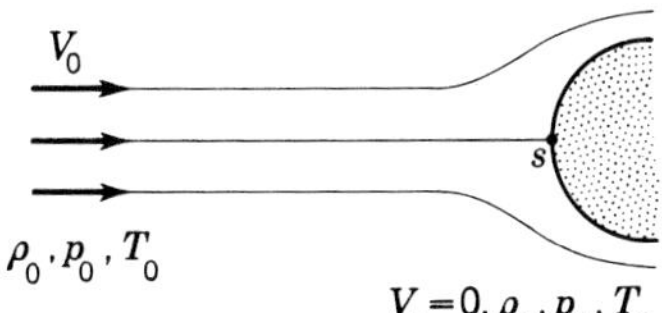

Figure 9.1 Stagnation point.

where c = sonic velocity in the undisturbed flow
V_0 = velocity in the undisturbed flow
p_0 = pressure in the undisturbed flow

Rearranging this equation, we get

$$\frac{p_s}{p_0} = \left[1 + \left(\frac{V_0}{c_0}\right)^2 (k-1)/2\right]^{k/(k-1)} \tag{9.14}$$

Substituting $k = \rho_0 c_0^2/p_0$ [Refer to Appendix 2, Eq. (A.5)] and expanding by the binomial theorem,

$$p_s = p_0 + \frac{1}{2}\rho_0 V_0^2\left(1 + \frac{V_0^2}{4c_0^2} + \cdots\right)$$

$$= p_0 + \gamma_0 \frac{V_0^2}{2g}\left(1 + \frac{1}{4}\mathbf{M}_0^2 + \cdots\right) \tag{9.15}$$

This equation is identical with Eq. (4.32), mentioned earlier. It is applicable only if $\mathbf{M} < 1$. In determining stagnation pressure, as can be seen from Eq. (9.15), the error from neglecting compressibility depends on the Mach number of the flow. At low Mach numbers the error is insignificant, but as $\mathbf{M}$ approaches unity, the error is sizable.

Illustrative Example 9.3 Find the stagnation pressure and temperature in nitrogen flowing at 600 fps if the pressure and temperature in the undisturbed flow field are 100 psia and 200°F, respectively. (See Appendix 3, Table A.5, for properties of gases.)

$$c = \sqrt{kRT} = \sqrt{(1.4)(1{,}773)(660)} = 1{,}280 \text{ fps}$$

$$\mathbf{M} = \frac{600}{1{,}280} = 0.469$$

$$\rho_0 = \frac{kp_0}{c_0^2} = \frac{1.4(100 \times 144)}{(1{,}280)^2} = 0.0123 \text{ slug/ft}^3$$

From Eq. (9.15)

$$p_s = 100(144) + \tfrac{1}{2}(0.0123)(600)^2[1 + \tfrac{1}{4}(0.469)^2]$$

$$= 14{,}400 + 2{,}214(1 + 0.055) = 16{,}740 \text{ lb/ft}^2 = 116 \text{ psia}$$

Applying Eq. (9.12)

$$\frac{c_p}{g}T_1 + \frac{V_1^2}{2g} = \frac{c_p}{g}T_s$$

$$\frac{6{,}210}{32.2}(660) + \frac{(600)^2}{64.4} = \frac{6{,}210}{32.2}T_s$$

$$T_s = 689°\text{R} = 229°\text{F}$$

9.5 ISENTROPIC FLOW

Frictionless adiabatic ($Q_H = 0$) flow is referred to as isentropic flow. Such flow does not occur in nature. However, flow through a nozzle or flow in a free stream of fluid over a reasonably short distance may be considered isentropic because there is very little heat transfer and fluid-friction effects are small. Equations for isentropic flow can be derived by substituting pv^k = constant in Eq. (9.11). The resulting equations for isentropic flow are

$$\frac{V_2^2 - V_1^2}{2g} = \frac{p_1}{\gamma_1}\frac{k}{k-1}\left[1 - \left(\frac{p_2}{p_1}\right)^{(k-1)/k}\right] = \frac{p_2}{\gamma_2}\frac{k}{k-1}\left[\left(\frac{p_1}{p_2}\right)^{(k-1)/k} - 1\right] \quad (9.16)$$

This equation may also be derived by integrating the Euler equation $dp/\gamma + V\,dV/g = 0$ along a stream tube, while noting that pv^k = constant.

The relation between the temperatures at two points along a streamline in isentropic flow can be derived by equating the expressions for $V_2^2 - V_1^2$ from Eqs. (9.11) and (9.16). The result is

$$T_2 = T_1 - \frac{p_1 g}{R\gamma_1}\left[1 - \left(\frac{p_2}{p_1}\right)^{(k-1)/k}\right] \quad (9.17)$$

Once again, temperatures and pressures are absolute.

9.6 EFFECT OF AREA VARIATION ON ONE-DIMENSIONAL COMPRESSIBLE FLOW

In steady flow the velocity of an *incompressible* fluid varies inversely with the cross-sectional area of the flow stream. This is not the case with a compressible fluid because variations in density will also influence the velocity. Moreover, the behavior of a compressible fluid, when there is a change in cross-sectional area, depends on whether the flow is *subsonic* ($\mathbf{M} < 1$) or *supersonic* ($\mathbf{M} > 1$). We shall now examine this phenomenon. In this discussion we shall confine our remarks to ideal flow.

The continuity equation (9.5*b*) may be written in differential form as

$$\frac{dA}{A} + \frac{d\rho}{\rho} + \frac{dV}{V} = 0 \quad (9.18)$$

Noting that $c^2 = dp/d\rho$ [Appendix 2, Eq. (A.3)], the Euler equation (9.8) for an *ideal* fluid may be expressed as

$$\frac{dp}{d\rho}\frac{d\rho}{\rho} + V\,dV = c^2\frac{d\rho}{\rho} + V\,dV = 0 \quad (9.19)$$

Combining the two preceding equations, replacing V/c with $\mathbf{M}$, and rearranging, we get

$$\frac{dA}{A} = (\mathbf{M}^2 - 1)\frac{dV}{V} \quad (9.20)$$

From this equation we can arrive at some significant conclusions as follows:

1. For subsonic flow ($\mathbf{M} < 1$):

   If $dV/V > 0$, $dA/A < 0$ (area must decrease for increase in velocity).
   If $dV/V < 0$, $dA/A > 0$ (area must increase for a decrease in velocity).

2. For supersonic flow ($\mathbf{M} > 1$):

   If $dV/V > 0$, $dA/A > 0$ (area must increase if velocity is to increase).
   If $dV/V < 0$, $dA/A < 0$ (area must decrease if velocity is to decrease).

3. For sonic flow ($\mathbf{M} = 1$):

$$\frac{dA}{A} = 0$$

Thus it is seen that subsonic and supersonic flows behave *oppositely* if there is an area variation. To accelerate a flow at subsonic velocity, a converging passage is required just as in the case of an incompressible flow. To accelerate a flow at supersonic velocity, however, a diverging passage is required.

For sonic velocity it is noted that $dA/A = 0$. This condition occurs at the throat of a converging or a converging-diverging passage. However, the flow will be sonic at the throat only if the pressure differential between the upstream region and the throat is large enough to accelerate the flow sufficiently. At modest pressure differentials the velocity at the throat will be subsonic ($\mathbf{M} < 1$). As the pressure differential between the upstream region and the throat is increased, the velocity at the throat will increase until sonic velocity ($\mathbf{M} = 1$) occurs there. With further increase in pressure differential the flow rate will increase (due to density increase) but the velocity at the throat will remain sonic. This is examined further in Secs. 9.7 and 9.8. Supersonic flow ($\mathbf{M} > 1$) will occur in the diverging section of a converging-diverging passage only if the flow at the throat is sonic. If the flow at the throat is subsonic, the flow in the diverging section will also be subsonic and the velocity will decrease in the diverging section. In Fig. 9.2

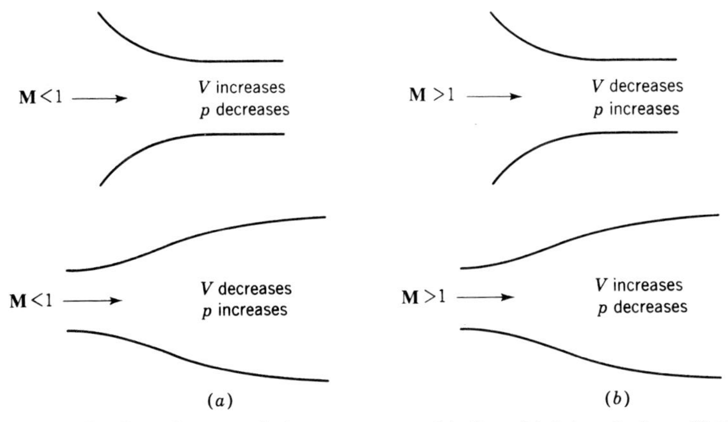

Figure 9.2 Effect of area variation on compressible flow. (*a*) Subsonic flow. (*b*) Supersonic flow.

is shown the behavior of subsonic and supersonic flow through converging and diverging passages.

9.7 COMPRESSIBLE FLOW THROUGH CONVERGING NOZZLE DISCHARGING FROM A LARGE TANK

Let us now consider one-dimensional flow of a compressible fluid through the converging nozzle of Fig. 9.3. Since there is little opportunity for heat transfer between sections 1 and 2 and since frictional effects are small, we will assume isentropic conditions. If the velocity of approach is negligible, Eq. (9.16) can be expressed as

$$\frac{V_2^2}{2g} = \frac{p_2}{\gamma_2}\frac{k}{k-1}\left[\left(\frac{p_1}{p_2}\right)^{(k-1)/k} - 1\right] \tag{9.21}$$

Noting that $c_2 = \sqrt{kgp_2/\gamma_2}$, the above equation can be rearranged to give

$$\left(\frac{V_2}{c_2}\right)^2 = \mathbf{M}_2^2 = \frac{2}{k-1}\left[\left(\frac{p_1}{p_2}\right)^{(k-1)/k} - 1\right] \tag{9.22}$$

Thus it is seen that the velocity of flow at the throat (section 2 of the figure) depends on the p_1/p_2 ratio. If there is a large enough pressure differential between the inside and outside of the tank, sonic velocity will occur at section 2. From the discussion of Sec. 9.6 it is recognized that supersonic flow is impossible in this situation. If the flow through the throat is subsonic, the pressure at the throat is identical with that outside the tank ($p_2 = p_3$). If the flow through the throat is sonic, the pressure at the throat may be equal to, but is generally greater than, that outside the tank ($p_2 \geqq p_3$).

Let us now assume the condition of sonic flow at the throat (that is, $\mathbf{M}_2 = 1.0$). Substituting $\mathbf{M}_2 = 1.0$ into Eq. (9.22) we get what is referred to as the *critical pressure ratio*[1]

$$\left(\frac{p_2}{p_1}\right)_c = \left(\frac{2}{k+1}\right)^{k/(k-1)} \tag{9.23}$$

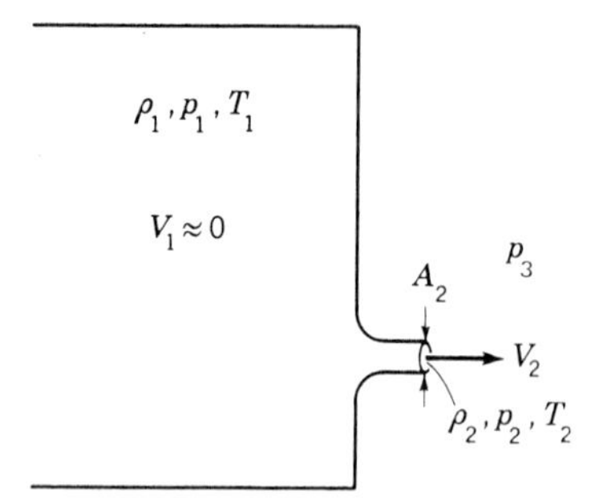

Figure 9.3 Flow through converging nozzle.

[1] In all equations here the pressure p is absolute pressure.

This critical pressure ratio exists whenever the pressure differential between p_1 and p_2 is great enough to create sonic velocity at the throat. If the flow through the throat is subsonic, then p_2/p_1 will be larger than the ratio given by Eq. (9.23). It should be noted that p_2/p_1 can never be smaller than $(p_2/p_1)_c$.

The rate of flow through the nozzle of Fig. 9.3 may be found by substituting V_2 from Eq. (9.21) into $G = \gamma_2 A_2 V_2$. Making use of the isentropic relation between the p's and γ's and rearranging, we get

Subsonic flow at throat:
$$G = A_2 \sqrt{2g \frac{k}{k-1} p_1 \gamma_1 \left[\left(\frac{p_2}{p_1} \right)^{2/k} - \left(\frac{p_2}{p_1} \right)^{(k+1)/k} \right]} \tag{9.24}$$

This expression is applicable so long as $p_2/p_1 > (p_2/p_1)_c$, as given by Eq. (9.23). Thus it applies to the condition where there is subsonic flow at the throat.

To determine the rate of flow G' through the nozzle when there is sonic flow at the throat, we substitute $p_2/p_1 = (p_2/p_1)_c = [2/(k+1)]^{k/(k-1)}$ into Eq. (9.24). The result is

Sonic flow at throat:
$$G' = A_2 \sqrt{p_1 \gamma_1 g k \left(\frac{2}{k+1} \right)^{(k+1)/(k-1)}} \tag{9.25}$$

Introducing Eq. (1.5), this equation may also be expressed as

Sonic flow at throat:
$$G' = g \frac{A_2 p_1}{\sqrt{T_1}} \sqrt{\frac{k}{R} \left(\frac{2}{k+1} \right)^{(k+1)/(k-1)}} \tag{9.26}$$

In this equation the expression under the second radical depends only on the properties of the gas. Thus a simple device for metering compressible flows is a converging nozzle at whose throat the sonic velocity is produced. It should be noted that with subsonic flow as p_1 is increased the value of p_2/p_1 becomes smaller and approaches $(p_2/p_1)_c$. At the point where p_2/p_1 first reaches the value of $(p_2/p_1)_c$ the flow at the throat changes from subsonic to sonic. At this point, a threshold point, Eq. (9.24) for subsonic flow and Eqs. (9.25) and (9.26) for sonic flow give the same result. As p_1 is increased beyond the threshold point, p_2/p_1 maintains the value of $(p_2/p_1)_c$ and the velocity at the throat remains sonic. However the flow rate increases directly with p_1 as can be seen from Eq. (9.26).

For air ($k = 1.4$) the value of $(p_2/p_1)_c = 0.528$. Isentropic conditions have been assumed in the preceding equations; hence the flows represent those for an ideal fluid. The flows for real fluids through converging nozzles are only slightly less than those given by these equations.

Illustrative Example 9.4 Air at 80°F flows from a large tank through a converging nozzle of 2.0-in exit diameter. The discharge is to an atmospheric pressure of 13.5 psia. Determine the flow through the nozzle for pressures within the tank of 5, 10, 15, and 20 psig. Assume isentropic conditions. Plot G as a function of p_1. Assume that the temperature within the tank is 80°F in all cases.

From Eq. (9.23) the critical pressure ratio for air is $(p_2/p_1)_c = 0.528$. If the flow at the throat is subsonic, $p_2/p_1 > (p_2/p_1)_c$. Thus for subsonic flow at the throat, $p_2/p_1 > 0.528$ and $p_3 = p_2$. So $p_3/p_1 > 0.528$ and $p_1 < p_3/0.528$.

Since $p_3 = 13.5$ psia, the flow at the throat will be subsonic if $p_1 < 25.6$ psia (12.1 psig) and sonic if $p_1 > 25.6$ psia (12.1 psig).

To find the flow rate for conditions where $p_1 < 12.1$ psig (subsonic flow at throat), we use Eq. (9.24). Substituting the appropriate value of p_1 into the equation and noting that for this condition $p_2 = p_3 = 13.5$ psia, we get

For $p_1 = 5$ psig (18.5 psia): $G = 1.20$ lb/s
For $p_1 = 10$ psig (23.5 psia): $G = 1.69$ lb/s

To find the flow rate for conditions where $p_1 > 12.1$ psig (sonic flow at the throat) we use Eq. (9.26). We get

For $p_1 = 15$ psig (28.5 psia) $G' = 2.04$ lb/s
For $p_1 = 20$ psig (33.5 psia) $G' = 2.40$ lb/s

Substituting $p_1 = 25.6$ psia in Eq. (9.24) for subsonic flow gives $G = 1.84$ lb/s as does Eq. (9.25) or Eq. (9.26). This is the threshold point at which the flow in the throat changes from subsonic to sonic. When $p_1 = 25.6$ psia the flow rate as found from Eq. (9.26) is

$$G'_{p=25.6\,\text{psia}} = \frac{32.2(0.0218)(144)(25.6)}{\sqrt{540}} \sqrt{\frac{(1.4)}{1715}\left(\frac{2}{2.4}\right)^{2.4/0.4}} = 1.84 \text{ lb/s}$$

As p_1 is increased beyond 12.1 psig, sonic flow prevails at the throat and the flow rate increases linearly with $(p_1)_{\text{abs}}$ as indicated by Eq. (9.26). The variation of the flow rate with p_1 is shown in the accompanying sketch. Other information concerning various aspects of this problem can be found by applying the equations that have been presented, namely, the gas law ($pv = RT$ or $\gamma = gp/RT$), the equation of state ($pv^k =$ constant or $p/\gamma^k =$ constant), continuity ($G = \gamma AV$), and the energy equation, Eq. (9.21). Applying these, for example, for the case where $p_1 = 5$ psig (18.5 psia) yields $\gamma_1 = 0.093$ lb/ft^3, $\gamma_2 = 0.075$ lb/ft^3, $V_2 = 741$ fps, $T_2 = 492$°R, and $p_2 = 13.57$ psia. Note that in this case $p_2 = p_3$.

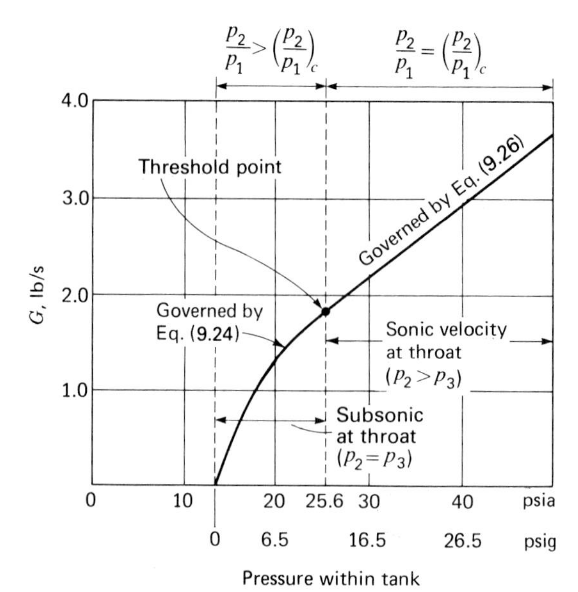

Illustrative Example 9.4

9.8 THE CONVERGING-DIVERGING NOZZLE DISCHARGING FROM A LARGE TANK

If a converging nozzle has attached to it a diverging section, it is possible to attain supersonic velocities in the diverging section. This will happen if sonic flow exists in the throat. In such a case the gas or vapor will continue to expand in the diverging section to lower pressures, and the velocity will continue to increase. The flow through such a converging-diverging nozzle is shown in Fig. 9.4. If there is not enough pressure differential to attain sonic velocity at the throat, the gas or vapor will behave in much the same manner as a liquid, with acceleration in the region up to the throat and deceleration in the diverging section beyond the throat. A plot of the pressure in the flow for such a condition is shown by the dashed lines *ABD* of Fig. 9.4.

Suppose in Fig. 9.4 that the pressure at 4 is gradually reduced while p_1 remains constant. In such a case $p_3 = p_4$, and the pressure at the throat (2) decreases while the velocity at the throat increases until the limiting sonic velocity is reached at the throat when the pressure plot is *ACE*. If the pressure at 4 is further reduced to *H*, the pressure plot is *ACFGH*, the jump from *F* to *G* being a pressure shock, or *normal shock wave*, similar to the hydraulic jump, or standing

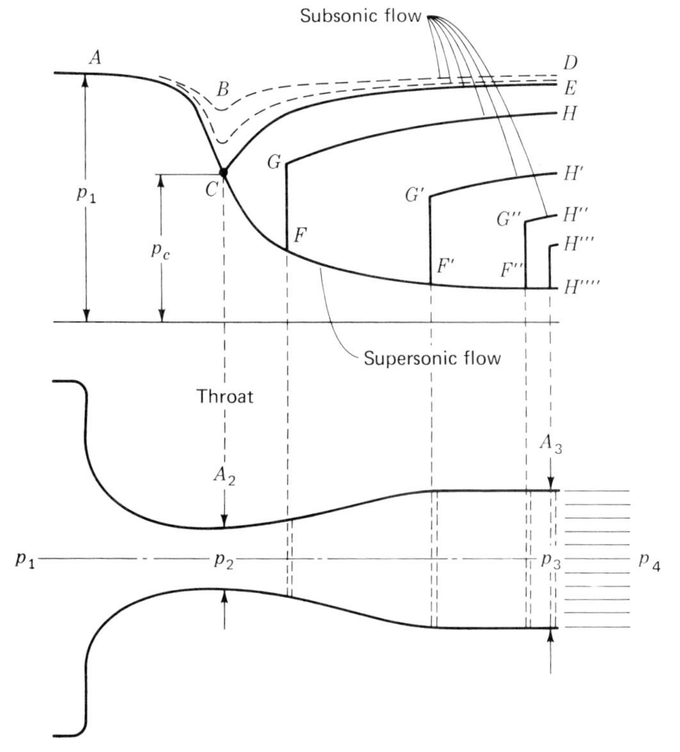

Figure 9.4 Flow of compressible fluid through converging-diverging nozzle.

wave, often seen in open channels conveying water. Through the shock wave the velocity is reduced abruptly from supersonic to subsonic, while at the same time the pressure jumps as shown by the lines FG, $F'G'$, and $F''G''$. The flow through a shock wave is not isentropic, since part of the kinetic energy is converted into heat.

Further reduction of the pressure at 4 causes the shock wave to move farther downstream until at some value given by H''' the shock wave is located at the downstream end of the nozzle. If the pressure at 4 is lowered below the level of H''', the shock wave occurs in the flow field downstream of the nozzle exit. Such flow fields are either two- or three-dimensional and cannot be described by the foregoing one-dimensional equations.

If the pressure at 4 is lowered to H'''', the flow will be supersonic throughout the entire region downstream from the throat, the velocity will increase continuously from 1 to its maximum value at 3 and the pressure will drop continuously from 1 to 3. As long as p_4 is above H'''', then $p_3 = p_4$; but if p_4 drops below H'''', then $p_3 > p_4$ and supersonic flow occurs through the entire length of tube.

If the pressure at 4 is above E in Fig. 9.4, the flow rate through the converging-diverging nozzle is given by Eq. (9.24). In this instance the p_2 of Eq. (9.24) must be replaced by the p_4 of Fig. 9.4. If the pressure at 4 is below E in Fig. 9.4, critical pressure, as defined by Eq. (9.23), will occur at the throat and the flow rate will be given by Eq. (9.25) or Eq. (9.26).

If p_1 is increased, the sonic velocity may be shown to remain unaltered, but since the density of the gas is increased, the rate of discharge will be greater. The converging nozzle and the converging-diverging nozzle are alike insofar as discharge capacity is concerned. The only difference is that with the converging-diverging nozzle, a supersonic velocity may be attained at discharge from the device, while with the converging nozzle, the sonic velocity is the maximum value possible.

Illustrative Example 9.5 Air discharges from a tank through a converging-diverging nozzle with a 2.0-in-diameter throat. Within the tank the pressure is 50 psia and the temperature is 80°F, while outside the tank the pressure is 13.5 psia. The nozzle is to operate with supersonic flow throughout its diverging section with a 13.5 psia pressure as its outlet. Find the required diameter of the nozzle outlet. Determine the flow rate and the velocities and temperatures at sections 2 and 3. Assume isentropic flow.

The pressure at the throat must be such that sonic velocity will occur there. Hence $p_2 = p_c = 0.528p_1 = 26.4$ psia.

The velocity at the outlet may be found from Eq. (9.16).

$$\frac{V_3^2}{2g} = \frac{RT_1}{g}\frac{k}{k-1}\left[1 - \left(\frac{p_3}{p_1}\right)^{(k-1)/k}\right] = \frac{1{,}715(540)}{32.2}\frac{1.4}{0.4}\left[1 - \left(\frac{13.5}{50}\right)^{0.4/1.4}\right] = 31{,}400 \text{ ft}$$

$$V_3 = 1{,}420 \text{ fps}$$

The flow rate is computed from Eq. (9.26)

$$G' = \frac{(32.2)0.0218(50 \times 144)}{\sqrt{540}}\sqrt{\frac{(1.4)}{1{,}715}\left(\frac{2}{2.4}\right)^{2.4/0.4}} = 3.58 \text{ lb/s}$$

The temperature at 3 may be determined by using Eq. (9.10):

$$V_3^2 - V_1^2 = 2c_p(T_1 - T_3)$$

$$(1{,}420)^2 = 2(6{,}000)(540 - T_3)$$

$$T_3 = 372°\text{R} = -88°\text{F}$$

From the perfect-gas law $p_3/\gamma_3 = RT_3/g$.

$$\gamma_3 = \frac{32.2(13.5 \times 144)}{1{,}715(372)} = 0.098 \text{ lb/ft}^3$$

Isentropic flow between 2 and 3 may be assumed, since the shock wave does not occur within that region. Thus $p_2/\gamma_2^{1.4} = p_3/\gamma_3^{1.4}$.

$$\frac{26.4}{\gamma_2^{1.4}} = \frac{13.5}{(0.098)^{1.4}}$$

$$\gamma_2 = 0.158 \text{ lb/ft}^3$$

The velocity at 2 may now be computed.

$$V_2 = \frac{G}{\gamma_2 A_2} = \frac{3.58}{0.158(0.0218)} = 1{,}040 \text{ fps}$$

The temperature at 2 results from application of Eq. (9.10):

$$(1{,}420)^2 - (1{,}040)^2 = 2(6{,}000)(T_2 - 372)$$

$$T_2 = 450°\text{R} = -10°\text{F}$$

The area at 3 is computed from

$$A_3 = \frac{G}{\gamma_3 V_3} = \frac{3.58}{0.098(1{,}420)} = 0.0257 \text{ ft}^2$$

Finally, $D_3 = 2.17$ in, the required outlet diameter.
Check for sonic velocity at throat:

$$c_2 = \sqrt{kRT_2} = \sqrt{1.4(1715 \times 450)} = 1040 \text{ fps} = V_2$$

With sonic velocity at the throat, if $D_3 < 2.17$ in, there will be supersonic flow throughout the tube and a shock wave will occur in the flow field downstream of the nozzle exit. If $D_3 > 2.17$ in, with sonic flow in the throat, in order to satisfy pressure conditions, a shock wave will occur in the tube somewhere between the throat and the nozzle exit.

9.9 ONE-DIMENSIONAL SHOCK WAVE

In Fig. 9.5 is shown a one-dimensional shock wave where the approaching supersonic flow changes to subsonic flow. This phenomenon is accompanied by a sudden rise in pressure, density, and temperature. Applying the impulse-momentum principle to the fluid in the shock wave, we get

$$\sum F_x = p_1 A_1 - p_2 A_2 = \frac{G}{g}(V_2 - V_1) \tag{9.27}$$

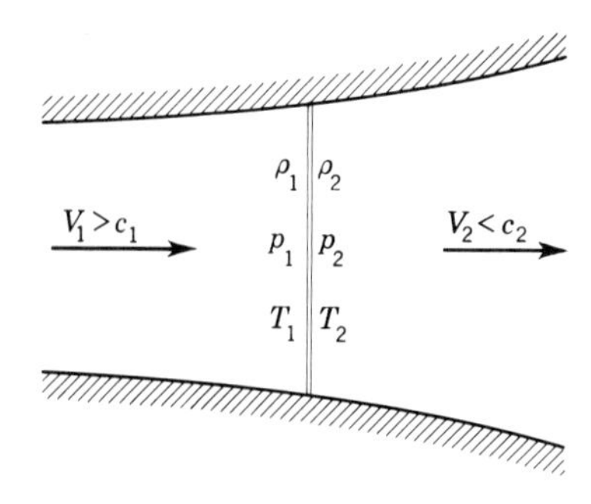

Figure 9.5 A one-dimensional normal shock wave.

Substituting the continuity conditions ($G = \gamma_1 A_1 V_1 = \gamma_2 A_2 V_2$) and noting that $A_1 = A_2$, we get

$$p_2 - p_1 = \frac{1}{g}(\gamma_1 V_1^2 - \gamma_2 V_2^2) \tag{9.28}$$

which is the pressure jump across the wave.

The flow across the shock wave may be considered adiabatic and can be expressed as

$$V_2^2 - V_1^2 = \frac{2k}{k-1}(p_1 v_1 - p_2 v_2) \tag{9.29}$$

This is identical with Eq. (9.11). It is suggested that the reader review the development of this equation, in Sec. 9.3.

Equations (9.28) and (9.29) may be solved simultaneously and rearranged algebraically to give some significant relationships. Several such relations are as follows:

$$\frac{p_2}{p_1} = \frac{2k\mathbf{M}_1^2 - (k-1)}{k+1} \tag{9.30}$$

$$\frac{V_2}{V_1} = \frac{(k-1)\mathbf{M}_1^2 + 2}{(k+1)\mathbf{M}_1^2} \tag{9.31}$$

and

$$\mathbf{M}_2^2 = \frac{2 + (k-1)\mathbf{M}_1^2}{2k\mathbf{M}_1^2 - (k-1)} \tag{9.32}$$

These equations permit one to find the physical properties of the flow on the two sides of the one-dimensional shock wave. These equations, of course, are applicable only if $\mathbf{M}_1 > 1$; that is, the oncoming flow must be supersonic. It will be seen that the shock wave is analogous to the hydraulic jump in open-channel flow (Sec. 11.19).

Illustrative Example 9.6 A normal shock wave occurs in the flow of air where $p_1 = 10$ psia (70 N/m^2, abs), $T_1 = 40°F$ (5°C), and $V_1 = 1400$ fps (425 m/s). Find p_2, V_2, and T_2.

$$\rho_1 = \frac{p_1}{RT_1} = \frac{10(144)}{1{,}715(460 + 40)} = 0.00168 \text{ slug/ft}^3$$

$$c_1 = \sqrt{kRT_1} = \sqrt{1.4 \times 1{,}715 \times (460 + 40)} = 1096 \text{ fps}$$

$$\mathbf{M}_1 = \frac{V_1}{c_1} = \frac{1{,}400}{1{,}096} = 1.28$$

From Eq. (9.30)

$$\frac{p_2}{p_1} = 1.75 \qquad p_2 = 17.5 \text{ psia}$$

From Eq. (9.31),

$$\frac{V_2}{V_1} = 0.675 \qquad V_2 = 945 \text{ fps}$$

$$\rho_1 V_1 = \rho_2 V_2 \qquad \rho_2 = \frac{0.00168}{0.675} = 0.00249 \text{ slug/ft}^3$$

$$pv = \frac{p}{\rho} = RT \qquad T_2 = \frac{p_2}{\rho_2 R} = 590°\text{R} = 130°\text{F}$$

In SI units:

$$\rho_1 = \frac{p_1}{RT_1} = \frac{70}{287(273 + 5)} = 8.8 \times 10^{-4} \text{ kg/m}^3$$

$$c = \sqrt{kRT_1} = \sqrt{1.4 \times 287 \times (273 + 5)} = 334 \text{ m/s}$$

$$\mathbf{M}_1 = \frac{V_1}{c_1} = \frac{425}{334} = 1.27$$

From Eq. (9.30)

$$\frac{p_2}{p_1} = 1.75 \qquad p_2 = 122.5 \text{ N/m}^2$$

From Eq. (9.31)

$$\frac{V_2}{V_1} = 0.675 \qquad V_2 = 286 \text{ m/s}$$

$$\rho_1 V_1 = \rho_2 V_2 \qquad \rho_2 = 8.8 \times \frac{10^{-4}}{0.675} = 1.3 \times 10^{-3} \text{ kg/m}^3$$

$$T_2 = \frac{p_2}{\rho_2 R} = \frac{122.5}{(1.3 \times 10^{-3})287} = 328 \text{ K} = 55°\text{C}$$

9.10 THE OBLIQUE SHOCK WAVE

When the velocity of a body through any fluid, whether a liquid or a gas, exceeds that of a sound wave in the same fluid, the flow conditions are entirely different from those for subsonic flow. Thus, instead of streamlines such as are shown in Fig. 3.11, the conditions are as shown in Fig. 9.6, which is a schlieren photograph[1] of supersonic flow past a sharp-nosed model in a wind tunnel. It could also represent a projectile in flight through still air. A conical compression or shock wave extends backward from the tip, as may be seen by the strong density gradient revealed as a bright shadow in the photograph. A streamline in the undisturbed fluid is unaffected by the solid boundary or by a moving projectile until it intersects a shock-wave front, when it is abruptly changed in direction, proceeding roughly parallel to the nose form. Where the conical nose is joined to the cylindrical body of the model, dark shadows may be seen, representing rarefaction waves. The streamlines are again changed in direction through this region, becoming parallel to the main flow again. A typical streamline has been superimposed on the photograph. In Fig. 9.7 the difference between a subsonic and supersonic flow pattern upstream of a body is shown.

The reason why streamlines are unaffected in front of a projectile is that the body travels faster than the disturbance can be transmitted ahead. This will be illustrated by reference to Fig. 9.8. Consider a point source of an infinitesimal disturbance moving at a supersonic velocity V through a fluid. At the instant

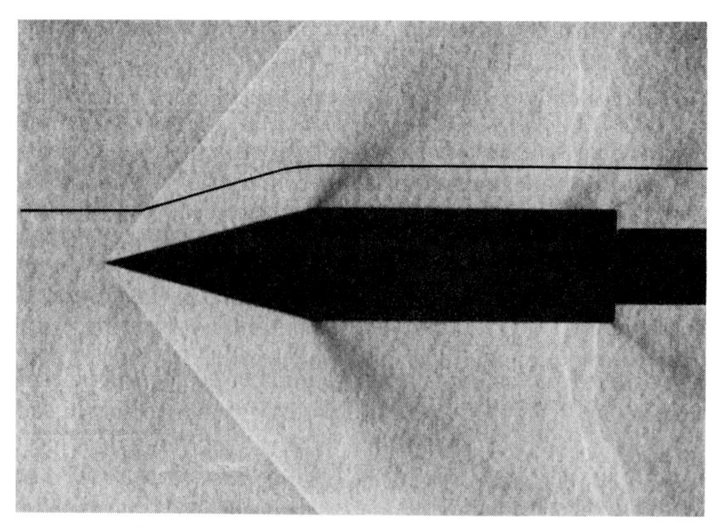

Figure 9.6 Schlieren photograph of head wave on 30° (total angle) cone at $\mathbf{M} = 1.88$. *(Photo by Guggenheim Aeronautical Laboratory, California Institute of Technology.)*

[1] The schlieren method employs the concept that the refraction of light is disturbed by a pressure (density) gradient.

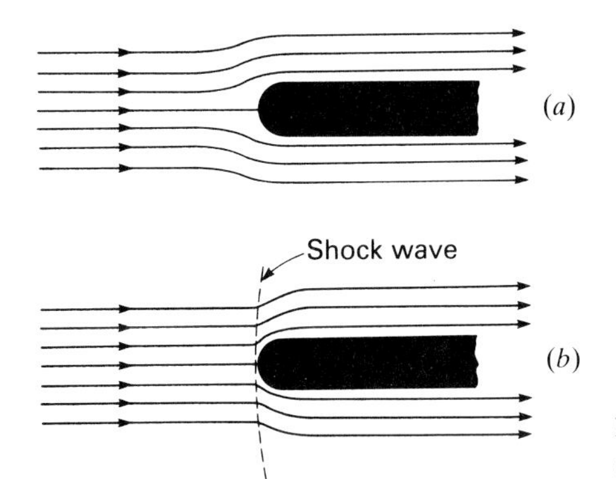

Figure 9.7 Flow pattern upstream of a body. (*a*) Subsonic flow. (*b*) Supersonic flow.

when this source passes through the point A_0, the disturbance commences to radiate in all directions with the velocity c of a sound wave in this medium. In successive instants the source passes through the points A_1, A_2, and A_3, the last of which may represent the position of the source at the instant of observation. While the source has covered the distance A_0A_3 with velocity V, the sound wave, traveling at the slower acoustic velocity c, has progressed only as far as radius A_0B_0. Similar termini of disturbances emanating from A_1 and A_2 form a straight-line envelope, which is the shock wave. The angle β is called the *Mach angle*, and it is seen that

$$\sin\beta = \frac{A_0B_0}{A_0A_3} = \frac{c}{V} = \frac{1}{\mathbf{M}} \tag{9.33}$$

where $\mathbf{M}$, the dimensionless velocity ratio V/c, is the Mach number as derived in Sec. 7.4.

In the case of the finite projectile of Fig. 9.6, the shock wave leaves the tip at an angle with the main flow which exceeds the Mach angle, on account of the

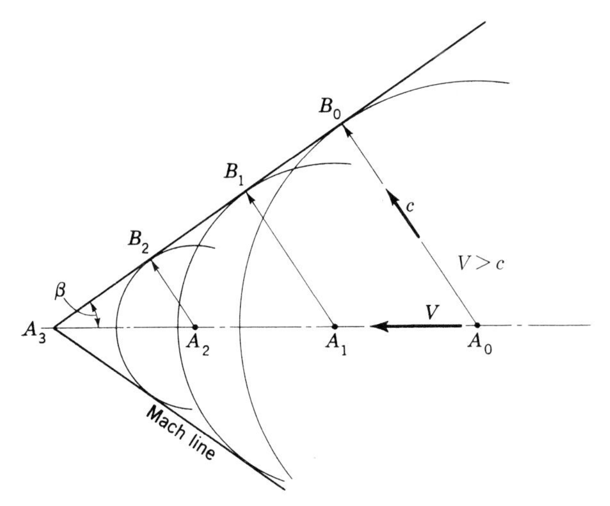

Figure 9.8 Schematic diagram of disturbance moving with supersonic velocity.

conical nose which follows the tip. Appropriate corrections[1] can be applied, however, and the shock-wave angle from such a sharp-nosed object can serve as an accurate means of measuring supersonic velocities. For a blunt-nosed object (Fig. 9.7) the shock wave is a combination of partly normal and partly oblique shock waves.

9.11 ISOTHERMAL FLOW

Most of the flows discussed thus far in this chapter are characterized by substantial changes in temperature. In isentropic flow (pv^k = constant) through a nozzle, for example, large changes in temperature occur because the variation in cross-sectional flow area causes substantial changes in V, p, ρ, and T.

Let us now examine some aspects of isothermal (T = constant) flow. In this case $pv = RT$ = constant so that p/γ = constant and $I_1 = I_2$. As a result $\hat{h}_1 = \hat{h}_2$ and Eq. (9.6) reduces to

$$\frac{V_1^2}{2g} + Q_H = \frac{V_2^2}{2g} \tag{9.34}$$

This equation is applicable to both real and ideal compressible fluids. For the case of isothermal compressible flow of a real fluid in a pipe of uniform diameter (Sec. 9.12) there is a drop in pressure along the pipe because of fluid friction, and thus a decrease in density. To maintain continuity the velocity must increase along the pipe and thus the kinetic energy also increases. Reference to Eq. (9.34) shows that, with $A_1 = A_2$, heat energy must be added to the fluid to maintain isothermal flow in a pipe of uniform diameter.

9.12 ISOTHERMAL FLOW IN A PIPE OF UNIFORM DIAMETER

To obtain some understanding of the characteristics of compressible flow with friction, let us consider isothermal flow in a pipe of uniform diameter. The flow of a gas through a long, uninsulated pipe in an isothermal environment may be assumed to approximate isothermal flow since the temperature rise created by fluid friction is very small. Applying the energy equation [Eq. (4.28)] in differential form to a short length dx of pipe, expressing the head loss due to pipe friction in the form of the Darcy-Weisbach equation [Eq. (8.13)], disregarding variations in z as we are dealing with a gas, and noting that $d(V^2/2g) = V\,dV/g$, we can write

$$-\frac{dp}{\gamma} = f\frac{dx}{D}\frac{V^2}{2g} + \frac{V\,dV}{g} \tag{9.35}$$

[1] Gilbrech, Donald A.: "Fluid Mechanics," p. 405, Wadsworth Publishing Co, Belmont, CA, 1965.

Recalling from Sec. 7.4 that the Mach number $\mathbf{M} = V/c$, where $c = \sqrt{kp/\rho} = \sqrt{kRT}$ = constant for isothermal flow, one can restate Eq. (9.35) as follows:

$$\frac{dp}{p} + k\mathbf{M}^2\left(\frac{d\mathbf{M}}{\mathbf{M}} + f\frac{dx}{2D}\right) = 0 \tag{9.36}$$

It has been observed that the f values given in Fig. 8.11 are applicable to compressible flow if $\mathbf{M} < 1$. For $\mathbf{M} > 1$, the values for f are about one-half of those of Fig. 8.11.

From continuity ρAV = constant, and hence ρV = constant, since we are dealing with a pipe of uniform diameter. Combining this with the perfect-gas law ($\rho = p/RT$), we get $pV = RT$ = constant for isothermal flow. Substituting this expression (pV = constant) into Eq. (9.36) and rearranging, we get for isothermal flow in a pipe of constant diameter,

$$\frac{dp}{p} = -\frac{dV}{V} = -\frac{d\mathbf{M}}{\mathbf{M}} = -\frac{k\mathbf{M}^2}{1 - k\mathbf{M}^2} f\frac{dx}{2D} \tag{9.37}$$

This equation shows that, when $\mathbf{M} < 1/\sqrt{k}$, the pressure will decrease in the direction of flow. On the other hand, if $\mathbf{M} > 1/\sqrt{k}$, the pressure increases in the direction of flow.

The weight flow rate $G = \gamma AV$, hence Reynolds number may be expressed as

$$\mathbf{R} = \frac{DV\rho}{\mu} = \frac{GD\rho}{A\gamma\mu} = \frac{GD}{\mu g A} \tag{9.38}$$

At usual pressures the viscosity μ of a gas depends only on temperature and is constant for a given temperature; therefore for steady isothermal flow (G = constant and T = constant), the Reynolds number (and therefore also f) is constant along the entire length of a uniform-diameter pipe for any given flow. From the perfect-gas law, $\gamma = \rho g = pg/RT$, and from continuity, $V = G/\gamma A$. Hence $V = GRT/pgA$. Substituting these values in Eq. (9.35) and rearranging, we obtain

$$-\left(\frac{2g^2A^2}{G^2RT}\right)p\,dp = \frac{f}{D}\,dx + \frac{2\,dV}{V}$$

For isothermal flow in a pipe of uniform diameter this equation is readily integrated for a length $L = x_2 - x_1$ to give

$$p_1^2 - p_2^2 = \frac{G^2RT}{g^2A^2}\left(f\frac{L}{D} + 2\ln\frac{p_1}{p_2}\right) \tag{9.39}$$

where p_1/p_2 has replaced V_2/V_1. That this substitution is valid can be seen by combining ρV = constant from continuity with $p/\rho = RT$ = constant for isothermal flow. Equation (9.39) may be used to find G if all other values are given, but if G and other values at (1) are given, p_2 may be found by successive approximation. In most cases the last term is small compared with fL/D and may be neglected as a first approximation. If it proves to be significant, a second solution, using an approximate value of p_1/p_2, may be made if greater accuracy is desired. It may be noted that p and A involve the *same area units*, and so, for numerical work in English units it is usually more convenient to use p in pounds per square inch and A in square inches.

There is a restriction to Eq. (9.39) similar to that discussed in Sec. 9.7 and inferred in Eq. (9.37). To illustrate, disregarding the logarithmic term, Eq. (9.39) can be expressed as $p_2^2 = p_1^2 - NL$, where N is a constant for the given conditions and L is the distance along the pipe from section 1 to section 2. According to the equation as L gets larger, p_2 will get smaller until it eventually drops to zero which, of course, is physically impossible. Actually what happens is that as L gets larger, p gets smaller as does ρ; and since $\rho V = \text{constant}$, V gets larger. However, there is a limit to how large V can get. This occurs when $\mathbf{M} = 1/\sqrt{k}$ as can be seen by examining Eq. (9.37). Thus Eq. (9.39) is applicable as long as $\mathbf{M} < 1/\sqrt{k}$. Another way of stating this is that for isothermal flow in a pipe of constant diameter there is a maximum length of pipe for which the given isothermal flow will proceed continuously. If the pipe exceeds this limiting length, there is a *choking* of the flow which limits the mass (or weight) flow rate.

Another way of expressing Eq. (9.39) is to divide it by p_1^2, substitute $\gamma_1 A_1 V_1$ for G, and note that $\mathbf{M}_1/\mathbf{M}_2 = p_2/p_1$ for this situation. Upon rearrangement, we get

$$\frac{\mathbf{M}_1^2}{\mathbf{M}_2^2} = 1 - k\mathbf{M}_1^2\left(2 \ln \frac{\mathbf{M}_2}{\mathbf{M}_1} + f\frac{L}{D}\right) \tag{9.40}$$

Equation (9.40) is a particularly useful form of the isothermal-flow equation as it can be handily employed to determine the maximum length of pipe such that the maximum possible velocity (limiting condition) will occur at the downstream end of the pipe. This is shown in the following illustrative example.

Illustrative Example 9.7 Air flows isothermally at 65°F through a horizontal 10- by 14-in rectangular duct at 100 lb/s. If the pressure at a section is 80 psia, find the pressure at a second section 500 ft downstream from the first. Assume the duct surface is very smooth; hence the lowest curve of Fig. 8.11 may be used to determine f.

First of all determine the applicability of Eq. (9.39)

$$R_h = \frac{A}{P} = \frac{140}{48} = 2.92 \text{ in} = 0.243 \text{ ft} \qquad \text{(Viscosity of air from Table A.2a)}$$

$$\mathbf{R}_1 = \frac{DV\rho}{\mu} = \frac{GD}{\mu g A} = \frac{G(4R)}{\mu g A} = \frac{100(4 \times 0.243)}{3.78 \times 10^{-7}(32.2)\frac{140}{144}} = 8.2 \times 10^6$$

$$\mathbf{R}_2 = \mathbf{R}_1 \text{ since} \qquad \rho_1 V_1 = \rho_2 V_2 \qquad \text{and} \qquad \mu_1 = \mu_2$$

From Fig. 8.11, $f = 0.0083$.

$$\gamma_1 = \frac{pg}{RT} = \frac{(80 \times 144)32.2}{1{,}715(460 + 65)} = 0.41 \text{ lb/ft}^3$$

$$V_1 = \frac{G}{\gamma_1 A} = \frac{100}{(0.41)\frac{140}{144}} = 250 \text{ fps}$$

$$c = \sqrt{kRT} = \sqrt{1.4 \times 1{,}715 \times 525} = 1{,}123 \text{ fps}$$

$$\mathbf{M}_1 = \frac{V_1}{c} = \frac{250}{1{,}123} = 0.222$$

The limiting value of $\mathbf{M}_1$ is $1/\sqrt{1.4} = 0.845$. Substituting into Eq. (9.40),

$$\frac{(0.222)^2}{(0.845)^2} = 1 - 1.4(0.222)^2\left[2 \ln \frac{0.845}{0.222} + 0.0083\frac{L}{4(0.243)}\right]$$

$$L = 1{,}260 \text{ ft}$$

Thus Eq. (9.39) applies for all values of $L < 1{,}260$ ft. Substituting $L = 500$ ft in Eq. (9.39) and neglecting the usually small logarithmic term,

$$(80 \times 144)^2 - p_2^2 = \frac{(100)^2(1{,}715)(525)}{(32.2)^2(140/144)^2}\left[0.0083 \times \frac{500}{4(0.243)}\right]$$

from which $p_2 = 67.1$ psia. Substituting this value of p_2 into Eq. (9.39) and considering the logarithmic term yields $p_2 = 66.6$ psia. Repeating the process again will give a more accurate answer.

Illustrative Example 9.8 For the case of Illustrative Example 9.7 with a duct length of 500 ft, compute the thermal energy (heat) that must be added to the fluid to maintain isothermal conditions.

Since the flow is isothermal, $p_1/\rho_1 = p_2/\rho_2 = RT =$ constant; $p_1 = 80$ psia and $p_2 = 66.6$ psia. Thus $\rho_1/\rho_2 = \frac{80}{66.6} = 1.20$ and $V_2/V_1 = 1.20$ since $\rho V =$ constant from continuity.

So $V_2 = 1.20(250) = 300$ fps; applying Eq. (9.34) $Q_H = (300)^2/64.4 - (250)^2/64.4 = 427$ ft·lb/lb of air. Since $G = 100$ lb/s, the rate at which heat must be added to the fluid is $100 \times 427 = 42{,}700$ ft·lb/s. Note, if $Q_H > 427$ ft·lb/lb of air, $T_2 > T_1$, and if $Q_H < 427$ ft·lb/lb of air, $T_2 < T_1$.

9.13 ADIABATIC FLOW IN A PIPE OF UNIFORM DIAMETER

The flow of fluids in well-insulated pipes is a case that approaches adiabatic flow (i.e., $Q_H = 0$). The usual applications of insulation are to pipes conveying steam or refrigerating fluids, such as ammonia vapor. In some situations such pipes are short, the pressure drops are relatively small, and the problem can be solved as if the fluid were incompressible. However, there are situations where the effects of compressibility must be considered.

Let us consider the case of steady adiabatic flow with friction in a pipe of uniform diameter. Substituting $V_1 = G/\gamma_1 A$ and $V_2 = G/\gamma_2 A$ into Eq. (9.11), we get

$$\frac{G^2}{\gamma_2^2 A^2} - \frac{G^2}{\gamma_1^2 A^2} = \frac{2k}{k-1}(p_1 v_1 - p_2 v_2) \tag{9.41}$$

where G is the weight rate of flow and A is the cross-sectional area of the pipe.

This equation can be rewritten as

$$\frac{G^2}{\gamma_1^2 A^2} + \frac{2k}{k-1} p_1 v_1 = \frac{G^2}{\gamma_2^2 A^2} + \frac{2k}{k-1} p_2 v_2 = X \tag{9.42}$$

where X is a constant evaluated from known conditions at section 1. Thus, in general

$$pv = \frac{p}{\rho} = \frac{k-1}{2k}\left(X - \frac{G^2}{\gamma^2 A^2}\right)$$

or

$$p = \frac{k-1}{2k}\left(X\rho - \frac{G^2}{g^2 A^2 \rho}\right) \tag{9.43}$$

Hence

$$dp = \frac{k-1}{2k}\left(X + \frac{G^2}{g^2 A^2 \rho^2}\right) d\rho$$

Multiplying both sides by ρ and integrating,

$$\int_1^2 \rho\, dp = \frac{k-1}{2k}\left[X\left(\frac{\rho_2^2 - \rho_1^2}{2}\right) + \frac{G^2}{g^2 A^2} \ln \frac{\rho_2}{\rho_1}\right] \tag{9.44}$$

The value of $\int_1^2 \rho\, dp$ may be obtained from Eq. (9.44) inasmuch as ρ may be found for any flow G and any value of p by Eq. (9.43). Practically, it will be better to assume values of ρ and find the corresponding values of p by Eq. (9.42).

Next let us rewrite Eq. (9.35). Multiplying each term by $\rho^2 g$ so that the first term becomes $\rho\, dp$, and substituting $\rho V = G/gA$ and $\rho\, dV = -V\, d\rho$ (from the differential of $\rho A V =$ constant), we obtain

$$\rho\, dp - \left(\frac{G}{gA}\right)^2 \frac{d\rho}{\rho} + \frac{f}{2D}\left(\frac{G}{gA}\right)^2 dx = 0$$

Integrating this (assuming f to be constant)[1] and rearranging with $x_2 - x_1 = L$,

$$f \frac{L}{D} \frac{1}{2}\left(\frac{G}{gA}\right)^2 = -\int_1^2 \rho\, dp - \left(\frac{G}{gA}\right)^2 \ln \frac{\rho_1}{\rho_2} \tag{9.45}$$

Since $\int_1^2 \rho\, dp$ may be evaluated by Eq. (9.44) it is possible to solve Eq. (9.45) to obtain a value of f if p_1 and p_2 have been measured for a known distance L, or the value of L may be found for any assumed values of p_1 and p_2 (or preferably ρ_1 and ρ_2) if f is given or assumed.

Thus it is possible through successive calculations to plot a curve such as that in Fig. 9.9 for any assumed flow and initial conditions where p_2 represents any pressure along the pipe at any distance x_2. However, as in the case of

[1] In adiabatic flow in a pipe of uniform diameter the Reynolds number is not constant along the pipe because of changes in viscosity caused by variations in temperature. Thus, since **R** varies, f must vary. For most situations, however, f may be assumed to have a constant value without introducing much error.

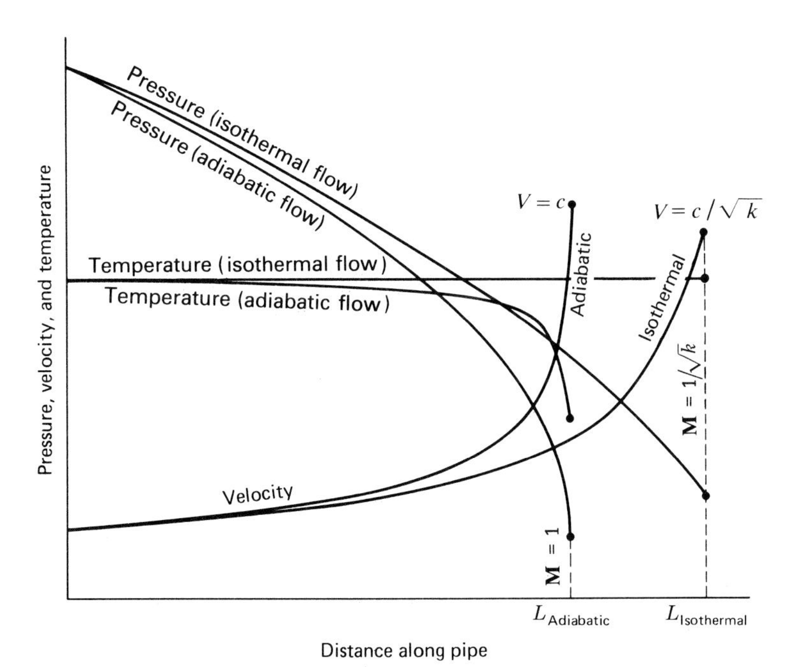

Figure 9.9 Subsonic flow of a compressible fluid in a pipe of constant diameter.

isothermal flow, there is a minimum value of p_2 where the velocity has attained its maximum value. It can be proved that in adiabatic flow, critical conditions occur when $\mathbf{M} = 1.0$.

The equations in this section [Eqs. (9.41) to (9.45)] apply equally well to supersonic flow. However the characteristics of the flow are reversed as shown in Table 9.1.

In both flows (subsonic and supersonic) the Mach number approaches a value of 1.0. In subsonic flow the Mach number gets larger as one moves along the pipe, and when it reaches a value of 1.0 choking of the flow occurs. In supersonic flow the Mach number gets smaller along the pipe, and when it reaches a value of 1.0 a shock wave occurs.

Table 9.1 Changes in the downstream direction for subsonic and supersonic adiabatic flow in a pipe of uniform diameter

Property	Subsonic flow	Supersonic flow
Velocity V	Increases	Decreases
Mach number $\mathbf{M}$	Increases	Decreases
Pressure p	Decreases	Increases
Temperature T	Decreases	Increases
Density ρ	Decreases	Increases

Illustrative Example 9.9 Air flows adiabatically through a 10- by 14-in rectangular duct at 100 lb/s. At a certain section the pressure is 80 psia and the temperature is 65°F (same data as Illustrative Example 9.7). Find the distance along the duct to the section where $\rho_2 = 0.80\rho_1$. Assume the duct surface is very smooth; hence the lowest curve of Fig. 8.11 may be used to determine f.

From Illustrative Example 9.7,

$$\rho_1 = \frac{\gamma_1}{g} = \frac{0.41}{32.2} = 0.0127 \text{ slug/ft}^3$$

$$V_1 = 250 \text{ fps} \qquad \mathbf{R}_1 = 8.2 \times 10^6 \qquad f = 0.0083$$

From Eq. (9.42),

$$X = \frac{100^2}{(0.41 \times 0.97)^2} + \frac{2(1.4)}{0.4}\left(\frac{80 \times 144}{0.0127}\right) = 6{,}412{,}000 \text{ ft}^2/\text{s}^2$$

Equation (9.43) yields

$$p_2 = \frac{0.4}{2.8}\left(6{,}412{,}000 \times 0.8 \times 0.0127 - \frac{100^2}{32.2^2(0.97)^2(0.8 \times 0.0127)}\right)$$

$$p_2 = 9160 \text{ lb/ft}^2\text{, abs} = 63.6 \text{ psia}$$

From Eq. (9.44), $\int_1^2 \rho\, dp = -26.9 \text{ lb}^2\cdot\text{s}^2/\text{ft}^4$. Putting this value in Eq. (9.45) and substituting the value for f yields

$$L = 563 \text{ ft}$$

Thus at a section 563 ft downstream from the first section, $\rho_2 = 0.8\rho_1 = 0.0102$ slug/ft³ and $p_2 = 63.6$ psia. Also,

$$V_2 = \frac{G}{\rho_2 g A} = \frac{100}{0.012 \times 32.2 \times 0.97} = 315 \text{ fps}$$

$$T_2 = \frac{p_2}{\rho_2 R} = \frac{63.6 \times 144}{0.0102 \times 1{,}715} = 523°\text{R} = 63°\text{F}$$

By assuming other values of ρ_2, one can get a complete picture of the flow at various sections along the pipe.

9.14 OTHER TYPES OF FLOW

The two types of compressible flow in pipelines that were just discussed are special cases which are often approximated in practice and which are amenable to mathematical treatment. For fluids which do not follow the perfect-gas laws, such as wet steam, for example, the preceding equations are only rough approximations. A more complicated thermodynamic treatment is necessary than is within the scope of this text.

In Fig. 9.9 are shown curves plotted to scale for the flow of air through a pipe of uniform diameter for both isothermal and adiabatic conditions, assuming the same initial values for each. Inspection of this diagram shows that for small pressure drops from the initial, such as up to $\Delta p/p_1 = 0.10$ (or $p_2/p_1 = 0.90$), there is very little difference between the two curves. Thus, for such a situation where $p_2/p_1 > 0.90$, adiabatic flow in a pipe can be analyzed as isothermal flow without introducing much error. The flow of gas in a pipe is rarely either isothermal or adiabatic. Isothermal flow requires that heat be transferred into the flowing fluid from the surrounding atmosphere at just the right rate, and this rate must increase along the length of the pipe. If the rate of heat transfer is less than this required amount, the performance curves will lie between the isothermal and the adiabatic curves in Fig. 9.9. Heat transfer is proportional to the temperature difference between the fluid and the surrounding atmosphere. If these temperatures are denoted as T_f and T_a, respectively, the heat transfer is some function of $T_a - T_f$. Isothermal flow is possible only if T_a is greater than T_f.

If the fluid in the pipe is very much colder than the surroundings, its absorption of heat might be such as to cause the pressure in the pipe to be higher than for isothermal flow. On the other hand, for example, air at a high temperature might be discharged directly from a compressor into a pipe so that T_f might be greater than T_a, which would cause heat to flow from the fluid in the pipe to the surrounding atmosphere. In this case the pressure along the pipe would decrease even faster that for adiabatic flow.

Because the energy equation for compressible fluids does not contain a term for friction and the momentum equation which contains a term for friction[1] does not include any term for heat transfer, it is seen that there is no simple analytical solution of such cases. An approximate approach to such problems is to divide the entire length of pipe into short reaches and employ the equations of incompressible flow using average values of density and velocity within each reach. This step-by-step method will give results that are approximately correct if the lengths of the reaches are made small enough. Small reach lengths are particularly important in the regions where the curves of Fig. 9.9 are sharply curved.

9.15 CONCLUDING REMARKS

In this chapter we have seen that in order to solve problems of compressible flow the equation of state of the gas must be combined with the energy equation and the continuity principle. Hence thermodynamics is commonly involved in compressible-flow problems and the expressions relating the various physical parameters are generally quite complicated. In the discussion we have restricted ourselves to one-dimensional flow, and have not considered multidimensional flow, which is of course an extremely important topic, especially when dealing with aircraft and missiles. The intent here was merely to provide an introduction to the flow of compressible fluids. Among the references cited at the end of the book will be found some excellent treatments of compressible flow.

[1] Friction is accounted for as one of the forces acting on the fluid element in the equation $\sum \mathbf{F} = \rho Q \Delta \mathbf{V}$.

PROBLEMS

9.1 Compute the change in enthalpy of 500 lb (2,200 N) of oxygen if its temperature is increased from 120°F (50°C) to 155°F (70°C).

9.2 Suppose the 500 lb (2,200 N) of oxygen of Prob. 9.1 were compressed isentropically to 80 percent of its original volume. Find the final temperature and pressure, the work required, and the change in enthalpy. Assume $T_1 = 120°F$ (50°C) and $p_1 = 200$ psia (1,400 kN/m², abs).

9.3 Using the data of Illustrative Example 9.2 compute $\Delta(p/\rho)$ and thus show that $\Delta h = \Delta i + \Delta(p/\rho)$.

9.4 Using the data of Illustrative Example 9.2, determine the work done in compressing the air by finding the area under a pressure-vs.-volume curve. Compute and tabulate volumes and pressures using volume increments which are 10 percent of the original volume.

9.5 Determine the sonic velocity in air at sea level and at elevations 5,000, 10,000, 20,000, and 30,000 ft. Assume standard atmosphere (Appendix 3, Table A.3).

9.6 Repeat Prob. 9.5 for sea level, 2,000 and 10,000 m, expressing the answers in SI units.

9.7 Find the stagnation pressure and temperature in air flowing at 88 fps if the pressure and temperature in the undisturbed field are 14.7 psia and 50°F respectively.

9.8 Air flows past an object at 600 fps. Determine the stagnation pressures and temperatures in the standard atmosphere at elevations of sea level, 5,000 and 30,000 ft.

9.9 Repeat Prob. 9.8 for an air speed of 200 m/s and elevations of sea level, 2,000 and 10,000 m, expressing the answers in SI units.

9.10 Air at 250 psia (1750 kN/m², abs) is moving at 500 fps (150 m/s) in a high-pressure wind tunnel at a temperature of 100°F (40°C). Find the stagnation pressure and temperature. Note the magnitude of the sonic velocity for the 250-psia 100°F (1750-kN/m² 40°C) air.

9.11 Show that Eq. (9.15) results from the binomial expansion of Eq. (9.14).

9.12 Air at a pressure of 150 psia (1000 kN/m², abs) and a temperature of 100°F (40°C) expands in a suitable nozzle to 15 psia (100 kN/m², abs). (*a*) If the flow is frictionless and adiabatic and the initial velocity is negligible, find the final velocity by Eq. (9.16). (*b*) Find the final temperature at the end of the expansion through use of Eq. (9.10).

9.13 Derive Eq. (9.16) for isentropic flow by integrating the Euler equation.

9.14 Carbon dioxide flows isentropically. At a point in the flow the velocity is 50 fps (15 m/s) and the temperature is 125°F (50°C). At a second point on the same streamline the temperature is 80°F (25°C). What is the velocity at the second point?

9.15 Refer to Prob. 9.14. If the pressure at the first point were 20 psia, determine the pressure and temperature on the nose of a streamlined object placed in the flow at that point.

9.16 Show in detail the development of Eq. (9.20) from Eqs. (9.19) and (9.18).

9.17 Verify that Eq. (9.23) results from the substitution of $\mathbf{M} = 1.0$ at the throat in Eq. (9.22).

9.18 Start with Eq. (9.21) and derive Eq. (9.24).

9.19 Differentiate Eq. (9.24) with respect to p_2/p_1 and set to zero to find the value of p_2/p_1 for which G is a maximum. The answer should correspond to Eq. (9.23).

9.20 Air flows at 150°F (65°C) from a large tank through a 1.5-in (4-cm)-diameter converging nozzle. Within the tank the pressure is 85 psia (600 kN/m², abs). Calculate the flow rate for external pressures of 10, 30, 50, and 70 psia (50, 200, 350, and 500 kN/m², abs). Assume isentropic conditions. Plot G as a function of p_3. Assume that the temperature within the tank is 150°F (65°C) in all cases. Compute also the temperature at the nozzle outlet for each condition.

9.21 Air flows at 25°C from a large tank through a 10-cm-diameter converging nozzle. Within the tank the pressure is 50 kN/m², abs. Calculate the flow rate for external pressures of 30, 20, and 10 kN/m², abs. Assume isentropic conditions. Plot G as a function of p_3. Assume that the temperature within the tank is 25°C in all cases. Compute also the temperature at the nozzle outlet for each condition.

9.22 Air within a tank at 120°F flows isentropically through a 2-in-diameter convergent nozzle into a 14.2-psia atmosphere. Find the flow rate for air pressures within the tank of 5, 10, 20, 40, and 50 psig.

9.23 Carbon dioxide within a tank at 40 psia (280 kN/m^2, abs) and 80°F (25°C) discharges through a convergent nozzle into a 14.2-psia (98 kN/m^2, abs) atmosphere. Find the velocity, pressure, and temperature at the nozzle outlet. Assume isentropic conditions.

9.24 In Prob. 9.23 if the pressure and temperature within the tank had been 20 psia (140 kN/m^2, abs) and 100°F (40°C), what would have been the velocity, pressure, and temperature at the nozzle outlet? Assume isentropic conditions.

9.25 Refer to Illustrative Example 9.4. If the pressure in the tank is 5 psig, confirm that $G = 1.20$ lb/s, $p_2 = 13.57$ psia, and $T_2 = 492°$R.

9.26 Air enters a converging-diverging nozzle at a pressure of 120 psia (830 kN/m^2, abs) and a temperature of 90°F (32°C). Neglecting the entrance velocity and assuming a frictionless process, find the Mach number at the cross section where the pressure is 35 psia (240 kN/m^2, abs).

9.27 Work Illustrative Example 9.5 with all data the same except for the pressure within the tank, which is 100 rather than 50 psia.

9.28 Air discharges from a large tank through a converging-diverging nozzle. The throat diameter is 3.0 in (7.5 cm), and the exit diameter is 4.0 in (10 cm). Within the tank the air pressure and temperature are 40 psia (290 kN/m^2, abs) and 150°F (65°C), respectively. Calculate the flow rate for external pressures of 39, 38, 36, and 30 psia, (280, 270, 250, and 200 kN/m^2, abs). Assume no friction.

9.29 Air is to flow through a converging-diverging nozzle at 18 lb/s. At the throat the pressure, temperature, and velocity are to be 20 psia, 100°F, and 500 fps, respectively. At outlet the velocity is to be 200 fps. Determine the throat diameter. Assume isentropic flow.

9.30 Air in a tank under a pressure of 140 psia (950 kN/m^2, abs) and 70°F (20°C) flows out into the atmosphere through a 1.00-in (2.5-cm)-diameter converging nozzle. (*a*) Find the flow rate. (*b*) If a diverging section with an outlet diameter of 1.50 in (4 cm) were attached to the converging nozzle, what then would be the flow rate? Neglect friction.

9.31 Repeat Prob. 9.30 for the case where the air within the tank is at 20 psia (140 kN/m^2, abs). Assume all other data to be the same.

9.32 Air discharges from a large tank through a converging-diverging nozzle with a 2.5-cm-diameter throat into the atmosphere. The gage pressure and temperature in the tank are 700 kN/m^2 and 40°C, respectively, the barometric pressure is 995 millibars. (*a*) Find the nozzle-tip diameter required for p_3 to be equal to the atmospheric pressure. For this case, what are the flow velocity, sonic velocity, and Mach number at the nozzle exit? (*b*) Determine the value of p_4 which will cause the shock wave to be located at the nozzle exit.

9.33 The pressure, velocity, and temperature just upstream of a normal shock wave in air are 10 psia, 2,200 fps, and 23°F. Determine the pressure, velocity, and temperature just downstream of the wave.

9.34 Just downstream of a normal shock wave the pressure, velocity, and temperature are 52 psia (360 kN/m^2, abs), 400 fps (110 m/s) and 120°F (50°C). Compute the Mach number upstream of the shock wave. Consider air and carbon dioxide.

9.35 Assuming the tip of the model in Fig. 9.6 to be a point source of infinitesimal disturbance, find the air velocity if the temperature is $-60°$F and $k = 1.4$. If the actual Mach number is 1.38, what is the percentage error involved in the preceding assumption?

9.36 A schlieren photograph of a bullet shows a Mach angle of 30°. The air is at a pressure of 14 psia (95 kN/m^2, abs) and 50°F (10°C). Find the approximate speed of the bullet.

9.37 Air flows isothermally in a long pipe. At one section the pressure is 90 psia (600 kN/m^2, abs), the temperature is 80°F (25°C), and the velocity is 100 fps (30 m/s). At a second section some distance from the first the pressure is 15 psia (100 kN/m^2, abs). Find the energy head loss due to friction, and determine the thermal energy that must have been added to or taken from the fluid between the two sections. The diameter of the pipe is constant.

9.38 Air flows isothermally through a long horizontal pipe of uniform diameter. At a section where the pressure is 100 psia (700 kN/m^2, abs), the velocity is 120 fps (35 m/s). Because of fluid friction the pressure at a distance point is 40 psia (280 kN/m^2, abs). (*a*) What is the increase in kinetic energy per pound of air? (*b*) What is the amount of thermal energy in Btu per pound (J/N) of air that must be transferred in order to maintain the temperature constant? (*c*) Is this heat transferred to the air in the pipe or removed from it? (*d*) If the temperature of the air is 100°F (40°C) and the diameter of the pipe is 3 in (7.5 cm) find the total heat transferred in Btu (joules) per hour.

9.39 Refer to Illustrative Example 9.7. Neglecting the logarithm term in Eq. (9.39), find the pressures at sections 100, 300, and 800 ft downstream of the section where the pressure is 80 psia. Plot the pressure as a function of distance along the pipe.

9.40 Carbon dioxide flows isothermally at 100°F (40°C) through a horizontal 6-in (15-cm)-diameter pipe. At this temperature $\mu = 4.0 \times 10^{-7}$ $lb \cdot s/ft^2$ (1.95×10^{-5} $N \cdot s/m^2$). The pressure changes from 150.0 to 140.0 psig (1000 to 930 kN/m^2, gage) in a 100-ft (30-m) length of pipe. Determine the flow rate if the atmospheric pressure is 14.5 psia (100 kN/m^2) and e for the pipe is 0.002 ft (0.60 mm).

9.41 Methane gas is to be pumped through a 24-in-diameter welded-steel pipe connecting two compressor stations 25 mi apart. At the upstream station the pressure is not to exceed 60 psia, and at the downstream station it is to be at least 20 psia. Determine the maximum possible rate of flow (in cubic feet per day at 60°F and 1 atm). Assume isothermal flow at 60°F.

9.42. Refer to Illustrative Example 9.9. Find the distance along the pipe to (*a*) where $\rho_2 = 0.9\rho_1$; (*b*) where $\rho_2 = 0.7\rho_1$; (*c*) where subsonic adiabatic flow ends. Compute the corresponding values of p, V, T, and $\mathbf{M}$ and plot the first three as a function of distance along the pipe.

9.43. Air flows adiabatically at 100 lb/s (450 N/s) in a 12-in (30-cm)-diameter horizontal pipe. At a certain section the pressure is 150 psia (1000 kN/m^2, abs) and the temperature 140°F (60°C). Determine the distance along the pipe to the section where $\rho_2 = 0.80\rho_1$. Assume $e/D = 0.0004$.

CHAPTER

TEN

FORCES ON IMMERSED BODIES

In this chapter the discussion relates primarily to fluid phenomena encountered in incompressible flow or in low-velocity compressible flow where the effects of compressibility are negligible. Near the end of the chapter in Sec. 10.14 there is a brief discussion of the effects of compressibility on drag and lift. These become important at Mach numbers above 0.7.

10.1 INTRODUCTION

A body which is wholly immersed in a homogeneous fluid may be subject to two kinds of forces arising from relative motion between the body and the fluid. These forces are termed the *drag* and the *lift*, depending on whether the force is parallel to the motion or at right angles to it, respectively. Fluid mechanics draws no distinction between two cases of relative motion, namely, when a body moves rectilinearly at constant speed through a stationary fluid or when a fluid travels at constant velocity past a stationary body. Thus it is possible to test airplane models in wind tunnels and torpedo models in water tunnels and predict with confidence the behavior of their prototypes when moving through still fluid. For instructional purposes it is somewhat simpler to fix our ideas on the stationary body in the moving fluid, while the practical result desired is more frequently associated with a body moving through still fluid.

In this chapter we shall first consider the drag, or resistance, forces. As we shall not be concerned with wave action at a free surface, gravity does not enter the problem and the forces involved are those due to inertia and viscosity.[1] The drag forces on a submerged body can be viewed as having two components: a *pressure drag* F_p and a *friction* or *surface, drag* F_f. The pressure drag, oftentimes referred to as the *form drag* because it depends largely on the form of the body, is equal to the integration of the components in the direction of motion of all the pressure forces exerted on the surface of the body. It may be expressed[2] as the dynamic component of the stagnation pressure [from Eq. (4.31)] acting on the projected area A of the body *normal to the flow* times a coefficient C_p which is dependent on the geometric form of the body and generally determined by experiment. Thus,

$$F_p = C_p \rho \frac{V^2}{2} A \tag{10.1}$$

The friction drag along a surface is equal to the integration of the components of the shear stress along the surface of the body in the direction of motion. For convenience, the friction drag is commonly expressed in the same general form as Eq. (10.1). Thus,

$$F_f = C_f \rho \frac{V^2}{2} BL \tag{10.2}$$

where C_f = friction-drag coefficient, dependent on viscosity, among other factors
L = length of surface parallel to flow
B = transverse width, conveniently approximated for irregular shapes by dividing total surface area by L

It is important to note that for a body such as a plate with both sides immersed in the fluid, Eq. (10.2) gives the drag for *one side* only.

From our experience with pipe flow we should expect that the friction drag would be more amenable to a theoretical approach than pressure drag. This is not necessarily the case. In previous chapters the *boundary layer* was described as a very thin layer of fluid adjacent to a surface, in which viscosity is important, while outside this layer the fluid can be considered as frictionless or ideal. This concept, originated by Prandtl, is one of the important advances in modern fluid mechanics. It means that the mathematical theory of ideal fluid flow, including the flow-net method studied in Chap. 3, can actually be used to determine the streamlines in the real fluid at a short distance away from a solid wall. The Bernoulli theorem may then be used to determine the normal pressures on the

[1] Actually, without viscosity there could be no drag force at all. The flow of a frictionless fluid about any body, as constructed mathematically or by the flow-net technique of Chap. 3, produces opposing stagnation points at the nose and tail of the body. The pressure distribution, as computed from the Bernoulli theorem and integrated over the entire body, always adds up to zero in the direction of the flow. This situation is known as *D'Alembert's paradox*.

[2] Note that Eq. (10.1) is of the same general form as the expression for the drag on a sphere that was developed by dimensional analysis in Sec. 7.7. The comparison indicates $C_p = \phi(\mathbf{R})$. If the effects of compressibility had been considered in Sec. 7.7, the comparison would have shown $C_p = \phi(\mathbf{R}, \mathbf{M})$.

surface, for such pressures are practically the same as those outside this thin layer.

The boundary layer may be entirely laminar, or it may be primarily turbulent with a viscous sublayer, as in Fig. 8.7. The boundary layer increases in thickness with distance from the leading edge of a surface, as shown in both Figs. 8.4 and 8.7. The important difference between the case at hand and that of pipe flow, however, is that in pipe flow the boundary layers from the opposite walls of the pipe merge together after a certain distance and the flow becomes "all boundary layer," while with airplanes, submarines, trains, etc., the boundary layer, even though it may reach a thickness of several inches, is still small compared with the dimensions of the "ideal fluid" outside the boundary layer in which the streamlines are determined by the presence of the body.

10.2 FRICTION DRAG OF BOUNDARY LAYER—INCOMPRESSIBLE FLOW

In Fig. 10.1 is shown the growth of a boundary layer along one side of a smooth plate in steady flow of an incompressible fluid. Let us consider the control volume shown in Fig. 10.2 which extends a distance δ from the plate, where δ is the thickness of the boundary layer at a distance x along the plate.[1] Along control surface AB the undisturbed velocity U exists. The pressure forces around the periphery of the control volume will cancel one another out since the undisturbed flow field pressure must exist along AB and DA, and the distance BC $(=\delta)$ is so small it will have a negligible effect on pressure variations.

Applying Eq. (6.7a), we get

$$\begin{aligned} -F_x = -\text{drag} &= \text{rate of momentum in } x \text{ direction leaving through } BC \\ &\quad + \text{rate of momentum in } x \text{ direction leaving through } AB \\ &\quad - \text{rate of momentum in } x \text{ direction entering through } DA \end{aligned} \tag{10.3}$$

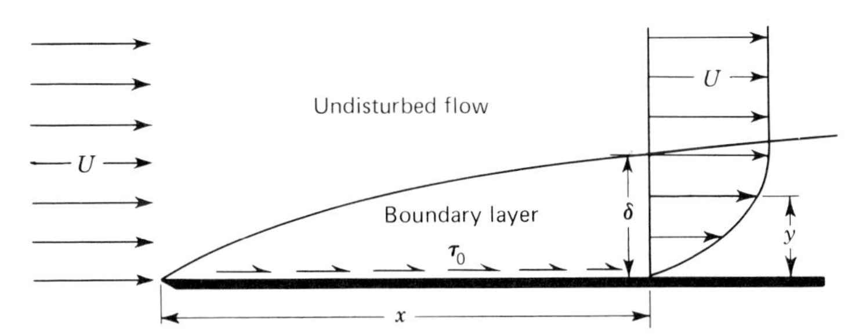

Figure 10.1 Growth of a boundary layer along a smooth plate (vertical scale exaggerated).

[1] Here we use δ to indicate the thickness of the boundary layer, usually defined as the distance from the boundary to the point where the velocity $u = 0.99U$. In this analysis, however, we will assume that $u = U$ at the edge of the boundary layer.

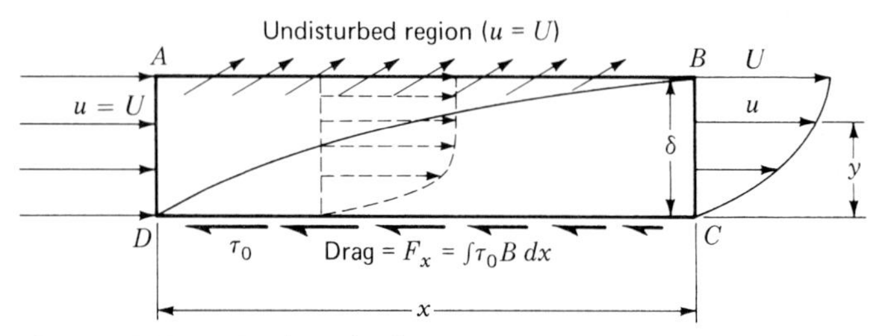

Figure 10.2 Control volume for flow over one side of a flat plate.

Since $Q_{BC} < Q_{DA}$, there is flow out of the control volume across control surface AB and $Q_{AB} = Q_{DA} - Q_{BC}$.

If the width of the plate is B, neglecting edge effects, the flows and momentums across the control surfaces can be expressed as follows:

Control surface	Flow rate	Rate of momentum in x direction
DA	$UB\delta$	$\rho(UB\delta)U$
BC	$B\int_0^\delta u\, dy$	$\rho B \int_0^\delta u^2\, dy$
AB	$UB\delta - B\int_0^\delta u\, dy$	$\rho\left(UB\delta - B\int_0^\delta u\, dy\right)U$

Substituting these momentum values in Eq. (10.3) gives

$$F_x = \rho B \int_0^\delta u(U - u)\, dy \tag{10.4}$$

where F_x is the total friction drag of the plate on the fluid *from the leading edge up to x* directed to the left, as shown on Fig. 10.2. Equal and opposite to this is the drag of the fluid on the plate.

It will now be assumed that the velocity profiles at various distances along the plate are similar to each other:

$$\frac{u}{U} = f\left(\frac{y}{\delta}\right) = f(\eta) \qquad \eta = \frac{y}{\delta}$$

There is experimental evidence that this assumption is valid if there is no pressure gradient along the surface and if the boundary layer does not change from laminar to turbulent within the region considered. Then substituting $u = Uf(\eta)$, $y = \delta\eta$ and $dy = \delta\, d\eta$ into Eq. (10.4) and noting that the limits of integration become 0 to 1, we get

$$F_x = \rho B U^2 \delta \int_0^1 f(\eta)[1 - f(\eta)]\, d\eta$$

which, for convenience, may be written

$$F_x = \rho B U^2 \, \delta \, \alpha \tag{10.5}$$

where α is a function of the boundary-layer velocity distribution only and is given by the indicated integral (not to be confused with the kinetic-energy factor α discussed in Chap. 4).

We next investigate the local wall shear stress τ_0 at distance x from the leading edge. From the definition of surface resistance, $dF_x = \tau_0 B \, dx$, or

$$\tau_0 = \frac{1}{B}\frac{dF_x}{dx} = \frac{1}{B}\frac{d}{dx}(\rho B U^2 \, \delta \, \alpha)$$

and as all terms in the expression for F_x are constant except δ,

$$\tau_0 = \rho U^2 \alpha \frac{d\delta}{dx} \tag{10.6}$$

This expression for the shear stress is valid for either laminar or turbulent flow in the boundary layer, but in this form it is not useful until the quantities α and $d\delta/dx$ are evaluated.

10.3 LAMINAR BOUNDARY LAYER FOR INCOMPRESSIBLE FLOW ALONG A SMOOTH FLAT PLATE

As in the case of laminar flow in pipes, we may examine the shear stress at the plate wall with the aid of the velocity gradient and the definition of viscosity.

$$\tau_0 = \mu\left(\frac{du}{dy}\right)_{y=0} = \frac{\mu}{\delta}\left(\frac{du}{d\eta}\right)_{\eta=0} = \frac{\mu U}{\delta}\left[\frac{df(\eta)}{d\eta}\right]_{\eta=0}$$

which may be abbreviated to

$$\tau_0 = \frac{\mu U \beta}{\delta} \tag{10.7}$$

where β, like α, is a dimensionless function of the velocity-distribution curve and is given by the expression in brackets.

Equations (10.6) and (10.7) are two independent expressions for τ_0. Equating them to one another results in a simple differential equation,

$$\delta \, d\delta = \frac{\mu\beta}{\rho U \alpha} dx$$

with solution

$$\frac{\delta^2}{2} = \frac{\mu\beta}{\rho U \alpha} x + C$$

where $C = 0$, since $\delta = 0$ at $x = 0$. Therefore

$$\delta = \sqrt{\frac{2\mu\beta x}{\rho U \alpha}} = \sqrt{\frac{2\beta}{\alpha}} \frac{x}{\sqrt{\mathbf{R}_x}} \tag{10.8}$$

where $\mathbf{R}_x = xU\rho/\mu$ may be called the *local* Reynolds number. It should be noted that $\mathbf{R}_x$ increases linearly in the downstream direction. Examination of the first expression of Eq. (10.8) shows that the thickness of the laminar boundary layer increases with distance from the leading edge; thus the shear stress [Eq. (10.7)] decreases as the layer grows along the plate.

To evaluate Eq. (10.8), we must know or assume the velocity profile in the laminar boundary layer. The velocity distribution may be closely represented by a parabola, as shown in Fig. 10.3. In dimensionless terms this curve becomes

$$\frac{u}{U} = f(\eta) = 2\eta - \eta^2 \tag{10.9}$$

The other velocity profile in Fig. 10.3 was derived by Blasius from the fundamental equations of viscous flow, with all factors considered, and has been closely checked by experiment.[1] This curve is based on the thickness δ being defined as that for which $u = 0.99\ U$.

As can be demonstrated by Prob. 10.2, the parabolic distribution will give numerical values for α and β of 0.133 and 2.0, respectively. The Blasius curve yields $\alpha = 0.135$ and $\beta = 1.63$, the principal difference lying in the milder slope of the velocity gradient at the wall. With the Blasius values substituted in Eq. (10.8) we obtain

$$\frac{\delta}{x} = \sqrt{\frac{2 \times 1.63}{0.135}}\frac{1}{\sqrt{\mathbf{R}_x}} = \frac{4.91}{\sqrt{\mathbf{R}_x}} \tag{10.10}$$

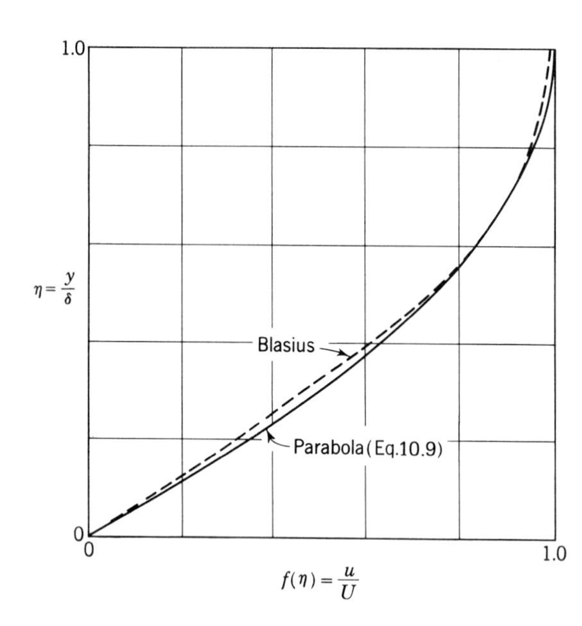

Figure 10.3 Velocity distribution in laminar boundary layer on flat plate. (Blasius curve adapted from Fig. 30 of *NACA Tech. Mem. 1217, 1949*.)

[1] See H. Schlichting, "Boundary Layer Theory," 4th ed., pp. 116–125, McGraw-Hill Book Co., New York, 1960.

If the value of δ from Eq. (10.10) is substituted in Eq. (10.7) with $\beta = 1.63$, there results for the shear stress

$$\tau_0 = 0.332 \frac{\mu U}{x} \sqrt{\mathbf{R}_x} \tag{10.11}$$

But we have another expression for shear stress, given in Eq. (8.11), $\tau_0 = c_f \rho U^2/2$.[1] Setting this equal to Eq. (10.11) allows a determination of the local friction coefficient

$$c_f = \frac{0.332\mu U \sqrt{\mathbf{R}_x}}{\rho x U^2/2} = \frac{0.664}{\sqrt{\mathbf{R}_x}} \tag{10.12}$$

If the boundary layer remains laminar over a length L of the plate, the total friction drag on one side of the plate is given by integrating Eq. (10.11):

$$F_f = B \int_0^L \tau_0 \, dx = 0.332B\sqrt{\rho\mu U^3} \int_0^L x^{-1/2} \, dx = 0.664B\sqrt{\rho\mu L U^3} \tag{10.13}$$

Comparing Eq. (10.13) with a standard friction-drag equation (10.2), and substituting U for the more general velocity V, it may be seen that for a laminar boundary layer

$$C_f = 1.328 \sqrt{\frac{\mu}{\rho L U}} = \frac{1.328}{\sqrt{\mathbf{R}}} \tag{10.14}$$

where it is noted that $\mathbf{R}$ is based on the characteristic length of the whole plate. The laminar boundary layer will remain laminar if undisturbed, up to a value of $\mathbf{R}_x$ of about 500,000. In this region the layer becomes turbulent increasing noticeably in thickness and displaying a marked change in velocity distribution.

Illustrative Example 10.1 Find the friction drag on one side of a smooth flat plate 6 in (15 cm) wide and 18 in (50 cm) long, placed longitudinally in a stream of crude oil ($s = 0.925$) at 60°F (20°C) flowing with undistributed velocity of 2 fps (60 cm/s).

From Fig. 1.3

$$\nu = 0.0010 \text{ ft}^2/\text{s}$$

Then, at $x = L$,

$$\mathbf{R} = \frac{LU}{\nu} = \frac{1.5 \times 2}{0.0010} = 3{,}000$$

well within the laminar range; that is, $\mathbf{R} < 500{,}000$.

From Eq. (10.14)

$$C_f = \frac{1.328}{\sqrt{\mathbf{R}}} = \frac{1.328}{\sqrt{3{,}000}} = 0.0242$$

[1] The reader will observe an apparent inconsistency between the notation used here and that used in Chap. 8. Thus, in pipe flow, the significant reference velocity is the mean velocity V in the pipe, while in flow over a plate, it is the *uniform* velocity U of the undisturbed fluid. Likewise, C_f has been employed in Chap. 8 to denote a friction coefficient for the fully developed boundary layer in a pipe, while c_f is used here to denote the local friction coefficient of the growing layer.

From Eq. (10.2)

$$F_f = 0.0242 \times 0.925 \times 1.94 \frac{2^2}{2} \frac{6 \times 18}{144} = 0.065 \text{ lb}$$

Find the thickness of the boundary layer and the shear stress at the trailing edge of the plate.
From Eq. (10.10)

$$\frac{\delta}{x} = \frac{4.91}{\sqrt{3{,}000}} = 0.0896$$

$$\delta = 0.0896 \times 1.5 = 0.1345 \text{ ft} = 1.61 \text{ in}$$

From Eq. (10.11), at $x = L$,

$$\tau_0 = 0.332 \frac{\nu \rho U}{L} \sqrt{\mathbf{R}} = 0.332 \times \frac{(0.0010 \times 0.925 \times 1.94)2}{1.5} \sqrt{3{,}000} = 0.0435 \text{ lb/ft}^2$$

In the given SI units:
From Fig. 1.3

$$\nu = 0.93 \times 10^{-4} \text{ m}^2/\text{s}$$

Then, at $x = L$,

$$\mathbf{R} = \frac{LU}{\nu} = \frac{(0.50 \text{ m})(0.60 \text{ m/s})}{0.93 \times 10^{-4}} = 3{,}220$$

well within the laminar range; that is $\mathbf{R} < 500{,}000$.
From Eq. (10.14)

$$C_f = \frac{1.328}{\sqrt{\mathbf{R}}} = 0.0234$$

From Eq. (10.2)

$$F_f = 0.0234 \times 0.925 \frac{10^3 \text{ kg}}{\text{m}^3} \times \frac{1}{2} (0.6 \text{ m/s})^2 (0.15 \times 0.50) \text{ m}^2$$

$$F_f = 0.29 \frac{\text{kg}\cdot\text{m}}{\text{s}^2} \times \frac{\text{N}}{\text{kg}\cdot\text{m/s}^2} = 0.29 \text{ N}$$

Find the thickness of the boundary layer and the shear stress at the trailing edge of the plate.
From Eq. (10.10)

$$\frac{\delta}{x} = \frac{4.91}{\sqrt{3{,}220}} = 0.086 \qquad \delta = 0.086 \times 50 \text{ cm} = 4.3 \text{ cm}$$

From Eq. (10.11), at $x = L$,

$$\tau_0 = 0.332 \frac{\mu U}{L} \sqrt{\mathbf{R}} = 0.332 \frac{\rho \nu U}{L} \sqrt{\mathbf{R}}$$

$$\tau_0 = 0.332 \frac{(0.925 \times 10^3)(0.93 \times 10^{-4})(0.6)}{0.50} \sqrt{3{,}220}$$

$$\tau_0 = 1.93 \text{ N/m}^2$$

10.4 TURBULENT BOUNDARY LAYER FOR INCOMPRESSIBLE FLOW ALONG A SMOOTH FLAT PLATE

Comparing the laminar and turbulent boundary layers in Fig. 10.4, the velocity distribution in the turbulent boundary layer shows a much steeper gradient near the wall and a flatter gradient throughout the remainder of the layer. As would

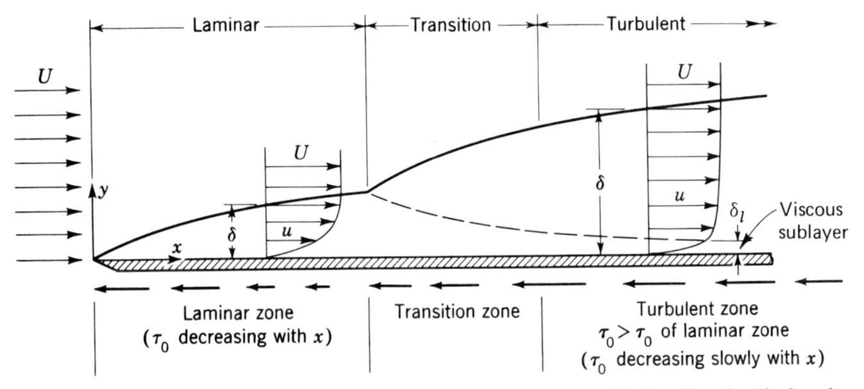

Figure 10.4 Laminar and turbulent boundary layers along a smooth flat plate (vertical scale greatly exaggerated).

be expected, then, the wall shear stress is greater in the turbulent boundary layer than in the laminar layer at the same Reynolds number. In this case, however, it is not practical to proceed along the lines of Eqs. (10.6) and (10.7), determining the shear stress from the velocity gradient at the wall. As an approximation we turn instead to turbulent flow in a circular pipe because of the wealth of experimental information on that subject compared with that of flow in a turbulent boundary layer along a smooth flat plate. We learned in Eq. (8.17) that the shear stress at the wall of a pipe is given by

$$\tau_0 = f\rho \frac{V^2}{8} \tag{10.15}$$

where V again denotes the average velocity in the pipe. Now we shall assume that the turbulent boundary layer occupies all the region between the wall and the center line of the pipe as Fig. 10.5. The radius of the pipe then becomes the thickness of the boundary layer, and by analogy, the velocity at the center of the pipe, here denoted by U, corresponds to the undisturbed velocity at the outer edge of the boundary layer. We may obtain an approximate relation between V and U by use of the pipe-factor equation (8.33). Taking a middle value of $f = 0.028$ and allowing for the 1 percent difference in velocity between the edge of the boundary layer and the free stream, we have

$$U = 1.235\ V \tag{10.16}$$

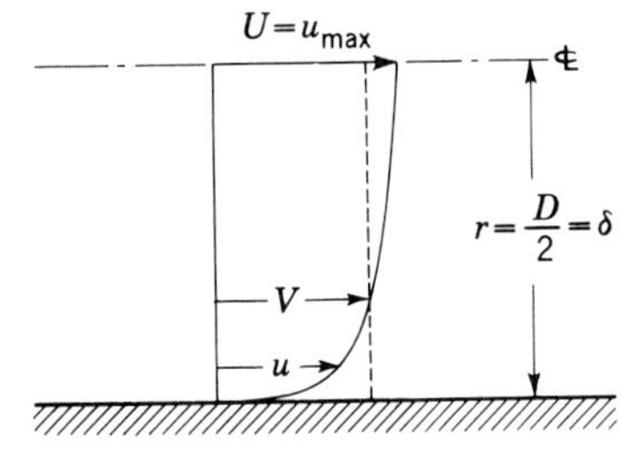

Figure 10.5 Flow in a pipe as a turbulent boundary layer.

To proceed further, we need a simple relation between f and $\mathbf{R}$ for turbulent pipe flow in a smooth pipe. The Blasius equation (8.37) provides a useful relationship. Substituting Eqs. (8.37) and (10.16) into (10.15) gives

$$\tau_0 = \frac{0.316}{(DV/\nu)^{1/4}} \frac{\rho V^2}{8} = \frac{0.316}{[(2\delta/\nu)(U/1.235)]^{1/4}} \frac{\rho}{8}\left(\frac{U}{1.235}\right)^2 = \frac{0.023\rho U^2}{(\delta U/\nu)^{1/4}} \tag{10.17}$$

If we now equate the two expressions for τ_0 [(10.6) and (10.17)], there results

$$\rho U^2 \alpha \frac{d\delta}{dx} = \frac{0.023\rho U^2}{(\delta U/\nu)^{1/4}} \tag{10.18}$$

Separating variables and integrating this expression (with the condition $\delta = 0$ at $x = 0$), there results

$$\delta = \left(\frac{0.0287}{\alpha}\right)^{4/5} \left(\frac{\nu}{Ux}\right)^{1/5} x \tag{10.19}$$

In using Eq. (10.6) it is assumed that we know the velocity distribution in the turbulent boundary layer. Of the many formulas for this distribution that have been proposed, the most convenient for our purpose is Eq. (8.38), the seventh-root law, which can be expressed as

$$u = Uf(\eta) = U(\eta)^{1/7} = U(y/\delta)^{1/7} \tag{10.20}$$

For the case of such a velocity distribution, $\alpha = 0.0972$. This is found by evaluating the integral definition of α (Sec. 10.2):

$$\alpha = \int_0^1 f(\eta)[1 - f(\eta)]\, d\eta$$

where $\eta = y/\delta$

$d\eta = dy/\delta$

$f(\eta) = (y/\delta)^{1/7}$

Substituting this value for α in Eq. (10.19) gives

$$\frac{\delta}{x} = \frac{0.377}{\mathbf{R}_x^{1/5}} \tag{10.21}$$

and substituting this value of δ in Eq. (10.17) yields

$$\tau_0 = 0.0587\rho \frac{U^2}{2}\left(\frac{\nu}{Ux}\right)^{1/5} \tag{10.22}$$

And since $\tau_0 = c_f \rho U^2/2$,

$$c_f = \frac{0.0587}{\mathbf{R}_x^{1/5}} \tag{10.23}$$

With this value for c_f (the *local friction coefficient*) we can express the total friction drag on one side of the plate as

$$F_f = B \int_0^L \tau_0 \, dx = 0.0735 \rho \frac{U^2}{2} \left(\frac{\nu}{UL} \right)^{1/5} BL \tag{10.24}$$

or in Eq. (10.2) the friction-drag coefficient for a turbulent boundary layer is

$$C_f = \frac{0.0735}{\mathbf{R}^{1/5}} \qquad \text{for} \qquad 500{,}000 < \mathbf{R} < 10^7 \tag{10.25}$$

where it is again noted that the characteristic length in $\mathbf{R}$ is L, the total length of the plate parallel to the flow. For Reynolds numbers above 10^7, Schlichting has proposed a modification of Eq. (10.25) which agrees better with experimental results,[1]

$$C_f = \frac{0.455}{(\log \mathbf{R})^{2.58}} \qquad \text{for} \qquad \mathbf{R} > 10^7 \tag{10.26}$$

For Reynolds numbers less than 10^7 this equation gives values for C_f that are very close to those given by Eq. (10.25) and consequently Eq. (10.26) is commonly employed over the entire range of Reynolds numbers above 500,000.

Illustrative Example 10.2 (*a*) Find the frictional drag on the top and sides of a box-shaped moving van 8 ft wide, 10 ft high, and 35 ft long, traveling at 60 mph through air ($\gamma = 0.0725$ lb/ft^3) at 50°F. Assume that the vehicle has a rounded nose so that the flow does not separate from the top and sides (see Fig. 10.12*b*). Assume also that even though the top and sides of the van are relatively smooth there is enough roughness so that for all practical purposes a turbulent boundary layer starts immediately at the leading edge (*b*) Find the thickness of the boundary layer and the shear stress at the trailing edge.

From Fig. 1.3, for air at 50°F, $\nu = 0.00015$ ft^2/s. Then

$$\mathbf{R} = \frac{LU}{\nu} = \frac{35 \times 88}{0.00015} = 20{,}530{,}000$$

As $\mathbf{R} > 10^7$, use Eq. (10.26):

$$C_f = \frac{0.455}{(7.31)^{2.58}} = 0.00268$$

Then, by Eq. (10.2),

$$F_f = 0.00268 \times \frac{0.0725}{32.2} \times \frac{(88)^2}{2} \times (10 + 8 + 10)35 = 22.9 \text{ lb}$$

Applying Eq. (10.21), we get

$$\delta_{35} = \frac{35 \times 0.377}{[(205.3)^{1/5} \times 10]} = 0.455 \text{ ft}$$

From Eq. (10.22)

$$(\tau_0)_{35} = 0.0587 \frac{0.0725}{32.2} \frac{88^2}{2} \left[\frac{0.00015}{88 \times 35} \right]^{1/5} = 0.0176 \text{ lb/ft}^2$$

[1] H. Schlichting, Boundary Layer Theory, Part II, *NACA Tech. Mem.* 1218, p. 39, 1949.

10.5 FRICTION DRAG FOR INCOMPRESSIBLE FLOW ALONG A SMOOTH FLAT PLATE WITH A TRANSITION REGIME

In the two preceding sections we have treated separately the resistance due to laminar and turbulent boundary layers on a smooth flat plate. When the plate is of such length that there is a transition from the laminar to the turbulent boundary layer on the plate surface, the friction drag may be computed as follows.

Let x_c in Fig. 10.6 be the distance from the leading edge to the point where the boundary layer becomes turbulent, which will normally occur at a value of $\mathbf{R}_x$ of about 500,000. The drag of the turbulent portion of the boundary layer may be approximated as the drag which would occur if a turbulent boundary layer extended along the whole plate, less the drag of a fictitious turbulent layer from the leading edge to x_c. Thus

$$F_{\text{turb}} \approx F_{\text{total turb}} - F_{\text{turb to } x_c}$$

When this is added to the drag from the laminar boundary layer up to x_c, we have, from Eqs. (10.2), (10.14), (10.26), and (10.25), for the total drag, assuming the plate is long enough[1] so that $\mathbf{R} > 10^7$,

$$F_f = \rho \frac{U^2}{2} B \left[\frac{1.328 x_c}{\sqrt{\mathbf{R}_c}} + \frac{0.455 L}{(\log \mathbf{R})^{2.58}} - \frac{0.0735 x_c}{\mathbf{R}_c^{1/5}} \right]$$

where $\mathbf{R}_c$ is based on the length x_c to the point of transition, while $\mathbf{R}$ is based on the total length L of the plate, as before. Next we observe that

$$\frac{\mathbf{R}_c}{\mathbf{R}} = \frac{x_c}{L} \qquad \text{or} \qquad x_c = \frac{\mathbf{R}_c}{\mathbf{R}} L$$

and thus

$$F_f = \rho \frac{U^2}{2} BL \left[1.328 \frac{\sqrt{\mathbf{R}_c}}{\mathbf{R}} + \frac{0.455}{(\log \mathbf{R})^{2.58}} - \frac{0.0735 \mathbf{R}_c^{4/5}}{\mathbf{R}} \right]$$

which, for $\mathbf{R}_c = 500{,}000$, reduces to

$$C_f = \frac{0.455}{(\log \mathbf{R})^{2.58}} - \frac{1{,}700}{\mathbf{R}} \tag{10.27}$$

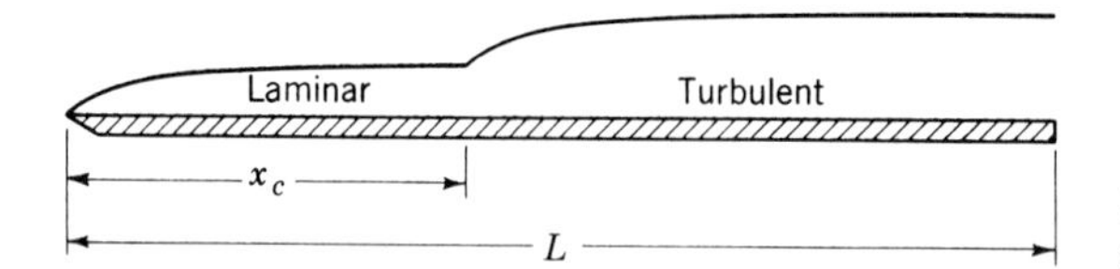

Figure 10.6 Boundary layers along a smooth flat plate of finite length.

[1] This expression for F_f is also quite accurate in the undisturbed transition regime for a shorter plate with $\mathbf{R} < 10^7$ because Eqs. (10.26) and (10.25) gives almost identical values for C_f when $\mathbf{R} < 10^7$.

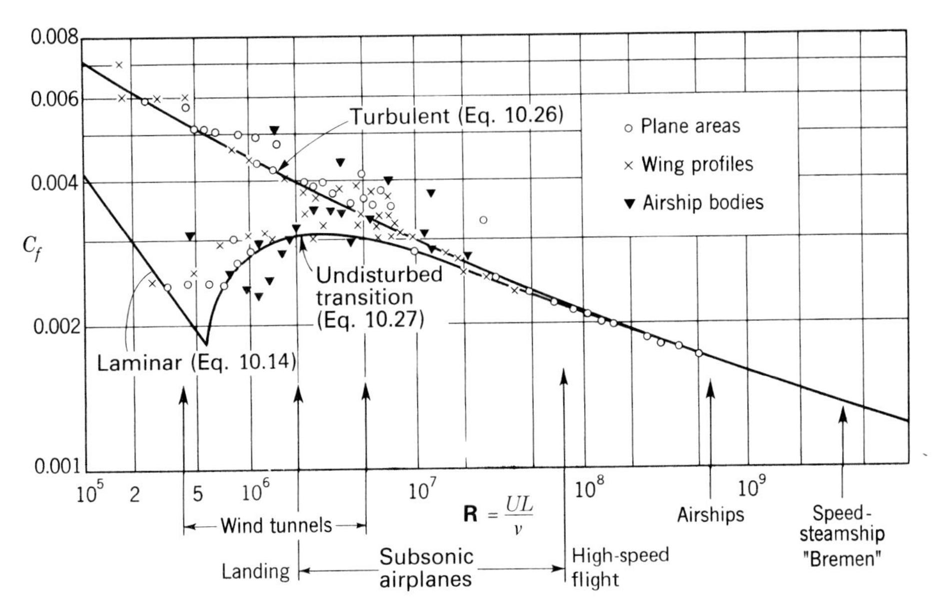

Figure 10.7 Drag coefficients for a smooth flat plate. (Adapted from *NACA Tech. Mem. 1218, p. 117, 1949.*)

Equations (10.14), (10.26), and (10.27) are plotted in Fig. 10.7, together with some comparison measurements and indicated ranges of applicability.

All the treatment of laminar and turbulent boundary layers has so far been based upon the surface of the immersed body being smooth. However, a local region of excess roughness can "trip" the laminar layer into becoming a turbulent layer at Reynolds numbers less than 500,000. The height of the roughness which will cause this tripping (*critical roughness*) is given approximately by[1]

$$e_c = \frac{15\nu}{\sqrt{\tau_0/\rho}} = 26\frac{\nu}{U}\mathbf{R}_x^{1/4} \tag{10.28}$$

where τ_0 is determined by Eq. (10.11).

We see that the height of the critical roughness depends on the distance from the leading edge. As the boundary layer grows along the plate, the roughness must be greater in order to upset the stability of the layer. It must be recalled that when $\mathbf{R}_x$ reaches a value in the neighborhood of 500,000, the laminar layer of itself becomes unstable, however smooth the surface, and changes to a turbulent boundary layer with a thin viscous sublayer lying close to the surface. As in the case of flow in pipes, the surface is considered hydraulically smooth if the effect of the roughness projections does not extend into this sublayer.

[1] I. Tani, J. Hama, and S. Mituisi, On the Permissible Roughness in the Laminar Boundary Layer, *Aeronaut. Res. Inst., Tokyo Imp. Univ., Rept.* 15, p. 419, 1940.

The thickness of the viscous sublayer is not a clearly determinable quantity, but it appears to be well agreed that the thickness of the strictly laminar region is given approximately by [see Eq. (8.29) for pipe flow]

$$\delta_s = \frac{3.5\nu}{\sqrt{\tau_0/\rho}} \tag{10.29}$$

while a transition layer extends to

$$\delta_t = \frac{60\nu}{\sqrt{\tau_0/\rho}} \tag{10.30}$$

In both these equations τ_0/ρ may be obtained from Eq. (10.22). If the roughness height is only of the order of δ_s and a little greater, the surface may still be considered smooth, but if the roughness height is greater than δ_t, the surface is truly rough and the drag is materially increased.

It may be remarked finally that a plate or wing which is to incur minimum drag must be very smooth near the leading edge, where the laminar layer or sublayer is thinnest, while greater roughness may be tolerated near the trailing edge. Since the wall shear is so much greater in a turbulent boundary layer than in a laminar one, anything that can be done to delay the breakdown of the laminar boundary layer will greatly reduce the frictional drag force on a body. The *laminar flow wing* for aircraft is one for which suction slots along the leading edge of the wing together with a smooth leading edge and a properly shaped wing profile help to maintain a favorable pressure gradient (Sec. 10.6) along the upper surface of the wing. This delays the breakdown of the laminar boundary layer, and thus such wings have much less drag than conventional ones.

Illustrative Example 10.3 A small submarine, which may be supposed to approximate a cylinder 10 ft in diameter and 50 ft long, travels submerged at 3 knots (5.06 fps) in sea water at 40°F. Find the friction drag assuming no separation from the sides.

Viscosity of sea water ≈ viscosity of fresh water

From Table A.1*a*, $\nu = 0.0000166$ ft²/s. Then

$$\mathbf{R} = \frac{50 \times 5.06}{0.0000166} = 1.52 \times 10^7$$

From Eq. 10.27 or Fig. 10.7

$$C_f = 0.0028$$

and

$$F_f = 0.0028 \times \frac{64}{32.2} \times \frac{(5.06)^2}{2} \times \pi \times 10 \times 50 = 112 \text{ lb}$$

Find the value of the critical roughness for a point 1 ft from the nose of the submarine.

At $x = 1$ ft

$$\mathbf{R}_x = \frac{5.06 \times 1}{0.0000166} = 305{,}000$$

By Eq. (10.28)

$$e_c = \frac{26 \times 0.0000166}{5.06}(305{,}000)^{1/4} = 0.0020 \text{ ft}$$

Find the height of roughness at the mid-section of the submarine which would class the surface as truly rough.

At $x = 25$ ft

$$\mathbf{R}_x = \frac{5.05 \times 25}{0.0000166} = 7.62 \times 10^6$$

By Eq. (10.22)

$$\tau_0/\rho = 0.0587 \times \frac{(5.06)^2}{2} \frac{1}{(76.2 \times 10^5)^{1/5}} = 0.0316 \text{ ft}^2/\text{s}^2$$

Then, by Eq. (10.30)

$$\delta_t = \frac{60 \times 0.0000166}{\sqrt{0.0316}} = 0.0056 \text{ ft}$$

10.6 BOUNDARY-LAYER SEPARATION AND PRESSURE DRAG

The motion of a thin stratum of fluid lying wholly inside the boundary layer is determined by three forces:

1. The forward pull of the outer free-moving fluid, transmitted through the laminar boundary layer by viscous shear and through the turbulent boundary layer by momentum transfer (Sec. 8.8).
2. The viscous retarding effect of the solid boundary which must, by definition, hold the fluid stratum immediately adjacent to it at rest.
3. The pressure gradient along the boundary. The stratum is accelerated by a pressure gradient whose pressure decreases in the direction of flow and is retarded by an adverse gradient.

The treatment of fluid resistance in the foregoing sections has been restricted to the drag of the boundary layer along a smooth flat plate located in an unconfined fluid, that is to say, in the *absence of a pressure gradient*. In the presence of a favorable pressure gradient the boundary layer is "held" in place. This is what occurs in the accelerated flow around the *forebody*, or upstream portion, of a cylinder, sphere, or other object, such as that of Fig. 3.11. If a particle enters the boundary layer near the forward stagnation point with a low velocity and high pressure, its velocity will increase as it flows into the lower pressure region along the side of the body. But there will be some retardation from wall friction (force 2 above) so that its total useful energy will be reduced by a corresponding conversion into thermal energy.

What happens next may best be explained by reference to Fig. 10.8. Let A represent a point in the region of accelerated flow with a normal velocity distribution in the boundary layer (either laminar or turbulent), while B is the point where the velocity outside the boundary layer reaches a maximum. Then C, D, and E are points downstream where the velocity outside the boundary layer decreases, resulting in an increase in pressure in accordance with ideal-flow theory. Thus the velocity of the layer close to the wall is reduced at C and finally brought to a stop at D. Now the increasing pressure calls for further retardation; but this is impossible, and so the boundary layer actually *separates* from the wall. At E there is a backflow next to the wall, driven in the direction of decreasing pressure—upstream in this case—and feeding fluid into the boundary layer which has left the wall at D.

Downstream from the point of separation the flow is characterized by irregular turbulent eddies, formed as the separated boundary layer becomes rolled up in the reversed flow. This condition generally extends for some distance downstream until the eddies are worn away by viscous attrition. The whole disturbed region is called the *turbulent wake* of the body (Fig. 10.9).

Because the eddies cannot convert their kinetic energy of rotation into an increased pressure, as the ideal-fluid theory would dictate, the pressure within the wake remains close to that at the separation point. Since this is always less than the pressure at the forward stagnation point, there results a net pressure difference tending to move the body with the flow, and this force is the pressure drag.

Although the laminar and turbulent boundary layers behave in essentially the same manner at a point of separation, the *location* of the separation point on a given curved surface will be very different for the two cases. In the laminar layer the transfer of momentum from the rapidly moving outer strata through the viscous-shear process to the inner strata is slow and ineffective. Consequently, the laminar boundary layer is "weak" and cannot long stick to the wall against an adverse pressure gradient. The transition to a turbulent boundary layer, on the other hand, brings a violent mixing of the faster-moving outer strata into the slower-moving inner strata, and vice versa. The mean velocity close to the boundary is greatly increased, as shown in Fig. 10.4. This added energy enables

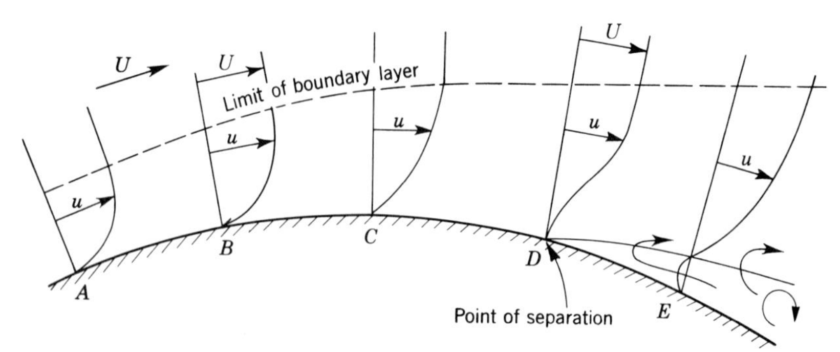

Figure 10.8 Growth and separation of boundary layer owing to increasing pressure gradient. Note that U has its maximum value at B and then gets smaller.

Figure 10.9 Turbulent wake behind a flat plate held normal to the flow.

the boundary layer better to withstand the adverse pressure gradient, with the result that *with a turbulent boundary layer the point of separation is moved downstream* to a region of higher pressure. An example of this is shown in Fig. 10.11.

10.7 DRAG ON THREE-DIMENSIONAL BODIES (INCOMPRESSIBLE FLOW)

The total drag on a body is the sum of the friction drag and the pressure drag.

$$F_D = F_f + F_p$$

In the case of a well-streamlined body, such as an airplane wing or the hull of a submarine, the friction drag is the major part of the total drag and may be estimated by the methods of the preceding articles on the boundary layer. Only rarely is it desired to compute the pressure drag separately from the friction drag. Usually, when the wake resistance becomes significant, one is interested in the total drag only. Indeed, it is customary to employ a single equation which gives the total drag.[1]

$$F_D = C_D \rho \frac{V^2}{2} A \tag{10.31}$$

in terms of an overall drag coefficient C_D, with the other quantities the same as in Eq. (10.1), except that in the case of the lifting vane (as an airplane wing) the area A is defined as the product of the span and the mean chord (Figs. 10.15 and 10.21). In such a case the area is neither strictly parallel to nor normal to the flow.

In the case of a body with sharp corners, such as the plate of Fig. 10.9 set normal to the flow, separation always occurs at the same point, and the wake extends across the full projected width of the body. This results in a relatively constant value of C_D, as may be seen from the plot for the flat disk in Fig. 10.10. If the body has curved sides, however, the location of the separation point will be determined by whether the boundary layer is laminar or turbulent. This location

[1] In the equations for total drag on submerged bodies, we revert to the use of V to designate a general reference velocity.

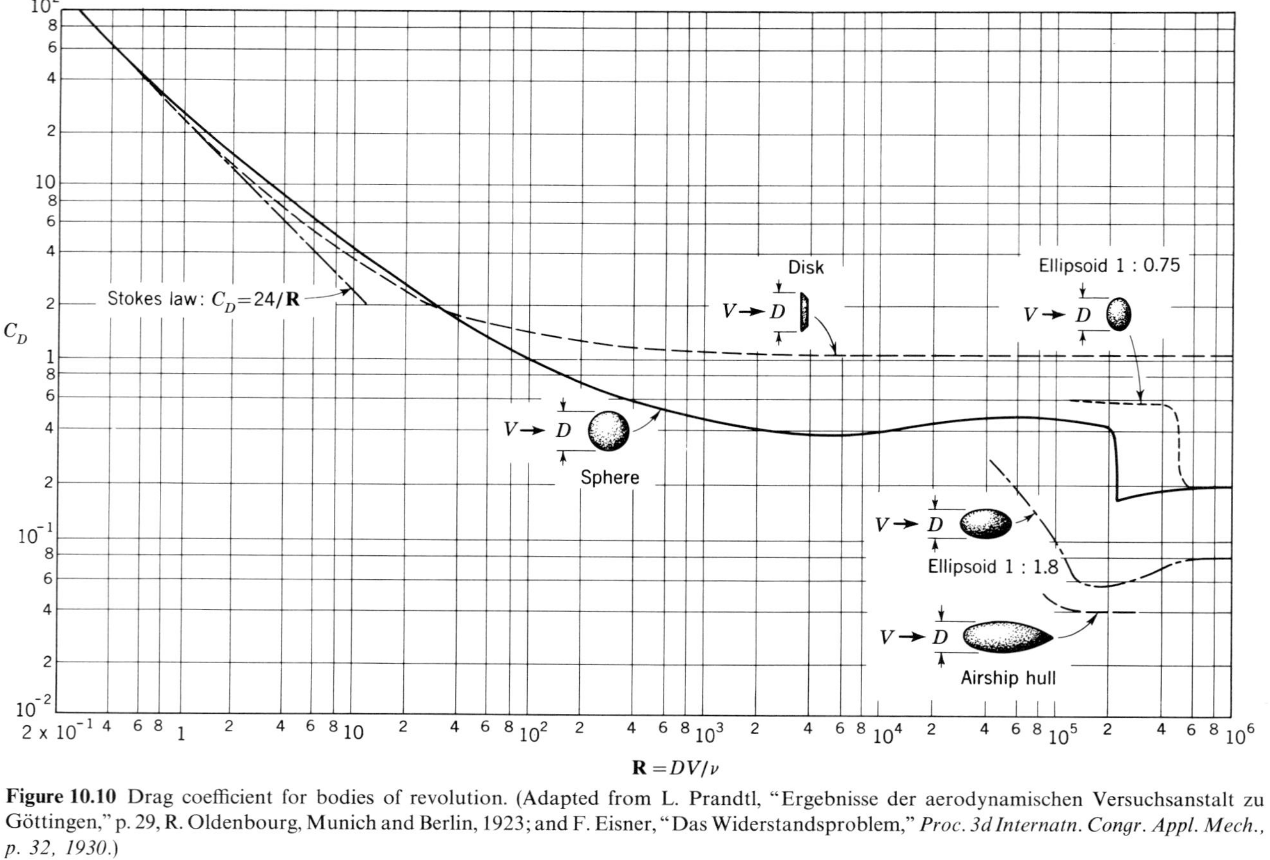

Figure 10.10 Drag coefficient for bodies of revolution. (Adapted from L. Prandtl, "Ergebnisse der aerodynamischen Versuchsanstalt zu Göttingen," p. 29, R. Oldenbourg, Munich and Berlin, 1923; and F. Eisner, "Das Widerstandsproblem," *Proc. 3d Internatn. Congr. Appl. Mech., p. 32, 1930.*)

in turn determines the size of the wake and the amount of the pressure drag.

The foregoing principles are vividly illustrated in the case of the flow around a sphere. For very low Reynolds numbers ($DV/\nu < 1$, in which D is the diameter of the sphere) the flow about the sphere is completely viscous and the friction drag is given by Stokes' law,

$$F_D = 3\pi\mu VD \qquad (10.32)$$

Equating this equation to Eq. (10.31), where A is defined as $\pi D^2/4$, the frontal area of the projected sphere, gives the result that $C_D = 24/\mathbf{R}$. The similarity between this case and the value of the friction factor for laminar flow in pipes is at once apparent. This regime of the flow about a sphere is shown as the straight line at the left of the log-log plot of C_D versus $\mathbf{R}$ in Fig. 10.10.

As $\mathbf{R}$ is increased beyond 1, the laminar boundary layer separates from the surface of the sphere, beginning first at the rear stagnation point, where the adverse pressure gradient is the strongest. The curve of C_D in Fig. 10.10 begins to level off as the pressure drag becomes of increasing importance and the drag becomes more proportional to V^2. With further increase in $\mathbf{R}$, the point of separation moves forward on the sphere, until at $\mathbf{R} \approx 1{,}000$ the point of separation becomes fairly stable at about 80° from the forward stagnation point.

For a considerable range of Reynolds numbers, conditions remain fairly stable, the laminar boundary layer separating from the forward half of the sphere and C_D remaining fairly constant at about 0.45. At a value of $\mathbf{R}$ of about 250,000 for the smooth sphere, however, the drag coefficient is suddenly reduced by about 50 per cent, as may be seen in Fig. 10.10. The reason for this lies in a change from a laminar to a turbulent boundary layer on the sphere. The point of separation is moved back to something like 115°, from the stagnation point, with a consequent decrease in the size of the wake and the pressure drag.

If the "level" of turbulence in the free stream is high, the transition from laminar to turbulent boundary layer will take place at lower Reynolds numbers. Because this phenomenon of shift in separation point is so well defined, the sphere is often used as a turbulence indicator. The Reynolds number producing a value of C_D of 0.3—which lies in the middle of the rapid-drop range—becomes an accurate measure of the turbulence.[1]

As was mentioned previously, the transition from a laminar to a turbulent boundary layer may also be prematurely induced by artificially roughening the surface over a local region. The two pictures of Fig. 10.11 clearly show the effectiveness of this procedure. By roughening the nose of the sphere the boundary layer is made turbulent and the separation point moved back. The added roughness and turbulent boundary layer cause an increase in friction drag, to be sure, but this is of secondary importance compared with the marked decrease in the size and effect of the wake. This explains the main reason why the surface of a golf ball is dimpled. A smooth-surfaced ball would have greater overall drag and would not travel as far when driven.

[1] H. Dryden, Reduction of Turbulence in Wind Tunnels, *NACA Tech. Report* 392, 1931.

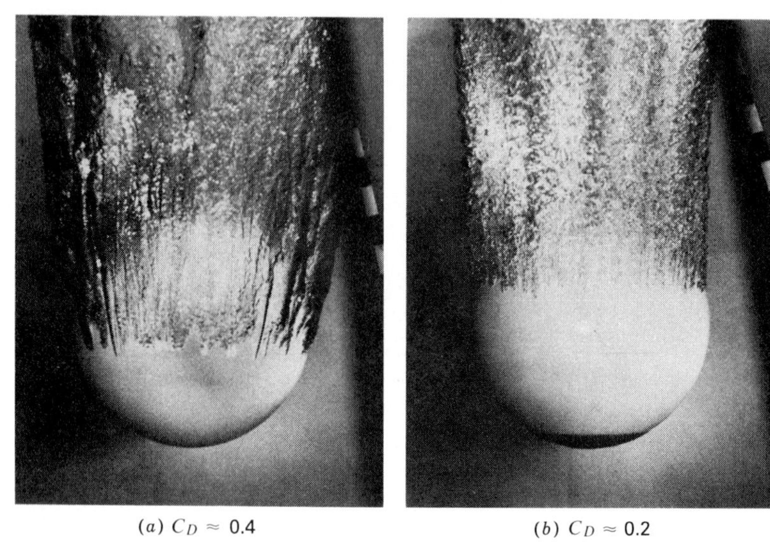

(a) $C_D \approx 0.4$ (b) $C_D \approx 0.2$

Figure 10.11 Shift in point of separation on 8.5-in-diameter sphere (bowling ball) at a velocity of approximately 25 fps in water. (a) Smooth sphere—laminar boundary layer. (b) Sphere with 4-in-diameter patch of sand grains cemented to nose—turbulent boundary layer. Reynolds numbers are the same. *(Photograph by U.S. Naval Ordnance Test Station, Pasadena Annex.)*

Plots of C_D versus **R** for various other three-dimensional shapes are also shown in Fig. 10.10. It may be pointed out here that the object of *streamlining* a body is to move the point of separation as far back as possible and thus to produce the minimum size of turbulent wake. This decreases the pressure drag, but by making the body longer so as to promote a gradual increase in pressure, the friction drag is increased. The optimum amount of streamlining, then, is that for which the sum of the friction and pressure drag is a minimum. Quite evidently, from what we have learned, attention in streamlining must be given to the rear end, or downstream part, of a body as well as to the front. The shape of the forebody is important principally to the extent that it governs the location of the separation point(s) on the afterbody. A rounded nose produces the least disturbance in the streamlines and is therefore the best for incompressible or compressible flow at subsonic velocities.[1] This is illustrated in Fig. 10.12 where flow about a blunt-nosed motor vehicle is compared to that about a rounded-nose vehicle.

Illustrative Example 10.4 Using the data of Illustrative Example 10.2 determine the total drag exerted by the air on the van. Assume that $C_D \approx 0.45$ (see Fig. 10.12).

$$F_D = C_D \rho \frac{V^2}{2} A = 0.45\left(\frac{0.0725}{32.2}\right)\frac{(88)^2}{2}(8 \times 10)$$

$$F_D = 314 \text{ lb}$$

[1] For supersonic flow a sharp-nosed body has less drag than a round-nosed body (Fig. 10.28).

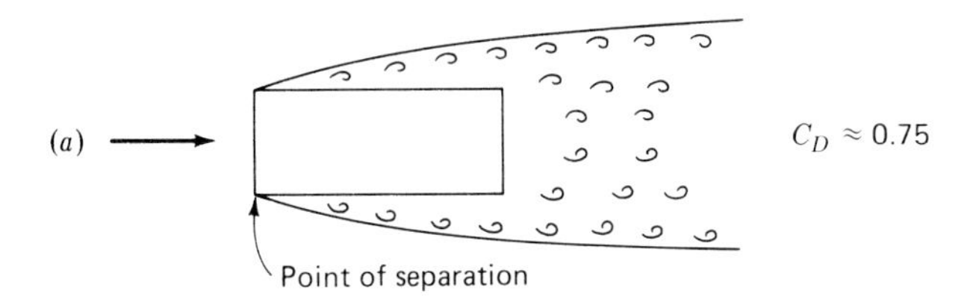

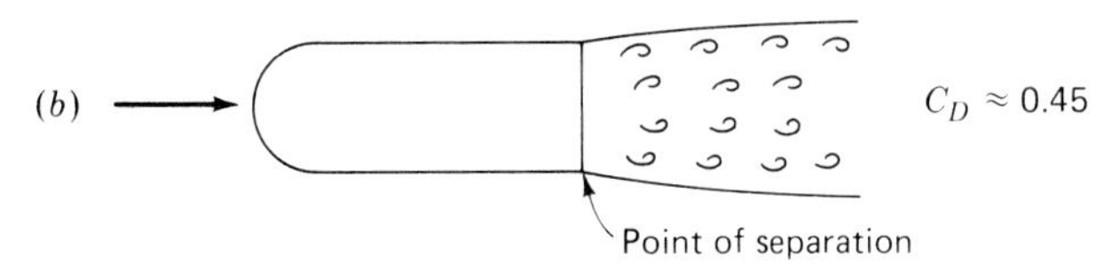

Figure 10.12 Plan view of flow about a motor vehicle (delivery van). (*a*) Blunt nose with separated flow along the entire side wall and a large drag coefficient $C_D = 0.75$. (*b*) Round nose with separation at the rear of the vehicle and smaller drag coefficient $C_D = 0.45$. (Adapted from H. Schlicting, *Boundary Layer Theory*, 4th ed., p. 34, McGraw-Hill Book Co., New York, N.Y., 1960).

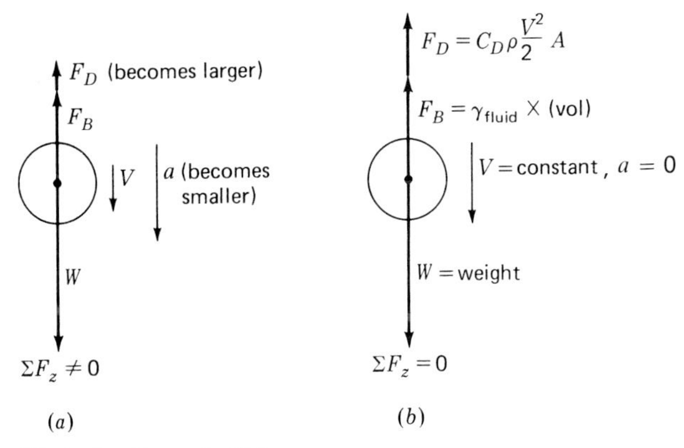

Illustrative Example 10.5

Thus the pressure drag = 314 − 23 = 291 lb; in this case the pressure drag is responsible for about 93 percent of the total drag while the friction drag comprises only 7 percent of the total.

Illustrative Example 10.5 Find the "free-fall" velocity of a 8.5-in-diameter sphere (bowling ball) weighing 16 lb when falling through the following fluids under the action of gravity: (*a*) through the standard atmosphere at sea level; (*b*) through the standard atmosphere at 10,000-ft elevation; (*c*) through water at 60°F; (*d*) through crude oil ($s = 0.925$) at 60°F.

When first released the sphere will accelerate (left-hand figure) because the forces acting on it are out of balance. This acceleration results in a buildup of velocity which causes an increase in the drag force. After a while the drag force will increase to the point where the forces acting on the sphere are in balance, as indicated in the right-hand figure. When that point is reached the sphere will attain a constant or terminal (free-fall) velocity. Thus for free-fall conditions,

$$\sum F_z = W - F_B - F_D = \text{mass} \times \text{acceleration} = 0$$

where W is the weight, F_B the buoyant force, and F_D the drag force. The buoyant force is equal to the unit weight of the fluid multiplied by the volume ($\pi D^3/6 = 0.186\ \text{ft}^3$) of the sphere. The given data are approximately as follows:

Fluid	γ lb/ft³	ρ slug/ft³	$\nu = \mu/\rho$ ft²/s	F_B lb
Air (sea level)	0.0765	0.00238	1.57×10^{-4}	0.0142
Air (10,000 ft)	0.0564	0.00176	2.01×10^{-4}	0.0105
Water, 60°F	62.4	1.94	1.22×10^{-5}	11.6
Oil, 60°F	57.7	1.79	0.001	10.7

The detailed analysis for the sphere falling through the standard sea-level atmosphere is as follows:

$$16 - 0.0142 - C_D \rho \frac{V^2}{2} A = 0$$

where

$$\rho = 0.00238\ \text{slugs/ft}^3 \qquad \text{and} \qquad A = \frac{\pi(8.5/12)^2}{4} = 0.394\ \text{ft}^2$$

or

$$15.986 = C_D(0.00238)\frac{V^2}{2}(0.394) = 0.00047\, C_D V^2$$

A trial-and-error solution is required. Let $C_D = 0.2$, then $V = 412$ fps.

$$\mathbf{R} = \frac{DV}{\nu} = \frac{(8.5/12)412}{1.57 \times 10^{-4}} = 1.86 \times 10^6$$

The values of C_D and **R** check Fig. 10.10; hence $C_D = 0.2$ and $V = 412$ fps.

Following a similar procedure for the other three fluids gives the following free-fall velocities:

Fluid	C_D	**R**	V_{fall}
Standard atmosphere at 10,000 ft	0.20	1.69×10^6	480 fps
Water at 60°F	0.19	453,000	7.4 fps*
Crude oil ($s = 0.925$) at 60°F	0.39	4,390	6.2 fps

* In this instance the Reynolds number is 453,000 which, for the case of a sphere, generally indicates a turbulent boundary layer (Fig. 10.10). This is very close to the point where the boundary layer changes from laminar to turbulent. If the water had been at a somewhat lower temperature and, hence, more viscous, a laminar boundary layer might have been present, in which case the free-fall velocity would have been only about 5.2 fps.

10.8 DRAG ON TWO-DIMENSIONAL BODIES (INCOMPRESSIBLE FLOW)

Two-dimensional bodies are also subject to friction and pressure drag. However, the flow about a two-dimensional body exhibits some peculiar properties which are not ordinarily found in the three-dimensional case of flow around a sphere. For example, with Reynolds numbers less than 1, the flow around a cylinder is completely viscous and the drag coefficient is given by the straight-line part of the curve at the left of Fig. 10.13. As the Reynolds number increases from 2 to about 30, the boundary layer separates symmetrically from the two sides of the cylinder and two weak, symmetrical standing eddies are formed. The equilibrium of the standing eddies is maintained by the flow from the separated boundary layer, and if the cylinder is of finite length the eddies increase in length with increase in velocity in order to dissipate their rotational energy to the free-streaming fluid.

At some limiting Reynolds number, usually about 60, depending on the shape of the cylinder (not necessarily circular), the width of the confining channel, and the turbulence in the stream, the eddies break off, having become too long to hang on, and wash downstream. This gives rise to the beginning of the so-called *Kármán vortex street.* Above this critical **R**, and visibly up to a value of about 120, the vortices are shed first from one side of the cylinder and then from the other. The result is a staggered double row of vortices in the wake of the object spaced approximately as shown in Fig. 10.14. This alternating shedding of vortices and the accompanying forces gives rise to the phenomenon of aerodynamic instability, of much importance in the design of tall smoke stacks and suspension bridges. It is also understood to account for the "singing" of wind blowing across wires. The frequency at which the vortices are shed has been given by G. F. Taylor and substantiated by Lord Rayleigh to be about

$$f \approx 0.20 \frac{V}{D}\left(1 - \frac{20}{\mathbf{R}}\right) \tag{10.33}$$

A serious condition may result if the natural frequency of vibration of the structure is equal or close to the frequency of the shedding of the vortices.

For Reynolds numbers above 120 or so it is difficult to perceive the vortex street, but the eddies continue to be shed alternately from each side up to a value of **R** of about 10,000. Beyond this the viscous forces become negligible, and it is not possible to say how the eddies form and leave the cylinder. As in the case of the sphere, the boundary layer for a circular cylinder becomes turbulent at a value of **R** of about 350,000. The corresponding sharp drop in C_D may be seen in Fig. 10.13.

Values of C_D for various other two-dimensional shapes are given in Fig. 10.13. As may be noted from the curve for the finite-cylinder, the resistance is decreased if three-dimensional flow can take place around the ends. This decrease in C_D occurs because the vortices can extend laterally into the flow field and permit dissipation of energy over a larger region.

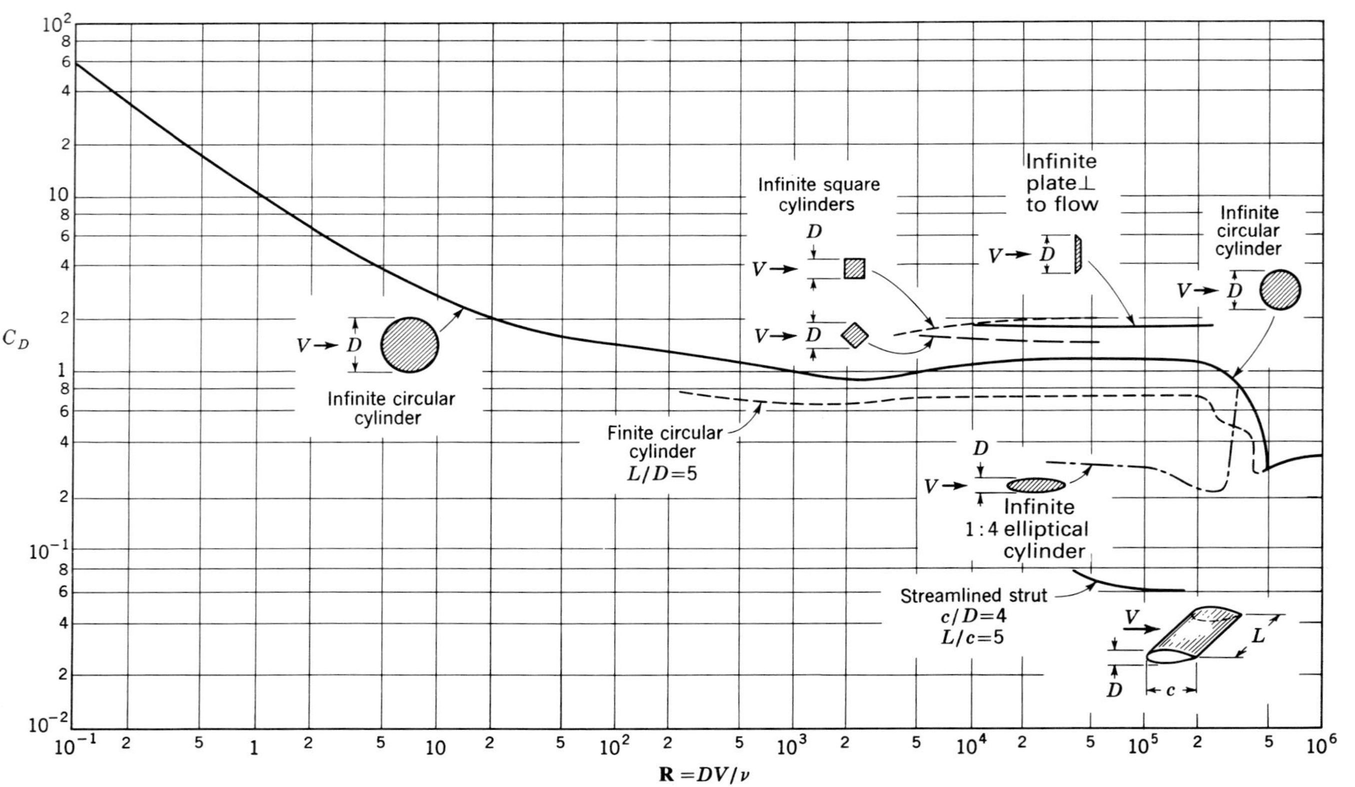

Figure 10.13 Drag coefficient for two-dimensional bodies. (Adapted from L. Prandtl, "Ergebnisse der aerodynamischen Versuchsanstalt zu Göttingen," p. 24, R. Oldenbourg, Munich and Berlin, 1923; F. Eisner, "Das Widerstandsproblem," *Proc. 3d Internatn. Congr. Appl. Mech.*, p. 32, 1930; A. F. Zahm, R. H. Smith, and G. C. Hill, "Point Drag and Total Drag of Navy Struts No. 1 Modified," *NACA Rept. 137*, p. 14, 1972; and W. F. Lindsey, "Drag of Cylinders of Simple Shapes," *NACA Rept. 619*, pp. 4–5, 1938.)

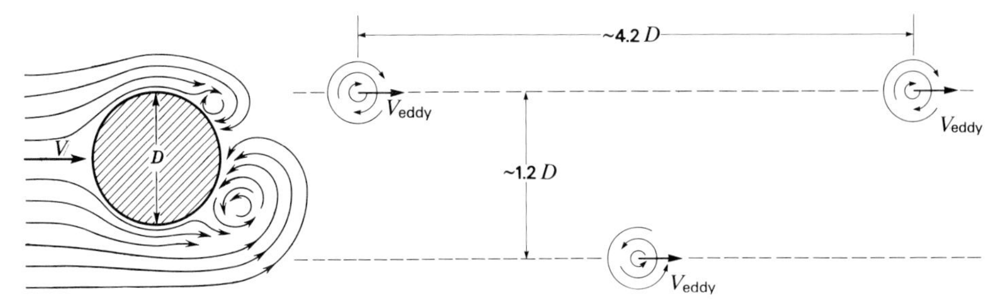

Figure 10.14 The Kármán vortex street following a cylinder. Velocity of eddies, $V_{eddy} < V$, velocity of the fluid.

Illustrative Example 10.6 What frequency of oscillation is produced by a 15 m/s wind at $-20°C$ blowing across a 2-mm-diameter wire at sea level?

From Appendix 3, Table A.2*b*:

$$\nu_{air} = 1.15 \times 10^{-5}\ m^2/s$$

$$\mathbf{R} = \frac{DV}{\nu} = \frac{2 \times 10^{-3}(15)}{1.15 \times 10^{-5}} = 2{,}600$$

Eq. (10.33),

$$f = 0.20\frac{15}{2 \times 10^{-3}}\left(1 - \frac{20}{2{,}600}\right) = 1{,}500\ \text{Hz}$$

10.9 LIFT AND CIRCULATION

At the start of this chapter we briefly mentioned the lift as a force which acts on an immersed body normal to the relative motion between the fluid and the body. The most commonly observed example of lift is that of the airplane wing supported in the air by this force. The elementary explanation for such a lift force is that the air velocity over the top of the wing is faster than the mean velocity,

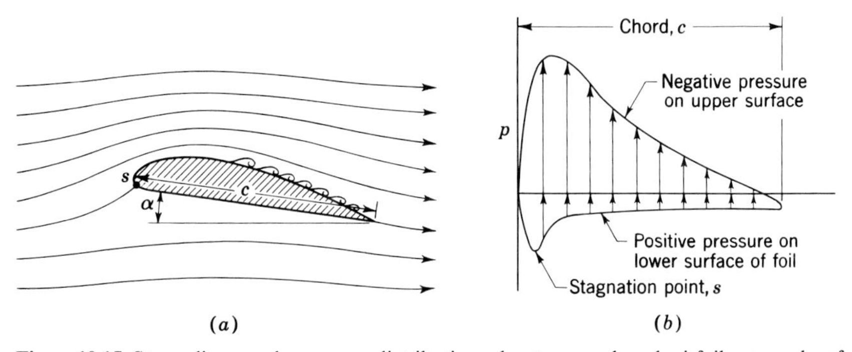

Figure 10.15 Streamlines and pressure distribution about a cambered airfoil, at angle of attack $\alpha = 8.6°$. (Data from L. Prandtl, "Ergebnisse der aerodynamischen Versuchsanstalt zu Göttingen," R. Oldenbourg, Munich and Berlin, 1923.)

while that along the underside is slower than the mean (Fig. 10.15*a*). The Bernoulli theorem then shows a lower pressure on the top and a higher pressure on the bottom (Fig. 10.15*b*), resulting in a net upward lift.

The increased velocity over the top of the wing of Fig. (10.15*a*) and the decreased velocity around the bottom of the wing can be explained by noting that a *circulation* (Sec. 5.3) is induced as the wing moves relative to the flow field. The strength of the circulation depends, in the real case, on the shape of the wing and its velocity and orientation with respect to the flow field. A schematic diagram of the situation is presented in Fig. 10.16.

The relationship between lift and circulation is one that has been studied exhaustively for years by many investigators. An understanding of this relationship is essential to the analysis of various aerodynamic and hydrodynamic problems. To illustrate the theory of lift, we shall consider the flow of an ideal fluid past a cylinder and assume that a circulation about the cylinder is imposed on the flow. First, though, let us consider the velocity field surrounding a free vortex (Fig. 10.17). The equation for this field was given in Sec. 4.20 as $vr = C$, a constant. The circulation can be readily computed by application of Eq. (5.8) if we choose the closed path as the circular streamline L_1 concentric with the center of the vortex. The velocity v_1 is constant around this path and tangent to it ($\cos \beta = 1$), and the line integral of dL is simply the circumference of the circle. Applying the same treatment to another concentric circle L_2, we get

$$\Gamma = v \oint_L dL = v_1(2\pi r_1) = v_2(2\pi r_2)$$

But from the free-vortex velocity field $v_1 r_1 = v_2 r_2 = C$, and hence

$$\Gamma = 2\pi C = 2\pi vr \tag{10.34}$$

which demonstrates that the circulation around two different curves, each completely enclosing the vortex center, is the same. It may be proved more rigorously that the circulation around *any path enclosing the vortex center* is given by $\Gamma = 2\pi C$. The circulation is seen to depend only on the vortex constant C, which is called the *strength* of the vortex.

A corollary states that the circulation around a path *not enclosing* the vortex center is zero. Let us take the path $EFGH$ of Fig. 10.17. Along the two radial lines EF and GH, $\cos \beta = 0$, while along the two circular arc segments, $\Gamma_{FG} = v_2 \phi r_2$ and $\Gamma_{HE} = -v_1 \phi r_1$, resulting in a net circulation of $\phi(v_2 r_2 - v_1 r_1) = 0$.

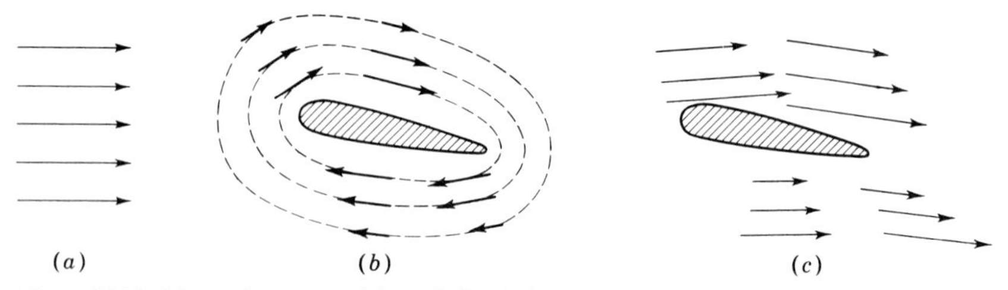

Figure 10.16 Schematic superposition of circulation on a uniform rectilinear flow field. (*a*) Uniform rectilinear flow field. (*b*) Circulation. (*c*) Net effect.

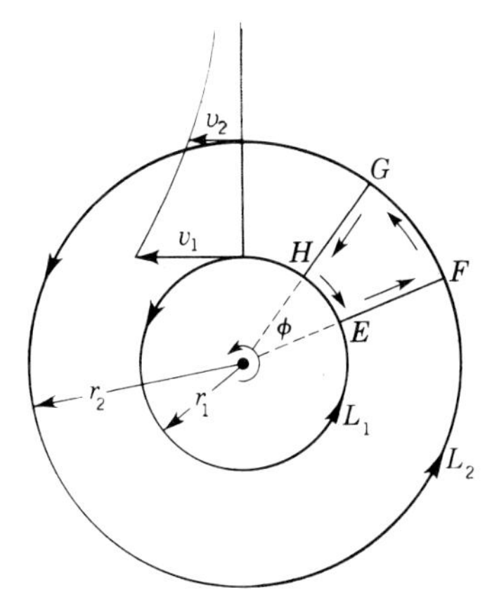

Figure 10.17 Circulation about a free vortex center.

10.10 IDEAL FLOW ABOUT A CYLINDER

Let us first consider uniform flow of an ideal fluid about a cylinder that is infinitely long. From classical hydrodynamics[1] it has been shown that with steady flow of uniform velocity U (Fig. 10.18*a*) the velocity v_{t_1} at the periphery ($r = R$) of the cylinder is given by

$$v_{t_1} = 2U \sin \theta \tag{10.35}$$

The pressure distribution on the cylinder may be computed by writing the Bernoulli theorem between a point at infinity in the free-streaming fluid and a point on the cylinder wall. Since the pressure distribution is completely symmetrical about the cylinder, there is no net lift or drag for this ideal case.

Putting this uniform flow aside for the moment, we next suppose a circulatory flow about the cylinder (Fig. 10.18*b*). Adopting the positive clockwise direction of circulation Γ, the peripheral velocity v_{t_2} on the surface of the cylinder due to circulation may be expressed as

$$v_{t_2} = \frac{\Gamma}{2\pi R} \tag{10.36}$$

where R is the radius of the cylinder. Thus in the flow field outside the cylinder, where $r > R$, $v_{t_2} = \Gamma/2\pi r$. This velocity distribution produces a pressure variation which is radially symmetrical, in accordance with the free-vortex theory. We see that the solid cylinder has replaced the vortex center in the circulation theory. If the reader wishes an explanation for the existence of the circulation, it may be supposed to arise from rotating the cylinder, which indeed may be demonstrated in a real fluid.

[1] Equation (10.35) can be developed by noting that $v_t = -\partial\psi/\partial r$ with v_t positive counterclockwise, where ψ is given by Eq. (5.18) in which $y = r \sin \theta$ and $m = 2\pi U R^2$ (Sec. 5.5).

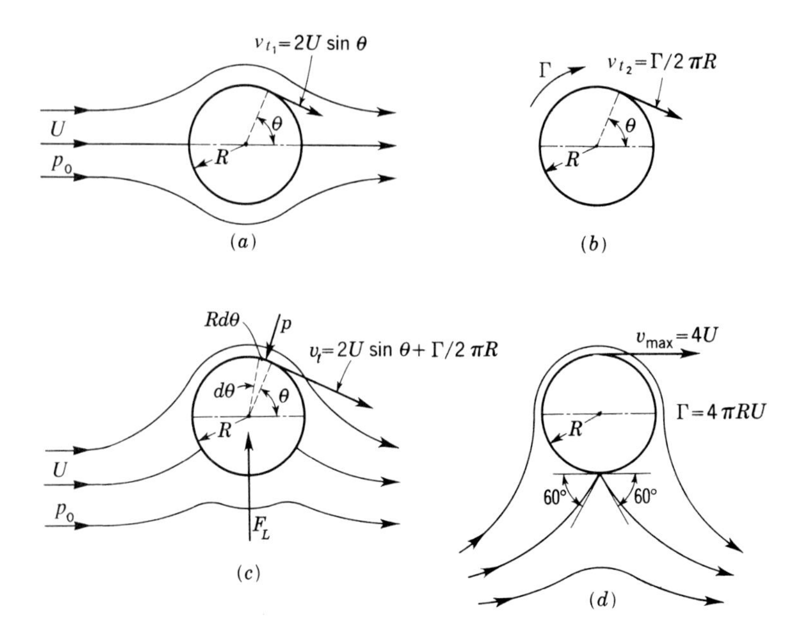

Figure 10.18 Circulation and lift from unsymmetrical flow about a cylinder.

Next let us *superpose* the circulatory flow onto the uniform motion, to form the unsymmetrical flow of Fig. 10.18*c*. The velocity at the periphery is the sum of the two contributions, or

$$v_t = 2U \sin\theta + \frac{\Gamma}{2\pi R} \tag{10.37}$$

The general equation for the pressure p at any point on the cirumference of the cylinder is obtained as follows:

$$\frac{p_0}{\gamma} + \frac{U^2}{2g} = \frac{p}{\gamma} + \frac{v_t^2}{2g}$$

where p_0 is the pressure at some distance away where the velocity is uniform. From these two equations

$$p - p_0 = \frac{\rho}{2}\left[U^2 - \left(2U \sin\theta + \frac{\Gamma}{2\pi R}\right)^2\right] \tag{10.38}$$

Since the elementary area per unit of length of the cylinder is $R\,d\theta$ and the lift F_L is the summation of all the components normal to the direction of U, the resulting value of F_L is obtained from

$$F_L = -B\int_0^{2\pi} (p - p_0)R \sin\theta \, d\theta$$

Substituting the expression for $p - p_0$ from Eq. (10.38) and integrating, this reduces to

$$F_L = \rho B U \Gamma \tag{10.39}$$

where F_L is the lift force and B is the length of the cylinder.

The existence of this transverse force on a rotating cylinder is known as the *Magnus effect*, after the man who first observed it in 1852. Equation (10.39)—the *Kutta–Joukowski theorem*—is known by the name of the two men who pioneered the quantitative investigation of the lift force shortly after the turn of the century. The great importance of this theorem is that it applies not only to the circular cylinder but to a cylinder of any shape, including in particular the lifting vane, or *airfoil*, as shown in Fig. 10.15.

It is clear from Fig. 10.18*c* that the stagnation points have shifted downward from the horizontal axis, but they are still symmetrical about the vertical axis. At the point of stagnation on the cylinder, v_t in Eq. (10.37) will be zero. Thus we have at the stagnation point

$$-2U \sin \theta = \frac{\Gamma}{2\pi R}$$

This shows that if we can measure the angle to the stagnation point and know the free-stream velocity, we may obtain the circulation from

$$\Gamma = -4\pi R U \sin \theta_s \tag{10.40}$$

where θ_s represents the angle between the horizontal diameter and the stagnation point in Fig. 10.18*c*. Figure 10.18*c* illustrates a case where $\Gamma < 4\pi R U$, that is, where $|\sin \theta| < 1$. For the case of $\Gamma = 4\pi R U$, $\sin \theta = -1$, and the two stagnation points meet together at the bottom of the cylinder as shown in Fig. 10.18*d*. The two streamlines make angles of 60° with the tangent to the cylinder and the maximum velocity in the flow for this case occurs at the top of the cylinder and is equal to

$$v_{\text{max}} = 2U + \frac{4\pi R U}{2\pi R} = 2U + 2U = 4U$$

Thus according to ideal-flow theory,[1] if the cylinder is rotated so that $v_t = 2U$ (that is, with $\omega = v_t/R = 2U/R$), the circulation thus produced will cause the stagnation point to occur at the bottom of the cylinder as in Fig. 10.18*d*. If the cylinder is rotated at still greater speed, the stagnation point is removed entirely from the cylinder surface, and a ring of fluid is dragged around with the cylinder.

[1] In the case of a real fluid, because of viscosity, the required velocity to bring the stagnation point to the bottom of the cylinder is about twice that indicated by ideal-flow theory.

Illustrative Example 10.7 A cylinder 4 ft in diameter and 25 ft long rotates at 90 rpm with its axis perpendicular to an airstream with a wind velocity of 120 fps. The specific weight of the air is 0.0765 lb/ft^3. Assuming no slip between the cylinder and the circulatory flow, find (*a*) the value of the circulation; (*b*) the transverse or lift force; and (*c*) the position of the stagnation points.

(*a*) Peripheral velocity

$$v_t = \frac{2\pi Rn}{60} = 2\pi \times 2 \times \frac{90}{60} = 18.84 \text{ fps}$$

From Eq. (10.34),

$$\Gamma = 2\pi R v_t = 2\pi \times 2 \times 18.84 = 237 \text{ ft}^2/\text{s}$$

(*b*) From Eq. (10.39),

$$F_L = \rho B U \Gamma = \frac{0.0765}{32.2} \times 25 \times 120 \times 237 = 1{,}690 \text{ lb}$$

(*c*) From Eq. (10.40),

$$\sin \theta_s = -\frac{\Gamma}{4\pi R U} = -\frac{237}{4\pi 2 \times 120} = -0.0786$$

Therefore

$$\theta_s = 184.5^\circ, \ 355.5^\circ$$

Actually, the real circulation produced by surface drag of the rotating cylinder would be only about one-half of that obtained above for the no-slip assumption.

10.11 LIFT OF AN AIRFOIL

The reader may well ask why so much attention is given to the flow about a cylinder when it is obvious that there are few practical applications of the lift on a cylinder.[1] The answer is that one of the most remarkable applications of mathematics to engineering is *conformal transformation*,[2] by which the flow about one body may be mapped into the flow about a body of different (though mathematically related) shape. Certain quantities, notably the circulation and relative position of the stagnation points, remain unchanged in the mapping. The importance of the circular cylinder, then, is that it can be mapped into a perfectly workable airfoil by the so-called *Joukowski transformation.* The position of the stagnation points is determined from the physical requirements of the flow about the airfoil, and these stagnation points, mapped back onto the cylinder, determine the circulation, by Eq. (10.40), and the lift, by Eq. (10.39).[3]

[1] In the early 1920s A. Flettner developed the "rotorship," which substituted motor-driven cylindrical rotors for sails. The ship was then driven by the Magnus effect but still required wind. A few trans-Atlantic crossings were made, but the rotorship was ultimately found to be uneconomical. See A. Fletner, "The Story of the Rotor," Willhoft, New York, 1926.

[2] For an excellent discussion of conformal transformation see H. R. Vallentine, "Applied Hydrodynamics," 2d ed., Butterworth & Co. (Publishers), Ltd., London, 1967.

[3] It must be understood that while the Joukowski profile is a workable lifting vane, the modern airfoil has undergone many modifications to improve its performance for various special purposes. The mapping theory described here applies exactly to the Joukowski foil and in principle to any lifting vane.

Let us examine the airfoil of Fig. 10.19. As fluid flows past the foil, there will be a tendency for stagnation points to form at the points of the foil, corresponding to the 0 and 180° points of the corresponding cylinder (Fig. 10.18*a*). Just where these points occur on the foil depends on the *angle of attack* α, or the attitude of the foil with respect to the oncoming flow, as shown in the figure. We shall assume a positive angle of attack in Fig. 10.19*a*, with corresponding initial stagnation points *a* and *b*. While the location of these stagnation points involves no difficulty in the case of the ideal fluid, we see at once that the condition at the trailing edge—with the air from the underside trying to flow around the sharp cusp of the foil—becomes a point of violent separation in real fluid flow.

This condition lasts no more than an instant, however, for stagnation point *b* is soon swept back to the trailing edge of the foil (Fig. 10.19*b*), where it stays. This stable position of the point *b* is necessary, according to the so-called *stagnation hypothesis* of N. Joukowski, in order to avoid an infinite velocity around the sharp cusp of the foil. Now this shift in the rear stagnation point of the foil corresponds to a shift in the rear stagnation point of the related cylinder to a negative angle, somewhat as shown in Fig. 10.18*c*. Vertical symmetry of the flow about the cylinder requires that the forward stagnation point move downward by the same angle. This in turn maps a new location of the forward stagnation point on the airfoil, and such a shift also takes place in the real flow. We see, then, that a circulation has become established about the airfoil, the magnitude of which is determined by the location of the stagnation points on the corresponding cylinder. The lift may then be determined analytically by Eq. (10.39).

Although the Joukowski hypothesis appears perfectly reasonable, we must investigate whether or not nature will actually perform this adjustment of the stagnation point to the cusp of the airfoil profile. Our acceptance of this hypothesis is complicated by the perfectly valid theorem of Thomson (Lord Kelvin), which states that "the circulation around a closed curve in the fluid does not change with time if one moves with the fluid." How is a circulation created around the airfoil where none existed before? The answer was first suggested by Prandtl and has been well substantiated with photographs. He showed that the initial separation point at the cusp caused a *starting vortex* to form, as shown in Fig. 10.20*a*. In order to satisfy Thomson's theorem, an equal and opposite circulation must automatically be generated around the foil (Fig. 10.20*b*). After this circulation has been established, the starting vortex breaks off and is left behind as the airplane moves forward, but just to satisfy Thomson's theorem, the starting vortex keeps whirling around (Fig. 10.20*c*) until it dies out from viscous

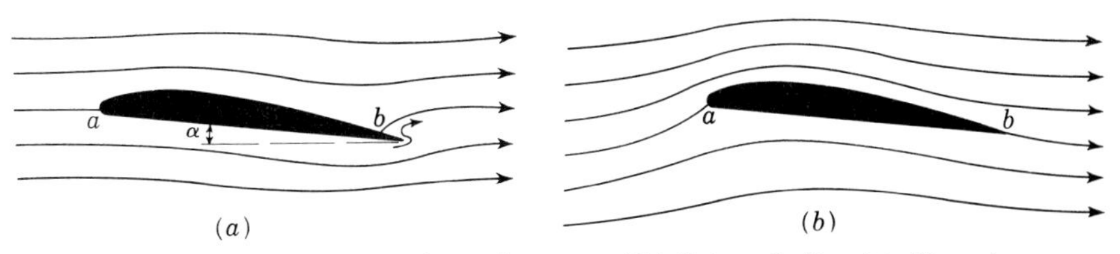

Figure 10.19 Adjustment of stagnation points to avoid infinite velocity at trailing edge.

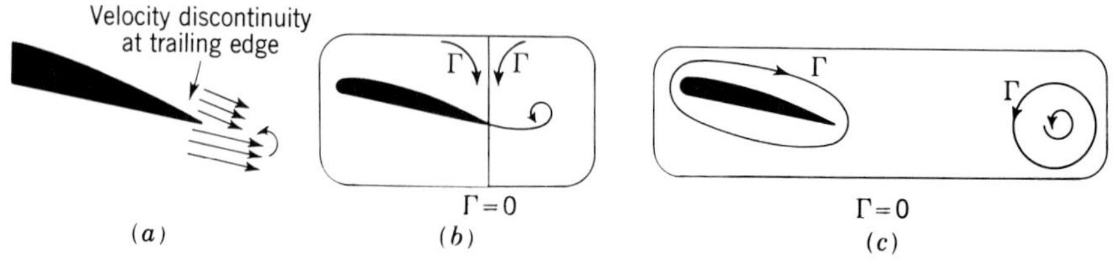

Figure 10.20 Life history of the starting vortex.

effects. The net circulation around a curve including the profile and this vortex is still zero. When their airfoil comes to a stop or changes its angle of attack, new vortices are formed to effect the necessary change in circulation.

10.12 INDUCED DRAG ON AIRFOIL OF FINITE LENGTH

The discussion of lift has so far been limited to strictly two-dimensional flow. When the foil or lifting vane is of finite length in a free fluid, however, there are end conditions which affect both the lift and the drag. Since the pressure on the underside of the vane is greater than that on the upper side, fluid will escape around the ends of the vane and there will be a general flow outward from the center to the ends along the bottom of the vane and inward from the ends to the center along the top. The movement of the fluid upward around the ends of the vane results in small *tip vortices* which are cast off from the wing tips. In theory, the Thomson theorem still holds, for the tip vortices are of equal and opposite magnitude. If the circulation is computed about a closed path passing along the axis of the airfoil and along the axes of the tip and starting vortices, as shown in Fig. 10.21, it will still add up to zero. Practically, of course, the circulation about the foil continues to exist, but the tip and starting vortices soon die out from viscous resistance.

The closed path consisting of the finite wing, the tip vortices, and the starting vortices of Fig. 10.21 constitutes a large vortex ring inside of which there is a

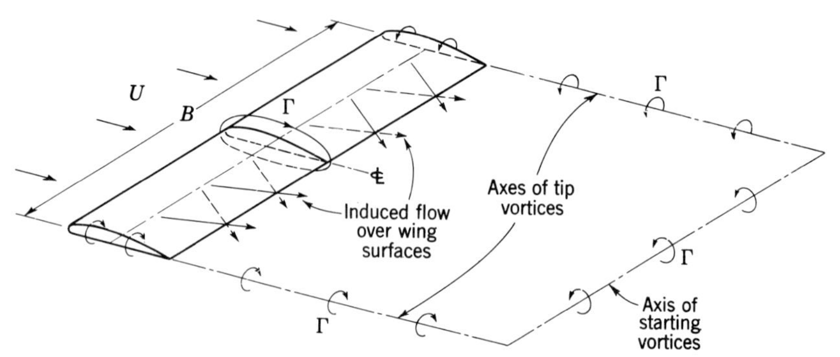

Figure 10.21 Wing of finite span.

downward velocity induced by the vortices. Prandtl showed this induced, or *downwash*, velocity U_i to be a constant if the wing is so constructed as to produce an elliptical distribution of lift over a given span. The downwash changes the direction of the flow in the vicinity of the foil from U to U_0 thus *decreasing the effective angle of attack* from α to α_0. The decrease in the effective angle of attack $\alpha_i = \alpha - \alpha_0 = \arctan(U_i/U)$, as shown in Fig. 10.22.

The wing may be analyzed on the basis of a foil of infinite length set in a stream of uniform velocity U_0, at angle of attack α_0. The lift F_{LO} generated from the circulation about the infinite foil must be normal to U_0. This force is seen to be resolved into two components, the true lift F_L normal to U and a component parallel to U called the *induced drag* F_{Di}. In conformity with our other drag terms, we represent the induced drag in the standard form

$$F_{Di} = C_{Di}\rho \frac{V^2}{2} A \tag{10.41}$$

It is now necessary to distinguish between the two- and three-dimensional cases of drag. The skin friction and pressure drag discussed earlier in this chapter will be lumped into the *profile drag* F_{DO}, which includes all drag forces acting on the profile of infinite length. The total drag on the finite span is then the sum of the profile and induced drags, or

$$F_D = F_{DO} + F_{Di} \tag{10.42}$$

As the angle α_i is small,

$$U_0 \approx U \qquad F_{LO} \approx F_L \qquad F_{Di} \approx \alpha_i F_L$$

It should be noted at this point that in addition to expressing the lift force by Eq. (10.39), it is convenient to express it as

$$F_L = C_L \rho \frac{V^2}{2} A \tag{10.43}$$

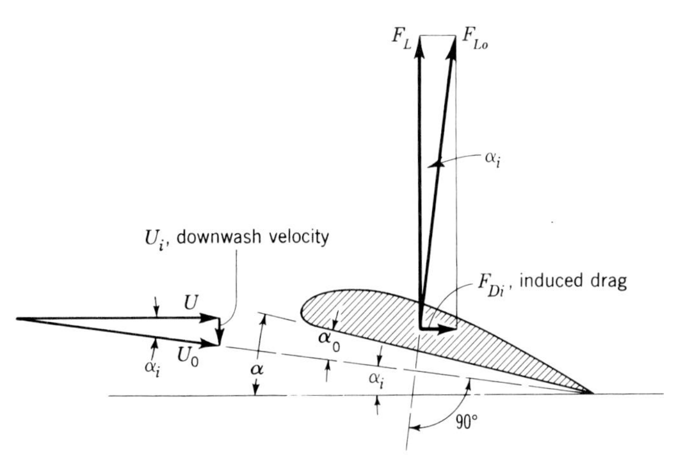

Figure 10.22 Definition sketch for induced drag.

where C_L is the lift coefficient whose value depends primarily on the angle of attack and the shape of the airfoil, and A is the projected area of the airfoil or body normal to the lift vector.

The computations for the elliptical distribution of lift are too complex to appear here, but they result in the simple relation

$$\frac{U_i}{U} = \alpha_i \text{ (radians)} = \frac{C_L}{\pi(B^2/A)} \tag{10.44}$$

where B is the span of the airfoil and A is its plane area. The quantity $B^2/A = B/c$ is referred to as the *aspect ratio* where c is the mean chord length.

From Eqs. (10.41), (10.42), and (10.43), together with the above expression for F_{Di}, we have

$$C_{Di} = \frac{C_L^2}{\pi(B^2/A)} \tag{10.45}$$

Dividing Eq. (10.42) by $\rho V^2 A/2$ and substituting equations similar to Eq. (10.41) and making use of Eq. (10.45) gives for the coefficient of total drag on a foil of finite length,

$$C_D = C_{D0} + C_{Di} = C_{D0} + \frac{C_L^2}{\pi(B^2/A)} \tag{10.46}$$

As would be expected, C_{Di} is seen to depend on the lift coefficient, i.e., the angle of attack α_0 and the *aspect ratio* B^2/A. For zero lift or infinite aspect ratio the induced drag would be zero. These equations are important in comparing data for an airfoil tested at one aspect ratio with data for another foil at a different aspect ratio.

The explanation of how the induced drag, occurring as it does in the ideal-fluid theory, fits into the D'Alembert paradox, which states that there is no drag on a body in ideal flow, is that the work done against the flow by the induced drag is conserved in the kinetic energy of the tip vortices cast from the ends of the foil. In a real fluid, evidence of the tip vortices may frequently be seen in the form of *vapor trails* extending for miles across the sky. The decreased temperature caused by the decreased pressure at the center of the vortex causes condensation of the moisture in the air.

10.13 LIFT AND DRAG DIAGRAMS

A wealth of data on the lift and drag of various airfoils has been obtained from wind-tunnel tests. The results of such tests may be presented graphically as plots of the lift and drag coefficients vs. the angle of attack. Since the efficiency of the airfoil is measured by the ratio of lift to drag, the value of C_L/C_D is generally plotted also. These three curves can be combined neatly into a single curve, suggested by Prandtl, known as a *polar diagram* (Fig. 10.23).

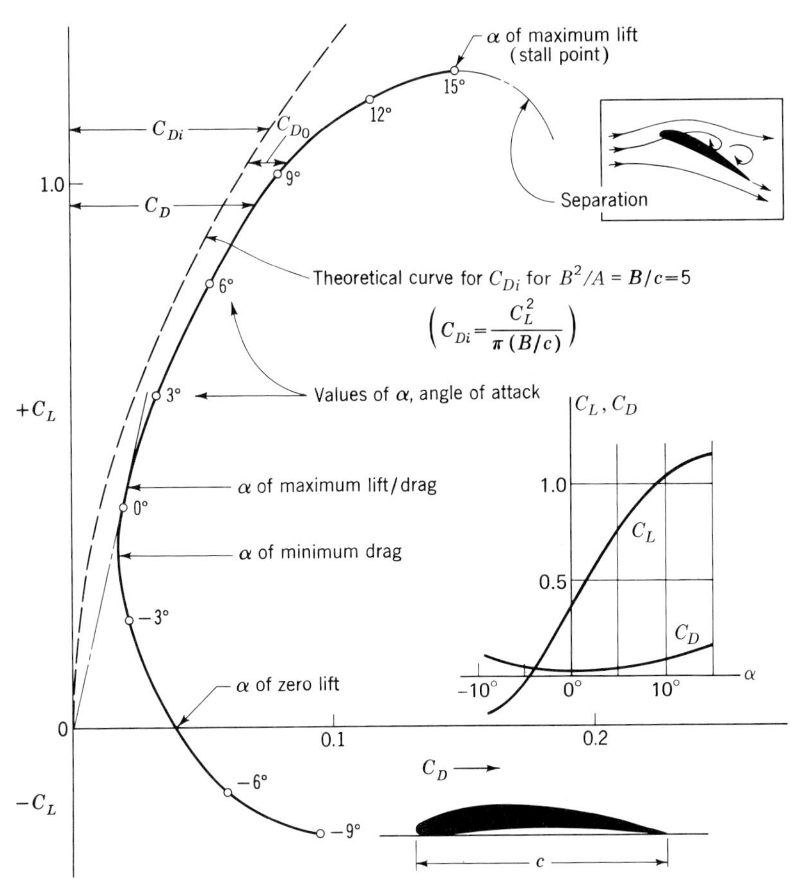

Figure 10.23 Polar diagram for wing of aspect ratio 5. (Curve from Prandtl–Tietjens, "Applied Hydro- and Aeromechanics," p. 152, McGraw-Hill Book Company, Inc., New York. 1934.)

The coordinates of the polar diagram are the lift and drag coefficients, while the angles of attack are represented by different points along the curve. The ratio of lift to drag is the slope of the line from the origin to the curve at any point. Evidently, the maximum value of the ratio occurs when this line is tangent to the curve. The lift is seen to increase with the angle of attack up to the *point of stall.* Beyond this point the boundary layer along the upper surface of the foil separates and creates a deep turbulent wake.

The polar diagram is notably instructive with regard to the drag coefficient, consisting of the coefficients of profile and induced drag as shown in Eq. (10.46). The dashed line in Fig. 10.23 is the parabola of Eq. (10.45). For an aspect ratio of 5, as shown, the induced drag is a major part of the total drag. For larger aspect ratios the parabola remains closer to the vertical axis, and the total drag is correspondingly decreased.

The polar diagram of a Clark Y airfoil, rectangular in plan, 6-ft chord by 36-ft span, is shown in Fig. 10.24. It will be observed that the angle of attack is

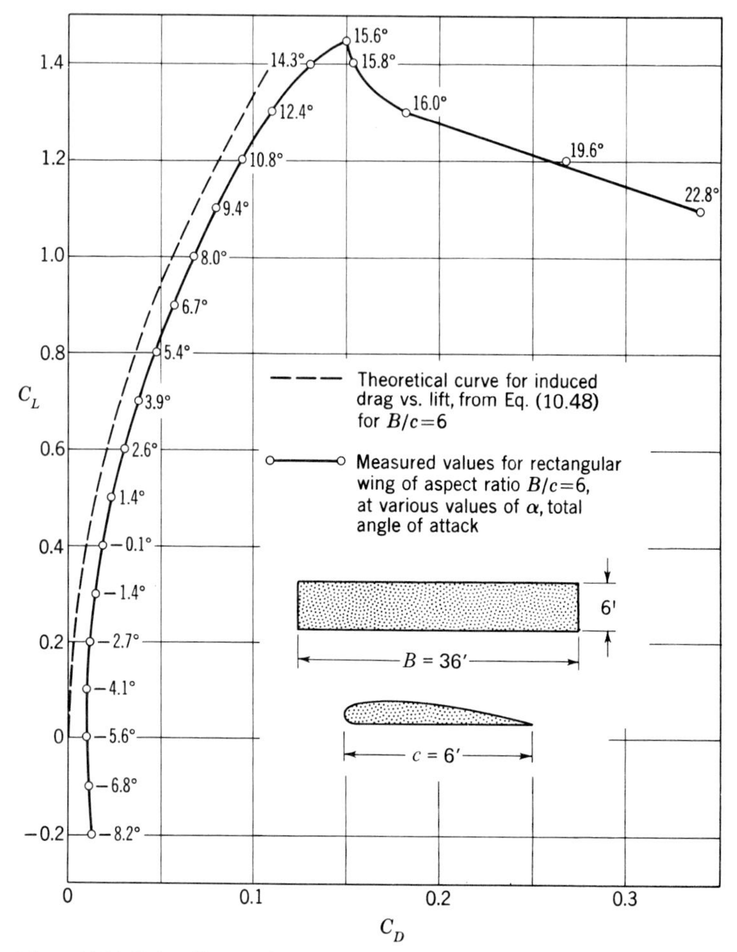

Figure 10.24 Polar diagram for rectangular Clark Y airfoil of 6-ft chord by 36-ft span. (Data from A. Silverstein, *NACA Rept. 502*, p. 15, 1934.)

read from a geometric reference which has little meaning by itself. The important reference angle is the angle of attack for zero lift, in this case $-5.6°$. In general, this is also the angle for minimum drag. The lift coefficient can be shown theoretically to be given by

$$C_L = 2\pi\eta\alpha_0' \tag{10.47}$$

where α_0' is the angle of attack (for the airfoil of infinite span) measured in radians from the attitude of no lift, and η is a correction factor for frictional effects, having a value of about 0.9 for modern airfoil sections.

It will be recalled from Sec. 10.12 that the induced-drag theory assumed an elliptical distribution of lift over the span of the finite airfoil. Such a distribution of lift is only an approximation, and for the rectangular airfoil the expressions for induced angle of attack and induced-drag coefficient given in Eqs. (10.44) and (10.45) must be corrected as follows:

$$\alpha_i \text{ (radians)} = \frac{C_L}{\pi(B/c)}(1+\tau) \tag{10.48}$$

and

$$C_{Di} = \frac{C_L^2}{\pi(B/c)}(1+\sigma) \tag{10.49}$$

where τ and σ are correction factors given in Fig. 10.25. Information on airfoils of shapes other than rectangular[1] and lift and drag on the fuselage, stabilizers, and control surfaces of aircraft[2] may be found in the literature.

Illustrative Example 10.8 For a rectangular Clark Y airfoil of 6-ft chord by 36-ft span, find the value of the friction coefficient η if the angle of attack $\alpha = 5.4°$ when the wing is moving at 300 fps through standard atmosphere at altitude 10,000 ft. Find the weight which the wing will carry and the horsepower required to drive it.

From Fig. 10.24, with $\alpha = 5.4°$, $C_L = 0.8$, $C_D = 0.047$. From Fig. 10.25, for $B/c = 6$, $\tau = 0.175$. From Eq. (10.48),

$$\alpha_i = \frac{0.8}{\pi(36/6)}(1+0.175) = 0.0498 \text{ rad} = 2.85°$$

From Fig. 10.22, $\alpha_0 = \alpha - \alpha_i = 5.40 - 2.85 = 2.55°$ and since the angle of zero lift is $-5.6°$,

$$\alpha_0' = 2.55 + 5.6 = 8.15° = 0.1422 \text{ rad}$$

From Eq. (10.47).

$$\eta = \frac{C_L}{2\pi\alpha_0'} = \frac{0.8}{2\pi \times 0.1422} = 0.896$$

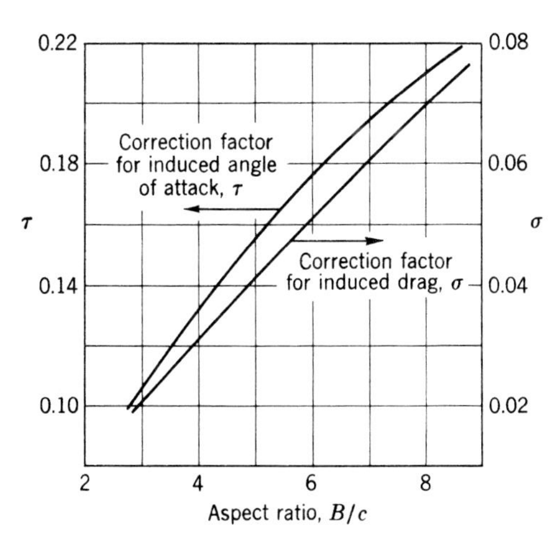

Figure 10.25 Correction factors for transforming rectangular airfoils from finite to infinite aspect ratio. (From A. Silverstein, *NACA Rept. 502*, Fig. 7, 1934.)

[1] Ira H. Abbott and Albert E. von Doenhoff, "Theory of Wing Sections Including Summary of Airfoil Data," Dover, New York, 1959.

[2] Hermann Schlichting and Erich Truckenbrodt, "Aerodynamics of the Airplane," translated by Heinrich J. Ramm, McGraw-Hill International Book Co., New York, 1979.

The wing will support a weight equal to the lift force,

$$F_L = C_L \rho \frac{V^2}{2}(B \times c)$$

From Appendix 3, Table A.2(*a*), at 10,000 ft, $\rho = 0.001756$ slug/ft^3.

$$F_L = 0.8 \times 0.001756 \times \frac{(300)^2}{2} \times 36 \times 6 = 13{,}650 \text{ lb}$$

while

$$F_D = \frac{0.047}{0.8} \times 13{,}680 = 802 \text{ lb}$$

$$\text{Horsepower required} = \frac{802 \times 300}{550} = 437 \text{ hp}$$

10.14 EFFECTS OF COMPRESSIBILITY ON DRAG AND LIFT

In Sec. 9.10 it was pointed out that when a body moves through a fluid at supersonic velocity a shock wave is formed. The wave pattern (Fig. 9.6) is the same whether the body is moving through the fluid or whether the fluid is moving past the body. Unlike subsonic flow (Fig. 3.11), in supersonic flow the streamlines in front of the body are unaffected because the body is moving faster than the disturbance can be transmitted ahead. This is demonstrated in Fig. 9.7.

With most bodies the drag coefficient tends to increase drastically at a Mach number of about 0.70. This is so because the body is encountering *transonic flow* phenomena, which means that at some place in the flow field supersonic flow is occurring. With a streamlined body the highest velocity in the flow field occurs at some point such as *b* in Fig. 10.26 near the body and away from its nose. The local Mach number at *b* will reach unity when the free-stream Mach number at *a* has a value of perhaps only 0.7 or 0.8. Thus a shock wave will form at *b*. Through the shock wave there is a sudden jump in pressure which causes an adverse pressure gradient in the boundary layer, resulting in separation and an increase in drag. Drag coefficients for several bodies as a function of the free-stream Mach number are given in Fig. 10.27. The increased drag is caused not only by the separation effects; a substantial amount of energy is dissipated in the shock wave. Skin friction also contributes to the drag, and at Mach numbers above 2 or 3 heating in the boundary layer from skin friction may be an important phenomenon. As the value of the Mach number increases beyond about 2,

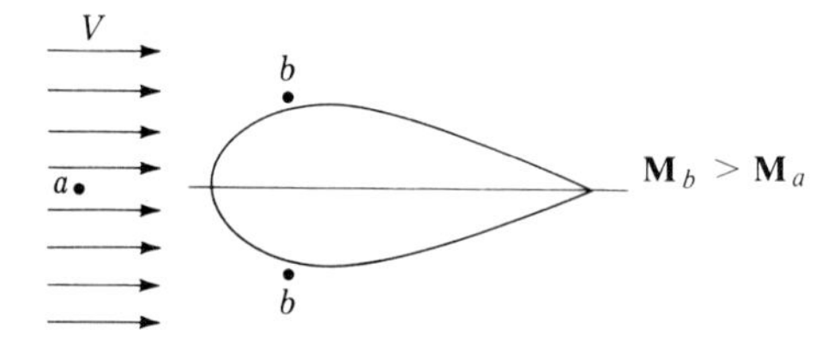

Figure 10.26 Local Mach number greater than freestream Mach number. When $\mathbf{M}_b = 1.0$, $\mathbf{M}_a \approx 0.7$.

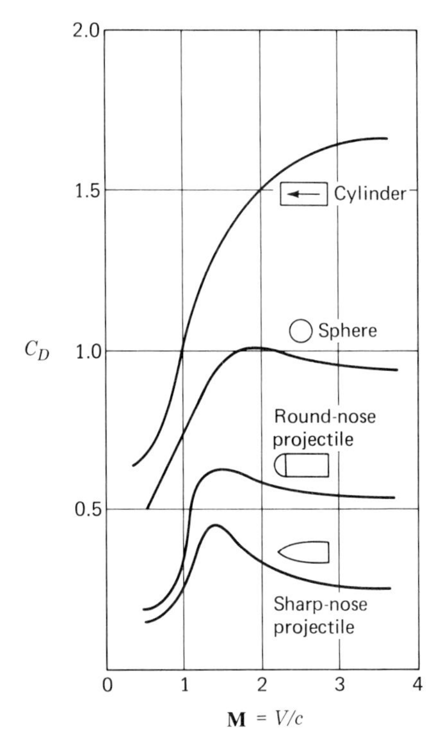

Figure 10.27 Drag coefficients as a function of Mach number.

for most bodies there is a drop in the value of the drag coefficient because of a shift in the point of separation.

Earlier it was mentioned that for streamlining against subsonic flow a rounded nose and a long, tapered afterbody generally result in minimum drag. *In supersonic flow the best nose form is a sharp point.* This tends to minimize the effect of the shock wave.

With respect to lift it has been found that for Mach numbers less than about 0.7 the lift coefficient for compressible fluids may be estimated by dividing the lift coefficient for incompressible flow by $\sqrt{1 - \mathbf{M}^2}$, where $\mathbf{M}$ is the Mach number of the free stream.[1] The reduced pressures on the top of the airfoil are responsible for this trend.

At $\mathbf{M} \approx 0.8$ there is a rather abrupt drop in the lift coefficient because the shock wave induced by the local supersonic flow creates high pressures on the top side of the airfoil which may result in *shock stall.* At somewhat higher values of the Mach number a shock wave forms on the bottom of the airfoil which tends to compensate for the preceding action and there is an increase in the value of the lift coefficient. Although there have been tremendous advances in theory, the best way to predict the drag and lift of a particular airfoil is by conducting model tests in a wind tunnel.

[1] This is known as the Prandtl-Glauert rule.

10.15 CONCLUDING REMARKS

There are some other aspects of drag that ought to be mentioned. One of these is drag under conditions of *supercavitation.* This occurs when bodies move at high speeds through liquids. It is particularly prevalent with blunt bodies. A large cavity is formed behind the body. This alters the pressure distribution because the limiting minimum absolute pressure in the cavity is the vapor pressure. Such problems require special treatment; experimental data provide the best information.

Another interesting drag problem is that of an object moving at the interface of two fluids of different density. A good example of this is that of a ship or boat moving through water. In such cases energy is expended in the generation of waves. The drag of the ship is caused primarily by skin friction and wave action; hence in a model test both the Reynolds and Froude criteria ought to be satisfied, but this is not practical. The modeling procedure usually employed involves determining the total drag of the model by testing it at the prototype Froude number. The frictional drag of the model is then estimated by using boundary-layer theory, as presented in Secs. 10.3 through 10.5, and subtracted from the total drag to get an estimate of the drag on the model caused by wave action. This is then translated by model laws to an estimate of the wave-action drag on the prototype to which is added the prototype friction drag as estimated by boundary-layer theory to give the total prototype drag.

Illustrative Example 10.9 A 20-cm-diameter round-nosed projectile whose drag coefficient is shown in Fig. 10.27 travels at 600 m/s through the standard atmosphere at an altitude of 6,000 m. Find the drag.

From Appendix 2, the sonic velocity is given by $c = \sqrt{kRT}$. From Appendix 3, Table A.3*b*, the temperature is $-24°$C, or 249 K. Then with $k = 1.4$ and $R \approx 287\ \text{m}^2/(\text{s}^2\cdot\text{K})$,

$$c = \sqrt{1.4 \times 287 \times 249} = 316\ \text{m/s}$$

and so

$$\mathbf{M} = \frac{600}{316} = 1.90$$

From Fig. 10.27, $C_D = 0.62$

$$\frac{p}{\rho} = RT \qquad \rho = \frac{p}{RT}$$

$$\rho = \frac{47.22\ \text{kN/m}^2\text{, abs}}{[287\ \text{m}^2/(\text{s}^2\cdot\text{K})]249\ \text{K}} = 0.00066\frac{\text{kN}\cdot\text{s}^2}{\text{m}^4} = 0.66\ \text{kg/m}^3$$

and

$$F_D = 0.62 \times 0.66 \times \frac{(600)^2}{2} \times \frac{\pi(0.20)^2}{4} = 2{,}310\ \text{N}$$

PROBLEMS

10.1 Commencing with the general equation for a parabola $u = ay^2 + by + c$, derive the velocity distribution of Eq. (10.9) in the dimensionless form shown.

10.2 For the parabolic velocity distribution of Eq. (10.9), derive the numerical values of α [Eq. (10.5)] and β [Eq. (10.7)] of 0.133 and 2.0, respectively.

10.3 Refer to the data of Illustrative Example 10.1. Find the shear stress at 6 in (16 cm) and 12 in (32 cm) back from the leading edge of the plate.

10.4 A smooth, flat plate 12 ft (3.6 m) wide and 2.5 ft (0.75 m) long parallel to the flow is immersed in water at 60°F (15°C) flowing at an undisturbed velocity of 2 fps (0.6 m/s). Find the thickness of the boundary layer and the shear stress (*a*) at the center and (*b*) at the trailing edge of the plate, assuming a laminar boundary layer over the whole plate. Also, find the total friction drag on one side of the plate.

10.5 For the critical Reynolds number of 500,000 for transition from laminar to turbulent flow in the boundary layer, find the corresponding critical Reynolds number for flow in a circular pipe. How does this compare with the value given in Chap. 8? (*Hint:* Consider the boundary-layer thickness to correspond to the radius of the pipe in laminar flow, while the undisturbed velocity U of the boundary-layer theory represents the centerline velocity u_{max} of the pipe flow.)

10.6 Derive Eq. (10.19) along the lines suggested in the text.

10.7 For the Prandtl seventh-root law given in Eq. (10.20) derive the value of $\alpha = 0.0972$ for the turbulent boundary layer.

10.8 Derive Eq. (10.22) from the information given.

10.9 Refer to the data of Illustrative Example 10.2. Find the shear stress on the sides of the van at 5, 15, and 25 ft back from the front edge of the sides.

10.10 Assume that the boundary layer of Prob. 10.4 is disturbed near the leading edge. Compute the corresponding quantities for the turbulent boundary layer covering the whole plate, and compare the results.

10.11 A streamlined train is 300 ft (90 m) long, with sides 9 ft (2.75 m) high and the top 9 ft (2.75 m) wide. Assuming the skin-friction drag on sides and top to be equal to that on one side of a flat plate 27 ft (8.25 m) wide and 300 ft (90 m) long, compute the power required to overcome the skin-friction drag when the train is traveling at 100 mph (45 m/s) through the ICAO standard atmosphere air at sea level.

10.12 It is well known that on the beach one can lie down to get out of the wind. Suppose the wind velocity 6 ft (2 m) above the beach is 20 fps (6 m/s). Approximately what would be the velocity at 0.5 ft (0.15 m) and at 1.0 ft (0.30 m) above ground level?

10.13 Compare the values of C_f as computed by Eqs. (10.25) and (10.26) for $\mathbf{R} = 10^7$.

10.14 Demonstrate the equality of the two expressions of Eq. (10.28).

10.15 A smooth, thin, flat plate with sharpened edges is 10 ft by 2 ft and is submerged in water the temperature of which is 60°F. If it moves through the water with a velocity of 1.22 fps in the direction of the 10-ft length, what is the total drag?

10.16 A harpoon is $\frac{3}{4}$ in (19 mm) in diameter and 5 ft (1.5 m) long, with a sharp tip. If this harpoon is launched in the water at 60°F (15°C) at a speed of 25 fps (7.5 m/s), find the friction drag. What will be the maximum thickness of the boundary layer?

10.17 An airplane wing having a chord length parallel to the flow of 6.5 ft (2 m) moves through standard atmospheric air at an altitude of 10,000 ft (4 km) and a speed of 250 mph (400 km/h). Find the critical roughness for a point one-tenth the chord length back from the leading edge. Find the surface drag on a section of this wing of 20-ft (6 m) span.

10.18 A flat plate 20 ft long is towed at 6 fps through a liquid ($\gamma = 50$ lb/ft^3, $\mu = 0.00026$ lb·s/ft^2). Determine the drag on the plate. Plot the boundary-layer profile, showing its thickness along the plate. Plot the local shear stress τ_0 as a function of x, and determine the area under this curve. Compare this value with the computed value of the drag. Assume that the boundary-layer changes from laminar to turbulent at a Reynolds number of 300,000. The plate is 1 ft wide.

10.19 Refer to the plate and data of Prob. 10.18. Make the necessary calculations to plot drag vs. velocity for velocities ranging from 0 to 50 fps.

10.20 Refer to the harpoon of Prob. 10.16. Determine the drag on the harpoon for velocities of 0, 10, 30, and 50 fps (0, 3, 9, and 15 m/s). Consider movement through (*a*) 60°F (15°C) water and (*b*) 60°F (15°C) air at sea level. In each case calculate the length x_c of the laminar zone. Plot curves of drag vs. velocity.

10.21 A steel sphere ($s = 7.8$) of diameter 0.25 in (6 mm) is released in a large tank of oil ($s = 0.825$). The sphere is observed to have a terminal velocity of 2.0 fpm (0.6 m/min). What is the viscosity of the oil?

10.22 What will be the terminal velocity of the sphere of Prob. 10.21 in 100°F water?

10.23 Compute the drag on a 15-in (38-cm)-diameter sphere from wind under sea-level conditions in a standard ICAO atmosphere. Plot drag vs. velocity for a range of velocities from 0 to 100 fps (30 m/s).

10.24 Repeat Prob. 10.23 for wind at 10,000-ft elevation in a standard ICAO atmosphere.

10.25 A metal ball of diameter 1.0 ft (30 cm) and weight 90 lb (400 N) is dropped from a boat into the ocean. Determine the maximum velocity the ball will achieve as it falls through the water. Properties of the ocean water: $\rho = 2.0$ slugs/ft^3 (1030 kg/m^3), $\mu = 3.3 \times 10^{-5}$ lb·s/ft^2 (0.0016 N·s/m^2).

10.26 Suppose a well-streamlined automobile has a body form corresponding roughly to the airship hull of Fig. 10.10, while a poorly streamlined car has a body approximating the 1:0.75 oblate ellipsoid, each with a diameter of 6 ft (2 m). Find the horsepower (kW) required to overcome air resistance in each of the two cases if the velocity is 60 mph (27 m/s) through standard air at sea level (Appendix 3, Table A.2).

10.27 It is desired to calculate approximately the pressure drag on the streamlined train of Prob. 10.11. As a rough approximation, assume that the nose and tail of the train are of the shape of the two halves of the prolate ellipsoid of Fig. 10.10, of 9 ft diameter. Find the drag on the ellipsoid (pressure drag on the train), and compare with the skin-friction drag on the train determined earlier.

10.28 A submerged 10 by 2 ft (3.0 by 0.6 m) flat plate is dragged through 60°F (15°C) water at 1.22 fps (0.4 m/s) with the flat side normal to the direction of motion. What is the approximate drag force? How does this compare with the drag force of 0.1598 lb (0.810 N) when pulled in the direction of the 10-ft (3-m) length? [*Hint:* Assume that the drag coefficient for the plate of finite length is in the same ratio to the coefficient for the infinite plate as is the ratio of coefficients for the finite cylinder ($L/D = 5$) and the infinite cylinder of Fig. 10.13, for the same Reynolds number.]

10.29 Compare the velocity of a 0.1-in (2.5-mm)-diameter spherical bubble of air rising through water with that of a 0.1-in (2.5-mm) drop of water falling through air. Assume standard air at sea level and water at the same temperature.

10.30 The drag coefficient for a hemispherical shell with the concave side upstream is approximately 1.33 if $\mathbf{R} > 10^3$. Find the diameter of a hemispherical parachute required to provide a fall velocity no greater than that caused by jumping from a height of 8 ft (2.5 m), if the total load is 200 lb (900 N). Assume standard air at sea level.

10.31 Find the rate of fall of a particle of sand ($s = 2.65$) in water at 60°F if the particle may be assumed spherical in shape and the diameter is (*a*) 0.1 mm; (*b*) 1.0 mm; (*c*) 10 mm. Express answers in centimeters per second.

10.32 A regulation football is approximately 6.78 in in diameter and weighs 14.5 oz. Its shape is not greatly different from the prolate ellipsoid of Fig. 10.10. Find the resistance when the ball is passed through still air (14.7 psia and 80°F) at a velocity of 40 fps. Neglect the effect of spin about the longitudinal axis. What is the deceleration at the beginning of the trajectory? Assuming no change in drag coefficient, find the percentage change in resistance if the air temperature is 20°F rather than 80°F.

10.33 What drag force is exerted at sea level by a 3-m-diameter braking parachute when the speed is 25 m/s? Assume $C_D = 1.20$. At what speed will the same braking force be exerted by this parachute at elevation 2,000 m? Assume C_D remains constant.

10.34 An eight-oar racing shell is traveling through 60°F water at a mean velocity of 12 mph. Each oar is 9 ft long, with a length of 6 ft from the oarlock to the center of the "spoon," which has a projected area of 120 in^2. Assume that all drag is due to the spoon, and that this drag is equal to that of a disk of equal area. (*a*) If the "stroke" is 32 per min and if each oarsman sweeps a right angle in one-fourth of his rowing cycle, what is the maximum thrust of the oars? It must be assumed that the shell moves at something less (say, 20 percent) than its mean velocity when the oars are in the water.

(*b*) The maximum velocity occurs during the backstroke when the oarsmen shift their weight toward the stern. Why? (*c*) The oarsman on his backstroke moves at half the angular velocity of his forward stroke, while the shell moves at perhaps 10 per cent above its mean velocity. Find the drag in 60°F air resulting from a "feathered" oar (turbulent boundary layer, Fig. 10.7) and an unfeathered one, in percentage of the forward thrust from part (*a*). What is the ratio of these two drag forces?

10.35 Find the bending moment at the base of a cylindrical radio antenna 0.30 in in diameter extended to 6 ft in length on an automobile traveling through standard air at 80 mph.

10.36 (*a*) Find the bending moment at the base of a cylindrical light post 40 cm in diameter and 10 m high when it is subject to a uniform wind velocity of 25 m/s at standard sea level. Neglect end effects. (*b*) Discuss the consequences of considering the atmospheric boundary layer above the surface of the earth.

10.37 (*a*) Repeat Prob. 10.36(*a*) for the case where the pole has a uniform taper from 40-cm-diameter at the base to 30 cm at its top. (*b*) If the pole has been tapered from 40 cm down to 20 cm, would the drag force on the pole be larger or smaller? Why?

10.38 Approximately what frequency of oscillation is produced when a 60-mph wind blows across a 0.125-in-diameter wire at (*a*) standard sea level; (*b*) standard atmosphere at 10,000-ft elevation?

10.39 If the mean velocity along the top of a wing having a 2-m chord is 150 km/hr and that along the bottom of the wing is 120 km/hr when the wing moves through still air ($\gamma = 0.072$ lb/ft^3) at 128 km/hr, estimate the lift per unit length of span. Give the answer in both English and SI units.

10.40 Commencing with the expression above Eq. (10.39) fill in the steps leading to Eq. (10.39). Take care to account for all changes in sign.

10.41 Calculate the value of the lift coefficient for the rotating cylinder in Illustrative Example 10.7, assuming the effective circulation to be half the theoretical. If the drag coefficient may be assumed to be unchanged by the rotation of the cylinder, find the total force of the wind on the rotor and its direction.

10.42 Assume the rotor of Illustrative Example 10.7 to be installed upright on a ship which is proceeding due north at 30 fps. The wind has an *absolute* velocity of 50 fps due east. If the drag coefficient of the cylinder is 1.0 and the "lines" of stagnation are separated by 120° on the rotor, find approximately the component of the total air force on the rotor in the direction of the ship's motion. Assume standard air.

10.43 Consider a cylinder of radius a in a stream of ideal fluid in which the undisturbed velocity and pressure are V and p_0 and the density is ρ. (*a*) Utilizing Eq. (10.35) and the Bernoulli theorem, evaluate the dimensionless pressure coefficient $(p - p_0)/(\rho V^2/2)$ for every 10° over the surface of one quadrant of the cylinder, and plot to scale, plotting the pressure radially from the cylinder surface. (*b*) What is the actual pressure in pounds per square inch (newtons per square meter) on the surface of a cylinder 1 ft (30 cm) in diameter, 70° from the forward stagnation point, if the cylinder is 20 ft (6 m) below the free surface of a stream of water at 60°F (15°C), flowing at 10 fps (3 m/s)?

10.44 Suppose a circulation of 20 ft^2/s (2 m^2/s) is superposed about the cylinder of Prob. 10.43. Find the location of the stagnation points and the lift for a length of 30 ft (9 m).

10.45 A double-stagnation point is observed to occur on a cylinder 3 ft (0.9 m) in diameter, rotating in a stream of standard air (sea level) having a velocity of 50 fps (15 m/s). Find the lift force per unit length of the cylinder. What is the lift coefficient?

10.46 There have often been arguments over the validity and extent of the curve of a pitched baseball. According to tests (*Life*, July 27, 1953), a pitched baseball was found to rotate at 1,400 rpm while traveling at 43 mph (69 km/h). The horizontal projection of the trajectory revealed a smooth curve of about 800 ft (245 m) radius. If the ball were 9 in (23 cm) in circumference and weighed 5 oz (1.4 N), find the transverse force required to produce the observed curvature. Assuming the shape of the ball to be roughly that of a cylinder having a diameter equal to the ball's diameter and a length of two-thirds its diameter, find the value of the circulation that would be required to produce the transverse force. Compare this with that obtained by assuming no slip at the equator of the ball. Assume standard air at sea level.

10.47 A wing with a 20-m span and 60-m^2 "plan-form" area moves horizontally through the standard atmosphere at 10,000 m with a velocity of 800 km/hr. If the wing supports 250,000 N, find (*a*) the required value of the lift coefficient: (*b*) the downwash velocity, assuming semielliptical distribution of lift over the span; (*c*) the induced drag.

10.48 A wing of 40-ft span and 320-ft^2 "plan-form" area moves horizontally through standard atmosphere at 10,000 ft with a velocity of 200 fps. If the wing supports 3,000 lb, find (*a*) the required value of the lift coefficient; (*b*) the downward velocity, assuming the elliptical distribution of lift over the span; (*c*) the induced drag.

10.49 If the plan form of the wing in Prob. 10.48 were rectangular, what would be the proper value of the induced angle of attack and induced drag?

10.50 An airplane having a Clark Y airfoil wing of 6-ft chord by 36-ft span, with polar diagram given in Fig. 10.24, weighs 1,500 lb. Neglecting the aerodynamic forces on the fuselage and tail, find the speed required to get the airplane off the ground. Find the horsepower required. Find the circulation about the wing and the strength of the starting vortex. Assume standard air at sea level and angle of attack for maximum ratio of lift to drag.

10.51 For the Clark Y airfoil of Fig. 10.24, evaluate the friction coefficient η of Eq. (10.47) for values of C_L of 0.6, 1.0, and 1.4.

10.52 A kite has a shape corresponding roughly to a rectangular airfoil of 2-ft chord and 4-ft span. When rigged as shown in the figure, the guideline exerts a tension of 11 lb when the wind velocity is 30 mph in standard air at 1,000 ft altitude. Evaluate C_L, C_{DO}, and the friction coefficient η.

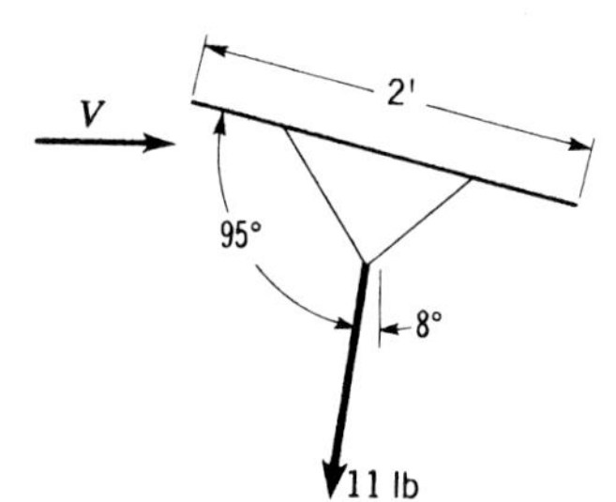

Problem 10.52

10.53 A rectangular airfoil with a 2-m chord and 12-m span has a drag coefficient of 0.062 and a lift coefficient of 0.94 at an angle of attack of 6.8°. What would be the corresponding lift coefficient, drag coefficient, and angle of attack for a wing having the same profile but with an aspect ratio of 7.5?

10.54 A sailplane weighing 400 lb (1800 N) including its load has a 4-ft (1.2 m)-chord by 24-ft (7.3 m)-span wing of the Clark Y section. Assuming that its characteristics are the same as those for the larger wing of the same aspect ratio shown in Fig. 10.24, find the angle of glide through standard air at 2,000 ft (600 m) which will produce the greatest horizontal distance range. Neglect air forces on the fuselage and tail. (*Note:* The aspect ratio of 6 is here chosen for convenience in working the problem with the available data. Actually, the sailplane may be constructed with an aspect ratio of about twice this, so as to reduce drag to minimum.)

10.55 At what angle of glide will the sailplane of Prob. 10.54 reach a minimum velocity of descent; in other words, at what angle will it remain in the air the longest time? (*Note:* A trial-and-error type of solution will be required here.)

10.56 If a supersonic jet aircraft traveling horizontally at 1,600 mph (2,500 km/h) passes overhead at an elevation of 5,000 ft (2 km), approximately how soon thereafter will the shock wave be felt at sea level?

10.57 If the round-nosed and sharp-nosed projectiles of Fig. 10.27 each represent a 900-lb (4000-N) bomb having a diameter of 20 in (50 cm), find their terminal velocities in standard air at sea level. Assume that the bombs travel nose first vertically downward.

10.58 Determine the rate of deceleration that will be experienced by the blunt-nosed projectile of Fig. 10.27 when it is moving (*a*) horizontally at 1,000 mph; (*b*) upward at an angle of 40° with the horizontal at a velocity of 1,000 mph. Assume standard sea-level atmosphere. The projectile has a diameter of 18 in and it weighs 600 lb.

CHAPTER

ELEVEN

STEADY FLOW IN OPEN CHANNELS

11.1 OPEN CHANNELS

An open channel is one in which the stream is not completely enclosed by solid boundaries and therefore has a free surface subjected only to atmospheric pressure. The flow in such a channel is caused not by some external head, but rather by the gravity component along the slope of the channel. Thus open-channel flow is often referred to as *free-surface flow* or *gravity flow*. This chapter will deal only with *steady* flow in open channels.

The principal types of open channel are natural streams and rivers; artificial canals; and sewers, tunnels, and pipelines not completely filled. Artificial canals may be built to convey water for purposes of water-power development, irrigation or city water supply, and drainage or flood control and for numerous other purposes. While there are examples of open channels carrying liquids other than water, there are few experimental data for such and the numerical coefficients given here apply only to water at natural temperatures.

The accurate solution of problems of flow in open channels is much more difficult than in the case of pressure pipes. Not only are reliable experimental data more difficult to secure, but there is a wider range of conditions than is met with in the case of pipes. Practically all pipes are round, but the cross sections of open channels may be of any shape, from circular to the irregular forms of natural streams. In pipes the degree of roughness ordinarily ranges from that of new smooth metal, on the one hand, to that of old corroded iron or steel pipes, on the other. But with open channels the surfaces vary from that of smooth timber or concrete to that of the rough or irregular beds of some rivers. Hence the choice of friction coefficients is attended by greater uncertainty in the case of open channels than in the case of pipes.

Uniform flow was described in Sec. 3.2 as it applies to hydraulic phenomena in general. In the case of open channels uniform flow means that the water cross section and depth remain constant over a certain *reach* of the channel and over time. This requires that the drop in potential energy due to the fall in elevation along the channel be exactly consumed by the energy dissipation through boundary friction and turbulence.

Uniform flow will eventually be established in any channel which continues sufficiently far with a constant slope and cross section. This may be stated in another way, as follows. For any channel of given roughness, cross section, and slope, there exists for a given flow rate one and only one water depth, y_0, at which the flow will be uniform. Thus, in Fig. 11.1, the flow is accelerating in the reach from A to C, becomes established as uniform flow from C to D, suffers a violent deceleration due to the change of slope between D and E, and finally approaches a new depth of uniform flow somewhere beyond E. There is acceleration in the reach from B to C because the gravity component along the slope is greater than the boundary shear resistance. As the flow accelerates, the boundary shear increases because of the increase in velocity, until at C the boundary shear resistance becomes equal to the gravity component along the slope. Beyond C there is no acceleration, the velocity is constant, and the flow is uniform. The depth in uniform flow is commonly referred to as the *normal depth*, y_0.

Open-channel flow is usually wholly rough, that is, it occurs at high Reynolds numbers. For open channels the Reynolds number is defined by $\mathbf{R} = R_h V/\nu$, where R_h is the hydraulic radius. Since $R_h = D/4$, the critical value of Reynolds number at which the changeover occurs from laminar flow to turbulent flow in open channels is 500, whereas in pressure conduits the critical value is 2,000.

In open-channel flow (Fig. 11.2) we refer to the slope of the *channel bed* S_0, the slope of the *water surface* S_w, and the slope of the *energy line* S. It is quite evident that in the case of flow in an open channel the hydraulic grade line coincides with the water surface provided there is no unusual curvature in the streamlines or stream tubes, for if a piezometer tube is attached to the side of the channel, the water will rise in it until its surface is level with that of the water in the channel at that point. Water depth y is always measured vertically and the

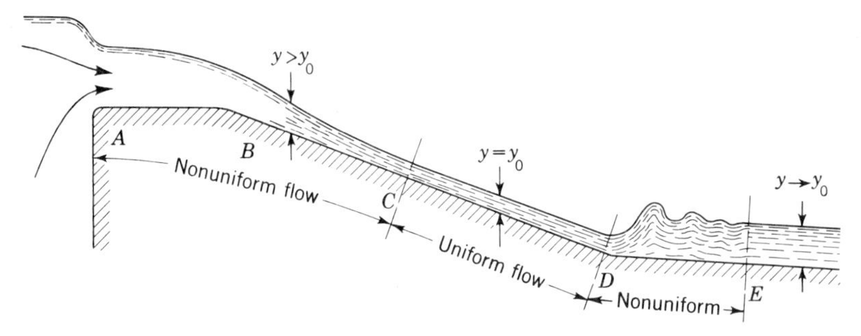

Figure 11.1 Steady flow down a chute or spillway.

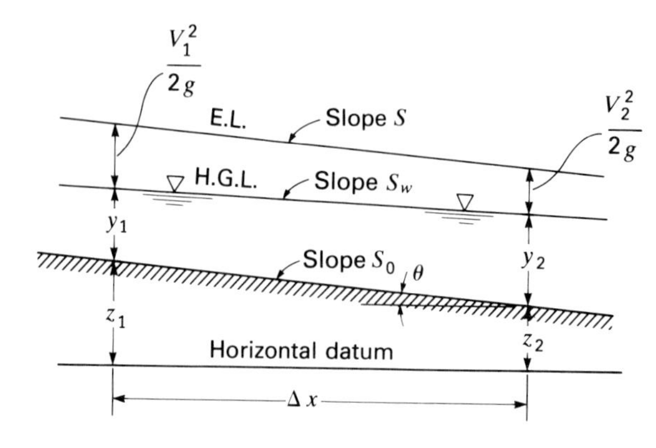

Figure 11.2 Open channel flow—definition sketch (L = distance along the channel bed between sections 1 and 2).

distance between sections is commonly defined by the horizontal distance Δx between them. The various slopes mentioned above are defined as follows:

$$S_0 = \frac{(z_1 - z_2)}{\Delta x} = -\frac{\Delta z}{\Delta x} \tag{11.1}$$

$$S_w = \frac{(z_1 + y_1) - (z_2 + y_2)}{\Delta x} = -\frac{\Delta(z + y)}{\Delta x} \tag{11.2}$$

$$S = \frac{h_L}{L} = \frac{(z_1 + y_1 + V_1^2/2g) - (z_2 + y_2 + V_2^2/2g)}{L} \tag{11.3}$$

where z, y, Δx and L are defined in Fig. 11.2. Note that S, the energy gradient, is defined as the head loss per length of flow path and that it is usually assumed that $\alpha = 1.0$. This assumption is reasonable when the flow depth is less than the channel width. Also, for all practical purposes, the angle θ between the channel bed and the horizontal is small; hence L, the distance along the channel bed between the two sections, is almost equal to Δx, the horizontal distance between the two sections.

11.2 UNIFORM FLOW—THE CHEZY FORMULA

In uniform flow (Fig. 11.3) the cross section through which flow occurs is constant along the channel and so also is the velocity. Thus, $y_1 = y_2$ and $V_1 = V_2$ and the channel bed, water surface, and energy line are parallel to one another. Also, $S_w = S_0 = -\Delta z/\Delta x = \tan\theta$, while $S = h_L/L = \sin\theta$ where θ is the angle the channel bed makes with the horizontal.

In most open channels (rivers, canals, and ditches) the bed slope is small (that is, $S_0 = |\Delta z/\Delta x| < 1/10$ or $\theta < 5.7°$). In such a case $\sin\theta \approx \tan\theta$ and $S_0 = S_w \approx S$. In this chapter it will be assumed that $\theta < 5.7°$ and thus that

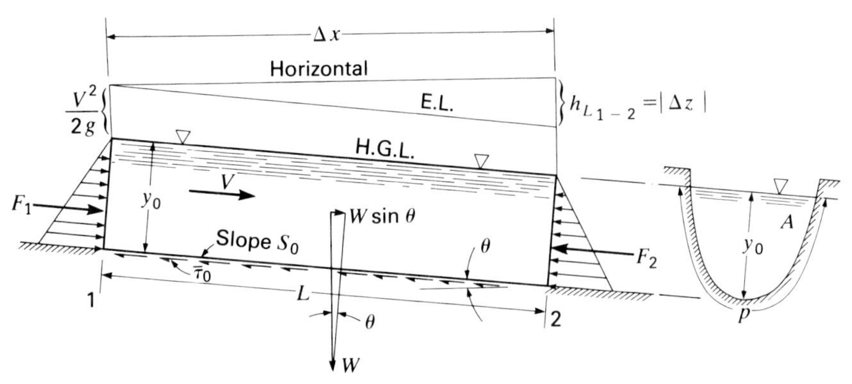

Figure 11.3 Uniform flow in open channel.

$S \approx S_0$. In channels where $\theta > 5.7°$ a distinction must be made in the difference between S and S_0. Moreover, on such slopes slug flow and air entrainment are a common occurrence.

In Sec. 8.4 a general equation for frictional resistance in a pressure conduit was developed. The same reasoning may now be applied to uniform flow with a free surface. Consider the short reach of length L along the channel between stations 1 and 2 in uniform flow with area A of the water section (Fig. 11.3). As the flow is neither accelerating nor decelerating, we may consider the body of water contained in the reach in static equilibrium. Summing forces along the channel, the hydrostatic-pressure forces F_1 and F_2 balance each other, since there is no change in the depth y between the stations. The only force in the direction of motion is the gravity component, and this must be resisted by the average boundary shear stress $\bar{\tau}_0$, acting over the area PL, where P is the wetted perimeter of the section. Thus

$$\gamma AL \sin \theta = \bar{\tau}_0 PL$$

But $\sin \theta = h_L/L = S$. Solving for $\bar{\tau}_0$, we have

$$\bar{\tau}_0 = \gamma \frac{A}{P} S = \gamma R_h S \tag{11.4}$$

where R_h is the hydraulic radius,[1] discussed in Sec. 8.3, and for small slopes (that is, $\theta < 5.7°$) S_0 may be taken as equal to S. Substituting the value of $\bar{\tau}_0$ from Eq. (8.11) and replacing S with S_0,

$$\bar{\tau}_0 = C_f \rho \frac{V^2}{2} = \gamma R_h S_0$$

[1] Strictly speaking A and P should be measured in a plane at right angles to L. However, since depth is measured vertically, values of A and P are calculated as those in the vertical plane. The resulting values of R_h are identical regardless of which way A and P are determined.

This may be solved for V in terms of either the friction coefficient C_f or the conventional friction factor f to give

$$V = \sqrt{\frac{2g}{C_f} R_h S_0} = \sqrt{\frac{8g}{f} R_h S_0} \tag{11.5}$$

In 1775 Chézy proposed that the velocity in an open channel varied as $\sqrt{R_h S_0}$, which led to the formula

$$V = C\sqrt{R_h S_0} \tag{11.6}$$

which is known by his name. It has been widely used both for open channels and for pipes under pressure. Comparing Eqs. (11.6) and (11.5), it is seen that $C = \sqrt{8g/f}$. Despite the simplicity of Eq. (11.6), it has the distinct drawback that C is not a pure number but has the dimensions $L^{1/2}T^{-1}$, requiring that values of C in metric units be converted before being used with English units in the rest of the formula.

As C and f are related, the same considerations that have been presented regarding the determination of a value for f in Chap. 8 apply also to C. For a small open channel with smooth sides, the problem of determining f or C is the same as that in the case of a pipe. But most channels are relatively large compared with pipes, thus giving Reynolds numbers which are higher than those commonly encountered in pipes. Also, open channels are frequently rougher than pipes, especially in the case of natural streams. A study of Fig. 8.11 reveals that, as the Reynolds number and the relative roughness both increase, the value of f becomes practically independent of **R** and depends only on the relative roughness.

11.3 MANNING FORMULA

One of the best as well as one of the most widely used formulas for open-channel flow is that of Robert Manning, who published it in 1890.[1] Manning found from many tests that the value of C varied approximately as $R_h^{1/6}$, and others observed that the proportionality factor was very close to the reciprocal of n, the coefficient of roughness in the classical Kutter formula.[2] This led to the formula which has since spread to all parts of the world. In metric units, the Manning formula is

$$V(\text{m/s}) = \frac{1}{n} R_h^{2/3} S_0^{1/2} \tag{11.7}$$

[1] Robert Manning, Flow of Water in Open Channels and Pipes, *Trans. Inst. Civil Engrs. (Ireland)*, vol. 20, 1890.

[2] The Kutter formula, for many years the most widely used of all open-channel formulas, is now of interest principally for its historical value and as an outstanding example of empirical hydraulics. This formula, which may be found in several handbooks, included terms to make C a function of S, based on some river-flow data which were later proved in error. The reader is referred to E. Ganguillet and W. R. Kutter, "Flows of Waters in Rivers and Other Channels," transl. by R. Hering and J. C. Trautwine, Jr., John Wiley & Sons, Inc., New York, 1869.

The dimensions[1] of n are seen to be $TL^{-1/3}$. To avoid converting the numerical value of n for use with English units, the formula itself is changed so as to leave the value of n unaffected. Thus, in English units, the Manning formula is

$$V(\text{fps}) = \frac{1.49}{n} R_h^{2/3} S_0^{1/2} \tag{11.8}$$

where 1.49 is the cube root of 3.28, the number of feet in a meter. Despite the dimensional difficulties of the Manning formula, which have long plagued those attempting to put all fluid mechanics on a rational dimensionless basis, it continues to be popular because it is simple to use and reasonably accurate. Representative values of n for various surfaces are given in Table 11.1.

Table 11.1 Values of n in Manning's formula
Prepared by R. E. Horton and others

	n	
Nature of surface	Min	Max
Neat cement surface	0.010	0.013
Wood-stave pipe	0.010	0.013
Plank flumes, planed	0.010	0.014
Vitrified sewer pipe	0.010	0.017
Metal flumes, smooth	0.011	0.015
Concrete, precast	0.011	0.013
Cement mortar surfaces	0.011	0.015
Plank flumes, unplaned	0.011	0.015
Common-clay drainage tile	0.011	0.017
Concrete, monolithic	0.012	0.016
Brick with cement mortar	0.012	0.017
Cast iron—new	0.013	0.017
Cement rubble surfaces	0.017	0.030
Riveted steel	0.017	0.020
Corrugated metal pipe	0.021	0.025
Canals and ditches, smooth earth	0.017	0.025
Metal flumes, corrugated	0.022	0.030
Canals:		
Dredged in earth, smooth	0.025	0.033
In rock cuts smooth	0.025	0.035
Rough beds and weeds on sides	0.025	0.040
Rock cuts, jagged and irregular	0.035	0.045
Natural streams:		
Smoothest	0.025	0.033
Roughest	0.045	0.060
Very weedy	0.075	0.150

[1] As it is unreasonable to suppose that the roughness coefficient should contain the dimension T, the Manning equation would be more properly adjusted so as to contain $\sqrt{g}$ within the constant in the numerator, thus yielding the dimension of $L^{1/6}$ for n.

In terms of flow rate Eqs. (11.8) and (11.7) may be expressed as

In English units:
$$Q(\text{cfs}) = \frac{1.49}{n} AR_h^{2/3} S_0^{1/2} \tag{11.9a}$$

In SI units:
$$Q(\text{m}^3/\text{s}) = \frac{1}{n} AR_h^{2/3} S_0^{1/2} \tag{11.9b}$$

It was mentioned in Sec. 8.11 that e is a measure of the absolute roughness of the inside of a pipe. The question naturally arises as to whether e and n may be functionally related to one another. Such a relation has been proposed by Powell (Fig. 11.4) on the basis of experimental data using the Prandtl-Kármán equation for rough pipes as a guide. In terms of the hydraulic radius Powell's relation is:

$$\frac{1}{\sqrt{f}} = 2 \log\left(14.8 \frac{R_h}{e}\right) \tag{11.10}$$

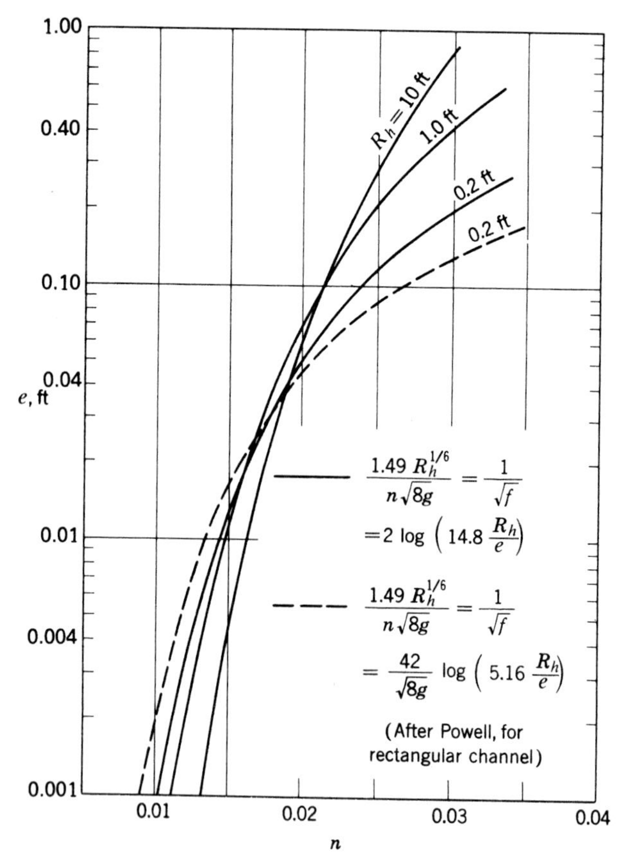

Figure 11.4 Correlation of n with absolute roughness.

Combining Eq. (11.5) with Eqs. (11.8) and (11.7), we get

In English units: $$n = 1.49R_h^{1/6}\sqrt{\frac{f}{8g}}$$

In SI units: $$n = R_h^{1/6}\sqrt{\frac{f}{8g}}$$

Substituting numerical values for g gives

In English units: $$n = 0.093f^{1/2}R_h^{1/6} \tag{11.11a}$$

In SI units: $$n = 0.113f^{1/2}R_h^{1/6} \tag{11.11b}$$

Thus we see that n is related to the friction factor, which depends on the relative roughness and Reynolds number and on the hydraulic radius which is indicative of the size of the channel.

Equating Eq. (11.11*a*) with Eq. (11.10) provides a correlation between e and n, which is plotted as the solid lines in Fig. 11.4 for three representative values of the hydraulic radius. The dashed line is the plot of another correlation proposed by Powell that gave a better fit to experimental data for small values of hydraulic radius.[1] The salient feature of these curves is that a large relative error in e results in only a small error in n. Another observation is that, if $e < 0.02$ ft, the value of n increases with increasing hydraulic radius.[2] For example, a conduit with $e = 0.001$ ft with a hydraulic radius of 1.0 ft will have an n of 0.011 while another conduit with the same surface roughness but with $R_h = 8.75$ ft will have an n of 0.013. On the other hand, for channels that are quite rough ($e > 0.10$ ft, for example) Manning's n gets smaller with increasing hydraulic radius.

11.4 SOLUTION OF UNIFORM FLOW PROBLEMS

Uniform flow problems usually involve the application of Manning's equation [Eqs. (11.9)]. The selection of an appropriate value for the Manning roughness factor n is critical to the accuracy of the results of a problem. When the channel surface is concrete or some other structural material, it is possible to select a reasonably accurate value for n, but for the case of a natural channel one must rely on judgment and experience, and in many instances the selected value may be quite inaccurate.

There are a number of different types of problems that are encountered when using Manning's equation. For example, to find the normal depth of flow for a particular flow rate in a given channel, a trial-and-error solution is required. On

[1] R. W. Powell, Resistance to Flow in Rough Channels, *Trans. Am. Geophys. Union*, vol. 31, no. 4., pp. 575–582, 1950.

[2] J. B. Franzini and P. S. Chisholm, Current Practice in Hydraulic Design of Conduits, *Water and Sewage Works*, vol. 110, pp. 342–345, Oct., 1963.

the other hand, the expected flow in a particular channel under given conditions can be solved for directly. Various types of sliderules, nomographs, tables,[1] and computer programs,[2] are available to serve as an aid to the solution of open-channel problems.

In applying Manning's equation to channel shapes such as Fig. 11.5, which simulates a river with overbank flow conditions, the usual procedure is to break the section into several parts, as indicated in the figure. It is assumed that there is no resistance along the dashed vertical line. Actually the flow in area A_2 tends to speed up the flow in area A_1, while the flow in A_1 tends to slow down the flow in area A_2. These two effects come very close to balancing out one another. If A/P for the total cross section had been computed by the usual method, that is, $R_h = (A_1 + A_2 + A_3)/(P_1 + P_2 + P_3)$, it would imply that the effect of boundary resistance is uniformly distributed over the flow cross section, which, of course, is not the case.

Another advantage of breaking the total section into parts is that possible variations in Manning's n can be taken into consideration. Thus, for the change shown in Fig. 11.5, in English units

$$Q = \frac{1.49}{n_1} A_1 R_{h_1}^{2/3} S_0^{1/2} + \frac{1.49}{n_2} A_2 R_{h_2}^{2/3} S_0^{1/2} + \cdots \tag{11.12}$$

where $R_{h_1} = A_1/P_1$, $R_{h_2} = A_2/P_2$, etc. The A and P are defined in Fig. 11.5. An equation of similar form can be written for SI units in which case the 1.49 becomes unity.

Illustrative Example 11.1 Find the depth for uniform flow in this channel when the flow rate is 225 cfs if $S_0 = 0.0006$ and n is assumed to be 0.016. Compute the corresponding value of e.

$$A = (10 + 2y)y$$

and

$$R_h = \frac{A}{P} = \frac{(10 + 2y)y}{10 + 2\sqrt{5}y}$$

Thus

$$Q = 225 = \frac{1.49}{0.016}(10 + 2y)y\left[\frac{(10 + 2y)y}{10 + 2\sqrt{5}y}\right]^{2/3}(0.0006)^{1/2}$$

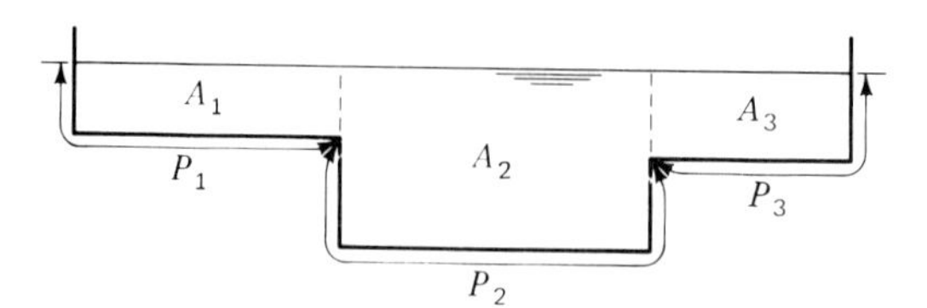

Figure 11.5

[1] E. F. Brater and H. W. King, "Handbook of Applied Hydraulics," 6th ed., McGraw-Hill Book Company, New York, 1976.

[2] T. E. Croley, "Hydrologic and Hydraulic Computations on Small Programmable Calculators," Iowa Institute of Hydraulic Research, University of Iowa, Iowa City, 1977.

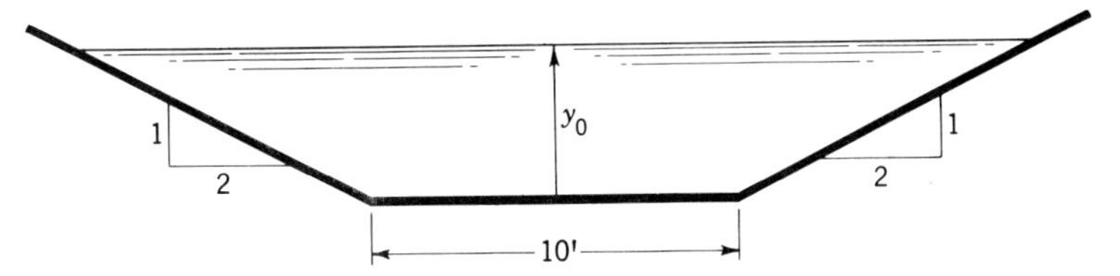

Illustrative Example 11.1

By trial, $y_0 = 3.4$ ft, uniform flow depth,

$$A = [10 + 2(3.4)]3.4 = 57.1 \text{ ft}^2$$

$$P = 10 + 2\sqrt{5}(3.4) = 25.2 \text{ ft}$$

$$R_h = \frac{A}{P} = \frac{57.1}{25.2} = 2.27 \text{ ft}$$

Rearranging Eq. (11.11*a*),

$$f = \frac{116n^2}{R_h^{1/3}} = 0.0226$$

Finally, using Eq. (11.10),

$$e = 0.0159 \text{ ft}$$

11.5 VELOCITY DISTRIBUTION IN OPEN CHANNELS

Vanoni[1] has demonstrated that the Prandtl universal logarithmic velocity-distribution law for pipes [Eq. (8.31)] also applies to a two-dimensional open channel, i.e., one of uniform depth which is infinitely wide. This equation may be written

$$\frac{u - u_{\max}}{\sqrt{gy_0 S}} = \frac{2.3}{K} \log \frac{y}{y_0}$$

where y_0 = depth of water in channel
u = velocity at a distance y from channel bed
K = von Kármán constant, having a value of about 0.40 for clear water[2]

This expression can be integrated over the depth to yield the more useful relation

$$u = V + \frac{1}{K}\sqrt{gy_0 S}\left(1 + 2.3 \log \frac{y}{y_0}\right) \tag{11.13}$$

which expresses the distribution law in terms of the mean velocity V. This equation is plotted in Fig. 11.6, together with velocity measurements that were made

[1] V. A. Vanoni, Velocity Distribution in Open Channels, *Civil Eng.*, vol. 11, pp. 356–357, 1941.
[2] For sediment-laden water the value of K may be as low as 0.2.

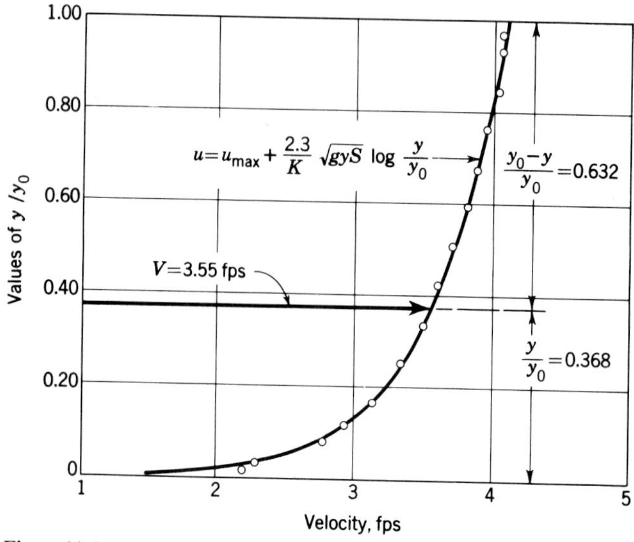

Figure 11.6 Velocity profile at center of a flume 2.77 ft wide for a flow 0.59 ft deep. (After Vanoni.)

on the center line of a rectangular flume 2.77 ft wide with a water depth of 0.59 ft. The filament whose velocity u is equal to V is seen to lie at a distance of $0.632y_0$ beneath the surface.

Velocity measurements made in a trapezoidal canal, reported by O'Brien,[1] gave the isovels (contours of equal velocity) shown in Fig. 11.7, with the accompanying values of the correction factors for kinetic energy and momentum. The point of maximum velocity is seen to lie beneath the surface, and the correction factors for kinetic energy and momentum are greater than in the corresponding case of pipe flow [Eqs. (8.35)]. Despite the added importance of these factors, however, the treatment in this chapter follows the earlier procedure of assuming

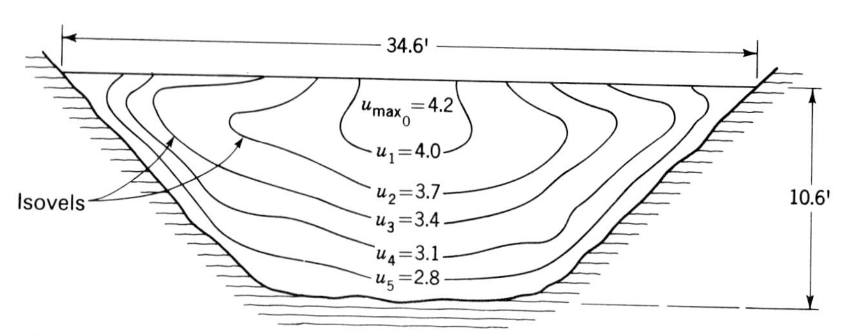

Figure 11.7 Velocity distribution in trapezoidal canal. $V = 3.32$ fps, $A = 230.5$ ft^2, $S_0 = 0.000057$, $\alpha = 1.105$, $\beta = 1.048$. (After O'Brien.)

[1] M. P. O'Brien and J. W. Johnson, Velocity Head Correction for Hydraulic Flow, *Eng. News-Record*, vol. 113, pp. 214–216, 1934.

the values of α and β to be unity, unless stated otherwise. Any thoroughgoing analysis would of course have to take account of their true values.

11.6 MOST EFFICIENT CROSS SECTION[1]

Any of the open-channel formulas given above show that, for a given slope and roughness, the velocity increases with the hydraulic radius. Therefore, for a given area of water cross section, the rate of discharge will be a maximum when R_h is a maximum, which is to say, when the wetted perimeter P is a minimum. Such a section is called the *most efficient cross section* for the given area. Or for a given rate of discharge, the cross-sectional area will be a minimum when the design is such as to make R_h a maximum (and thus P a minimum). This section would be the most efficient cross section for the given rate of discharge.

Of all geometric figures the circle has the least perimeter for a given area. The hydraulic radius of a semicircle is the same as that of a circle. Hence a semicircular open channel will discharge more water than one of any other shape, assuming that the area, slope, and surface roughness are the same. Semicircular open channels are often built of pressed steel and other forms of metal, but for other types of construction such a shape is impractical. For wooden flumes the rectangular shape is usually employed. Canals excavated in earth must have a trapezoidal cross section, with side slopes less than the *angle of response* of the saturated bank material. Thus there are other factors besides hydraulic efficiency which determine the best cross section.

The shape of the most efficient trapezoidal cross section can be determined by expressing the wetted perimeter P as a function of the section area A, the depth y, and the angle of side slope. By differentiating P with respect to the depth y, while holding A and the angle of side slope constant, it can be shown that the hydraulic radius of the most efficient cross section is one for which $R_h = y/2$. This corresponds to a rectangle whose depth is one-half the width. It also indicates that the most efficient trapezoid is the half-hexagon and the most efficient triangle is the one that has a total vertex angle of 90°.

11.7 CIRCULAR SECTIONS NOT FLOWING FULL

In circular pipes, flow frequently occurs at partial depth. The maximum rate of discharge in such a section occurs at slightly less than full depth, as may be shown by reference to Fig. 11.8. Thus

$$A = \frac{D^2}{4}(\theta - \sin\theta\cos\theta) = \frac{D^2}{4}(\theta - \tfrac{1}{2}\sin 2\theta)$$

$$P = D\theta$$

[1] The discussion here relates to *hydraulic* efficiency. The channel section with the greatest hydraulic efficiency is rarely the one of greatest economic efficiency because of cost of excavation, right-of-way, etc.

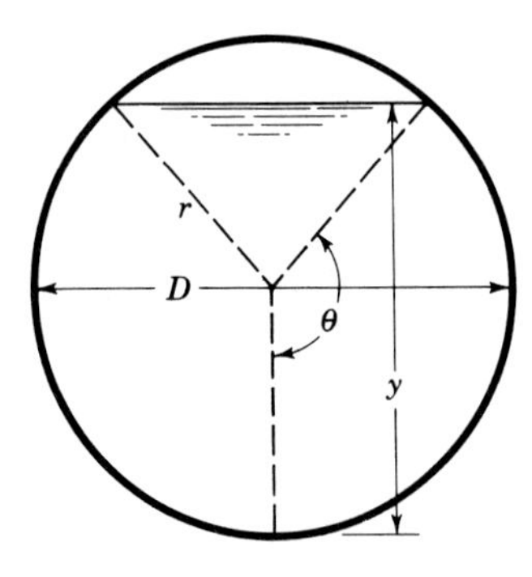

Figure 11.8 Circular section not full.

where θ is expressed in radians. This gives

$$R_h = \frac{A}{P} = \frac{D}{4}\left(1 - \frac{\sin\theta\cos\theta}{\theta}\right) = \frac{D}{4}\left(1 - \frac{\sin 2\theta}{2\theta}\right)$$

For the maximum rate of discharge, the Manning formula indicates that $AR_h^{2/3}$ must be a maximum.

Substituting the preceding expressions for A and R_h into $AR_h^{2/3}$ and differentiating with respect to θ, setting equal to zero and solving for θ gives $\theta = 151.2°$, which corresponds to $y = 0.938D$ for the condition of maximum discharge. By differentiating $R_h^{2/3}$ the maximum velocity is found to occur at $0.81D$.[1] Despite the foregoing analysis, circular sections are usually designed to carry the design capacity when flowing full, since the conditions producing maximum flow frequently include sufficient backwater to place the conduit under slight pressure.

The simplest way to handle the problem of a partially full circular section is to compute the velocity or flow rate for the pipe-full condition and adjust to partly full conditions by using a chart such as Fig. 11.9. On this chart the effect of the variation of Manning's n with depth is taken into consideration.

The rectangle, trapezoid, and circle are the simplest geometric shapes from the standpoint of hydraulics, but other forms of cross section are often used, either because they have certain advantages in construction or because they are desirable from other standpoints. Thus oval- or egg-shaped sections are common for sewers and similar channels where there may be large fluctuations in the rate of discharge. It is desirable that the velocity, when a small quantity is flowing, be kept high enough to prevent the deposit of sediment, and when the conduit is full, the velocity should not be so high as to cause excessive wear of the pipe wall. For example, in sanitary sewers it is desirable to maintain a velocity of at least 2 ft/s to prevent deposition of sediment, while in concrete channels conveying storm water, velocities in excess of 10 or 12 ft/s may result in excessive abrasion of the channel sides and bottom.

[1] The above derivation is based upon a roughness coefficient which remains constant as the depth changes. Actually, the value of n has been shown to increase by as much as 28 per cent from the full to about one-quarter full depth, where it appears to be a maximum. This effect causes the actual maximum discharge and velocity to occur at water depths of about 0.97 and 0.94 full depth, respectively as indicated in Fig. 11.9. See Design and Construction of Sanitary and Storm Sewers, *ASCE Manual Eng. Practice*, vol. 37, pp. 86–88, 1976.

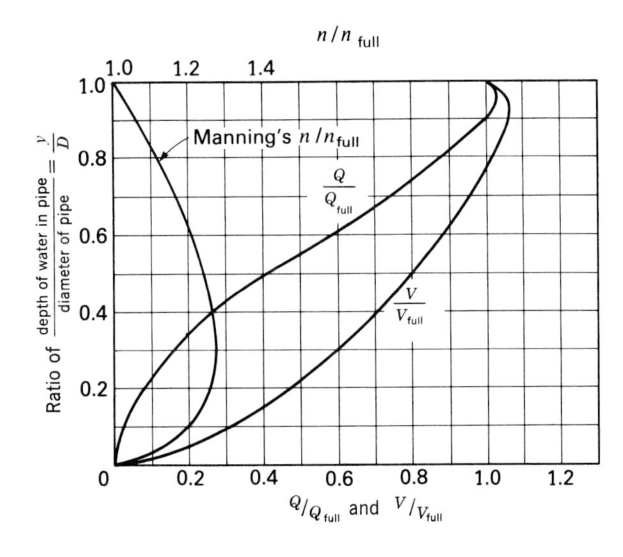

Figure 11.9 Hydraulic characteristics of circular pipe flowing partly full (n variable with depth).

11.8 LAMINAR FLOW IN OPEN CHANNELS

Laminar flow in open channels is sometimes encountered in industrial processes where a very viscous liquid is flowing in a trough or similar conveyance structure. More commonly, however, laminar flow occurs as *sheet flow*, a thin sheet of flowing liquid, such as that in drainage from sidewalks, streets, and airport runways. Sheet flow is essentially two-dimensional and can be analyzed in that fashion. Consider two-dimensional uniform laminar flow at depth y_0 as shown in Fig. 11.10. The forces acting on a unit width of the liquid represented by the shaded area include the hydrostatic forces which cancel out, the gravity force[1] component $\gamma(y_0 - y)LS$ and the shear force $\tau \times L$ exerted along the lower boundary of the shaded liquid by the liquid below it. Since the flow is uniform there is no acceleration and thus these two forces must balance one another. Hence

$$\gamma(y_0 - y)LS = \tau \times L$$

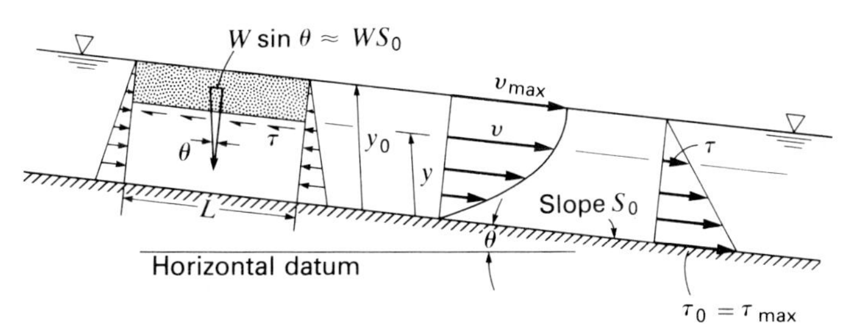

Figure 11.10 Laminar flow in open channel of infinite width and uniform depth showing the velocity profile and the shear stress distribution.

[1] Depth y is always measured vertically. Thus $\gamma(y_0 - y)L$ is not precisely the weight of the shaded volume of liquid. However, for the small slopes usually encountered it is a good approximation.

where τ is the shear stress in the fluid at distance y above the channel bed. From this expression it can be seen that the shear stress must vary linearly from zero at the liquid surface to a maximum value at the channel bed.

Since the flow is laminar we can replace τ with $\mu\, dv/dy$ (Sec. 1.11). Making this substitution we get

$$\gamma(y_0 - y)S = \mu \frac{dv}{dy}$$

Separating variables and integrating, noting that $v = 0$ when $y = 0$, gives

$$dv = \frac{\gamma}{\mu}(y_0 - y)S\, dy$$

and

$$v = \frac{\gamma}{\mu}\left(y_0 y - \frac{y^2}{2}\right)S = \frac{g}{\nu}\left(y_0 y - \frac{y^2}{2}\right)S \tag{11.14}$$

This is the equation of a parabola; thus the velocity profile is parabolic as it was for laminar flow in a pipe (Sec. 8.6).

We can now integrate Eq. (11.14) over the depth from $y = 0$ to $y = y_0$ to obtain an expression for q, the flow rate per unit width

$$dq = \int_0^{y_0} v\, dy$$

$$q = \frac{g}{\nu}(y_0^3/3)S \tag{11.15}$$

From Eqs. (11.14) and (11.15) it can be shown that the average velocity V for this case is equal to $\frac{2}{3}v_{max}$. In contrast, for laminar flow in a pipe flowing full it was shown (Sec. 8.6) that $V = \frac{1}{2}v_{max}$. The total flow Q through any width B will be $q \times B$.

Equation (11.15) shows that if the flow is laminar the flow rate is independent of the roughness. However, in situations where there are significant irregularities in the surface over which the liquid flows, the flow may be disturbed such that it is not everywhere laminar. Consequently Eqs. (11.14) and (11.15) must be applied with caution.

11.9 SPECIFIC ENERGY AND ALTERNATE DEPTHS OF FLOW—WIDE RECTANGULAR CHANNELS

The specific energy E at a particular section is defined as the energy head referred to the channel bed as datum. Thus

$$E = y + \frac{V^2}{2g} \tag{11.16}$$

If the channel is of uniform depth and relatively wide, the flow near the center of the channel will be unaffected by the side boundaries of the channel and the flow q per unit width b can be expressed as $q = Q/b$. The average velocity $V = Q/A = qb/by = q/y$ and Eq. (11.16) can be expressed as

$$E = y + \frac{1}{2g}\left(\frac{q^2}{y^2}\right) \tag{11.17}$$

Let us consider how E will vary with y if q remains constant. Physically, this situation would occur if the slope of a wide rectangular channel could be changed with the flow rate remaining constant. Manning's equation [Eq. (11.9)] indicates that on a steep slope, with a given flow rate, the normal depth of flow y_0 will be relatively small in contrast to a larger depth on a flatter slope. That this is so can be seen by writing the Manning equation in terms of the flow q per unit width noting that $A = by$ and $R_h = A/P = by/(2b + y) \approx y$ if $b \gg y$. Thus, for a wide rectangular channel with uniform flow ($y = y_0$), in English units

$$q = \frac{Q}{b} = \frac{1.49}{nb} AR_h^{2/3}S_0^{1/2} = \frac{1.49}{n}\frac{by_0}{b} y_0^{2/3}S_0^{1/2} = \frac{1.49}{n} y_0^{5/3}S_0^{1/2} \tag{11.18}$$

For the case of constant q, Eq. (11.17) can be restated as follows:

$$(E - y)y^2 = \frac{q^2}{2g} = \text{constant} \tag{11.19}$$

A plot of E vs. y is a parabola with asymptotes $(E - y) = 0$ (that is, $E = y$) and $y = 0$. Such a curve, shown in Figure 11.11, is known as the *specific energy*

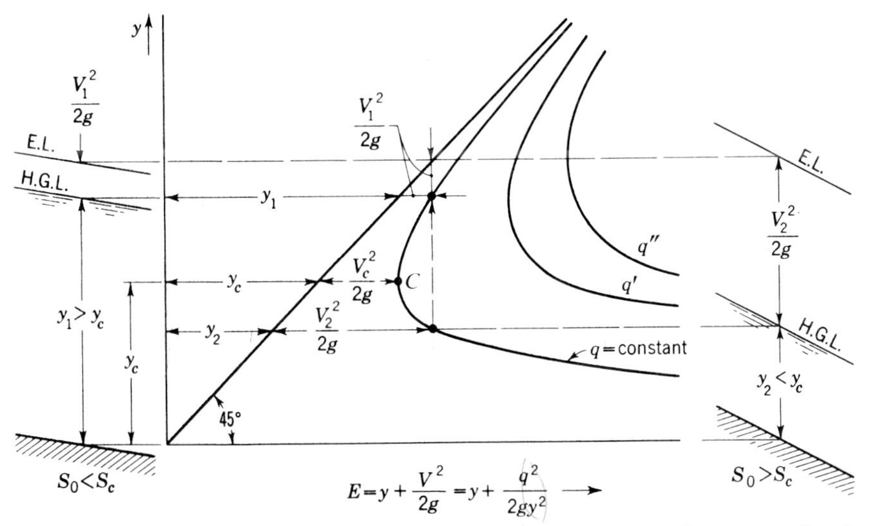

Figure 11.11 Specific-energy diagram for three constant rates of discharge in a rectangular channel. (Bed slopes are greatly exaggerated.)

diagram. Actually each different value of q will give a different curve, as shown in Fig. 11.11. For a particular q there are two possible values of y for a given value of E. These are known as *alternate depths.* Equation (11.19) is a cubic equation with three roots, the third root being negative has no physical meaning. The two alternate depths represent two totally different flow regimes—slow and deep on the upper portion of the curve and fast and shallow on the lower portion of the curve. Point C represents the dividing point between the two regimes of flow. At C, for a given q, the value of E *is a minimum* and the flow at this point is referred to as *critical flow.* The depth of flow at that point is the *critical depth* y_c and the velocity is the *critical velocity* V_c. A relation for critical depth in a wide rectangular channel can be found by differentiating E of Eq. (11.17) with respect to y to find the value of y for which E is a minimum. Thus

$$\frac{dE}{dy} = 1 - \frac{q^2}{gy_c^3} = 0$$

from which

$$q^2 = gy_c^3$$

Substituting $q = Vy = V_c y_c$ gives

$$V_c^2 = gy_c \qquad \text{and} \qquad V_c = \sqrt{gy_c} \tag{11.20}$$

where the subscript c indicates critical flow conditions (minimum specific energy for a given q).

Equation (11.20), applicable to wide open channels, may also be expressed as

$$y_c = \frac{V_c^2}{g} = \left(\frac{q^2}{g}\right)^{1/3} \tag{11.21}$$

From Eq. (11.20)

$$\frac{V_c^2}{2g} = \frac{1}{2} y_c \tag{11.22}$$

Hence

$$E_c = E_{\text{min}} = y_c + \frac{V_c^2}{2g} = \frac{3}{2} y_c \tag{11.23}$$

and

$$y_c = \frac{2}{3} E_c = \frac{2}{3} E_{\text{min}} \tag{11.24}$$

Another approach to alternate depths is to solve Eq. (11.19) for q and note the variation in q for changing values of y for a constant value of E. Physically this situation is encountered when water flows from a large reservoir of constant surface elevation over a high, frictionless, broad-crested weir provided with a movable sluice gate near its downstream end (Fig. 11.12*a*). As the gate is opened the flow rate increases until the opening becomes just large enough for critical depth to occur. With further opening of the gate there is no increase in flow rate. As long as the water impinges on the gate (Fig. 11.12*a*), the flow is subcritical (a') upstream of the gate and supercritical (a'') downstream of the gate.

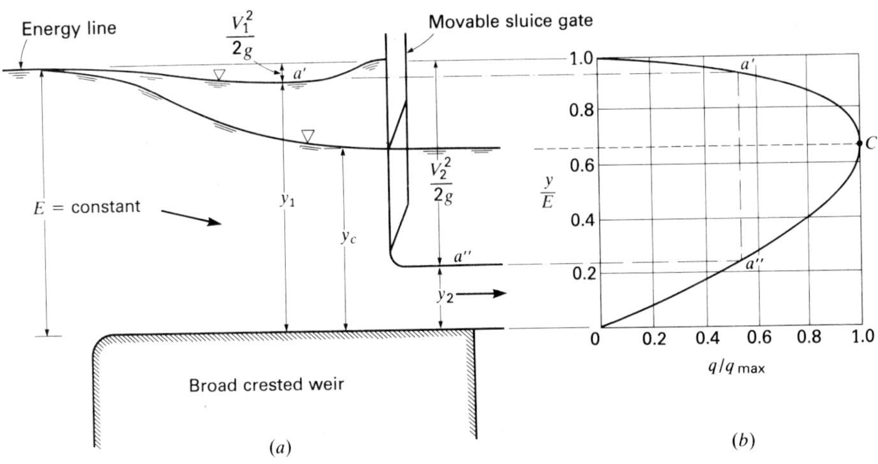

Figure 11.12 Variation of depth y and discharge q per unit width for constant specific energy E. (a) Side-view of flow over broad-crested weir. (b) Dimensionless discharge curve.

Rewriting Eq. (11.19) gives

$$q = y\sqrt{2g(E - y)} \tag{11.25}$$

This is the equation of the curve which is shown in dimensionless form in Fig. 11.12b. It is seen that maximum discharge for a given specific energy occurs when the depth is between $0.6E$ and $0.7E$. This may be established more exactly by differentiating Eq. (11.25) with respect to y and equating to zero. Thus

$$\frac{dq}{dy} = \sqrt{2g}\left(\sqrt{E - y} - \frac{1}{2}\frac{y}{\sqrt{E - y}}\right) = 0$$

from which

$$y_c = \tfrac{2}{3}E \tag{11.26}$$

where y_c is called the *critical depth* for the given specific energy. This equation is identical to Eq. (11.24). Thus, there is a maximum value of q for a given E as indicated by point C of Fig. 11.12b. This curve is often referred to as the discharge curve. The flow depicted by the upper portion of the curve has characteristics similar to those of the upper portion of that in Fig. 11.11. Similarly, the flow depicted by the lower portion of the curve has characteristics similar to the lower portion of that of Fig. 11.11. The significance of these two regimes of flow is discussed in Sec. 11.11. Point C of Fig. 11.12b represents critical flow conditions.

The maximum rate of discharge in a wide rectangular channel for a given value of E may be determined by substituting E from Eq. (11.26) into Eq. (11.25).

$$q_{\max} = y_c\sqrt{2g(E_c - y_c)} = \sqrt{gy_c^3} \tag{11.27}$$

In Fig. (11.11) it is seen that when the flow is near critical, a small change in specific energy results in a large change in depth. Hence flow at or near critical depth is unstable and there will be an undulating water surface. Because of this it is undesirable to design channels with slopes near the critical.

We summarize the foregoing discussion as axioms of open-channel flow related to conditions at a given section in a *wide rectangular channel*:

1. A flow condition, i.e., a certain rate of discharge flowing at a certain depth, is completely specified by any two of the variables y, q, V, and E, except the combination q and E, which yields in general two alternate stages of flow.
2. For any value of E there exists a critical depth, given by Eq. (11.24), for which the flow is a maximum.
3. For any value of q there exists a critical depth, given by Eq. (11.21), for which the specific energy is a minimum.
4. When flow occurs at critical depth, both Eq. (11.20) and (11.24) are satisfied and the velocity head is one-half the depth.
5. For any flow condition other than critical, there exists an alternate stage at which the same rate of discharge is carried by the same specific energy. The alternate depth may be found from either the specific-energy diagram (Fig. 11.11) or the discharge curve (Fig. 11.12), by extending a vertical line to the alternate limb of the curve. Analytically, the alternate depth is found by solving Eq. (11.17).

11.10 CELERITY OF GRAVITY WAVES AND CRITICAL VELOCITY

Consider a small wave of height Δy traveling at a velocity (or *celerity*) c to the left across the surface of stationary water whose depth is y (Fig. 11.13*a*). Let us now replace this with an equivalent steady flow to the right having a velocity $V_1 = -c$ (Fig. 11.13*b*). In this situation the wave is standing still with respect to the observer. Applying the principle of continuity to Fig. 11.13*b* we have $V_1 y = V_2(y + \Delta y)$ where V_2 is the average velocity of flow past section 2 in Fig. 11.13*b*.

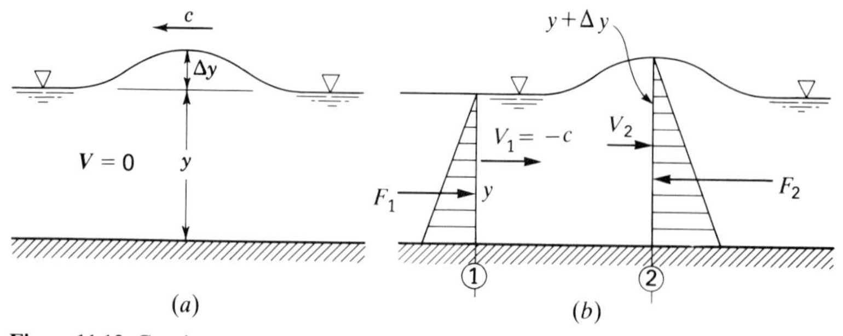

Figure 11.13 Gravity wave of small amplitude. (*a*) As seen by a stationary observer. (*b*) As seen by an observer moving with the wave.

Applying the impulse-momentum principle, $F = \rho Q(\Delta V)$, to the control volume between sections 1 and 2 for a unit width of channel gives

$$\gamma \frac{y^2}{2} - \gamma \frac{(y + \Delta y)^2}{2} = \frac{\gamma}{g}(yV_1)(V_2 - V_1)$$

Substituting the expression for V_2 from continuity and letting $c = -V_1$ in the above gives

$$c = \sqrt{g(y + \Delta y)\left(\frac{y + \frac{1}{2}\Delta y}{y}\right)} \approx \sqrt{g(y + \Delta y)} \approx \sqrt{gy} \tag{11.28}$$

The latter expressions apply if the surface disturbance is relatively small, that is, $\Delta y \ll y$. It is seen that the celerity of a wave will increase as the depth of the water increases. But this is the celerity relative to the water. If the water is flowing, the absolute speed of travel of the wave will be the resultant of the two velocities. Thus

$$c_{\text{abs}} = c \pm V \tag{11.29}$$

When water is flowing at critical depth y_c, the flow velocity will be V_c $(=\sqrt{gy_c})$ and the flow velocity and wave celerity will be almost equal in magnitude provided $\Delta y/y$ is small. This means that when the surface is disturbed from any cause, the wave so produced *cannot travel upstream.* This also applies to depths less than y_c for then V will be greater than V_c. When the depth is greater than y_c, the velocity of flow is less than V_c and the wave celerity is greater than the flow velocity. Under these conditions any surface disturbance will be able to travel upstream against the flow. Hence the entire stream picture is dependent on whether the stream velocity is smaller or greater than the critical velocity. The situation is analogous to that of the sonic wave as described in Sec. 9.10. A *standing wave* will exist upstream of a bridge pier, for example, if V is greater than V_c. The wave will be at such a direction that $\sin \beta = c/V = \sqrt{gy}/V$, where β is the angle between the direction of flow and the wave front, as shown in Fig. 9.8.

11.11 SUBCRITICAL AND SUPERCRITICAL FLOW

In Sec. 11.9 reference was made to the upper and lower portions of the specific-energy diagram and the discharge curve. The upper portion of those curves represents *subcritical* (also referred to as tranquil or upper-stage) flow while the lower portion of the curves represents *supercritical* (also referred to as rapid or lower-stage) flow. At the point between the upper and lower portions of the curves the flow is critical (Sec. 11.9) and the slope required to give uniform flow at critical depth is known as the critical slope S_c. An expression for critical slope in a wide rectangular channel results when Eq. (11.21) for critical flow is set equal to Eq. (11.18) as follows:

$$S_c = \frac{gn^2}{(1.49)^2 y_c^{1/3}} \tag{11.30}$$

If $S_0 > S_c$ the slope is known as a *steep slope*. Normal depth for uniform flow on such a slope will be less than critical depth and hence the flow will be supercritical. In contrast, if $S < S_c$ the normal depth will be greater than critical and the flow is subcritical. Such a slope is referred to as a *mild slope*. Reference to Eq. (11.30) indicates that the hydraulic steepness of a channel slope is determined by more than its elevation gradient. A steep slope for a channel with a smooth lining could be a mild slope for the same flow with a rough lining. Even for a given channel with a given boundary roughness, the slope may be mild for a low rate of discharge and steep for a higher one.

It may be recalled that the Froude number (Sec. 7.4) is defined as $V/\sqrt{gL}$. If the depth of flow is used to represent the significant length parameter in the Froude number (that is, ($\mathbf{F} = V/\sqrt{gy}$), we find by comparing this with Eq. (11.20) that the flow is critical if $\mathbf{F} = 1.0$, the flow is subcritical if $\mathbf{F} < 1.0$, and the flow is supercritical if $\mathbf{F} > 1.0$.

Previously it was mentioned that with supercritical flow a surface disturbance will not propagate upstream but will form a standing wave because $V > c$. Also, it can be shown from several of the previously stated equations that $V^2/2g > y/2$ for supercritical flow and $V^2/2g < y/2$ for subcritical flow. A brief résumé of some of the characteristics of subcritical, critical, and supercritical flow is given in Table 11.2. Another important characteristic of supercritical flow, which is not true of subcritical flow, is that it can be followed by a hydraulic jump (Sec. 11.18).

Table 11.2 Characteristics of subcritical, critical, and supercritical flow in rectangular channel

	Subcritical	Critical	Supercritical
Depth of flow	$y > y_c$	$y = y_c = \left(\frac{q^2}{g}\right)^{1/3}$	$y < y_c$
Velocity of flow	$V < V_c$	$V = V_c = \sqrt{gy}$	$V > V_c$
Slope for uniform flow	Mild slope $S_0 < S_c$	Critical slope $S = S_c$ [Eq. (11.30)]	Steep slope $S_0 > S_c$
Froude number	$\mathbf{F} < 1.0$	$\mathbf{F} = V/\sqrt{gy} = 1.0$	$\mathbf{F} > 1.0$
Disturbance	Will propagate upstream	Will hold fast	Will form standing wave with $\sin\beta = c/V$
Other feature	$\frac{V^2}{2g} < \frac{1}{2}y$	$\frac{V^2}{2g} = \frac{1}{2}y$	$\frac{V^2}{2g} > \frac{1}{2}y$

11.12 CRITICAL DEPTH IN NONRECTANGULAR CHANNELS

For nonrectangular channels the Froude number is defined by $\mathbf{F} = V/\sqrt{gy_h}$ where y_h is the *hydraulic depth* defined by $y_h = A/B$ where A is the cross-sectional flow area and B is the width of the flow area at the water surface.

Let us now consider an irregularly shaped flow section (Fig. 11.14) of area A carrying a flow Q. Thus Eq. (11.17) becomes

$$E = y + \frac{Q^2}{2gA^2} \tag{11.31}$$

Differentiating with respect to y and setting to zero to find y_c, that is, the value of y for which E is a minimum,

$$\frac{dE}{dy} = 1 - \frac{Q^2}{2g}\left(\frac{2}{A^3}\frac{dA}{dy}\right) = 0$$

As A may or may not be a reasonable function of y, it is helpful to observe that $dA = B\,dy$, and thus $dA/dy = B$, the *width of the water surface.* Substituting this in the above expression results in

$$\frac{Q^2}{g} = \left(\frac{A^3}{B}\right)_{y=y_c} \tag{11.32}$$

as the equation which must be satisfied for critical flow. For a given cross section the right-hand side is a function of y only. A trial-and-error solution is generally required to find the value y_c of y which satisfies Eq. (11.32)

We may next solve for V_c, the critical velocity in the irregular channel, by observing that $Q = A_c V_c$. Substituting this in Eq. (11.32) yields

$$\frac{V_c^2}{g} = \frac{A_c}{B_c} \qquad \text{or} \qquad V_c = \sqrt{\frac{gA_c}{B_c}} = \sqrt{g(y_h)_c} \tag{11.33}$$

If the channel is rectangular, $A_c = B_c y_c$, and the above is seen to reduce to Eq. (11.20) and Eq. (11.32) reduces to Eq. (11.21).

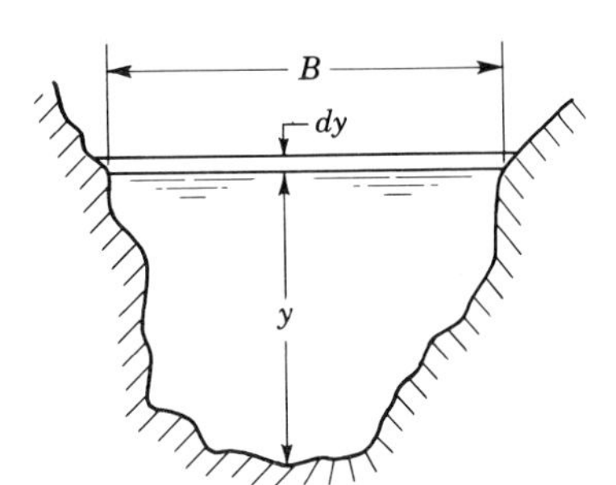

Figure 11.14

It has already been pointed out that the cross section most commonly encountered in open-channel hydraulics is not rectangular but trapezoidal. As repeated trial-and-error solutions of Eq. (11.32) become very tedious, the practicing hydraulic engineer avails himself of numerous tables, curves, and computer programs which have been prepared for finding the critical depth in trapezoidal channels of any bottom width and side slopes.[1]

Illustrative Example 11.2 In the accompanying figure, water flows uniformly at a steady rate of 14.0 cfs in a very long triangular flume which has side slopes 1:1. The bottom of the flume is on a slope of 0.006, and $n = 0.012$. (*a*) Is the flow subcritical or supercritical? (*b*) Find the relation between $V_c^2/2g$ and y_c for this channel.

$$A = \tfrac{1}{2}(2y)y = y^2$$

$$P = 2\sqrt{2}y = 2.83y$$

$$R_h = \frac{A}{P} = 0.354y$$

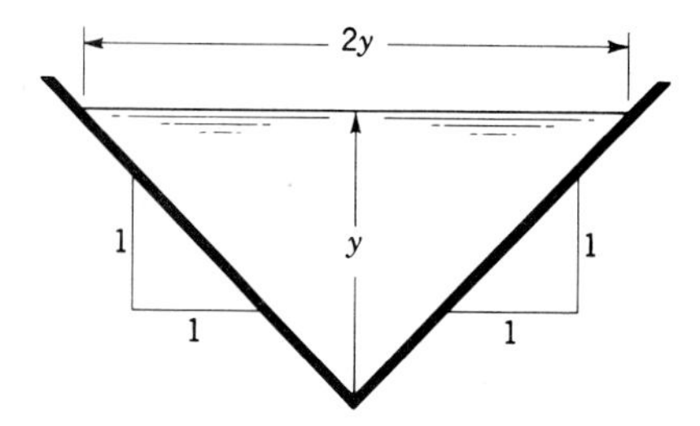

Illustrative Example 11.2

From Eq. (11.9*a*)

$$14 = \frac{1.49}{0.012} y_0^2 (0.354 y_0)^{2/3} (0.006)^{1/2}$$

from which $y_0 = 1.495$ ft, uniform flow depth,

$$\frac{Q^2}{g} = \frac{A^3}{B}$$

$$\frac{(14)^2}{32.2} = \frac{(y_c^2)^3}{2y_c}$$

and

$$y_c = 1.65 \text{ ft}$$

[1] S. Kolupaila, Universal Diagram Gives Critical Depth in Trapezoidal Channel, *Civil Eng.*, vol. 20, p. 785, 1950; discussion of this, with simplified single curve, by C. G. Edson in *ibid.*, vol. 21, p. 159, 1951. See also tables 8.4 to 8.8 in E. F. Brater and H. W. King, "Handbook of Applied Hydraulics," 6th ed., McGraw-Hill Book Company, New York, 1976.

Since y_c is greater than uniform flow depth, the flow is supercritical. If the data in this problem had been given in SI units rather than in English units, the procedure for solution would have been the same except that Eq. (11.9*b*) would have been used instead of Eq. (11.9*a*).

(*b*) The relation between $V^2/2g$ and y for critical flow in a flow section of any shape may be found from Eq. (11.33). For this triangular section with a vertex of 90°,

$$\frac{V_c^2}{g} = \frac{A_c}{B_c} = \frac{y^2}{2y} = \frac{y}{2}$$

and

$$\frac{V_c^2}{2g} = \frac{y_c}{4}$$

Consequently we see that the relation between $V^2/2g$ and y for critical flow conditions depends on the geometry of the flow section. If the vertex angle of the triangle had been different, the relation would have been different.

11.13 OCCURRENCE OF CRITICAL DEPTH

When flow changes from subcritical to supercritical or vice versa, the depth must pass through critical depth. In the former, the phenomenon gives rise to what is known as a *control section.* In the latter a hydraulic jump (Sec. 11.19) usually occurs. In Fig. 11.15 is depicted a situation where the flow changes from subcritical to supercritical. Upstream of the break in slope there is a *mild* slope, the flow is subcritical, and $y_{0_1} > y_c$. Downstream of the break there is a *steep* slope, the flow is supercritical, and $y_{0_2} < y_c$. At the break in slope the depth passes through critical depth. This point in the stream is referred to as a control section since the depth at the break controls the depth upstream. A similar situation occurs when water from a reservoir enters a canal in which the uniform depth is less than critical. In such an instance (Fig. 11.16), the depth passes through critical depth in the vicinity of the entrance to the canal. Once again, this section is known as a control section. By measuring the depth at a control section, one can compute a reasonably accurate value of Q by application of Eq. (11.21) for rectangular channels or Eq. (11.32) for nonrectangular channels.

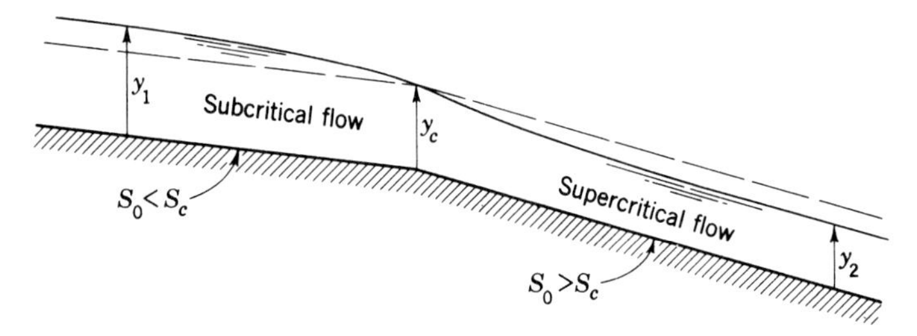

Figure 11.15 Change in flow from subcritical to supercritical at a break in slope.

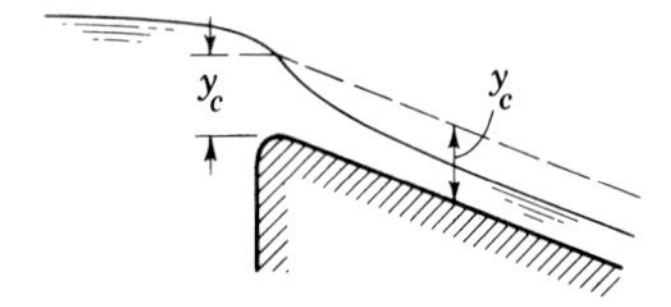

Figure 11.16 Hydraulic drop entering a steep slope.

Another instance where critical depth occurs is that of a free outfall (Fig. 11.17*a*) with subcritical flow in the channel prior to the outfall. Since friction produces a constant diminution in energy in the direction of flow, it is obvious that at the point of outfall the total energy must be less than at any point upstream. As critical depth is the value for which the specific energy is a minimum, one would expect critical depth to occur at the outfall. However, the value for the critical depth is derived on the assumption that the water is flowing in straight lines. In the free outfall gravity creates a curvature of the streamlines, with the result that the depth at the brink is less than critical depth. It has been found by experiment that the depth at the brink $y_b \approx 0.72 y_c$. Also, critical depth generally occurs upstream of the brink a distance of somewhere between $3y_c$ and $10y_c$. If the flow is supercritical, there is no drop-down curve (Fig. 11.17*b*).

Critical depth may occur in a channel if the bottom is humped or if the sidewalls are moved in to form a contracted section. In such cases critical depth will not always occur (Illustrative Example 11.3). Generally, the head loss through such a contraction is very small and may usually be neglected without introducing a sizable error.

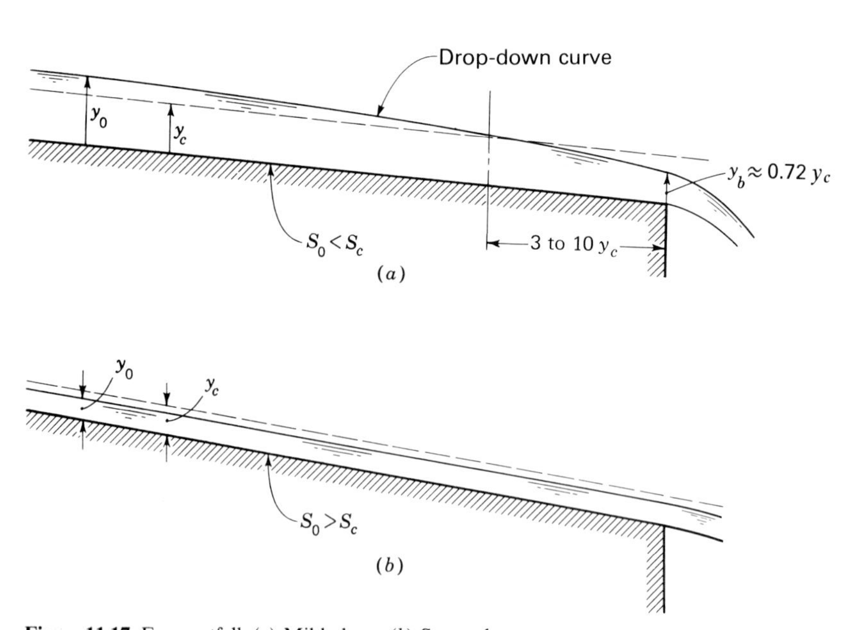

Figure 11.17 Free outfall. (*a*) Mild slope. (*b*) Steep slope.

11.14 HUMPS AND CONTRACTIONS

Let us examine the case of a hump in a rectangular channel (Fig. 11.18). We will neglect head loss. In Fig. 11.18*a* is shown a small hump. If the flow upstream of the hump is subcritical there will be a slight depression in the water surface over the hump. Inspection of the subcritical (upper) portion of the specific-energy diagram (Fig. 11.11) substantiates this statement. The hump (locally raised bed) causes a drop in specific energy E. A decrease in E is accompanied by a decrease in y which exceeds the decrease in E, hence there is a depression in the water surface over the hump. If the height of the hump is increased with no change in q there will be a further drop in the water surface over the hump (Fig. 11.18*b*) for the same reason. Further increases in hump height create further depression of the water surface over the hump until finally the depth on the hump becomes critical (Fig. 11.18*c*). The minimum height of the hump that causes critical depth is here referred to as the *critical hump height*. If the hump is made still higher (Fig. 11.18*d*), critical depth is maintained on the hump and the depth upstream of the hump is increased. This phenomenon is referred to as *damming action*. Downstream of the hump the flow will be supercritical. In such a situation, the hump, if it has a flat top, becomes a broad-crested weir (Sec. 12.13).

Flow through a contraction behaves in a manner similar to that of flow over a hump. If the flow is subcritical a small contraction will cause a slight depression in the water surface through the contraction. Further contraction creates further depression in the water surface until critical depth occurs in the contraction. That this is so can be seen by examining the subcritical (upper) portion of Fig. 11.12. In this case as q gets larger because of the contraction, y gets smaller. Upon further contraction the depth remains critical in the contracted section. However the depth increases because q, the flow per unit width, gets larger [Eq. (11.21)]. Thus damming action occurs. The maximum width (i.e., minimum contraction) at which critical depth occurs in the contracted section is referred to here as the *critical contracted width*.

With supercritical flow, humps and contractions behave differently than they do with subcritical flow according to the trends indicated by the supercritical (lower) portions of Figs. 11.11 and 11.12. Thus, if the flow is supercritical, the water depth at the hump or contraction will increase as the hump height is made

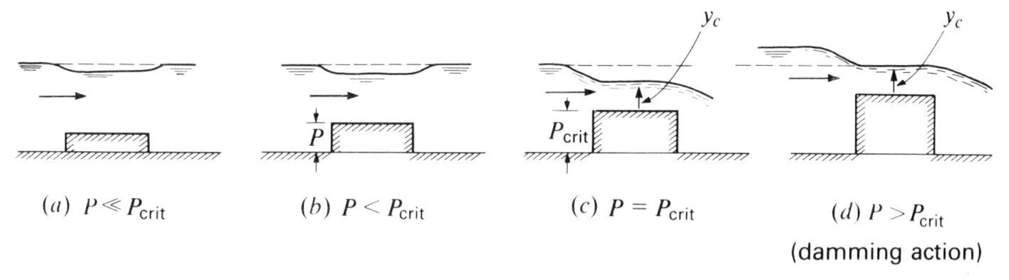

Figure 11.18 Subcritical flow over hump in a rectangular channel. (Flow rate is the same in all four cases. P = hump height; P_{crit} = critical hump height.)

larger or as the amount of contraction is increased until critical depth is reached. Beyond that, further increases in hump height or increases in the amount of contraction cause damming action.

A hump may be combined with a contraction to give a section not unlike a venturi meter (Sec. 12.7). The same principles discussed above for the hump and the contraction may be applied for such a situation. In so doing it must be noted that q is changed by the contraction when dealing with the hump.

Illustrative Example 11.3 In the accompanying figure, uniform flow of water occurs at 27 cfs in a 4-ft wide rectangular flume at a depth of 2.00 ft. (*a*) Is the flow subcritical or supercritical? (*b*) If a hump of height $\Delta z = 0.30$ ft is placed in the bottom of the flume, calculate the water depth on the hump. Neglect head loss in flow over the hump. (*c*) If the hump height is raised to $\Delta z = 0.60$ ft, what then are the water depths upstream and downstream of the hump? Once again neglect head loss over the hump.

(*a*) First find critical depth. From Eq. (11.21)

$$y_c = \left(\frac{q^2}{g}\right)^{1/3} = \left[\frac{(\frac{27}{4})^2}{32.2}\right]^{1/3} = 1.12 \text{ ft}$$

Since the normal depth (2.00 ft) is greater than the critical depth, the flow is subcritical and the channel slope is mild.

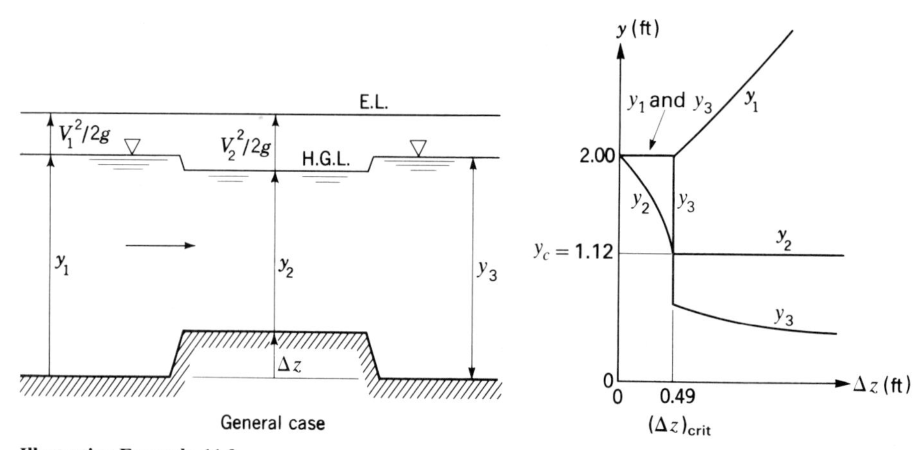

Illustrative Example 11.3

(*b*) Find the critical hump height. Write the energy equation between sections 1 and 2, assume critical flow on the hump and apply continuity.

$$2.00 + \frac{V_1^2}{2g} = (\Delta z)_{\text{crit}} + 1.12 + \frac{V_2^2}{2g} \qquad (a)$$

$$V_2 = \frac{27}{4 \times 1.12} = 6.03 \text{ fps} \qquad (b)$$

$$V_1 = \frac{27}{4 \times 2} = 3.38 \text{ fps} \qquad (c)$$

Substituting (*b*) and (*c*) in (*a*) gives

$$(\Delta z)_{\text{crit}} = 0.49 \text{ ft}$$

Thus the minimum-height hump that will produce critical depth on the hump is 0.49 ft.

Since the actual hump height, $\Delta z = 0.30$ ft, is less than the critical hump height, 0.49 ft, critical flow does not occur on the hump and there is no damming action.

Let us now find the depth y_2 on the hump.

Energy:
$$2.00 + \frac{V_1^2}{2g} = 0.30 + y_2 + \frac{V_2^2}{2g} \qquad (d)$$

Continuity:
$$(4 \times 2)V_1 = 4y_2 V_2 = 27 \text{ cfs} \qquad (e)$$

Eliminating V_1 and V_2 from Eqs. (*d*) and (*e*) gives three roots for y_2; $y_2 = 1.60$ ft, 0.82 ft, or a negative answer that has no physical meaning. Since the hump height is less than $(\Delta z)_{\text{crit}}$, the flow on the hump must be subcritical (that is, $y_2 > y_c$). Hence $y_2 = 1.60$ ft and the drop in the water surface on the hump $= 2.00 - (0.30 + 1.60) = 0.10$ ft.

(*c*) In this case the hump height $\Delta z = 0.60$ ft which is greater than the critical hump height. Hence critical depth ($y_c = 1.12$ ft) will occur on the hump. Writing the energy equation for this case, we have

$$y_1 + \frac{V_1^2}{2g} = 0.60 + 1.12 + \frac{V_2^2}{2g} \qquad (f)$$

From continuity,
$$(4 \times y_1)V_1 = 27 \text{ cfs} \qquad (g)$$

and, for critical flow,
$$\frac{V_2^2}{2g} = \tfrac{1}{2}y_2 = 0.56 \text{ ft} \qquad (h)$$

Combining Eqs. (*f*), (*g*), and (*h*) gives

$$y_1 + \frac{(27/4)^2}{(2g)y_1^2} = 2.28$$

from which $y_1 = 2.12$ ft, 0.66 ft, or a negative answer which has no physical meaning. In this case, damming action occurs and the depth y_1, upstream of the hump is 2.12 ft. On the hump the depth passes through critical depth of 1.12 ft and just downstream of the hump the depth will be 0.66 ft. The depth will then increase in the downstream direction following an M_2 water surface profile (Fig. 11.20) until a hydraulic jump (Sec. 11.19) occurs to return the depth to the normal uniform flow depth of 2.00 ft.

This higher type of hump is commonly built of concrete and is used for flow measurement purposes. Such structures are discussed under broad-crested weirs, in Sec. 12.13. The right-hand sketch of this example shows the relation between the hump height Δz, and the water depths y_1, y_2, and y_3 for the condition where $Q = 27$ cfs.

In the preceding example, if the flow had been supercritical rather than subcritical, a reverse situation would have occurred. With supercritical flow, the depth on a low hump will be slightly greater than the depth upstream of the hump. As the hump height is increased, the depth on the hump will increase until critical depth is reached. With further increase in the hump height, critical depth will prevail on the hump and damming action will occur upstream. That this is so can be reasoned out by reference to Fig. 11.11.

Similar situations take place in contractions and expansions with contractions behaving like humps and expansions behaving oppositely. Thus the water surface will be depressed through a contraction if the flow is subcritical until the contraction is so narrow that critical depth occurs in the contracted section. Upon further contraction, critical depth continues to occur in the contracted section, but damming action (with a rise in the water surface) takes place upstream.

11.15 NONUNIFORM, OR VARIED, FLOW

As a rule, uniform flow is found only in artificial channels of constant shape and slope, although even under these conditions the flow for some distance may be nonuniform, as shown in Fig. 11.1. But with a natural stream the slope of the bed and the shape and size of the cross section usually vary to such an extent that true uniform flow is rare. Hence the application of the equations given in Secs. 11.2 and 11.3 to natural streams can be expected to yield results that are only approximations to the truth. In order to apply these equations at all, the stream must be divided into lengths within which the conditions are approximately the same.

In the case of artificial channels which are free from the irregularities found in natural streams, it is possible to apply analytical methods to the various problems of nonuniform flow. In many instances, however, the formulas developed are merely approximations, and we must often resort to trial solutions and even purely empirical methods.[1]

In the case of pressure conduits, we have dealt with uniform and nonuniform flow without drawing much distinction between them. This can be done because in a closed pipe the area of the water section, and hence the mean velocity, is fixed at every point. But in an open channel these conditions are not fixed, and the stream adjusts itself to the size of cross section that the energy gradient (i.e., slope of the energy line) requires.

In an open stream on a falling grade, without friction, the effect of gravity is to tend to produce a flow with a continually increasing velocity along the path, as in the case of a freely falling body. However, the gravity force is opposed by frictional resistance. The frictional force increases with velocity, while gravity is constant; so eventually the two will be in balance, and uniform flow will occur. When the two forces are not in balance, the flow is nonuniform.

There are two types of nonuniform flow. In one the changing conditions extend over a long distance, and this may be called *gradually varied flow*. In the other the change may take place very abruptly and the transition is thus confined to a short distance. This may be designated as a *local nonuniform phenomenon* or *rapidly varied flow*. Gradually varied flow can occur with either subcritical or supercritical flow, but the transition from one condition to the other is ordinarily abrupt, as between D and E in Fig. 11.1. Other cases of local nonuniform flow

[1] For the treatment of many types of flow see Ven Te Chow, "Open-channel Hydraulics," McGraw-Hill Book Company, New York, 1959.

occur at the entrance and exit of a channel, at changes in cross section, at bends, and at obstructions such as dams, weirs, or bridge piers.

11.16 ENERGY EQUATION FOR GRADUALLY VARIED FLOW

The principal forces involved in flow in an open channel are inertia, gravity, hydrostatic force due to change in depth, and friction. The first three represent the useful kinetic and potential energies of the liquid, while the fourth dissipates useful energy into the useless kinetic energy of turbulence and eventually into heat because of the action of viscosity. Referring to Fig. 11.19, the total energy of the elementary volume of liquid shown is proportional to

$$H = z + y + \alpha \frac{V^2}{2g} \tag{11.34}$$

where $z + y$ is the potential energy head above the arbitrary datum, and $\alpha V^2/2g$ is the kinetic energy head, V being the mean velocity in the section. Each term of the equation represents energy in foot-pounds per pound of fluid (or newton-meters per newton of fluid in SI units). Once again, as in Secs. 11.1 and 11.2, we define L as the distance along the channel bed between any two sections and Δx as the horizontal distance between the sections. For all practical purposes these two distances can be considered as being equal to one another.

The value of α will generally be found to be higher in open channels than in pipes, as was explained in Sec. 11.5. It may range from 1.05 to 1.40, and in the case of a channel with an obstruction the value of α just upstream may be as high as 2.00 or even more. As the value of α is not known unless the velocity distribution is determined, it is often omitted from the equations, but an effort should be made to employ it if a high degree of accuracy is necessary. In the numerical problems in this chapter, α is assumed to be unity.

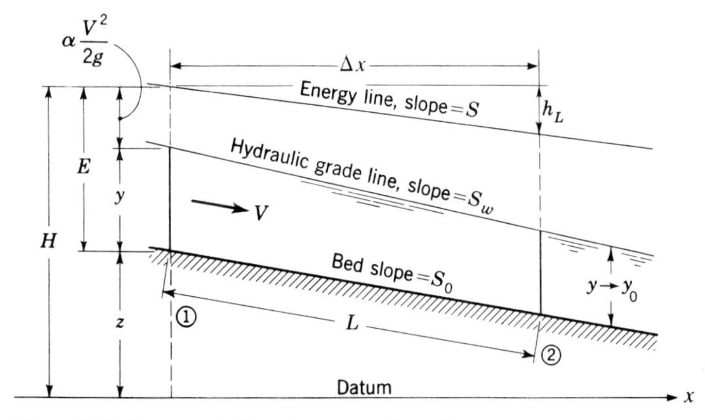

Figure 11.19 Energy relations for nonuniform flow.

Differentiating Eq. (11.34) with respect to x, the horizontal distance along the channel, the rate of energy dissipation is found to be (with $\alpha = 1$)

$$\frac{dH}{dx} = \frac{dz}{dx} + \frac{dy}{dx} + \frac{1}{2g}\frac{d(V^2)}{dx} \tag{11.35}$$

The energy gradient $S = -dH/dL \approx -dH/dx$, while the slope of the channel bed is $S_0 = -dz/dx$, and the slope of the hydraulic grade line or water surface is given by $S_w = -dz/dx - dy/dx$.

The energy equation for steady flow between two sections (1) and (2) of Fig. 11.19 a distance Δx apart is

$$z_1 + y_1 + \alpha_1 \frac{V_1^2}{2g} = z_2 + y_2 + \alpha_2 \frac{V_2^2}{2g} + h_L \tag{11.36}$$

As $z_1 - z_2 = S_0(\Delta x)$ and $h_L \approx S(\Delta x)$, the energy equation may also be written in the form (with $\alpha_1 = \alpha_2 = 1$)

$$y_1 + \frac{V_1^2}{2g} = y_2 + \frac{V_2^2}{2g} + (S - S_0)(\Delta x) \tag{11.37}$$

An approximate analysis of gradually varied, nonuniform flow can be achieved by considering a length of stream as consisting of a number of successive reaches, in each of which uniform flow occurs. Greater accuracy results from smaller depth variations in each reach. The Manning formula [Eq. (11.8)] is applied to average conditions in each reach to provide an estimate of the value of S for that reach as follows:

In English units:
$$S = \left(\frac{nV_m}{1.49R_m^{2/3}}\right)^2 \tag{11.38a}$$

In SI units:
$$S = \left(\frac{nV_m}{R_m^{2/3}}\right)^2 \tag{11.38b}$$

where V_m and R_m are the means of the respective values at the two ends of the reach. With this value of S, with S_0 and n known, and with the depth and velocity at one end of the reach given, the length Δx to the end corresponding to the other depth can be computed from Eq. (11.37), rearranged as follows:

$$\Delta x = \frac{(y_1 + V_1^2/2g) - (y_2 + V_2^2/2g)}{S - S_0} \tag{11.39}$$

In practice, the depth range of interest is divided into small increments, usually equal, which define reaches whose lengths can be found by using Eq. (11.39).

Illustrative Example 11.4 At a certain section in a very smooth 6-ft-wide rectangular channel the depth is 3.00 ft when the flow rate is 160 cfs. Compute the distance to the section where the depth is 3.20 ft if $S_0 = 0.0020$ and $n = 0.012$.

The calculations are shown in the following table. The total distance is calculated to be 73 ft. The accuracy could be improved by taking more steps. In computing $S - S_0$, a slight error in the calculated value of S will introduce a sizeable error in the calculated value of x. Thus it is important that S be calculated as accurately as possible.

y, ft	A, ft^2	P $(6 + 2y)$, ft	R_h, ft	V, fps	$V^2/2g$, ft	$y + \dfrac{V^2}{2g}$	Numerator $\Delta\left(y + \dfrac{V^2}{2g}\right)$	V_{avg}, fps	R_{avg}, ft	S Eq. (11.38)	Denominator $S - S_0$	Δx, ft Eq. (11.39)
3.00	18.00	12.00	1.500	8.89	1.227	4.227						
							0.022	8.74	1.512	0.00284	0.00084	26
3.10	18.60	12.20	1.525	8.60	1.149	4.249						
							0.029	8.47	1.536	0.00262	0.00062	47
3.20	19.20	12.40	1.548	8.33	1.078	4.278						

$\sum(\Delta x) = 73$

If the data for this problem has been given in SI units rather than in English units, the procedure for solution would have been the same except that Eqs. (11.38*b*) and (11.7) would have been used rather than Eqs. (11.38*a*) and (11.8). Because of the extensive calculations required, this type of problem is commonly solved through use of a computer. Section 11.17 will indicate that depth ranges for this procedure must not cross the normal or critical depths. Therefore, as a precaution, y_0 and y_c should first be computed from Eqs. (11.9) and (11.21).

11.17 WATER-SURFACE PROFILES IN GRADUALLY VARIED FLOW

As there are some 12 different circumstances giving rise to as many different fundamental types of varied flow, it is helpful to have a logical scheme of type classification. In general, any problem of varied flow, no matter how complex it may appear, with the stream passing over dams, under sluice gates, down steep chutes, on the level, or even on an upgrade, can be broken down into reaches such that the flow within any reach is either uniform or falls within one of the given nonuniform classifications. The stream is then analyzed one reach at a time, proceeding from one reach to the next until the desired result is obtained.

The following treatment is based, for simplicity, on channels of rectangular section. The section will be considered sufficiently wide and shallow so that we may confine our attention to a section 1 ft wide through which the velocity is essentially uniform. It is important to bear in mind that the following development is based on a constant value of q, the discharge per unit width, and upon one value of n, the roughness coefficient.

Commencing with the last term of Eq. (11.35), we may observe that since $V = q/y$,

$$\frac{1}{2g}\frac{d(V^2)}{dx} = \frac{1}{2g}\frac{d}{dx}\left(\frac{q^2}{y^2}\right) = -\frac{q^2}{g}\frac{1}{y^3}\frac{dy}{dx}$$

Substituting this, plus the S and S_0 terms defined earlier, into Eq. (11.35), yields

$$-S = -S_0 + \frac{dy}{dx}\left(1 - \frac{q^2}{gy^2}\right)$$

or

$$\frac{dy}{dx} = \frac{S_0 - S}{1 - q^2/gy^3} = \frac{S_0 - S}{1 - V^2/gy} = \frac{S_0 - S}{1 - \mathbf{F}^2} \tag{11.40}$$

Evidently, if the value of dy/dx as determined by Eq. (11.40) is positive, the water depth will be increasing along the channel; if negative, it will be decreasing. Let us examine the numerator and denominator of Eq. (11.40).

Looking first at the numerator, S may be considered as the energy gradient [such as would be obtained from Eq. (11.38)] which would carry the given discharge at depth y with uniform flow. From Eq. (11.38*a*), noting that $R_h = y$ for a very wide rectangular channel and that $V = q/y$, we get

$$S = \frac{V^2 n^2}{(1.49)^2 R_h^{4/3}} = \frac{q^2 n^2}{(1.49)^2 y^{10/3}}$$

Equation (11.9*a*) modified for uniform flow in a very wide rectangular channel may be expressed in similar form as

$$S_0 = \frac{q^2 n^2}{(1.49)^2 y_0^{10/3}}$$

Comparing these two expressions we see that $S/S_0 = (y_0/y)^{10/3}$. Consequently, for constant q and n, when $y > y_0$, $S < S_0$, and the numerator is positive. Conversely, when $y < y_0$, $S > S_0$, and the numerator is negative.

To investigate the denominator of Eq. (11.40) we observe that, if $\mathbf{F} = 1$, $dy/dx = \infty$; if $\mathbf{F} > 1$, the denominator is negative; and if $\mathbf{F} < 1$, the denominator is positive.

The foregoing analyses have been combined graphically into a series of water-surface profiles [Fig. (11.20)]. The surface profiles are classified according to slope and depth as follows. If S_0 is positive, the bed slope is termed *mild* (M) when $y_0 > y_c$, *critical* (C) when $y_0 = y_c$, and *steep* (S) $y_0 < y_c$; if $S_0 = 0$, the channel is *horizontal* (H); and if S_0 is negative, the bed slope is called *adverse* (A). If the stream surface lies above both the normal (uniform flow) and critical depth lines, it is of type 1; if between these lines, it is of type 2; and if below both lines, it is of type 3. The 12 forms of surface curvature are labeled accordingly in Fig. 11.20.

It must be pointed out that the scale of the drawings in Fig. 11.20 is greatly reduced in the horizontal direction. The problems at the end of this chapter demonstrate that gradually varied flow generally extends over many hundreds of feet, and if plotted to an undistorted scale, the rate of change in depth would be scarcely discernible. It may be noted further that, since even a hydraulically steep slope varies but a few degrees from the horizontal, it makes little difference whether the depth y is measured vertically (as shown) or perpendicular to the bed.

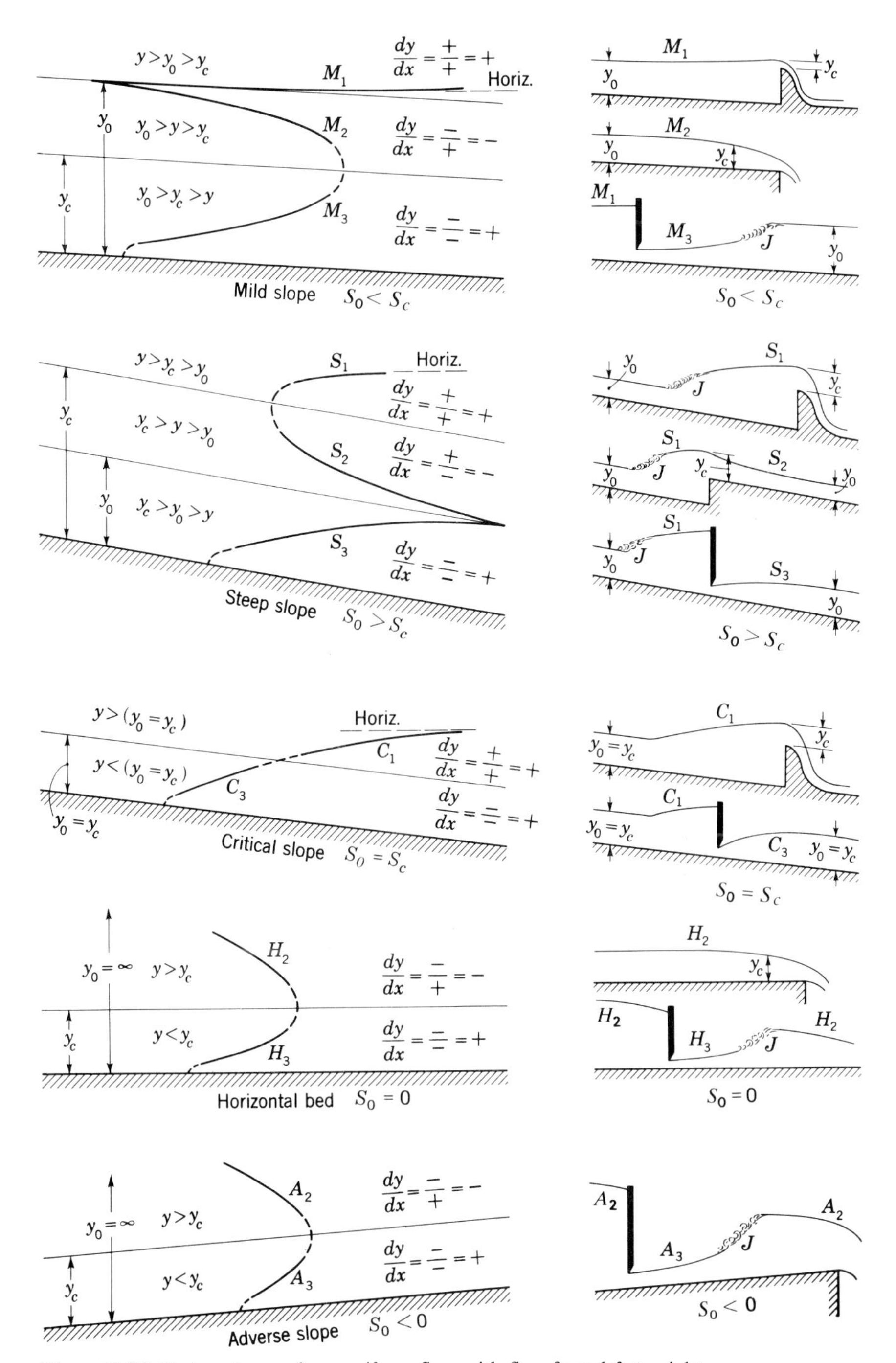

Figure 11.20 Various types of nonuniform flow with flow from left to right.

It will be observed that some of the curves of Fig. 11.20 are concave upward while others are concave downward. Although the mathematical proof for this is not given, the physical explanation is not hard to find. In the case of the type 1 curves, the surface must approach a horizontal asymptote as the velocity is progressively slowed down because of the increasing depth. Likewise, all curves which approach the normal or uniform depth line must approach it asymptotically, because uniform flow will prevail at sections sufficiently remote from disturbances, as pointed out in Sec. 11.1. Theoretically the curves which cross the critical-depth line must do so vertically, as the denominator of Eq. (11.40) becomes zero in this case. The critical-slope curves, for which $y_0 = y_c$, constitute exceptions to both the foregoing statements, since it is not possible for a curve to be both tangent and perpendicular to the critical-uniform depth line.

To the right of each water-surface profile is shown a representative example of how this particular curve can occur "in nature." Many of the examples show a rapid change from a depth below the critical to a depth above the critical. This is a local phenomenon, known as the hydraulic jump, which is discussed in detail in Sec. 11.19.

The qualitative analysis of water-surface profiles has been restricted to rectangular sections of large width. The curve forms of Fig. 11.20 are, however, applicable to any channel of uniform cross section if y_0 is the depth for uniform flow and y_c is the depth which satisfies Eq. (11.32). The surface profiles can even be used qualitatively in the analysis of natural stream surfaces as well, provided that local variations in slope, shape, and roughness of cross section, etc., are taken into account. The step-by-step method presented in Sec. 11.16 for the solution of nonuniform-flow problems is not restricted to uniform channels and is therefore suited to water-surface-profile computations for any stream whatever.

11.18 EXAMPLES OF WATER-SURFACE PROFILES

The M_1 Curve

The most common case of gradually varied flow is where the depth is already above the critical and is increased still further by a dam, as indicated in Fig. 11.21. Referring to the specific energy diagram of Fig. 11.11, this case is found on the upper limb of the diagram, for here also, as the depth increases, the velocity diminishes without any abrupt transitions, so that a smooth surface curve is

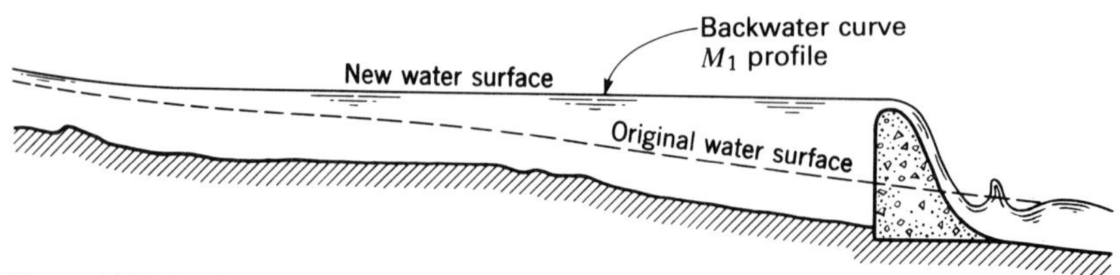

Figure 11.21 Backwater curve in a natural stream.

obtained. In the case of flow in an artificial channel with a constant bed slope, the water-surface curve would be asymptotic at infinity to the surface for uniform flow, as we noted before. But the problems that are usually of more important interest are those concerned with the effect of a dam on a natural stream and the extent to which it raises the water surface at various points upstream. The resulting water-surface profile in such a case is commonly known as a *backwater curve.*

For an artificial channel where the conditions are uniform, save for the variation in water depth, the problem may be solved by use of Eqs. (11.38) and (11.39). Usually, the solution commences at the dam, where conditions are assumed to be known, and the lengths of successive reaches upstream, corresponding to assumed increments of depth, are computed. A tabular type of solution (Illustrative Example 11.4) is the most helpful, with column headings corresponding to the various elements of Eqs. (11.38) and (11.39), the last column being $\Sigma\, \Delta x$ which sums up the length from the dam to the point in question. It is important, if accuracy is desired, to keep the depth increment small within any reach; a depth change of 10 per cent or less is fairly satisfactory. The smaller the depth increment used in this step-by-step procedure, the greater the accuracy of the final result. This type of problem where successive calculations are required can advantageously be solved through use of a computer.

For a natural stream, such as that shown in Fig. 12.10, the solution is not so direct, because the form and dimensions of a cross section cannot be assumed and then the distance to its location computed. As there are various slopes and cross sections at different distances upstream, the value of Δx in Eq. (11.37) must be assumed, and then the depth of stream at this section can be computed by trials, as in Prob. 11.61. The solution is then pursued in similar fashion on a reach by reach basis. The accuracy of the results depends on the selection of a proper value for Manning's n, which is difficult when dealing with natural streams. For this reason and because of irregularities in the flow cross sections, the refinements of Eq. (11.37) are not always justified and it is often satisfactory to assume uniform flow by applying Eq. (11.9) to each successive reach.

The M_2 Curve

This curve, representing accelerated subcritical flow on a slope which is flatter than critical, exists, like the M_1 curve, because of a control condition *downstream.* In this case, however, the control is not an obstruction but the removal of the hydrostatic resistance of the water downstream, as in the case of the free overfall shown in Fig. 11.17*a*. As in the M_1 curve, the surface will approach the depth for uniform flow at an infinite distance upstream. Practically, because of slight wave action and other irregularities, the distinction between the M_2, or *drop-down,* curve and the curve for uniform flow disappears within a finite distance.

The M_3 Curve

This occurs because of an upstream control, as the sluice gate shown in Fig. 11.20. The bed slope is not sufficient to sustain lower-stage flow, and at a certain point determined by energy and momentum relations, the surface will pass through an hydraulic jump unless this is made unnecessary by the existence of a free overfall before the M_3 curve reaches critical depth.

The *S* Curves

These may be analyzed in much the same fashion as the *M* curves, having due regard for downstream control in the case of subcritical flow and for upstream control for supercritical flow. Thus a dam or an obstruction on a steep slope produces an S_1 curve (Fig. 11.20) which approaches the horizontal asymptotically but cannot so approach the uniform depth line, which lies below the critical depth. Therefore this curve must be preceded by a hydraulic jump. The S_2 curve shows accelerated lower-stage flow, smoothly approaching uniform depth. Such a curve will occur whenever a steep channel receives flow at critical depth, as from an obstruction (as shown) or reservoir. The sluice gate on a steep channel will produce the S_2 curve, which also approaches smoothly the uniform depth line.

The *C* Curves

These curves, with the anomalous condition ($dy/dx = \infty$) at $y_0 = y_c$ are not of frequent occurrence.

The *H* and the *A* Curves

These curves have in common the fact that there is no condition of uniform flow possible. The H_2 and A_2 drop-down curves are similar to the M_2 curve, but even more noticeable. The value of $y_b = 0.72y_c$ given in Sec. 11.13 applies strictly only to the H_2 curve, but is approximately true for the M_2 curve also. The sluice gate on the horizontal and adverse slopes produces H_3 and A_3 curves which are like the M_3 curve, but they do not exist for as great a distance as the M_3 curve before a hydraulic jump occurs. Of course, it is not possible to have a channel of any appreciable length carry water on a horizontal grade, and even less so on an adverse grade.

Other Examples

Some other interesting water-surface profiles occur when the slope of a channel of uniform section changes abruptly from a mild to a milder slope or to a less mild slope. In this case the flow is everywhere subcritical. Similar water-surface profiles occur when a channel on a constant slope that is mild throughout its entire length has an abrupt change in width to an either narrower or wider channel. These possibilities are depicted in Fig. 11.22.

Other water-surface profiles include those that occur when the slope of a channel changes abruptly from steep to either steeper or less steep. In this case the flow is supercritical. Similar profiles occur when a channel on a constant slope that is steep throughout its entire length has an abrupt change in width to

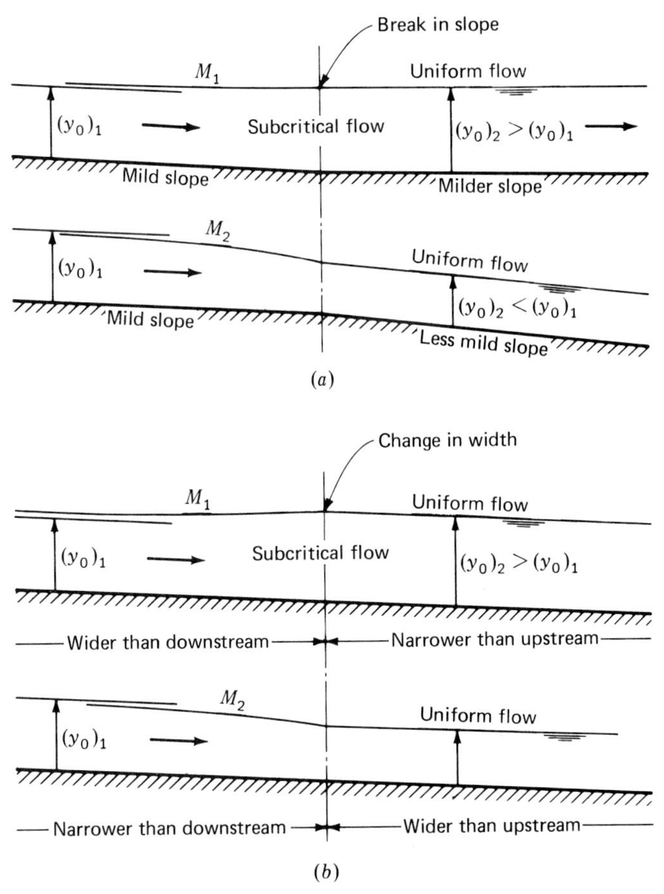

Figure 11.22 Subcritical flow water-surface profiles. (*a*) Constant section with change in slope. (*b*) Constant mild slope with change in width.

an either wider or narrower channel. As an exercise it is suggested that the reader sketch profiles similar to Fig. 11.22 for the steep-slope situations. In these cases it will be found that, with steep slopes (supercritical flow), uniform flow occurs upstream of the change in either slope or width, while with mild slopes (subcritical flow), uniform flow prevails downstream of the change.

11.19 THE HYDRAULIC JUMP

By far the most important of the local nonuniform-flow phenomena is that which occurs when supercritical flow has its velocity reduced to subcritical. We have seen in the surface profiles of Fig. 11.20 that there is no ordinary means of changing from supercritical to subcritical flow with a smooth transition, because theory calls for a vertical slope of the water surface. The result, then, is a marked discontinuity in the surface, characterized by a steep upward slope of the profile, broken throughout with violent turbulence, and known universally as the *hydraulic jump.*

The specific reason for the occurrence of the hydraulic jump can perhaps best be explained by reference to the M_3 curve of Fig. 11.20. Downstream of the sluice gate the flow decelerates because the slope is not great enough to maintain supercritical flow. The specific energy decreases as the depth increases (proceeding to the left along the lower limb of the specific-energy diagram, Fig. 11.11). Were this condition to progress until the flow reached critical depth, an increase in specific energy would be required as the depth increased from the critical to the uniform flow depth downstream. But this is a physical impossibility. Therefore the jump forms before critical depth is reached.

The hydraulic jump can also occur from an upstream condition of uniform supercritical flow to a nonuniform S_1 curve downstream when there is an obstruction on a steep slope, as illustrated in Fig. 11.20, or again from a nonuniform upstream condition to a nonuniform downstream condition, as illustrated by the H_3, H_2 or the A_3, A_2 combinations. An M_3, M_2 combination is also possible. In addition to the foregoing cases, where the channel bed continues at a uniform slope, a jump will form when the slope changes from steep to mild, as on the apron at the base of the spillway, illustrated in Fig. 11.23. This is an excellent example of the jump serving a useful purpose, for it dissipates much of the destructive energy of the high-velocity water, thereby reducing downstream erosion.

The equation relating the depths before and after the hydraulic jump will be derived for the case of a horizontal channel bottom (the H_3, H_2 combination of Fig. 11.20). For channels on a gradual slope (i.e., less than about 3°) the gravity component of the weight is relatively small and may be neglected without introducing significant error. The friction forces acting are negligible because of the short length of channel involved and therefore the only significant forces are hydrostatic forces. Applying Newton's second law [Eq. (6.7*a*)] to the element of fluid contained between sections 1 and 2 of Fig. 11.23 we have

$$\sum F_x = \gamma h_{c_1} A_1 - \gamma h_{c_2} A_2 = \frac{\gamma}{g} Q(V_2 - V_1)$$

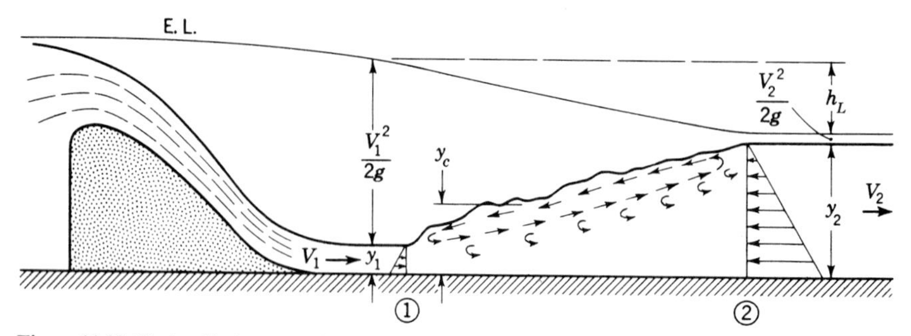

Figure 11.23 Hydraulic jump on horizontal bed following a spillway; horizontal scale foreshortened between sections 1 and 2 approximately $2\frac{1}{2}$:1.

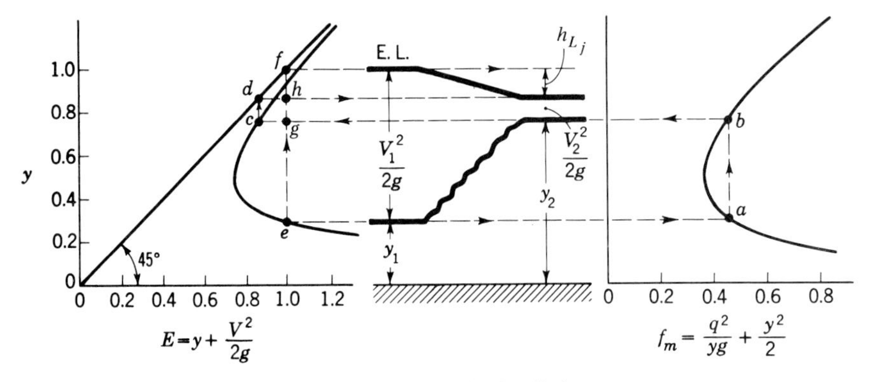

Figure 11.24 Energy and momentum relations in hydraulic jump.

which can be reordered to give

$$\frac{\gamma}{g}QV_1 + \gamma h_{c_1}A_1 = \frac{\gamma}{g}QV_2 + \gamma h_{c_2}A_2 \tag{11.41}$$

This states that the momentum plus the pressure force on the cross-sectional area is constant, or dividing by γ and observing that $V = Q/A$,

$$F_m = \frac{Q^2}{Ag} + Ah_c = \text{constant} \tag{11.42}$$

This equation applies to any shape of cross section.

In the case of a rectangular channel, this reduces for a unit width to:

$$f_m = \frac{q^2}{y_1 g} + \frac{y_1^2}{2} = \frac{q^2}{y_2 g} + \frac{y_2^2}{2} \tag{11.43}$$

A curve of values of f_m for different values of y is plotted to the right of the specific-energy diagram shown in Fig. 11.24. Both curves are plotted for the condition of 2 cfs/ft of width. As the loss of energy in the jump does not affect the "force" quantity f_m, the latter is the same after the jump as before, and therefore any vertical line intersecting the f_m curve serves to locate two *conjugate depths* y_1 and y_2. These depths represent possible combinations of depth that could occur before and after the jump.

Thus, in Fig. 11.24, the line for the initial water level y_1 intersects the f_m curve at a as shown, giving the value of f_m, which must be the same after the jump. The vertical line ab then fixes the value of y_2. This depth is then transposed to the specific-energy diagram to determine the value cd of $V_2^2/2g$. The value of $V_1^2/2g$ is the vertical distance ef, and the head loss h_{L_j} caused by the jump is the drop in energy from 1 to 2. Or

$$h_{L_j} = \left(y_1 + \frac{V_1^2}{2g}\right) - \left(y_2 + \frac{V_2^2}{2g}\right) \tag{11.44}$$

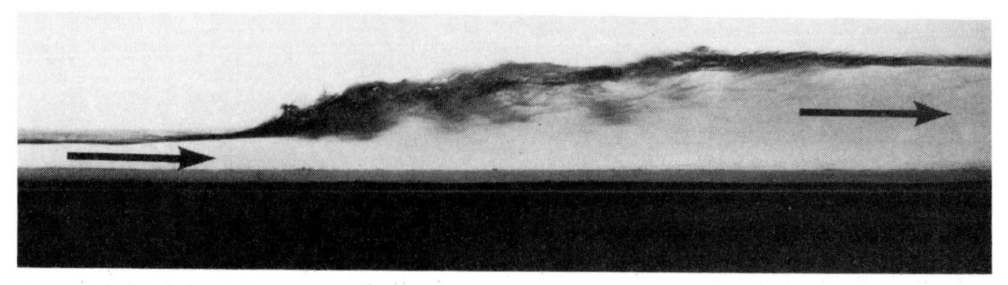

Figure 11.25 Hydraulic jump. *(Photograph by Hydrodynamics Laboratory, California Institute of Technology.)*

On Fig. 11.24 the head loss in the hydraulic jump is given by either the horizontal distance cg or the vertical distance fh.

When the rate of flow and the depth before or after the jump are given, it is seen that Eq. (11.43) becomes a cubic equation when solving for the other depth. This may readily be reduced to a quadratic, however, by observing that $y_2^2 - y_1^2 = (y_2 + y_1)(y_2 - y_1)$,

$$\frac{q^2}{g} = y_1 y_2 \frac{y_1 + y_2}{2} \tag{11.45}$$

Solving the above equation by the quadratic formula gives

$$y_2 = \frac{y_1}{2}\left(-1 + \sqrt{1 + \frac{8q^2}{gy_1^3}}\right) \tag{11.46a}$$

or

$$y_1 = \frac{y_2}{2}\left(-1 + \sqrt{1 + \frac{8q^2}{gy_2^3}}\right) \tag{11.46b}$$

These equations relate the depths before and after hydraulic jump (i.e., the conjugate depths) in a rectangular channel. They give good results if the channel slope is less than about 0.05. For steeper channel slopes the effect of the gravity component of the weight of liquid between sections 1 and 2 of Fig. 11.23 must be considered.

Although the length of jump is difficult to determine, a good approximation for length of jump is about $5y_2$. This relation may be seen to be approximately true by examination of Fig. 11.25, a photograph of a hydraulic jump in a horizontal channel, caused by a sluice gate upstream.

It can be shown that for a given flow rate in a rectangular channel the minimum value of f_m [Eq. (11.43)] occurs at the same depth as the minimum value of E. Differentiating f_m with respect to y and equating to zero gives

$$f_m = \frac{q^2}{yg} + \frac{y^2}{2}$$

$$\frac{df_m}{dy} = -\frac{q^2}{gy^2} + \frac{2y}{2} = 0$$

and

$$y = \left(\frac{q^2}{g}\right)^{1/3}$$

This expression is identical to Eq. (11.21). Thus we have shown that for a given q the minimum value of f_m occurs at the same depth as does the minimum value of E. This is indicated on Fig. 11.24.

Since the flow must be supercritical in order for a jump to occur, $\mathbf{F}_1$, the Froude number of the flow just upstream of the jump must be greater than 1.0. If $\mathbf{F}_1$ is only slightly larger than 1.0 a poorly formed, *undular jump* will occur. At higher values of $\mathbf{F}_1$ the jumps are more pronounced and the turbulence within them and the head loss through them is correspondingly greater.

11.20 LOCATION OF HYDRAULIC JUMP

The problem of determining the location of a hydraulic jump involves a combined application of the principles discussed in Secs. 11.18 and 11.19. Examples of the location of a hydraulic jump are shown in Fig. 11.26. In case (1) the jump occurs downstream of the break in slope, while in case (2) the jump is located upstream of the break. The reasons for this are illustrated by the following example.

Illustrative Example 11.5 Analyze the water-surface profile in a long rectangular channel with concrete lining ($n = 0.013$). The channel is 10 ft wide, the flow rate is 400 cfs, and there is an abrupt change in channel slope from 0.0150 to 0.0016. Find also the horsepower loss in the jump.

$$400 = \frac{1.49}{0.013}(10y_{0_1})\left(\frac{10y_{0_1}}{10 + 2y_{0_1}}\right)^{2/3}(0.015)^{1/2}$$

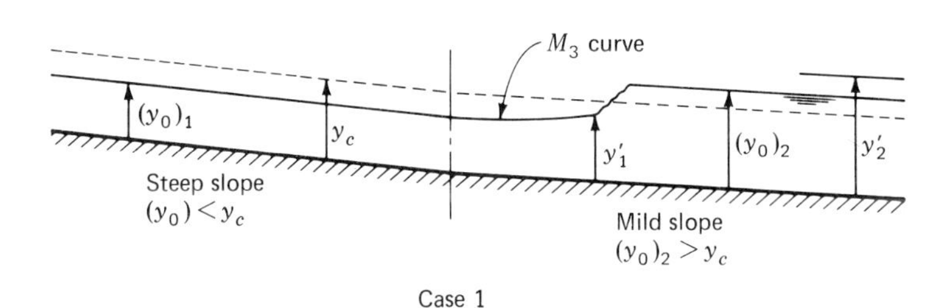

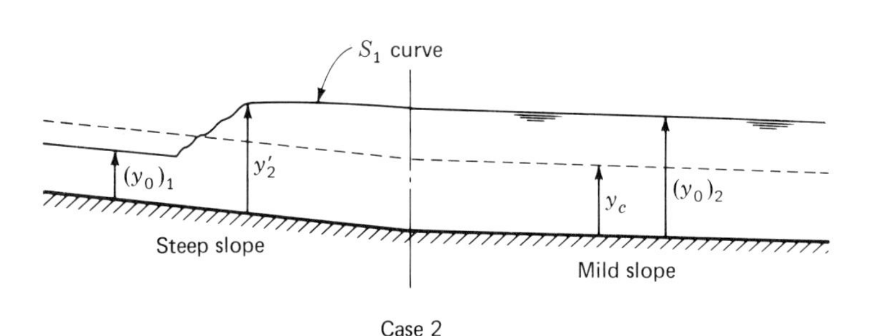

Figure 11.26 Examples of location of a hydraulic jump. [*Note:* y'_2 is the conjugate depth of $(y_0)_1$ and y'_1 is the conjugate depth of $(y_0)_2$.]

By trial, $y_{0_1} = 2.17$ ft (normal depth on upper slope)

Using a similar procedure, the normal depth y_{0_2} on the lower slope is found to be 4.80 ft.

$$y_c = \left(\frac{q^2}{g}\right)^{1/3} = \left[\frac{(\frac{400}{10})^2}{32.2}\right]^{1/3} = 3.68 \text{ ft}$$

Thus flow is supercritical ($y_{0_1} < y_c$) before break in slope and subcritical ($y_{0_2} > y_c$) after break, so a hydraulic jump must occur.

Applying Eq. (11.46*a*) to determine the depth conjugate to the 2.17-ft (upper-slope) normal depth, we get

$$y_2' = \frac{2.17}{2}\left\{-1 + \left[1 + \frac{8(40)^2}{32.2(2.17)^3}\right]^{1/2}\right\} = 5.75 \text{ ft}$$

Therefore the depth conjugate to the upper-slope normal depth of 2.17 ft is 5.75 ft. This jump cannot occur because the normal depth y_{0_2} on the lower slope is less than 5.75 ft.

Applying Eq. (11.46*b*) to determine the depth conjugate to the 4.80-ft (lower-slope) normal depth, we get

$$y_1' = \frac{4.80}{2}\left\{-1 + \left[1 + \frac{8(40)^2}{32.2(4.8)^3}\right]\right\} = 2.76 \text{ ft}$$

The lower conjugate depth of 2.76 ft will occur downstream of the break in slope. Thus the condition here is similar to that depicted in Fig. 11.26*a*. The location of the jump (i.e., its distance below the break in slope) may be found by applying Eq. (11.39):

$$\Delta x = \frac{E_1 - E_2}{S - S_0}$$

$$E_1 = 2.17 + \frac{(400/21.7)^2}{64.4} = 7.45 \text{ ft}$$

$$E_2 = 2.76 + \frac{(400/27.6)^2}{64.4} = 6.02 \text{ ft}$$

$$V_m = \frac{1}{2}\left(\frac{400}{21.7} + \frac{400}{27.6}\right) = 16.46 \text{ fps}$$

$$R_m = \frac{1}{2}\left(\frac{21.7}{14.34} + \frac{27.6}{15.52}\right) = 1.645 \text{ ft}$$

From Eq. (11.38*a*)

$$S = \left[\frac{(0.013)(16.45)}{1.49(1.645)^{2/3}}\right]^2 = 0.01060$$

Finally,

$$\Delta x = \frac{7.45 - 6.02}{0.0106 - 0.0016} = 160 \text{ ft}$$

Thus depth on the upper slope is 2.17 ft; downstream of the break the depth increases gradually (M_3 curve) to 2.76 ft over a distance of approximately 160 ft; then a hydraulic jump occurs from a depth of 2.76 ft to 4.80 ft; downstream of the jump the depth remains constant (i.e., normal) at 4.80 ft.

$$\text{HP loss} = \frac{\gamma Q h_{L_j}}{550} \qquad \text{where} \qquad h_{L_j} = \Delta E$$

Before jump:

$$E_1' = 2.76 + \frac{(400/27.6)^2}{64.4} = 6.02 \text{ ft}$$

After jump: $$E_{0_2} = 4.80 + \frac{(400/48.0)^2}{64.4} = 5.88 \text{ ft}$$

Hence $$\text{HP loss} = \frac{62.4(400)(6.02 - 5.88)}{550} = 6.35$$

In the two examples shown in Fig. 11.26 the flow is at normal depth either before or after the hydraulic jump. These situations are straightforward and are easy to handle as shown in the preceding illustrative example. There are instances, however, where the flow is not normal either before or after the jump. Such a situation will occur, for example, when water flows out from under a sluice gate and there is an overflow weir downstream of the gate. If the channel is on a steep slope an hydraulic jump will occur between an S_3 and an S_1 water surface profile while if the channel is on a mild slope the jump will occur between an M_3 and an M_1 profile (Fig. 11.20). To find the location of such a jump, varied flow calculations must be made in the downstream direction from the sluice gate and in the upstream direction from the overflow weir, and the location of the jump is found by trial. Using Eq. (11.46*a*) the conjugate depth y_2' of the depth y_1 at a particular station on the first water-surface profile is calculated. This is repeated for different values of y_1 at different stations on the first profile until the calculated conjugate depth y_2' corresponds to the depth at that same station on the second water-surface profile. The solution can also be achieved by calculating conjugate depths y_1' of various depths y_2 using Eq. (11.46*b*). Graphically plotting these various curves greatly aids the solution process.

11.21 FLOW AROUND CHANNEL BENDS

When a body moves along a curved path of radius r at constant speed it has a normal acceleration $a_n = V^2/r$ toward the center of the curve, and hence the body must be acted on by a force in that direction. In Fig. 11.27 it may be seen that this force comes from the unbalanced pressure forces due to the difference in liquid levels between the outer and inner banks of the channel. Assuming that the velocity V across the rectangular section is uniform and that $r \gg B$, we can write, for a unit length of channel, $\sum F_n = ma_n$,

$$\frac{\gamma y_2^2}{2} - \frac{\gamma y_1^2}{2} = \frac{\gamma}{g} B\left(\frac{y_1 + y_2}{2}\right)\frac{V^2}{r}$$

which by algebraic transformation can be written as

$$\Delta y = y_2 - y_1 = \frac{V^2 B}{gr} \tag{11.47}$$

where B is the top width of the water surface as shown in Fig. 11.27. It can be shown that Eq. (11.47) applies to any shape of cross section. If the effect of velocity distribution and variations in curvature across the stream are considered, the difference in water depths between the outer and inner banks may be as

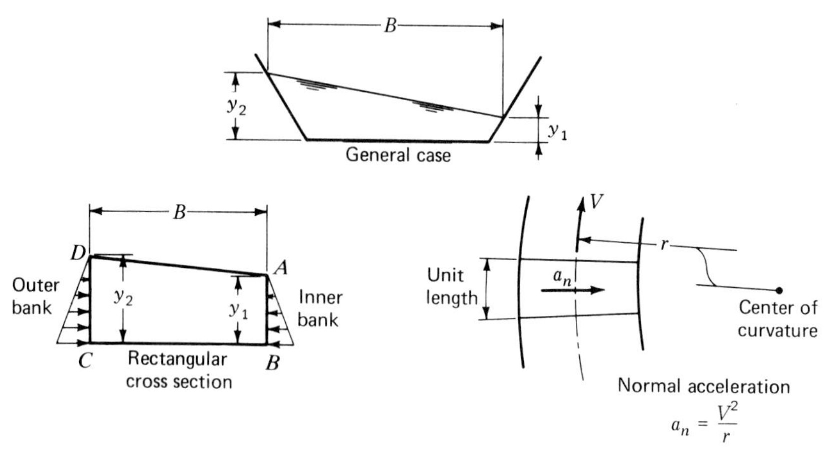

Figure 11.27 Flow in an open-channel bend.

much as 20 per cent more than that given by Eq. (11.47). If the actual velocity distribution across the stream is known, the width may be divided into sections and the difference in elevation computed for each section. The total difference in surface elevation across the stream is the sum of the differences for the individual sections.

With supercritical flow the complicating factor is the effect of disturbance waves, generated by the very start of the curve. These waves, one from the outside wall and one from the inside, traverse the channel, making an angle β with the original direction of flow, as discussed in Sec. 11.10. The result is a criss-cross wave pattern. The water surface along the outside wall will rise from the beginning of the curve, reaching a maximum at the point where the wave from the inside wall reaches the outside wall. The wave is then reflected back to the inside wall and so on around the bend. Hence, in supercritical flow around a bend the increase in water depth on the outer wall is equal to that created by centrifugal effects plus or minus the height of the waves. The height of these waves is approximately $\Delta y/2$ [Eq. (11.47)]. Thus, the water-depth increase on the outer wall varies between zero and Δy, and the depth decrease on the inner wall varies similarly.

Several schemes to lessen the surface rise from wave effect have been investigated.[1] The bed of the channel may be banked so that all elements of the flow are acted upon simultaneously, which is not possible when the turning force comes from the wall only. As in a banked-railway curve, this requires a transition section with a gradually increasing superelevation preceding the main curve. Another method is to introduce a counterdisturbance to offset the disturbance wave caused by the curve. Such a counterdisturbance can be provided by a

[1] R. T. Knapp, Design of Channel Curves for Supercritical Flow, *Trans. ASCE*, vol. 116, p. 296, 1951.

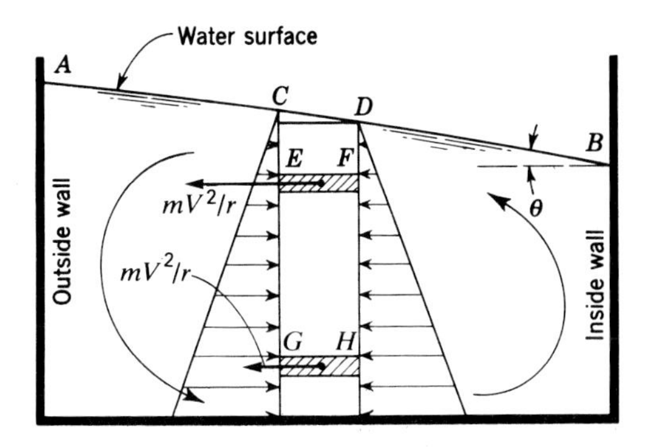

Figure 11.28 Schematic sketch of flow around a bend in a rectangular channel curving to the right, looking downstream, with spiral flow counterclockwise.

section of curved channel of twice the radius of the main section, by a spiral transition curve, or by diagonal sills on the channel bed, all preceding the main curve.

Flow around a bend in an open channel is complicated by the development of a secondary flow as the liquid travels around the curve. The development of secondary flow in the bends of pressure conduits was discussed in Sec. 8.21.

The flow around a channel bend is called *spiral flow* because superposition of the secondary flow on the forward motion of the liquid causes the liquid to follow a path like a corkscrew (or spiral). The existence of spiral flow is readily explained by reference to Fig. 11.28. The water surface is superelevated at the outside wall. The element EF is subjected to a centrifugal force mV^2/r which is balanced by an increased hydrostatic force on the left side, due to the superelevation of the water surface at C above that at D. The element GH is subjected to the same net hydrostatic force inward, but the centrifugal force outward is much less because the velocity is decreased by friction near the bottom. This results in a cross flow inward along the bottom of the channel, which is balanced by an outward flow near the water surface; hence the spiral. This spiral flow is largely responsible for the commonly observed erosion of the outside bank of a river bend, with consequent deposition and building up of a sand bar near the inside bank.

11.22 TRANSITIONS

Special transition sections are often used to join channels of different size and shape in order to avoid undesirable flow conditions and to minimize head loss. If the flow is subcritical, a straightline transition (Fig. 11.29) with an angle of about 12.5° is fairly satisfactory and will result in a head loss of about one-tenth of the difference in velocity heads for accelerating flow and three-tenths of the difference in velocity heads for decelerating flow. Without the transition, i.e., with an abrupt change in section with square corners, the corresponding head losses are about 0.5 and 1.0 times the difference in velocity heads. At Froude numbers between about 0.5 and 1.0, complex warped transitions are advisable. In supercritical flow

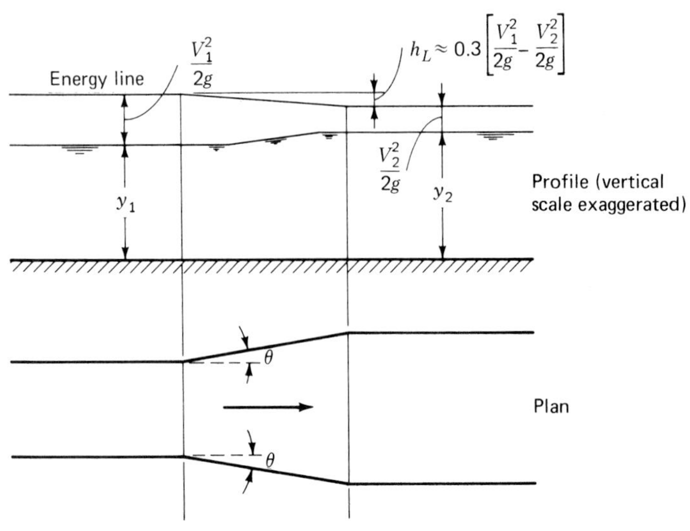

Figure 11.29 Simple open-channel transition for decelerating flow.

(**F** > 1), surface waves are formed and special procedures are required for transition design.[1]

At channel entrance from a reservoir or from a larger channel the head loss for a square-edged entrance is about 0.5 times the velocity head. By rounding the entrance (Illustrative Example 11.6) the head loss can be reduced to slightly less than 0.2 times the velocity head.

Illustrative Example 11.6 Consider a rectangular flume 4.5 m wide, built of unplaned planks ($n = 0.014$), leading from a reservoir in which the water surface is maintained constant at a height of 1.8 m above the bed of the flume at entrance (see accompanying figure). The flume is on a slope of 0.001. The depth 300 m downstream from the head end of the flume is 1.20 m. Assuming an entrance loss of $0.2V_1^2/2g$, find the flow rate for the given conditions.

For a first approximate answer we shall consider the entire flume as one reach. The equations to be satisfied are

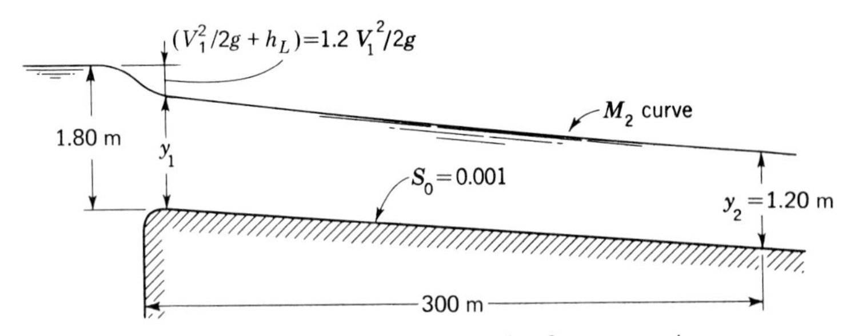

Illustrative Example 11.6 Mild-slope flume leading from reservoir.

[1] Arthur T. Ippen, Design of Channel Contractions, *Trans. ASCE*, vol. 116, pp. 326–346, 1951.

Energy at entrance:

$$y_1 + \frac{1.2V_1^2}{2g} = 1.80 \qquad (a)$$

Energy equation (11.37) for the entire reach:

$$y_1 + \frac{V_1^2}{2g} = 1.20 + \frac{V_2^2}{2g} + (S - 0.001)L \qquad (b)$$

where S is given by Eq. (11.38*b*):

$$S = \left(\frac{nV_m}{R_m^{2/3}}\right)^2 \qquad (c)$$

The procedure is to make successive trials of the upstream depth y_1. This determines corresponding values of V_1, q, V_2, V_m, R_m, and S. The trials are repeated until the value of Δx from Eq. (*b*) is close to 300 m. The solution is conveniently set in tabular form as follows:

Trial y_1, m	V_1, Eq. (*a*), m/s	$q = y_1V_1$, m^2/s	$V_2 = q/1.20$, m/s	V_m, m/s	R_{h_1}, m	R_{h_2}, m	R_{h_m}, m	S, Eq. (*c*)	Δx, Eq. (*b*), m
1.50	2.22	3.33	2.78	2.50	0.90	0.78	0.89	0.00143	358
1.48	2.29	3.39	2.82	2.56	0.89	0.78	0.835	0.00163	226

Thus $y_1 \approx 1.49$ m and the flow rate $Q = qB \approx 3.36 \times 4.5 \approx 15.1\ m^3/s$. The accuracy of the result, of course, depends on one's ability to select the correct value for Manning's n. If n was assumed to be 0.015, for example, rather than 0.014, the result would have been quite different. Also, a more accurate result can be obtained by dividing the flume into reaches in which the depth change is about 10 percent of the depth.

11.23 HYDRAULICS OF CULVERTS

A culvert is a conduit passing under a road or highway. In section, culverts may be circular, rectangular, or oval. Culverts may operate with either a submerged entrance (Fig. 11.30) or a free entrance (Fig. 11.31).

In the case of a submerged entrance there are three possible regimes of flow as indicated in Fig. 11.30. Under conditions (*a*) and (*b*) of the figure the culvert is said to be flowing under *outlet control*, while condition (*c*) represents *entrance control*. In (*a*) the outlet is submerged, possibly because of inadequate channel capacity downstream or backwater from an intersecting stream. In (*b*) the normal depth y_0 of the flow is greater than the culvert height D and the culvert flows full.[1] The same equation is applicable to both conditions (*a*) and (*b*), namely

$$\Delta h = h_e + h_f + h_v \qquad (11.48)$$

[1] If there is a contraction of flow at the culvert entrance, reexpansion will require about six diameters. Hence a very short culvert may not flow full even though $y_0 \geq D$.

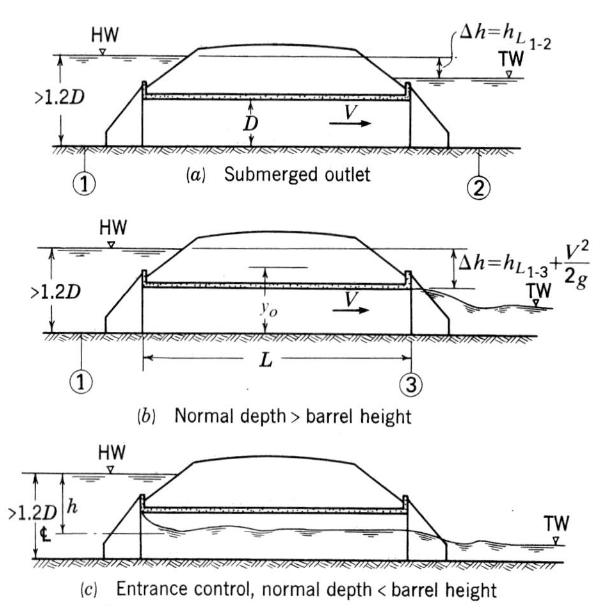

Figure 11.30 Flow conditions in culverts with submerged entrance.

where Δh is defined[1] in Figs. 11.30*a* and 11.30*b*, h_e is the entrance head loss, h_f is the friction head loss in the culvert barrel, and h_v is the velocity head loss at submerged discharge under condition (*a*) or the residual velocity head at discharge under condition (*b*).

Entrance loss is a function of the velocity head in the culvert, while friction loss may be computed using Manning's equation [Eq. (11.9)]. Thus, in English units,[2]

$$\Delta h = k_e \frac{V^2}{2g} + \frac{n^2 V^2 L}{2.22 R^{4/3}} + \frac{V^2}{2g} \tag{11.49}$$

This expression in English units can be reduced to

$$\Delta h = \left(k_e + \frac{29 n^2 L}{R^{4/3}} + 1 \right) \frac{V^2}{2g} \tag{11.50}$$

The entrance coefficient k_e is about 0.5 for a square-edged entrance and about 0.05 if the entrance is well rounded. If the outlet is submerged, the head loss may

[1] The energy grade line at the entrance and exit is above the actual water surface by the velocity head of the approaching or leaving water. With water ponded at the entrance or exit this velocity head is usually negligible but should be included in computations if it is of significant magnitude.

[2] In metric units the 2.22 disappears from the denominator of the second term on the right-hand side of Eq. (11.49) and the constant 29 in the numerator of the second term in parentheses of Eq. (11.50) becomes 19.62.

be reduced somewhat by flaring the culvert outlet so that the outlet velocity is reduced and some of the velocity head recovered. Tests show that the flare angle should not exceed about 6° for maximum effectiveness.

If normal depth in the culvert is less than the barrel height, with the inlet submerged and the outlet free, the condition (*c*) illustrated in Fig. 11.30*c* will normally result. This culvert is said to be flowing under *entrance control*, i.e., the entrance will not admit water fast enough to fill the barrel, and the discharge is determined by the entrance conditions. The inlet functions like an orifice for which

$$Q = C_d A \sqrt{2gh} \tag{11.51}$$

where h is the head on the center of the orifice[1] and C_d is the orifice coefficient of discharge. The head required for a given flow Q is therefore

$$h = \frac{1}{C_d^2} \frac{Q^2}{2gA^2} \tag{11.52}$$

For a sharp-edged entrance without suppression of the contraction $C_d = 0.62$, while for a well-rounded entrance C_d approaches unity. If the culvert is set with its invert at stream-bed level, the contraction is suppressed at the bottom. Flared wingwalls may also cause partial suppression of the side contractions. Because of the wide variety of entrance conditions which may be encountered, it is impracticable to cite appropriate values of C_d, and for a specific design these must be determined from model tests or tests of similar entrances.

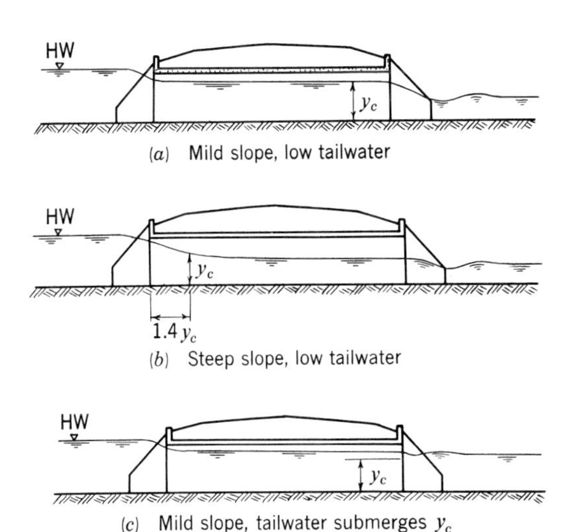

Figure 11.31 Flow conditions in culverts with free entrance.

[1] Equation (11.51) applies for an orifice on which the head h to the center of the orifice is large compared with the orifice height D. When the headwater depth is $1.2D$ (that is, $h = 0.7D$), an error of 2 percent results. In the light of other uncertainties in the design this error can be ignored.

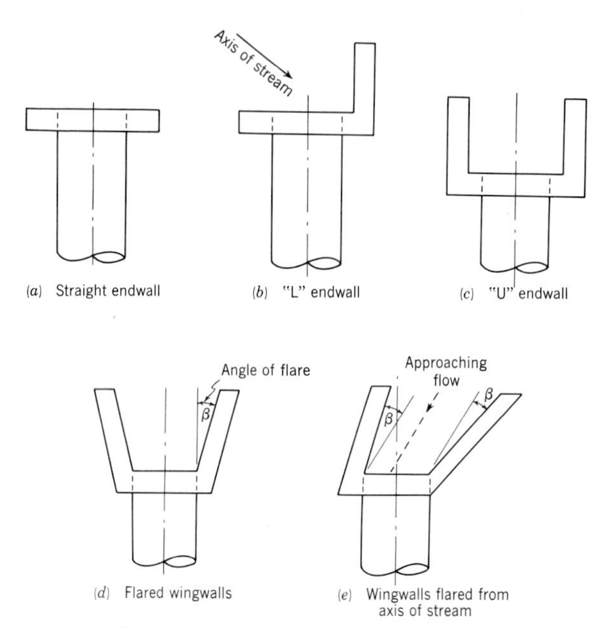

Figure 11.32 Culvert endwalls and wingwalls.

Some box culverts may be designed so that the top of the box forms the roadway. In this case the headwater should not submerge the inlet, and one of the flow conditions of Fig. 11.31 (free entrance) will exist. In cases (*a*) and (*b*) critical depth in the barrel controls the headwater elevation, while in case (*c*) the tailwater elevation is the control. In all cases the headwater elevation may be computed using the principles of open-channel flow discussed in this chapter with an allowance for entrance and exit losses.

When the culvert is on a steep slope [case (*b*)], critical depth will occur at about $1.4y_c$ downstream from the entrance. The water surface will impinge on the headwall when the headwater depth is about $1.2D$ if y_c is $0.8D$ or more. Since it would be inefficient to design a culvert with y_c much less than $0.8D$, a headwater depth of $1.2D$ is approximately the boundary between free-entrance conditions (Fig. 11.30) and submerged-entrance conditions (Fig. 11.31).

The geometry of the entrance structure is an important aspect of culvert design. Entrance structures (Fig. 11.32) serve to protect the embankment from erosion and, if properly designed, may improve the hydraulic characteristics of the culvert. The straight endwall (*a*) is used for small culverts on flat slopes when the axis of the stream coincides with culvert axis. If an abrupt change in flow direction is necessary, the L endwall (*b*) is used. The U-shaped endwall (*c*) is the least efficient form from the hydraulic viewpoint and has the sole advantage of economy of construction. Where flows are large, the flared wingwall (*d*) is preferable. The flare should, however, be made from the axis of the approaching stream (*e*) rather than from the culvert axis. Some hydraulic advantage is gained by warping wingwalls into a smooth transition, but the gain is not usually sufficient to offset the cost of the complex forming required for such warped surfaces.

The purpose of the culvert outlet is to protect the downstream slope of the fill from erosion and prevent undercutting of the culvert barrel. Where the discharge velocity is low or the channel below the outlet is not subject to erosion, a straight endwall or a U-shaped endwall may be quite sufficient. At higher velocities, lateral scour of the embankment or channel banks may result from eddies at the end of the walls, especially when the culvert is much narrower than the outlet channel. With moderate velocities, flaring of outlet wingwalls is helpful, but the flare angle β must be small enough so that the stream from the culvert will adhere to the walls of the transition.

PROBLEMS

11.1 For the channel of Illustrative Example 11.1 compute the "open-channel Reynolds number" assuming that water at 50°F is flowing. Refer to Fig. 8.11 to verify whether or not the flow is wholly rough. Determine e from Fig. 8.11 and compare it with the value computed in the example.

11.2 Assuming the values of f versus $\mathbf{R}$ and e/D given for pipes in Fig. 8.11 to apply to open channels as well, find the rate of discharge of water at 60°F in a 100-in-diameter smooth concrete pipe flowing half full, if the pipe is laid on a grade of 2 ft/mi. Note that D should be replaced by $4R$.

11.3 For the channel of Illustrative Example 11.1, compute the flow rate for depths of 1, 3, 5, and 7 ft. Plot a curve of Q versus y.

11.4 The figure shows a cross section of a canal forming a portion of the Colorado River Aqueduct, which is to carry 1,600 cfs (44 m^3/s). The canal is lined with concrete, for which n is assumed to be 0.014. (*a*) What must be the grade of the canal, and what will be the drop in elevation per mile (kilometer)? (*b*) If the flow in the canal were to decrease to 800 cfs (22 m^3/s), all other data, including the slope, being the same, what would be the depth of the water?

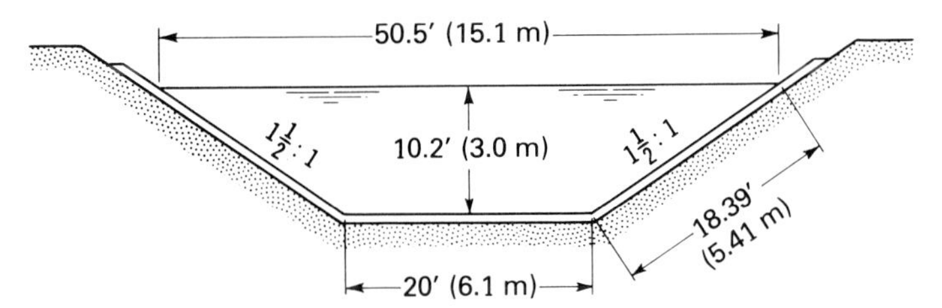

Problem 11.4

11.5 In Prob. 11.4 find the corresponding value of e and compare it with values previously given for concrete pipe. Does it fall in the range given?

11.6 What would be the capacity of the canal of Prob. 11.4 if the grade were to be 1.2 ft/mi?

11.7 Water flows uniformly in a 2-m-wide rectangular channel at a depth of 45 cm. The channel slope is 0.002 and $n = 0.014$. Find the flow rate in m^3/s.

11.8 At what depth will water flow in a 3-m-wide rectangular channel if $n = 0.017$, $S = 0.00085$, and $Q = 4$ m^3/s?

11.9 The figure shows a tunnel section on the Colorado River Aqueduct. The area of the water cross section is 191 ft^2, and the wetted perimeter is 39.1 ft. The flow is 1,600 cfs. If $n = 0.013$ for its concrete lining, find the slope.

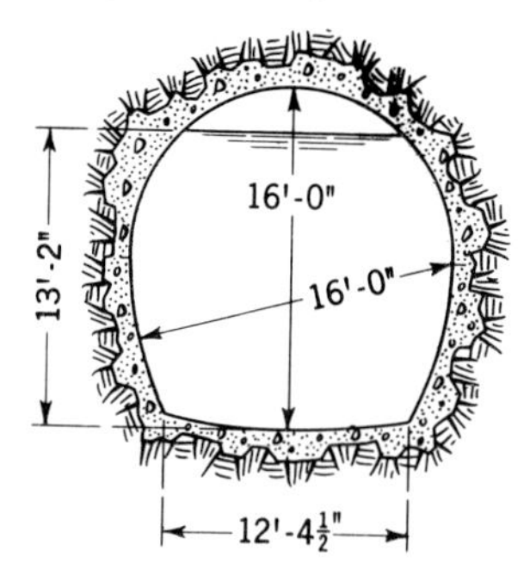

Problem 11.9

11.10 A monolithic concrete inverted siphon on the Colorado River Aqueduct is circular in cross section and is 16 ft (4.88 m) in diameter. Obviously, it is completely filled with water, unlike the case of Prob. 11.9. (*a*) If $n = 0.013$, find the slope of the hydraulic grade line for a flow of 1,600 cfs (45 m^3/s). (*b*) Solve the same problem using the methods of Chap. 8 and assuming a mean value of e from Table 8.1 for concrete pipe. Compare the result with that of part (*a*).

11.11 A 30-in-diameter pipe is known to have a Manning's n of 0.021. What is Manning's n for a 96-in-diameter pipe having exactly the same e as the smaller pipe?

11.12 If $a = 2$ m, $b = 5$ m, $d = 3$ m, $w = 25$ m, and $n = 0.014$, what slope is required so that the flow will be 30 m^3/s when the depth of flow is 2.50 m?

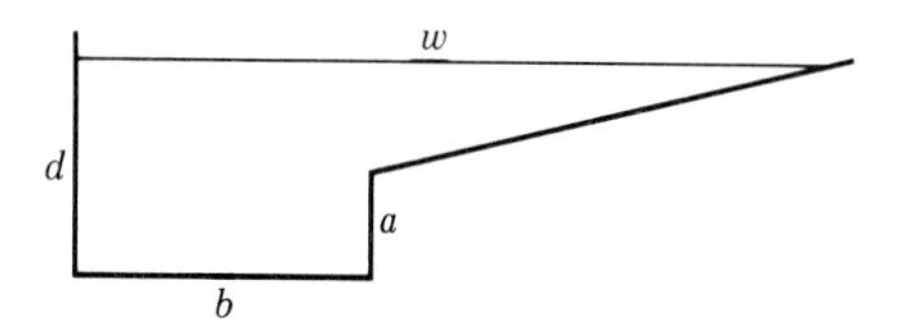

Problem 11.12

11.13 Water flows in a rectangular flume 5 ft (1.5 m) wide made of unplaned timber ($n = 0.013$). Find the necessary channel slope if the water flows uniformly at a depth of 2 ft (0.6 m) and at 15 fps (4.5 m/s).

11.14 Refer to Fig. 11.5. Suppose the widths of A_1, A_2, and A_3 are 100, 30, and 200 ft (30, 10, and 60 m) and the total depths are 2, 10, and 3 ft (0.5, 3.0, and 1.0 m). Compute the flow rate if $S = 0.0016$, $n_1 = n_3 = 0.04$, and $n_2 = 0.025$.

11.15 Refer to the figure for Prob. 11.12. Find the flow rate at water depths of 1, 2, 3, 4, and 5 ft if $n = 0.020$ and $S = 0.0015$. The dimensions are as follows: $a = 3$ ft, $b = 6$ ft, $d = 5$ ft, and $w = 36$ ft.

11.16 In the figure, with uniform flow in the wide open channel, $a = 2.80$ ft. Find b if $n = 0.020$.

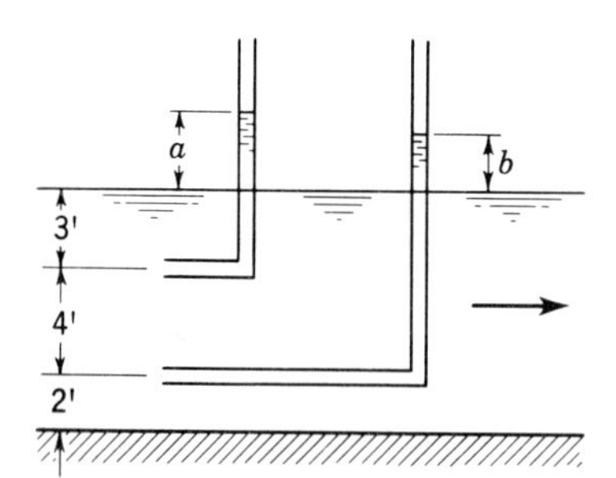

Problem 11.16

11.17 Using Eq. (11.13), determine the depth below the surface (clear water) at which the velocity is equal to the mean velocity. Also find the average of the velocities at 0.2 and 0.8 depths. Let $y = 4$ ft, $S = 0.001$, and $n = 0.025$.

11.18 Water flows uniformly in a very wide rectangular channel at a depth y_0 of 1.5 m. $S = 0.006$ and $n = 0.015$. Calculate the velocities at y-values of 0.15, 0.3, 0.6, 0.9, and 1.5 m and plot the velocity profile. Note the value of the maximum velocity at the water surface.

11.19 Consider a variety of rectangular sections all of which have a cross-sectional area of 20 ft^2. Plot the hydraulic radii versus channel widths for a range of channel widths from 2 to 20 ft and note the depth : width ratio when R_h is maximum.

11.20 Set up a general expression for the wetted perimeter P of a trapezoidal channel in terms of the cross-sectional area A, depth y, and angle of side slope ϕ. Then differentiate P with respect to y with A and ϕ held constant. From this prove that $R_h = y/2$ for the section of greatest hydraulic efficiency (i.e., smallest P for a given A).

11.21 Using the results of Prob. 11.20 prove that the most efficient triangular section is the one with a 90° vertex angle.

11.22 The amount of water to be carried by a canal excavated in smooth earth ($n = 0.030$) is 370 cfs (10 m^3/s). It has side slopes of 2:1, and the depth of water y is to be 5 ft (1.5 m) or less (see figure). If the slope is 2.5 ft/mi (45 cm/km), what must be the width at the bottom? How does this compare with the most efficient trapezoidal section for these side slopes?

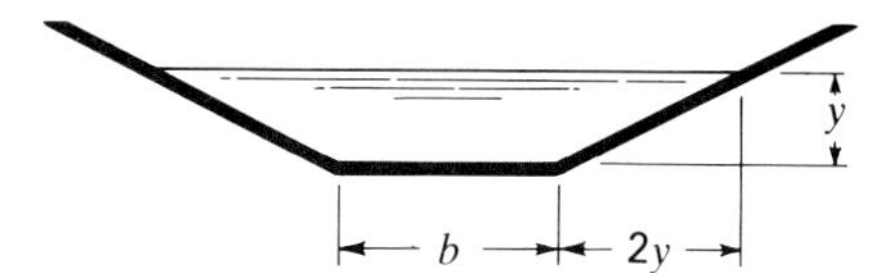

Problem 11.22

11.23 Refer to the figure for Prob. 11.22. If the discharge in the canal ($n = 0.030$) is to be 200 cfs (5 m^3/s) while the depth y is 5 ft (1.5 m) and the velocity is not to exceed 150 ft/min (45 m/min), what must be the width at the bottom and the drop in elevation per mile (km)? Compare this with the bottom width for maximum efficiency.

11.24 A rectangular flume of planed timber ($n = 0.012$) slopes 1 ft per 1,000 ft (1 m per km). (*a*) Compute the rate of discharge if the width is 6 ft (2 m) and the depth of water is 3 ft (1 m). (*b*) What would be the rate of discharge if the width were 3 ft (1 m) and the depth of water 6 ft (2 m)? (*c*) Which of the two forms would have the greater capacity and which would require less lumber?

11.25 What diameter of semicircular channel will have the same capacity as a rectangular channel of width 8 ft and depth 3 ft. Assume S and n are the same for both channels. Compare the lengths of the wetted perimeters.

11.26 Prove that the value of θ given in Sec. 11.7 for the point of maximum discharge is correct. After differentiating, a trial-and-error type of solution will be found most practical here.

11.27 Water flows uniformly in a circular concrete pipe ($n = 0.014$) of diameter 10 ft at a depth of 4 ft. Using Fig. 11.9, determine the flow rate and the average velocity of flow. $S = 0.0003$.

11.28 At what depth will water flow at 0.25 m^3/s in a 100-cm-diameter concrete pipe on a slope of 0.004? (*a*) Assume $n = 0.013$. (*b*) Repeat with $n = 0.015$ and compare results.

11.29 Evaluate the friction factor f for laminar flow in terms of the Reynolds number, and compare with Eq. (8.23) for pipe flow. (*Note:* Recall that for a wide channel the hydraulic radius is approximately equal to the depth.)

11.30 Eastern lubricating oil (SAE 30) at 90°F (30°C) flows down a flat plate 10 ft (3 m) wide. What is the maximum rate of discharge at which laminar flow may be ensured, assuming that the critical Reynolds number is 500? What should be the slope of the plate to secure a depth of 6 in (15 cm) at this flow rate?

11.31 In Sec. 11.8 the velocity distribution in laminar sheet flow was found to be $u = (gS/2\nu)y_0^2[1 - (y/y_0)^2]$, where y in this case is the variable distance downward from the surface. Evaluate α, and compare it with the result of Prob. 4.1.

11.32 At what rate (cfs/ft of width) will 60°F water flow in a wide rectangular channel on a slope of 0.00015 if the depth is 0.01 ft? Assume laminar flow and justify this assumption by computing Reynolds number.

11.33 At what rate (L/s per meter of width) will water at 15°C flow in a wide, smooth, rectangular channel on a slope of 0.0003, if the depth is 8.0 mm? Assume laminar flow and justify the assumption by computing Reynolds number.

11.34 Consider a wide rectangular channel on a given slope. With what power of the discharge does the depth vary? With what power of the discharge does the critical depth vary? As the flow increases, does the Froude number increase or decrease? Assume Manning equation, with constant value of n.

11.35 Differentiate Eq. (11.16) to obtain the expression for y_c given in Eq. (11.21).

11.36 Water flows with a velocity of 4 fps and at a depth of 2 ft in a wide rectangular channel. Is the flow subcritical or supercritical? Find the alternate depth for the same discharge and specific energy by two methods: (*a*) by direct solution of Eq. (11.17); (*b*) by use of Fig. 11.12.

11.37 Water flows down a wide rectangular channel of concrete ($n = 0.014$) laid on a slope of 0.002. Find the depth and rate of flow in SI units for critical conditions in this channel.

11.38 Water is released from a sluice gate in a rectangular channel 5 ft (1.5 m) wide such that the depth is 2 ft (0.6 m) and the velocity is 15 fps (4.5 m/s). Find (*a*) the critical depth for this specific energy; (*b*) the critical depth for this rate of discharge; (*c*) the type of flow and the alternate depth by either direct solution or the discharge curve.

11.39 A flow of 100 cfs is carried in a rectangular channel 10 ft wide at a depth of 1.2 ft. If the channel is made of smooth concrete ($n = 0.012$), find the slope necessary to sustain uniform flow at this depth. What roughness coefficient would be required to produce uniform critical flow for the given rate of discharge on this slope?

11.40 A long straight rectangular channel 10 ft (3 m) wide is observed to have a wavy water surface at a depth of about 6 ft (2 m). Estimate the rate of discharge.

11.41 A thin rod is placed vertically in a stream which is 3 ft deep, and the resulting small disturbance wave is observed to make an angle of about 55° with the axis of the stream. Find the approximate velocity of the stream.

11.42 At a point in a shallow lake, the wave from a passing boat is observed to rise 1 ft (30 cm) above the undisturbed water surface. The observed speed of the wave is 10 mph (16 km/h). Find the approximate depth of the lake at this point. Compute it three ways and comment on them.

11.43 A rectangular channel 10 ft wide carries a flow of 200 cfs. Find the critical depth and the critical velocity for this flow. Find also the critical slope if $n = 0.020$.

11.44 A flow of 10 cfs (0.28 m^3/s) of water is carried in a 90° triangular flume built of planed timber ($n = 0.011$). Find the critical depth and the critical slope.

11.45 A trapezoidal canal with side slopes of 2:1 has a bottom width of 10 ft (3 m) and carries a flow of 600 cfs (20 m^3/s). (*a*) Find the critical depth and critical velocity. (*b*) If the canal is lined with brick ($n = 0.015$), find the critical slope for the same rate of discharge.

11.46 For a circular conduit with a diameter of 10 ft, compute the specific energy for a flow of 100 cfs at depths of 1, 3, 5, and 8 ft assuming $\alpha = 1.0$. At what depth is E the least? Check to see if Eq. (11.32) is satisfied at this depth.

11.47 A circular conduit of well-laid (smooth) brickwork when flowing half full is to carry 400 cfs at a velocity of 10 fps. (*a*) What will be the necessary fall per mile? (*b*) Will the flow be subcritical or supercritical?

11.48 This figure and table describe the cross section of an open channel for which $S_0 = 0.02$ and $n = 0.015$. The sketch is drawn to the scale shown. When the flow rate is 100 cfs, find (*a*) the depth for uniform flow and (*b*) the critical depth. *Hint:* The cross-sectional area may be found by planimetry or counting squares and the wetted perimeter by use of dividers.

y ft	x_L ft	x_R ft
0	0	0
1	1.5	1.1
2	2.6	2.0
3	3.25	3.3
4	3.6	5.5
5	3.85	8.2

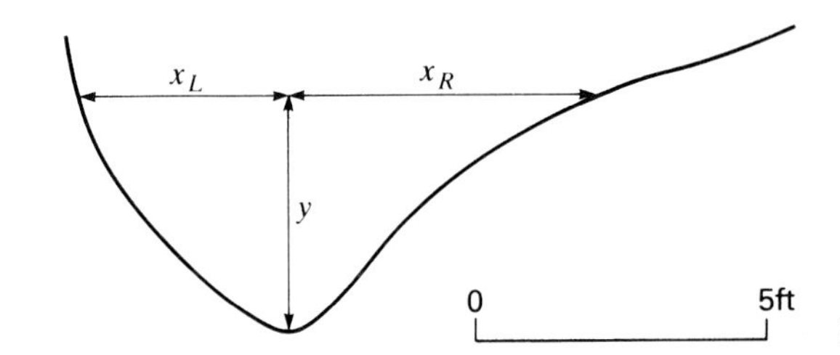

Problem 11.48

11.49 Refer to the figure, table, and hint for Prob. 11.48, replacing feet dimensions with meters. Let the slope be 0.007 with $n = 0.015$. When the flow rate is 50 m^3/s, find (*a*) the depth for uniform flow; (*b*) the critical depth.

11.50 A rectangular channel 2 m wide carries 2 m^3/s of water in subcritical uniform flow at a depth of 0.8 m. A frictionless hump is to be installed across the bed. Find the critical hump height (i.e., the minimum hump which causes y_c on it).

11.51 Work Illustrative Example 11.3 ($y_0 = 2.0$ ft) for the case where the flow rate is 50 cfs.

11.52 A flow of 1.8 m^3/s is carried in a rectangular channel 1.6 m wide at a depth of 0.9 m. Will critical depth occur at a section where (*a*) a frictionless hump 15 cm high is installed across the bed, (*b*) a frictionless sidewall constriction (with no hump) reduces the channel width to 1.3 m, and (*c*) the hump and the sidewall constriction are installed together? Show calculations.

11.53 Work all parts of Illustrative Example 11.3 ($y_0 = 2.0$ ft) for the case where the flow rate is 16 cfs.

11.54 A rectangular channel 4 ft (1.2 m) wide carries 40 cfs (1.1 m^3/s) of water in uniform flow at a depth of 2.80 ft (0.85 m). If a bridge pier 1 ft (0.3 m) wide is placed in the middle of this channel, find the local change in the water-surface elevation. What is the minimum width of constricted channel which will not cause a rise in water surface upstream?

11.55 A rectangular channel 10 ft wide carries 20 cfs in uniform flow at a depth of 0.90 ft. Find the local change in water-surface elevation caused by an obstruction 0.20 ft high on the floor of the channel.

11.56 Suppose that the channel of Prob. 11.55 is so sloped that uniform flow of 20 cfs occurs at a depth of 0.30 ft. Find the local change in water-surface elevation caused by the 0.20-ft-high obstruction.

11.57 Suppose that the depth of uniform flow in the channel of Prob. 11.54 is 0.90 ft (27 cm). Find the change in water-surface elevation caused by the bridge pier. The flow rate is 40 cfs (1.1 m^3/s).

11.58 Fifty cubic feet per second (1.4 m^3/s) of water flows uniformly in a channel of width 6 ft (1.8 m) at a depth of 2.5 ft (0.75 m). What is the change in water-surface elevation at a section contracted to a 4-ft (1.2 m) width with a 0.2-ft (6 cm) depression in the bottom?

11.59 A rectangular flume of planed timber ($n = 0.012$) is 5 ft (1.5 m) wide and carries 60 cfs (1.7 m^3/s) of water. The bed slope is 0.0006, and at a certain section the depth is 3 ft (0.9 m). Find the distance (in one reach) to the section where the depth is 2.5 ft (0.75 m). Is the distance upstream or downstream?

11.60 Suppose that the slope of the flume in Prob. 11.59 is now changed so that, with the same flow, the depth varies from 4 ft (1.2 m) at one section to 3 ft (0.9 m) at a section 1,000 ft (300 m) downstream. Find the new bed slope of the flume, using one reach. Sketch the flume, the energy grade line, and the water surface to assure that the answer is reasonable.

11.61 Suppose that the slope of the flume in Prob. 11.59 is now increased to 0.01. With the same flow as before, find the depth 50 ft (15 m) downstream from a section where the flow is 1.5 ft (45 cm) deep. Use only one reach. Is the flow subcritical or supercritical? [*Note:* In this case it will not be possible to make a direct solution from Eq. (11.25). A trial-and-error solution may best be set in the form of a table with the headings y_2, V_2, V_m, A_2, P_2, R_2, R_m, $R_m^{2/3}$, S, etc.]

11.62 Repeat Prob. 11.61, but increase the 50-ft (15-m) distance to 500 ft (150 m).

11.63 A test on a rectangular glass flume 10 in (25 cm) wide yielded the following data on a reach of 30-ft (9-m) length: with still water, $z_1 - z_2 = 0.009$ ft (2.7 mm); with a measured flow of 0.1516 cfs (4.3 L/s), $y_1 = 0.361$ ft (11.02 cm), $y_2 = 0.366$ ft (11.17 cm). Find the value of the roughness coefficient n using only one reach.

11.64 A rectangular flume 10 ft (3 m) wide is built of planed timber ($n = 0.012$) on a bed slope of 0.2 ft per 1,000 ft (20 cm per km), ending in a free overfall. If the measured depth at the fall is 1.82 ft (0.55 m), find (*a*) the rate of flow; (*b*) the distance upstream from the fall to where the depth is 4 ft (1.2 m). [*Note:* Assume that critical depth occurs at a distance of $4y_c$ upstream from the fall, and employ reaches with end depths of 2.7, 3.0, 3.4, and 4.0 ft (0.82, 0.9, 1.0, and 1.2 m).]

11.65 A wide rectangular channel dredged in earth ($n = 0.035$) is laid on a slope of 10 ft/mi and carries a flow of 100 cfs/ft of width. Find the water depth 2 mi upstream of a section where the depth is 28.9 ft. (*a*) Compute, using a single reach, and (*b*) compare the result with that obtained using three equal reaches.

11.66 The slope of a stream of rectangular cross section is $S_0 = 0.0002$, the width is 160 ft, and the value of the Chézy C is 78.3 $ft^{1/2}/s$. Find the depth for a uniform flow of 88.55 cfs per foot of width of the stream. If a dam raises the water level so that at a certain distance upstream the increase is 5 ft, how far from this latter section will the increase by only 1 ft? Use reaches with 1-ft depth increments.

11.67 A portion of an outfall sewer is approximately a circular conduit 5 ft in diameter and with a slope of 1 ft in 1,100 ft. It is of brick, for which $n = 0.013$. (*a*) What would be its maximum capacity for uniform flow? (*b*) If it discharges 120 cfs with a depth at the end of 2.90 ft, how far back from the end must it become a pressure conduit? Proceeding from the mouth upstream, find by tabular solution the length of sewer which is not flowing full. Use three reaches with equal depth increments. (*c*) Find the pressure at the top and bottom of the pipe at the section that is 600 ft upstream from the point at which the sewer no longer flows full. Sketch the energy line and hydraulic grade line from this point to the end of the sewer.

11.68 For the channel of Prob. 11.48 find the distance between a section where the depth is 3.0 ft to another where the depth is 2.5 ft. Which section is upstream? Use only one reach. See the hint of Prob. 11.48.

11.69 When the flow in a certain natural stream is 7,600 cfs, it is required to find the elevation of the water surface at different sections upstream from a certain initial point. A survey of the channel shows that conditions are fairly similar for a length of 1,500 ft upstream from the initial point, and then beyond that there is another stretch of 2,200 ft, and so on. Assuming a rise in the water surface in the distance of 1,500 ft to be 0.20 ft, a study of the stream bed shows the average values of the area and wetted perimeter to be as given in the table below. The computed head loss, based on average velocity and hydraulic radius, is seen to be 0.283 ft, which is greater than that assumed. Hence assume a larger value, and repeat. Complete the following table, and find the probable rise in elevation in the first 1,500 ft. In a similar manner the rises in other lengths may be computed, and the sum of all of them up to the desired point will give the elevation at that point above the initial. Assume $n = 0.036$.

Assumed rise SL, ft	A average, ft^2	P average, ft	R average, ft	V average, fps	$SL = LV^2/C^2R$, ft
0.20	3,100	350	8.86	2.45	0.283
0.25	3,180	359	8.86		
0.26	3,190	360	8.86		
0.27	3,220	363	8.86		
0.28	3,230	364	8.86		

11.70 Classify the water-surface profile of Prob. 11.59 as one of the forms shown in Fig. 11.20. Show all necessary calculations.

11.71 Repeat Prob. 11.70 when the channel slope is -0.00040.

11.72 The flow in a 15-ft-wide rectangular channel which has a constant bottom slope is 1,400 cfs. A computation using Manning's equation indicates that the normal depth is 6.0 ft. At a certain section the depth of flow in the channel is 2.8 ft. Will the depth increase, decrease, or remain the same as one proceeds downstream from this section? Sketch a physical situation where this type of flow will occur.

11.73 Classify the water-surface profile of Prob. 11.61 as one of the forms shown in Fig. 11.20. Show all necessary calculations. Sketch the profile.

11.74 Repeat Prob. 11.73 for the channel of Prob. 11.63 if its $n = 0.012$.

11.75 A trapezoidal canal dredged in smooth earth ($n = 0.030$) has a bottom width of 15 ft, side slopes of 1:1, and a bed slope $S_0 = 0.0003$. With a flow of 800 cfs, $y_c = 4.05$ ft, and $y_0 = 10.8$ ft. Find the length of M_2 curve extending from a free overfall back to where the depth is 10 ft. Use reaches with end depths of 6, 8, and 10 ft.

11.76 A rectangular channel 10 ft wide carries 100 cfs in uniform flow at a depth of 1.67 ft. Suppose that an obstruction such as a submerged weir is placed across the channel, rising to a height of 6 in above the bottom. (*a*) Will this obstruction cause an hydraulic jump to form upstream? Why? (*b*) Find the water depth over the obstruction, and classify the surface profile, if possible, to be found upstream from the weir. Sketch the resulting water surface profile and energy line, showing y_c and y_0.

11.77 Suppose that the slope and roughness of the channel in Prob. 11.76 are such that uniform flow of 100 cfs occurs at 1.00 ft. Consider an obstruction rising 4 in above the bottom of the channel. Will an hydraulic jump form upstream? As in Prob. 11.76, classify the surface profile found just upstream from the obstruction.

11.78 Analyze the water-surface profile in a long rectangular channel ($n = 0.013$). The channel is 10 ft wide, the flow rate is 400 cfs, and there is an abrupt change in slope from 0.0016 to 0.0150. Make a sketch showing normal depths, critical depths, and water surface profile types. Refer to Illustrative Example 11.5 for information on normal depths and critical depth.

11.79 Repeat Prob. 11.78 for the case where the flow rate is 150 cfs.

11.80 Repeat Prob. 11.79 for the case where the slope change is from 0.0016 to 0.0006. Compute the approximate distance upstream from the break to the point where normal depth occurs, using one reach.

11.81 In a 6-ft-wide rectangular channel ($S_0 = 0.002$, $n = 0.013$) water flows at 250 cfs. A low dam (broad-crested weir) placed in the channel raises the water to a depth of 8.9 ft. Analyze the water-surface profile upstream from the dam.

11.82 Solve Prob. 11.81 if the channel slope is (*a*) 0.0005, (*b*) 0.0008, and (*c*) 0.005.

11.83 In a rectangular channel 10 ft (3 m) wide with a flow of 200 cfs (5.65 m^3/s) the depth is 1 ft (0.3 m). If a hydraulic jump is produced, what will be the depth after it? What will be the loss of energy?

11.84 The hydraulic jump may be used as a crude flowmeter. Suppose that in a horizontal rectangular channel 5 ft wide the observed depths before and after a hydraulic jump are 0.66 and 3.00 ft, respectively. Find the rate of flow and the head loss.

11.85 Repeat Prob. 11.83 for the case where the channel bed slopes at 10 degrees. For this slope, jump length $\approx 4y_2$. Assume friction force = 400 lb/ft (6 kN/m) of width. Also find horsepower (kW) loss.

11.86 The tidal bore, which carries the tide into the estuary of a large river, is an example of an abrupt translatory wave, or moving hydraulic jump. Suppose such a bore is observed to rise to a height of 12 ft above the normal low-tide river depth of 8 ft. The speed of travel of the bore upstream is observed to be 15 mph. Find the approximate velocity of the undisturbed river. Does this represent subcritical or supercritical flow? (*Note:* The theory developed in Sec. 11.19 is based on the hydraulic jump in a fixed position. In the case of a moving jump, all kinematic terms must be considered relative to the moving wave as a frame of reference.)

11.87 An hydraulic jump occurs in a triangular flume having side slopes 1:1. The flow rate is 15 cfs ($0.45\ m^3/s$) and the depth before jump is 1.0 ft (0.3 m). Find the depth after jump and the power loss in the jump.

11.88 An hydraulic jump occurs in a 15-ft (5-m)-wide rectangular channel carrying 200 cfs ($6\ m^3/s$) on a slope of 0.005. The depth after jump is 4.5 ft (1.4 m). (*a*) What must be the depth before jump? (*b*) What are the losses of energy and power in the jump?

11.89 A very wide rectangular channel with bed slope $S_0 = 0.0003$ and roughness $n = 0.020$ carries a steady flow of 50 cfs/ft of width. If a sluice gate (Fig. 12.32) is so adjusted as to produce a minimum depth of 1.5 ft in the channel, determine whether an hydraulic jump will form downstream, and if so, find (using one reach) the distance from the gate to the jump.

11.90 A rectangular channel 10 ft wide carries 300 cfs in uniform flow at a depth of 4 ft. By how much should the outside wall be elevated above the inside wall for a bend of 40-ft radius to the center line of the channel?

11.91 Repeat Prob. 11.90 for the same conditions except that the normal depth is 2 ft.

11.92 A rectangular channel 4 m wide carries $6\ m^3/s$ in uniform flow at a depth of 1.5 m. What will be the maximum difference in water-surface elevations between the inside and outside of a circular bend in this channel if the radius of the bend is 25 m?

11.93 Refer to Fig. 11.29. A rectangular channel changes in width from 4 to 6 ft. Measurements indicate that $y_1 = 2.50$ ft and $Q = 50$ cfs. Determine the depth y_2 by (*a*) neglecting head loss; (*b*) considering the head loss to be given as shown on the figure.

11.94 A rectangular flume of planed timber ($n = 0.012$) 20 ft wide, 1,000 ft long, with horizontal bed leads from a reservoir in which the still-water surface is 10 ft above the flume bed. Assume that the depth of the downstream end of the flume is fixed at 8 ft by some control section downstream. Allowing an entrance loss of 0.2 × velocity head, find the flow rate in the flume using one reach.

11.95 Find the flow rate in the flume of Prob. 11.94 if it ends in a free fall, all other conditions remaining the same. The critical depth may be supposed to occur at about $6y_c$ back from the fall. Thus the length of the reach is $1{,}000 - 6y_c$, and $y_2 = y_c$. As a first trial, y_c may be given a reasonable value, say 3 ft.

11.96 What is the capacity of a 4-by-4-ft concrete box culvert ($n = 0.013$) with a rounded entrance ($k_e = 0.05$, $C_d = 0.95$) if the culvert slope is 0.005, the length 120 ft, and the headwater level 6 ft above the culvert invert? Assume (*a*) free outlet conditions, (*b*) tailwater elevation 1 ft above top of box at outlet, and (*c*) tailwater elevation 2 ft above top of box at outlet. Neglect headwater and tailwater velocity heads.

11.97 Repeat Prob. 11.96*a* for the cases where the culvert slope is (*a*) 0.04 and (*b*) 0.08.

11.98 A culvert under a road must carry $4.3\ m^3/s$. (*a*) The culvert length will be 30 m and the slope will be 0.003. If the maximum permissible headwater level is 3.6 m above the culvert invert, what size of corrugated-pipe culvert ($n = 0.025$) would you select? The outlet will discharge freely. Neglect velocity of approach. Assume square-edged entrance with $k_e = 0.5$, $C_d = 0.65$. (*b*) Repeat for a culvert length of 100 m.

11.99 A 120-ft-long corrugated metal pipe ($n = 0.022$) of 30-in diameter is tested in a laboratory. The headwater is maintained at a level which is 5 ft above the pipe invert at entrance. Assume a square-edged entrance with $k_e = 0.5$, $C_d = 0.68$, and neglect headwater and tailwater velocity heads. Compute values of Q for S_0 values of 0.0, 0.01, 0.03, and 0.08. Whenever $y_0 > D$ assume that condition (*b*) of Fig. 11.30 prevails. As the slope is increased, at what slope does the flow change to condition (*c*) of Fig. 11.30?

CHAPTER

TWELVE

FLUID MEASUREMENTS

Both the engineer and the scientific investigator are often faced with the problem of measuring various fluid properties such as density, viscosity, and surface tension. Also, measurements are often required of various fluid phenomena, as pressure, velocity, and flow rate. In this chapter only the principles and theory of such measurements will be discussed. Detailed information on the various measuring devices may be found elsewhere in the literature.[1] This chapter draws on many concepts presented in previous chapters, hence it serves as a good review.

12.1 MEASUREMENT OF FLUID PROPERTIES

The measurement of *liquid density* is most commonly determined by weighing a known volume of the liquid. Other techniques include hydrostatic weighing, where a nonporous solid object of known volume is weighed (*a*) in air and then (*b*) in the liquid whose density is to be determined. The *hydrometer* is a variation of this technique. The densities are calculated from fluid statics. Another, though not very accurate, technique of determining liquid density is achieved by placing two immiscible liquids in a U tube, one of known density, the other of unknown density. From fluid statics the unknown density may be found. These various techniques of measuring liquid density are illustrated in Probs. 12.1 through 12.5.

[1] See, for example, "Fluid Meters: Their Theory and Application," 6th ed., American Society of Mechanical Engineers, New York, 1971.

The measurement of *viscosity* is generally made with a device known as a *viscometer*. Various types of viscometers are available. They all depend on the creation of laminar-flow conditions. We shall confine our discussion to the measurement of the viscosity of liquids. Since viscosity varies considerably with temperature, it is essential that the fluid be at a constant temperature when a measurement is being made. This is generally accomplished by immersing the device in a constant-temperature bath.

Several types of *rotational viscometers* are available. These generally consist of two concentric cylinders that are rotated with respect to one another. The narrow space between them is filled with liquid whose viscosity is to be measured. The rate of rotation under the influence of a given torque is indicative of the viscosity of the liquid. One difficulty with this type of viscometer is that mechanical friction must be accounted for, and this is difficult to deal with accurately.

The *tube-type viscometer* is perhaps the most reliable. Figure 12.1 shows the Saybolt viscometer. In this device the liquid is originally at M, with the bottom of the tube plugged. The plug is removed, and the time required for a certain volume of liquid to pass through the tube is a measure of the kinematic viscosity of the liquid. In this device the flow is unsteady and the tube is of such small diameter that the flow may be assumed to be laminar. Substituting the expression for V for laminar flow in tubes [Eq. (8.22)] into the continuity expression [Eq. (3.3)] gives

$$Q = \frac{\pi D^4 \gamma h_L}{128 \mu L} \tag{12.1}$$

As an approximation, let h_L be the average imposed head during the flow period and let $Q \approx V_L/t$, where V_L is the volume of liquid that flows out of the tube in

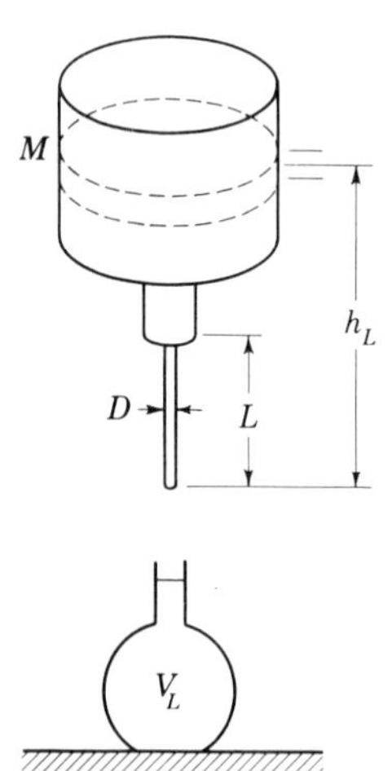

Figure 12.1 Falling-head tube-type viscometer.

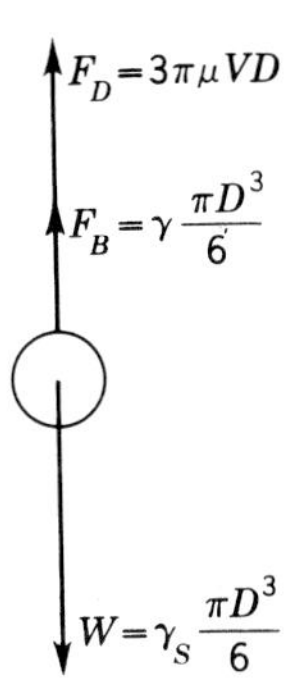

Figure 12.2 Free-body diagram of sphere falling at terminal velocity.

time t. Substituting $Q = V_L/t$ into Eq. (12.1), we get

$$\nu = \frac{\pi D^4 h_L}{128 V_L L} gt \tag{12.2}$$

Since D, L, V_L, and h_L are constants of the device, $\nu = Kgt$, and the kinematic viscosity is seen to be proportional to the measured time. Equation (12.2) gives good results if the tube is relatively long. However, for a short tube, as with the Saybolt viscometer, a correction factor[1] must be applied if the tube is too short for the establishment of laminar flow (Sec. 8.7).

There are several other types of tube viscometers, but they are all based on the same principle. Some come with a set of tubes of various diameters so that measurements can be made on liquids with a wide range of viscosities in a convenient time period. Because the dimensions of such fine tubes cannot be perfectly duplicated, each tube is individually calibrated by measuring the time for a liquid of known viscosity at a given temperature to discharge the standard volume.

A third type of viscometer is the *falling-sphere* type. In such a device the liquid is placed in a tall transparent cylinder and a sphere of known weight and diameter is dropped in it. If the sphere is small enough, Stokes's law (Sec. 10.7) will prevail and the fall velocity of the sphere will be approximately inversely proportional to the absolute viscosity of the liquid. That this is so may be seen by examining the free-body diagram of such a falling sphere (Fig. 12.2). The forces acting include gravity, buoyancy, and drag. Stokes' law states that if $DV/\nu < 1$, the drag force on a sphere is given by $F_D = 3\pi\mu VD$, where V is the velocity of the sphere and D is its diameter. When the sphere is dropped in a liquid, it will quickly accelerate to terminal velocity, at which time $\sum F_z = 0$. Thus

$$W - F_B - F_D = \gamma_s \frac{\pi D^3}{6} - \gamma \frac{\pi D^3}{6} - 3\pi\mu VD = 0$$

[1] Exact procedures for viscosity determinations are available in the standards of the American Society for Testing Materials.

where γ_s and γ represents the specific weight of the sphere and liquid, respectively. Solving the above equation, we get

$$\mu = \frac{D^2(\gamma_s - \gamma)}{18V} \tag{12.3}$$

In the preceding development it was assumed that the sphere was dropped into a liquid of infinite extent. In actuality, the liquid will be contained in a tube and a *wall effect* will influence the drag force and hence the fall velocity. It has been found that wall effect[1] can be expressed approximately as

$$\frac{V}{V_t} \approx 1 + \frac{9D}{4D_t} + \left(\frac{9D}{4D_t}\right)^2 \tag{12.4}$$

where D_t is the tube diameter, and V_t represents the fall velocity in the tube. Equation (12.4) is reliable only if $D/D_t < \frac{1}{3}$.

Other fluid properties such as surface tension, elasticity, vapor pressure, specific heats at constant pressure and constant temperature, and gas constant are commonly determined by physicists, and the techniques for their measurement will not be discussed here.

12.2 MEASUREMENT OF STATIC PRESSURE

To get an accurate measurement of static pressure in a flowing fluid, it is important that the measuring device fit the streamlines perfectly so as to create no disturbance to the flow. In a straight reach of conduit the static pressure is ordinarily measured by attaching to a piezometer a pressure gage or a U-tube manometer. The piezometer opening in the side of the conduit should be normal to and flush with the surface. Any projection, such as (*c*) in Fig. 12.3, will result in

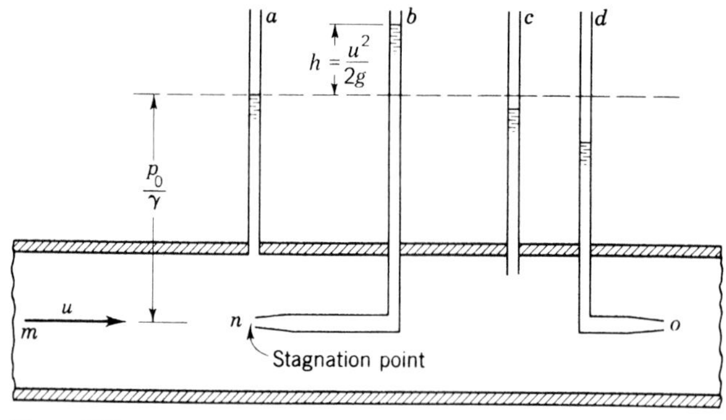

Figure 12.3

[1] J. S. McNown, H. M. Lee, M. B. McPherson, and S. M. Engez, Influence of Boundary Proximity on the Drag of Spheres, *Proc. 7th Intern. Congr. Appl. Mech.*, 1948.

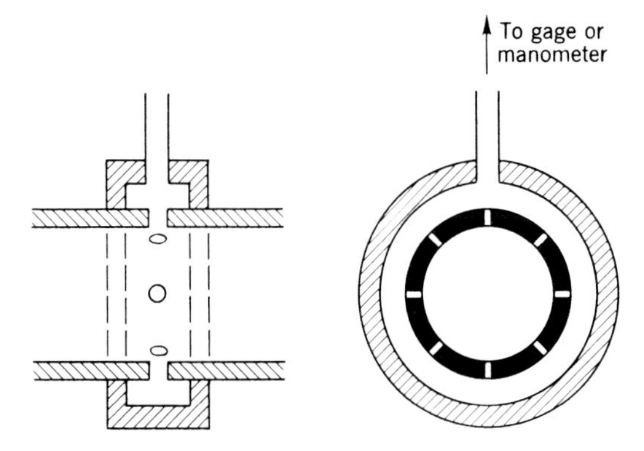

Figure 12.4 Piezometer ring.

error. It has been found, for example, that a projection of 0.10 in (2.5 mm) will cause a 16 percent change in the local velocity head. In this case the recorded pressure is depressed below the pressure in the undisturbed fluid because the disturbance of the streamline pattern increases the velocity, hence decreasing the pressure according to the Bernoulli equation.

When measuring the static pressure in a pipe, it is desirable to have two or more openings around the periphery of the section to account for possible imperfections of the wall. For this purpose a *piezometer ring* (Fig. 12.4) is used.

To measure the static pressure in a flow field, the *static tube* (Fig. 12.5) is used. In this device the pressure is transmitted to a gage or manometer through piezometric holes that are evenly spaced around the circumference of the tube. This device will give good results if it is perfectly aligned with the flow. Actually, the mean velocity past the piezometer holes will be slightly larger than that of the undisturbed flow field; hence the pressure at the holes will generally be somewhat below the pressure of the undisturbed fluid. This error can be minimized by making the diameter of the tube as small as possible. If the direction of the flow is unknown for two-dimensional flows, a *direction-finding tube* (Fig. 12.6) may be used. This device is a cylindrical tube having two piezometer holes located as shown. Each piezometer is connected to its own measuring device. The tube may be rotated until each tube shows the same reading. Then, from symmetry, one can determine the direction of flow. It has been found that if the piezometer openings are located as shown, the recorded pressures will correspond very closely to those of the undisturbed flow.

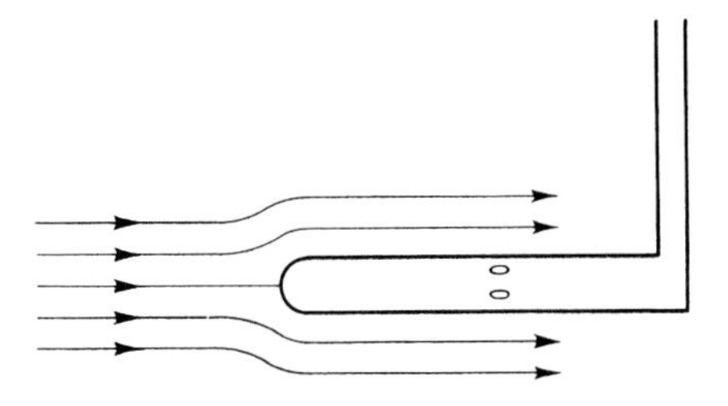

Figure 12.5 Static tube.

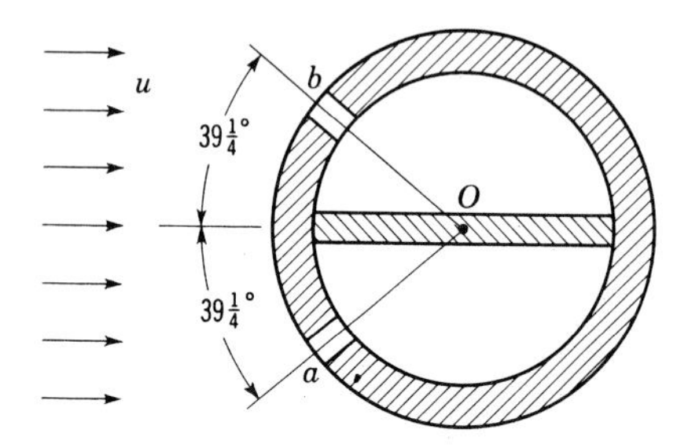

Figure 12.6 Direction-finding tube.

To obtain a reading of the fluid pressure a piezometer tube may be connected to a bourdon gage (Sec. 2.6) or to a pressure transducer (Fig. 2.9). The latter is sometimes connected to a strip-chart recorder or the pressure reading may be displayed on a panel in digital form.

12.3 MEASUREMENT OF VELOCITY WITH PITOT TUBES

One means of measuring the local velocity u in a flowing fluid is the pitot tube, named after Henri Pitot, who used a bent glass tube in 1730 to measure velocities in the River Seine. In Sec. 4.16 it is shown that the pressure at the forward stagnation point of a stationary body in a flowing fluid is $p_s = p_0 + 1/2\rho u^2$, where p_0 and u are the pressure and velocity, respectively, in the undisturbed flow upstream from the body. If $p_s - p_0$ can be measured, the velocity at a point is determined by this relation. The stagnation pressure can be measured by a tube facing upstream, such as (*b*) in Fig. 12.3. For a liquid jet or open stream with parallel streamlines, only this single tube is necessary, since the height h to which the liquid rises in the tube above the surrounding free surface is equal to the velocity head in the stream approaching the tip of the tube.

For a closed conduit under pressure it is necessary to measure the static pressure also, as shown by tube (*a*) in Fig. 12.3, and to subtract this from the total pitot reading to secure the differential head h. The differential pressure may be measured with any suitable manometer arrangement. The formula for the pitot tube for incompressible flow may be derived by writing the energy equation between points m and n of Fig. 12.3,

$$\frac{p_0}{\gamma} + \frac{u^2}{2g} = \frac{p_s}{\gamma} \tag{12.5}$$

from which

$$u^2 = 2g\left(\frac{p_s}{\gamma} - \frac{p_0}{\gamma}\right)$$

and finally

$$u = \sqrt{2g\left(\frac{p_s}{\gamma} - \frac{p_0}{\gamma}\right)} \tag{12.6}$$

This equation gives the ideal velocity of flow[1] at the point in the stream where the pitot tube is located. In actuality the right-hand side of this equation must be multiplied by a factor varying from 0.98 to 0.995 to give the true velocity. This is so because the directional velocity fluctuations of turbulence cause a pitot tube to read a value somewhat higher than the temporal mean axial component of velocity.

Where conditions are such that it is impractical to measure static pressure at the wall, a combined pitot-static tube, as in Fig. 12.7, may be used. The static pressure is measured through two or more holes drilled through an outer tube into an annular space. Rarely are the piezometer holes located in precisely the correct position to indicate the true value of p_0/γ. Hence Eq. (12.6) is modified as follows:

$$u = C_I\sqrt{2g\left(\frac{p_s}{\gamma} - \frac{p_0}{\gamma}\right)} \tag{12.7}$$

where C_I, a coefficient of instrument, is introduced to account for this discrepancy. Either English units or SI units may be used with this equation since C_I is dimensionless. However, when a coefficient possesses dimensions [see Eq. (12.26), for example], an equation developed for English units must be modified for application to SI units, and vice versa. A particular type of pitot-static tube with a blunt nose, the *Prandtl tube*, is designed so that $C_I = 1$. For other pitot-static tubes, coefficient C_I must be determined by calibration in the laboratory.

Another instrument, the *pitometer*, consists of two tubes, one pointing upstream and the other downstream, such as tubes (*b*) and (*d*) of Fig. 12.3. The reading for tube (*d*) will be considerably below the level of the static head. The equation applicable to a pitometer is identical to Eq. (12.7), except that p_0/γ is replaced by the pressure head sensed by the downstream tube.

Most of these devices will give reasonably accurate results even if the tube is as much as $\pm 15°$ out of alignment with the direction of flow.

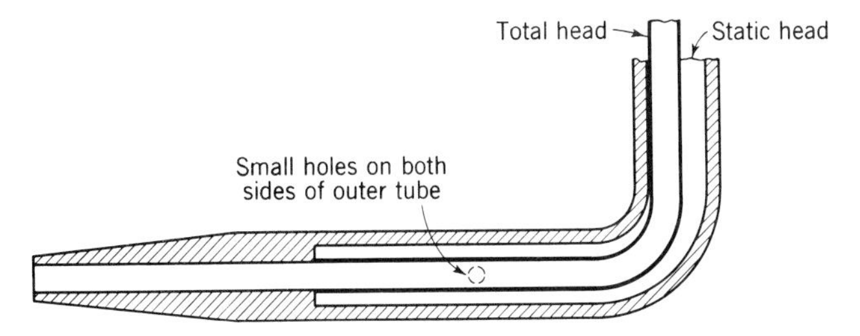

Figure 12.7 Pitot-static tube.

[1] Equations (12.5), (12.6) and (12.7) as well as those presented in Secs. 12.6 through 12.9 apply strictly to incompressible fluids. However these equations will all give very good results when applied to compressible fluids if $\mathbf{M} < 0.1$. At high values of $\mathbf{M}$ the effects of compressibility must be considered as discussed in Sec. 12.10.

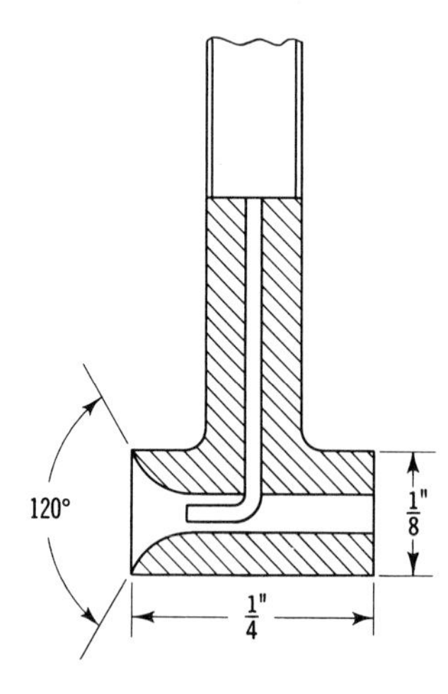

Figure 12.8 Kiel probe.

Still greater insensitivity to angularity may be obtained by guiding the flow past the pitot tube by means of a shroud, as shown in Fig. 12.8. Such an arrangement, called a *Kiel probe,* is used extensively in aeronautics. The stagnation-pressure measurement with this device is accurate to within 1 percent of the dynamic pressure for yaw angles up to $\pm 54°$. A disadvantage is that the static pressure must be measured independently.

The direction-finding tube (Fig. 12.6) may be used to determine velocity. The procedure is to orient it properly so that both piezometers give the same reading. This reading is the static head. Then turn the tube through $39\frac{1}{4}°$ to obtain the stagnation pressure head. The difference in the two readings is the velocity head. This device has been used extensively in wind tunnels and in the investigations of hydraulic machinery.

Two types of pitot tubes used in experimental work are the *Stanton tube* and the *Preston tube.* These are used to provide data for the determination of boundary shear stress. They are very small tubes that measure velocities very close to the boundary. The boundary shear stresses are calculated by substituting the measured velocities into the theory of velocity profiles in circular pipes. It is assumed that this theory holds for flat boundaries as well as curved boundaries in the region very close to the boundary.

12.4 MEASUREMENT OF VELOCITY BY OTHER METHODS

Other methods for measuring local velocity will be discussed in this section.

Current Meter and Rotating Anemometer

These two instruments, which are the same in principle, determine the velocity as a function of the speed at which a series of cups or vanes rotate about an axis either parallel or normal to the flow. The instrument used in water is called a *current meter*, and when designed for use in air, it is called an *anemometer*. As the force exerted depends upon the density of the fluid as well as upon its velocity, the anemometer must be so made as to operate with less friction than the current meter.

If the meter is made with cups which move in a circular path about an axis perpendicular to the flow, it always rotates in the same direction and at the same rate regardless of the direction of the velocity, whether positive or negative, and it even rotates when the velocity is at right angles to its plane of rotation. Thus this type is not suitable where there are eddies or other irregularities in the flow. If the meter is constructed of vanes rotating about an axis parallel to the flow, resembling a propeller, it will register the component of velocity along its axis, especially if it is surrounded by a shielding cylinder. It will rotate in an opposite direction for negative flow and is thus a more dependable type of meter.

Hot-Wire and Hot-Film Anemometer

The *hot-wire anemometer* measures the instantaneous velocity at a point. It consists of a small sensing element that is placed in the flow field at the point where the velocity is to be measured. The sensing element is a short thin wire, which is generally of platinum or tungsten, connected to a suitable electronic circuit. The operation depends on the fact that the electrical resistance of a wire is a function of its temperature; that the temperature, in turn, depends upon the heat transfer to the surrounding fluid; and that the rate of heat transfer increases with increasing velocity of flow past the wire.

In one type of hot-wire anemometer the wire is maintained at a constant temperature by a variable voltage which changes the current through the wire. Thus, when an increase in velocity tends to cool the wire, a balancing device creates an increase in voltage to increase the current through the wire. This tends to heat up the wire to counteract the cooling and thus maintain it at constant temperature. The voltage provides a measure of the velocity of the fluid. The hot-wire anemometer is a very sensitive instrument particularly adapted to the measurement of turbulent velocity fluctuations as in Fig. 3.6. A *hot-film anemometer*, though similar to the hot-wire, is more rugged in that its sensing element consists of a metal film laid over a glass rod and provided with a protective coating.

Float Measurements

A crude technique for estimating the average velocity of flow in a river or stream is to observe the velocity at which a float will travel down a stream. To get good results the reach of stream should be straight and uniform with a minimum of surface disturbances. The average velocity of flow V will generally be about (0.85 ± 0.05) times the float velocity.

Photographic and Optical Methods

The camera is one of the most valuable tools in a fluid-mechanics research laboratory. In studying the motion of water, for example, a series of small spheres consisting of a mixture of benzene and carbon tetrachloride adjusted to the same specific gravity as the water can be introduced into the flow through suitable nozzles. When illuminated from the direction of the camera, these spheres will stand out in a picture. If successive exposures are taken on the same film, the velocities and accelerations of the particles can be determined. A similar technique involves the use of hydrogen bubbles generated through use of a fine wire which serves as the negative electrode of a dc electric circuit. By pulsing the voltage across the wire the water is electrolyzed thus releasing hydrogen bubbles. Short uninsulated sections of wire will permit the bubbles to be emitted at fixed points along the wire. This, when combined with intermittent pulsing, will aid in flow visualization.

In the study of compressible fluids many techniques have been devised to measure optically the variations in density, as given by the *interferometer*, or the rate at which density changes in space, as determined in the *shadowgraph* and *schlieren* methods.[1] From such measurements of density and density gradient it is possible to locate shock waves. Although of great importance, these photographic methods are too complex to warrant further description here.

Other Methods

Other devices for measuring velocity of flow include magnetic flowmeters, acoustic flowmeters and laser-Doppler anemometers. Magnetic flowmeters are used to measure velocity of flow in liquids. The liquid serves as a conductor and develops a voltage as it travels through a magnetic field. With proper calibration, this device can be used to measure the average velocity in pipes. Small magnetic flowmeters can be used to measure local velocities in a flowing liquid. However, their accuracy drops off in the vicinity of boundaries.

Acoustic meters and laser-Doppler anemometers depend on the effect of the moving fluid on waves, the former on sound waves and the latter on light waves. These devices are expensive and are used primarily for research. One of their advantages is that they can be employed so as not to disturb the flow.

12.5 MEASUREMENT OF DISCHARGE

There are various ways of measuring discharge. In a pipe, for example, the velocity may be determined at various radii using a pitot-static tube or a pitot

[1] For an excellent discussion of optical methods used in the study of fluid flow, see Irving Shames, "Mechanics of Fluids," appendix A.7, pp. 528–535, McGraw-Hill Book Company, New York, 1962. Another good reference is H. Liepmann and A. Roshko, "Elements of Gas Dynamics," pp. 153–170, John Wiley & Sons, Inc., New York, 1957.

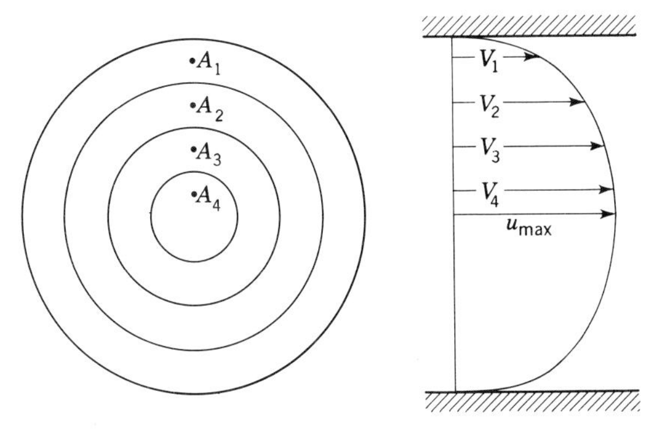

Figure 12.9 Determination of pipe discharge

$$Q = \sum A_i V_i = A_1 V_1 + A_2 V_2 + \cdots.$$

tube in combination with a wall piezometer. The cross section of a pipe may then be considered as a series of concentric rings, each with a known velocity. The flow through these rings is summed up, as in Fig. 12.9, to determine the total flow rate.

To determine the flow in a river or stream, a similar technique is used. The stream is divided into a number of convenient sections, and the average velocity in each section is measured. A pitot tube could be used for such measurements, but a current meter is more commonly used. It has been found that the average velocity occurs at about 0.6 × depth (Sec. 11.5), so the velocity is generally measured at that level. Another widely used method is to take the average of the velocities at 0.2 × depth and 0.8 × depth. This procedure for determining stream discharge is shown in Fig. 12.10. A crude estimate of the flow in a river or stream can be made by multiplying (0.85 × float velocity) times the area of the average cross section in the reach of stream over which the float measurement was made.

Devices for the direct measurement of discharge can be divided into two categories, those which measure by weight or positive displacement a certain quantity of fluid and those which employ some aspect of fluid mechanics. An

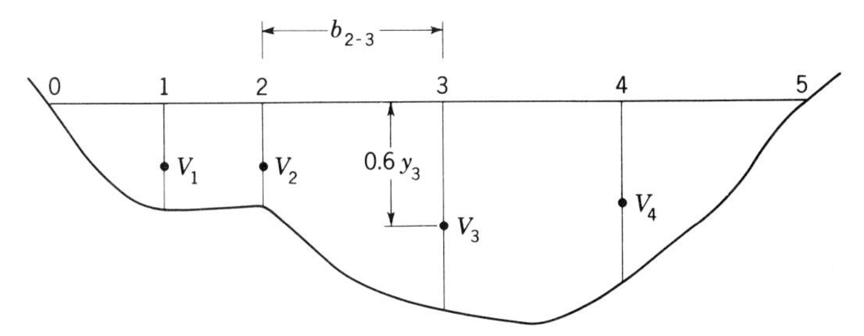

Figure 12.10 Determination of discharge in a stream.

$$Q_{2\text{-}3} = \left(\frac{y_2 + y_3}{2}\right) b_{2\text{-}3} \frac{V_2 + V_3}{2}$$

$$Q_{\text{total}} = \sum Q$$

example of the first type of device is the household water meter in which a nutating disk oscillates in a chamber. On each oscillation a known quantity of water passes through the meter. The second type of flow-measuring device, dependent on basic principles of fluid mechanics in combination with empirical data, will be discussed in the following sections.

12.6 ORIFICES, NOZZLES, AND TUBES

Among the devices used for the measurement of discharge are orifices and nozzles. Tubes are rarely so used but are included here because their theory is the same and experiments upon tubes provide information relating to entrance losses from reservoirs into pipelines. An *orifice* is an opening (usually circular) in the wall of a tank or in a plate normal to the axis of a pipe, the plate being either at the end of the pipe or in some intermediate location. An orifice is characterized by the fact that the thickness of the wall or plate is very small relative to the size of the opening. A *standard orifice* is one with a sharp edge as in Fig. 12.11*a* or an absolutely square shoulder as in Fig. 12.11*b* so that there is only a line contact with the fluid. Those shown in Fig. 12.11*c* and *d* are not standard because the flow through them is affected by the thickness of the plate, the roughness of the surface, and for (*d*) the radius of curvature. Hence such orifices should be calibrated if high accuracy is desired.

A *nozzle* is a converging tube, as in Fig. 12.12, if it is used for liquids; but for a gas or a vapor a nozzle may first converge and then diverge (Sec. 9.8) to produce supersonic flow. In addition to possible use as a flow measuring device a nozzle has other important uses, such as providing a high-velocity stream for fire fighting or for power in a steam turbine or a Pelton water wheel.[1]

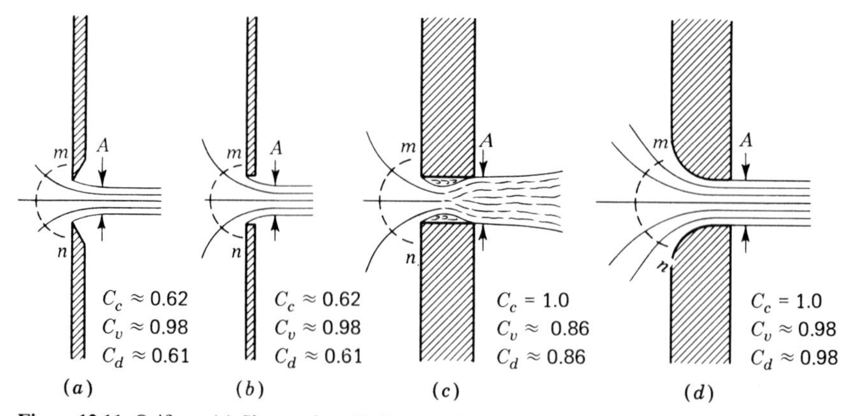

Figure 12.11 Orifices. (*a*) Sharp-edge. (*b*) Square shoulder. (*c*) Thick-plate, square edge. (*d*) Rounded.

[1] Pelton water wheel is the common name for an impulse turbine (Sec. 15.2).

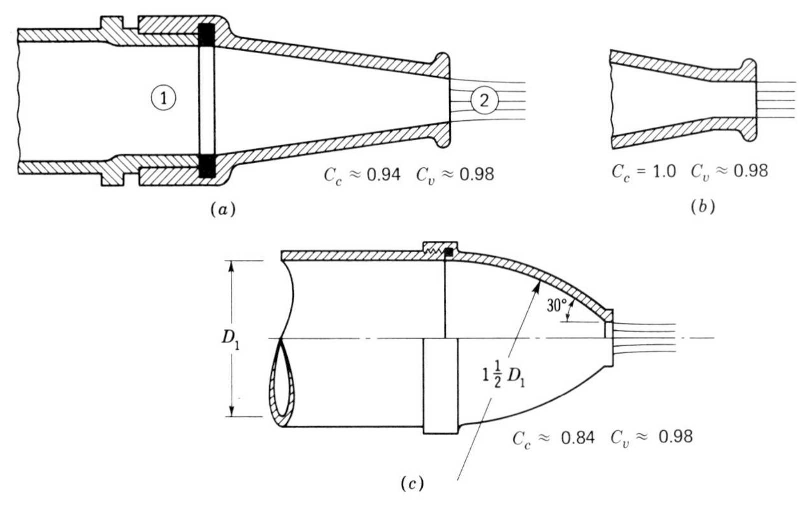

Figure 12.12 Nozzles. (*a*) Conical. (*b*) Straight-tip. (*c*) Fire.

A *tube* is a short pipe whose length is not more than two or three diameters. There is no sharp distinction between a tube and the thick-walled orifices of Fig. 12.11*c* and *d*. A tube may be of uniform diameter, or it may diverge.

A jet is a stream issuing from an orifice, nozzle, or tube. It is not enclosed by solid boundary walls but is surrounded by a fluid whose velocity is less than its own. The two fluids may be different or they may be of the same kind. A free jet is a stream of liquid surrounded by a gas and is therefore directly under the influence of gravity. A submerged jet is a stream of any fluid surrounded by a fluid of the same type, that is, a gas jet discharging into a gas or a liquid jet discharging into a liquid. A submerged jet is buoyed up by the surrounding fluid and is not directly under the action of gravity.

Jet Contraction

Where the streamlines converge in approaching an orifice, as shown in Fig. 12.13, they continue to converge beyond the upstream section of the orifice until they reach the section *xy* where they become parallel. Commonly this section is about $0.5D_o$ from the upstream edge of the opening, where D_o is the diameter of the orifice. The section *xy* is then a section of minimum area and is called the *vena contracta*. Beyond the vena contracta the streamlines commonly diverge because of frictional effects.[1] In Fig. 12.11*c* the minimum section is referred to as a

[1] Of course, if a free jet is discharged vertically downward, the acceleration due to gravity will cause its velocity to increase and the area to decrease continuously, so that there may be no apparent section of minimum area. In such special cases, the vena contracta should be taken as the place where marked contraction ceases and before the place where gravity has increased the velocity to any appreciable extent above the true jet velocity.

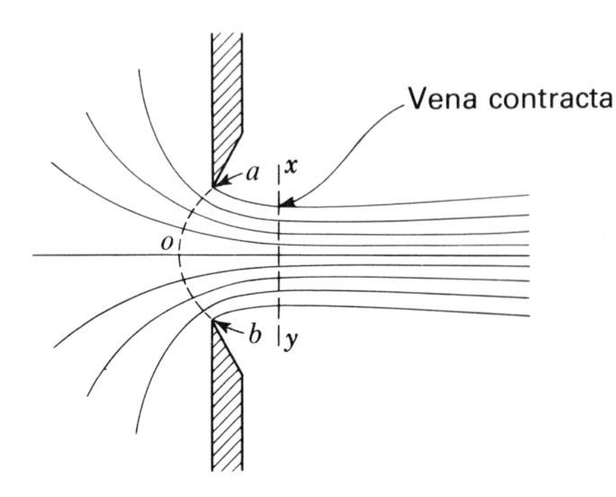

Figure 12.13 Jet contraction.

submerged vena contracta as it is surrounded by its own fluid. In Fig. 12.11*d* there is no vena contracta as the rounded entry to the opening permits the streamlines to gradually converge to the cross-sectional area of the orifice.

Jet Velocity and Pressure

Jet velocity is defined as the average velocity at the vena contracta in Fig. 12.11*a* and *b* and at the downstream edge of the orifices in Fig. 12.11*c* and *d*. The velocity at these sections is practically constant across the section except for a small annular region around the outside (Fig. 12.14*b*). In all four of the jets of Fig. 12.11 the pressure is practically constant across the diameter of the jet wherever the streamlines are parallel, and this pressure must be equal to that in the medium surrounding the jet at that section. At sections *mn* in Fig. 12.11 where the streamlines are curved, the effective cross-sectional area of the flow (at right angles to streamlines) is greater than at the minimum section and hence, the average velocities at sections *mn* are considerably less than the jet velocities. The same is true of section *aob* of Fig. 12.13. In Fig. 12.14*a* the velocity and pressure distributions at section *aob* of Fig. 12.13 are shown. These variations are the result of the curvature of the streamlines and centrifugal effects (Sec. 4.16).

Coefficient of Contraction C_c

The ratio of the area A of a jet (Fig. 12.11), to the area A_o of the orifice or other opening, is called the *coefficient of contraction*. Thus $A = C_c A_o$.

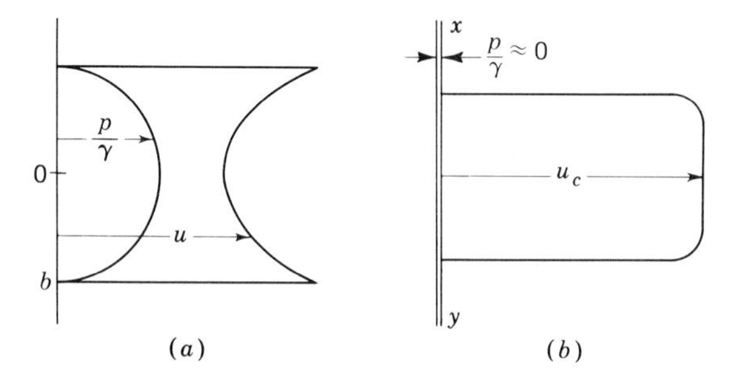

Figure 12.14 Pressure and velocity variation in jet. (*a*) At section *aob* of Fig. 12.13. (*b*) At vena contracta (section *xy*) in Fig. 12.13.

Coefficient of Velocity C_v

The velocity that would be attained in the jet if friction did not exist may be termed the ideal velocity V_i.[1] It is practically the value of u_c in Fig. 12.14. Because of friction the actual average velocity V is less than the ideal velocity, and the ratio V/V_i is called the *coefficient of velocity*. Thus $V = C_v V_i$.

Coefficient of Discharge C_d

The ratio of the actual rate of discharge Q to the ideal rate of discharge Q_i (the flow that would occur if there were no friction and no contraction) is defined as the coefficient of discharge. Thus $Q = C_d Q_i$. By observing that $Q = AV$ and $Q_i = A_o V_i$, it is seen that $C_d = C_c C_v$.

Determining the Coefficients

The coefficient of contraction can be determined by using outside calipers to measure the jet diameter at the vena contracta and then comparing the jet area with the orifice area. The contraction coefficient is very sensitive to small variations in the edge of the orifice or in the upstream face of the plate. Thus slightly rounding the edge of the orifice in Fig. 12.11*b* or roughening the orifice plate will increase the contraction coefficient materially.

The average velocity V of a free jet may be determined by a velocity traverse of the jet with a fine pitot tube or it may be obtained by measuring the flow rate and dividing by the cross-sectional area of the jet. The velocity may also be computed approximately from the coordinates of the trajectory of the jet, as discussed in Sec. 4.17. The ideal velocity V_i is computed by the Bernoulli theorem. Thus C_v for an orifice, nozzle, or tube may be computed by dividing V by V_i.

The coefficient of discharge is the one that can most readily be obtained and with a high degree of accuracy. It is also the one that is of the most practical value. For a liquid the actual Q can be determined by some standard method such as a volume or a weight measurement over a known time. For a gas one can note the change in pressure and temperature in a container of known volume from which the gas may flow. Obviously, if any two of the coefficients are measured, the third can be computed from them. Thus, in equation form

Ideal flow rate: $$Q_i = A_i V_i = A_o\sqrt{2g(\Delta H)} \tag{12.8}$$

Actual flow rate: $$Q = AV = C_c A_o (C_v \sqrt{2g(\Delta H)}) \tag{12.9}$$

and $$C_d = \frac{Q}{Q_i} = C_c C_v \tag{12.10}$$

[1] This is frequently called the *theoretical velocity*, but the authors feel this is a misuse of the word "theoretical." Any correct theory should allow for the fact that friction exists and affects the result. Otherwise it is not correct theory but merely an incorrect hypothesis.

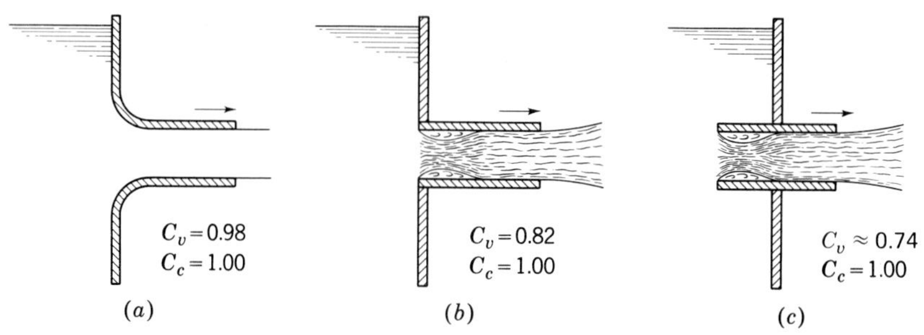

Figure 12.15 Coefficients for tubes.

where ΔH is the total difference in energy head between the upstream section and the minimum section of the jet (section A of Fig 12.11). It should be recalled that the total energy head $H = z + p/\gamma + V^2/2g$. If the flow is from a tank, the velocity of approach is negligible and may be neglected. If the discharge is to the atmosphere (free jet), the downstream pressure head is zero, whereas if the jet is submerged, the downstream pressure head is equal to the depth of submergence[1] (Fig. 12.17) in the case of a liquid or to the pressure head surrounding the jet in the case of a gas.

Typical values of the coefficients for orifices, nozzles, and tubes are as indicated in Figs. 12.11,[2] 12.12, and 12.15 respectively. It is apparent from Fig. 12.15 that rounding the entrance to a tube increases the coefficient of velocity. Any device that provides a uniform diameter for a long enough distance before exit, such as the tubes of Fig. 12.15 or the nozzle tip of Fig. 12.12*b*, will usually create a $C_c = 1.0$. Although this increases the size of the jet from the given area, it also tends to produce more friction.

If the geometry of the orifice, nozzle or tube is standard such as those of Figs. 12.11, 12.12, and 12.15, the coefficients should be very close to the values indicated on the figures. However, the best way to determine the coefficients of a device, particularly one of unusual shape, is by experiment in the laboratory. Also, one can make a fair estimate of the contraction by sketching the flow net. If one wishes to estimate the coefficient of discharge of an orifice, nozzle, or tube it is usually best to estimate velocity and contraction coefficients separately and calculate the discharge coefficient from them.

Borda Tube

Tubes (*b*) and (*c*) in Fig. 12.15 are shown as flowing full, and because of the turbulence, the jets issuing from them will have a "broomy" appearance. Because

[1] If the jet discharges into a different liquid, the depth of submergence must be converted to equivalent depth of the *flowing liquid.*

[2] Surface tension may become important when orifices operate under low heads. The coefficient of contraction of small, sharp-edged, and square-edged orifices such as those of Fig. 12.11*a* and 12.11*b* have values of C_c as high as 0.72 rather than the usual 0.62 when operating under heads less than about 0.5 ft.

of the contraction of the jet at entrance to these tubes the local velocity in the central portion of the stream will be higher than that at exit from the tubes, and hence the pressure will be lower. If the pressure is lowered to that of the vapor pressure of the liquid, the streamlines will then no longer follow the walls. In such a case tube (*b*) of Fig. 12.15 becomes equivalent to orifice (*b*) in Fig. 12.11, while tube (*c*) of Fig. 12.15 behaves as shown in Fig. 12.16. If its length is less than its diameter, the reentrant tube is called a *Borda mouthpiece*. Because of the greater curvature of the streamlines for a reentrant tube, the velocity coefficient is lower (Fig. 12.15*c*) than for any other type of entry, if the tube flows full. But if the jet springs clear as in Fig. 12.16, the velocity coefficient is as high as for a sharp-edged orifice.

The Borda mouthpiece is of interest because it is one device for which the contraction coefficient can be very simply calculated. For all other orifices and tubes there is a reduction of the pressure on the walls adjacent to the opening, but the exact pressure values are unknown. But for the reentrant tube, the velocity along the wall of the tank is almost zero at all points, and hence the pressure is essentially hydrostatic. In the case of a Borda tube the only unbalanced pressure is that on an equal area A_0 opposite to the tube (Fig. 12.16), and its value is $\gamma h A_o$. The time rate of change of momentum due to the flow out of the tube is $\rho Q V = \gamma A V^2/g$, where A is the area of the jet. Equating force to time rate of change of momentum, $\gamma h A_o = \gamma A V^2/g$, and thus, $V^2 = ghA_o/A$. Ideally, $V^2 = 2gh$, and thus, ideally, $C_c = A/A_o = 0.5$. The actual values of the coefficients for a Borda tube are $C_c = 0.52$, $C_v = 0.98$, and $C_d = 0.51$.

Head Loss

The relationship between the head loss and the coefficient of velocity of an orifice, nozzle, or tube may be found by comparing the ideal energy equation with the actual (or real) energy equation between points 1 and 2 in Fig. 12.12*a*.

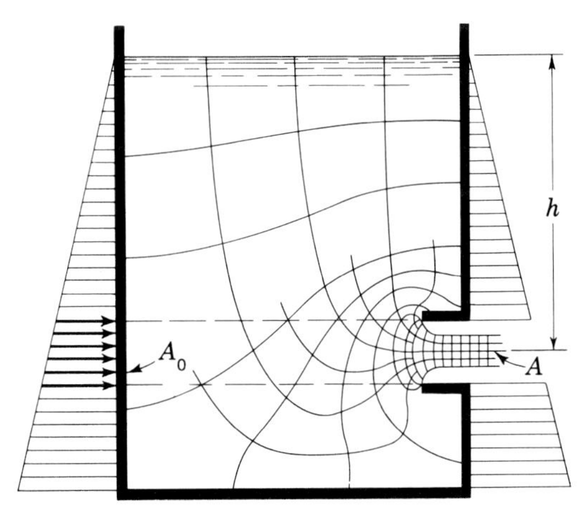

Figure 12.16 Borda tube.

The ideal energy equation is

$$z_1 + \frac{p_1}{\gamma} + \frac{V_1^2}{2g} = z_2 + \frac{p_2}{\gamma} + \frac{V_2^2}{2g}$$

In the case of a free jet, $p_2 = 0$ while for the most general case of a submerged jet $p_2 \neq 0$. From continuity $A_1 V_1 = A_2 V_2$, hence we can write

$$z_1 + \frac{p_1}{\gamma} + \left(\frac{A_2}{A_1}\right)^2 \frac{V_2^2}{2g} = z_2 + \frac{p_2}{\gamma} + \frac{V_2^2}{2g}$$

which leads to

$$(V_2)_{\text{ideal}} = \frac{1}{\sqrt{1 - (A_2/A_1)^2}} \sqrt{2g\left[\left(z_1 + \frac{p_1}{\gamma}\right) - \left(z_2 + \frac{p_2}{\gamma}\right)\right]} \tag{12.11}$$

The real energy equation accounts for head loss and is expressed as

$$z_1 + \frac{p_1}{\gamma} + \frac{V_1^2}{2g} - h_{L_{1-2}} = z_2 + \frac{p_2}{\gamma} + \frac{V_2^2}{2g}$$

which leads to

$$(V_2)_{\text{actual}} = \frac{1}{\sqrt{1 - (A_2/A_1)^2}} \sqrt{2g\left[\left(z_1 + \frac{p_1}{\gamma}\right) - \left(z_2 + \frac{p_2}{\gamma}\right) - h_{L_{1-2}}\right]} \tag{12.12}$$

Remembering that $V_{\text{actual}} = C_v V_{\text{ideal}}$, and combining this with the above expressions for V_{ideal} and V_{actual} gives

$$h_{L_{1-2}} = \left(\frac{1}{C_v^2} - 1\right)\left[1 - \left(\frac{A_2}{A_1}\right)^2\right]\frac{V_2^2}{2g} \tag{12.13}$$

This equation is perfectly general; it expresses the head loss between a section upstream of an orifice and the jet (section A in Fig. 12.11) or between sections 1 and 2 in Fig. 12.12*a*, etc. If the orifice or nozzle takes off directly from a tank where $A_1 \gg A_2$, then the velocity of approach is negligible and Eq. (12.13) reduces to

$$h_{L_{1-2}} = \left(\frac{1}{C_v^2} - 1\right)\frac{V_2^2}{2g} \tag{12.14}$$

For the tubes of Fig. 12.15 with $C_v = 0.98$, 0.82, and 0.74, Eq. (12.14) yields $h_L = 0.04V_2^2/2g$, $0.5V_2^2/2g$, and $0.8V_2^2/2g$ respectively. These correspond to the values for minor loss at entrance shown in Fig. 8.14.

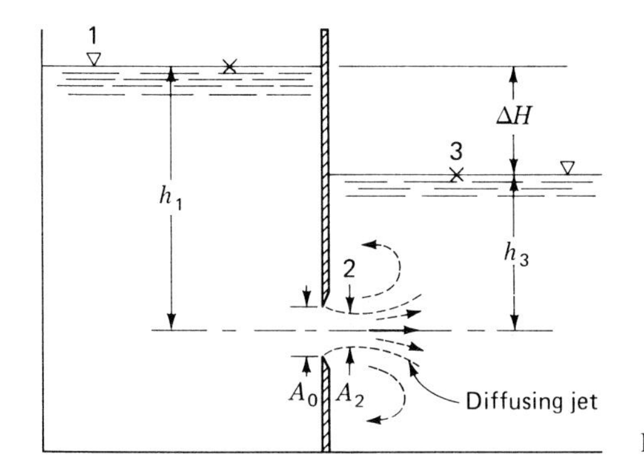

Figure 12.17 Submerged jet.

Submerged Jet

For the case of a submerged jet, as shown in Fig. 12.17, the ideal energy equation is written between 1 and 2, realizing that the pressure head on the jet at 2 is equal to h_3. Thus

$$h_1 = h_3 + \frac{V_i^2}{2g}$$

or

$$V_i = \sqrt{2g(h_1 - h_3)} = \sqrt{2g(\Delta H)}$$

where V_i is the ideal velocity at the vena contracta of the submerged jet and ΔH is the net head differential expressed in terms of the flowing liquid. Hence $Q = C_c C_v A_o \sqrt{2g(\Delta H)}$ as in Eq. (12.9).

For a submerged orifice, nozzle, or tube the coefficients are practically the same as for a free jet, except that, for heads less than 10 ft and for very small openings, the discharge coefficient may be slightly less. It is of interest to observe that, if the energy equation is written between 1 and 3, the result is $h_{L_{1-3}} = h_1 - h_3 = \Delta H$. Actually, the head loss in this case is that of Eq. (12.14) plus that of a submerged discharge, as described in Sec. 8.17. Hence, for submerged orifices, nozzles, and tubes,

$$h_{L_{1-3}} = \left(\frac{1}{C_v^2} - 1\right)\frac{V_2^2}{2g} + \frac{V_2^2}{2g}$$

where $V_2 = C_v V_i$, the velocity at the vena contracta.

Illustrative Example 12.1 A 2-in circular orifice (not standard) at the end of a 3-in-diameter pipe discharges into the atmosphere a measured flow of 0.60 cfs of water when the pressure in the pipe is

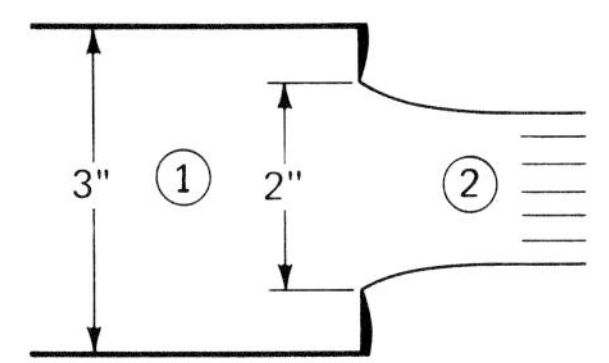

Illustrative Example 12.1

10.0 psi. The jet velocity is determined by a pitot tube to be 39.2 fps. Find the values of the coefficients C_v, C_c, C_d. Find also the head loss for inlet to throat.

Define the inlet as section 1 and the throat as section 2.

$$\frac{p_1}{\gamma} = 10\left(\frac{144}{62.4}\right) = 23.1 \text{ ft}$$

$$V_1 = \frac{Q}{A_1} = \frac{0.60}{0.0491} = 12.22 \text{ fps} \qquad \frac{V_1^2}{2g} = 2.32 \text{ ft}$$

Express the ideal energy equation from 1 to 2 to determine the ideal velocity at 2

$$\frac{p_1}{\gamma} + \frac{V_1^2}{2g} = \frac{V_2^2}{2g}$$

$$23.1 + 2.3 = \frac{V_2^2}{2g} \qquad (V_2)_{\text{ideal}} = 40.4 \text{ fps}$$

$$C_v = \frac{V}{V_i} = \frac{39.2}{40.4} = 0.97$$

Area of jet

$$A_2 = \frac{Q}{V} = \frac{0.60}{39.2} = 0.0153 \text{ ft}^2$$

$$C_c = \frac{A_2}{A_o} = \frac{0.0153}{0.0218} = 0.70$$

Hence,

$$C_d = C_c C_v = 0.68$$

From Eq. (12.13)

$$h_{L_{1-2}} = \left[\frac{1}{(0.97)^2} - 1\right]\left[1 - \left(\frac{2}{3}\right)^4\right]\frac{V_2^2}{2g} = 0.051\frac{V_2^2}{2g}$$

$$= 0.051\frac{(39.2)^2}{64.4} = 1.22 \text{ ft}$$

As a check, determine the actual velocity at 2 by expressing the real energy equation from 1 to 2.

$$\frac{p_1}{\gamma} + \frac{V_2^2}{2y} - h_{L_{1-2}} = \frac{V_2^2}{2g}$$

$$23.1 + 2.3 - 1.2 = \frac{V_2^2}{2g} \qquad (V_2)_{\text{actual}} = 39.5 \text{ fps}$$

which is a good check. A better check would result if all numbers were carried out to more places.

12.7 VENTURI TUBE

The converging tube is an efficient device for converting pressure head to velocity head, while the diverging tube converts velocity head to pressure head. The two may be combined to form a venturi tube, named after Venturi, an Italian, who investigated its principle about 1791. It was applied to the measurement of water by Clemens Herschel in 1886. As shown in Fig. 12.18, it consists of a tube with a constricted *throat* which produces an increased velocity accompanied by a reduction in pressure, followed by a gradually diverging portion in which the velocity

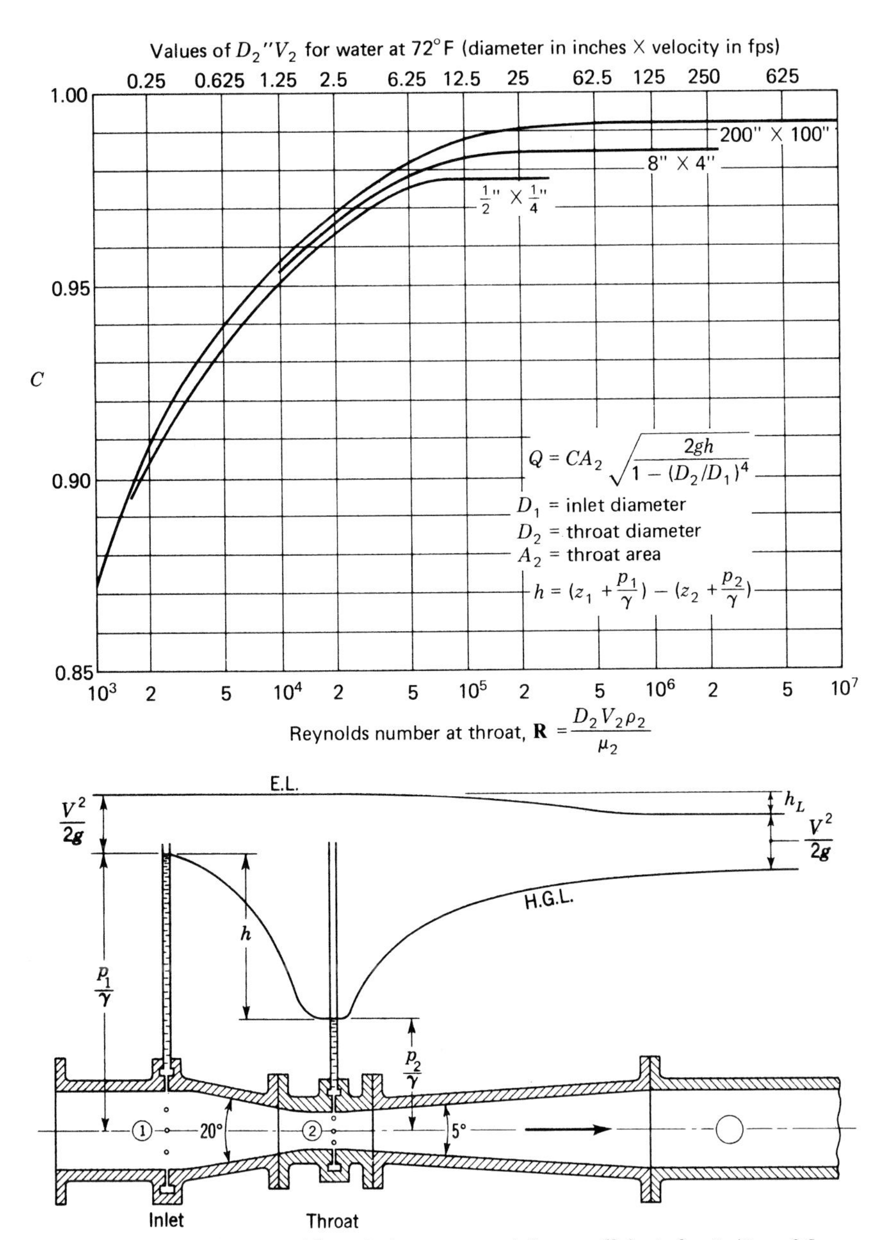

Figure 12.18 Venturi meter with conical entrance and flow coefficients for $D_2/D_1 = 0.5$.

is transformed back into pressure with slight friction loss. As there is a definite relation between the pressure differential and the rate of flow, the tube may be made to serve as a metering device. The venturi meter is used for measuring the rate of flow of both compressible and incompressible fluids.[1] In this section we

[1] As mentioned previously, if $\mathbf{M} < 0.1$, a compressible fluid can be dealt with as if it were incompressible without introducing much error.

shall consider the application of the venturi meter to incompressible fluids. In Sec. 12.10 the application of the venturi meter to compressible fluids will be discussed.

Writing the Bernoulli equation between sections 1 and 2 of Fig. 12.18, we have, for the ideal case,

$$\frac{p_1}{\gamma} + z_1 + \frac{V_1^2}{2g} = \frac{p_2}{\gamma} + z_2 + \frac{V_2^2}{2g}$$

Substituting the continuity equation, $V_1 = (A_2/A_1)V_2$, we get for the ideal throat velocity

$$(V_2)_{\text{ideal}} = \sqrt{\frac{1}{1-(A_2/A_1)^2}}\sqrt{2g\left[\left(\frac{p_1}{\gamma}+z_1\right)-\left(\frac{p_2}{\gamma}+z_2\right)\right]}$$

As there is some friction loss between (1) and (2), the true velocity is slightly less than the value given by this expression. Hence we may introduce a discharge coefficient C, so that the flow is given by

$$Q = A_2V_2 = \frac{CA_2}{\sqrt{1-(D_2/D_1)^4}}\sqrt{2g\left[\left(\frac{p_1}{\gamma}+z_1\right)-\left(\frac{p_2}{\gamma}+z_2\right)\right]} \tag{12.15}$$

In this situation the discharge coefficient C is identical to the velocity coefficient C_v since the coefficient of contraction $C_c = 1.0$. In the preceding equation it should be noted by reference to Eq. (2.12) that if a differential manometer is used with piezometric connections at sections (1) and (2)

$$\left(\frac{p_1}{\gamma}+z_1\right)-\left(\frac{p_2}{\gamma}+z_2\right) = MR\left(\frac{s_m}{s_f}-1\right) \tag{12.16}$$

where MR is the manometer reading, and s_m and s_f are the specific gravities of the manometer and flowing fluids, respectively.

The venturi tube provides an accurate means for measuring flow in pipelines. With a suitable recording device the flow rate can be integrated so as to give the total quantity of flow. Aside from the installation cost, the only disadvantage of the venturi meter is that it introduces a permanent frictional resistance in the pipeline. This loss is practically all in the diverging part from (2) to (3) (Fig. 4.10) and is ordinarily from $0.1h$ to $0.2h$, where h is the static-head differential between the upstream section and the throat, as indicated in Fig. 12.18.

Values of D_2/D_1 may vary from $\frac{1}{4}$ to $\frac{3}{4}$, but a common ratio is $\frac{1}{2}$. A small ratio gives increased accuracy of the gage reading, but is accompanied by a higher friction loss and may produce an undesirably low pressure at the throat, sufficient in some cases to cause liberation of dissolved air or even vaporization of the liquid at this point. This phenomenon, called *cavitation*, has been described in Sec. 4.8. The angles of convergence and divergence indicated in Fig. 12.18 are considered optimum, though somewhat larger angles are sometimes used to reduce the length and cost of the tube.

For accuracy in use, the venturi meter should be preceded by a straight pipe whose length is at least 5 to 10 pipe diameters. The approach section becomes more important as the diameter ratio increases, and the required length of straight pipe depends on the conditions preceding it. Thus the vortex formed from two short-radius elbows in planes at right angles, for example, is not eliminated within 30 pipe diameters. Such a condition can be alleviated by the installation of straightening vanes preceding the meter.[1] The pressure differential should be obtained from piezometer rings (Fig. 12.4) surrounding the pipe, with a number of suitable openings in the two sections. In fact, these openings are sometimes replaced by very narrow slots extending most of the way around the circumference.

Unless specific information is available for a given venturi tube, the value of C may be assumed to be about 0.99 for large tubes and about 0.97 or 0.98 for small ones, provided the flow is such as to give reasonably high Reynolds numbers. A roughening of the surface of the converging section from age or scale deposit will reduce the coefficient slightly. Venturi tubes in service for many years have shown a decrease in C of the order of 1 to 2 per cent.[2] Dimensional analysis of a venturi tube indicates that the coefficient C should be a function of Reynolds number and of the geometric parameters D_1 and D_2. Values of venturi-tube coefficients are shown in Fig. 12.18. This diagram is for a diameter ratio of $D_2/D_1 = 0.5$, but it is reasonably valid for smaller ratios also. For best results a venturi meter should be calibrated by conducting a series of tests in which the flow rate is measured over a wide range of Reynolds numbers.

Occasionally, the precise calibration of a venturi tube has given a value of C greater than 1. Such an abnormal result is sometimes due to improper piezometer openings. But another explanation is that the α's at sections 1 and 2 are such that this is so.

12.8 FLOW NOZZLE

If the diverging discharge cone of a venturi tube is omitted, the result is a *flow nozzle* of the type shown in Fig. 12.19. This is simpler than the venturi tube and can be installed between the flanges of a pipeline. It will answer the same purpose, though at the expense of an increased friction loss in the pipe. Although the venturi-meter equation [Eq. (12.15)] can be employed for the flow nozzle, it is more convenient and customary to include the correction for velocity of approach with the coefficient of discharge, so that

$$Q = KA_2\sqrt{2g\left[\left(\frac{p_1}{\gamma} + z_1\right) - \left(\frac{p_2}{\gamma} + z_2\right)\right]} \tag{12.17}$$

[1] W. S. Pardoe, The Effect of Installation on the Coefficients of Venturi Meters, *Trans. ASME*, vol. 65, p. 337, 1945.

[2] C. M. Allen and L. J. Hooper, Venturi and Weir Measurements, *Mech. Eng.*, June, 1935, p. 369. W. S. Pardoe, *Mech. Eng.*, January, 1936, p. 60.

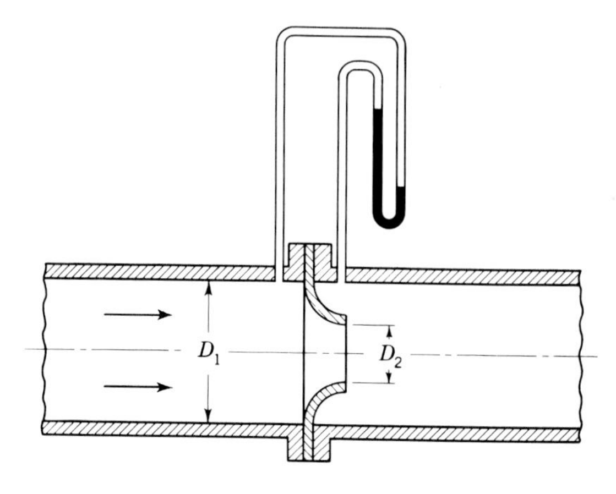

Figure 12.19 Flow nozzle.

where K is called the *flow coefficient* and A_2 is the area of the nozzle throat. Comparison with Eq. (12.15) establishes the relation

$$K = \frac{C}{\sqrt{1 - (D_2/D_1)^4}} \tag{12.18}$$

Although there are many designs of flow nozzles, the ISA (International Standards Association) nozzle (Fig. 12.20) has become an accepted standard form in many countries. Values of K for various diameter ratios of the ISA nozzle are shown in Fig. 12.21 as a function of Reynolds number. Note that in this case the

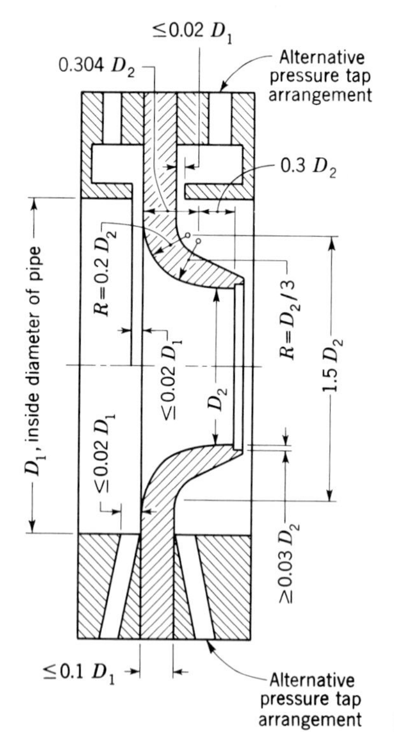

Figure 12.20 ISA flow nozzle.

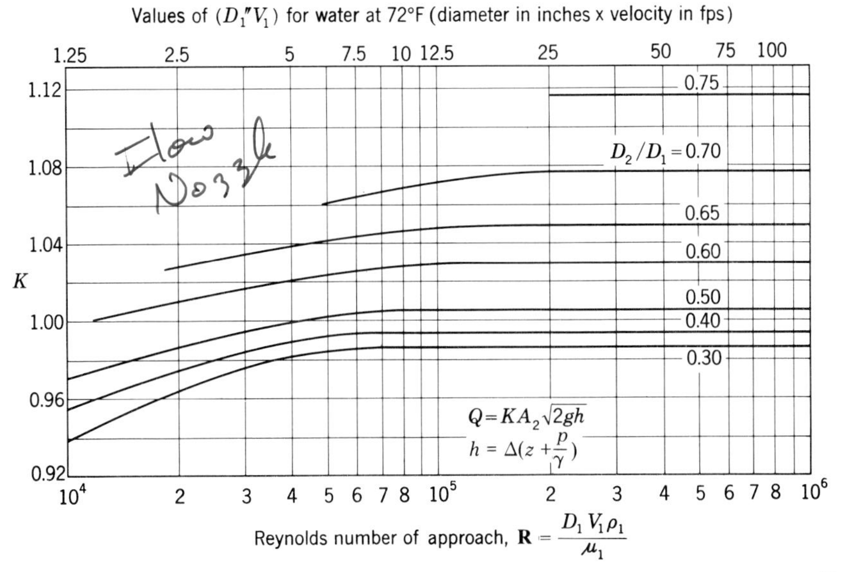

Figure 12.21 Flow coefficients for ISA nozzle. (Adapted from ASME *Flow Measurement*, 1959.)

Reynolds number is computed for the approach pipe rather than for the nozzle throat, which is a convenience since **R** in the pipe is frequently needed for other computations also.

As shown in Fig. 12.21, many of the values of K are greater than unity, which is a result of including the correction for approach velocity with the conventional coefficient of discharge. There have been many attempts to design a nozzle for which the velocity-of-approach correction would just cancel the discharge coefficient, leaving a value of the flow coefficient equal to unity. Detailed information on these so-called *long-radius* nozzles may be found in the ASME publications on fluid meters and flow measurement.[1]

As in the case of the venturi meter, the flow nozzle should be preceded by at least 10 diameters of straight pipe for accurate measurement. Two alternative arrangements for the pressure taps are shown in Fig. 12.20.

12.9 ORIFICE METER

An orifice in a pipeline, as in Fig. 12.22, may be used as a meter in the same manner as the venturi tube or the flow nozzle. It may also be placed on the end of the pipe so as to discharge a free jet. The flow rate through an orifice meter is commonly expressed as

$$Q = KA_o\sqrt{2g\left[\left(\frac{p_1}{\gamma} + z_1\right) - \left(\frac{p_2}{\gamma} + z_2\right)\right]} \tag{12.19}$$

[1] Flow Measurement by Means of Standardized Nozzles and Orifice Plates, pt. 5, chap. 4, in *ASME Power Test Codes on Instruments and Apparatus*, 1959.

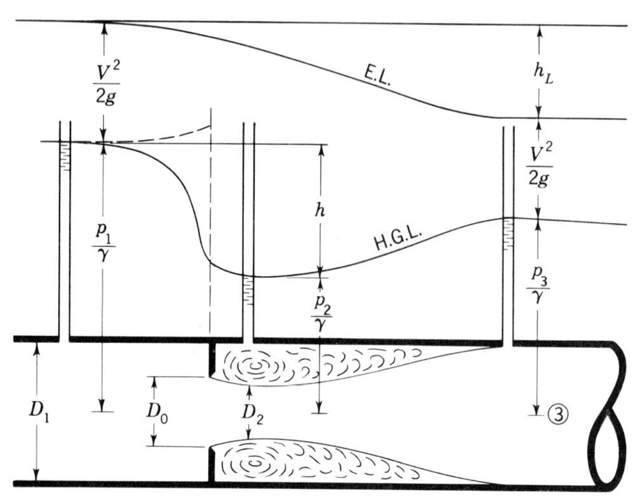

Figure 12.22 Thin-plate orifice in a pipe. (Scale distorted: the region of eddying turbulence will usually extend 4 to 8 × D_1 dowstream depending upon the Reynolds number.)

This is the same form as Eq. (12.17) except that A_2 is replaced by A_o, the cross-sectional area of the orifice opening. Typical values of K for a standard orifice meter are given in Fig. 12.23. The variation of K with Reynolds number is quite different than the trend of the flow coefficients for venturi tubes and flow nozzles. At high Reynolds numbers K is essentially constant, but as the Reynolds number is lowered, an increase in the value of K for the orifice is noted with maximum value of K occurring at Reynolds numbers between 200 and 600, depending on the D_o/D_1 ratio of the orifice. The lowering of the Reynolds number increases viscous action which causes a decrease in C_v and an increase in C_c. The latter apparently predominates over the former until C_c reaches a maximum value of about 1.0. With a further decrease in Reynolds number K then becomes smaller because C_v continues to decrease.

The difference between an orifice meter and a venturi tube or flow nozzle is that for both of the latter there is no contraction, so that A_2 is also the area of the throat and is fixed, while for the orifice, A_2 is the area of the jet and is a variable and is less than A_o, the area of the orifice. For the venturi tube or flow nozzle the discharge coefficient is practically a velocity coefficient, while for the orifice it is much more affected by variations in C_c than it is by variations in C_v.

The pressure differential may be measured between a point about one pipe diameter upstream of the orifice and the vena contracta, approximately one-half the pipe diameter downstream. The distance to the vena contracta is not constant, but decreases as D_o/D_1 increases. The differential can also be measured between the two corners on each side of the orifice plate. These *flange taps* have the advantage that the orifice meter is self-contained; the plate may be slipped into a pipeline without the necessity of making piezometer connections in the pipe.

The orifice has merit as a measuring device for it may be installed in a pipeline with a minimum of trouble and expense. Its principal disadvantage is the

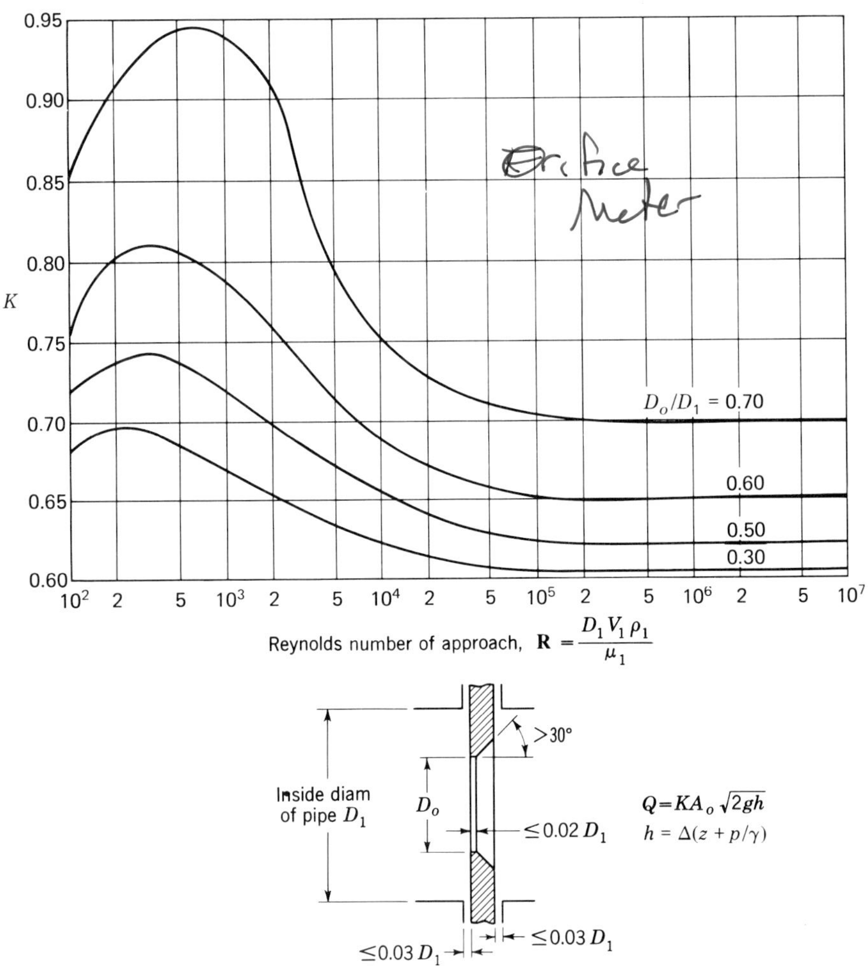

Figure 12.23 VDI orifice meter and flow coefficients for flange taps. (*Adapted from NACA Tech. Mem.* 952.)

greater frictional resistance offered by it as compared with the venturi tube or flow nozzle.[1]

Illustrative Example 12.2 A 2-in ISA flow nozzle is installed in a 3-in pipe carrying water at 72°F. If a water-air manometer shows a differential of 2 in, find the flow.

This is a trial-and-error type of solution. First assume a reasonable value of K. From Fig. 12.21, for $D_2/D_1 = 0.67$, for the level part of the curve, $K = 1.06$. Then from Eq. (12.17),

$$Q = KA_2\sqrt{2g\left[\left(\frac{p_1}{\gamma} + z_1\right) - \left(\frac{p_2}{\gamma} + z_2\right)\right]}$$

where
$$A_2 = \frac{\pi}{4} \times \frac{2^2}{144} = 0.0218 \text{ ft}^2$$

[1] For further information on orifices, see "Fluid Meters: Their Theory and Application," 6th ed., American Society of Mechanical Engineers, New York, 1971.

and
$$\Delta\left(\frac{p}{\gamma}+z\right)=\frac{2}{12}=0.167\text{ ft}$$

Thus
$$Q=1.06\times 0.0218\sqrt{64.4\times 0.167}=0.0757\text{ cfs}$$

With this first determination of Q,
$$V_1=\frac{Q}{A}=\frac{0.0757}{0.0492}=1.54\text{ fps}$$

Then
$$D_1''V_1=3\times 1.54=4.62$$

From Fig. 12.21, $K = 1.04$ and
$$Q=\frac{1.04}{1.06}\times 0.0757=0.0743\text{ cfs}$$

No further correction is necessary.

12.10 FLOW MEASUREMENT OF COMPRESSIBLE FLUIDS

Strictly speaking, most of the equations that have been presented in the preceding part of this chapter apply only to incompressible fluids, but practically, they may be used for all liquids and even for gases and vapors where the pressure differential is small relative to the total pressure. As this is the condition usually encountered in the metering of all fluids, even compressible ones, the preceding treatment has extensive application. However, there are conditions in metering fluids where compressibility must be considered.

As in the case of incompressible fluids, equations may be derived for ideal frictionless flow and then a coefficient introduced to obtain a correct result. The ideal condition that will be imposed on the compressible fluid is that the flow be isentropic, i.e., frictionless adiabatic process (no transfer of heat—Sec. 1.8). The latter is practically true for metering devices, as the time for the fluid to pass through is so short that very little heat transfer can take place. An expression applicable to *pitot tubes* for subsonic flow of compressible fluids can be derived by introducing the conditions at the upstream tip of the tube (that is, $V_2 = 0$ and $p_2 = p_s$) in Eq. (9.16) and substituting the first expression for R from Eq. (9.4). Doing so gives for pitot tubes

$$\frac{V_1^2}{2}=c_pT_1\left[\left(\frac{p_2}{p_1}\right)^{(k-1)/k}-1\right]=c_pT_2\left[1-\left(\frac{p_1}{p_2}\right)^{(k-1)/k}\right] \tag{12.20}$$

The static pressure p_1 may be obtained from the side openings of the pitot tube or from a regular piezometer, and the stagnation pressure p_s $(= p_2)$ is indicated by the pitot tube itself. A coefficient must be applied if the side openings do not measure the true static pressure. Equation (12.20) does not apply to supersonic conditions because a shock wave would form upstream of the stagnation point. In such a case a special analysis considering the effect of the shock wave is required.

To develop an expression applicable to compressible flow through *venturi tubes* we take Eq. (9.16) and combine it with continuity $(G = \gamma_1 A_1 V_1 = \gamma_2 A_2 V_2)$ to get

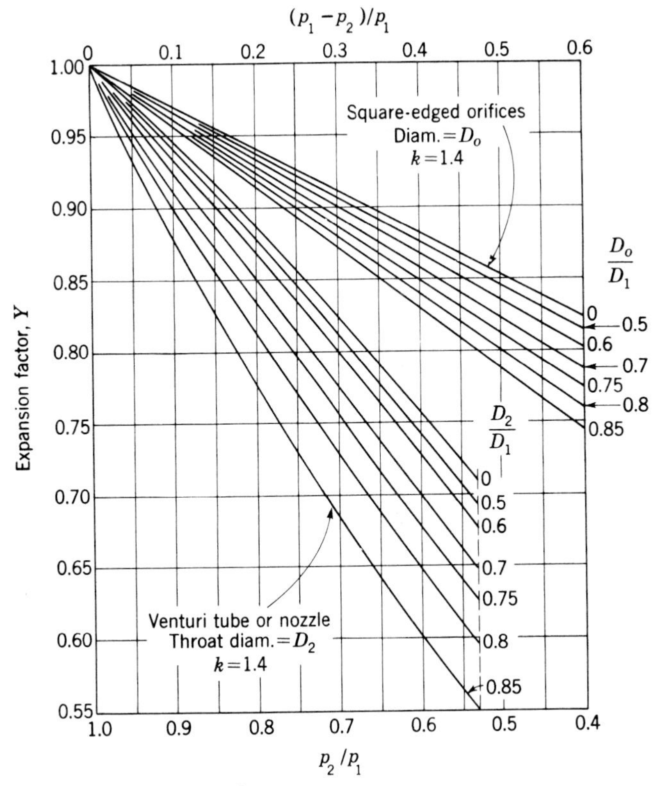

Figure 12.24 Expansion factors.

$$G_{\text{ideal}} = A_2\sqrt{2g\frac{k}{k-1}p_1\gamma_1\left(\frac{p_2}{p_1}\right)^{2/k}\frac{1-(p_2/p_1)^{(k-1)/k}}{1-(A_2/A_1)^2(p_2/p_1)^{2/k}}} \tag{12.21}$$

This equation can be transformed into an equation for the actual weight rate of flow through venturi tubes by introducing the discharge coefficient C (Fig. 12.18) and an expansion factor Y.[1] The resulting equation is

$$G = CYA_2\sqrt{2g\gamma_1\frac{p_1-p_2}{1-(D_2/D_1)^4}} \tag{12.22}$$

where C has the same value as for an incompressible fluid at the same Reynolds number and γ_1 may be replaced by p_1/RT_1 if desired. Values of Y for $k = 1.4$ are plotted in Fig. 12.24.

In Fig. 12.24 it may be observed that for the venturi meter no values for Y are shown for p_2/p_1 ratios less than 0.528. This is so because, for air and other

[1] For a venturi or nozzle throat where $C_c = 1$,

$$Y = \sqrt{\frac{[k/(k-1)](p_2/p_1)^{2/k}[1-(p_2/p_1)^{(k-1)/k}]}{1-(p_2/p_1)}}\sqrt{\frac{1-(D_2/D_1)^4}{1-(D_2/D_1)^4(p_2/p_1)^{2/k}}}$$

gases having adiabatic constant $k = 1.4$, the p_2/p_1 ratio will always be greater than 0.528 if the flow is subsonic, as was pointed out in Sec. 9.7.

Equation (12.22) is directly applicable to the flow of compressible fluids through venturi tubes where $C_c = 1.0$, provided the flow is subsonic. The equation can also be used for flow nozzles and orifice meters, though for flow nozzles C should be replaced by $K\sqrt{1-(D_2/D_1)^4}$ [from Eq. (12.18)], so that Fig. 12.21 can be used directly. For orifice meters the C of Eq. (12.22) should be replaced by $K\sqrt{1-(D_o/D_1)^4}$, D_2 should be replaced by D_o and A_2 should be replaced by A_o where D_o is the diameter of the orifice opening and A_o its area.

For compressible fluids the C_c of an orifice meter depends on the p_2/p_1 ratio; hence Y varies in a different manner than in the case of a venturi. Values of Y for orifice meters are shown in Fig. 12.24. In the case of an orifice meter the maximum jet velocity is the sonic velocity c, but this does not impose a limit on the rate of discharge because the jet area continues to increase with decreasing values of p_2/p_1. For this reason the values of Y for the orifice are extended in Fig. 12.24 to lower values of p_2/p_1.

The general case of flow measurement under supersonic conditions will not be discussed in this text. If supersonic flow occurs in a converging or converging-diverging nozzle attached to the end of a pipe or to a tank, Equations (9.24), (9.25), and (9.26) may be employed to compute ideal flow rates where the velocity of approach is negligible. This can be transformed into actual flow rates by introducing a proper flow coefficient.

Illustrative Example 12.3 Determine the weight flow rate when air at 20°C and 700 kN/m², abs flows through a venturi meter if the pressure at the throat of the meter is 400 kN/m², abs. The diameters at inlet and throat are 25 and 12.5 cm respectively. Assume that $C = 0.985$.

Substitute the given data in Eq. (12.22), obtaining the value of Y from Fig. 12.24. For $p_2/p_1 = \frac{400}{700} = 0.57$ and $D_2/D_1 = 0.50$, $Y \approx 0.72$. Thus,

$$G = 0.985 \times 0.72 \times \frac{\pi(0.125)^2}{4}\sqrt{2(9.81)\gamma_1 \frac{700-400}{1-(0.5)^4}}$$

We find γ_1 from $pv = RT$ or $\gamma = pg/RT$,

$$\gamma_1 = \frac{(700\ \text{kN/m}^2)(9.81\ \text{m/s}^2)}{[287\ \text{m}^2/(\text{s}^2\cdot\text{K})](273+20)\ \text{K}} = 0.0817\ \text{kN/m}^3$$

Substituting this value for γ_1 into the first expression gives

$$G = 0.1971\ \text{kN/s} = 197.1\ \text{N/s}$$

If the relation between C and $\mathbf{R}_2$ for this meter is known, the value of $\mathbf{R}_2$ for the computed value of G can be determined. If the assumed value of C does not correspond with this value of $\mathbf{R}_2$, a slight adjustment in the value of C can be made to give a more accurate answer.

12.11 RECTANGULAR WEIRS

The weir has long been a standard device for the measurement of water in an open channel. In its simplest form the water flows over the top of a plate as shown in Fig. 12.25. The rate of flow is determined by measuring the height H

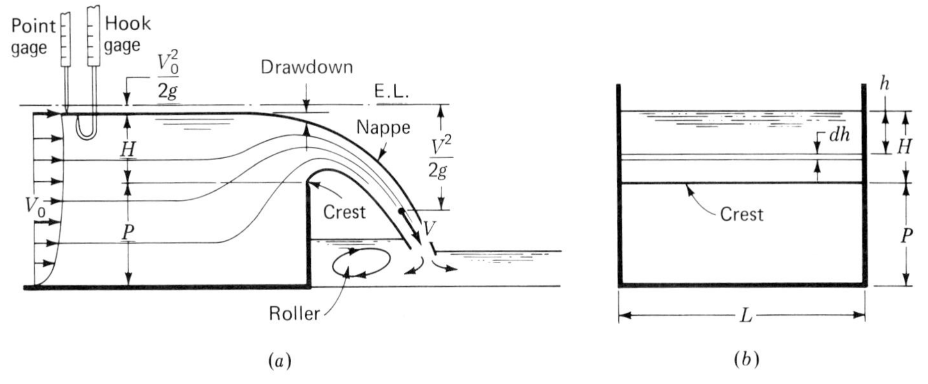

Figure 12.25 Flow over sharp-crested weir. (*a*) Side view. (*b*) Looking upstream.

(head), relative to the crest, at a distance upstream from the crest at least four times the maximum head that is to be employed. The amount of drawdown at the crest is typically about $0.15H$.

The upstream face of the weir plate should be smooth, and the plate should be strictly vertical. The crest should have a sharp, square upstream edge, and a bevel on the downstream side, so that the nappe springs clear, making a line contact for all but the very lowest heads. If it does not spring clear, the flow cannot be considered as true weir flow and the experimentally determined coefficients do not apply. The velocity at any point in the nappe is related to the energy line as shown in Fig. 12.25. The approach channel should be long enough so that normal velocity distribution exists and the surface should be as free of waves as possible.

Suppressed Rectangular Weir

This type of weir is as wide as the channel, and the width of the nappe is the same as the length of the crest. As there are no contractions of the stream at the sides, it is said that end contractions are *suppressed*. It is essential that the sides of the channel upstream be smooth and regular. It is common to extend the sides of the channel downstream beyond the crest so that the nappe is confined laterally. The flowing water tends to entrain air from this enclosed space under the nappe, and unless this space is adequately ventilated, there will be a partial vacuum and perhaps all the air may eventually be swept out. The water will then cling to the downstream face of the plate, and the discharge will be greater for a given head than when the space is vented. Therefore venting of a suppressed weir is necessary if the standard formulas are to be applied.

To derive the flow equation for a rectangular weir having a crest of length L, consider an elementary area $dA = L\,dh$ in the plane of the crest, as shown in Fig. 12.25. This elementary area is in effect an horizontal slot of length L and height dh. Neglecting velocity of approach, the ideal velocity of flow through this area will be equal to $\sqrt{2gh}$. The apparent flow through this area is

$$dQ = L\,dh\sqrt{2gh} = L\sqrt{2g}\,h^{1/2}\,dh$$

and this is to be integrated over the whole area, i.e., from $h = 0$ to $h = H$.
Performing this integration, we obtain an ideal Q_i which is

$$Q_i = \sqrt{2g}\, L \int_0^H h^{1/2} = \tfrac{2}{3}\sqrt{2g}\, LH^{3/2}$$

The actual flow over the weir will be less than the ideal flow because the effective flow area is considerably smaller than $L \times H$ due to drawdown from the top and contraction of the nappe from the crest below. Introducing a coefficient of discharge C_d to account for this,

$$Q = C_d \tfrac{2}{3}\sqrt{2g}\, LH^{3/2} \tag{12.23}$$

Dimensional analysis of weir flow leads to some interesting conclusions that provide a basis for an understanding of the factors that influence the coefficient of discharge. The physical variables that influence the flow Q over the weir of Fig. 12.25 include L, H, P, g, μ, σ, and ρ. Using the Buckingham Π theorem (Sec. 7.7), and without going into the details, the following results:

$$Q = \phi\left(\mathbf{W}, \mathbf{R}, \frac{P}{H}\right) L\sqrt{g}\, H^{3/2}$$

Thus, comparing this expression with Eq. (12.23), we conclude that C_d depends on **W**, **R**, and P/H. It has been found that P/H is the most important of these (Probs. 12.62 and 12.68). The Weber number **W**, which accounts for surface-tension effects, is important only at low heads. In the flow of water over weirs the Reynolds number is generally quite high, so viscous effects are generally insignificant. If one were to calibrate a weir for the flow of oil, however, **R** would undoubtedly affect C_d substantially. Typical values of C_d for sharp-crested weirs with water flowing range from about 0.62 for $H/P = 0.10$ to about 0.75 for $H/P = 2.0$.

Small-scale but precise experiments covering a wide range of conditions led Rehbock of the Karlsruhe Hydraulic Laboratory in Germany to the following expression for C_d in Eq. (12.23).

$$C_d = 0.605 + \frac{1}{305H} + 0.08\frac{H}{P} \tag{12.24}$$

This equation was obtained by fitting a curve to the plotted values of C_d for a great many experiments and is purely empirical. Capillarity is accounted for by the second term, while velocity of approach (assumed to be uniform) is responsible for the last term. Rehbock's formula has been found to be accurate within 0.5 per cent for values of P from 0.33 to 3.3 ft and for values of H from 0.08 to 2.0 ft with the ratio H/P not greater than 1.0. It is even valid for greater ratios than 1.0 if the bottom of the discharge channel is lower than that of the approach channel so that backwater does not affect the head.

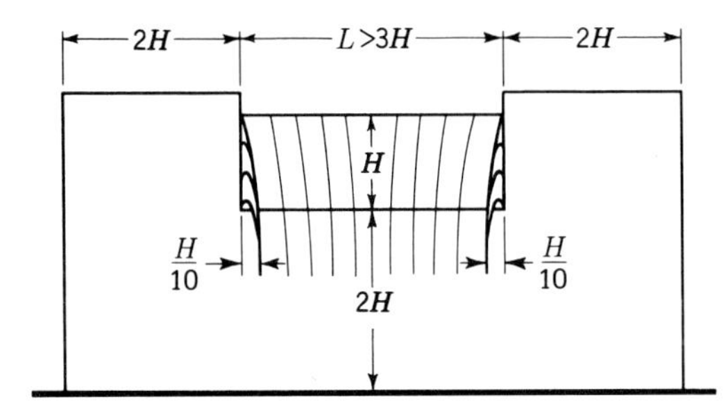

Figure 12.26 Limiting proportions of standard contracted weirs.

It is convenient to express Eq. (12.23) as

$$Q = C_W L H^{3/2} \tag{12.25}$$

where C_W, the *weir coefficient*,[1] replaces $C_d \frac{2}{3}\sqrt{2g}$.

Using a value of 0.62 for C_d in Eq. (12.23), we can write

$$Q \approx \begin{cases} 3.33 L H^{3/2} & \text{in English units} \\ 1.84 L H^{3/2} & \text{in SI units} \end{cases} \tag{12.26}$$

These equations give good results if $H/P < 0.4$, which is well within the usual operating range. If the velocity of approach V_0 is appreciable, a correction must be applied to the preceding equations either by changing the form of the equation or, more commonly, by changing the value of the coefficient.

Rectangular Weir with End Contractions

When the length L of the crest of a rectangular weir is less than the width of the channel, there will be a lateral contraction of the nappe so that its width is less than L. Experiments by Francis[2] indicated that under the conditions depicted in Fig. 12.26 the effect of each side contraction is to reduce the effective width of the nappe by $0.1H$. Hence for such a situation the flow rate may be computed by employing any of the three preceding equations and substituting $(L - 0.1nH)$ for L, where n is the number of end contractions, normally 2 but sometimes 1.

Cipolletti Weir

In order to avoid correcting for end contractions a Cipolletti weir is often used. It has a trapezoidal shape with side slopes of four vertical on one horizontal. The additional area adds approximately enough to the effective width of the stream to offset the lateral contractions.

[1] Since C_W is not dimensionless, its value in English units is different from that in SI units, as indicated in Eq. (12.26).

[2] James B. Francis, "Lowell Hydraulic Experiments," 5th ed., D. Van Nostrand Company, Inc., Princeton, N.J., 1909.

Illustrative Example 12.4 Flow is occurring in a rectangular channel at a velocity of 3 fps and depth of 1.0 ft. Neglecting the effect of velocity of approach and employing Eq. (12.26), determine the height of a sharp-crested suppressed weir that must be installed to raise the water depth upstream of the weir to 4 ft.

$$L = \text{length of weir crest} = \text{width of channel}$$

$$Q = AV = LyV = L(1)(3) = 3.33LH^{3/2}$$

$$H^{3/2} = \frac{3.0}{3.33} = 0.90 \qquad H = 0.93 \text{ ft}$$

$$P = \text{height of weir} = 4.00 - 0.93 = 3.07 \text{ ft}$$

12.12 TRIANGULAR, OR V-NOTCH, WEIR

For relatively small flows the rectangular weir must be very narrow and thus of limited maximum capacity, or else the value of H will be so small that the nappe will not spring clear but will cling to the plate. For such a case the triangular weir has the advantage that it can function for a very small flow and also measure reasonably large flows as well. The vertex angle is usually between 10 and 90° but rarely larger.

In Fig. 12.27 is a triangular weir with a vertex angle θ. The ideal rate of discharge through an elementary area dA is $dQ = \sqrt{2gh}\,dA$. Now $dA = 2x\,dh$, and $x/(H - h) = \tan\theta/2$. Substituting in the foregoing, and introducing a coefficient of discharge C_d, the following result is obtained for the entire notch:

$$Q = C_d 2\sqrt{2g}\tan\frac{\theta}{2}\int_0^H (H - h)h^{1/2}\,dh$$

Integrating between limits and reducing, the fundamental equation for all triangular weirs is obtained:

$$Q = C_d \frac{8}{15}\sqrt{2g}\tan\frac{\theta}{2}H^{5/2} \tag{12.27}$$

For a given angle θ and assuming C_d is constant, this may be reduced to

$$Q = KH^{5/2} \tag{12.28}$$

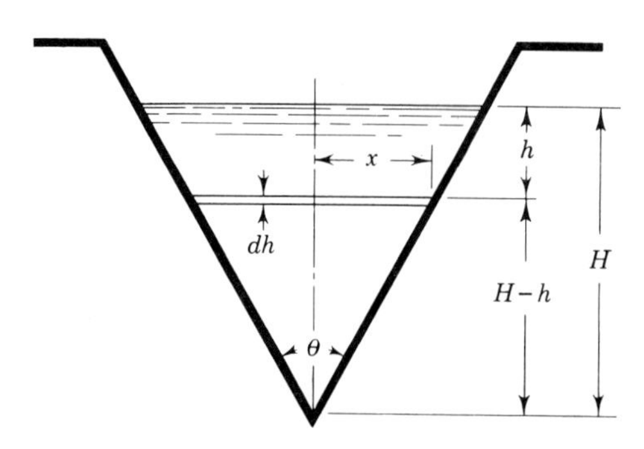

Figure 12.27 Triangular weir.

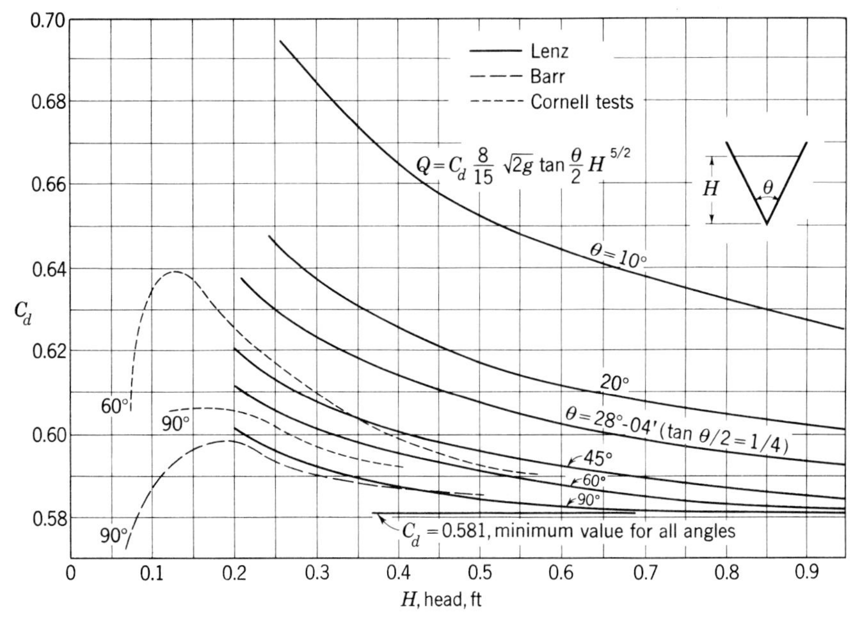

Figure 12.28 Coefficients for triangular weirs.

The value of the constant K in English units will be different from that in SI units.

In Fig. 12.28 are presented experimental values for C_d for water flowing over V-notch weirs with central angles varying from 10 to 90°. The solid lines represent tests by Lenz;[1] the dotted lines are from data taken at Cornell University;[2] the dashed line represents a 90° weir with a fine sharp edge, reported by Barr.[3] The rise in C_d at heads less than 0.5 ft is due to incomplete contraction. At lower heads the frictional effects reduce the coefficient. At very low heads, when the nappe clings to the weir plate, the phenomenon can no longer be classed as weir flow and Eqs. (12.27) and (12.28) are inapplicable.

12.13 BROAD-CRESTED WEIR[4] AND FREE OUTFALL

Another type of weir is the broad-crested weir (Fig. 12.29), which is usually built of concrete. One of its advantages is that it is rugged and can stand up well under field conditions.

[1] Arno T. Lenz, Viscosity and Surface Tension Effects on V-notch Weir Coefficients, *Trans. ASCE*, vol. 108, pp. 759–802, 1943.

[2] *Eng. News*, vol. 73, p. 636, 1915.

[3] James Barr, Experiments upon the Flow of Water over Triangular Notches, *Engineering*, Apr. 8–15, 1910.

[4] When the flow is supercritical, the presence of surface waves causes broad-crested weirs to be rather impractical for use as metering devices.

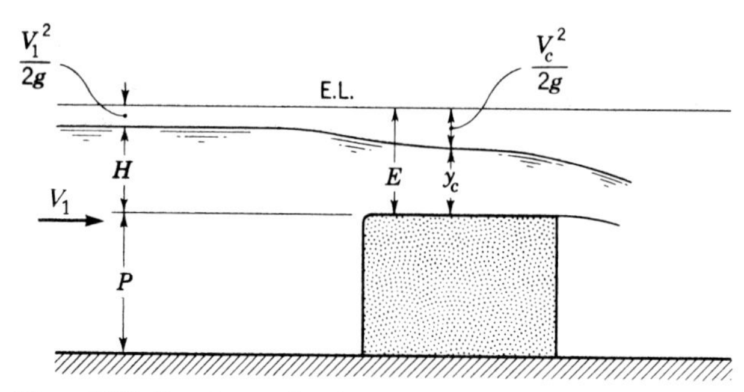

Figure 12.29 Broad-crested weir.

The broad-crested weir, as mentioned in Illustrative Example 11.3, is a critical-depth meter; that is, if the weir is high enough, critical depth occurs on the crest of the weir. In Eq. (11.20) it was shown that for a rectangular channel $V_c = \sqrt{gy_c}$, while Eq. (11.24) stated that when the flow is critical $y_c = \frac{2}{3}E$. Employing these relations, we can write for the flow over a broad-crested weir,

$$Q = AV = (Ly_c)\sqrt{gy_c} = L\sqrt{g}\,y_c^{3/2} = L\sqrt{g}(\tfrac{2}{3})^{3/2}E^{3/2} \tag{12.29}$$

Let us now substitute this expression into Eq. (12.23), which is applicable to broad-crested weirs as well as sharp-crested ones, since both have rectangular flow cross sections. This yields

$$C_d = \frac{1}{\sqrt{3}}\left(\frac{E}{H}\right)^{3/2} \tag{12.30}$$

For very high weirs (that is, P/H large) the velocity of approach becomes small, so that $E \to H$ (Fig. 12.29) and thus $E/H \to 1$ and $C_d \to 1/\sqrt{3} = 0.577$. With a lower weir and the same flow rate the velocity of approach becomes larger and hence E/H increases in magnitude, resulting in larger values of C_d for lower broad-crested weirs. The foregoing discussion, of course, assumes critical flow (Sec. 11.9) on the weir. The actual value of C_d for a broad-crested weir depends on the length of the weir and whether or not the upstream corner of the weir is rounded.[1]

Another method for determining the flow rate in a rectangular channel is to measure the depth of flow y_b at the brink of a free outfall (Fig. 11.17). Substituting $y_c = y_b/0.72$ in Eq. (11.21) permits an approximate determination of the rate of flow per foot of channel width. This method can be used only if the flow is subcritical.

12.14 OVERFLOW SPILLWAY

An overflow spillway is a section of dam designed to permit water to pass over its crest. Overflow spillways are widely used on gravity, arch, and buttress dams.

[1] E. F. Brater and H. W. King: "Handbook of Applied Hydraulics," 6th ed, McGraw-Hill Book Company, New York, N.Y., 1976.

The ideal spillway should take the form of the underside of the nappe of a sharp-crested weir (Fig. 12.30*a*) when the flow rate corresponds to the maximum design capacity of the spillway (Fig. 12.30*b*). Figure 12.30*c* shows an *ogee spillway* which closely approximates the ideal.[1] The reverse curve on the downstream face of the spillway should be smooth and gradual. A radius of about one-fourth of the spillway height has proved satisfactory.

The discharge of an overflow spillway is given by the weir equation[2]

$$Q = C_w L H^{3/2} \tag{12.31}$$

where Q = discharge, cfs or m³/s
C_w = coefficient
L = length of the crest, ft or m
H = head on the spillway (vertical distance from the crest of the spillway to the reservoir level), ft or m

The coefficient C_w varies with the design and head. For the standard overflow crest of Fig. 12.30*c* the variation of C_w is given in Fig. 12.31. Experimental models are often used to determine spillway coefficients. End contractions on a spillway reduce the effective length below the actual length L. Square-cornered piers disturb the flow considerably and reduce the effective length by the width of the piers plus about $0.2H$ for each pier. Streamlining the piers or flaring the

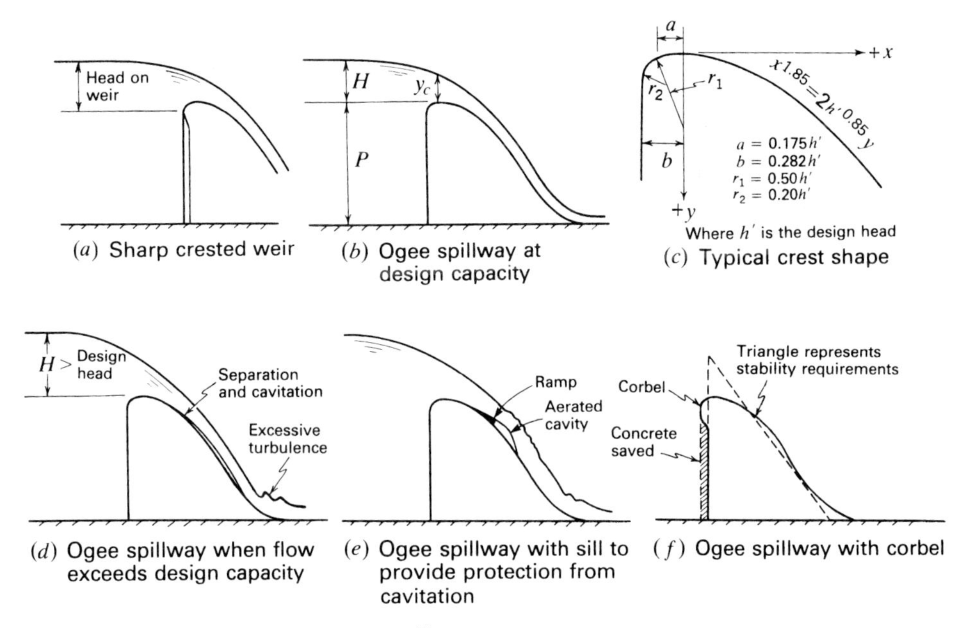

Figure 12.30 Characteristics of an ogee spillway.

[1] Hydraulic Models as an Aid to the Development of Design Criteria, U.S. Waterways Expt. Sta., Bull. 37, Vicksburg, Miss., 1951.

[2] See Eq. (12.25).

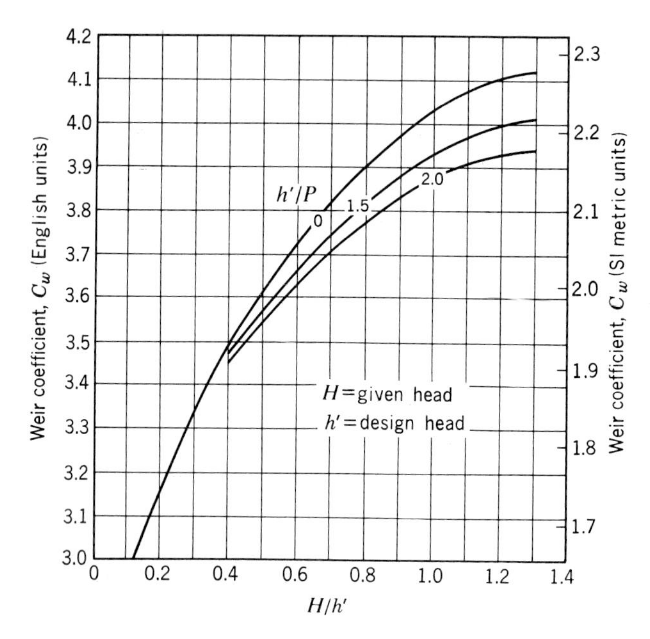

Figure 12.31 Variation of discharge coefficient with head for an ogee-spillway crest such as shown in Fig. 12.30.

spillway entrance minimizes the flow disturbance. If the cross-sectional area of the reservoir just upstream from the spillway is less than five times the area of flow over the spillway, the approach velocity will increase the discharge a noticeable amount. The effect of approach velocity can be accounted for by the equation

$$Q = C_w L \left(H + \frac{V_0^2}{2g} \right)^{3/2} \tag{12.32}$$

where V_0 is the approach velocity.

On high spillways if the overflowing water breaks contact with the spillway surface, a vacuum will form at the point of separation (Fig. 12.30*d*) and cavitation may occur. Cavitation and vibration from the alternate making and breaking of contact between the water and the face of the spillway may result in serious structural damage. Recent developments in the design of overflow spillways show that a ramp of proper shape and size when properly located (Fig. 12.30*e*) will direct the water away from the spillway surface to form a cavity.[1] To be effective, air must be freely admitted to the cavity. The result is that air is entrained in the water, the water bulks up, and when it returns to the spillway surface there is no problem with cavitation. On very high spillways these ramps may be used in tandem.

[1] K. Zagustin and N. Castillejo, Model-Prototype Correlation for Flow Aeration in Guru Dam Spillway, *Proceedings International Association for Hydraulic Research*, vol. 3, 1983.

There are a number of other types of spillways including chute, side-channel, and shaft spillways.[1]

12.15 SLUICE GATE

The sluice gate shown in Fig. 12.32 is a device used to control the passage of water in an open channel. When properly calibrated, it may also serve as a means of flow measurement. As the lower edge of the gate opening is flush with the floor of the channel, contraction of the bottom side of the issuing stream is entirely suppressed. Side contractions will of course depend on the extent to which the opening spans the width of the channel. The complete contraction on the top side, however, because of the larger velocity components parallel to the face of the gate, will offset the suppressed bottom contraction, resulting in a coefficient of contraction nearly the same as for a slot with contractions at top and bottom.

Flow through a sluice gate differs fundamentally from flow through a slot in that the jet is not free but guided by a horizontal floor. Consequently, the final jet pressure is not atmospheric, but distributed hydrostatically in the vertical section. Writing the energy equation with respect to the stream bed as datum from point 1 to point 2 in the free-flow case (Fig. 12.32*a*) and neglecting head loss,

$$\frac{V_1^2}{2g} + y_1 = \frac{V_2^2}{2g} + y_2$$

from which, introducing continuity,

$$V_{2_i} = \frac{1}{\sqrt{1 - (A_2/A_1)^2}} \sqrt{2g(y_1 - y_2)} \qquad (12.33)$$

The actual flow rate $Q = C_d Q_i = C_c C_v (A V_{2_i})$, where $A = aB$ is the area of the gate opening and B is the width of the opening.

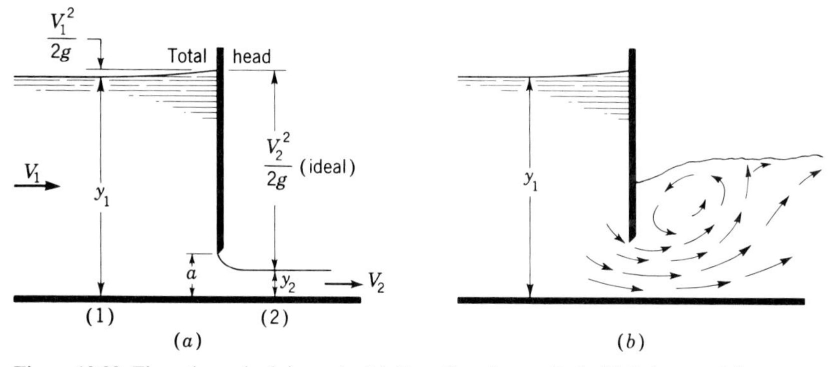

Figure 12.32 Flow through sluice gate. (*a*) Free flow [$y_2 = C_c a$]. (*b*) Submerged flow.

[1] For a brief discussion of the different types of spillways see: R. K. Linsley and J. B. Franzini, Water Resources Engineering, 3d ed., McGraw-Hill Book Co., New York, 1979.

Absorbing the effects of flow contraction, friction, velocity of approach, and the downstream depth y_2 into an experimental flow coefficient, a simple discharge equation for flow under a sluice gate results:

$$Q = K_s A \sqrt{2gy_1} \tag{12.34}$$

where K_s is defined as the *sluice coefficient*.[1]

Values of K_s depend on a and y_1 (Fig. 12.32*a*) and are usually between 0.55 and 0.60 for free flow, but are materially reduced when the flow conditions downstream are such as to produce submerged flow, as shown in Fig. 12.32*b*.

12.16 MEASUREMENT OF LIQUID SURFACE ELEVATION

The elevation of the surface of a liquid at rest may be determined through use of a piezometer column, manometer, or pressure gage (Sec. 2.6). These will also give accurate results when applied to a stationary liquid contained in a tank that is moving provided the tank is not undergoing an accelerating motion (Sec. 2.11). Staff gages, such as those used at reservoirs, provide approximate liquid surface elevation data.

Various methods are used to measure the surface elevation of moving liquids. To determine the head H on a weir the elevation difference between the crest of the weir and the liquid surface must be measured. In the field the elevation of the liquid surface is often determined through use of a stilling well connected by a pipe to the main liquid body. A float in the well is used to actuate a clock-driven liquid-level recorder so that a continuous record of the liquid-surface elevation is obtained. In the laboratory a *hook gage* or *point gage* (Fig. 12.25) is commonly employed for liquid-surface level determinations. The point gage is particularly suitable for fast-moving liquids where a hook gage would create a local disturbance in the liquid surface. In all liquid-surface level determinations care should be taken to make the measurements in regions where there is no curvature of streamlines; otherwise centrifugal effects will give a false reading of the piezometric head.

Other methods for determining the elevation of a liquid surface include the use of the sonic devices, electrical gages, and bubblers. A sonic device is mounted at some convenient location above the liquid surface and the time required for a sound pulse to travel vertically downward to the liquid surface and return is indicative of the relationship between the elevation of the liquid surface and the device. Electrical gages include those with capacitive sensors and those with resistive sensors. In the resistive type two parallel bare-wire conductors are partially immersed in the liquid as a component of an electrical system. The electrical resistance between them is a function of the liquid depth. With proper

[1] Values of discharge coefficients for sluice and other types of gates may be found in Hunter Rouse (ed.), "Engineering Hydraulics," pp. 536–543, John Wiley & Sons, Inc., New York, 1950.

circuitry and calibration the device can be used to provide data on liquid-surface elevation. The sonic devices and the electrical gages are equally applicable to liquids at rest or in motion. The bubbler is used primarily for liquids at rest. In the bubbler system, the minimum pressure required to drive a gas into a liquid (i.e., form bubbles) at a depth is a measure of the depth of the liquid. Knowing the elevation at which the bubbles are emitted permits determination of the elevation of the liquid surface.

12.17 OTHER METHODS OF MEASURING DISCHARGE

In addition to the foregoing "standard" devices for measuring the flow of fluids, there are a number of supplementary devices less amenable to exact theoretical analysis but worthy of brief mention. One of the simplest for measuring flow in a pipeline is the *elbow meter*, which consists of nothing more than piezometer taps at the inner and outer walls of a 90° elbow in the line. The pressure difference, due to the centrifugal effects at the bend, will vary approximately as the velocity head in the pipe. Like other meters, the elbow should have sections of a straight pipe upstream and downstream and should be calibrated in place.[1]

The *rotameter* (Fig. 12.33) consists of a vertical glass tube that is slightly tapered, in which the metering *float* is suspended by the upward motion of the fluid around it. Directional notches cut in the float keep it rotating and thus free of wall friction. The rate of flow determines the equilibrium height of the float, and the tube is graduated to read the flow directly. The rotameter is also used for gas flow, but the weight of the float and the graduation must be changed accordingly.

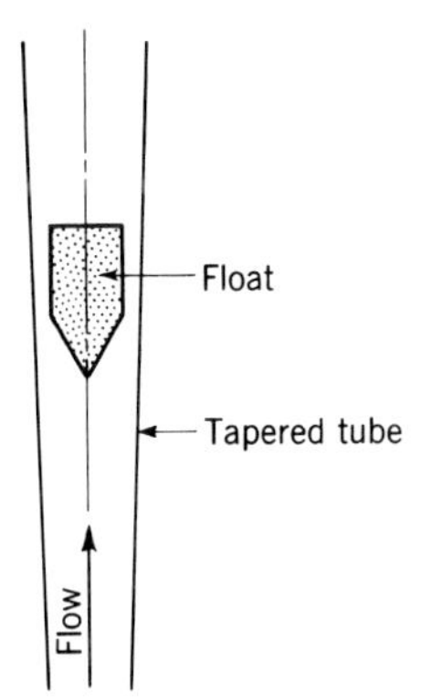

Figure 12.33 Rotameter type of flow meter.

[1] W. M. Lansford, The Use of an Elbow in a Pipe Line for Determining the Rate of Flow in the Pipe, *Univ. Ill. Eng. Expt. Sta. Bull.* 289, December, 1936.

Other techniques for measuring flow rate include the *salt-velocity* method. In this method a charge of concentrated salt is injected into the flow at an upstream station. Its arrival at a downward station is detected from conductivity measurements. In flowing between the two stations, the salt disperses and its arrival at the downstream station is spread out over a considerable period of time. The time of travel between the two stations is taken as the time from the instant of injection at the upstream station to the time at which the centroid of the imposed conductivity-time curve passes the downstream station. Knowing the travel time and the distance, the velocity may be determined, and then by multiplying by the cross-sectional area, we get the flow rate.

PROBLEMS

12.1 A small object weighs 1.32 lb in air and 1.02 lb in a liquid. The volume of this object is known to be 0.0060 ft^3. What is the density of the liquid?

12.2 A small object weighs 14.00 N in air and 9.80 N in a liquid. The volume of this object is known to be 292 cm^3. What is the density of the liquid?

12.3 A hydrometer is made in the form of a $\frac{1}{4}$-in (1.0-cm)-diameter cylinder of length 10 in (25 cm). Attached to the end of the cylinder is a 1-in-diameter sphere. The entire device weighs 14.0 g. What range of specific gravities can be measured with this device?

12.4 To what depth will the hydrometer of Prob. 12.3 sink when placed in a liquid having a density of 1.74 $slugs/ft^3$?

12.5 Carbon tetrachloride ($s = 1.59$) is placed in an open U tube. A liquid is poured into one of the legs of the tube. A liquid column 15.4 in (38.5 cm) high balances a carbon tetrachloride column 10.0 in (25.0 cm) high. What is the specific weight of the liquid? The text states that this method will give only approximate values. Why is this so?

12.6 A rotational viscometer is constructed of two concentric cylinders of height 10.0 in (25.0 cm). The OD of the inner cylinder is 3.950 in (9.90 cm), and the ID of the outer cylinder is 4.050 in (10.10 cm). When a torque of 5.0 ft·lb (7.0 N·m) is applied to the outer cylinder, it was found to rotate at 1 revolution per 3.5 s. Find the viscosity of the fluid. Neglect mechanical friction.

12.7 A tube viscometer similar to the one of Fig. 12.1 has a tube diameter of 0.0422 in and a tube length of 3.05 in. The vertical distance from the liquid surface in the reservoir to the tube outlet changed from 9.50 to 9.00 in during a run. The flow volume was 50 cm^3, and the time was 126.4 s. Find the kinematic viscosity of the liquid.

12.8 Fifty cubic centimeters of water at 80°F flows through a tube-type viscometer in 50.5 s. An equal volume of oil at 60°F flows through the same viscometer in 800 s. Find the absolute viscosity of the oil if $s = 0.86$.

12.9 A liquid ($\rho = 880$ kg/m^3) flows through a glass tube of diameter 2.0 mm and length 4.5 m under a head of 50 cm at a steady rate of 30 cm^3/min. Find the absolute viscosity and the kinematic viscosity of the liquid and confirm by calculation that the flow is laminar. Express the answers in stokes and poises.

12.10 Water at 50°F (10°C) flows through a tube-type viscometer in 100.0 s. How long will it take 90°F (38°C) water to pass through the same viscometer?

12.11 A 0.25-in-diameter lead sphere ($s = 11.4$) falls through an oil ($s = 0.86$) at a constant velocity of 0.150 fps. The oil is contained in a 2.25-in-diameter tube. Find the viscosity of the oil. Check $\mathbf{R}$ to see if it is less than 1.0.

12.12 A 16-mm-diameter glass bead ($s = 2.60$) falls through a liquid ($s = 1.59$) at a constant velocity of 14.5 cm/min. The liquid is contained in a 10-cm-diameter tube. Find the absolute viscosity and kinematic viscosity of the liquid.

12.13 A 4-in (10-cm)-diameter tube contains oil ($s = 0.9$) having a viscosity of 0.005 lb·s/ft^2 (0.25 N·s/m^2). Find the maximum size of steel sphere ($s = 7.8$) that will satisfy Stokes' law. What will be the fall velocity of this sphere?

12.14 A 0.10-in-diameter sphere has a fall velocity of 0.005 fps when a certain liquid is contained in a 1.0-in-diameter tube. Compute the fall velocities in tubes of diameter 0.50, 2.0, 4.0, and 10.0 in. Plot fall velocity vs. tube diameter.

12.15 In the figure, pressure gage A reads 10.0 psi, while pressure gage B reads 11.0 psi. Find the velocity if 50°F air is flowing. Atmospheric pressure is 26.8 in Hg. Assume $C_I = 1.0$ and neglect compressibility effects.

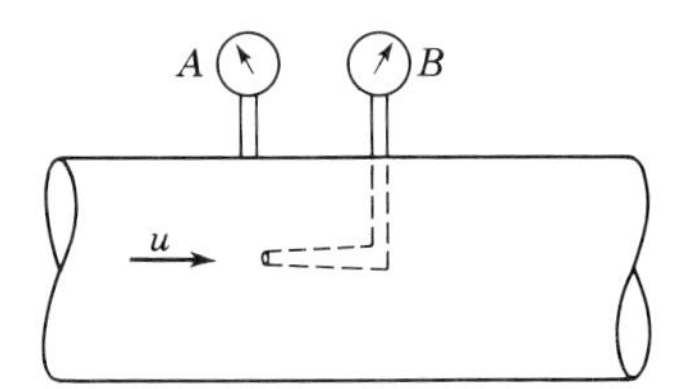

Problem 12.15

12.16 In Prob. 12.15, if the two pressure gages were replaced by a differential manometer containing water, what would be the reading on the manometer?

12.17 In the figure for Prob. 12.15 kerosene ($s = 0.81$) is flowing. The pressure gages at A and B read 65 and 140 N/m^2. Find the velocity u assuming $C_I = 1.0$.

12.18 The pitometer in the figure is connected to a mercury manometer, and the reading is 4.0 in (10.0 cm). The velocity is known to be 11.8 fps (3.6 m/s). If carbon tetracholoride ($s = 1.59$) is flowing what is C_I for the instrument?

Problem 12.18

12.19 In the sketch for Prob. 12.18 suppose air at 50°F is flowing. The pitometer is attached to a manometer containing a liquid ($s = 0.85$). Plot the velocity u versus the manometer reading assuming $C_I = 0.92$. Assume the air is at standard atmospheric pressure.

12.20 A pitot tube is placed in a pipe carrying water at 60°F (15°C). The pitot tube and a wall piezometer tube are connected to a water-mercury manometer which registers a differential of 3 in (7.5 cm). Assuming $C_I = 0.99$, what is the velocity approaching the tube?

12.21 Suppose that the fluids of Prob. 12.20 are reversed so that mercury is flowing in the pipe and water is the gage fluid (with the manometer now inverted). With the same gage differential, what would be the velocity of the mercury?

12.22 A Prandtl tube is placed on the center line of a smooth 12-in-diameter pipe in which 80°F water is flowing. The reading on a differential manometer attached to this Prandtl tube is 10 in of carbon tetrachloride ($s = 1.59$). Find the flow rate.

12.23 A pitot-static tube for which $C_I = 0.98$ is connected to an inverted U tube containing oil ($s = 0.85$). Water is flowing. What is the velocity if the manometer reading is 4.0 in (10 cm)?

12.24 In Fig. 12.9 let the pipe diameter be 24 in (64 cm) and suppose the flow is laminar, that is, $u = u_{max} - kr^2$. Divide the circle into concentric rings with radii 3, 6, 9, and 12 in (8, 16, 24, and 32 cm), and compute the flow rate by the method of Fig. 12.9 by taking the velocities at radii of 1.5, 4.5, 7.5, and 10.5 in (4, 12, 20, and 28 cm) as representative of the rings. Use a value of 10 fps (3 m/s) for u_{max}. Compare the result with that obtained by integration.

12.25 Water issues from a circular orifice under a head of 40 ft. The diameter of the orifice is 4 in. If the discharge is found to be 479 ft^3 in 3 min, what is the coefficient of discharge? If the diameter at the vena contracta is measured to be 3.15 in, what is the coefficient of contraction and what is the coefficient of velocity?

12.26 A jet discharges from an orifice in a vertical plane under a head of 12 ft (3.65 m). The diameter of the orifice is 1.5 in (3.75 cm), and the measured discharge is 0.206 cfs (6.0 L/s). The coordinates of the center line of the jet are 11.54 ft (3.46 m) horizontally from the vena contracta and 3.0 ft (0.9 m) below the center of the orifice. Find the coefficients of discharge, velocity, and contraction.

12.27 The velocity of water in a 4-in (10-cm)-diameter pipe is 10 fps (3 m/s). At the end of the pipe is a nozzle whose velocity coefficient is 0.98. If the pressure in the pipe is 8 psi (55 kN/m^2), what is the velocity in the jet? What is the diameter of the jet? What is the rate of discharge? What is the head loss?

12.28 A jet of water 3 in in diameter is discharged through a nozzle whose velocity coefficient is 0.96. If the pressure in the pipe is 12 psi and the pipe diameter is 8 in and if it is assumed that there is no contraction of the jet, what is the velocity at the tip of the nozzle? What is the rate of discharge?

12.29 The nozzle in the figure throws a stream of water vertically upward such that the power available in the jet at point 2 is 3.42 hp (2.55 kW). If the pressure at the base of the nozzle, point 1, is 21.0 psi (145 kPa), find (*a*) the theoretical height to which the jet will rise; (*b*) the coefficient of velocity; (*c*) the head loss between points 1 and 2; (*d*) the theoretical diameter of the jet at a point 20 ft (6 m) above point 2.

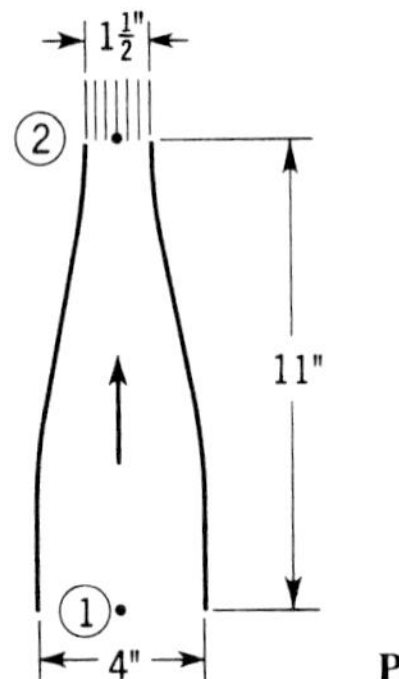

Problem 12.29

12.30 The loss of head due to friction in an orifice, nozzle, or tube may be expressed as $h_L = kV^2/2g$, where V is the actual velocity of the jet. (*a*) Compute k for the three tubes in Fig. 12.15. (*b*) If the tubes discharge water under a head of 5 ft (1.5 m), compute the loss of head in each case.

12.31 The diverging tube shown in the figure discharges water when $h = 5$ ft. The area A is twice area A_o. Neglecting all friction losses, find (*a*) velocity at throat; (*b*) pressure head at throat.

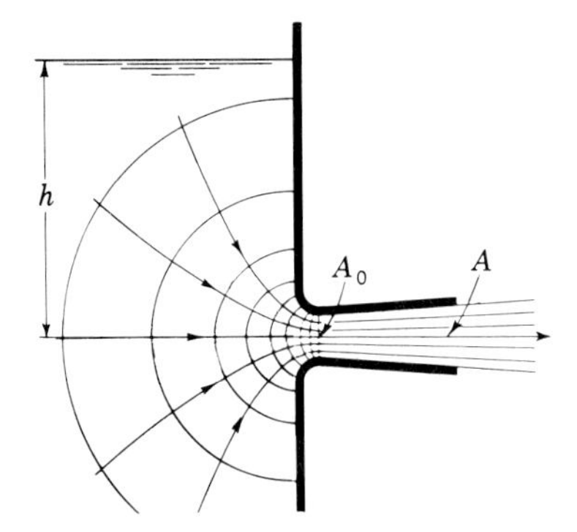

Problem 12.31

12.32 If the barometric pressure is 14.7 psia and the water temperature is 80°F, what is the maximum value of h at which the tube will flow full, all other data being the same as in Prob. 12.31? What will happen if the value of h is made greater than this?

12.33 For a rounded entrance and tube flowing full as in the sketch for Prob. 12.31, $C_c = 1.0$ both for the throat and the exit, and thus $C_v = C_d$ for both sections. For the throat, assume the value of C_v as given for (*a*) in Fig. 12.15, and assume that for the tube as a whole the discharge coefficient applied to the exit end is 0.70. If $h = 5$ ft, find the velocity at the throat and the pressure head at the throat, and compare with Prob. 12.31.

12.34 Suppose that the diverging tube shown in the figure for Prob. 12.31 is discharging water when $h = 2.5$ m. The area A is $1.8 \times A_o$. Neglecting all friction losses, find (*a*) the velocity at the throat; (*b*) the pressure head at the throat.

12.35 If the tube of Prob. 12.34 is operating at standard atmospheric conditions at 2,000-m elevation, what would be the maximum value of h at which the tube will flow full?

12.36 Find the maximum theoretical head at which the Borda tube of Fig. 12.16 will flow full if the liquid is water at 80°F (25°C) and the barometer reads 28.4 in (720 mm) Hg. Assume $C_d = 0.72$ for the tube flowing full.

12.37 In the figure, the pitot tube in a water jet at elevation 60 ft registers a pressure of 16.5 psi. The orifice at the bottom of the large open tank has a diameter of 1.00 in. Find C_c and C_v of the orifice. Neglect air resistance. The flow rate is 0.12 cfs.

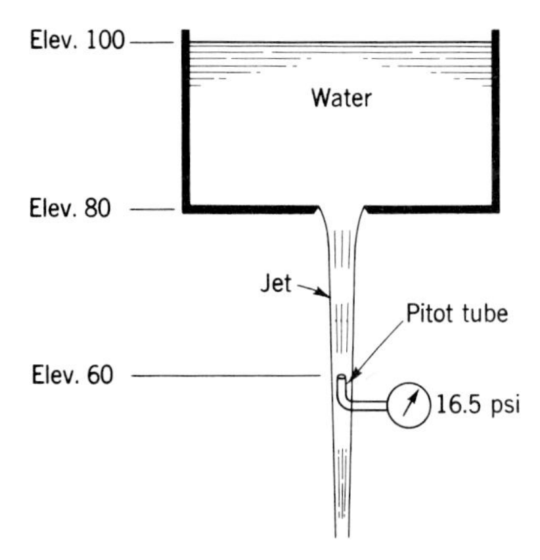

Problem 12.37

12.38 Water flows from one tank to an adjacent tank through a 50-mm sharp-edged orifice. The head of water on one side of the orifice is 2.0 m and that on the other is 0.4 m. Assuming $C_c = 0.62$ and $C_v = 0.95$, calculate the flow rate.

12.39 Find the flow rate of 72°F water for this venturi tube if a mercury manometer reads $y = 4$ in (10 cm) for the case where $D_1 = 8$ in (20 cm), $D_2 = 4$ in (10 cm), and $\Delta z = 1.5$ ft (0.45 m). Assume the discharge coefficients of Fig. 12.18 are applicable. In addition, find the head loss from inlet to throat. Also, determine the total head loss across the meter. Assume a diverging cone angle of 10°.

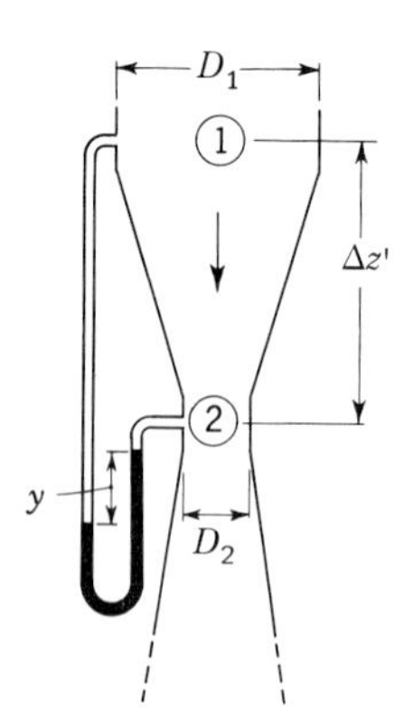

Problem 12.39

12.40 Repeat Prob. 12.39 for the case of a horizontal venturi tube (i.e., $\Delta z = 0$) with all other data the same.

12.41 Refer to Prob. 12.39. What would be the flow rate if the manometer fluid was carbon tetrachloride ($s = 1.59$) with all other data remaining the same?

12.42 Find the flow rate of water for the venturi tube shown in the figure for Prob. 12.39 if $D_1 = 80$ cm, $D_2 = 40$ cm, $\Delta z = 200$ cm, and $y = 15$ cm of mercury. Assume Fig. 12.18 is applicable.

12.43 In the figure for Prob. 12.39 suppose $D_1 = 2$ in, $D_2 = 1$ in, and $\Delta z = 6$ in. Oil ($s = 0.90$) with a kinematic viscosity of 0.0005 ft^2/s is flowing. Determine the manometer reading y if mercury is used as the manometer fluid. Assume the C values of Fig. 12.18 are applicable. The rate of flow is 0.10 cfs.

12.44 Suppose a 10-cm ISA flow nozzle (Figs. 12.20 and 12.21) is used in a 20-cm pipe to measure the flow of 40°C water. What would be the reading on a mercury manometer for the following flow rates: (*a*) 1.5 L/s; (*b*) 15 L/s; (*c*) 150 L/s?

12.45 A 6-in ISA flow nozzle is used to measure the flow of crude oil ($s = 0.855$) at 15°F. If a mercury manometer shows a reading of 3.5 in, what is the flow? Assume $D_2/D_1 = 0.70$.

12.46 Repeat Prob. 12.44 for a VDI orifice.

12.47 Repeat Prob. 12.45 for a VDI orifice.

12.48 A 6-in pipe carries 1.5 cfs of water at 72°F. Find the differential head and head loss across the following types of meters: (*a*) a 6- by 3-in VDI orifice; (*b*) a 6- by 3-in ISA flow nozzle; (*c*) a 6- by 3-in venturi meter.

12.49 Assume that air at 70°F (20°C) and 100 psia (700 kN/m^2, abs) flows through a venturi tube and that the pressure at the throat is 60 psia (420 kN/m^2, abs). The inlet area is 0.60 ft^2 (0.060 m^2), and the throat area is 0.15 ft^2 (0.015 m^2). The tube coefficient is 0.98, and k for air is 1.4. (*a*) Find the rate of discharge using Eq. (12.21). (*b*) Evaluate Y from Fig. 12.24, and use it to find the rate of discharge.

12.50 What would be the value of Y and the rate of discharge for a square-edged orifice for the same data as in Prob. 12.49, assuming $C = 0.60$?

12.51 What is the value of the throat velocity in Prob. 12.49?

12.52 Air flows through a 15- by 7.5-cm venturi meter. At inlet the air temperature is 15°C and the pressure is 140 kN/m^2. Determine the flow rate if a mercury manometer reads 15 cm. Assume an atmospheric pressure of 101.3 kN/m^2, abs.

12.53 Natural gas, for which $k = 1.3$ and $R = 3{,}100$ ft·lb/(slug·°R), flows through a venturi tube with pipe and throat diameters of 12 and 6 in, respectively. The initial pressure of the gas is 150 psia, and its temperature is 60°F. If the meter coefficient is 0.98, find the rate of flow for a throat pressure of 100 psia.

12.54 Helium, for which $k = 1.66$, and $R = 12{,}400$ ft·lb/(slug·°R), is in a tank under a pressure of 50 psia and a temperature of 80°F. It flows out through an orifice $\frac{1}{2}$ in in diameter. For such an orifice, $C_v = 0.98$, and $C_c = 0.62$ for liquids. Find the rate of flow if the pressure into which the gas discharges is 40 psia. Assume $Y = 0.95$.

12.55 For the data in Prob. 12.54 find the rate of discharge if $p_2 = p_c$.

12.56 Air is in a tank under a pressure of 200 psia (1400 kN/m^2, abs) and a temperature of 100°F (40°C). It flows out through an orifice having an area of 1.5 in^2 (10 cm^2) into a space where the pressure is 80 psia (550 kN/m^2, abs). Compute the rate of discharge assuming $C_d = 0.60$.

12.57 Using the same data as in Prob. 12.56, what would be the flow if the air discharged into a space where the pressure is 15 psia (105 kN/m^2, abs)?

12.58 Air in a tank at 1500 kN/m^2, abs and 40°C flows out through a 5.0-cm-diameter orifice into a space where the pressue 500 kN/m^2, abs. Compute the rate of discharge assuming $C_d = 0.60$. Refer to Sec. 9.7. Repeat for external pressures of 750, 1,000, and 1,250 kN/m^2, abs.

12.59 For air at $p_1 = 100$ psia (700 kN/m^2, abs) and a temperature of 70°F (20°C), find the critical pressure and the corresponding throat velocity in a suitable nozzle, neglecting the velocity of approach. What will the values be if $D_2/D_1 = 0.80$?

12.60 A rectangular sharp-crested weir 3.0 ft (0.9 m) high extends across a rectangular channel which is 8 ft (2.4 m) wide. When the head is 1.200 ft (36 cm) find the rate of discharge by neglecting the velocity of approach.

12.61 Suppose the rectangular weir of Prob. 12.60 is contracted at both ends. Find the rate of discharge for a head of 1.20 ft (36 cm) by the Francis formula. What would be the maximum value of H for which the Francis formula could be used?

12.62 Plot a family of curves of C_d versus P/H with H as a parameter. Use the Rehbock formula. These curves give a complete picture of the variation of C_d for sharp-crested rectangular weirs. Include P/H values of 0.5, 1.0, 2.0, 5.0, 10.0 and H values of 0.2, 1.0, and 5.0 ft.

12.63 (*a*) What is the rate of discharge of water over a 45° triangular weir when the head is 0.5 ft? (*b*) With the same head, what would be the increase in discharge obtained by doubling the notch angle, i.e., for a 90° weir? (Use curves of C_d versus H.) (*c*) What would be the head for discharge of 2.0 cfs of water over a 60° triangular weir?

12.64 For the Cipolletti weir, derive the slope ($\frac{1}{4}$:1) of the sides of the trapezoid by setting the reduction in discharge due to contraction equal to the increase in discharge due to the triangular area added.

12.65 Develop in general terms an expression for the percent of error in Q over a triangular weir if there is a small error in the measurement of the vertex angle. Assume there is no error in the weir coefficient. Compute the percent error in Q if there is a 2° error in the measurement of the total vertex angle of a triangular weir having a total vertex angle of 75°.

12.66 A broad-crested weir rises 1.0 ft (0.3 m) above the bottom of a horizontal channel. With a measured head of 2.0 ft (0.6 m) above the crest, what is the rate of discharge per unit width?

12.67 A broad-crested weir of height 2.00 ft (0.6 m) in a channel 5.00 ft (1.5 m) wide has a flow over it of 9.50 cfs (0.27 m^3/s). What is the water depth, just upstream of the weir?

12.68 Using Eq. (12.30), plot C_d versus P/H for broad-crested weirs. Include P/H values of 0.5, 1.0, 2.0, 5.0, 10.0 and H values of 0.2, 1.0, and 5.0 ft.

12.69 All the weir crests discussed in this chapter produce flow rates which vary as the head to some power greater than 1. In certain cases, such as in the outlet of a constant-velocity sedimentation chamber, it is desirable to employ a weir form in which Q varies directly with H. The *proportional-flow weir* is set flush with the bottom of the channel, as shown in the figure, while the sides taper inward, following the hyperbola $x\sqrt{y} = k$, a constant. Commencing with the head $h = H - y$, on the element of area $dA = 2x\,dy = 2(k/\sqrt{y})\,dy$, prove that the discharge equation for such a weir may be written as $Q = C_d \pi k \sqrt{2gH}$, and evaluate k in terms of the width B and the velocity V in the rectangular approach channel.

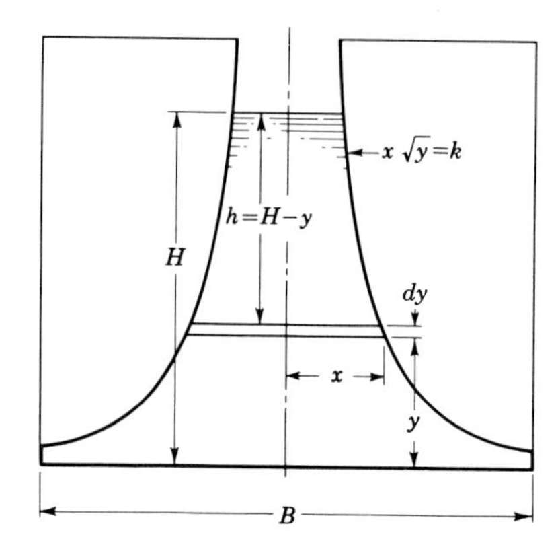

Problem 12.69

12.70 A 60° V-notch weir and a rectangular weir with end contactions having a crest length of 2 ft (0.6 m) are both used to measure a flow rate of approximately 0.25 cfs (7 L/s). Assuming C_d is known precisely for both weirs, compute the percentage of error in Q that would result from an error of 0.02 ft (6 mm) in the respective head measurements.

12.71 A rectangular channel 6 ft (1.8 m) wide contains a sluice gate which extends across the width of the channel. If the gate produces free flow when it is open 0.4 ft (0.12 m) with an upstream depth of 3.5 ft (1.05 m) find the rate of discharge, assuming $C_d = 0.60$ and $C_c = 0.62$. Evaluate K_s.

12.72 Refer to Illustrative Example 4.8. If $C_v = 0.98$, what is the flow rate? If $C_c = 0.62$, what is the height of the opening? Find K_s.

CHAPTER

THIRTEEN

UNSTEADY-FLOW PROBLEMS

13.1 INTRODUCTION

This text deals mostly with steady flow, since the majority of cases of engineering interest are of this nature.[1] However, there are a number of cases of unsteady flow that are very important, some of which are discussed in this chapter. It has been explained that turbulent flow is unsteady in the strictest sense of the word, but if the mean temporal values are constant over a period of time, it is called mean steady flow. Attention is here directed to cases where the mean temporal values continuously vary.

There are two main types of unsteady flow that will be considered here. The first is where the water level in a reservoir or pressure tank is steadily rising or falling, so that the rate of flow varies continuously, but where change takes place slowly. The second is where the velocity in a pipeline is changed rapidly by the fast closing or opening of a valve.

In the first case, of slow change, the flow is subject to the same forces as have previously been considered. Fast changes, of the second type, require the consideration of elastic forces.

Unsteady flow also includes such topics as oscillations in connected reservoirs and in U tubes and such phenomena as tidal motion and flood waves in open channels. Likewise, the field of machinery regulation by servomechanisms is intimately connected with unsteady motion. However, none of these topics will be considered here.

[1] Where the unsteadiness is not too rapid, unsteady flow can usually be approximated by assuming the flow is steady at different rates over successive time-periods of short duration.

13.2 DISCHARGE WITH VARYING HEAD

When flow occurs under varying head, the rate of discharge will continuously vary. Let us consider the situation depicted in Fig. 13.1 in which V_L represents the volume of liquid contained in the tank at a particular instant of time. There is inflow at the rate Q_i and outflow at rate Q_o. The change in volume during a small time interval dt can be expressed as

$$dV_L = Q_i dt - Q_o\, dt$$

If A_s = area of the surface of the volume while dz is the change in level of the surface, then $dV_L = A_s\, dz$. Equating these two expressions for dV_L,

$$A_s\, dz = Q_i\, dt - Q_o\, dt \tag{13.1}$$

Either Q_i or Q_o or both may be variable. The outflow Q_o is usually a function of z. For example, if liquid is discharged through an orifice or a pipe of area A under a differential head z, $Q_o = C_d A\sqrt{2gz}$, where C_d is a numerical discharge coefficient and z is a variable. If the liquid flows out over a weir or a spillway of length L, $Q_o = CLh^{3/2}$, where C is the appropriate coefficient and h is the head on the weir or spillway (Sec. 12.11). In either case z or h is the variable height of the liquid surface above the appropriate datum. The inflow Q_i commonly varies with time, however such problems will not be considered here. We will consider only the cases where $Q_i = 0$ or where Q_i = constant.

Rewriting Eq. (13.1) and integrating gives an expression for t, the time for the water level to change from z_1 to z_2. Thus

$$t = \int_{z_1}^{z_2} \frac{A_s\, dz}{Q_i - Q_o} \tag{13.2}$$

The right-hand side of this expression can be integrated if Q_i is zero or constant and if A_s and Q_o can be expressed as functions of z. In the case of natural reservoirs, the surface area cannot be expressed as a simple mathematical function of z but values of it may be obtained from a topographic map. In such a

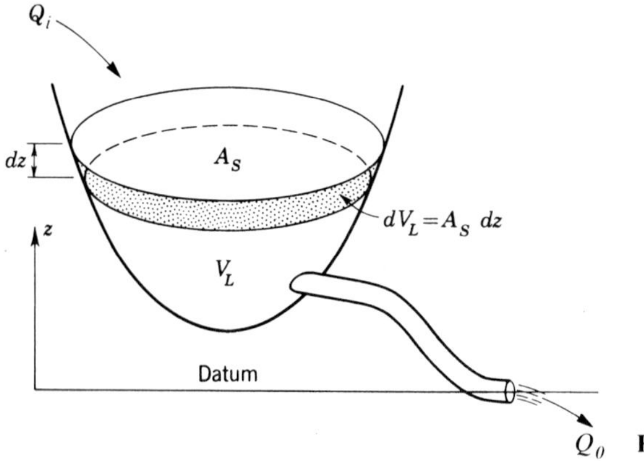

Figure 13.1

case, Eq. (13.2) may be solved graphically by plotting values of $A_s/(Q_i - Q_o)$ against simultaneous values of z. The area under such a curve to some scale is the numerical value of the integral.

It may be observed that instantaneous values for Q_0 have been expressed in the same manner as for steady flow. This is not strictly correct, since for unsteady flow the energy equation should also include an acceleration head [Eq. (13.6)]. The introduction of such a term renders the solution much more difficult. In cases where the value of z does not vary rapidly, no appreciable error will result if this acceleration term is disregarded. Therefore the equations will be written as for steady flow.

Illustrative Example 13.1 The open wedge-shaped tank in the accompanying figure has a length of 15 ft perpendicular to the sketch. It is drained with a 3-in diameter pipe of length 10 ft whose discharge end is at elevation zero. The coefficient of loss at pipe entrance is 0.50, the total of the bend loss coefficients is 0.20, and f for the pipe is 0.018. Find the time required to lower the water surface in the tank from elevation 8 to 5 ft. Neglect the possible change of f with **R**, and assume that the acceleration effects in the pipe are negligible.

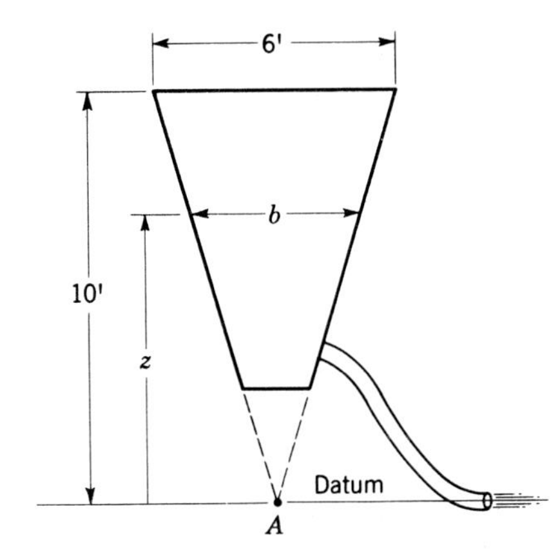

Illustrative Example 13.1

Energy equation from water surface to jet at discharge:

$$z - \left[0.5 + 0.2 + 0.018\left(\frac{10}{0.25}\right)\right]\frac{V^2}{2g} = \frac{V^2}{2g}$$

$$z - 1.42\frac{V^2}{2g} = \frac{V^2}{2g}$$

$$V = 5.16z^{1/2}$$

$$Q_o = AV = \frac{\pi}{4}(0.25)^2 5.16z^{1/2} = 0.253z^{1/2}$$

The area of the water surface may be expressed as

$$A_s = 15b = 15Kz$$

At the top of the tank, $A_s = 15 \times 6 = 15K(10)$, $K = 0.6$. Thus

$$A_s = 15(0.6)z = 9z$$

Applying Eq. (13.2),

$$t = \int_8^5 \frac{9z\,dz}{0 - 0.253z^{1/2}} = -\frac{9}{0.253}\int_8^5 z^{1/2}\,dz$$

$$= -35.5[\tfrac{2}{3}z^{3/2}]_8^5 = 271 \text{ s}$$

Note that if the pipe had discharged at an elevation other than zero, the integral would have been different, because the head on the pipe would then have been $z + h$, where h is the vertical distance of the discharge end of the pipe below (h positive) or above (h negative) point A of the figure.

13.3 UNSTEADY FLOW OF INCOMPRESSIBLE FLUIDS IN PIPES

When the flow in a pipe is unsteady, the energy equation has a term, the *accelerative head* $(L/g)(dV/dt)$, which accounts for the effect of the acceleration of the fluid. Let us refer back to Sec. 4.11 where the energy equation for one-dimensional steady flow of a real fluid was developed. We shall follow the same procedure that was used there by writing $\sum F = ma$; however in this situation, with unsteady flow, at a particular point on the streamline at a particular instant of time, we express the acceleration as $V(dV/ds) + dV/dt$. This comes from the general expression for acceleration in unsteady flow [Eq. (3.20)]. Applying $\sum F = ma$ to the cylindrical fluid element of Fig. 4.4, we get for unsteady flow

$$-dp\,dA - \rho g\,dA\,dz - \tau(2\pi r)\,ds = \rho\,ds\,dA\left(V\frac{dV}{ds} + \frac{dV}{dt}\right)$$

In this case $dA = \pi r^2$. Making this substitution for dA and dividing through by $-\rho\pi r^2$ gives

$$\frac{dp}{\rho} + V\,dV + g\,dz + ds\frac{dV}{dt} = -\frac{2\tau\,ds}{\rho r} \tag{13.3}$$

This equation is similar to Eq. (4.27), except that it has an extra term. This extra term $ds(dV/dt)$ accounts for the effect of acceleration caused by the unsteadiness of the flow.

Equation (13.3) may also be expressed as

$$\frac{dp}{\gamma} + d\frac{V^2}{2g} + dz + \frac{ds}{g}\frac{dV}{dt} = -\frac{2\tau\,ds}{\gamma r} \tag{13.4}$$

This equation applies to unsteady flow of both compressible and incompressible real fluids. However, once again an equation of state relating γ to p and T must be introduced before integration if we are dealing with a compressible fluid. For an incompressible fluid (γ = constant), we can integrate directly.

Integrating from some section 1 to another section 2, where the distance between them is L, we get

$$\frac{p_2}{\gamma} - \frac{p_1}{\gamma} + \frac{V_2^2}{2g} - \frac{V_1^2}{2g} + z_2 - z_1 + \frac{L}{g}\frac{dV}{dt} = -\frac{2\tau L}{\gamma r}$$

or

$$\left(\frac{p_1}{\gamma} + \frac{V_1^2}{2g} + z_1\right) - \frac{2\tau L}{\gamma r} = \left(\frac{p_2}{\gamma} + \frac{V_2^2}{2g} + z_2\right) + \frac{L}{g}\frac{dV}{dt} \tag{13.5}$$

For the case of a circular pipe of radius r_0 we recognize the term $2\tau L/\gamma r$ as representing the head loss in the pipe over the length L. This can be seen by examining Eq. (8.5) and noting that the hydraulic radius $R_h = r_0/2$ and $\tau = \tau_0$ when $r = r_0$. Thus, substituting h_L for $2\tau L/\gamma r$ in Eq. (13.5), we get the general energy equation applicable to incompressible unsteady flow,

$$\left(\frac{p_1}{\gamma} + \frac{V_1^2}{2g} + z_1\right) - h_L = \left(\frac{p_2}{\gamma} + \frac{V_2^2}{2g} + z_2\right) + \frac{L}{g}\frac{dV}{dt} \tag{13.6}$$

where $(L/g)(dV/dt)$ represents the accelerative head. In this equation h_L represents the head loss between sections 1 and 2 while L is the distance between the sections. It is presumed that the head loss at any instant is equal to the steady-flow head loss for the flow rate at that instant. Experimental evidence indicates that this presumption is reasonably valid.

If the pipe consists of two or more pipes in series, an $(L/g)(dV/dt)$ term for each pipe should appear in the equation just as there would be a separate term for the head loss in each pipe. To clarify the discussion further, the simple case of unsteady flow of an incompressible fluid in a horizontal pipe is shown in Fig. 13.2. The left-hand sketch shows the steady-flow case, while unsteady flow is depicted in the two right-hand sketches. The analysis below the sketches indicates that, with the same instantaneous flow rates, the pressure is depressed at section 2 if the acceleration is positive or increased if it is negative.

Illustrative Example 13.2 Although the unrealistic assumptions of instantaneous change in pump speed and head are made in this example, it will serve to illustrate application of Eq. (13.6). When the centrifugal pump in the accompanying figure is rotating at 1,650 rpm, the steady-flow rate is 1,600 gpm. Let us suppose that the pump speed can be increased instantaneously to 2,000 rpm. Determine the flow rate as a function of time. Assume that the head developed by the pump is proportional to the square of the rotative speed. Writing the unsteady-flow energy equation,

$$50 - 0.5\frac{V_1^2}{2g} - f_1\frac{L_1}{D_1}\frac{V_1^2}{2g} + h_p - f_2\frac{L_2}{D_2}\frac{V_2^2}{2g} = \frac{V_2^2}{2g} + \frac{L_1}{g}\frac{dV_1}{dt} + \frac{L_2}{g}\frac{dV_2}{dt}$$

where the subscripts 1 and 2 refer to the 10- and 6-in diameter pipes, respectively. Note that the accelerative head for each pipe depends on the respective L and dV/dt values.

From continuity,

$$V_1 = \frac{A_2 V_2}{A_1} = \left(\frac{6}{10}\right)^2 V_2 = 0.36V_2$$

Hence

$$\frac{dV_1}{dt} = \frac{A_2}{A_1}\frac{dV_2}{dt} = 0.36\frac{dV_2}{dt}$$

Thus

$$50 - 0.5\frac{(0.36V_2)^2}{2g} - 0.030\left(\frac{200}{10/12}\right)\frac{(0.36V_2)^2}{2g} + h_p - 0.020\left(\frac{750}{6/12}\right)\frac{V_2^2}{2g} = \frac{V_2^2}{2g} + \frac{200}{g}(0.36)\frac{dV_2}{dt} + \frac{750}{g}\frac{dV_2}{dt}$$

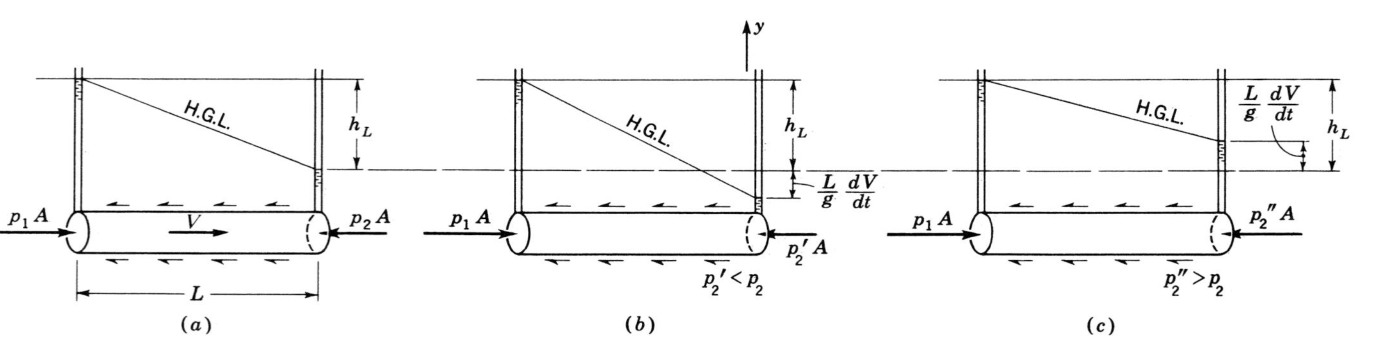

Figure 13.2 Steady and unsteady flow of incompressible fluid in a horizontal pipe. (Flow is instantaneously equal in all three pipes.) (*a*) Steady flow ($dV/dt = 0$). (*b*) Unsteady flow (dV/dt is positive). (*c*) Unsteady flow (dV/dt is negative).

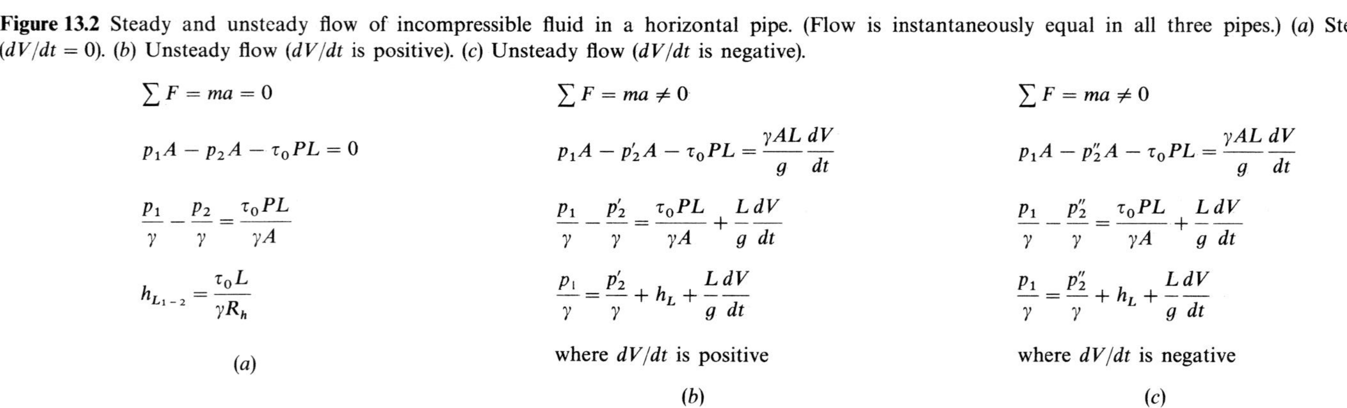

(a)

$$\sum F = ma = 0$$

$$p_1A - p_2A - \tau_0 PL = 0$$

$$\frac{p_1}{\gamma} - \frac{p_2}{\gamma} = \frac{\tau_0 PL}{\gamma A}$$

$$h_{L_{1-2}} = \frac{\tau_0 L}{\gamma R_h}$$

(b)

$$\sum F = ma \neq 0$$

$$p_1A - p_2'A - \tau_0 PL = \frac{\gamma AL}{g}\frac{dV}{dt}$$

$$\frac{p_1}{\gamma} - \frac{p_2'}{\gamma} = \frac{\tau_0 PL}{\gamma A} + \frac{L}{g}\frac{dV}{dt}$$

$$\frac{p_1}{\gamma} = \frac{p_2'}{\gamma} + h_L + \frac{L}{g}\frac{dV}{dt}$$

where dV/dt is positive

(c)

$$\sum F = ma \neq 0$$

$$p_1A - p_2''A - \tau_0 PL = \frac{\gamma AL}{g}\frac{dV}{dt}$$

$$\frac{p_1}{\gamma} - \frac{p_2''}{\gamma} = \frac{\tau_0 PL}{\gamma A} + \frac{L}{g}\frac{dV}{dt}$$

$$\frac{p_1}{\gamma} = \frac{p_2''}{\gamma} + h_L + \frac{L}{g}\frac{dV}{dt}$$

where dV/dt is negative

P = wetted perimeter; R_h = hydraulic radius.

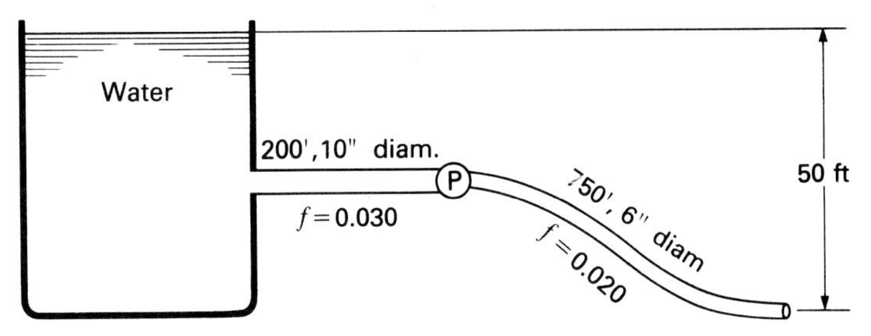

Illustrative Example 13.2

Evaluating and combining terms,

$$50 + h_p = 32.0\frac{V_2^2}{2g} + \frac{822}{g}\frac{dV_2}{dt} \tag{a}$$

With the original steady-flow conditions ($dV/dt = 0$),

$$V_2 = \frac{Q}{A_2} = \frac{1{,}600/449}{0.1963} = 18.15 \text{ fps}$$

and

$$h_p = 32\frac{V_2^2}{2g} - 50 = 113.7 \text{ ft}$$

After the speed is increased to 2,000 rpm

$$h_p = 113.7\left(\frac{2{,}000}{1{,}650}\right)^2 = 167.0 \text{ ft}$$

Substituting into (a),

$$50 + 167 = 32\frac{V_2^2}{2g} + \frac{822}{g}\frac{dV_2}{dt}$$

Expressing the foregoing in terms of Q,

$$217 = 12.89Q^2 + 130.0\frac{dQ}{dt} \tag{b}$$

Solving for dt and integrating, noting that at $t = 0$, $Q = 3.57$ cfs (1,600 gpm)

$$\int_0^t dt = 130\int_{3.57}^{Q}\frac{dQ}{217 - 12.89Q^2}$$

$$t = 1.229 \ln\frac{4.10 + Q}{4.10 - Q} - 3.27$$

$$e^{0.814t + 2.66} = \frac{4.10 + Q}{4.10 - Q}$$

Finally

$$Q = 4.10\frac{e^{0.814t + 2.66} - 1}{e^{0.814t + 2.66} + 1}$$

Note that as t gets larger, Q approaches 4.10 cfs (1,840 gpm), the steady-state flow rate for the condition where $h_p = 169$ ft.

It should be noted that the speed of a pump cannot be changed instantaneously fron one value to another, as was assumed in this example. To solve this problem correctly the operating characteristics of the pump and motor and the moment of inertia of the rotating system would have to be known.

13.4 ESTABLISHMENT OF STEADY FLOW

Determining the time for the flow to become steady in a pipeline when a valve is suddenly opened at the end of the pipe can be accomplished through application of Eq. (13.6). Immediately after the valve is opened (Fig. 13.3), the head H is available to accelerate the flow. Thus flow commences, but as the velocity increases the accelerating head is reduced by fluid friction and minor losses. Let us assume the total head loss h_L can be expressed as $kV^2/2g$, where k is constant, although it may vary somewhat with velocity unless the pipe is very rough. Writing Eq. (13.6) between sections 1 and 2 in Fig. 13.3 gives

$$H - k\frac{V^2}{2g} = \frac{V^2}{2g} + \frac{L}{g}\frac{dV}{dt}; \qquad k = k_L + f\frac{L}{D}$$

Let us define the steady-flow velocity by V_0. Noting that for steady flow $(dV/dt) = 0$, we get

$$V_0 = \sqrt{\frac{2gH}{(1 + k)}}$$

Substituting the value of H from this expression into the above energy equation gives

$$dt = \left(\frac{2L}{1 + k}\right)\frac{dV}{V_0^2 - V^2}$$

Integrating and noting that the constant of integration = zero, since $V = 0$ at $t = 0$ and $\ln V_0/V_0 = 0$, we get

$$t = \frac{L}{(1 + k)V_0}\ln\frac{V_0 + V}{V_0 - V} \tag{13.7}$$

This equation indicates that V approaches V_0 asymptotically and that equilibrium will be attained only after an infinite time (Fig. 13.3*b*), but it must be

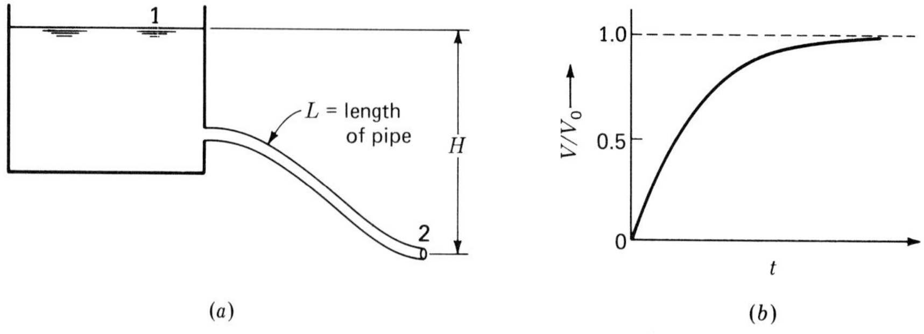

Figure 13.3 Establishment of steady flow (V_0 = velocity at steady flow).

remembered that this is an idealized case. In reality there will be elastic waves and damping, so that true equilibrium will be reached in a finite time. Also, Eq. (13.7) applies to a submerged discharge as well as a free discharge since $V^2/2g$ represents the head loss at submerged discharge in one case and the residual velocity head in the other.

Illustrative Example 13.3 Two large water reservoirs are connected to one another with a 10-cm-diameter pipe ($f = 0.02$) of length 15 m. The water-surface elevation difference between the reservoirs is 2.0 m. A valve in the pipe, initially closed, is suddenly opened. Determine the times required for the flow to reach $\frac{1}{4}$, $\frac{1}{2}$, and $\frac{3}{4}$ of the steady-state flow rate. Assume the water-surface elevations remain constant. Repeat for pipe lengths of 150 m and 1,500 m with all other data remaining the same. In the first case $L/D = 15/0.10 = 150$, hence minor losses are significant (Sec. 8.22). Assume square-edged entrance.

Square-edge entrance: $k_L = 0.5$

$$k = k_L + f\frac{L}{D} = 0.5 + 0.02\frac{15}{0.10} = 3.5$$

For steady flow:

$$V_0 = \sqrt{\frac{2gH}{(1+k)}} = \sqrt{\frac{2(9.81)2}{1+3.5}} = 2.95 \text{ m/s}$$

For unsteady flow use Eq. (13.7):

$$t = \frac{L}{(1+k)V_0}\ln\frac{V_0+V}{V_0-V} = \frac{15}{4.5(2.95)}\ln\frac{2.95+V}{2.95-V}$$

$$t = 1.129 \ln\frac{2.95+V}{2.95-V}$$

For $Q = \frac{1}{4}Q_0$ substitute $V = \frac{1}{4}V_0$, etc.:

Q	V, m/s	$\frac{2.95+V}{2.95-V}$	ln	t, s
$0.25Q_0$	0.74	1.667	0.511	0.577
$0.50Q_0$	1.48	3.00	1.099	1.240
$0.75Q_0$	2.21	7.00	1.946	2.197

For the other two lengths the results are as follows:

Q	$L = 150$ m	$L = 1{,}500$ m
$0.25Q_0$	2.18 s	7.04 s
$0.50Q_0$	4.69 s	15.15 s
$0.75Q_0$	8.30 s	26.84 s

13.5 VELOCITY OF PRESSURE WAVE IN PIPES

Unsteady phenomena, with rapid changes taking place, frequently involve the transmission of pressure in waves or surges. As shown in Appendix 2, the velocity of a pressure (sonic) wave is

$$c = \sqrt{\frac{g}{\gamma} E_v} = \sqrt{\frac{E_v}{\rho}} \tag{13.8}$$

where E_v is the volume modulus of the medium. For water, a typical value of E_v is 300,000 psi (2.07×10^6 kN/m²), and thus the velocity of a pressure wave in water is $c = 4{,}720$ fps (1,440 m/s). But for water in an elastic pipe, this value is modified by the stretching of the pipe walls, and as shown in Appendix 2, E_v is replaced by K, such that

$$K = \frac{E_v}{1 + (D/t)(E_v/E)}$$

where D and t are the diameter and wall thickness of the pipe, respectively, and E is the modulus of elasticity of the pipe material. As the ratios D/t and E_v/E are dimensionless, any consistent units may be used in each.

The velocity of a pressure wave in an elastic pipe is then

$$c_P = \sqrt{\frac{g}{\gamma} K} = c \sqrt{\frac{1}{1 + \dfrac{D}{t}\dfrac{E_v}{E}}} \tag{13.9}$$

Values[1] of the modulus of elasticity E for steel, cast iron, and concrete are about 30,000,000, 15,000,000 and 3,000,000 psi, respectively. Values of the volume modulus E_v for various liquids are given in Appendix 3, Table A.4.

For normal pipe dimensions the velocity of a pressure wave in a water pipe usually ranges between 2,000 and 4,000 fps (600 and 1,200 m/s), but it will always be less than 4,720 fps (1,440 m/s), the velocity of a pressure wave in water.

13.6 WATER HAMMER

In the preceding unsteady-flow cases in this chapter, the changes of velocity were presumed to take place slowly. But if the velocity of a liquid in a pipeline is abruptly decreased by a valve movement, the phenomenon encountered is called *water hammer*. This is a very important problem in the case of hydroelectric plants, where the flow of water must be rapidly varied in proportion to the load changes on the turbine. Water hammer occurs in liquid-flow pressure systems

[1] Corresponding values of E for steel, cast iron, and concrete in SI units are 207×10^6, 103×10^6, and 20.7×10^6 kN/m², respectively.

whenever a valve is closed. The terminology "water hammer" is perhaps misleading since the phenomenon can occur in any liquid.

Instantaneous Closure

Although it is physically impossible to close a valve instantaneously, such a concept is useful as an introduction to the study of real cases. For convenience let us start off by considering steady flow in a horizontal pipe (Fig. 13.4*a*) with a partly open valve. Then let us assume that the valve at N is closed instantaneously. The lamina of liquid next to the valve will be compressed by the rest of the column of liquid flowing against it. At the same time the walls of the pipe

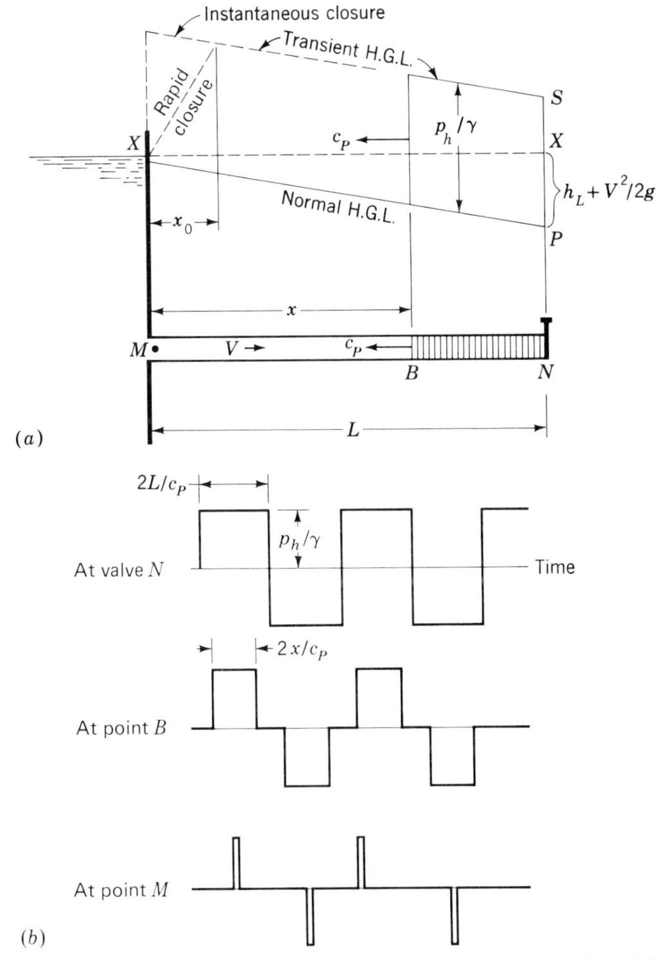

Figure 13.4 Water hammer with pipe friction and damping neglected. (*a*) Valve at end of pipeline is abruptly closed and the pressure wave has traveled part way up the pipe. (*b*) Water-hammer pressure heads at N, B, and M as a function of time for instantaneous valve closure.

surrounding this lamina will be stretched by the excess pressure produced. The next upstream lamina will then be brought to rest, and so on. The liquid in the pipe does not behave as a rigid incompressible body but the phenomenon is affected by the elasticity of both the liquid and the pipe. The cessation of flow and the resulting pressure increase move upstream along the pipe as a wave with the velocity c_P as given by Eq. (13.9).

After a short interval of time the liquid column BN will have been brought to rest, while the liquid in the length MB will still be flowing with its initial velocity and initial pressure. When the pressure wave finally reaches the inlet at M, the entire mass in the length L will be at rest but will be under an excess pressure throughout. During travel of the pressure wave from N to M there will be a transient hydraulic grade line parallel to the original steady flow grade line XP but at a height p_h/γ above it, where p_h represents the water hammer pressure.

It is impossible for a pressure to exist at M that is greater than that due to depth MX, and so when the pressure wave arrives at M, the pressure at M drops instantly to the value it would have for zero flow. But the entire pipe is now under an excess pressure; so the liquid in it is compressed, and the pipe walls are stretched. Then some liquid starts to flow back into the reservoir, and a wave of pressure unloading travels along the pipe from M to N. Assuming there is no damping, at the instant this unloading wave reaches N, the entire mass of liquid will be under the normal pressure indicated by the line XP, but the liquid is still flowing back into the reservoir. This reverse velocity will produce a drop in pressure at N that ideally will be as far below the normal, steady-flow pressure as the pressure an instant before was above it. Then a wave of rarefaction travels back up the pipe from N to M. Ideally, there would be a series of pressure waves traveling back and forth over the length of the pipe and alternating equally between high and low pressures. Actually, because of damping due to fluid friction and imperfect elasticity of liquid and pipe, the total pressure at any point in the pipe will fluctuate back and forth heading gradually toward the pressure for the no-flow condition indicated by XX in Fig. 13.4*a*.

The time for a round trip of the pressure wave from N to M and back again is

$$T_r = 2\frac{L}{c_P} \tag{13.10}$$

where L is the length of pipe, and so for an instantaneous valve closure the excess pressure remains constant for this length of time, before it is affected by the return of the unloading pressure wave; and in like manner the pressure defect during the period of rarefaction remains constant for the same length of time. At a distance x from the inlet, such as at B, the time for a round trip of a pressure wave is only $2x/c_P$, and hence at that point the time duration of the excess or deficient pressure will be $2x/c_P$, as shown in Fig. 13.4*b*. At the inlet M, where $x = 0$, the excess pressure occurs for only an instant.

In Fig. 13.5 is shown a close-up in the vicinity of the valve. If the valve is closed abruptly, a pressure wave travels up the pipe with a celerity c_P. In a short

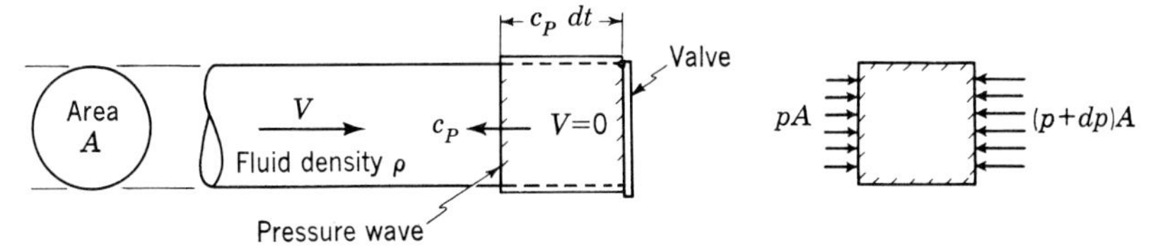

Figure 13.5 Definition sketch for analysis of water hammer in pipes.

interval of time dt an element of liquid of length $c_P\,dt$ is brought to rest. Applying Newton's second law, $F\,dt = M\,dV$, and neglecting friction,

$$[pA - (p + dp)A]\,dt = (\rho A c_P\,dt)\,dV$$

$$-dp = \rho c_P\,dV$$

or

$$\Delta p = -\rho c_P(\Delta V)$$

which indicates the change in pressure Δp that results from an instantaneous change in velocity ΔV.

In the case of instantaneous and complete closure of a valve the velocity is reduced from V to zero, that is, $\Delta V = -V$; Δp then represents the increase in pressure due to valve closure, so the water hammer pressure $p_h = \Delta p$. Thus for instantaneous valve closure

$$p_h = \rho c_P V \tag{13.11}$$

It will be observed that the pressure increase is independent of the length of the pipe and depends solely upon the celerity of the pressure wave in the pipe and the change in the velocity of the water. The total pressure at the valve immediately after closure is $p_h + p$, where p is the pressure in the pipe just upstream of the valve prior to closure.

Consider now conditions at the valve as affected by both pipe friction and damping. In Fig. 13.4*a*, when the pressure wave from N has reached B, the water in BN will be at rest and for zero flow the hydraulic grade line should be a horizontal line. There is thus a tendency for the grade line to flatten out for the portion BN. Hence, instead of the transient gradient having the slope imposed by friction, as shown in the figure, it will approach a horizontal line starting from the transient value at B. Thus the pressure head at N will be raised to a slightly higher value than NS shortly after the valve closure.

This slight increase in pressure head at the valve over the theoretical value $c_P V/g$ has been borne out by tests. In Fig. 13.6 the line *ab* is shown as sloping upward because of the adverse pressure gradient, and for the same reason *ef* may slope slightly downward, as all conditions are now reversed. Also, because of damping, the waves will be of decreasing amplitude until the final equilibrium pressure is reached.

All of the preceding analysis assumes that the wave of rarefaction will not cause the minimum pressure at any point to drop down to or below the vapor pressure. If it should do so, the water would separate and produce a discontinuity.

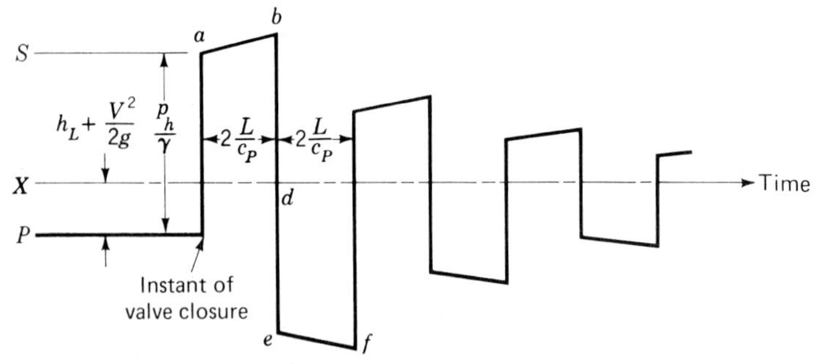

Figure 13.6 Pressure history at valve with instantaneous closure, considering pipe friction and damping.

Rapid Closure ($t_c < 2L/c_P$)

It is physically impossible for a valve to be closed instantaneously; so we shall now consider the real case where the valve is closed in a finite time t_c which is more than zero but less than $2L/c_P$. In Fig. 13.7 are shown actual pressure recordings for such a case. The slope of the curve during the time t_c depends entirely upon the operation of the valve and its varying effect upon the velocity in the pipe. But the maximum pressure rise is still the same as for instantaneous closure. The only differences are that it endures for a shorter period of time and the vertical lines of Fig. 13.6 are changed to the sloping lines of Fig. 13.7. If the time of valve closure were exactly $2L/c_P$, the maximum pressure rise at the valve would still be the same, but the curves in Fig. 13.7 would all end in sharp points for both maximum and minimum values, since the time duration of maximum pressure would be reduced to zero.[1]

No matter how rapid the valve closure may be, so long as it is not the idealized instantaneous case, there will be some distance from the intake, such as

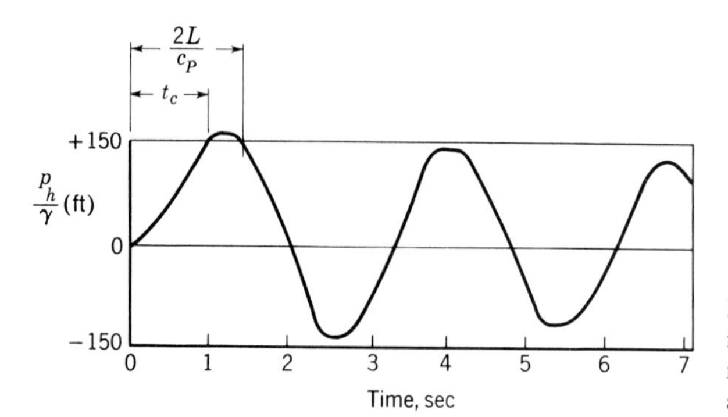

Figure 13.7 Rapid valve closure in time t_c less than $2L/c_p$. Actual measurement of pressure changes at valve.

[1] Figures 13.7 and 13.8 are from water-hammer studies made by the Southern California Edison Co. on an experimental pipe with the following data: $L = 3{,}060$ ft, internal diameter $= 2.06$ in, $c_P = 4{,}371$ fps, $V = 1.11$ fps, $c_P V/g = 151$ ft, $2L/c_P = 1.40$ s, static head $= 306.7$ ft, head before valve closure $= 301.6$ ft, $h_L = 5.1$ ft. In Fig. 13.7 the time of closure $= 1$ s, and it will be noted that the actual rise in pressure head is slightly more than 151 ft. In Fig. 13.8 the time of closure $= 3$ s.

x_0 in Fig. 13.4*a*, within which the valve closure time is more than $2x_0/c_P$. Thus, in any real case, the maximum pressure rise cannot extend all the way to the reservoir intake. In the actual case, the maximum pressure rise will be constant at the instantaneous value p_h for a distance from the valve up to this point a distance x_0 from the intake. From this point the excess pressure will diminish to zero at the intake. This is shown as a uniform rate of decrease in Fig. 13.4*a*.

Slow Closure ($t_c > 2L/c_P$)

The preceding discussion has assumed a closure so rapid (or a pipe so long) that there is an insufficient time for a pressure wave to make the round trip before the valve is closed. Slow closure will be defined as one in which the time of valve movement is greater than $2L/c_P$. In this case the maximum pressure rise will be less than in the preceding because the wave of pressure unloading will reach the valve before the valve is completely closed. This will prevent any further increase in pressure.

Thus in Fig. 13.8 the pressure would continue to rise if it were not for the fact that at a time $2L/c_P$ a return unloading pressure wave reaches the valve and stops the pressure rise at a value of about 53 ft as contrasted with nearly three times that value in Fig. 13.7. Subsequent pressure changes, as elastic waves travel back and forth, are very complex and require a detailed step-by-step analysis that is beyond the scope of this text. In brief, the method consists in assuming the valve movement to take place in a series of steps each of which produces a pressure Δp proportional to each ΔV.[1]

Tests have shown that for slow valve closure, i.e., in a time greater than $2L/c_P$, the excess pressure produced decreases uniformly from the value at the valve to zero at the intake. The maximum water-hammer pressure p'_h developed by gradual closure of a valve when $t_c > 2L/c_P$ is given approximately by

$$p'_h \approx \frac{2L/c_P}{t_c} p_h = \frac{2Lp_h}{c_P t_c} = \frac{2LV\rho}{t_c} \tag{13.12}$$

where t_c is the time of closure.

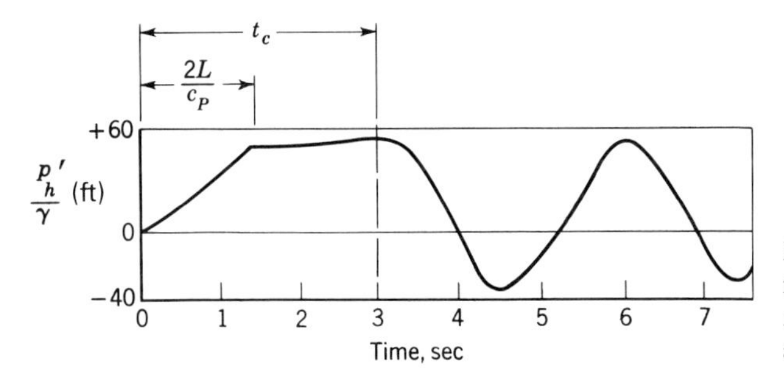

Figure 13.8 Slow valve closure in time t_c greater than $2L/c_p$. Actual measurement of pressure changes at valve.

[1] For details of computing successive pressures for slow valve closure and for further explanation of much of this condensed treatment, see John Parmakian, "Waterhammer Analysis," Prentice-Hall, Inc., Englewood Cliffs, N.J., 1955.

A pipe can be protected from the effects of high water-hammer pressure through the use of slow-closing valves, the use of automatic relief valves which permit water to escape when the pressure exceeds a certain value, or through application of surge chambers as explained in the following section.

Illustrative Example 13.4 In Fig. 13.4*a* the elasticity and dimensions of the pipe are such that the celerity of the pressure wave is 3,200 fps. Suppose the pipe has a length of 2,000 ft and a diameter of 4 ft. The flow rate is initially 30 cfs. Water is flowing. Find (*a*) the water-hammer pressure for instantaneous valve closure; (*b*) the approximate water-hammer pressure at the valve if it is closed in 4.0 s; (*c*) the water-hammer pressure at the valve if it is manipulated so that the flow rate drops almost instantly from 30 to 10 cfs; (*d*) the maximum water-hammer pressure at a point in the pipe 300 ft from the reservoir if a 1.0-s valve closure reduces the flow rate from 10 cfs to zero.

$$V = \frac{Q}{A} = \frac{30}{\pi 2^2} = 2.39 \text{ fps}$$

(*a*) $p_h = \rho c_P V = 1.94(3{,}200)(2.39) = 14{,}840 \text{ lb/ft}^2 = 103.0 \text{ psi}$

(*b*) $p'_h \approx \dfrac{4{,}000/3{,}200}{4.0} p_h = \dfrac{1.25}{4.00}(103.0) = 32.2 \text{ psi}$

(*c*) For this case of partial closure Eq. (13.11) may be written as $\Delta p_h = -\rho c_P(\Delta V)$.

$$\Delta V = \frac{10 - 30}{\pi 2^2} = -1.592 \text{ fps}$$

$$\Delta p_h = -1.94(3{,}200)(-1.592) = 9{,}880 \text{ lb/ft}^2 = 68.6 \text{ psi}$$

(*d*) If $2x_0/c_P = 1.0$ s, $x_0 = 1{,}600$ ft, so that full water-hammer pressure will be developed in the pipe only in the region that is farther than 1,600 ft from the reservoir.

For this case, at valve

$$p_h = 1.94(3{,}200)\frac{2.39}{3} = 4{,}940 \text{ lb/ft}^2 = 34.3 \text{ psi}$$

At point 300 ft from reservoir:

$$p_h = 34.3\frac{300}{1{,}600} = 6.43 \text{ psi}$$

13.7 SURGE TANKS[1]

In a hydroelectric plant the flow of water to a turbine must be decreased very rapidly whenever there is a sudden drop in load. This rapid decrease in flow will result in high water-hammer pressures and may result in the need for a very strong and hence expensive pipe. There are several ways to handle a situation of this sort; one is by use of a *surge tank*, or *surge chamber*. A simple surge tank is a vertical standpipe connected to the pipeline as shown in Fig. 13.9. With steady

[1] Analysis of surge tank phenomena is commonly conducted using computer programs employing numerical methods to solve the differential equations.

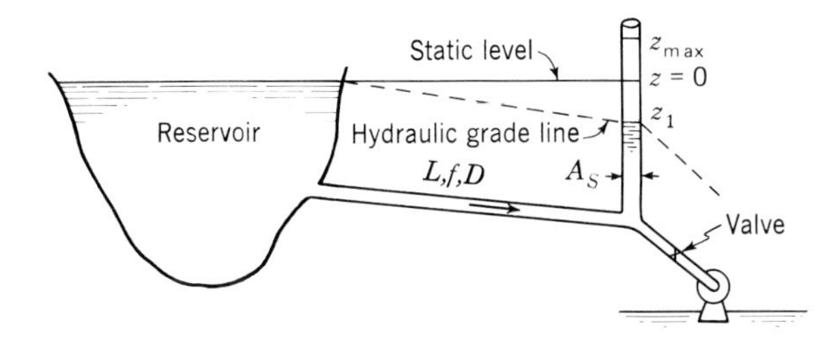

Figure 13.9 Definition sketch for surge-tank analysis.

flow in the pipe, the water level z_1 in the surge tank is below the static (no flow) level ($z = 0$). When the valve is suddenly closed, water rises in the surge tank. The water surface in the tank will then fluctuate up and down until damped out by fluid friction. The section of pipe upstream of the surge tank is in effect afforded protection from the high water-hammer pressures that would exist on valve closure if there were no tank.

An approximate analysis for this simple surge tank may be acquired by writing the energy equation for unsteady flow between the surface of the reservoir and the water surface in the surge tank under the condition where the valve is closed. Neglecting fluid friction, velocity head, and inertial effects in the tank, neglecting pipe entrance and tank entrance losses and neglecting velocity head in the pipe, we get for flow out of the reservoir

$$0 - f\frac{L}{D}\frac{V^2}{2g} = z + \frac{L}{g}\frac{dV}{dt} \tag{13.13}$$

In this equation z represents the level of the water surface in the surge tank measured positively upward from the static water level where $z = 0$ and $(L/g)(dV)/(dt)$ represents the accelerative head in the pipe between the reservoir and the surge tank.

With the valve completely closed, the continuity equation is

$$AV = A_s\frac{dz}{dt} \tag{13.14}$$

where A and A_s are the cross-sectional areas of the pipe and surge tank, respectively. Combining Eqs. (13.13) and (13.14) and noting that $(dV/dt) = (dV/dz)(dz/dt)$, integrating and solving for V yields

$$V^2 = \frac{2gAD^2}{LA_s f^2}\left(1 - \frac{fA_s}{AD}z\right) - Ce^{-(fA_s/AD)z} \tag{13.15}$$

which expresses the relation between velocity in the pipe and water-surface level in the tank over the interval from valve closure to the top of the first surge. Equation (13.15) may be used to estimate the maximum height of surge z_{max} by finding the constant of integration C for steady-state conditions at the instant of closure ($z = z_1$) and then solving for z_{max} when $V = 0$. Since the derivation neglected fluid friction, velocity head, and inertial effects in the surge tank as well as velocity head in the pipe and minor losses at pipe entrance and surge tank junction and assumed instantaneous valve closure, the value of z_{max} as computed by Eq. (12.15) will be larger than the true value and thus the results provide a conservative estimate for preliminary design of simple surge tanks.

Surge tanks are usually open at the top and of sufficient height so that they will not overflow. In some instances they are permitted to overflow if no damage will result. There are many types of surge tanks. Some have a restriction to entry; others have a closed top so that there is an air cushion within the tank during operation.

The surge tank, in addition to providing protection against water-hammer pressures, fulfills another desirable function. That is, in the event of a sudden demand for increased flow, it can quickly provide some excess water, while the entire mass of water in a long pipeline is being accelerated. The acceleration of masses of liquids in pipelines was discussed in Sec. 13.4.

PROBLEMS

13.1 (*a*) Suppose a ship lock has vertical sides and that water enters or discharges through a conduit area A such that $Q = C_d A\sqrt{2gz}$, where z is the variable difference in level between the water surface in the lock and that outside. Prove that for the water level in the lock to change from z_1 to z_2 the time is

$$t = \left(\frac{2A_s}{C_d A\sqrt{2g}}\right)(z_1^{1/2} - z_2^{1/2})$$

(*Note:* If the lock is being filled, the signs must be reversed.) (*b*) Suppose the lock is 300 ft (90 m) long by 90 ft (27.5 m) wide, and water enters through a conduit for which the discharge coefficient is 0.50. If the water surface in the lock is initially 36 ft (11 m) below the level of the surface of the water upstream, how large must the conduit be if the lock is to be filled in 5 min?

13.2 (*a*) Suppose a reservoir has vertical sides and that initially there is a steady flow into it such that the height of the surface above the level of a spillway ($Q = C_W LH^{3/2}$) is z_1. If the inflow is suddenly cut off, prove that the time required for the water level to fall from z_1 to z_2 is $t = (2A_s/C_W L) \times (1/\sqrt{z_2} - 1/\sqrt{z_1})$. (*Note:* $z = H$.) (*b*) How long will it take theoretically for the outflow to cease entirely? What factors make this theoretical answer unrealistic? (*c*) The crest of the overflow spillway is 100 ft (30 m) long, and the value of C_W (Eq. 12.25) is 3.45 (1.91). For the range of levels here considered the area of the water surface is constant and is 700,000 ft^2 (65,000 m^2). Initially, there is a flow into the reservoir at such a rate that the height of the water surface above that of the spillway crest is stabilized at 3 ft (90 cm), and then the inflow is suddenly diverted. Find the length of time for the water surface to fall to a height of 1 ft (30 cm) above the level of the spillway.

13.3 The crest of the overflow spillway of a reservoir is 40 ft long, and the value of C_W (Eq. 12.25) is 3.50. The area of the water surface is assumed constant at 600,000 ft^2 for the range of heights here considered. Initially, the water surface is 3 ft below the level of the spillway crest. If suddenly there is turned into this reservoir a flow of 500 cfs, what will be the height of water in the reservoir for equilibrium? How long a time will be required for this height to be reached? How long a time will be required for the water surface to reach a height of 2 ft above the level of the spillway? (*Note:* This last can be solved by integration after substituting x^3 for $z^{3/2}$ and consulting integral tables. However, it will be easier to solve it graphically either by plotting and actually measuring the area under the curve or by computing the latter by some method, such as Simpson's rule.)

13.4 Water enters a reservoir at such a rate that the height of water above the level of the spillway crest is 3 ft. The spillway ($Q = C_W LH^{3/2}$) is 100 ft long, and the value of C_W is 3.45. The area of the water surface for various water levels is as follows:

z, ft	A_s, ft^2
3.00	860,000
2.50	830,000
2.00	720,000
1.50	590,000
1.25	535,000
1.00	480,000

If the inflow is suddenly reduced to 150 cfs, what will be the height of water for equilibrium? How long will it take, theoretically, for equilibrium to be attained? How long will it take for the level to drop from 3 to 1 ft above that of the spillway?

13.5 Work Prob. 13.4 using the same numbers but changing ft to m, ft^2 to m^2, and cfs to m^3/s.

13.6 The figure shows a tank with vertical sides containing a liquid with a surface area A_s. The liquid discharges through an orifice under a head z which varies from the initial height h to 0 as the tank empties down to the orifice level. Neglecting friction losses, what is the cumulative kinetic energy of the jet during the time required for the liquid surface to drop from h to 0? How does this kinetic-energy summation compare with the total energy of the mass of fluid initially in the tank above the orifice level?

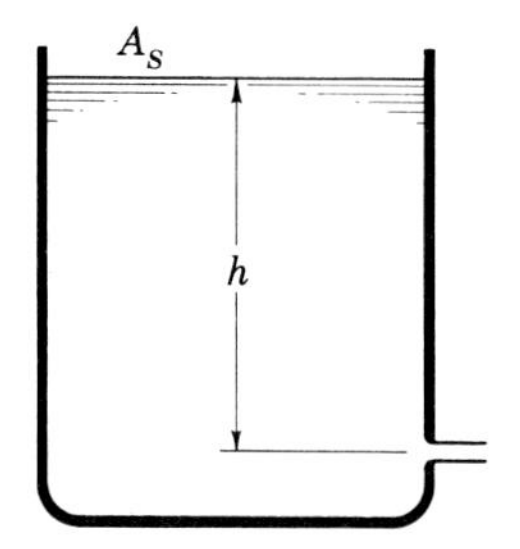

Problem 13.6

13.7 Suppose that in Prob. 13.6 the value of A_s is 10 ft^2 (0.9 m^2), h is 16 ft (5 m), and the jet diameter is 4 in (10 cm). How long will it take the tank to empty down to the orifice level? Plot a graph of h vs. time, using increments of 4 ft (1 m).

13.8 The figure shows a tank whose shape is the frustum of a cone with a 2-ft^2 orifice in the bottom. Assume $C_d = 0.62$. If the water level outside of the tank is constant at section 2, how long will it take the water level in the tank to drop from section 1 to section 2? (*Note:* Diameter of tank $= Ky$, and $y = z + h_2$, where z is the variable distance between surface levels.)

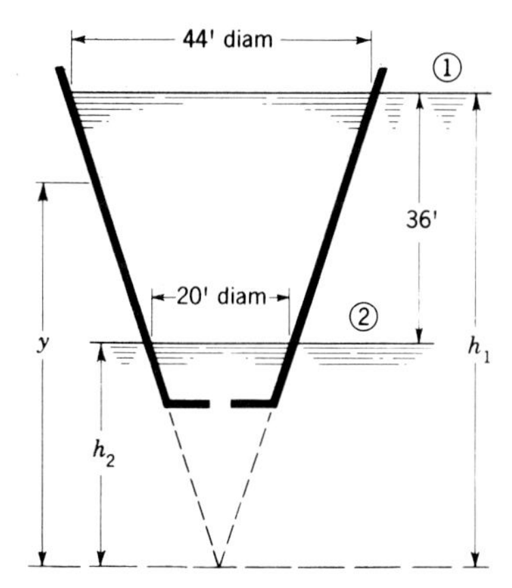

Problem 13.8

13.9 If in the figure for Prob. 13.8 the water surface outside the tank is constant at section 1 and the tank is initially empty, how long will it take for the water level in the tank to rise from section 2 to section 1? Assume a 2-ft^2 orifice with $C_d = 0.62$.

13.10 The tank in the figure has vertical sides and is 5 ft (1.5 m) in the dimension normal to the plane of the paper. It is divided by a vertical plate in which is a submerged orifice 0.5 ft^2 (0.045 m^2) in area. Assume $C_d = 0.65$. How long a time will be required for the two water surfaces to equalize?

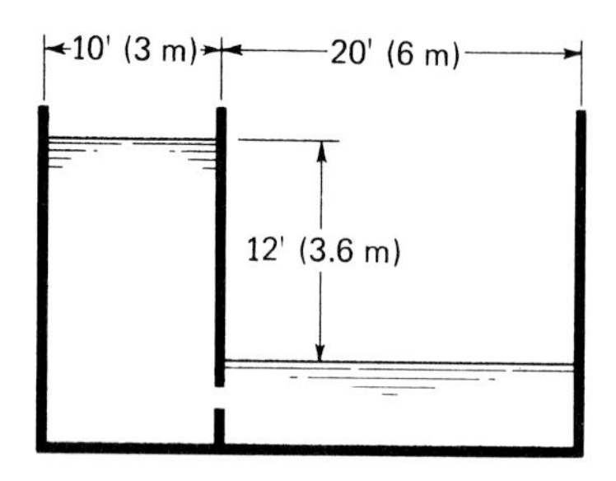

Problem 13.10

13.11 A 1-in (25-mm)-diameter smooth brass pipe 1,000 ft (300 m) long drains an open cylindrical tank which contains oil having $\rho = 1.8$ slugs/ft^3 (950 kg/m^3), $\mu = 0.0006$ lb·s/ft^2 (0.03 N·s/m^2). The pipe discharges at elevation 100 ft (30 m). Find the time required for the oil level to drop from elevation 120 (36) to elevation 108 ft (32.5 m) if the tank is 4 ft (1.2 m) in diameter.

13.12 Verify that the neglect of the $(L/g)(dV/dt)$ term was justified in Illustrative Example 13.1 by finding its value when $z = 5$ ft.

13.13 A large reservoir is being drained with a pipe system as shown in the figure. Initially, when the pump is rotating at 200 rpm, the flow rate is 6.3 cfs (0.18 m^3/s). If the pump speed is increased instantaneously to 250 rpm, determine the flow rate as a function of time. Assume that the head h_p developed by the pump is proportional to the square of the rotative speed; that is, $h_p \propto n^2$.

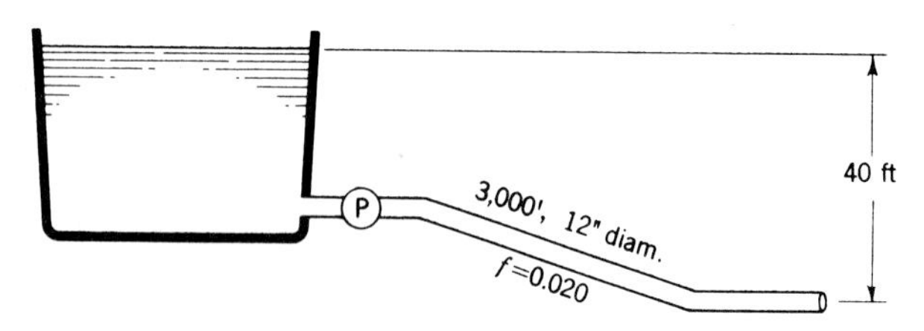

Problem 13.13

13.14 (*a*) Repeat Prob. 13.13 with all data the same except use a 24-in (60-cm)-diameter pipe rather than a 12-in (30-cm) pipe. (*b*) Repeat also for the case of a 10-inch (25-cm)-diameter pipe.

13.15 Refer to Illustrative Example 13.2 and make a plot of flow rate versus time.

13.16 Work Illustrative Example 13.2 for the case where the pipe lengths are 400 and 1,500 ft rather than 200 and 750 ft. All other data are the same.

13.17 Suppose that in Illustrative Example 13.2 the pump speed had been reduced instantaneously from 1,650 to 1,150 rpm. What then would have been the rate of deceleration of the flow immediately after the change in pump speed?

13.18 Attached to the tank in the figure is a flexible 1-in (25-mm)-diameter hose ($f = 0.015$) of length 200 ft (60 m). The tank is hoisted in such a manner that $h = 20 + 3t$, where h is the head in feet ($h = 6 + 0.9t$, where h is the head in meters) and t is the time in seconds. (*a*) Find as accurately as you can the flow rate at $t = 10$ s. (*b*) Suppose h were decreasing at the same rate. What, then, would be the flow rate when $h = 50$ ft (15 m)?

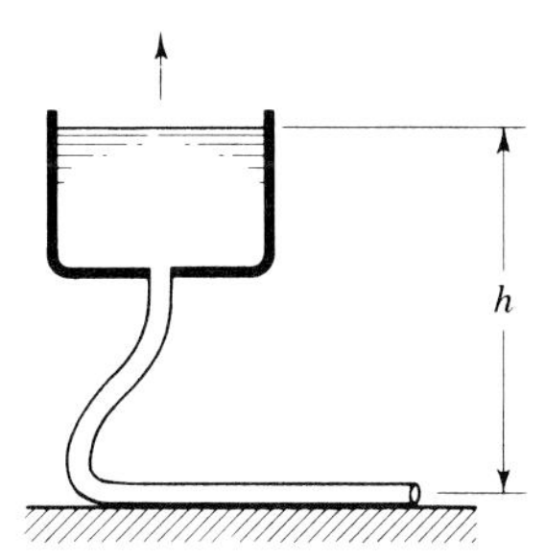

Problem 13.18

13.19 Repeat Prob. 13.18 for the case of a 4-in (10-cm)-diameter hose with all other data to remain the same.

13.20 In Prob. 13.18 suppose h was changed instantaneously from 20 to 50 ft (6 to 15 m). Under these conditions find the flow rate at $t = 10$ s.

13.21 A 4-in-diameter pipe of length 3,000 ft drains a reservoir. The elevation difference between the reservoir water surface and the pipe outlet is 100 ft. The pipe entrance is flush. Initially there is no flow since there is a plug at the pipe outlet. The plug is then removed. Plot Q versus t, assuming $f = 0.020$.

13.22 Repeat Prob. 13.21 for the case where the pipe length is 300 ft rather than 3,000 ft.

13.23 A 15-cm-diameter pipe of length 500 m drains a reservoir. The elevation difference between the reservoir water surface and the pipe outlet is 60 m. Initially there is no flow because a valve at the pipe outlet is closed. The valve is suddenly opened; plot Q versus t assuming $f = 0.03$.

13.24 A large water reservoir is drained by a pipeline that consists of 200 ft (60 m) of 6-in (15-cm)-diameter pipe ($f = 0.030$) followed by 500 ft (150 m) of 10-in (25-cm)-diameter pipe ($f = 0.020$). The point of discharge is 100 ft (30 m) below the elevation of the reservoir water surface. A valve at the discharge end of the pipe is initially closed. It is then quickly opened. Derive an equation similar to Eq. (13.7) applicable to this situation, and plot flow rate versus time. Neglect minor losses.

13.25 A 10-in-diameter pipe ($f = 0.020$) of length 300 ft is connected to a reservoir. Entrance losses are negligible. At the discharge end of the pipe is a nozzle that produces a 4-in-diameter jet. The elevation difference between the jet and the water surface in the reservoir is 40 ft. The nozzle has a coefficient of velocity of 0.95. Initially, there is a tight-fitting plug in the nozzle, which is then removed. For this situation derive an equation similar to Eq. (13.7) and plot flow rate vs. time. Assume that the liquid level in the reservoir does not drop.

13.26 An open tank containing oil ($s = 0.85$, $\mu = 0.0005$ lb·s/ft^2) is connected to a 2-in-diameter smooth pipe of length 3,000 ft. The elevation drop from the liquid surface in the tank to the point of discharge is 15 ft. A valve on the discharge end of the pipe, initially closed, is then opened. Plot the ensuing flow rate vs. time.

13.27 An open tank containing oil ($s = 0.82$, $v = 0.002$ m^2/s) is connected to a 10-cm-diameter pipe of length 400 m. The elevation drop from the oil surface in the tank to the pipe outlet is 2.5 m. A valve at the end of the pipe, initially closed, is suddenly opened. Plot the ensuing flow rate as a function of time.

13.28 A vertical 1-in (25-mm)-diameter pipe 10 ft (3 m) long is full of oil of specific gravity 0.88 and viscosity 0.004 lb·s/ft^2 (0.2 N·s/m^2). Find the time required to drain the pipe after a plug is removed from the lower end. Assume that the head loss is given by the equation of established laminar flow and that surface-tension effects are negligible.

13.29 Find the celerity of a pressure wave in benzene (Appendix 3, Table A.4) contained in a 6-in (15-cm)-diameter steel pipe having a wall thickness of 0.285 in (7.2 mm).

13.30 (*a*) What is the celerity of a pressure wave in a 5-ft (1.5-m)-diameter water pipe with 0.5-in steel (10-cm concrete) walls? (*b*) If the pipe is 4,000 ft (1,200 m) long, what is the time required for a pressure wave to make the round trip from the valve? (*c*) If the initial water velocity is 8 fps (2.5 m/s), what will be the rise in pressure at the valve if the time of closure is less than the time of a round trip? (*d*) If the valve is closed at such a rate that the velocity in the pipe decreases uniformly with respect to time and closure is completed in a time $t_c = 5L/c_P$, approximately what will be the pressure head at the valve when the first pressure unloading wave reaches the valve?

13.31 For the situation depicted in Illustrative Example 13.4 find the water-hammer pressure at the valve if a flow of 80 cfs is reduced to 25 cfs in 3.0 s. Under these conditions what would be the maximum water-hammer pressures at points 500 and 1,500 ft from the reservoir?

13.32 Water is flowing through a 30-cm-diameter welded-steel pipe of length 2,000 m that drains a reservoir under a head of 40 m. The pipe has a thickness of 8 mm. (*a*) If a valve at the end of the pipe is closed in 10 s, approximately what water-hammer pressure will be developed? (*b*) If the steady-state flow is instantaneously reduced to one-half its original value, what water-hammer pressure would you expect?

13.33 In the figure, the total length of pipe is 10,000 ft, its diameter is 36 in, and its thickness is $\frac{3}{4}$ in. Assume $E = 30{,}000{,}000$ psi and $E_v = 300{,}000$ psi. If the initial velocity for steady flow is 10 fps and the valve at G is partially closed so as to reduce the flow to half of the initial velocity in 4 s, find (*a*) maximum pressure rise from the water hammer; (*b*) the location of the point of maximum total pressure.

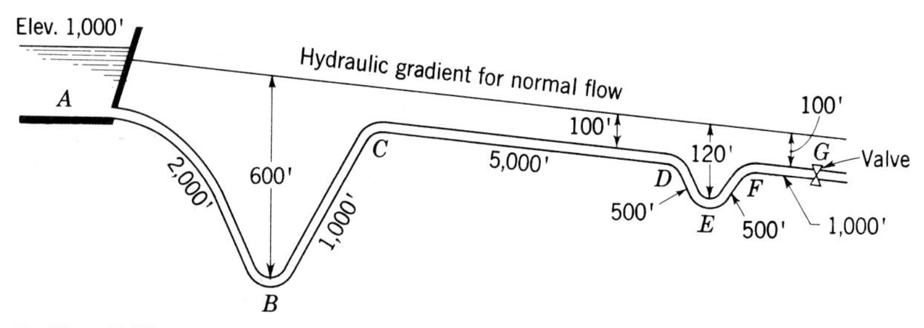

Problem 13.33

13.34 Refer to the figure for Prob. 13.33, but take all the dimensions given in feet to be in meters instead. This 10-km-long pipe has a diameter of 1.5 m and a wall thickness of 25 mm. Assume $E = 200$ GPa and $E_v = 2$ GPa. The initial steady flow velocity is 6 m/s. The valve at G is then partially closed so as to reduce the velocity to 1 m/s in 13 s. Find (*a*) the maximum pressure rise due to water hammer, and (*b*) the location of the point of maximum total pressure.

13.35 Using Eqs. (13.11) and (13.12) and the data for Figs. 13.7 and 13.8 as given in the footnote, compute the water-hammer pressure for each case and compare the answers with the actual measurements. Also, for the given data, compute f.

13.36 Derive Eq. (13.15).

13.37 A 36-in steel pipe 3,000 ft long supplies water to a small power plant. What height would be required for a simple surge tank 6 ft in diameter situated 50 ft upstream from the valve at a point where the center line of the pipe is 120 ft below the water surface in the reservoir if the tank is to protect against instantaneous closure of a valve at the plant? The valve is 150 ft below reservoir level, and the discharge is 150 cfs. Take $f = 0.015$. The surge tank is not to overflow. Neglect all velocity heads and minor losses; in the surge tank (only) neglect fluid friction and inertial effects.

13.38 Repeat Prob. 13.37 for the case where the surge tank is to have a diameter of 10 ft.

13.39 Using the data of Prob. 13.37, find the diameter of surge tank that will produce a surge requiring a tank height of 175 ft.

13.40 A 90-cm steel pipe ($f = 0.015$) of length 1,000 m supplies water to a small power plant. What height would be required for a simple surge tank 3 m in diameter situated 10 m upstream from the valve at a point where the centerline of the pipe is 40 m below the water surface in the reservoir? Assume instantaneous closure of the valve. The valve is 50 m below reservoir level and the flow is 3.2 m^3/s. The surge tank is not to overflow. Neglect all velocity heads and minor losses; in the surge tank (only) neglect fluid friction and inertial effects.

13.41 A 1.25-m-diameter penstock ($f = 0.018$) 950 m long carries water at Q m^3/s from a reservoir to a power plant. When the outlet valve is closed instantaneously, water rises in a 2-m-diameter surge tank immediately adjacent to the outlet valve. Determine the maximum allowable initial discharge Q so that the resulting surge will not rise more than 10 m above the reservoir water surface. Neglect all velocity heads and minor losses; in the surge tank (only) neglect fluid friction and inertial effects. (*Note:* Trial and error can be avoided.)

CHAPTER

FOURTEEN

SIMILARITY LAWS AND FACTORS FOR TURBOMACHINES

In this and the next three chapters[1] our discussion will be confined to turbomachines (i.e., those that rotate) conveying constant-density fluids. Chapters 15 and 16 deal with hydraulic turbines while centrifugal and axial-flow pumps are considered in Chap. 17. In Secs. 6.9, 6.10, and 6.11 there is a discussion of the torque developed in rotating machinery and of flow in rotating channels; it is suggested that the reader review that material before proceeding further. Other types of fluid machinery, none of which will be considered in this text, include steam turbines, blowers, compressors, and positive-displacement pumps such as rotary pumps and piston-in-a-cylinder reciprocating pumps.

14.1 EFFICIENCY DEFINITIONS

The efficiency of a turbine or pump can be broken down into three components: *volumetric* efficiency, *hydraulic* efficiency, and *mechanical* efficiency.

Turbine

The overall efficiency η of a turbine is defined as

$$\eta = \frac{\text{power delivered to the shaft (brake power)}}{\text{power taken from the water (work power)}} = \frac{T\omega}{\gamma Q h} \tag{14.1}$$

[1] English units of measurement are used exclusively in Chaps. 14 through 17. If SI units are given, they should be converted to English units before using an equation if the equation is not dimensionally correct.

where T is the torque delivered to the shaft by the turbine, ω is the rotative speed in radians per second, Q is the flow rate, and h is the net head on the turbine (Secs. 15.4 and 16.7).

The volumetric efficiency η_v refers to the possible loss of efficiency through leakage around the outside of the *rotor* or rotating element.[1] In other words, not all of the fluid flowing is necessarily effective in the energy transfer process. In the case of a turbine, let Q_L represent this leakage while Q represents the net flow passing through the turbine. Then $Q - Q_L$ represents the flow that is effectively acting on the rotor. Consequently, the volumetric efficiency is

$$\eta_v = \frac{Q - Q_L}{Q} \tag{14.2}$$

Ordinarily this leakage is a very small percentage of the flow and for some machines it does not exist, but under unfavorable conditions it may be important.

The hydraulic efficiency η_h of a turbine is the ratio of $\gamma(Q - Q_L)h''$, the power transferred from the water to the rotor, to $\gamma(Q - Q_L)h$, the available power in the fluid that effectively flows through the rotor. In Sec. 6.10 it was shown that the head utilized by the rotor, $h'' = (u_1 V_1 \cos \alpha_1 - u_2 V_2 \cos \alpha_2)/g$. This can also be expressed as $h'' = h - h_f$, where h_f is the fluid-friction head loss in flow through the turbine including loss at the exit. Thus the hydraulic efficiency of a turbine is

$$\eta_h = \frac{h - h_f}{h} = \frac{h''}{h} \tag{14.3}$$

The mechanical efficiency η_m of a turbine is the ratio of the power available at the shaft to that exerted by the water on the rotor. Thus,

$$\eta_m = \frac{T\omega}{(T + T_f)\omega} = \frac{bp}{bp + fp} \tag{14.4}$$

where T_f represents the torque required to overcome mechanical friction. The term bp represents the *brake power*,[2] or power available at the shaft of the machine, while fp $(= T_f\omega)$ represents the power used up in overcoming mechanical friction, which includes friction in the bearings and stuffing boxes and disk friction between the sides of the rotor and the fluid in the adjacent casing. The mechanical efficiency of hydraulic turbines is usually relatively high, about 95 to 98 per cent.

The *total*, or *overall*, efficiency can be found by noting that $bp + fp$, the power that is transferred from the water to the rotor, can be expressed as

[1] The rotor of a turbine is referred to as the *runner* while that of a pump is called the *impeller*.

[2] Brake power (and therefore brake horsepower) represents the power transmitted by the shaft of a turbine *or* pump. This must be distinguished from the work power (and therefore work horsepower) which represents the power *taken from* the liquid by a turbine *or delivered to* the liquid by a pump.

$\gamma(Q - Q_L) \times (h - h_f)$. Substituting this into Eq. (14.4) and comparing with Eq. (14.1) gives for the overall efficiency of a turbine,

$$\eta = \eta_v \eta_h \eta_m \tag{14.5}$$

Pump

For a pump the efficiencies are analogous to those for a turbine but they are essentially inverted. If there is leakage at a rate Q_L back from the high-pressure side to the low-pressure side of a pump, there is a loss of energy because work is done upon the fluid that has leaked. For a pump the volumetric efficiency is

$$\eta_v = \frac{Q}{Q + Q_L} \tag{14.6}$$

where Q represents the flow actually delivered. The *hydraulic efficiency* of a pump is

$$\eta_h = \frac{h}{h''} \tag{14.7}$$

where h represents the net head delivered to the fluid by the pump while h'' is the head transferred from the rotor to the fluid. In this case $h = h'' - h_f$, where h_f is the hydraulic head loss. The *mechanical efficiency* of a pump is

$$\eta_m = \frac{bp - fp}{bp} \tag{14.8}$$

where bp is the power at the pump shaft $(= T\omega)$ while fp represents the power lost to mechanical friction in the bearings and stuffing boxes as well as in the disk friction.

The *total*, or *overall*, efficiency of a pump can be found by noting that $bp - fp$, the power that is transmitted from the rotor to the water, can be expressed as $\gamma(Q + Q_L)(h + h_f)$. Relating this to the preceding equations gives for the overall efficiency of a pump,

$$\eta = \frac{\text{power delivered to the fluid (work power)}}{\text{power put into the shaft (brake power)}} = \frac{\gamma Q h}{T\omega} = \eta_v \eta_h \eta_m \tag{14.9}$$

These same equations apply to compressors, blowers, and fans, but if there is an appreciable change in the density of the fluid, some modifications may be necessary. In Fig. 17.13 the various losses of power (and, hence, of energy) are shown for the case of a pump. Details on the nature of the hydraulic losses are presented for turbines in Sec. 16.11 and for pumps in Sec. 17.8.

14.2 SIMILARITY LAWS

Similarity laws permit the prediction of the performance of a prototype machine from the test of a scaled model. These laws also permit prediction of the performance of a given machine under different conditions of operation from those under which it may have been tested. Similarity laws are based on the concept that two geometrically similar machines with similar velocity diagrams at entrance to and exit from the rotating element are *homologous*. This means that their streamline patterns will be geometrically similar, i.e., that their behaviors will bear a resemblance to one another.

Similarity laws can be derived by dimensional analysis. The most significant variables[1] affecting the operation of a turbomachine are the head h, the discharge Q, the rotative speed n, the diameter of the rotor D, and the acceleration of gravity g. Thus, from the Buckingham Π-theorem (Sec. 7.7), since there are five dimensional variables and two fundamental dimensions (L and T), there will be three dimensionless groups. We have

$$f(h, Q, n, D, g) = 0$$

Upon grouping these variables into dimensionless quantities, we get

$$f'\left(\frac{Q}{nD^3}, \frac{g}{n^2D}, \frac{h}{D}\right) = 0$$

Laboratory tests on turbomachines have demonstrated that the second dimensionless quantity is inversely proportional to the third. These can be combined to give

$$\frac{g}{n^2D} = K\frac{D}{h} \qquad \text{and} \qquad K = \frac{gh}{n^2D^2}$$

Thus

$$f''\left(\frac{Q}{nD^3}, \frac{gh}{n^2D^2}\right) = 0 \tag{14.10}$$

Next we shall examine the turbine and the pump individually to confirm the preceding relationships.

Turbine

In Fig. 14.1 is shown the profile of a turbine runner, and in Fig. 6.10 is shown a section of the same runner in a plane at right angles to the runner shaft. A turbine runner may be regarded as a special form of orifice in that the flow through it is proportional to an area times a velocity and the latter is some function of $\sqrt{2gh}$. The net circumferential area is $A_c = f\pi DB$, where f is the fractional part of the area that is free space, the vanes taking up the rest. The value of f is usually about 0.95. For a series of homologous runners $B/D = m =$ constant, and therefore $A_c = f\pi mD^2$.

The radial component of the velocity entering the runner is (Fig. 6.10) $V_r = V_1 \sin \alpha_1 = C_r\sqrt{2gh}$, where C_r is a factor determined by test and influenced

[1] If it is desired to relate the operation of one turbomachine to another with different fluids in each, then kinematic viscosity would be a significant variable.

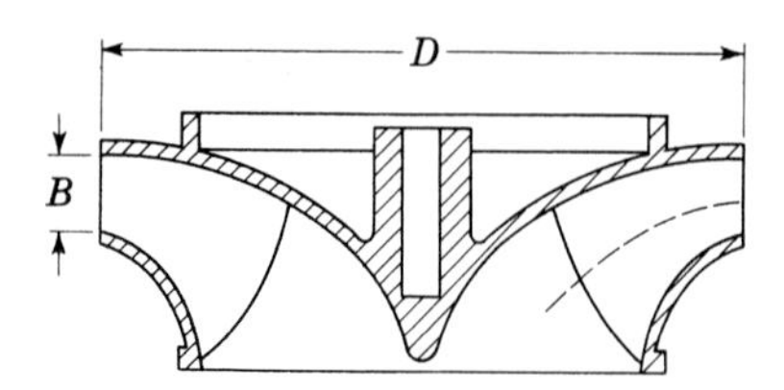

Figure 14.1 Profile of turbine runner or pump impeller.

by the type of runner, though it is substantially constant for a series of homologous runners. As

$$Q = A_c V_r = (f\pi m C_r \sqrt{2g}) D^2 \sqrt{h}$$

the expression within the parentheses may be replaced by K_q so that

$$Q = K_q D^2 \sqrt{h} \tag{14.11}$$

where K_q is a constant for a series of homologous runners. Note that K_q has the same dimensions as $\sqrt{g}$.[1]

From Eq. (6.21) torque is a function of ρQ, r_1, r_2, V_1, and V_2. Now r_1 and r_2 are proportional to D, and V_1 and V_2 are both functions of $\sqrt{h}$. Hence torque is proportional to ρQrV which is proportional to $\rho(D^2\sqrt{h})D\sqrt{h}$, or

$$T = K_t \gamma D^3 h \tag{14.12}$$

where K_t is also constant for a series of homologous runners.

The rotative speed is $\omega = u/r$, where ω is expressed in radians per second, though for practical engineering purposes it is customary to use revolutions per minute for most hydraulic machinery. As

$$n = \frac{60\omega}{2\pi} = \frac{60u_1}{\pi D}$$

and u_1 is proportional to $\sqrt{h}$,

$$n = K_n \frac{\sqrt{h}}{D} \tag{14.13}$$

Power $P = T\omega$, and as $\omega = 2\pi n/60$, it follows that

$$P = K_p \gamma D^2 h^{3/2} \tag{14.14}$$

Since $P = \eta\gamma Qh$, Eq. (14.14) could also have been obtained by inserting in this expression the value of Q from Eq. (14.11).

[1] We are assuming here that g is a constant and hence its effect drops out. Strictly speaking, however, the g should be retained. For example, Eqs. (14.11) to (14.18) must be modified to account for variation in g when comparing the performance of a turbomachine on the earth with one on the moon.

Pump

The same Fig. 14.1 will serve to illustrate the case of a pump, but for the centrifugal pump the flow is outward. There is, however, a practical difference in usage. For a turbine we are usually interested in its operation under a certain head which is fixed by nature. For a pump we are usually interested in its operation at a certain rotative speed, determined by the motor which drives it. For this reason, when dealing with pumps, it is convenient to have n in the equations.

From Eq. (14.13)

$$h = K_1 D^2 n^2 \tag{14.15}$$

and substituting this expression for h in Eqs. (14.11), (14.12), and (14.14), we obtain

$$Q = K_2 D^3 n \tag{14.16}$$

$$T = K_3 \gamma D^5 n^2 \tag{14.17}$$

$$P = K_4 \gamma D^5 n^3 \tag{14.18}$$

For any one design of a turbine or a pump these constants can be evaluated, preferably from test data, and then used for a series of different sizes, provided they are all homologous. Also, note that the relations predicted in Eq. (14.10) were confirmed in Eqs. (14.16) and (14.15).

14.3 RESTRICTION ON USE OF SIMILARITY LAWS

Similarity laws are of great practical value, but care must be exercised when applying them. Thus, in comparing two machines of different sizes, the two must be homologous and the variation in the values of h, D, and n should not be too large. For example, a machine which operates satisfactorily at low speeds may cavitate at high speeds. The values of K in each of Eqs. (14.11) to (14.18) change somewhat as h, D, and n are varied because the efficiencies of homologous machines are not identical. Large machines are usually more efficient than smaller ones because their flow passages are larger. Also, efficiency usually increases with speed of rotation because power output varies with the cube of the speed while mechanical losses increase only as the square of the speed.

Typical performance curves for homologous turbines and pumps are shown in Figs. 14.2 and 14.3. In Fig. 14.2 is shown the discharge through a turbine at different rotative speeds under a constant head, which we shall assume to be h_2. There is no simple or accurate theory which will determine the shape of this curve so it must be established by test. Suppose we have the complete curve from experiment for the case where $h = h_2$ and we wish to compute the corresponding curve for some other head, such as h_1. This may be done by the use of Eqs. (14.11) and (14.13) together, but not by one of them alone. Thus, if we use the former equation to give $Q_1 = Q_2\sqrt{h_1/h_2}$, this value will be found only at a speed

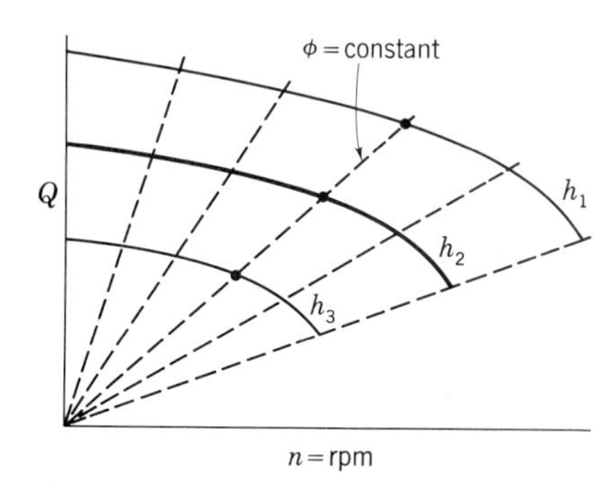

Figure 14.2 Performance of a typical hydraulic turbine at various constant heads.

$n_1 = n_2\sqrt{h_1/h_2}$. As both Q and n are seen to vary as $\sqrt{h}$, it follows that corresponding points are on straight lines through the origin. These lines (Fig. 14.2) represent lines of constant ϕ (defined in Sec. 14.4).

Consider Fig. 14.3, which shows the relation between h and Q for a typical centrifugal pump running at some constant speed n_1. Again, this curve is one which cannot be calculated by any simple or accurate theory and is determined experimentally. At some other speed n_2, Eq. (14.15) gives us $h_2 = h_1(n_2/n_1)^2$, but at the same time Eq. (14.16) gives $Q_2 = Q_1(n_2/n_1)$. Since h varies as n^2 and Q varies as n, corresponding values of h and Q for homologous pumps lie along parabolas passing through the origin. These parabolas represent lines of constant ϕ. To compare the performance of two homologous machines, the values of ϕ must be the same.

14.4 PERIPHERAL-VELOCITY FACTOR

For a turbine runner (Fig. 6.10) or a pump impeller (Fig. 6.11), the ratio of its peripheral velocity to $\sqrt{2gh}$ is referred to as the *peripheral-velocity factor*, denoted by ϕ. Thus, for a turbine,

$$u_1 = \phi\sqrt{2gh} \tag{14.19}$$

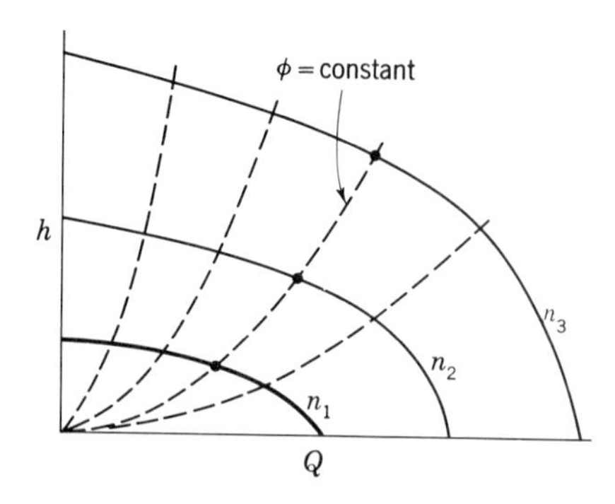

Figure 14.3 Performance of a typical centrifugal pump at various constant speeds.

while for a centrifugal pump,

$$u_2 = \phi\sqrt{2gh} \qquad \text{or} \qquad h = \frac{1}{\phi^2}\frac{u_2^2}{2g} \tag{14.20}$$

inasmuch as u_1 and u_2 are the designations of the peripheral speeds of these respective machines. For an axial-flow machine it is the vane-tip speed u_t that is significant. In the case of an impulse wheel (Chap. 15), u is defined as the speed of the bucket at the centerline of the nozzle. Omitting subscripts,

$$\omega = \frac{u}{r} = \frac{\phi\sqrt{2gh}}{r}$$

but for practical engineering use,

$$n = \frac{60u}{\pi D} = \frac{60\phi\sqrt{2gh}}{\pi D} \tag{14.21}$$

which may be reduced to the convenient form

$$nD = 153.2\phi\sqrt{h} \tag{14.22}$$

For any machine its peripheral velocity might be any value from zero up to some maximum under a given head, and ϕ would consequently vary through as wide a range. But the speed which is of most practical significance is that at which the efficiency is a maximum. The value of this dimensionless factor for this particular speed may be designated as ϕ_e, but frequently the subscript is omitted. It is the numerical value of ϕ_e that is usually inserted into Eq. (14.22).

The numerical value of ϕ_e depends upon the type of the machine, but for a series of homologous machines it is a constant. Its numerical value in a specific case may be estimated by theory, but practically it is determined by test.

14.5 SPECIFIC SPEED

Turbines

A most useful practical factor for turbines is the expression commonly known as *specific speed*. Using the expression for Q from Sec. 14.2 we can express the brake horsepower (horsepower delivered to the turbine shaft) as

$$\text{bhp} = \frac{\eta\gamma Qh}{550} = \left(\frac{\eta\gamma f\pi\sqrt{2g}}{550}\,mC_r\right)D^2h^{3/2}$$

where η, the overall efficiency of the turbine, is introduced to account for the difference between the power available in the flowing water and the power output of the turbine shaft.

Substituting in this equation the value of D^2 as obtained from Eq. (14.21), we obtain

$$\text{bhp} = \left[\frac{\eta\gamma f(2g)^{3/2}(60)^2}{550\pi}\phi^2 m C_r\right]\frac{h^{5/3}}{n^2}$$

It is seen that, in addition to numerical constants, the expression in brackets contains certain design factors. If the expression within the brackets is represented by n_s^2, then, taking the square root of this equation and rearranging, we obtain a quantity known as the *specific speed*, which is

$$n_s = \frac{n\sqrt{\text{bhp}}}{h^{5/4}} \tag{14.23}$$

Although under a given head any value of rotative speed and corresponding brake horsepower might be used, the only ones that will yield a significant value for n_s are those for which the efficiency is a maximum. Hence *the value of n to be used should be the most efficient speed for the given head*, and the brake horsepower should be that for maximum efficiency at that speed.

The only justification for the terminology "specific speed" is that if the turbine is made of such a size as to develop 1 bhp under 1 ft head, then n_s would be the revolutions per minute, but the dimensions of n_s are really $F^{1/2}L^{-3/4}T^{-3/2}$. An inspection of the terms in the brackets will show that n_s is a function of the design factors $\phi\sqrt{mC_r}$, and therefore its value depends on the design of the turbine. In fact, it might better be called the *type characteristic*, or some similar name, because it does indicate the type of turbine. In Fig. 15.5 is shown the runner of an impulse wheel with a specific speed $n_s = 5.6$ while in Fig. 16.2 are section views of the runners of reaction turbines with specific speeds of 21.3, 80, and 160.[1]

Pumps

A specific-speed factor is equally useful for pumps, but it will appear in a different form. In the case of a turbine, we are primarily interested in the power it will deliver, whereas in the case of a pump or similar machine, we are primarily interested in the quantity rate at which it will deliver fluid. Substituting the value of D^2 from Eq. (14.21) into the expression $Q = A_c V_r$ (Sec. 14.2) we obtain

$$Q = \left[\frac{f(2g)^{3/2}(60)^2}{\pi}\phi^2 m C_r\right]\frac{h^{3/2}}{n^2}$$

As in the case of the turbine, the expression in brackets contains not only numerical constants but also design factors, and if this expression is represented by n_s^2, we obtain a *specific speed*

$$n_s = \frac{n\sqrt{Q}}{h^{3/4}} \tag{14.24}$$

[1] The specific speed of turbines is sometimes computed using metric units, i.e., kW for brake horsepower and meters for head. In that case, $(n_s)_{\text{metric}} = 3.8 \times (n_s)_{\text{English}}$.

For a given pump this expression might have any value from zero at no flow to infinity at maximum Q and zero head, but the only value that is significant is that at the point of maximum efficiency, that is, $n = n_e$, the speed for maximum efficiency.

In the case of pumps handling liquids, it is customary to use cubic feet per second for large flows and to use gallons per minute for most usual capacities. Employing this common unit, the specific speed of pumps is usually expressed as

$$N_s = \frac{n\sqrt{\text{gpm}}}{h^{3/4}} \tag{14.25}$$

where $N_s = 21.2n_s$. Similarly to the turbine, the specific speed for a pump is the revolutions per minute for a pump of such a size as to deliver unit volume of fluid per unit time at a head of 1 ft, although the actual dimensions are $L^{3/4}T^{-3/2}$. The real value of the expression is that it indicates the type of pump. In Fig. 17.18 are section views of the impellers of centrifugal and axial-flow pumps over a range of values of N_s from 600 to 12,000.[1]

For a multistage pump the value of h to be used in computing specific speed is the head per stage. For a double-suction pump the specific speed is sometimes based on the total capacity, but in general it is preferable to compute it by using one-half of the total capacity on the basis that a double-suction impeller is the equivalent of two single-suction impellers placed back to back.

Illustrative Example 14.1 At its optimum point of operation a given centrifugal pump with an impeller diameter of 50 cm delivers 3.2 m³/s of water against a head of 25 m when rotating at 1,450 rpm.

(*a*) If its efficiency is 82 per cent, what is the brake power of the driving shaft?

(*b*) If a homologous pump with an impeller diameter of 80 cm is rotating at 1,200 rpm, what would be the discharge, head, and shaft power? Assume both pumps operate at the same efficiency.

(*c*) Compute the specific speed of both pumps.

SOLUTION

(*a*) $$\eta = \frac{\gamma Q h}{T\omega} \quad \text{or} \quad bp = T\omega = \frac{\gamma Q h}{\eta}$$

$$bp = \frac{(9.81\ \text{kN/m}^3)(3.2\ \text{m}^3/\text{s})(25\ \text{m})}{0.82} = 957\ \text{kN}\cdot\text{m/s} = 957\ \text{kW} = 1{,}280\ \text{hp}$$

(*b*) Eq. (14.15):

$$h \propto D^2 n^2 \qquad h_2 = h_1\left(\frac{80}{50}\right)^2\left(\frac{1{,}200}{1{,}450}\right)^2 = 43.8\ \text{m}$$

Eq. (14.16):

$$Q \propto D^3 n \qquad Q_2 = Q_1\left(\frac{80}{50}\right)^3\left(\frac{1{,}200}{1{,}450}\right) = 10.85\ \text{m}^3/\text{s}$$

$$bp = \frac{\gamma Q h}{\eta} = \frac{9.81(10.85)(43.8)}{0.82} = 5{,}690\ \text{kW}$$

[1] The specific speed of centrifugal pumps is sometimes computed by using metric units, i.e., m³/s for gpm and meters for head. In that case, $(n_s)_{\text{metric}} = 0.0194 \times (N_s)_{\text{English}}$.

As a check, from Eq. (14.18), $P \propto D^5 n^3$,

$$(bp)_2 = (bp)_1 \left(\frac{80}{50}\right)^5 \left(\frac{1{,}200}{1{,}450}\right)^3 = 957 \times 10.49 \times 0.567 = 5{,}690 \text{ kW}$$

(*c*) Converting to English units:

$$h_1 = 82.0 \text{ ft} \qquad Q_1 = 50{,}700 \text{ gpm} \qquad h_2 = 143.8 \text{ ft} \qquad Q_2 = 171{,}800 \text{ gpm}$$

$$N_{s_1} = \frac{1{,}450\sqrt{50{,}700}}{82.0^{3/4}} = 11{,}980 \qquad N_{s_2} = \frac{1{,}200\sqrt{171{,}800}}{143.8^{3/4}} = 11{,}980$$

Since the pumps are homologous we expect them to have the same specific speed.

PROBLEMS

14.1 At a hydroelectric plant the difference in elevation between the surface of the water at intake and at a tailrace is 600 ft. When the flow is 80 cfs, the friction loss in the penstock is 60 ft and the head utilized by the turbine is 460 ft. The mechanical friction in the turbine is 100 hp, and the leakage loss is 3 cfs. Find (*a*) hydraulic efficiency; (*b*) volumetric efficiency; (*c*) power delivered to runner; (*d*) power delivered to shaft (brake horsepower); (*e*) mechanical efficiency; (*f*) overall efficiency.

14.2 A turbine rotating at 200 rpm and operating under a net head of 150 m delivers a torque of 772 kN·m to its shaft when the flow rate is 12 m^3/s. What is the efficiency of the turbine?

14.3 A turbine runs at 150 rpm, discharges 200 cfs, and develops 1,600 bhp under a net head of 81 ft. (*a*) What is its efficiency? (*b*) What would be the revolutions per minute, Q, and brake horsepower of the same turbine under a net head of 162 ft for homologous conditions?

14.4 If a turbine homologous to that in Prob. 14.3 has a runner of twice the diameter, what would be the revolutions per minute, Q, and brake horsepower under the same head of 81 ft?

14.5 All dimensions of pump A are one-third as large as the corresponding dimensions of pump B. When operating at 300 rpm, B delivers 100 gpm (6 L/s) of water against a head of 50 ft (15 m). Assuming the same efficiency: (*a*) What will be the speed and capacity of A when it develops a head of 50 ft (15 m); (*b*) what will be the speed and head of A when it delivers 100 gpm (6 L/s); (*c*) what will be the head and capacity of A when it operates at 300 rpm?

14.6 A model centrifugal pump is made with a scale ratio of 1:10. The model was tested at 3,600 rpm and delivered 3 cfs at a head of 125 ft with an efficiency of 90 percent. Assuming the prototype to have an efficiency of 91 percent and to develop the same head, what will be its speed, capacity, and horsepower required? The liquid pumped is water.

14.7 A model centrifugal pump has a scale ratio of 1:15. The model was tested at 3200 rpm and delivered 0.10 m^3/s of water at a head of 40 m with an efficiency of 86 percent. Assuming the prototype has an efficiency of 88 percent, what will be its speed, capacity, and power requirement at a head of 50 m?

14.8 An 18-in-diameter centrifugal-pump runner discharges 25 cfs at a head of 100 ft when running at 1,200 rpm. (*a*) If its efficiency is 85 percent, what is the brake horsepower? (*b*) If the same pump were run at 1,800 rpm, what would be h, Q, and brake horsepower for homologous conditions?

14.9 An axial-flow pump delivers 300 L/s at a head of 6.0 m when rotating at 2,000 rpm. (*a*) If its efficiency is 80 percent, how many kilowatts of power must the shaft deliver to the pump? (*b*) If this same pump were operated at 2,400 rpm, what would be h, Q, and the power delivered by the shaft for homologous conditions?

14.10 What head will the pump of Prob. 14.8 develop if it is operating on the moon at 1200 rpm and delivering 25 cfs?

14.11 If ϕ for the runner in Prob. 14.3 is 0.72, what is the diameter of the runner? What, then, are the values of K_q, K_t, K_n, and K_p for that series?

14.12 What are the values of the torque exerted by the runners in Probs. 14.3 and 14.4? What is the specific speed for the runners of Probs. 14.3 and 14.4?

14.13 (*a*) At peak efficiency a Pelton waterwheel under a net head of 2,350 ft delivers 17,500 bhp at 450 rpm. What is its specific speed? (*b*) At peak efficiency a reaction turbine under a head of 92 ft delivers 35,000 bhp at 100 rpm. What is its specific speed? (*c*) If $\phi = 0.46$ for the former and 0.75 for the latter, what is the diameter of each runner?

14.14 For the pump in Prob. 14.8, what are the values of K_1, K_2, and K_4? What is the specific speed?

CHAPTER

FIFTEEN

IMPULSE TURBINES

15.1 DEFINITION[1]

An impulse turbine, whether for water, steam, or gas, is one in which the total drop in pressure of the fluid takes place in one or more stationary nozzles and there is no change in pressure of the fluid as it flows through the rotating wheel. As there is no pressure variation in flow over the buckets or vanes, the fluid does not fill the passageway between one bucket or vane and the next. Customarily, only a portion of the circumference of the wheel is acted upon by the fluid at any one instant.

The energy of the fluid entering the rotor is in the form of kinetic energy of the jets. In flowing over the buckets or vanes of the runner this kinetic energy is transformed into mechanical work delivered to the shaft, a part is dissipated in fluid friction, and there is some residual kinetic energy of the fluid leaving the rotor.

The foregoing statements and the theory presented in this text apply to all types of turbines, but steam and gas turbines involve thermodynamic principles in addition, and so their complete theory is outside the scope of this book. Because

[1] The reader should refer back to Chap. 14 for discussion of similarity laws, specific speed, and other characteristics of turbines.

of practical features, the mechanical construction of hydraulic turbines differs widely from that of steam or gas turbines, and physically they bear little resemblance to each other. In this text the descriptive material will be confined to the hydraulic turbine.

15.2 THE HYDRAULIC IMPULSE TURBINE

Several types of hydraulic impulse turbines have been produced in the past, but the only one that has survived is the Pelton wheel (Fig. 15.1), so called in honor of Lester A. Pelton (1829–1908), who contributed much to its development in the early gold-mining days in California. Pelton was granted a patent in 1880 on an improved type of bucket, its principal feature being a *splitter* in the middle, since before that time the buckets were merely cups. Later, W. A. Doble brought out the *ellipsoidal* bucket, which is the basis of the modern forms.[1]

15.3 SETTINGS

Impulse turbines are usually set with the shaft horizontal, and there is usually only one jet on a wheel. If an electric generator is driven by one wheel, the generator is mounted between two bearings, with the wheel outside. It is then called a *single-overhung* unit. Often a single generator is driven by two wheels, as in Fig. 15.1, and this is a double-overhung unit.

Occasionally, two or more jets may be employed on one wheel to increase the power of a horizontal-shaft wheel, but more commonly the multijet arrangement is used with a vertical-shaft turbine, as in Fig. 15.2.

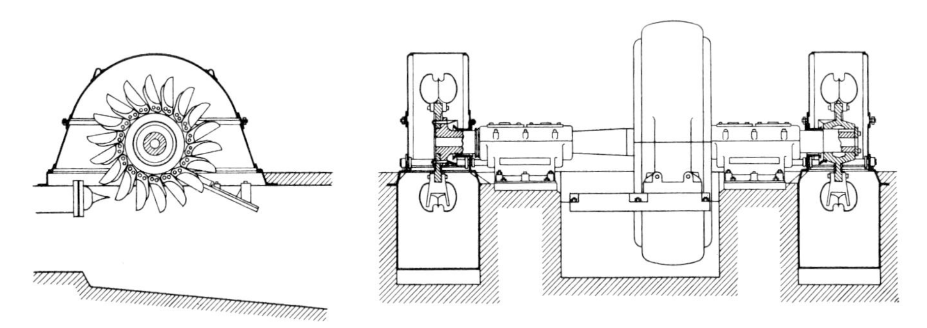

Figure 15.1 Double-overhung impulse wheel.

[1] The term *Pelton* is commonly used to designate a type and does not necessarily imply that the wheel was built by the company bearing that name. In fact, it is the designation commonly used by many European manufacturers.

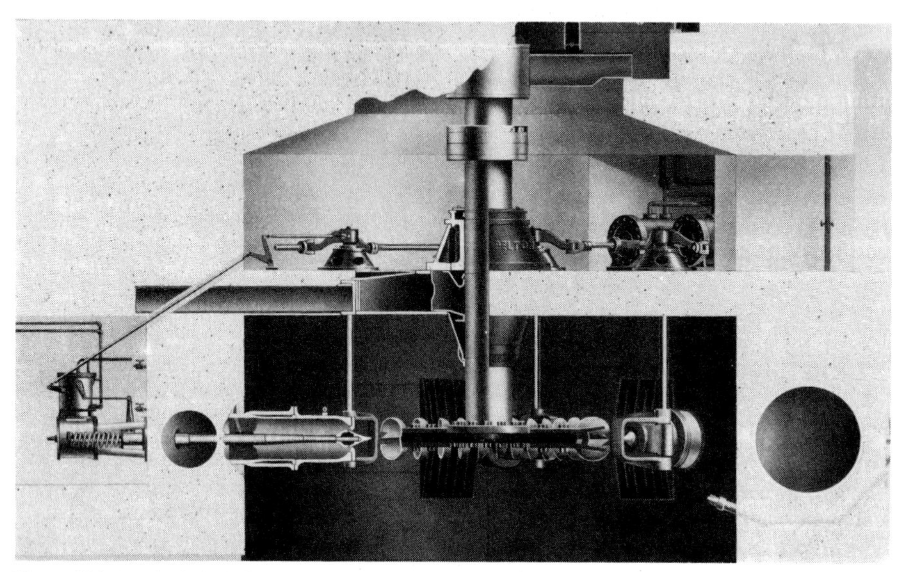

Figure 15.2 Vertical-shaft impulse turbine with six nozzles at Bridge River plant in British Columbia. Gross head = 1,226 ft, net head = 1,118 ft, 62,000 hp, n = 300 rpm, pitch diameter = 95 in.

15.4 HEAD ON IMPULSE TURBINE

The gross head for a power plant is the difference in elevation between headwater and tailwater, or $Y + z$ in Fig. 15.3. The pressure within the case of an impulse wheel is atmospheric, and consequently this is the pressure at which the jet is discharged. Thus the portion z is unavailable; so the gross head on the wheel itself is Y only. This is also called the *static* head. It is impractical to set the wheel too near the surface of the tailwater because it might then be submerged with any rise in level of the latter. However, as Pelton wheels are usually installed under high heads, the percentage loss due to this setting is small. Thus, if the head is 2,000 ft and z is 10 ft, the loss is only 0.5 percent.

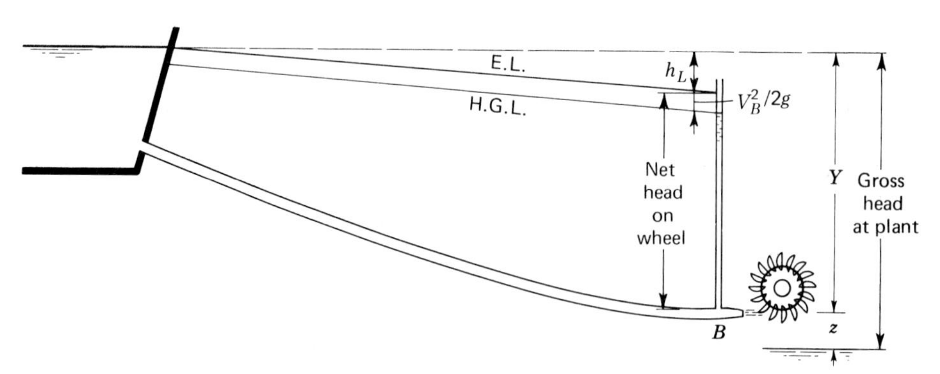

Figure 15.3

The net, or effective head h on the wheel, i.e., the head at the entrance to (or base of) the nozzle, is the static head minus the pipe friction losses. Since the nozzle is considered an integral part of the turbine, the net head is that at B in Fig. 15.3; so the effective, or net, head is

$$h = \frac{p_B}{\gamma} + \frac{V_B^2}{2g} \tag{15.1}$$

The energy, or head, supplied at the nozzle is expended in four ways. A small amount is lost in fluid friction in the nozzle [Eq. (12.13)], a portion is expended in fluid friction over the buckets, kinetic energy is carried away in the water discharged from the buckets, and the rest is delivered to the buckets. Thus

$$h = \left(\frac{1}{C_v^2} - 1\right)\left[1 - \left(\frac{A_2}{A_1}\right)^2\right]\frac{V_1^2}{2g} + k\frac{v_2^2}{2g} + \frac{V_2^2}{2g} + h'' \tag{15.2}$$

where C_v = coefficient of velocity of the nozzle (Sec. 12.6)
A_2 = cross-sectional flow area of the jet
A_1 = cross-sectional flow area of the pipe upstream of the nozzle
V_1 = jet velocity
v_2 = velocity of water relative to bucket at exit from bucket
V_2 = absolute velocity of water leaving the bucket
h'' = energy head delivered to the buckets

Typical values of k, the *bucket friction loss coefficient*, vary from 0.2 to about 0.6. The greater part of the energy delivered to the buckets is transferred to the shaft, but some of it is used in overcoming mechanical friction in the bearings and in windage loss (air friction).

15.5 NOZZLES

In any turbine, in order to maintain a constant speed of rotation, it is necessary that the flow rate be varied in proportion to the load on the machine; and for the impulse wheel this is done by varying the size of the jet. This is accomplished by varying the position of the needle in the needle nozzle. The shape of both nozzle tip and needle should be such as to cause a minimum friction loss for all positions of the needle and also such as to avoid cavitation damage to the needle at any position.

An important feature in attaining high efficiency in an impulse wheel is that the jet be uniform, the ideal being to have all particles of water moving in parallel lines with equal velocities and with no spreading out of the jet. Air friction retards the water on the outside of the jet, and the needle causes the velocity in the center to be slightly reduced. Careful design of both nozzle tip and needle will minimize these effects, and a gain in turbine efficiency of several percent has been made by improved nozzle design producing better jets. Values of C_v for needle nozzles vary from about 0.95 when partly closed to permit one-half of maximum flow to 0.99 at the fully opened position.

For a given pipeline there is a unique jet diameter that will deliver maximum power to a jet. Refer to Fig. 15.3 and note that the power of the jet issuing from the nozzle may be expressed as

$$P_{\text{jet}} = \gamma Q \frac{V_j^2}{2g} \tag{15.3}$$

where V_j is the jet velocity [equal to V_1 of Eq. (15.2)]. As the size of the nozzle opening is increased, the flow rate Q gets larger while the jet velocity V_j gets smaller. Hence, from Eq. (15.3) and from the preceding statement, we must conclude that there is some intermediate size of nozzle opening (and, hence, of jet diameter) that will provide maximum power to the jet. This is best illustrated by an example.

Illustrative Example 15.1 A 6-in-diameter pipe ($f = 0.020$) of length 1,000 ft delivers water from a reservoir with a water-surface elevation of 500 ft to a nozzle at elevation 300 ft. The jet from the nozzle is used to drive a small impulse turbine. If the head loss through the nozzle can be expressed as $0.04\, V_j^2/2g$, find the jet diameter that will result in maximum power in the jet. Neglect the head loss at entrance to the pipe from the reservoir. Evaluate the power in the jet.

Energy equation:

$$500 - 0.02 \frac{1{,}000}{0.5} \frac{V_p^2}{2g} - 0.04 \frac{V_j^2}{2g} = 300 + \frac{V_j^2}{2g}$$

If we define the pipe diameter and velocity as D_p and V_p and the jet diameter and velocity as D_j and V_j, from continuity we get:

$$A_p V_p = A_j V_j \qquad D_p^2 V_p = D_j^2 V_j$$

Since the pipe diameter $D_p = 0.50$ ft.

$$0.25 V_p = D_j^2 V_j \qquad \text{and} \qquad V_p = 4 D_j^2 V_j$$

Substituting this expression for V_p in the energy equation gives

$$200 = \frac{V_j^2}{2g}(1.04 + 640 D_j^4)$$

Assuming different values for D_j, we can compute corresponding values of V_j and Q, and then the jet power can be computed using Eq. (15.3). The results are as follows:

D_j, in	D_j, ft	V_j, fps	A_j, ft^2	$Q = A_j V_j$, cfs	P_{jet}, hp
1.0	0.083	111	0.0054	0.60	12.8
1.5	0.125	105	0.0122	1.28	24.2
2.0	0.167	91	0.0218	2.00	29.8
2.5	0.208	76	0.0338	2.57	26.2
3.0	0.250	60	0.0491	2.94	18.8
4.0	0.333	38	0.0873	3.29	8.4
6.0	0.500	18	0.197	3.49	1.9

Thus a 2-in-diameter jet is the optimum; it will have about 30 hp.

An alternate procedure for solving this problem is to set up an algebraic expression for the power of the jet, P_{jet}, as a function of the jet diameter, D_j, and differentiate P_{jet} with respect to D_j and equate to zero to find the value of D_j for which P_{jet} is a maximum.

15.6 SPEED REGULATION

The rotative speed of an impulse turbine is maintained constant through use of a governor. When the load on a turbine drops the wheel tends to speed up; this affects the governor which, in turn, actuates a mechanism to reduce the power of the jet that impinges on the buckets. In most designs this is accomplished by moving the needle to reduce the flow in the delivery pipe.[1] This may result in serious water-hammer pressures. There are several ways in which this problem may be avoided.

One means of solving this problem is the use of a jet deflector. This is a plate which can be moved rapidly by the governor to deflect a portion of the jet so that a decreased amount of water actually hits the buckets. Then the needle can be slowly moved to decrease the flow in the pipe gradually. As this is being done, the deflector is simultaneously withdrawn from the jet.

Another means is the use of an auxiliary needle nozzle below the main one. As the governor rapidly moves the needle to reduce the flow of water to the buckets, at the same time it opens the auxiliary nozzle, which discharges a jet that misses the wheel altogether. Thus there need be no change in flow in the pipe. However, to save wasting water, a delaying mechanism will slowly close the auxiliary nozzle.

Still a third arrangement is the deflecting needle nozzle, where the entire nozzle is movable about a ball-and-socket joint at its base. The governor can then swing the nozzle so that the jet is directed downward from the horizontal, only a portion of it striking the buckets. The nozzle is restored to its initial position as the needle is slowly closed.

It is seen that all these governing arrangements are such as to reduce the flow of water to the impulse wheel rapidly and yet bring about a reduction of flow in the pipeline more slowly and with a minimum waste of water. However, in case of a sudden demand for more power, the governor will move the needle rapidly to provide a bigger nozzle opening, but it will take time for the flow in a long pipe to be accelerated. The only possible aid in this case is to have a big surge tank (Sec. 13.7) located as close to the power plant as topography permits. Fortunately, while there may be sudden decreases in load, as when a circuit breaker opens, increases in load usually come on more gradually.

[1] If the nozzle is operating at an opening larger than that for maximum jet power, the nozzle will have to be opened wider, thus increasing the flow but decreasing the power of the jet as indicated in Illustrative Example 15.1.

15.7 WHEEL CONSTRUCTION

Figure 15.4 shows an impulse runner where the buckets are bolted to a rim, but runners are also cast in one piece, as shown in Fig. 15.5.

In both cases there is a notch to permit the bucket to attain a position more nearly tangent to the direction of the jet before the bucket lip intercepts the jet. The jet then strikes a splitter, which divides it so that equal quantities flow out each side, thus eliminating end thrust on the shaft.

The faces of the buckets are surfaces of double curvature more or less ellipsoidal in shape and are smooth-ground. The buckets are made of bronze or steel. The height and width of the bucket should each be 2.5 to 4 times the jet diameter; otherwise bucket efficiency will suffer. The exact proportions depend on the ratio of wheel diameter to jet diameter.

Figure 15.4 Runner at Big Creek-2A plant of Southern California Edison Co. Static head = 2,418 ft, net head = 2,200 ft, n = 250 rpm, pitch diameter = 162 in. (This was one of the original runners for 50-cycle generation. All have been replaced by 300-rpm runners for 60-cycle generation.)

Figure 15.5 Integral case runner. Specific speed $n_s = 5.6$.

15.8 ACTION OF JET

In Fig. 15.6 it is seen that a bucket initially intercepts the jet at a and, as the bucket moves toward b, a growing portion of the jet is intercepted. When the tip of the bucket reaches b, the entire jet will be intercepted unless the succeeding bucket has moved into the path of the jet. At first the direction of u, the bucket velocity, is downward relative to the direction of the jet; when the bucket splitter is directly below the axis of rotation (position c), u and V_1 will be in the same straight line, and subsequently the direction of u will be upward relative to the direction of the jet. In other words, u varies in direction during the time when the water is entering the bucket, and the angle α varies from α' to α''. Hence, to construct a single triangle for the entrance velocities, it is necessary to use an angle α_1, which is an average value. The average value of this angle will decrease as the wheel speed increases from zero. The velocity triangle shown in Fig. 15.6 is for average entrance relations when the wheel is running at the proper speed for maximum efficiency.

Under normal operating conditions all water that issues from the nozzle will act upon the buckets, for whatever water does not act on the first bucket will act on the second bucket and so on. The photograph of Fig. 15.7 shows that the water is acting upon several buckets at the same time. Although the amount of water per unit time striking a single moving bucket is G', as in Sec. 6.6, the total

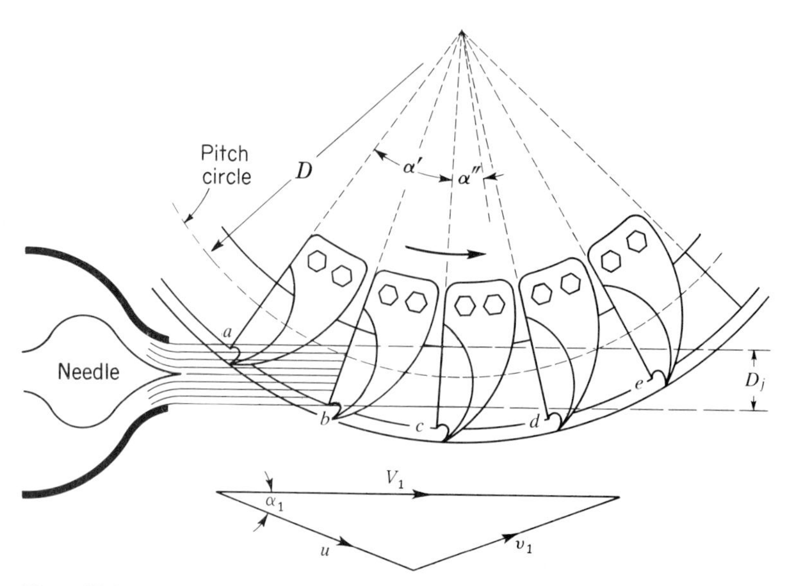

Figure 15.6

Figure 15.7 Instantaneous photograph of a 12-in Pelton wheel at 1,125 rpm, $\phi = 0.8$. (*Courtesy of California Institute of Technology, Mechanical Engineering Laboratory.*)

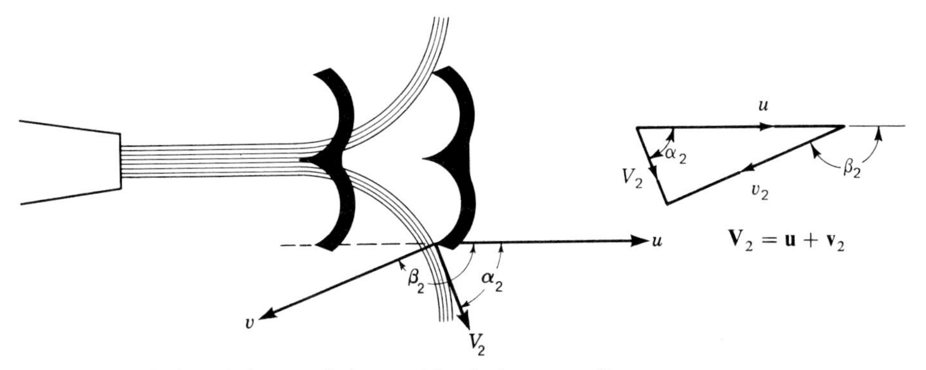

Figure 15.8 Velocity relations at discharge with velocity vector diagram.

discharge from the nozzle G acts upon the wheel as a unit, because, unlike the single object, the wheel as a whole is not moving away from the nozzle. However, if the wheel runs at a speed that is in excess of the proper speed, some of the water may go straight through without ever catching up with a bucket before the latter swings up above the line of action.

The velocity relations at discharge, as shown in Fig. 15.8, must also represent average values. The view here shown is in a plane at right angles to that of Fig. 15.6 and represents the case where the bucket is directly below the axis of rotation and the bucket velocity u and the jet velocity V_1 are in the same straight line.

Inspection of Fig. 15.6 shows that, when the bucket is entering the jet at a, the radius of the wheel at which the water enters is greater than that at which it leaves, whereas, when the bucket has traveled to points d or e where it is moving upward, the water is leaving the bucket at a maximum radius. Hence initially, r_1 is greater than r_2, and later it is just the reverse. As an average it may be assumed that $r_1 = r_2$, which is a close approximation to reality.

15.9 WHEEL DIAMETER

The diameter D that is used in equations and calculations for the impulse wheel is the diameter of the pitch circle. This, as shown in Fig. 15.6, is the circle to which the center line of the jet is tangent.

The ratio of the pitch diameter to the diameter of the jet may be very large, and there is no theoretical upper limit. An extreme case in practice is that of a Pelton wheel which was direct-connected to a reciprocating air compressor. Because of the high head, the bucket velocity was 138.5 fps, but the compressor required a rotative speed as low as 80 rpm. Thus the diameter of the wheel was 33 ft. The exact size of the jet is not known to the authors, but since the buckets were only 5 by 7 in, the diameter of the jet could not have been very far from 1.5 in. Thus $D/D_j = 264$. In Valais, Switzerland, are some Pelton wheels 11.67 ft in diameter with jets 1.5-in diameter, giving a ratio of 93.4.

On the other hand, there are limiting minimum values of this ratio, which depends in part upon the spacing of the buckets. For good efficiency with bolted buckets a ratio of 12 is very good; this means that the diameter of the wheel in feet equals the diameter of the jet in inches. However, with some sacrifice in efficiency, a value as low as 9 seems to be the practical lower limit for bolted buckets, but a ratio as low as 6 may be used for the integral type of Fig. 15.5.

15.10 TORQUE AND POWER FOR IDEAL CASES

Case 1

The ideal case first considered will be obtained by assuming $\alpha_1 = 0°$ and $\beta_2 = 180°$, although such values are physically impossible in a real wheel. However, these assumptions permit a convenient analysis, for all velocities are in the same straight line, and it is not necessary to solve velocity triangles. Assume, further, that $r_1 = r_2 = D/2$, that there is no fluid friction in flow over the buckets, and that all the water discharged by the nozzle acts upon the buckets. For these ideal conditions (Fig. 15.9) the simple relations are $v_1 = V_1 - u$, $v_2 = v_1$, $V_2 = u - v_2 = 2u - V_1$, and

$$\Delta V = V_2 - V_1 = 2(u - V_1)$$

The tangential force acting on the water is $\rho Q(\Delta V)$; hence for this ideal case the tangential force acting on the wheel is

$$F_u = \rho Q(-\Delta V) = 2\rho Q(V_1 - u) \tag{15.4}$$

and the torque acting on the wheel is

$$T = F_u \times D/2 = \rho QD(V_1 - u) \tag{15.5}$$

This is the equation of a straight line as in Fig. 15.10, and it shows that the torque is a maximum at zero speed and decreases to zero when $u = V_1$. At the point of zero torque, $\phi = C_v$ since $u = \phi\sqrt{2gh}$ (Sec. 14.4) and $V_1 = C_v\sqrt{2gh}$.

Instead of plotting either revolutions per minute or peripheral velocity, the dimensionless factor ϕ is used to express the speed, since the numerical values of this factor will apply with only slight variations for any size of wheel under any head. Dimensionless factors might also be used in Fig. 15.10 for torque, but it is more convenient to reduce these values for the 24-in wheel to their values for 1 ft

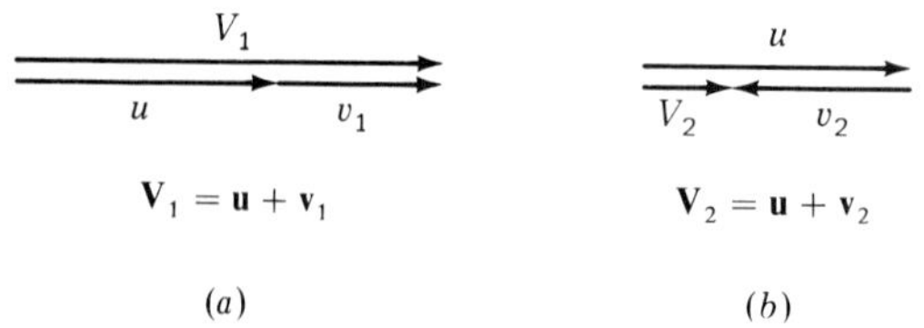

V_1 / u / v_1

$\mathbf{V}_1 = \mathbf{u} + \mathbf{v}_1$

(a)

u / V_2 / v_2

$\mathbf{V}_2 = \mathbf{u} + \mathbf{v}_2$

(b)

Figure 15.9 Velocity vector diagrams for idealized situation (Case 1). (a) At entrance to bucket. (b) At exit from bucket.

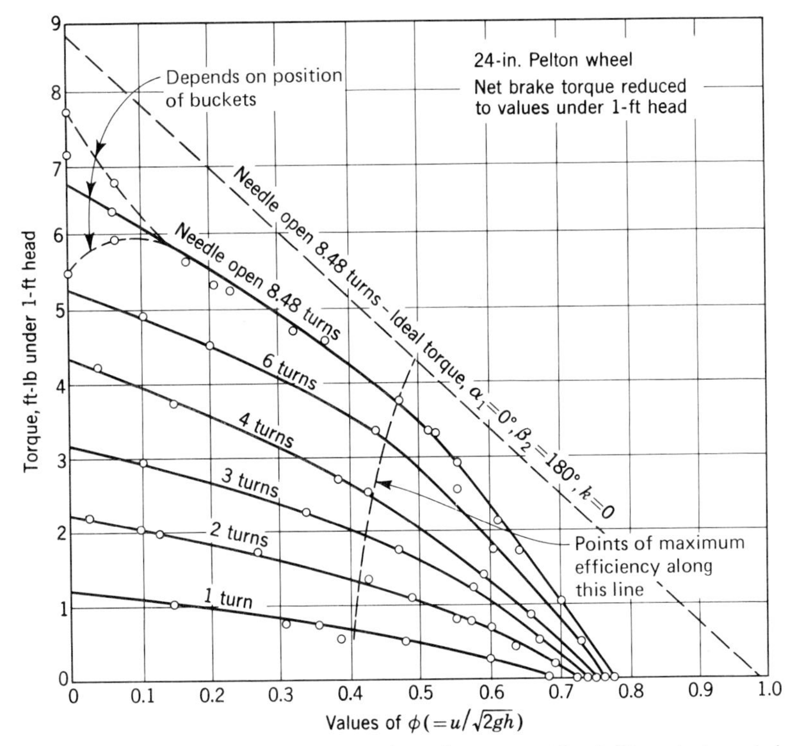

Figure 15.10 Relation between torque and speed at constant head. (From tests made by F. G. Switzer and R. L. Daugherty.)

head. The similarity laws (Sec. 14.2) then enable values to be computed from these for any size of homologous wheel under any head.

An equation for power is obtained by multiplying F_u by u; this is the equation of a parabola, such as shown in Fig. 15.11. The power will be zero when the wheel speed is zero and also when the torque is zero. In this ideal case the only loss is that due to the kinetic energy at discharge from the buckets. As $V_2 = 2u - V_1$ for this case where $\beta_2 = 180°$, it is seen that the kinetic energy at discharge is zero when $u = V_1/2$. Hence, for this speed ratio, i.e., when $\phi = C_v/2$, the power of the wheel is a maximum for this idealized case and is equal to the power of the jet.

Inasmuch as the power input to the wheel is independent of the wheel speed and is therefore constant, the efficiency curve has the same shape as the power curve.

Case 2

A closer approach to reality may be realized by using the factor k to take account of friction loss in flow through the buckets, as in Eq. (15.2), and also considering the true bucket angle β_2, which is usually about 165°. In this case

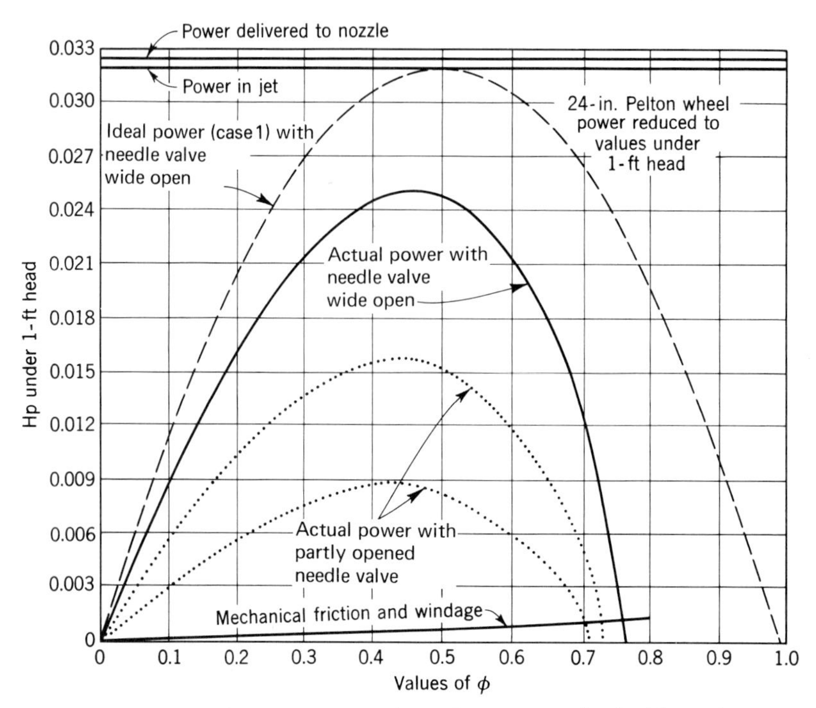

Figure 15.11 Relation between power and speed at constant head with maximum nozzle opening. (From tests made by F. G. Switzer and R. L. Daugherty.)

we are still assuming $\alpha_1 = 0°$. Using the velocities of the water relative to the buckets, the energy equation can be expressed as $v_1^2/2g - kv_2^2/2g = v_2^2/2g$. Hence, $v_2 = v_1/\sqrt{1+k}$. From Fig. 15.12 we observe that $V_2 \cos \alpha_2 = u + v_2 \cos \beta_2 = u + v_1 \cos \beta_2/\sqrt{1+k}$, from which, noting that $v_1 = V_1 - u$ and $\Delta V = V_2 \cos \alpha_2 - V_1$, we get for the tangential force acting on the buckets,

$$F_u = \rho Q\left(1 - \frac{\cos \beta_2}{\sqrt{1+k}}\right)(V_1 - u) \tag{15.6}$$

which is still the equation of a straight line if k is considered as constant. This approach is still somewhat idealized since α_1 is taken as zero along with other simplifying assumptions. However, the value of the torque at all speeds is lower than that given by application of Eq. (15.5) and thus Eq. (15.6) when multiplied by $D/2$ gives a torque nearer to the true value.

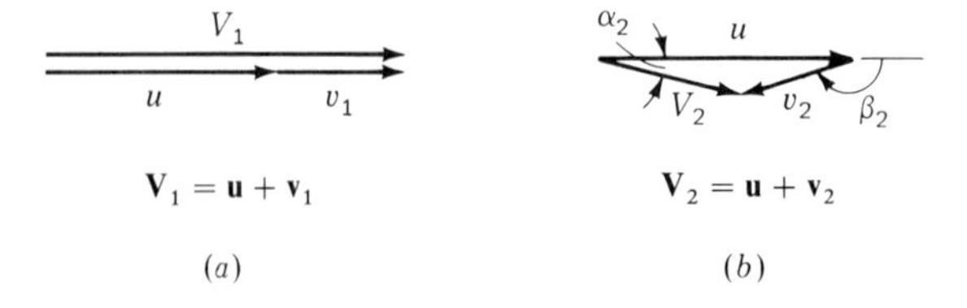

Figure 15.12 Velocity vector diagrams for the idealized *real* situation (Case 2). (*a*) At entrance to bucket. (*b*) At exit from bucket.

For this case, where $k \neq 0$ and neglecting mechanical friction and windage, the power and efficiency curves would still be parabolas, with their maximum values at $u = V_1/2$ or $\phi = C_v/2$, but both these maximum values will be lower than in case 1, because fluid friction in flow over the buckets is now being considered.

Although α_1 is not zero, as here assumed, an approximate expression for hydraulic efficiency may be obtained by multiplying Eq. (15.6) by u and dividing by γQh. If V_1 and u are replaced by $C_v\sqrt{2gh}$ and $\phi\sqrt{2gh}$, respectively, the result is

$$\eta_h = 2\left(1 - \frac{\cos \beta_2}{\sqrt{1 + k}}\right)(C_v - \phi)\phi \tag{15.7}$$

The special significance of this equation is that it shows that the hydraulic efficiency is independent of the head and depends only upon dimensionless quantities.

15.11 ACTUAL TORQUE AND POWER[1]

In Fig. 15.10 are shown the brake-torque curves for a Pelton wheel with different nozzle openings, the maximum being when the needle, the position of which is controlled by a screw thread, has been opened 8.48 revolutions from its closed position. Referring to Fig. 15.6 and the accompanying discussion, it is obvious that, when the wheel is prevented from rotating, the torque exerted by the jet upon it will vary within certain limits, depending upon the position of the bucket or buckets upon which the jet is acting. This variation is shown for the wide-open nozzle only in Fig. 15.10, but it exists for all the other nozzle openings. When the wheel was permitted to run at a slow speed, the observed fluctuation was less but was still very definite, as shown for $\phi = 0.07$, but when the wheel was run at higher speeds, the brake torque became constant. It should be made clear that the torque exerted by the fluid on the wheel fluctuates somewhat at all speeds, but what is recorded and plotted in Fig. 15.10 is the torque measured by a brake. The inertia of the rotating parts at the higher speeds produces an almost uniform torque output.

The difference between the ideal torque shown in Fig. 15.10 and the actual brake torque for the wide-open nozzle is due in part to the effect of k and β_2 in case 2 of Sec. 15.10. Also, an important factor is the value of the average angle α_1, especially at very low values of ϕ. As the wheel speed increases, the average value of α_1 decreases, but is probably never zero.

[1] The power of a turbine may be defined in three ways: the *rated power* is that guaranteed by the manufacturer; the *maximum power* is usually a little more than the rated power; and the *normal power* is that for maximum efficiency. Turbines are usually operated close to normal power.

It is observed that the torque curves of Fig. 15.10 are very nearly straight lines for the lower speeds, but at higher speeds the torque decreases more rapidly and becomes zero at a maximum peripheral velocity much less than V_1, that is, for ϕ much less than 1.0. This is due to two factors, one of them being that at speeds above the design value some of the water fails to complete its action upon the buckets or even to overtake them at all, as explained in Sec. 15.8. Another factor is that at these higher speeds the back of the bucket hits the water that has previously been intercepted by it and throws it up around the case. Thus the back of the bucket is doing work upon some of the water. In addition, there is some normal mechanical friction and windage as shown in Fig. 15.11. Hence, for all these reasons, the actual brake-torque curve differs from the purely ideal.

There are some very practical observations to be made from an inspection of the curves of Figs. 15.10 and 15.11. While the value of ϕ for maximum efficiency is not $C_v/2$, as for the ideal cases, it is only a very little less than that, usually 0.43 to 0.47. Also, it is seen that ϕ for maximum efficiency decreases slightly for smaller nozzle openings. It is also seen that the torque at zero speed is nearly twice the torque at *normal speed*, the latter being the speed at which the efficiency is a maximum. But the important feature is that the maximum speed possible, or the runaway speed, is only about 70 percent more than the normal speed. If the wheel and the generator are designed to withstand a speed only that much above normal, no damage can result if some failure of the governor permits the wheel to run away.

From the torque-speed relations of Fig. 15.10 the power-speed curves can be plotted; that for the maximum nozzle opening is shown as a solid line in Fig. 15.11. Power curves for smaller nozzle openings have lower values, and their maximum points are found at lower values of ϕ, as indicated by the dotted lines in Fig. 15.11. Because the actual brake-torque curve is not a straight line, the actual power curve is not a parabola, the right-hand side of the curve being steeper than the left-hand side.

The power consumed in mechanical friction in the bearings and in windage was determined by a separate test and is shown in Fig. 15.11. For this particular wheel at normal speed the mechanical friction and windage are about 1.5 percent of the maximum brake horsepower. Under normal operating conditions the volumetric efficiency η_v of an impulse wheel is traditionally assumed to be 1.0.

For a constant nozzle opening and a constant head, the tangential force or the torque varies as ΔV_u, which is the tangential component of the vector change of absolute velocity. As shown by the photographs in Fig. 15.13 and the velocity diagrams below each, the value of ΔV_u is a maximum when the wheel is at rest and is a minimum at runaway speed. The velocity diagrams show also that the value of V_2 decreases as the wheel speed increases from zero until it reaches some minimum value and then increases again. Hence, at some intermediate speed, the kinetic energy lost at discharge from the buckets is a minimum. There is no simple or exact theory to determine this point, but as a close approximation, the discharge velocity will be close to minimum when $u = v_2$, and consequently α_2 will then be only a little less than 90°. However, the efficiency is a maximum

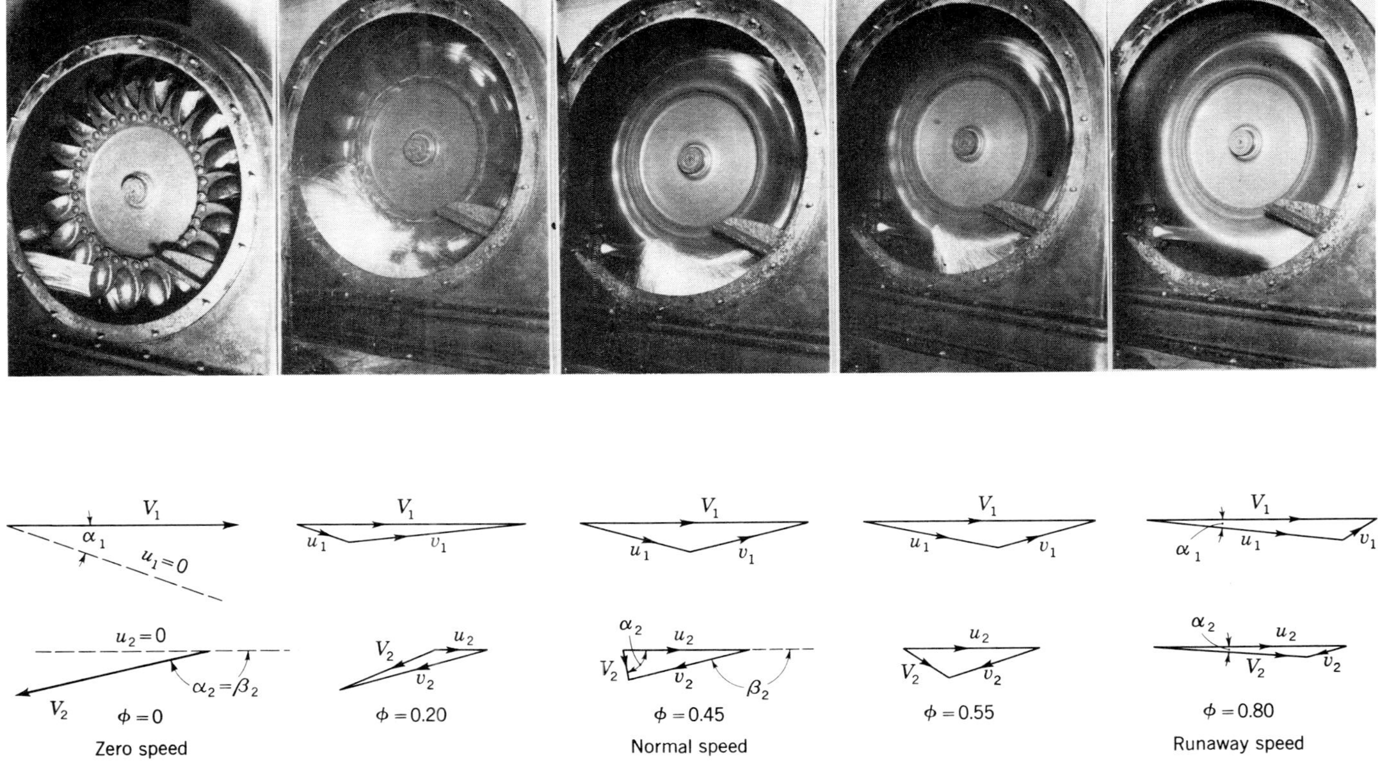

Figure 15.13 A 42-in Pelton wheel at different speeds under the same head. (Photographs at Cornell University by R. L. Daugherty.)

when the sum of *all* losses is a minimum. Because the relative velocity of flow over the buckets decreases with increasing wheel speed, the fluid-friction losses in the buckets decrease with speed. But in Fig. 15.11 it is seen that mechanical-friction and windage losses increase with speed. The summation of these three losses reaches a minimum at a speed slightly greater than the one for which the discharge loss is a minimum. Hence the maximum overall efficiency is usually not far from the speed where $\alpha_2 = 90°$.

Illustrative Example 15.2 Find the torque and power transferred to the buckets of an impulse wheel with $\alpha_1 = 0°$, $\beta_2 = 160°$, $k = 0.44$, $\phi = 0.46$, $C_v = 0.98$, a jet diameter of 10 in, and a pitch diameter of 10 ft. Find also, the hydraulic efficiency, and, expressed as a percentage of the total head, find: the head loss in bucket friction, the energy head loss at discharge, and the head loss in the nozzle.

For operation under the purely artificial value of 1 ft net head, $u = \phi\sqrt{2g} = 8.02\phi = 3.68$ fps, $V_1 = 0.98\sqrt{2g} = 7.86$ fps, and

$$\gamma Q = \gamma A V = 62.4 \times 0.545 \times 7.86 = 267 \text{ lb/s}$$

Hence, multiplying Eq. (15.6) by r, the expression for the torque exerted on the wheel by the water is

$$T = r \times F_u = 5\left(\frac{267}{32.2}\right)\left(1 + \frac{0.940}{1.2}\right)8.02(C_v - \phi)$$

With $C_v = 0.98$ and $\phi = 0.46$, $T = 309$ ft·lb. The power transferred from the water to the buckets is

$$F_u u = T\omega = T\frac{u}{r} = 309\left(\frac{0.46 \times 8.02}{5}\right) = 228 \text{ ft·lb/s}$$

The power input is $\gamma Q h = 267 \times 1 = 267$ ft·lb/s. The hydraulic efficiency is $\frac{228}{267} = 0.85$. This could have been determined directly from Eq. (15.7). Note that the answers given here are for $h = 1$ ft. For other values of h, the flow rate, torque, and power can be found by adjusting the values for $h = 1$ ft according to the similarity laws discussed in Sec. 14.2.

Under 1 ft head, $v_1 = V_1 - u = 7.86 - 3.68 = 4.18$ fps and $v_2 = v_1/\sqrt{1 + k} = 4.18/\sqrt{1.44} = 3.48$ fps. Hence the head loss in bucket friction is $0.44(3.48)^2/2g = 0.083$ ft, or 8.3 percent.

$$V_2 \cos\alpha_2 = u + v_2 \cos\beta_2 = 3.68 + 3.48 \cos 160° = 3.68 - 3.27 = 0.41 \text{ fps}$$

$$V_2 \sin\alpha_2 = v_2 \sin\beta_2 = 3.48 \sin 160° = 1.19 \text{ fps}$$

Hence $\cot\alpha_2 = 0.41/1.19 = 0.344$ or $\alpha_2 = 71°$, and $V_2 = 1.19/0.945 = 1.26$ fps from which the energy head loss at discharge is $1.26^2/2g = 0.025$ ft, or 2.5 percent. The head loss in the nozzle is approximately $(1/C_v^2 - 1)V_1^2/2g$, about 4.0 percent; so the total hydraulic loss is $8.3 + 2.5 + 4.0 \approx 14.8$ percent, which gives a close check on the computed hydraulic efficiency of 85 percent.

15.12 EFFICIENCY VARIATION WITH SIZE AND HEAD

Ideally, the efficiency of homologous impulse wheels is independent of size and head. However, the largest of a series of homologous wheels will have a slightly higher efficiency because mechanical-friction and windage losses do not increase at the same rate as the hydraulic properties. For the same reason a given wheel of any size will generally operate at a slightly higher efficiency at higher heads.

However, there is a factor that may operate to decrease the efficiency with an increase in head. The absolute velocity of the water discharged from the buckets will increase as $\sqrt{h}$, and the rebound velocity from the vertical sides of the setting (Fig. 15.1) will vary in proportion. If this rebound velocity is high enough, some of the water reflected from the walls will hit the wheel. This will result in an increased loss of the windage type. Hence, for high-head installations, the width of the chamber in which the lower part of the wheel resides should be made great enough to prevent this.

15.13 OPERATION AT CONSTANT SPEED

A turbine is usually operated at a constant rotative speed, and this is necessarily some synchronous speed if it drives an ac generator. In the United States, 60-cycle current is most common,[1] and under such conditions the rotative speed of the turbine in revolutions per minute is given by $n = 7{,}200/N$, where N is the number of poles in the generator and must be an even integer. Most generators have from 12 to 96 poles.

The foregoing discussion shows that there is one speed ratio that is most efficient for any one nozzle opening but that this optimum speed ratio varies slightly with the nozzle opening, as shown in Fig. 15.10. Also, the static, or gross, head varies with changing level in the reservoir at intake and the net head varies not only because of this but with the flow through the pipe. Thus, at the San Francisquito Plant 1 of the City of Los Angeles, the head on the plant varies from 940 ft static to 830 ft at full load. Not only will ϕ vary with wheel speed, but even if the speed of rotation is constant, it will vary if the head changes. Hence the maximum efficiency of a turbine at some constant speed might not be the maximum efficiency of which the turbine is capable.

In Fig. 15.14 is shown an efficiency curve for an impulse turbine at a constant rotative speed with varying load. If this speed were at some other value, but for the same head, the curve would be different. The figure shows that for an impulse wheel the efficiency curve is relatively flat over a wide load range. Thus

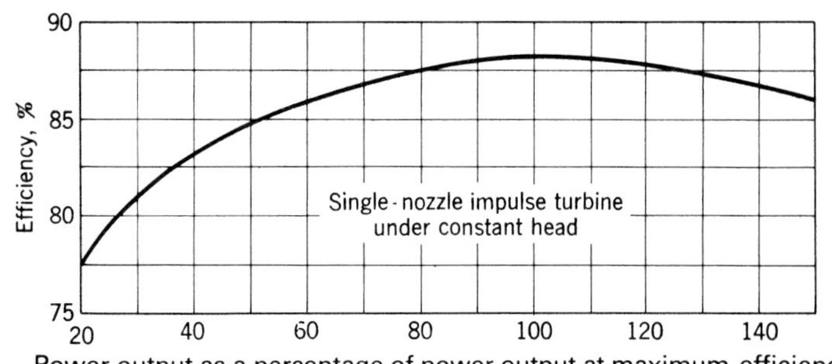

Figure 15.14 Variation of efficiency with power output of an impulse turbine at constant speed.

[1] In some places 50-cycle current is used, in which case $n = 6{,}000/N$.

this type of turbine is desirable for a variable load, especially if the load is light over long periods of time.

15.14 SPECIFIC SPEED

In Sec. 14.5 the expression for the specific speed n_s of a turbine was derived. In English units, it is expressed as

$$n_s = \frac{n_e\sqrt{\text{bhp}}}{h^{5/4}} \tag{15.8}$$

where n_e is defined as the rotative speed in revolutions per minute at operating conditions of highest possible efficiency, bhp is the brake horsepower delivered to the shaft of the turbine under these conditions, and h is the net operating head.[1] For multijet impulse turbines the specific speed is based on the brake horsepower per jet.

It is instructive to see what factors determine the numerical value of the specific speed for the Pelton wheel. Thus (from Sec. 14.4)

$$n = \frac{60u}{\pi D} = 60\phi\frac{\sqrt{2gh}}{\pi D}$$

$Q = V_1(\pi D_j^2/4)$, where D_j is the jet diameter, and bhp $= \gamma Qh\eta/550$. Making these various substitutions and reducing,

$$n_s = \frac{60(2g)^{3/4}\phi_e\sqrt{\gamma C_v \eta}}{2\sqrt{\pi 550}}\frac{D_j}{D} \tag{15.9}$$

As the factors ϕ_e, C_v, and η should vary only slightly, it is seen that the value of the specific speed depends principally upon the ratio of the wheel to the jet diameter. It was mentioned in Sec. 15.9 that there is no definite physical limit to the maximum value of the ratio D/D_j, but a large value results in a relatively large and therefore more costly wheel, and the efficiency suffers because of the proportionally large bearing friction and windage loss.

However, if the ratio D/D_j becomes too small, the size of the buckets becomes unreasonable in relation to the wheel diameter; it is physically impossible to space the buckets close enough together; and some of the water from the nozzle will not act upon the wheel at all, as explained in Sec. 15.8. But even before this point is reached, the efficiency of the wheel will suffer because of the increased departure from the tangential action, which is the ideal.

Thus, as seen in Fig. 15.15, the impulse wheel attains its best efficiency at a normal specific speed of about 4.5, and the upper limit for a single nozzle wheel is about 6 or 7, for beyond that value the efficiency drops too much. It should be emphasized that these values are for single-nozzle wheels only. If two or more

[1] Sometimes the term *rated* specific speed is used. This refers to a guaranteed bhp that can be achieved; it is greater than at normal operating conditions but is at a lower efficiency.

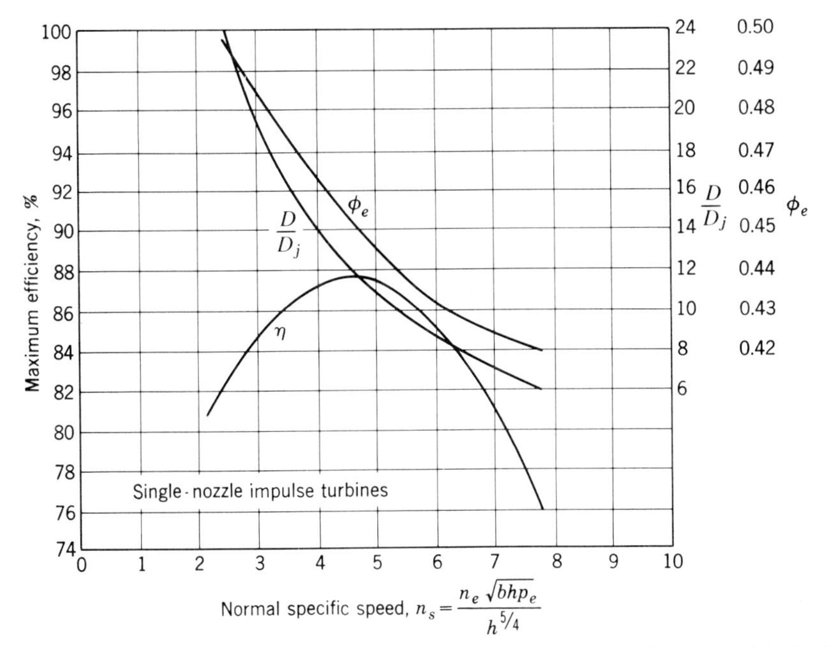

Figure 15.15 Plotted from test values given by R. S. Quick, Problems Encountered in the Design and Operation of Impulse Turbines, *Trans. ASME*, vol. 62, no. 1, Jan. 1940.

nozzles are used on one wheel, the specific-speed values here mentioned are to be multiplied by the square root of the number of nozzles.

The efficiency curve shown in Fig. 15.15 should be understood to indicate a trend rather than absolute values in all cases. Wheels of poor design or small size, operating under low head or other unfavorable conditions, will have efficiencies lower than indicated by this single curve.

Illustrative Example 15.3 (*a*) A turbine is to operate at 400 rpm under a net head of 1,320 ft. If a single 6-in-diameter water jet is used, find the specific speed of this machine assuming $C_v = 0.98$, $\phi = 0.45$, and $\eta = 0.85$. Find also the required pitch diameter of the wheel.

$$V = C_v\sqrt{2gh} = 0.98 \times \sqrt{64.4 \times 1{,}320} = 286 \text{ fps}$$

$$Q = AV = 0.196 \times 286 = 56.0 \text{ cfs}$$

$$\text{bhp} = \eta\frac{\gamma Q h}{550} = 0.85\left(\frac{62.4 \times 56 \times 1{,}320}{550}\right) = 7{,}130$$

$$n_s = \frac{n_e\sqrt{\text{bhp}}}{h^{5/4}} = \frac{400\sqrt{7{,}130}}{(1{,}320)^{5/4}} = 4.24$$

$$u = \phi\sqrt{2gh} = 0.45 \times \sqrt{64.4 \times 1{,}320} = 131.2 \text{ fps}$$

$$n = 400 \text{ rpm} = \frac{u(60)}{\pi D} = \frac{131.2 \times 60}{\pi D}$$

$$D = 6.26 \text{ ft} = 75.2 \text{ in}$$

(*b*) In lieu of the single impulse wheel of the preceding example, suppose that three identical single-nozzle wheels are to be used, operating under the same head of 1,320 ft. The total flow rate is to be 56.0 cfs. Determine the required specific speed of these turbines, their pitch diameter, the jet diameter, and the operating speed. Once again, assume $C_v = 0.98$, $\phi = 0.45$, and $\eta = 0.85$.

$$Q = \tfrac{56}{3} = 18.7 \text{ cfs}$$

$$\text{bhp} = \frac{7{,}130}{3} = 2{,}376$$

$$n_s = \frac{n_e\sqrt{2{,}376}}{(1{,}320^{5/4})} = n_e \times 0.00613$$

$$n = \frac{131.2(60)}{\pi D} = \frac{2{,}500}{D}$$

From the two preceding expressions it is apparent that the required n_s depends on the operating speed, as does D. Hence there are a number of possible answers. If we let the operating speed be 400 rpm (18-pole generator for 60-cycle electricity),

$$n_s = 400 \times 0.00613 = 2.45$$

$$D = \frac{2{,}500}{n} = 6.26 \text{ ft} = 75.2 \text{ in}$$

Thus three $n_s = 2.45$ wheels of pitch diameter 75 in operating at 400 rpm would suffice. These wheels would have relatively small buckets. At such a low specific speed the optimum efficiency of impulse wheels is usually less than 0.85. An alternative solution, for example, would be to use an operating speed of 600 rpm (12-pole generator for 60-cycle electricity). For this case.

$$n_s = 600 \times 0.00613 = 3.68$$

$$D = \frac{2{,}500}{n} = 4.18 \text{ ft} = 50.1 \text{ in}$$

Thus three $n_s = 3.68$ wheels, of pitch diameter 50.1 in, operating at 600 rpm would suffice.

Let us now determine the required jet diameter for these 50.1-in-diameter wheels operating at 600 rpm.

$$Q = AV = \frac{\pi D_j^2}{4} 286 = 18.7 \text{ cfs}$$

from which

$$D_j = 0.288 \text{ ft} = 3.46 \text{ in}$$

As a rough check, refer to Fig. 15.15, which shows that $D/D_j = 15$ for $n_s = 3.68$. This confirms the preceding calculations where $D/D_j = 50.1/3.46 = 14.5$.

Another solution would be to employ two identical single-nozzle turbines with $n_s = 4.5$ and pitch diameter 50 inches operating at 600 rpm. In this case a nozzle diameter of 4.24 in would be required.

15.15 INSTALLATIONS

The impulse wheel is especially adapted for use under high heads. For heads above 1,500 ft it is the only type that can be considered, and for many years it was the only one ever used for heads above 800 ft. Among some interesting installations are the following.

At Dixence, Switzerland, a double-overhung unit operates under a gross head of 5,735 ft and a net head of 5,330 ft, giving a jet velocity of approximately 580 fps. Each single-nozzle wheel delivers 25,000 hp, or 50,000 hp for the unit, and runs at 500 rpm. The jets are 3.71 in in diameter, and the pitch diameter of the wheels is 10.89 ft.

At Reisseck, Austria, an impulse wheel delivers 31,000 hp at 750 rpm under a gross head of 5,800 ft.

The Fully plant in Valais, Switzerland, operates under a gross head of 5,410 ft and a net head of 4,830 ft. There are four wheels of 3,000 hp each running at 500 rpm. The jets are 1.5 in in diameter, and the wheels are 11.67 ft in diameter.

A multiple-nozzle wheel at Pragnières, France, delivers 100,000 hp at 428 rpm under a gross head of 3,920 ft.

In North America there are a number of plants operating under heads of between 2,000 and 2,500 ft.

At the Big Creek-2A plant of the Southern California Edison Co. is an Allis–Chalmers double-overhung single-jet impulse turbine 137 in in diameter with an 8.4-in jet on each wheel and a similar Pelton Water Wheel Co. impulse turbine 135.25 in in diameter with an 8.5-in jet on each wheel. Each unit runs at 300 rpm, and the maximum static head with a full reservoir is 2,418 ft. With both units carrying full load, the net head is 2,200 ft, and the maximum output is 65,100 bhp for the Allis–Chalmers unit and 67,300 bhp for the Pelton unit. At maximum efficiency under 2,200 ft head, the former delivers 50,000 bhp, and the latter 31,000 bhp.

The Bucks Creek plant of the Feather River Power Co. has a gross head of 2,562 ft and a net head of 2,350 ft. The double-overhung unit has a capacity of 35,000 hp and runs at 450 rpm.

At the Balch plant of the former San Joaquin Light and Power Co., now the Pacific Gas and Electric Co., is an Allis–Chalmers double-overhung unit with wheels 115 in in diameter with a 7.5-in jet on each wheel. Its runs at 360 rpm and develops 49,000 hp under a net head of 2,243 ft.

The Kitimat plant in British Columbia of the Aluminum Company of Canada has three four-nozzle vertical-shaft units that operate at 327 rpm under a net head of 2,500 ft. Each unit is rated at 140,000 hp.

One of the highest head plants in South America is at Ros Molles in Chile where an impulse wheel develops 11,500 hp at 1,000 rpm under a head of 3,460 ft. At Cementos El Cairo in Colombia are two single-overhung impulse turbines operating under a head of 1,900 ft. Each unit develops 3,000 hp at 900 rpm.

One of the largest jets in the world is 14 in in diameter from a nozzle with an orifice diameter of 16.75 in. This is in an Allis-Chalmers double-overhung unit in the San Francisquito Plant 1 of the City of Los Angeles. Each wheel is 176 in in diameter, and the speed is 171.6 rpm. The maximum static head on the plant is 940 ft, but when the entire plant is running, the net head is 830 ft. In a test of this one unit under a net head of 870 ft, each nozzle discharged 250 cfs and the unit delivered 40,100 hp with an efficiency of 81 percent. The maximum efficiency was obtained at approximately 20,000 hp and was 86 percent.

PROBLEMS

15.1 Repeat Illustrative Example 15.1 for the case where the length of the pipe is 10,000 ft. All other data to remain the same.

15.2 Repeat Illustrative Example 15.1 for the case where the pipe diameter is 12 in. All other data to remain the same.

15.3 A 36-in pipeline ($f = 0.020$) of length 10,000 ft connects a reservoir whose water surface elevation is 1,800 ft to a nozzle at elevation of 1,000 ft. The jet from the nozzle is used to drive an impulse turbine. If the head loss through the nozzle is expressible as $0.04V_j^2/2g$, determine the jet diameter that will give a jet of maximum horsepower. Evaluate this horsepower.

15.4 A series of vanes is acted on by a 3-in water jet having a velocity of 100 fps, $\alpha_1 = \beta_1 = 0°$. Find the required blade angle β_2 in order that the force acting on the vanes in the direction of the jet is 200 lb. Neglect friction. Solve for vane velocities of 85, 50, and 0 fps.

15.5 (*a*) Repeat Prob. 15.4 but include friction by assuming $v_2 = 0.9v_1$. (*b*) Take the case of $u = 50$ fps and compute k using each of the following equations:

$$\gamma Q \frac{V_1^2}{2g} = Fu + \gamma Q \frac{V_2^2}{2g} + \gamma Q \frac{kv_2^2}{2g} \tag{1}$$

$$\frac{v_1^2}{2g} - k\frac{v_2^2}{2g} = \frac{v_2^2}{2g} \tag{2}$$

How do the values of k compare? Briefly describe what the two equations represent.

15.6 Under a net head of 1 ft the Pelton wheel for Fig. 15.11 discharges 0.286 cfs at full nozzle opening and the maximum power is 0.025 bhp for a value of $\phi = 0.465$. The corresponding brake torque is 3.69 ft·lb as shown in Fig. 15.10. Assuming that the similarity laws apply precisely, determine the discharge, torque, power, and rotative speed of this wheel when it operates under a head of 1,600 ft.

15.7 A wheel and nozzle similar to that of Figs. 15.10 and 15.11 with a pitch diameter of 12 ft is used under a net of 1,600 ft. What is the torque, power, and rpm at point of best efficiency for full nozzle opening?

15.8 At the bucket speed of 200 fps in Prob. 15.12, what is the value of the head lost in friction in the buckets, and what is the head lost in the kinetic energy at discharge?

15.9 Using the data of Prob. 15.12, find values of (*a*) head lost in bucket friction; (*b*) velocity head at discharge; (*c*) total head lost, for bucket speeds of 180, 185, 190, 195, 200, 205 fps. (This is best solved by tabulation.) At what bucket speed is the discharge loss a minimum? At what bucket speed is the total head loss a minimum?

15.10 A small impulse wheel 2.5 ft in diameter is driven by the jet of Illustrative Example 15.1 ($\alpha_1 = 0°, \beta_2 = 160°$). Assuming $v_2 = 0.8v_1$, compute the horsepower output and hydraulic efficiency of the turbine for $\phi = 0.2$, 0.3, 0.4, 0.5, 0.6, and 0.8 using jet diameters of 1, 2, and 3 in.

15.11 In Prob. 15.16 find the power lost in hydraulic friction in the buckets. Find the value of V_2, and determine the power carried away in the water discharged from the buckets.

15.12 (*a*) A Pelton wheel is 6 ft in diameter and is acted upon by a jet that is 6 in in diameter. The velocity of the jet is 400 fps. The bucket angle is 165°, and k may be assumed to be 0.21. Assuming ideal case 2, find the torque in foot-pounds for bucket speeds of 0, 100, 200, 300, and 400 fps. (*b*) If the velocity coefficient of the nozzle is 0.98, approximately what are the horsepower and efficiency for the same five speeds? In this case $\phi_e = 0.49$.

15.13 For the Pelton wheel of Fig. 15.10 with the nozzle wide open and $Q = 0.286$ cfs, find the torque for 1 ft head for zero speed ($\phi = 0$) and for $\phi = 0.49$, assuming the ideal case 2 with $C_v = 0.98$, $\beta_2 = 165°$, and $k = 0.7$.

15.14 Solve Prob. 15.13 for a wheel that is 8 ft in diameter and with the jet also four times the diameter of that in Fig. 15.10.

15.15 Find the torque, power, efficiency, and rotative speed of the impulse wheel of Illustrative Example 15.2 if it were to operate under a head of 60 ft. Assume for this case $\phi = 0.49$ rather than 0.46.

15.16 A nozzle having a velocity coefficient of 0.98 discharges a jet 6 in in diameter under a head of 900 ft. As a simplifying assumption take $\alpha_1 = 0°$. The wheel diameter is 8 ft, $\beta_2 = 165°$, and it may be assumed that $k = 0.5$. The mechanical efficiency of the wheel is 97 percent. What is the hydraulic efficiency? What is the gross efficiency? Assume $\phi = 0.46$.

15.17 (*a*) A 24-in laboratory Pelton wheel was tested under a head of 65.5 ft. With the nozzle open 6 turns of the needle, the net brake load at 275 rpm was 40 lb at a brake arm of 5.25 ft and the discharge was 1.897 cfs. Find values of brake horsepower and efficiency under operating conditions. What would be the values of torque and brake horsepower under 1 ft head? (*b*) At 275 rpm the bearing-friction and windage losses were found to be 0.2 hp. What percentage is this of the brake horsepower? What is the value of the mechanical efficiency? What is the value of the hydraulic efficiency?

15.18 If the Pelton wheel of Prob. 15.17 were run at double the speed, or 550 rpm, the head should be $4 \times 65.5 = 262$ ft for the same value of ϕ. At this higher head the power input will be $4^{3/2}$, or 8 times the value found in Prob. 15.17; the hydraulic losses will also be in the same proportion, and so the hydraulic efficiency will remain unchanged. A special test showed that at 550 rpm the bearing friction and windage losses were 0.8 hp.

What would then be the power input to the shaft? What will be the brake horsepower? What percentage is 0.8 hp of the brake horsepower? What are the values of the mechanical efficiency and of the total efficiency?

15.19 Find the approximate hydraulic efficiency of an impulse wheel for which the nozzle velocity coefficient is 0.97 and the bucket angle is 160°, if $\phi = 0.46$ and $k = 0.1$.

15.20 Assume a general case for any type of impulse turbine where $x = r_2/r_1$; prove that the equation for torque is

$$T = \rho Q r_1 \left(V_1 \cos \alpha_1 - x^2 u_1 - \frac{x \cos \beta_2}{\sqrt{1+k}} \sqrt{V_1^2 + x^2 u_1^2 - 2u_1 V_1 \cos \alpha_1} \right)$$

15.21 A nozzle having a velocity coefficient of 0.98 discharges a jet 7 in in diameter under a net head of 1,600 ft. This jet acts upon a wheel with the following dimensions: $D = 7$ ft, $\alpha_1 = 15°$, $\beta_2 = 160°$, and it is assumed $k = 0.6$. Find the tangential force exerted upon the buckets when $\phi = 0.45$.

15.22 Compute the specific speed for the turbine of Fig. 15.2.

15.23 Compute the specific speed for P.G. & E.'s units at the Balch plant (see Sec. 15.15).

15.24 A double-overhung impulse-turbine installation is to develop 20,000 hp at 257 rpm under a net head of 1,120 ft. Determine n_s, wheel-pitch diameter, and approximate jet diameter. Repeat for the following cases: (*a*) single wheel with single nozzle; (*b*) single wheel with four nozzles.

15.25 A multinozzle impulse turbine is to be designed to develop 60,000 hp at 300 rpm under a head of 1,200 ft. How many nozzles should this turbine have? Specify the approximate wheel diameter for this design. How many single-jet machines ought one to use to satisfy these requirements? Specify the jet diameters in both instances assuming $C_v = 0.96$.

15.26 A test of the impulse turbine shown in Fig. 15.2 gave a maximum efficiency of 92 percent at 35,000 bhp and 89.95 percent at the rated 62,000 bhp. The net head is 1,118 ft at full load, and the speed is 300 rpm. The static head is 1,226 ft. What is the net head at the load for maximum efficiency? What are the normal and the rated values of specific speed per jet?

15.27 Refer to Illustrative Example 15.3. Suppose a two-nozzle single-wheel installation were designed to operate under a head of 1,320 ft with a total flow (for both nozzles) of 56 cfs. Determine the required specific speed of this turbine, its pitch diameter, and jet diameter for rotative speeds of 300, 400, and 600 rpm. C_v, ϕ, and η remain unchanged.

15.28 A six-jet impulse turbine operating at 300 rpm develops 60,000 hp under a net head of 1,060 ft. The runner has a diameter of 6.0 ft. How large a homologous runner would be needed for a single-jet machine operating under the same head and developing the same horsepower?

15.29 An impulse turbine ($n_s \approx 5$) develops 100,000 hp under a head of 2,000 ft. (*a*) For 60-cycle electricity calculate the turbine speed (rpm), wheel diameter (ft), and number of poles in the generator. (*b*) Solve the problem for a six-nozzle unit using the same n_s, bhp, and head. In both instances assume $\phi = 0.45$.

15.30 It is desired to develop 15,000 bhp under a head of 1,000 ft. Make any necessary assumptions, and estimate the diameter of the wheel required and the rotative speed.

15.31 A single-nozzle impulse wheel is required to develop 20,000 bhp under a net head of 1,190 ft and is to run at 225 rpm. What should be the approximate diameter of the wheel?

15.32 The pressure of the water at the base of a nozzle of a Pelton wheel is 700 psi, and the velocity at that same point is 20 fps. The jet diameter is 8 in, and the velocity coefficient of the nozzle is 0.98. (*a*) If the efficiency of the wheel is 86 percent, find the brake horsepower. (*b*) If the pitch diameter of the wheel is 8 ft, what should be the normal operating speed? (*c*) What is an approximate value for the runaway speed? (*d*) What is the torque exerted on the wheel at normal operating speed? (*e*) What is an approximate value of the torque at zero speed?

15.33 The discharges under 65.5 ft head for the nozzle openings shown in Fig. 15.10 were 0.397, 0.773, 1.114, 1.414, 1.896, and 2.315 cfs. At a constant speed of 275 rpm, or $\phi = 0.443$, the net brake scale readings at a brake arm of 5.25 ft were 6.8, 14.9, 22.0, 28.9, 40.0, and 48.0 lb, respectively. When there was no load at 275 rpm, the discharge was 0.110 cfs. Compute and plot power input and efficiency vs. brake horsepower.

15.34 When the speed of the wheel in Prob. 15.33 was raised to 300 rpm, the discharge at no load was 0.125 cfs; for the same head of 65.5 ft the net scale readings were 5.9, 12.9, 19.8, 25.5, 36.0, and 43.8 lb, respectively. Compute and plot power input and efficiency vs. brake horsepower.

CHAPTER

SIXTEEN

REACTION TURBINES

16.1 DEFINITION[1]

A reaction turbine is one in which the major portion of the pressure drop takes place in the rotating wheel. As a consequence the proportions must be such that the fluid fills all the runner passages completely. This makes it necessary that the fluid be admitted to the rotor around its entire circumference. Since the entire circumference of the reaction turbine is in action, its rotor need not be as large as that of an impulse wheel for the same power.

16.2 EVOLUTION OF THE REACTION TURBINE

The first reaction turbine known is the steam turbine of Hero in Egypt about 120 B.C. It may never even have been built, but the drawings for it are still in existence and show a spherical vessel in which steam was generated and discharged through two small nozzles in a tangential direction. The reaction of these jets would cause the device to rotate.

In the hydraulic field the rotating lawn sprinkler is an elementary reaction turbine. As stated in Sec. 6.12, the addition of more arms to permit a greater flow, so as to produce a net power output, developed a power machine known as Barker's mill. A continuing increase in the number of arms terminated in a complete wheel with passages separated by vanes, but the device was not very

[1] The reader should refer back to Chap. 14 for discussion of similarity laws, specific speed, and other characteristics of turbines.

efficient, until in 1826, a Frenchman by the name of Fourneyron added stationary guide vanes in the central portion. These guide vanes gave the water a definite tangential component, thereby imparting angular momentum to the fluid entering the rotor. This outward-flow turbine was efficient, but the mechanical construction was not good because the rotating element was on the outside with the fixed guide vanes near the axis.

The inward-flow turbine permits a better mechanical construction since the rotor and shaft form a compact unit in the center, while the stationary guide vanes are on the outside. Several crude inward-flow turbines were constructed around 1838, but the first to be well designed was built in 1849 by the eminent hydraulic engineer James B. Francis, who made an accurate test of this turbine.[1] All inward-flow reaction hydraulic turbines are known as *Francis turbines*, both in this country and in Europe, even though they have developed into very different forms from the original.

The design of the original Francis turbine is shown in Fig. 16.1. It is a purely radial-flow turbine with both entrance and discharge edges of the runner vanes parallel to the axis of rotation, so that the radii at entrance are the same for all streamlines, as they are at exit. The inner diameter of the runner was almost as large as the outer diameter.

To make a more compact runner, the inner diameter was reduced and the water was discharged with a velocity having an axial as well as a radial component, as in Fig. 16.2*a*. Carrying this a step further, the dimensions of the runner parallel to the shaft were increased, resulting in the *mixed-flow runner* of Fig. 16.2*b*, also called a *Francis runner*. In this runner all flow lines have both axial and radial components throughout. The velocity at exit near the *band*, or *shroud ring*, may even have a slight outward component at high specific speeds. Inasmuch as the different streamlines vary so much from *crown* to band, it is obvious that the application of any simple theory to this type of runner is impossible.

The specific speeds of Francis runners range from 10 to 110, but more commonly from 20 to 80. In order to obtain both speed and power under very

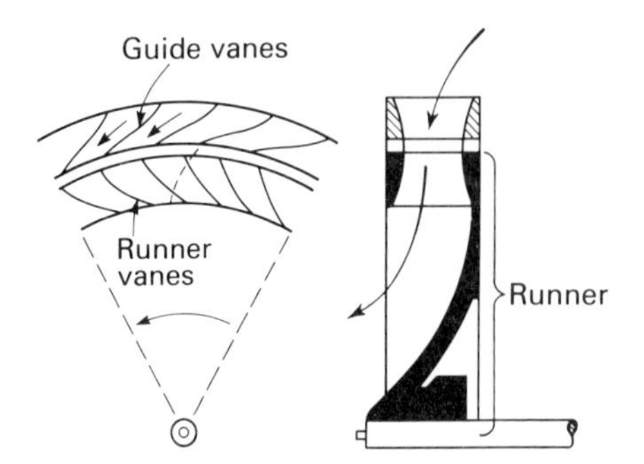

Figure 16.1 Original Francis turbine.

[1] J. B. Francis, "Lowell Hydraulic Experiments," 5th ed., D. Van Nostrand Company, Inc., Princeton, N.J., 1909.

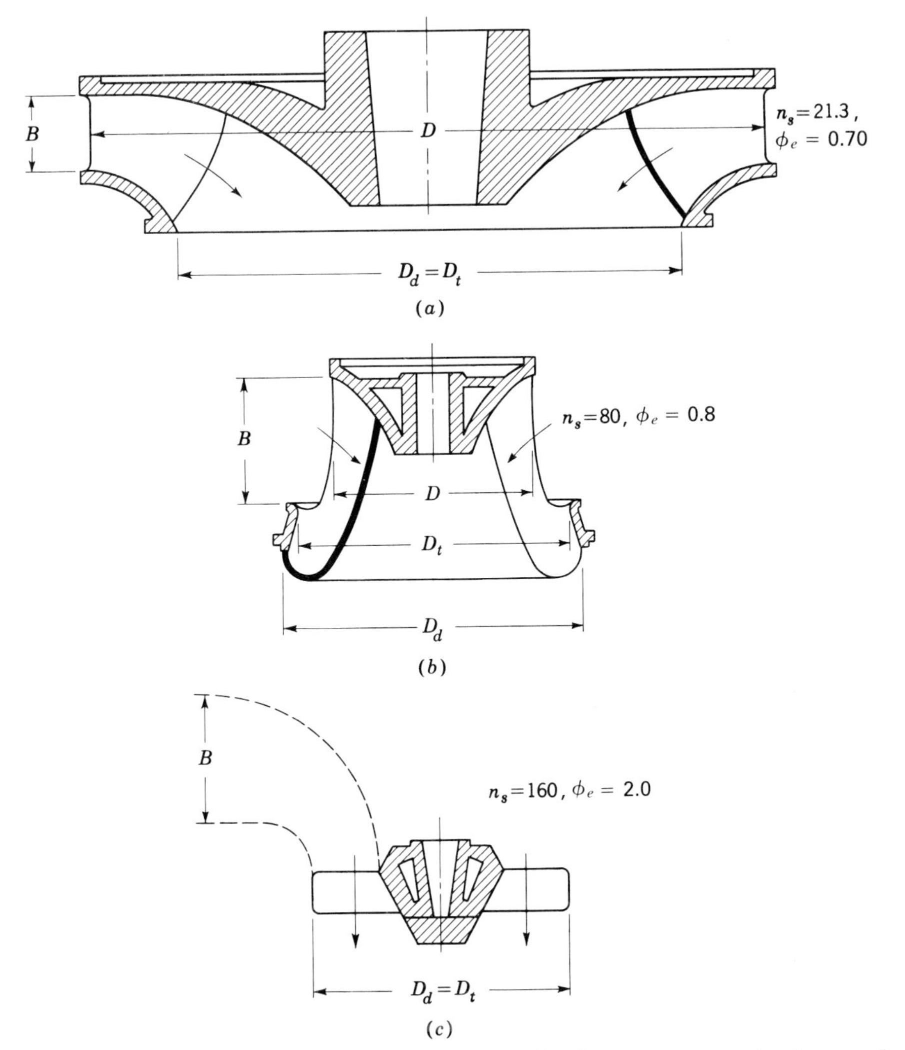

Figure 16.2 Relative sizes of different types of runner for the same power under the same head. (*a*) Radial-flow Francis. (*b*) Mixed-flow Francis. (*c*) Axial-flow (propeller).

low heads, the *axial-flow*, or *propeller*, type of runner shown in Fig. 16.2*c* is employed. With this design specific speeds are from 100 to 250.

The axial-flow, or propeller, type of runner may have fixed blades, or the pitch of the blades may be adjusted to vary the area between them in proportion to the load. This operation is usually accomplished through the actuation of a governor. This type of axial-flow turbine with adjustable blades is called a *Kaplan turbine*. It has a much higher efficiency on part load then the fixed-blade propeller.

Since 1974, because of the escalating cost of oil, there has been renewed interest in hydroelectric development, particularly small units. Reverse running pumps can be successfully used as turbines for such purposes.

16.3 SIGNIFICANCE OF SPECIFIC SPEED

Figure 16.2 shows the decrease in the size of a Francis runner for the same power under the same head as the specific speed is increased. It also shows that there is a definite relation between the profile of the runner, or the type of runner, and the value of the specific speed. Thus the numerical value of the latter at once fixes the type or design of the runner.

For a small power under a high head a Pelton wheel with its low specific speed is used to keep the rotative speed down to a suitable value, while for a large power under a low head a reaction turbine with its higher specific speed is used to raise the rotative speed to a suitable value. Since $n_e\sqrt{\text{bhp}} = n_s h^{5/4}$ (Eq. 15.8), it is apparent that for a given specific speed and a fixed head any number of combinations of rotative speed and power are possible. If the power is low, the speed can be high, or vice versa. It is also obvious that if both speed and power are high, the specific speed must be high if the head is low.

Illustrative Example 16.1 (*a*) In Sec. 15.15 an impulse wheel is described, running at 500 rpm under a head of 5,330 ft and producing 25,000 hp. The specific speed of this turbine is 1.74. Investigate the possibility of using a reaction turbine at this installation.

Suppose we use the turbine of Fig. 16.2*a*, for which $n_s = 21.3$. Applying Eq. (15.8), we find $n_e = 6{,}120$ rpm, which is impractical. Moreover, it would not be economically feasible to construct a casing designed to withstand such high pressures.

(*b*) At Keokuk on the Mississippi River are reaction turbines with runners 16 ft in diameter and 12 ft high. They run at 57.7 rpm and develop 10,000 hp under a net head of 32 ft and thus have a specific speed of 76. Investigate the possibility of using an impulse wheel at this installation.

Suppose we assume an impulse wheel with $n_s = 5$. Applying Eq. (15.8), we find $n_e = 3.8$ rpm, which is impractical. It should be noted, however, that an impulse wheel is applicable at a 32-ft head if the power to be developed is quite low. For example, with a rotative speed of 100 rpm, an impulse wheel with $n_s = 5$ will develop 14.5 hp.

(*c*) How might one produce 22,500 hp at 600 rpm under a head of 81 ft?

Applying Eq. (15.8), we find $n_s = 370$, which is impossible. Either the power may be divided among several units or a single unit may be used if the speed is reduced to a lower value. One possible solution is to use four units of 5,625 hp, each having a specific speed of 185. Another solution is to use a single unit operating at 240 rpm with a specific speed of 148. From a practical viewpoint it is advantageous to have more than one unit at any installation so that power can be developed even when a unit is shut down for repairs.

16.4 CONSTRUCTION OF REACTION TURBINES

A Francis runner is shown in Fig. 16.3, and a Kaplan runner in Fig. 16.7. Francis runners are cast in one piece unless the size is too large for shipment, in which case they are cast in sections and bolted together. The blades are warped surfaces, and sometimes a die is made and the blades are formed from sheet steel. These blades are then welded to the crown and to the shroud ring.

Axial runners with fixed blades may be cast in one piece, or again the blades may be formed by a die and then welded or bolted to the hub. Of course, Kaplan runners must be made with separate, movable blades.

Figure 16.3 Francis runner at Niagara Falls. $h = 214$ ft, $n = 107$ rpm, $\eta = 93.8$ percent at 72,500 hp, diameter = 176 in, overall diameter at band = $183\frac{3}{8}$ in. *(Courtesy of Allis-Chalmers Mfg. Co.)*

Francis runners are surrounded by pivoted guide vanes,[1] as shown in Figs. 6.10 and 16.4. The water is greatly accelerated in the guide-vane passages and given a definite tangential-velocity component as it enters the runner. The governor regulates the flow rate by rotating these vanes about their pivots so as to vary the area between them. This also has the effect of varying the angle α_1 from practically 0° up to values of 15 to 40° for maximum gate opening, depending upon the specific speed of the turbine. At maximum efficiency the value of α_1 is usually 10 to 15 degrees. The value of V_1, however, is not much affected by the change in the guide-vane angle.

The water rotates as a free vortex in the space between the ends of the guide vanes and the entrance edges of the turbine runner. Guide vanes for a propeller turbine (Fig. 16.7) are placed in the same way as for a Francis runner, but there is considerable distance between them and the propeller. In the case of the Kaplan turbine, the governor moves the guide vanes about their pivots and at the same time changes the angle of the blades by means of a mechanism in the hub.

For large turbines there is also a *stay ring*, or *speed ring*, outside the guide-vane assembly. This contains stay vanes which are fixed in position and whose primary function is to serve as columns to aid in supporting the weight of the generator above. The vanes should be so shaped as to conform to the natural streamlines, which are spiral in character. The water velocity increases in passing through this assembly because of the decreasing cross-sectional area.

[1] This type of assembly is commonly referred to as *wicket gates*.

Figure 16.4 Guide-vane assembly.

The guide-vane assembly, or the stay-vane assembly if there is one, is surrounded in turn by a spiral case (*scroll case*) such as shown in Fig. 16.5, which maintains a uniform velocity around the turbine circumference. For high heads the case is of metal, but for low heads it may be merely formed in the concrete. It is preferably circular in cross section as in Figs. 16.5 and 16.6, but if formed in concrete, it may be rectangular as in Fig. 16.7.

Figure 16.5 Scroll case. (*Courtesy of The James Leffel and Co.*)

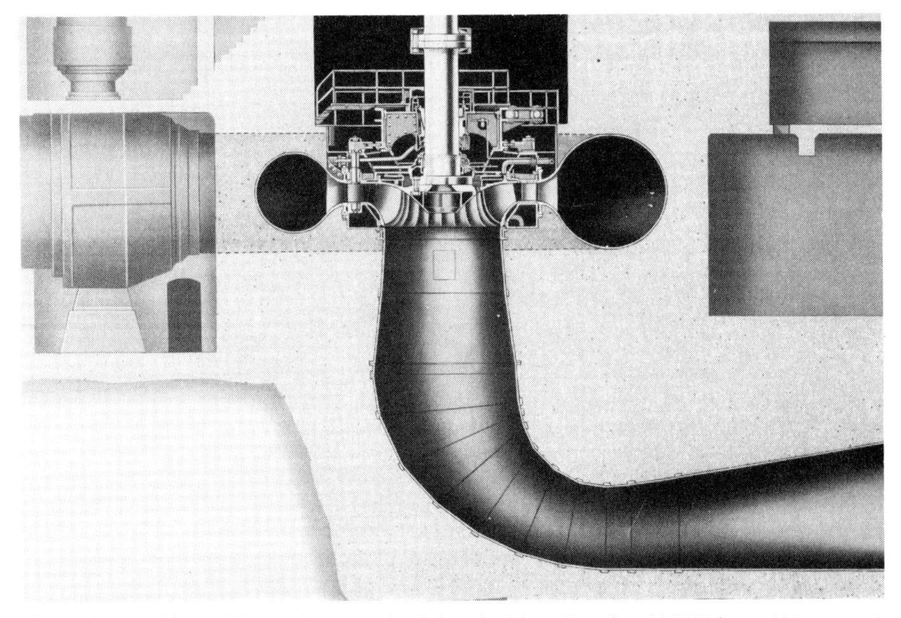

Figure 16.6 Turbine at Hoover Dam on the Colorado River. Rated at 115,000 hp at 180 rpm under a head of 480 ft. Runner diameter = 171 in. (*Courtesy of Allis-Chalmers Mfg. Co.*)

Figure 16.7 Kaplan turbine at Watts Bar Dam. 42,000 hp at 94.7 rpm under a head of 52 ft.

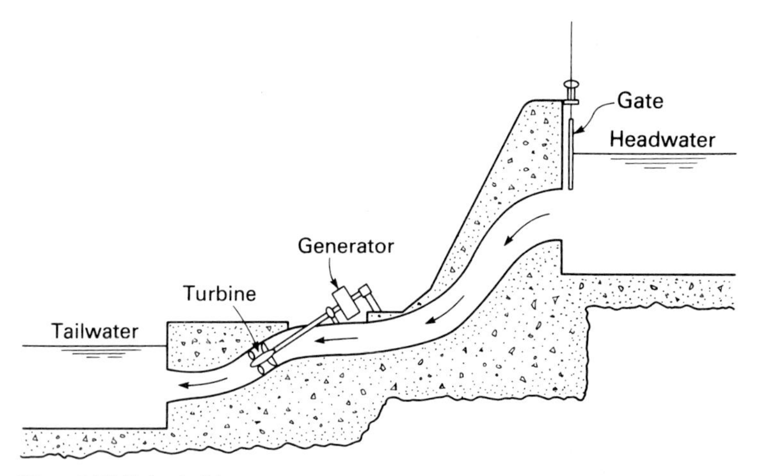

Figure 16.8 Tube turbine.

With the recent emphasis on small-scale hydroelectric power development, several manufacturers are now supplying standardized packaged units (turbine and generator) in various sizes. Previously nearly all turbines and generators were custom made. A widely used packaged unit is the *tube turbine* (Fig. 16.8). The tube turbine may have fixed or movable blades and is commonly used for heads in the 10- to 60-foot range. The tube turbine is an axial-flow hydraulic turbine with a generator located outside the passageway by extending the turbine shaft past a slight bend in the passageway. The turbines may be located upstream or downstream of the generator in any shaft position from vertical to horizontal.

16.5 SETTINGS

Some reaction turbines are set with a horizontal shaft, especially in small sizes or for very high heads. But the majority of reaction turbines are set with vertical shafts, as shown in Figs. 16.6, 16.7, 16.9 and 16.10. One advantage of the vertical setting is that the draft tube is then more efficient.

16.6 DRAFT TUBES

The draft tube is an integral part of a reaction turbine, and its design criteria should be specified by the turbine manufacturer. It has two functions. One is to enable the turbine to be set above the tailwater level without losing any head thereby. A reduced pressure is produced at the upper end of the draft tube, which compensates for the height at which the turbine runner is set. Within limits the turbine can be set at different elevations without altering the net head. By its use there is an unbroken stream of liquid from headwater to tailwater.

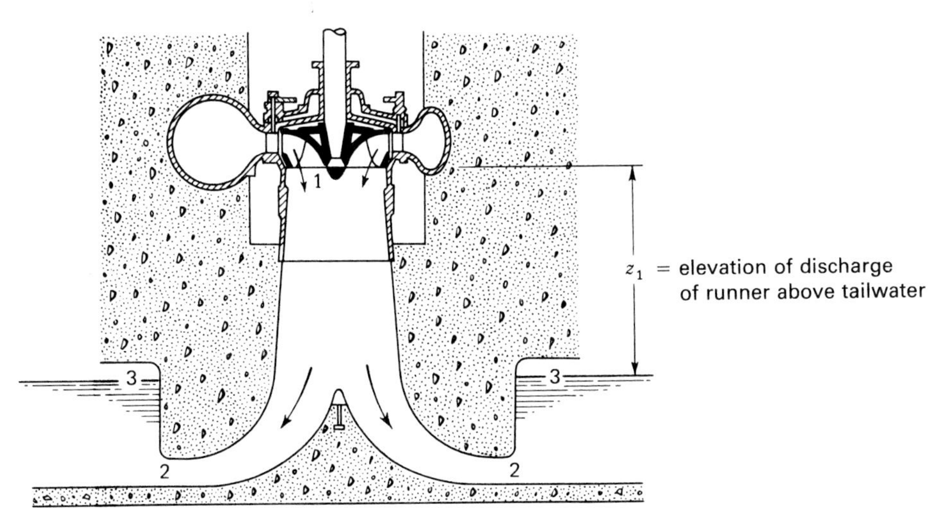

Figure 16.9 Moody spreading draft tube.

The second function of the draft tube is to reduce the head loss at submerged discharge to thereby increase the net head available to the turbine runner. This is accomplished by using a gradually diverging tube whose cross-sectional area at discharge is considerably larger than the cross-sectional area at entrance to the tube. Applying the energy equation to the draft tube of Fig. 16.9 and letting z_1 signify the elevation of the entrance (or top) of the draft tube above the surface of the water in the tailrace, the absolute pressure head at that section is given by

$$\frac{(p_1)_{\text{abs}}}{\gamma} = \frac{p_{\text{atm}}}{\gamma} - z_1 - \frac{V_1^2}{2g} + h_L + \frac{V_2^2}{2g} \tag{16.1}$$

where p_{atm} is the atmospheric pressure, h_L is the friction head loss in the diverging tube, and $V_2^2/2g$ is the submerged-discharge kinetic-energy head loss at exit from

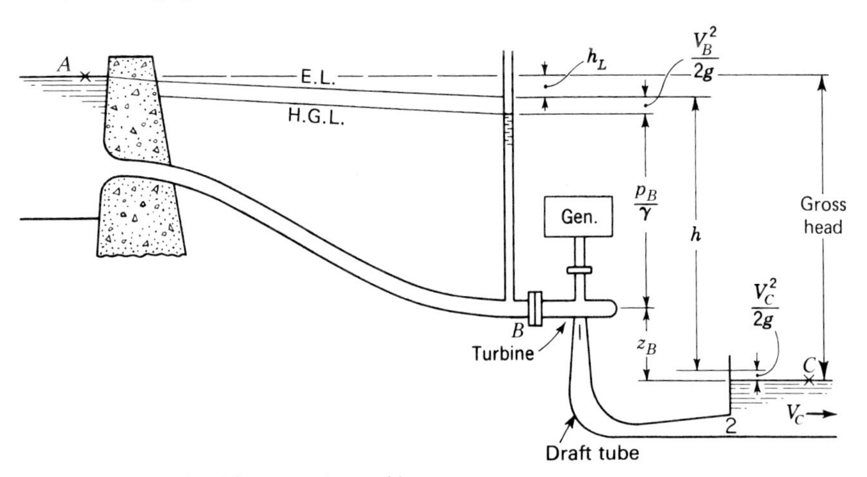

Figure 16.10 Net head h on reaction turbine.

the tube. The latter can be reduced by increasing the cross-sectional area A_2 at exit from the tube. However there is a practical upper limit to this because of tube length, since the angle of divergence of the tube must be kept reasonably small to prevent or at least minimize separation of the flow from the tube wall. The friction head loss in the tube can be estimated through application of Eq. (8.51) using the loss coefficients of Fig. 8.20. It should be noted that there is an upper limit on the tolerable value of z_1, as discussed in Sec. 16.13.

Many different designs of draft tubes have been developed to turn the water through 90° with the least loss of energy. Among them is the Moody spreading draft tube shown in Fig. 16.9. In some cases the central cone is extended up to meet the runner so as to form a solid core in the entire central portion of the tube. If the water leaves the turbine with any whirl, there will be a free vortex in the tube, and it has been shown that, as the radius of a free vortex approaches zero, the whirl component approaches infinity. Since this is physically impossible, the central core of a free vortex cannot follow the free-vortex law, and this is conducive to eddy losses, which are avoided by the solid core. In some cases, as at Hoover Dam, a little air is bled into the central section of the draft tube to provide smoother operation with less vibration, but not enough air is admitted to impair the vacuum materially.

Figures 16.6 and 16.9 show draft tubes where the principal reduction in velocity is in the vertical conical portion and the 90° turn at the bottom has been made with a flattened cross section, but one of increasing cross-sectional area. This design has been found to give good efficiency and is extensively used. There are still other forms, a simple design being a vertical diverging tube. In a few installations a nondiverging draft tube has been used.

16.7 NET HEAD

For a reaction turbine the net head h is the difference between the energy level just upstream of the turbine and that of the tailrace. Thus in Fig. 16.10 the net head on the turbine is $h = H_B - H_C$, or

$$h = \left(z_B + \frac{p_B}{\gamma} + \frac{V_B^2}{2g} \right) - \frac{V_C^2}{2g} \tag{16.2}$$

where V_C is the velocity in the tailrace. By comparing Eq. (16.2) with Eq. (15.1) it is apparent that, for the same setting, the net head on a reaction turbine will be greater than that on a Pelton wheel. The difference is of small importance in a high-head plant, but it is important for a low-head plant.

The draft tube is considered an integral part of a reaction turbine, thus the head h' that is effectively available to act on the runner of a reaction turbine is

$$h' = h - k' \frac{(V_1 - V_2)^2}{2g} + \frac{(V_2 - V_C)^2}{2g} \tag{16.3}$$

where h is the net head as defined in the foregoing and the other two terms refer to the friction head loss in the draft tube and the loss at submerged discharge from the tube. The head h'' (Sec. 14.1) that is actually extracted from the water by the runner is smaller than h' by an amount equal to the hydraulic friction losses and shock losses in the scroll case, guide vanes, and runner, as discussed in Sec. 16.11.

16.8 OPERATION AT DIFFERENT SPEEDS

The characteristics of a typical reaction turbine at various speeds under a constant-head and a constant-gate opening are shown in Fig. 16.11. These are similar to those for the impulse turbine in Figs. 15.10 and 15.11, with certain exceptions. The flow is no longer independent of the runner speed because of the unbroken flow from headwater to tailwater, and any change within the runner will affect the flow. In Sec. 6.12 it was seen that, for the rotating lawn sprinkler, the rate of flow increased with the rotative speed. This would also be true for any outward-flow reaction turbine.

For the inward-flow Francis turbine the centrifugal action decreases the flow with increasing wheel speed, as shown in Fig. 16.11. The next important difference is that the value of ϕ_e (Sec. 14.4) for maximum efficiency is not less than 0.5, as in the Pelton wheel, but greater. In this specific case it is about 0.8, but in

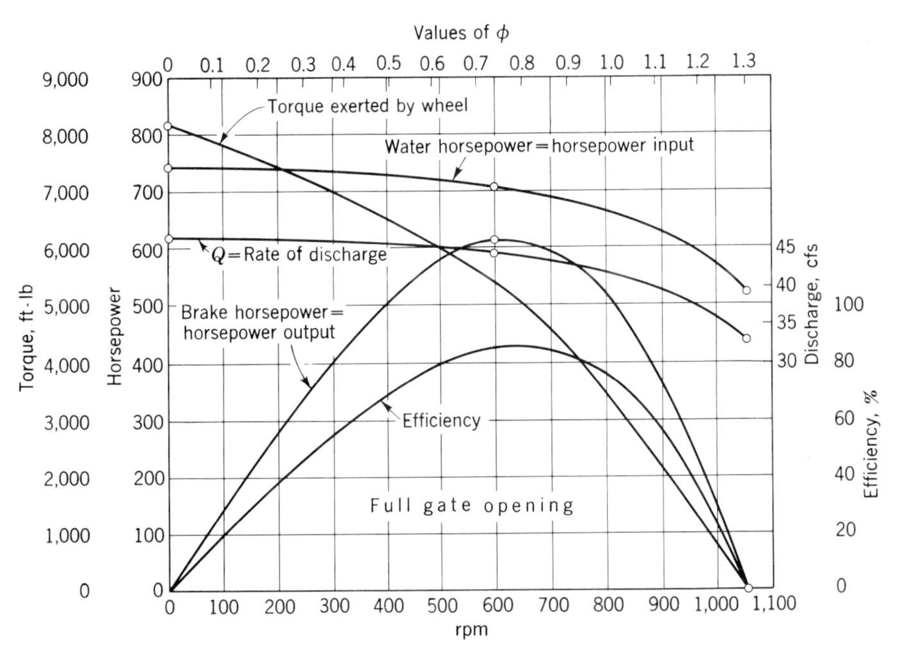

Figure 16.11 A certain Francis turbine operating at variable speed under constant head with constant gate opening. Runner diameter = 27 in, head = 140.5 ft.

general it ranges between 0.7 and 0.85 for Francis turbines and between 1.4 and 2.0 for axial-flow turbines. The maximum possible value of ϕ (Sec. 14.4) for reaction turbines is greater than 1 while for impulse wheels it is slightly less than 1. Inasmuch as the power input is not a constant, the maximum efficiency for a reaction turbine occurs at a speed slightly higher than that of maximum power output.

16.9 OPERATION AT CONSTANT SPEED

As with the impulse turbine (Sec. 15.13) the reaction turbine is also usually operated at a constant rotative speed. With 60-cycle current the rotative speed of the turbine in revolutions per minute is given by $n = 7{,}200/N$ where N is the number of poles in the generator. Constant rotative speed is achieved by varying the gate opening with load. It will here be assumed that the net head is constant, although generally it decreases slightly with increased load because the pipe friction varies approximately as Q^2. Also, the static head may change from time to time, for the level of both headwater and tailwater may vary. This is important for low-head plants, where in time of flood the tailwater level usually rises more than the headwater level. This decrease in head may cause a serious reduction in the power that can then be generated.

Figure 16.12 shows the performance of a certain reaction turbine at constant speed. The efficiency curve is not as flat as that for the impulse wheel in Fig. 15.14. Figure 16.13 shows typical efficiency curves for various types of turbines.

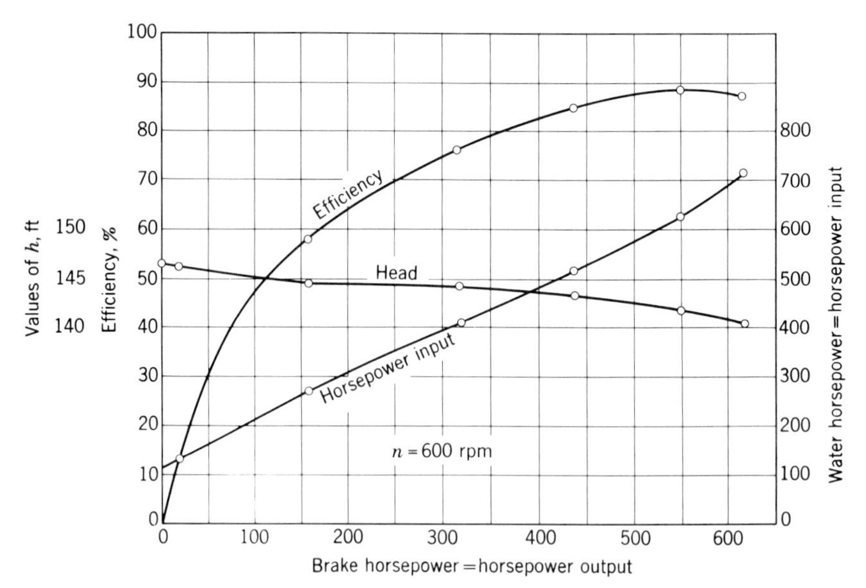

Figure 16.12 A certain Francis turbine operating at constant speed and variable gate opening. Runner diameter = 27 in.

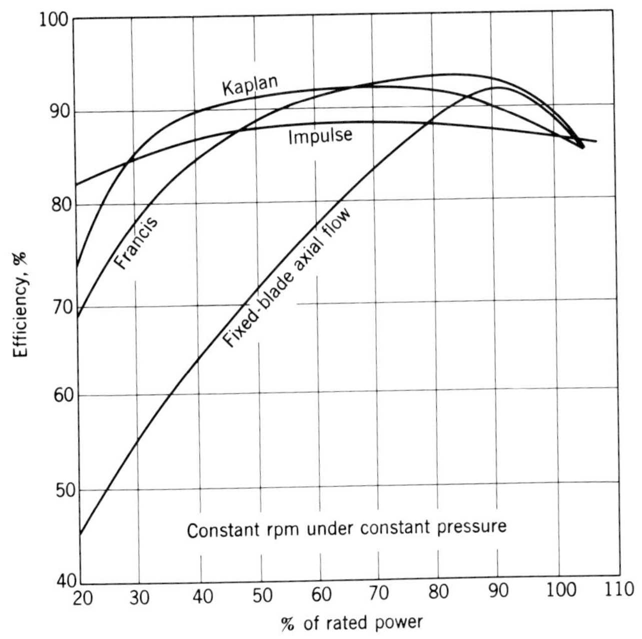

Figure 16.13 Typical efficiency curves for the various types of turbines at constant speed under constant head.

The impulse turbine and the Kaplan turbine both have very flat efficiency curves. The power at maximum efficiency for both is much lower than the rated power (see footnote, Sec. 15.11). The Francis turbine has a high maximum efficiency but a poorer part-load efficiency than either the impulse wheel or the Kaplan turbine. The fixed-blade propeller turbine has a high efficiency at the maximum point but a very low part-load efficiency. The normal power of the Francis turbine is closer to the rated power than it is with either the impulse turbine or the Kaplan turbine, while the normal power of the fixed-blade propeller turbine is very close to the rated power.

For the reaction turbine operating at constant speed the requirement that the runner passages be completely filled means that the relative velocity through the fixed areas must decrease as the load is decreased. That is, the continuity equation $Q = A_1V_1 = a_1v_1 = a_2v_2$ must be satisfied. At part load the area A_1 is decreased by the movement of the guide vanes about their pivots. This also decreases the angle α_1 (Fig. 6.10) and Q is reduced substantially. The net result is that the velocity diagrams at entrance and exit are changed. With reduced load, at entrance the flow does not enter the runner tangentially to the blades, resulting in a *shock loss*, and at exit, V_2 may increase, resulting in an increase in the kinetic energy loss at discharge from the runner. Also, an increased whirl at discharge causes the water to flow through the draft tube with spiral streamlines, which decreases the draft-tube efficiency. In addition, the quantity of water leaking past the seal rings will not be diminished even though Q is less, thus reducing the volumetric efficiency. For these reasons the efficiency of a reaction turbine tends to be less on light load than that of a Pelton wheel, although it may be more efficient at normal load.

16.10 EFFICIENCY

The trend of maximum efficiency of turbines as a function of specific speed is shown in Fig. 16.14. These are optimum values and apply to large turbines. Small turbines, no matter how well designed or constructed, cannot be expected to yield values as high as these.

One reason for the difference between large and small turbines is that of relative leakage. For a large turbine the leakage loss is very small, being of the order of 1 per cent. For a small runner the clearance distances in the seal rings cannot be reduced in proportion to other dimensions, and thus the leakage loss becomes a larger percentage value. Also, for the same average velocity, the fluid friction in flow through the small passages is greater than that through the larger passages, because of greater relative roughness and steeper velocity gradients resulting in larger shear stress at the boundary.

The effect of size on turbine efficiency η is of importance in transferring test results on small models to their prototypes. For both Francis turbines and propeller turbines this can be done by the *Moody step-up formula*, which is

$$\frac{1-\eta_1}{1-\eta_2} = \left(\frac{D_2}{D_1}\right)^{1/5} \tag{16.4}$$

This applies, of course, only to homologous machines. It has some theoretical basis and has been found in practice to give satisfactory results.

Equation (16.4) does not apply to Pelton wheels since it is assumed that their efficiencies are nearly independent of size. This is logical because they have no leakage losses to make a difference. Thus, although a large reaction turbine may

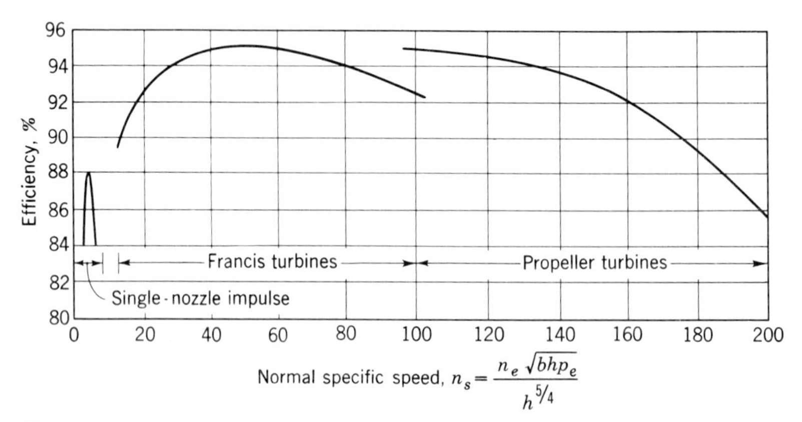

Figure 16.14 Optimum values of turbine efficiency.

be more efficient that a Pelton wheel, a small one may be less efficient. It is impossible to give an absolute value of size below which a reaction turbine would be less efficient than an impulse turbine, but a rough approximation would be that if the diameter of the reaction runner is less than about 20 in, its maximum efficiency may be less than that of a Pelton wheel.

It will be observed in Fig. 16.14 that the most favorable specific speed for a Francis turbine is around 50 and that the efficiency is lower at both extremes. A Francis runner of low specific speed will have a large diameter D and a narrow width B for a given power. Disk friction loss due to the drag exerted by the water in the spaces between the runner and the case varies approximately as D^5, and so this loss is proportionally large. Also, there is increased fluid friction in the long and narrow runner passages characteristic of the low-specific-speed Francis runner. At higher specific speeds these effects diminish in importance and the efficiency increases. But at very high specific speeds the fluid friction is greater because of the higher relative velocity through the runner. Also, the kinetic energy lost at discharge is greater. Therefore the maximum efficiency decreases, as n_s exceeds about 50; the propeller turbine becomes more desirable at specific speeds above 90.

16.11 THEORY[1]

The primary object in this chapter is to explain the operating characteristics of reaction turbines and to point out some features in their design. An understanding of these is of value to many engineers. However, very few engineers will ever have occasion to design the detail of a runner vane, and that topic will not be discussed here.

The energy losses in a reaction turbine may be very simply described as the so-called *shock loss* at entrance to the runner if the relative velocity of the water leaving the guide vanes is abruptly changed in either magnitude or direction or both when it enters the runner; fluid friction in the casing, through the guide-vane passages and through the runner passages; kinetic energy loss due to the absolute velocity head of the water at discharge from the runner, of which up to 80 percent might be regained in the most efficient draft tube; and mechanical friction of the bearings and stuffing boxes, as well as disk friction. All these losses vary in different ways, and it is not possible to have all of them a minimum at the same point.

In order to avoid shock loss at entrance, it is necessary that the runner-vane angle, which will be designated by β'_1 and which is fixed by construction, should be the same as β_1, determined by the velocity triangle. The latter will vary with

[1] It is suggested that at this point the reader review Sec. 6.9 giving particular attention to Fig. 6.10.

the operating conditions. The relations as determined by the velocity triangle (Fig. 6.10) are

$$V_1 \sin \alpha_1 = v_1 \sin \beta_1$$

$$V_1 \cos \alpha_1 = u_1 + v_1 \cos \beta_1$$

Eliminating v_1 between these two equations,

$$u_1 = \frac{\sin (\beta_1 - \alpha_1)}{\sin \beta_1} V_1$$

and if β_1 be assigned a fixed value of β'_1, this is the relation between u_1 and V_1 for which there is no shock loss. If, on the other hand, u_1 and V_1 are given, the equation may be put into a more convenient form as

$$\cot \beta_1 = \frac{V_1 \cos \alpha_1 - u_1}{V_1 \sin \alpha_1}$$

which, upon letting $\beta_1 = \beta'_1$, determines the value of the vane angle β'_1 for no shock loss. Employing $V_1 = C_1\sqrt{2gh}$ and $u_1 = \phi\sqrt{2gh}$, the above equations can be expressed in dimensionless forms as

$$\phi = \frac{\sin (\beta_1 - \alpha_1)}{\sin \beta_1} C_1 \tag{16.5}$$

and

$$\cot \beta_1 = \frac{C_1 \cos \alpha_1 - \phi}{C_1 \sin \alpha_1} \tag{16.6}$$

Values of the runner vane angle β'_1 smaller than 90° have been found to cause cavitation at the inlet to give poor efficiency; so the angle is generally made 90° or more. A value of β'_1 in the vicinity of 95 to 100° is fairly common.

The hydraulic efficiency of turbines is $\eta_h = h''/h$. From Eq. (6.22) we have $h'' = (u_1 V_1 \cos \alpha_1 - u_2 V_2 \cos \alpha_2)/g$. Defining $V_1 = C_1\sqrt{2gh}$, $V_2 = C_2\sqrt{2gh}$, and $u_1 = \phi\sqrt{2gh}$, we get for the hydraulic efficiency of a turbine,

$$\eta_h = \frac{h''}{h} = \frac{u_1 V_1 \cos \alpha_1 - u_2 V_2 \cos \alpha_2}{gh} = 2\phi\left(C_1 \cos \alpha_1 - \frac{r_2}{r_1} C_2 \cos \alpha_2\right) \tag{16.7}$$

For maximum efficiency, α_2 will be close to 90°, for then the value of V_2, and hence the loss of kinetic energy at discharge from the runner, will be a minimum. Experimental evidence indicates that α_2 for maximum efficiency varies from 85° for low-specific-speed Francis turbines to about 75° for high-specific-speed ones. As a simplifying assumption for this discussion let us assume $\alpha_2 = 90°$. Equation (16.7) then becomes

$$\eta_h \approx 2\phi_e C_1 \cos \alpha_1 \tag{16.8}$$

From this equation it is seen that ϕ_e and C_1 are inversely proportional to each other. For the Pelton wheel, ϕ is a little less than 0.5 and $C_1 = C_v$, the nozzle velocity coefficient, is nearly 1. For the reaction turbine, where less than half of the net head is converted into the kinetic energy leaving the guide vanes and entering the runner, the value of C_1 must be of the order of 0.6, and therefore ϕ_e is correspondingly high. As the specific speed increases, values of C_1 decrease and those of ϕ_e increase. As previously mentioned, values of ϕ_e for Francis turbines range from 0.70 to 0.85 and for axial-flow turbines from 1.4 to 2.0 while ϕ_e for impulse turbines generally lie between 0.43 and 0.48.

16.12 TURBINE PROPORTIONS AND FACTORS

Figures 16.2 and 16.15 show the nomenclature used for reaction turbines. In a high-specific-speed Francis runner the diameter at the entrance varies from crown to band and the nominal diameter D is taken at the mid-height of the guide vanes. For our purpose much use will be made of this dimension. For the designer, however, an important dimension is the *throat diameter* D_t, which is less than D for low-specific-speed Francis runners and greater than D for high-specific-speed Francis runners. The diameter D_d of the upper end of the draft tube is slightly less than D_t for low-specific-speed Francis runners and slightly greater than D_t for high-specific-speed Francis runners. For axial-flow runners $D = D_d = D_t$ (Fig. 16.2*c*).

Values of certain runner proportions and factors are shown in Fig. 16.16. All quantities which may vary with either load or speed are here shown as the values for the point of maximum efficiency. The exact form of a runner profile and the values of these factors will vary from one manufacturer to the next and are developed as a result of experience. However, the values shown are representative and do show the trend of each as a function of specific speed.

The theory as presented in this chapter represents the capacity of a runner as a function of D, but the quantity which determines capacity is the discharge area of the runner. This area is determined on the drafting board and is carefully checked in manufacture.

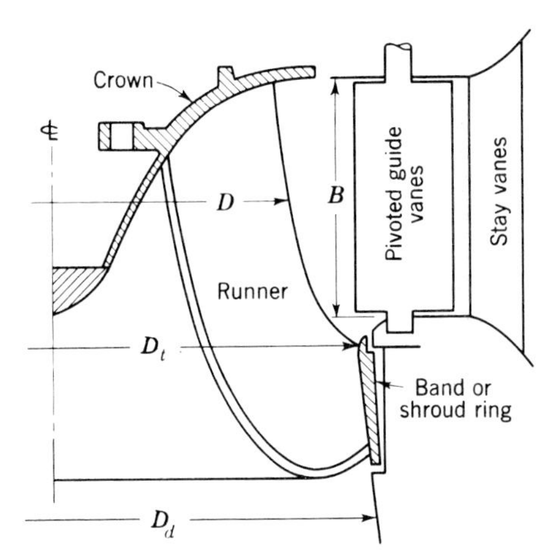

Figure 16.15 Section through a Francis turbine showing guide vanes and runner.

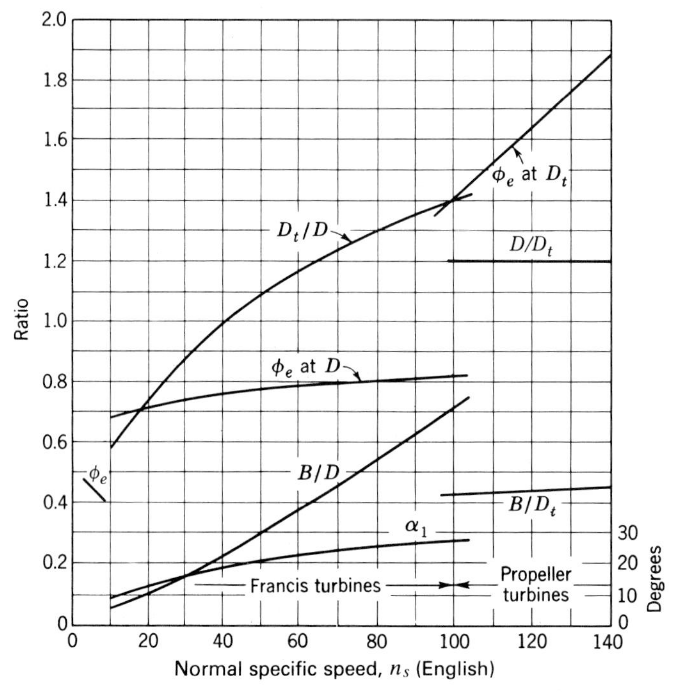

Figure 16.16 Characteristics of turbines as a function of specific speed.

The circumferential area of the runner is

$$A_c = 0.95\pi BD = 0.95\pi\left(\frac{B}{D}\right)D^2$$

where 0.95 is a factor arbitrarily inserted to correct for the area occupied by the runner vanes. In a specific design the precise value could be obtained. The ratio B/D is a function of specific speed (Fig. 16.16).

The radial component of the velocity at entrance is $V_r = C_r\sqrt{2gh}$ and $Q = V_r A_c$. Since $\text{bhp} = \eta\gamma Qh/550$ and also $\text{bhp} = n_s^2 h^{5/2}/n^2$, where $n = 153.2\phi\sqrt{h}/D$ from Eq. (14.22), it is possible to employ all these relations to obtain

$$C_r = \frac{n_s^2}{63{,}800\phi_e^2(B/D)\eta} \tag{16.9}$$

if γ is assumed as 62.4 lb/ft^3. (For any other specific weight γ', multiply 63,800 by γ'/γ.) Then, by employing values of ϕ_e and B/D from Fig. 16.16 a value of ηC_r may be obtained for any specific speed. It is seen that in any given case the value of C_r depends upon the turbine efficiency, which is proper, because, for a given power for a given specific speed, the quantity Q required will depend upon the turbine efficiency. The value of the efficiency η is not necessarily that shown in Fig. 16.14, for those in the figure are merely typical values.

Also, $V_r = V_1 \sin \alpha_1$, and it is therefore clear that, as C_r and V_r depend upon the turbine efficiency, so also must α_1 depend upon turbine efficiency. That is, in order to provide a greater flow for a turbine of lower efficiency, the gates must be opened to a larger angle to provide a larger area for the flow. Hence the values of α_1 shown in Fig. 16.16 are only approximations.

The foregoing discussion and equations apply to the Francis turbine. For the axial-flow turbine, D is replaced by D_t and the value of ϕ_e becomes that for D_t, the maximum diameter of the turbine blade. If the diameter of the guide-vane-tip circle is D_o, Eq. (16.9) is replaced by

$$C_r = \frac{n_s^2 D_t^2}{63{,}800\phi_e^2 B D_o \eta} \tag{16.10}$$

Often D_o is slightly greater than D_t, but in many cases the two are practically identical. If the latter is the case, then Eq. (16.10) reduces to the same form as Eq. (16.9).

Illustrative Example 16.2 (*a*) Find typical values for the blade angles α_1 and β_1' of a Francis turbine having a specific speed $n_s = 20$ and an hydraulic efficiency $\eta_h = 0.94$.

From Figs. 16.14 and 16.16, $\eta = 0.925$, $B/D = 0.10$ and $\phi_e = 0.72$. Assume $\alpha_2 = 90°$. Then, from Eq. (16.9)

$$C_r = \frac{20^2}{63{,}800(0.72)^2(0.10)(0.925)} = 0.1307$$

From Eq. (16.8), $C_1 \cos \alpha_1 = \eta_h/2\phi_e$. Thus

$$C_1 \cos \alpha_1 = \frac{0.94}{2 \times 0.72} = 0.653$$

Since $\quad V_r = V_1 \sin \alpha_1 = (C_1\sqrt{2gh}) \sin \alpha_1$ and $V_r = C_r\sqrt{2gh}$,

it follows that $\quad C_1 \sin \alpha_1 = C_r$

Hence $$\tan \alpha_1 = \frac{C_1 \sin \alpha_1}{C_1 \cos \alpha_1} = \frac{C_r}{C_1 \cos \alpha_1} = \frac{0.1307}{0.653} = 0.200$$

from which $\alpha_1 = 11°20'$. Hence $C_1 = C_r/\sin \alpha_1 = 0.1307/0.1964 = 0.666$

and $$\cot \beta_1 = \frac{C_1 \cos \alpha_1 - \phi}{C_1 \sin \alpha_1} = \frac{0.653 - 0.72}{0.1307} = -0.514$$

from which $\beta_1 = 117°13'$. This should also be the blade angle β_1'.

(*b*) If the turbine of (*a*) above is to be used to develop 3,600 bhp under a head of 402 ft at 600 rpm, determine approximate values of B and D and of V_r and Q.

Eq. (15.8): $$n_s = \frac{600\sqrt{3{,}600}}{(402)^{5/4}} = 20 \quad \text{as in } (a)$$

Eq. (14.22): $$D = \frac{153.2\phi_e\sqrt{h}}{n} = \frac{153.2 \times 0.72 \times \sqrt{402}}{600} = 3.69 \text{ ft}$$

From Fig. 16.16, $B = 0.10 \times 3.69 = 0.369$ ft, and $D_t = 0.735 \times 3.69 = 2.71$ ft. Thus

$$A_c = 0.95\pi DB = 4.06 \text{ ft}^2.$$

$V_r = C_r\sqrt{2gh} = 0.1307 \times \sqrt{64.4 \times 402} = 21.0$ ft/s, and $Q = AV = 4.06 \times 21.0 = 85.4$ cfs.

16.13 CAVITATION IN TURBINES

Cavitation (Sec. 4.8) is undesirable because it results in pitting (Fig. 16.17), mechanical vibration, and loss of efficiency. In reaction turbines, the most likely place for the occurrence of cavitation is on the back sides of the runner blades near their trailing edges. Cavitation may be avoided by designing, installing, and operating a turbine in such a manner that at no point will the local absolute pressure drop to the vapor pressure of the water. The most critical factor in the installation of reaction turbines is the vertical distance from the runner to the tailwater (*draft head*).

In comparing the cavitation characteristics of turbines it is convenient to define a *cavitation parameter* σ as

$$\sigma = \frac{p_{\text{atm}}/\gamma - p_v/\gamma - z_B}{h} \tag{16.11}$$

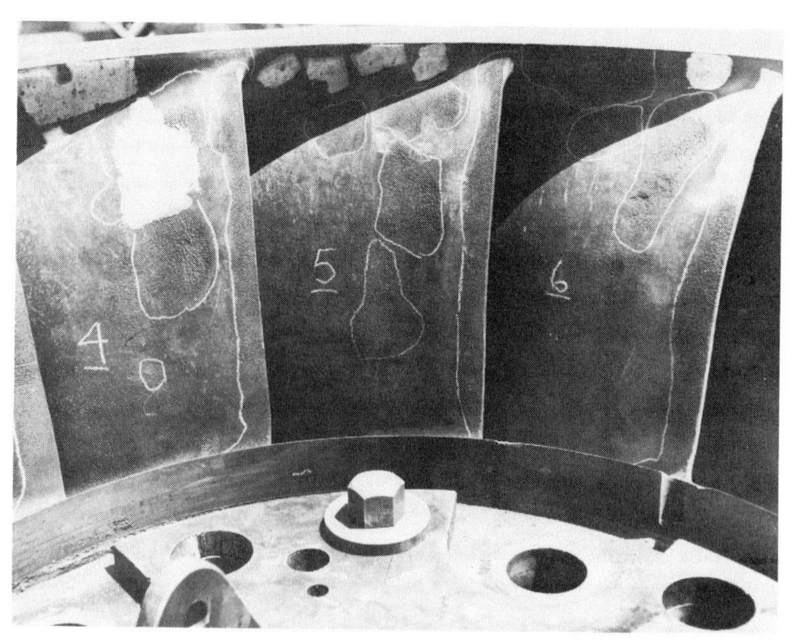

Figure 16.17 Cavitation pitting of Francis wheel and scroll case at Mammoth Pool Powerhouse after $2\frac{1}{2}$ yr of operation. Conditions of service were relatively severe. Turbine rating: 88,000 hp at effective head of 950 ft; operating speed 360 rpm. The bright shiny areas are stainless-steel welds that have withstood cavitation pitting for over a year. (*Courtesy of Southern California Edison Co.*)

where z_B and h are as defined in Fig. 16.10. The term $p_{atm}/\gamma - p_v/\gamma$ represents the height to which water will rise in a water barometer. At sea level with 70°F water, $p_{atm}/\gamma - p_v/\gamma = 33.1$ ft. At higher elevations and at higher water temperatures it is smaller than 33.1 ft. The minimum values of σ at which cavitation occurs is σ_c. Its value can be determined experimentally for a given turbine by noting the operating conditions under which cavitation first occurs as evidenced by the presence of noise, vibration, or loss of efficiency.

From Eq. (16.11) we see that the maximum permissible elevation of the turbine runner above tailwater is given by

$$z_B = p_{atm}/\gamma - p_v/\gamma - \sigma_c h \tag{16.12}$$

Typical values of σ_c are as follows:

	Francis turbines					Propeller turbines		
n_s	20	40	60	80	100	100	150	200
σ_c	0.025	0.10	0.23	0.40	0.64	0.43	0.73	1.5

Inspection of these values shows that a turbine of high specific speed must be set much lower than one of low specific speed. In fact, for a high net head h, it might be necessary to set a high specific-speed turbine below the level of the tailwater surface (i.e., with negative draft head). This is a factor which restricts the use of propeller turbines to the low head range, which is fortunately the condition for which they are best suited in other ways also.

Figure 16.18 shows recommended limits of safe specific speed of turbines for various heads and settings based on experience at existing power plants.

Illustrative Example 16.3 Find the maximum permissible head under which a Francis turbine ($n_s = 70$) can operate if it is set 10 ft above tailwater at an elevation of 5,000 ft with water temperature at 60°F.

At 5,000 ft elevation $$\frac{p_{atm}}{\gamma} = \frac{12.2 \times 144}{62.4} = 28.2 \text{ ft} \qquad \text{(Table A.3}a\text{)}$$

At 60°F, $$\frac{p_v}{\gamma} = \frac{0.26 \times 144}{62.4} = 0.6 \text{ ft} \qquad \text{(Table A.1}a\text{)}$$

From the above table, $$\sigma_c = 0.31$$

From Eq. (16.12),

$$10 = 28.2 - 0.6 - 0.31(h)$$

$h = 57$ ft (maximum permissible head to assure against cavitation)

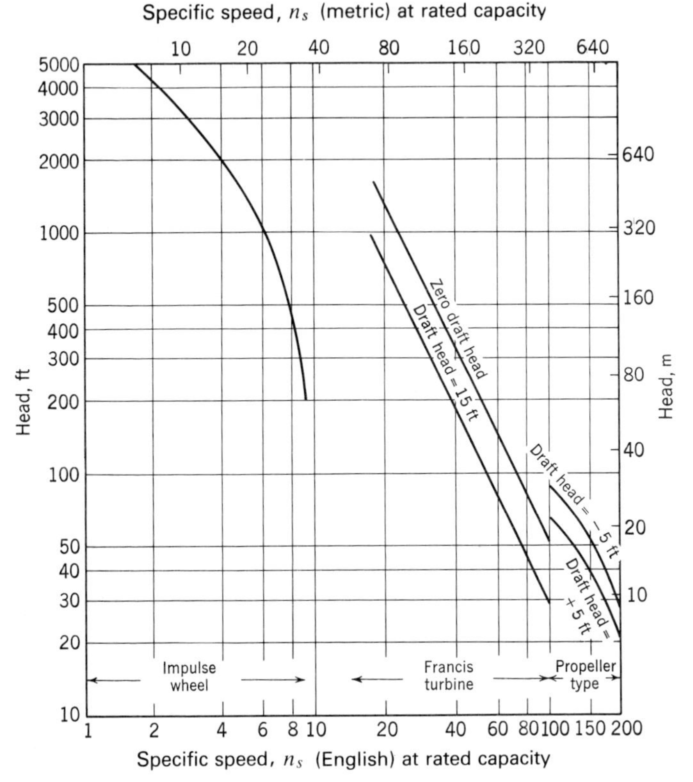

Figure 16.18 Recommended limits of specific speed for turbines under various effective heads at sea level with water temperature at 80°F. (After Moody.)

16.14 SELECTION OF TURBINES

Inspection of Eq. (14.23) indicates that at high heads for a given speed and power output, a low-specific speed machine such as an impulse wheel is required. On the other hand, an axial-flow turbine with a high n_s is indicated for low heads. An impulse turbine may, however, be suitable for a low-head installation if the flow rate (or power requirement) is small, but often under such conditions the required size of the impulse wheel is prohibitive. Impulse wheels have been used for heads as low as 50 ft if the capacity is small, but they are more commonly employed for heads greater than 500 or 1,000 ft. The limiting head for Francis turbines is about 1,500 ft because of possible cavitation and the difficulty of building casings to withstand such high pressures. By choosing a high speed of operation and, hence, a high-specific-speed turbine, the runner size and thus first cost are reduced. However, there is some loss of efficiency at high specific speeds.

In the selection of a turbine or turbines at a given installation, options are available with respect to the number and type (n_s) of turbines. Generally it is considered good practice to have at least two turbines at an installation so that the plant can continue operation while one of the turbines is shut down for repairs or inspection. The head h is determined primarily by topography, and the flow Q by the hydrology of the watershed and characteristics of the reservoir. Some of the factors influencing the choice of turbines are apparent in the following example.

Illustrative Example 16.4 Two or more identical turbines are to be selected for an installation where the net head is 350 ft and the total flow is to be 600 cfs. Select turbines for this installation assuming 90 percent efficiency.

The total available power (Eq. 4.16) is

$$\frac{\gamma Q h \eta}{550} = \frac{62.4 \times 600 \times 350 \times 0.90}{550} = 21{,}400 \text{ hp}$$

Assume two turbines at an operating speed of 75 rpm (96-pole generators for 60-cycle electricity).

$$n_s = \frac{75\sqrt{21{,}400/2}}{350^{5/4}} = 5.12$$

Thus, if the operating speed is 75 rpm, use two impulse turbines with $n_s = 5.12$ in which case $\phi_e \approx 0.45$. The required wheel diameter of these turbines is found from Eq. (14.22):

$$D = \frac{153.3\sqrt{350} \times 0.45}{75} = 17.2 \text{ ft}$$

A wheel diameter of 17.2 ft is quite large; a smaller size is possible by increasing the rotative speed. If $n = 100$ rpm, $n_s = (5.15 \times 100)/75 = 6.9$ and $D = (17.2 \times 75)/100 = 12.9$ ft. Other combinations of n_s and D could be used with other speeds; however, in accordance with Fig. 15.15, n_s should be less than about 7.0 to ensure high efficiency if impulse wheels are selected. Another possible solution is four identical turbines with $n_s = 5.8$ and $D = 10.7$ ft operating at 120 rpm.

Finally, let us explore the possibility of using Francis turbines. Assume two Francis turbines operating at 600 rpm (12-pole generator for 60-cycle electricity).

$$n_s = \frac{600\sqrt{21{,}400/2}}{350^{5/4}} = 41$$

According to Fig. 16.18 these turbines will be safe against cavitation only if they are set at zero or negative draft head (i.e., with the runner at the same elevation as the tailwater or below the tailwater).

To provide greater safeguard against cavitation we might select a lower-specific-speed machine, but then its efficiency may not be so good as indicated by Fig. 16.14. A good choice would be two Francis turbines with $n_s = 30.8$ operating at 450 rpm. The required diameter of these turbines would be about 4.8 ft assuming $\phi = 0.75$. There are actually an infinite number of alternatives. The things to watch out for are: (*a*) freedom from cavitation (Fig. 16.18); (*b*) reasonably high efficiency (Fig. 16.14); and (*c*) size not too large [Eq. (14.22)]. Flexibility of choice is achieved through variation in the number of units (and hence brake horsepower per unit) and in the operating speed. Variation in the draft-head setting also provides some flexibility.

16.15 PUMP TURBINE

In recent years the *pump-turbine* hydraulic machine has been developed. It is very similar in design and construction to the Francis turbine. When water enters the rotor at the periphery and flows inward the machine acts as a turbine. With water entering at the center (or *eye*) and flowing outward, the machine acts as a pump. The direction of rotation is, of course, opposite in the two cases. The pump turbine is connected to a motor generator which acts as either a motor or generator depending on the direction of rotation.

The pump turbine is used at pumped-storage hydroelectric plants which pump water from a lower reservoir to an upper reservoir during off-peak load periods so that water is available to drive the machine as a turbine during the time that peak power generation is needed.

An example of a pump turbine is that at the Kisenyama Pumped Storage Project of the Kansai Electric Company in Japan. There are two identical pump turbines at that installation. Under the normal range of operating conditions each machine has the following characteristics,

As a turbine ($n = 225$ rpm):
  Develops 322,000 hp at maximum net head of 722 ft.
  Develops 238,000 hp at minimum net head of 607 ft.
As a pump ($n = 225$ rpm):
  Delivers 3,880 cfs at minimum net head of 649 ft.
  Delivers 3,040 cfs at maximum net head of 755 ft.

When operating as a turbine these machines have a specific speed $n_s = 35$ in the English system of units. As a pump, in the English system, these machines have $n_s = 98$ and $N_s = 2{,}080$ (Sec. 17.5).

16.16 INSTALLATIONS

The reaction turbine is especially adapted for use under moderately low heads and even for high heads up to a possible maximum of 1,500 ft for large powers, if the water does not carry silt that would cut the clearance rings and produce excessive leakage loss. The propeller type of reaction turbine is particularly suitable for the combination of low head with large power. Its cavitation characteristics limit it to relatively low head use.

Among some interesting installations of reaction turbines are the following. At Fionnay, Switzerland, a Francis turbine operates under a head of 1,490 ft and delivers 63,200 hp at 750 rpm. In Austria a head of 1,430 ft is used for a 77,700-hp turbine at 500 rpm. In Norway a head of 1,360 ft is used to develop 69,000 hp at 500 rpm. In Italy a reaction turbine running at 1,000 rpm develops 20,140 hp under a head of 1,320 ft. At Oak Grove, Ore., a Francis turbine under a head of 850 ft delivers 35,000 hp at 514 rpm. These are all low-specific-speed reaction turbines.

Examples of higher specific speeds are a Francis turbine in France, which delivers 154,000 hp at 187.5 rpm under a head of 336 ft and another one, also in France, which runs under a head of 233 ft and develops 135,000 hp at 150 rpm. At Conowingo on the Susquehanna River 54,000-hp units run at 81.8 rpm under a head of 89 ft. The Conowingo runners are 18 ft in maximum diameter. Another example of large reaction turbines are the four vertical-shaft Francis turbines that Mitsubishi recently manufactured for installation at the Kootenay Canal Power Station in British Columbia, Canada. These are each designed to develop 196,000 hp when operating at 128.6 rpm under a net head of 268 ft. At Grand Coulee Dam on the Columbia River in the State of Washington there are six large Francis turbines ($n_s = 55$). Each unit develops over 800,000 hp under a head of 285 feet at 72 rpm. Each runner has a diameter of 32 feet and weighs 550 tons. Water is delivered to each unit at a rate as high as 30,000 cfs in a 40-ft diameter penstock. The turbines will operate at heads ranging from 220 to 355 feet.

Kaplan turbines are represented by two units at Camargos in Brazil. Each unit develops 35,700 hp under a head of 88.6 feet at 150 rpm. A Kaplan turbine in Italy runs at 600 rpm under a head of 141 feet and delivers 7,500 hp. At Bonneville, on the Columbia River, are units which have runners 280 in in diameter and with only five blades. They deliver 66,000 hp at 75 rpm at 50 ft head. At Safe Harbor on the Susquehanna River are six units with runners 220 in in diameter running at 109.1 rpm under a head of 53 ft and delivering 42,500 hp each. At Wheeler Dam in Alabama a fixed-blade propeller unit delivers 45,000 hp at 85.7 rpm under a head of 48 ft. At Rock River in Illinois a Kaplan turbine runs under a head of only 7 ft. It has a runner 138 in in diameter and develops 800 hp at 80 rpm.

PROBLEMS

16.1 Refer to Illustrative Example 16.1*c*. Suppose two units are to be used. Select several different specific-speed–operating-speed combinations that would satisfy the requirements.

16.2 In the test of a Francis turbine the pressure at the flange at the entrance to the spiral-turbine case where the diameter is 30 in was read by a mercury manometer. At a flow of 44.5 cfs the manometer differential reading in the U tube was 9.541 ft Hg, the top of the lower mercury column being 9.730 ft above the surface of the water in the tailrace. Neglecting the small velocity head in the tailrace, find the net head on the turbine.

16.3 A small Francis runner ($n_s = 30$, $D = 2$ ft) is tested and found to have an efficiency of 0.893 when operating under optimum conditions. Approximately what would be the maximum efficiency of a homologous runner ($n_s = 30$) having a diameter of 6 ft.

16.4 If a turbine homologous to that in Prob. 16.9 were made with a runner diameter of 135 in, what would be its probable efficiency under the same head?

16.5 A 1:8 model of a 12-ft-diameter turbine is operated at 600 rpm under a net head of 54.0 ft. Under this mode of operation the brake horsepower and Q of the model were observed to be 332 and 62 cfs, respectively. (*a*) From the above data compute the specific speed of the model and the value of ϕ. (*b*) Calculate the efficiency and shaft torque of the model. (*c*) What would be the efficiency of the 12-ft-diameter prototype? (*d*) The prototype is to operate at 144 rpm under a net head of 200 ft. Find the horsepower output of the prototype and the flow rate.

16.6 A radial-flow reaction turbine has the following characteristics: $A_c = a_1 = 40$ in^2, $a_2 = 28$ in^2, $Q = 1{,}600$ gpm, $\eta = 0.82$, $r_1 = 10$ in, $r_2 = 5$ in, $\beta'_1 = 140°$, $\beta_2 = 155°$. Assume shockless entrance and assume $\alpha_2 = 90°$. Find the rotative speed under these conditions. Find also the torque, the brake horsepower, the net head, and the value of C_1. Assume $\phi_e = 0.75$.

16.7 It is desired to install a single turbine that will develop 4,200 hp under a head of 247 ft. If a turbine with $n_s \approx 25$ were selected, what rotative speed would you suggest for 50-cycle electricity? How many poles do you recommend for the generator? Using Fig. 16.16 and Eq. (14.22) specify the values of ϕ_e, D, D_t, B, and α_1.

16.8 This Francis turbine is to produce 9,000 hp at 300 rpm under a net head of 150 ft. If the mechanical efficiency is 96 percent and the overall efficiency 84 percent, what guide vane angle should be used? Do not use any plotted data in the solution of this problem. Assume 2 percent leakage loss, $\alpha_2 = 90°$, $r_1 = 2$ ft, and vane thickness is negligible.

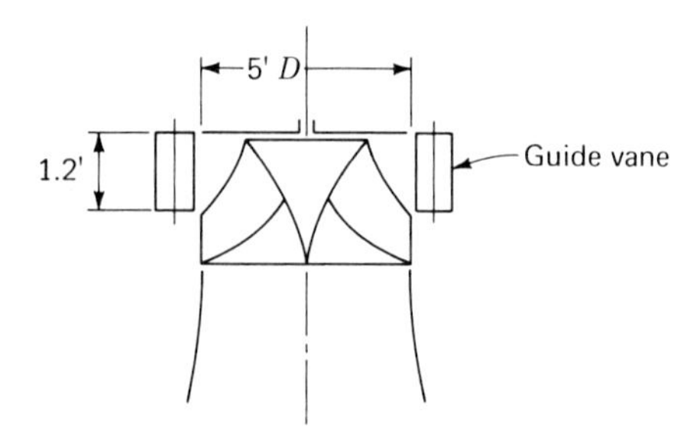

Problem 16.8

16.9 The Francis turbine for which the test curves are shown in Figs. 16.11 and 16.12 has a 27-in-diameter runner and a maximum efficiency of 88 percent when discharging 38.8 cfs and developing 550 bhp at 600 rpm under a net head of 141.8 ft. Compute n_s, ϕ_e, and C_r, assuming $B/D = 0.15$.

16.10 A Kaplan turbine is to run at 75 rpm and develop 66,000 hp at a head of 50 ft. Assume that for this particular design $\phi_e = 1.61$ at D_t and $B/D_t = 0.403$. (These values are not identical with those in Fig. 16.16 because of a difference in the practice of different companies.) Compute the maximum diameter of the runner and the height of the guide vanes, assuming the overall efficiency is 92 percent.

16.11 In addition to the data of Prob. 16.9, the mechanical-friction losses in the Francis turbine were measured and found to be 2.7 hp. Assuming that the leakage is 1 percent of the measured discharge and that $\alpha_2 = 90°$, find values of η_h, α_1, and β_1.

16.12 The Grand Coulee turbines have runner diameters of 197 in. The height of the guide vanes is 34.375 in. The diameter of the throat of the runner and also the diameter of the draft tube adjacent to the runner are 172 in. Each turbine is rated at 150,000 hp under a head of 330 ft at 120 rpm. At this power the efficiency is 89 percent and the absolute velocity of the water entering the runner is 77.2 fps. Compute (*a*) rated specific speed; (*b*) ϕ; (*c*) C_1; (*d*) C_r; (*e*) α_1 and β_1 for this full gate opening.

16.13 The turbines of Prob. 16.12 have a maximum efficiency of 93 percent at 125,000 hp under a head of 330 ft at 120 rpm. The axial component of velocity at the top of the draft tube is 23 fps, and if it is assumed that at this normal load the angle of whirl in the draft tube is 7°, which is good practice, some assumptions will lead to an estimate of $(r_2/r_1)C_2 \cos\alpha_2 = 0.013$. Compute (*a*) normal speed; (*b*) C_r; (*c*) $C_1 \cos\alpha_1$, assuming hydraulic efficiency = 94 percent; (*d*) α_1 and β_1 for normal power; (*e*) C_1.

16.14 Francis gave the following dimensions for the turbine runner (Fig. 16.1) designed and tested by him: $D = 9.338$ ft, $D_2 = 7.987$ ft, $B = 0.9990$ ft, $B_2 = 1.2300$ ft, minimum distance between runner vanes at exit = 0.1384 ft, minimum distance between guide vanes at exit = 0.1467 ft, 40 runner vanes made of $\frac{1}{4}$-in iron plate, 40 guide vanes made of $\frac{3}{16}$-in iron plate. (To avoid pulsating flow, the number of runner vanes should not be the same as the number of guide vanes or any multiple of them.) From these data and scaling the drawings, it is estimated that approximately $\alpha_1 = 13°$, $\beta_2 = 168°$, $\beta'_1 = 115°$, and $a_2 = 6.65$ ft^2 where a_2 is the effective flow area perpendicular to v_2. At the most efficient speed Francis reported the test data as $h = 13.378$ ft, $Q = 113$ cfs, $n = 40.3$ rpm, bhp = 136.6, and $\eta = 0.797$, from which $\phi = 0.672$, $u_1 = 19.7$ fps, $u_2 = 16.85$ fps, and $\sqrt{2gh} = 29.3$ fps. Compute (*a*) specific speed, (*b*) C_r, (*c*) C_1, (*d*) β_1, and compare with the vane angle β'_1, as measured.

16.15 The turbine of Prob. 16.9 has a horizontal shaft, and at the time of the test the center line of the shaft was 12.67 ft above the surface of the water in the tailrace. The discharge edge of the runner at its highest point is 0.83 ft above the centerline of the shaft. If the temperature of the water were as high as 80°F (vapor pressure = 0.5 psia) and the barometer pressure were 14.6 psia, what would be the value of the cavitation factor σ?

16.16 If it is assumed that the critical value of the cavitation factor for the turbine in Prob. 16.9 is 0.06 and if the other data of Prob. 16.15 are used, what would be the maximum allowable height of the centerline of the shaft above the tailwater surface?

16.17 In Prob. 16.16, assume all values the same except that the net head on the turbine is 400 ft. What would then be the maximum allowable height of the centerline of the shaft above the tailwater level?

16.18 Find the maximum permissible head under which an axial-flow turbine ($n_s = 160$) can operate if it is set 5 ft below tailwater. The installation is at elevation 3,150 ft, and the water temperature is 65°F.

16.19 Would you expect problems with cavitation in the prototype of Prob. 16.5(*d*) if it were set 5 ft above tailwater elevation?

16.20 What is the least number of identical turbines that can be used at a powerhouse where the available head is 1,200 ft and $Q = 1650$ cfs? Assume turbine efficiency is 90 percent and speed of operation 138.5 rpm. Specify the size and specific speed of the units.

16.21 A single hydraulic turbine is to be selected for a power site with a net head of 100 ft. The turbine is to produce 25,000 hp at maximum efficiency. What speed (rpm) and diameter should this turbine have if (*a*) a Francis turbine is selected? (*b*) a propeller turbine is selected? What are the highest "settings" (above or below tailwater) which should be recommended for these machines for them to run cavitation-free at their points of maximum efficiency?

16.22 For 50-cycle electricity how many poles would you recommend for a generator which is connected to a turbine operating under a design head of 3,000 ft with a flow of 80 cfs? Assume turbine efficiencies as given in Fig. 16.14 and be sure the turbine is free of cavitation.

16.23 Select two, four, and six identical turbines for an installation where $h = 400$ ft and total $Q = 300$ cfs. Develop 60-cycle electricity using either 36- or 72-pole generators. Be sure your selection is safe from cavitation. Assume the turbine efficiency is 90 percent.

16.24 A turbine is to be installed where the net available head is 185 ft, and the available flow will average 900 cfs. What type of turbine would you recommend? Specify the operating speed and number of generator poles for 60-cycle electricity if a turbine with the highest tolerable specific speed that will safeguard against cavitation is selected. Assume the turbine is set 5 ft above tailwater. Assume turbine efficiency is 90 percent. Approximately what size of runner is required?

16.25 (*a*) A turbine is to be installed at a point where the available head is 175 ft and the available flow will average 1,000 cfs. What type of turbine would you recommend? Specify the operating speed and number of generator poles for 60-cycle electricity if a turbine with the highest tolerable specific speed to safeguard against cavitation is selected. Assume static draft head of 10 ft and 90 percent turbine efficiency. Approximately what size of turbine runner is required? (*b*) For the same conditions select a set of two identical turbines to be operated in parallel. Specify the speed and size of the units.

16.26 A 24-in pipeline ($f = 0.020$) 15,000 ft long connects two reservoirs whose water-surface elevations differ by 300 ft. Compute the maximum horsepower that can be expected from a reaction turbine placed in this line. What is the maximum possible discharge. Investigate operating conditions over the full discharge range. Neglect minor losses and assume no cavitation. Let the turbine efficiency be 100 percent.

16.27 (*a*) The top of the draft tube for the turbine in Prob. 16.9 is 24.5 in in diameter, where it joins the flange of the elbow discharge from the runner, the latter being on a horizontal shaft. The top of the draft tube at this flange connection is 11.0 ft above the surface of the water in the tailrace. The discharge end of the draft tube is 42 in in diameter, and the velocity in the tailrace may be considered negligible. The total loss in the draft tube is $0.15\ V_2^2/2g$ plus the discharge loss of $V_3^2/2g$, where subscripts 2 and 3 refer to the top and bottom ends of the draft tube, respectively. When the flow is 38.8 cfs, what is the pressure at the top of the draft tube? (*b*) Suppose the draft tube were of uniform diameter, what would then be the pressure at the top of the tube? How much head is saved by the diverging tube? Assume the draft tube has a length of 18 ft and $f = 0.020$.

16.28 In the figure is shown a spiral case for a large vertical-shaft turbine. To assist in supporting the weight of a generator on the floor above the turbine, columns are inserted in the casing in the form of stay vanes in a casting known as a speed ring. These vanes should conform to natural streamlines. (Guide vanes, which do direct the course of the water, are inside the speed ring, and the runner is in the very center. These details are not shown.) In the figure let $r_1 = 18$ ft, $r_2 = 8$ ft, $r_3 = 6$ ft, $A_1 = 200$ ft^2, $B_2 = 3$ ft, $B_3 = 2.5$ ft, $\alpha_1 = 40°$. If the water enters the turbine case at (1) with a velocity of 8 fps, find the tangential and radial components of velocity at entrance to and exit from the speed ring. What should be the directions of the stay vanes at entrance and at exit?

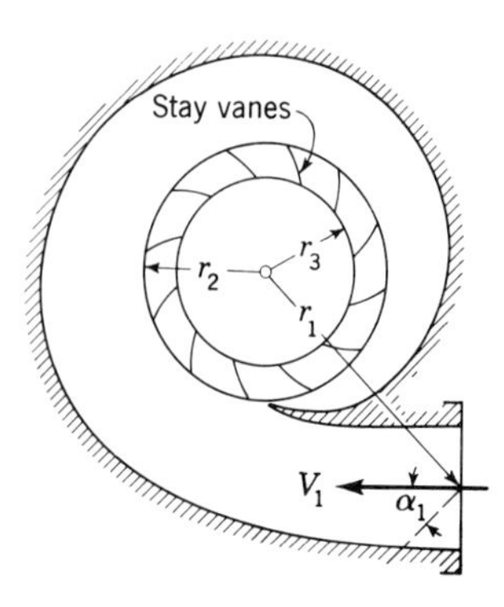

Problem 16.28

16.29 From the data given in Prob. 16.14, compute (*a*) v_2; (*b*) $v_2 \cos \beta_2$; (*c*) α_2; (*d*) V_2; (*e*) percent of the head lost in the kinetic energy at discharge from the runner. (This turbine was submerged below tailwater level and had no draft tube.)

16.30 In the test of the turbine of Prob. 16.14 at zero speed, $h = 13.565$ ft and $Q = 110.3$ cfs. (*a*) Determine the magnitude and direction of the absolute velocity at discharge from the runner. (*b*) What percentage of the head was the unit kinetic energy at discharge from the runner? (*c*) How was the rest of the head expended?

16.31 In the test of the turbine of Prob. 16.14 at runaway speed, $h = 13.596$ ft, $Q = 99.8$ cfs, $u_1 = 37.7$ fps, $u_2 = 32.3$ fps. (*a*) Determine the magnitude and direction of the absolute velocity at discharge from the runner? (*b*) What percentage of the head was the unit kinetic energy at discharge from the runner? (*c*) How was the rest of the head expended? (*d*) What was the maximum value of ϕ?

16.32 Water enters a rotating wheel with a relative velocity of 200 fps; $r_1 = 4.0$ ft, and $n = 420$ rpm. There is no pressure drop in flow over the vanes. Assume $k = 0.2$. Find the relative velocity at discharge if (*a*) $r_2 = 3.0$ ft; (*b*) $r_2 = 5.0$ ft.

16.33 Water enters a rotating wheel which is so proportioned that the passages are completely filled. $Q = 400$ cfs, $a_1 = 10$ ft^2, $a_2 = 8$ ft^2, $r_1 = 1.5$ ft, $r_2 = 1.0$ ft, and $n = 540$ rpm. Assume $k = 0.2$. Find the drop in pressure head between entrance and exit.

16.34 When operating at optimum efficiency of 90 percent the flow through a small radial reaction turbine, Fig. 6.10, is 4.0 cfs. The head on the machine is 29.6 ft. Its dimensions are: $r_1 = 0.8$ ft, $r_2 = 0.2$ ft, $\beta_1 = 60°$, $\beta_2 = 140°$, and vane height $Z = 0.4$ ft. Determine the specific speed of this turbine. Also compute ϕ and compare it with typical values given in the text.

16.35 Compute the specific speeds for the various turbines for which data are given in Sec. 16.16.

16.36 Determine the approximate values of the specific speeds of the Kisenyama pump turbines when operating as a turbine.

16.37 It is desired to develop 300,000 hp under a head of 49 ft and to operate at 60 rpm. (*a*) If turbines with a specific speed of approximately 150 are to be used, how many units will be required? (*b*) If Francis turbines with a specific speed of 80 were to be used, how many units would be required.

16.38 A 12-ft-diameter reaction turbine is to be operated at 100 rpm under a net head of 96 ft. A 1:8 model of this turbine is built and tested in the laboratory. If the model is operated at 450 rpm, under what net head should it be tested to simulate normal operating conditions in the prototype?

16.39 (*a*) At a plant of the Utah Power and Light Co. is a turbine runner 76 in in diameter which develops 8,500 hp at 300 rpm under a head of 440 ft. (*b*) At Niagara Falls a turbine runner 176 in in diameter develops 72,500 hp at 107 rpm under a head of 214 ft. (*c*) At Cedar Rapids a turbine runner 143 in in diameter develops 10,800 hp at 55.6 rpm under a head of 32 ft. (*d*) At Rock River in Illinois a turbine runner 138 in in maximum diameter develops 800 hp at 80 rpm under a head of 7 ft. For each case compute the specific speed and the value of ϕ.

CHAPTER

SEVENTEEN

CENTRIFUGAL AND AXIAL-FLOW PUMPS

17.1 DEFINITIONS[1]

The centrifugal pump is so called because the pressure increase within its rotor due to centrifugal action is an important factor in its operation. In brief, it consists of an impeller rotating within a case, as in Fig. 17.1. Fluid enters the impeller in the center portion, called the *eye*, flows outwardly, and is discharged around the entire circumference into a casing. During flow through the rotating impeller the fluid receives energy from the vanes, resulting in an increase in both pressure and absolute velocity. Since a large part of the energy of the fluid leaving the impeller is kinetic, it is necessary to reduce the absolute velocity and transform the larger portion of this velocity head into pressure head. This is accomplished in the volute casing surrounding the impeller (Fig. 17.1) or in flow through diffuser vanes (Fig. 17.2). The velocity vectors at entrance to and exit from the vanes of a radial-flow impeller are shown in Fig 6.11.

As with the reaction turbine, the demand for greater capacity, without increasing the diameter to obtain it, resulted in an increase in the dimensions parallel to the shaft. This in turn required an increase in the eye diameter to accommodate the larger flow and a corresponding change in the vanes at entrance, resulting in the mixed-flow impeller whose specific speed is higher than that of a radial-flow impeller.

[1] The reader should refer back to Chap. 14 for discussion of similarity laws, specific speed, and other characteristics of pumps.

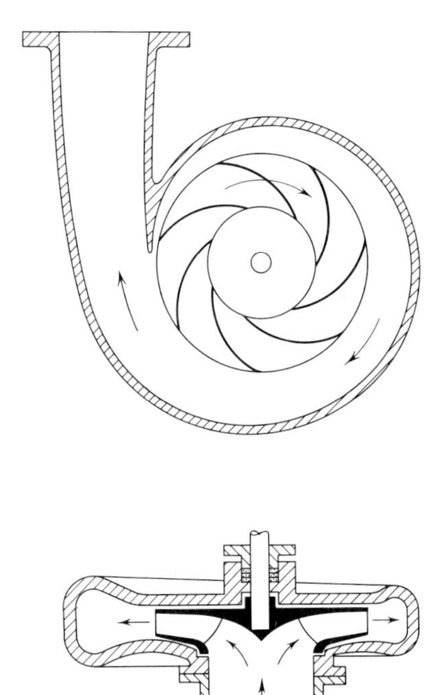

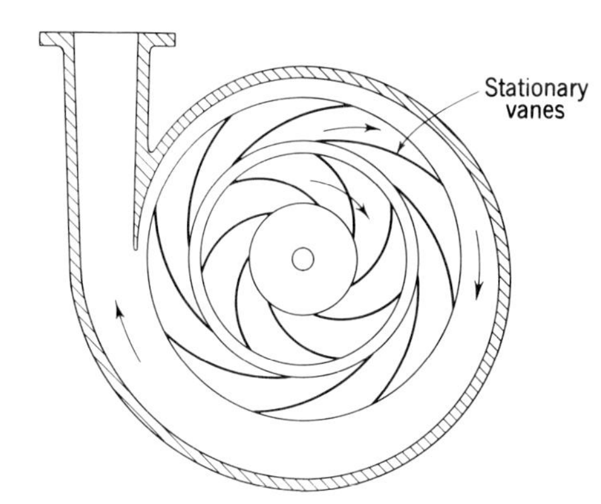

Figure 17.1 Volute centrifugal pump.

A still further increase in specific speed is obtained with the propeller, or axial-flow, pump (Fig. 6.13*b*). In this type there is no change in radius of a given streamline, and hence centrifugal action plays no part. However, the theorem of angular momentum applies alike to all types.

The principles of this chapter apply equally to fans and blowers as well as to centrifugal pumps provided there is only a small change in density of the air or other gas.

Figure 17.2 Diffuser (or turbine) pump. This type of pump is not typical of modern practice and would be found only in large pumps where the diffuser vanes are needed for structural reasons. In modern turbine pumps the diffuser vanes are three-dimensional as in Fig. 17.4 and cannot readily be shown in a drawing.

17.2 CLASSIFICATION

Centrifugal pumps are divided into two general classes: (1) volute pumps and (2) diffuser, or turbine, pumps. In the former the impeller is surrounded by a spiral case, as in Fig. 17.1, the outer boundary of which may be a curve called a *volute.* The absolute velocity of the fluid leaving the impeller is reduced in the volute casing, with a resultant increase in pressure. In the diffuser pump, shown in Fig. 17.2, the impeller is surrounded by diffuser vanes which provide gradually enlarging passages to effect a gradual reduction in velocity. Because of the superficial resemblance to a reaction turbine, this type is often called a *turbine pump.* However, it is still a centrifugal pump. These diffusion vanes are usually fixed or immovable, but in a very few instances they have been pivoted like the guide vanes in a turbine in order that the angle might be changed to conform to conditions with different rates of flow.

Centrifugal pumps are also divided into single-suction pumps, as in Fig. 17.3, and double-suction pumps. The latter have the advantage of symmetry, which ideally should eliminate end thrust. They also provide a larger inlet area with lower intake velocities than would be possible with a single-suction pump of the same outside diameter of the impeller.

All types of pumps may be single-stage or multistage. With the latter, two or more identical impellers are arranged in series, usually on a vertical shaft. The

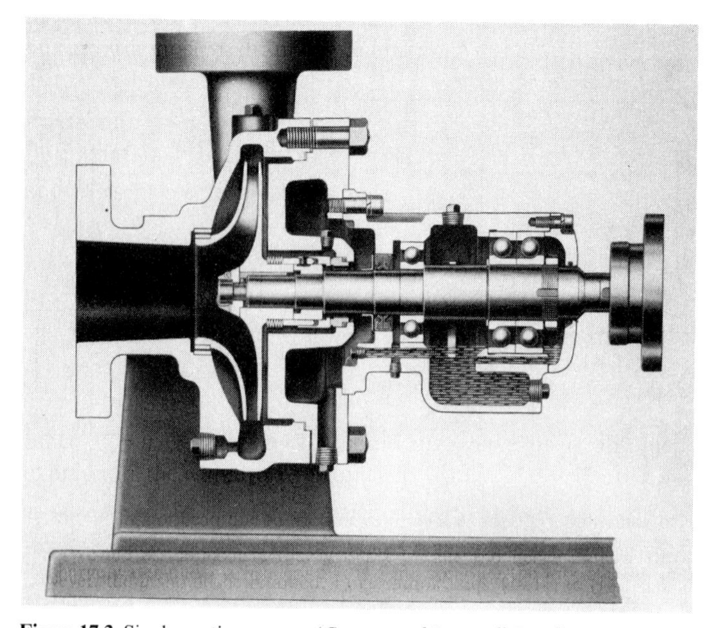

Figure 17.3 Single-suction pump. (*Courtesy of Ingersoll-Rand.*)

quantity of flow is the same as for one alone, but the total head developed by the unit is the product of the head of one stage times the number of stages.

A very special type is the deep-well pump of Fig. 17.4. Since this must be installed in a well casing of limited size, the total diameter of the pump assembly must be relatively small, and thus the impellers are even smaller in diameter. Because of the small diameter of the impeller, the head developed is not very great in one stage, and so for a deep well it is necessary to have a number of stages in order to lift the water to the desired height.

Since the casings, or *bowls*, of the deep-well pump are usually concentric and are not volutes and the water must be led from the discharge from one impeller into the eye of the next, it is customary to employ diffusion vanes in the intervening passages.

Figure 17.5 shows the impeller for Fig. 17.4. It is of the mixed-flow type and is also an *open*, or *unshrouded, impeller*. The stationary casing forms one boundary wall for the rotor passage, which necessitates the vanes having a small

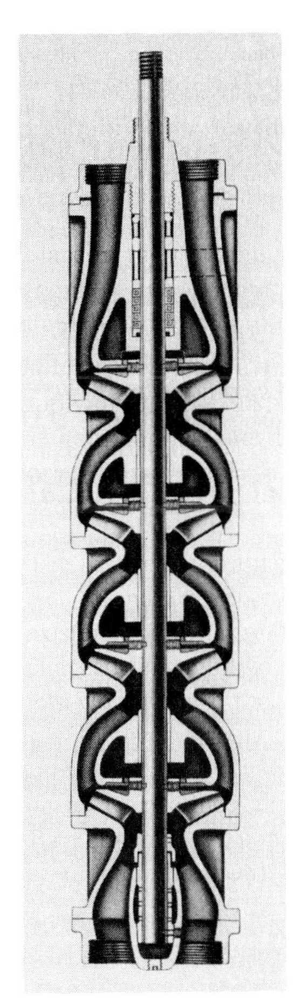

Figure 17.4 Deep-well multistage mixed-flow turbine pump. (*Courtesy of Byron Jackson Company.*)

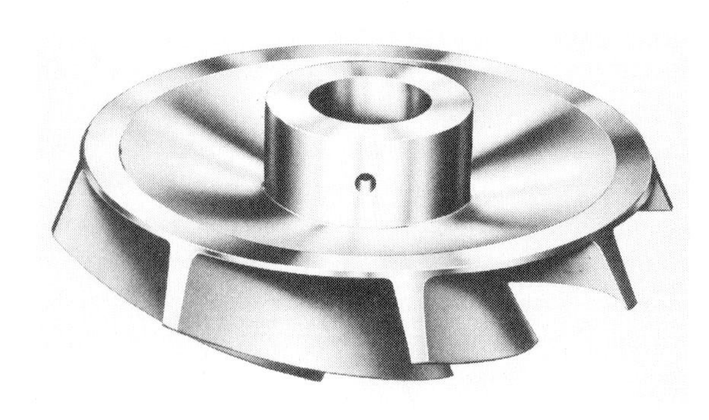

Figure 17.5 Open, or unshrouded, impeller for pump of Fig. 17.4. (*Courtesy of Byron Jackson Company.*)

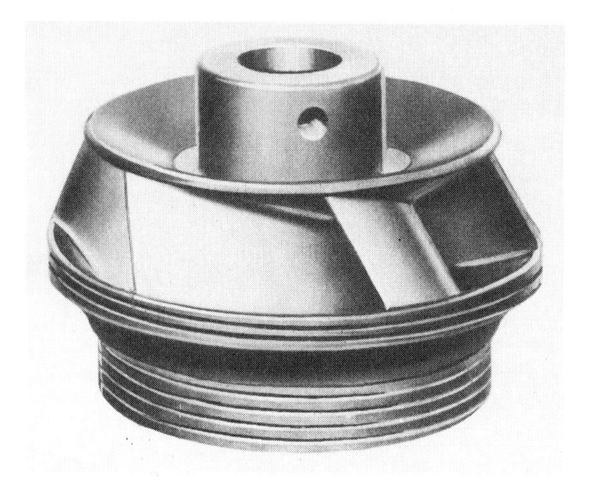

Figure 17.6 Shrouded mixed-flow impeller for deep-well pump. (*Courtesy of Byron Jackson Company.*)

clearance with the casing. By contrast Fig. 17.6 shows a *shrouded impeller* for another deep-well pump, and in this the rotor passages are completely enclosed as in Fig. 17.3. Open impellers are used where the material being pumped is likely to clog the passages of a shrouded impeller.

17.3 SIZE AND RATING OF PUMPS

It is customary to express the size of a pump by the internal diameter of the flange at discharge, probably because this indicates the size of the discharge pipe that might be used. This gives no indication as to the size of the impeller, which is in contrast with turbine practice, where the impeller diameter is usually given.

A pump is rated by its capacity (i.e., discharge) and head at the point of maximum efficiency for a given rotative speed. Of course, both of these values depend upon the speed that is specified. These values will be referred to as the *normal* capacity and the *normal* head for that speed.

17.4 HEAD DELIVERED

The mode of operation of a pump depends on the system in which it is operating. The *pump characteristic curve* (Fig. 17.7) shows the relation between the head developed by the pump and its rate of discharge when the pump is operating at a given rotative speed. If the pump is delivering fluid through a piping system with a static lift Δz, the head that the pump must develop is equal to the static lift plus the total head loss (proportional approximately to Q^2). The *system characteristic curve* shows the relation between the required pumping head and the flow rate in the pipe line. The actual pump-operating head and flow rate are determined by the intersection of the two curves.

The particular values of h and Q determined by this intersection may or may not be those for the maximum efficiency of the particular pump. If they are not, this means that the pump is not exactly suited to the specific conditions. Further discussion of the behavior of pumps and their relationship to the systems in which they operate is presented in Sec. 17.13.

In the test of a pump the head is determined by measuring the pressures on both the suction and discharge sides of the pump, computing the velocities by dividing the measured discharge by the respective cross-sectional areas, and noting the difference in elevation between the suction and discharge sides. The net head h delivered by the pump to the fluid is

$$h = H_d - H_s = \left(\frac{p_d}{\gamma} + \frac{V_d^2}{2g} + z_d\right) - \left(\frac{p_s}{\gamma} + \frac{V_s^2}{2g} + z_s\right) \tag{17.1}$$

where the subscripts d and s refer to the discharge and suction sides of the pump, as shown in Fig. 17.8. If the discharge and suction pipes are the same size, the velocity heads cancel out, but frequently the intake pipe is larger than the discharge pipe. It should be noted that h, the head put into the fluid by the pump, was previously referred to as h_p in Sec. 4.6.

The official test code provides that the head on a pump be the difference between the total energy heads at the intake and discharge flanges. However,

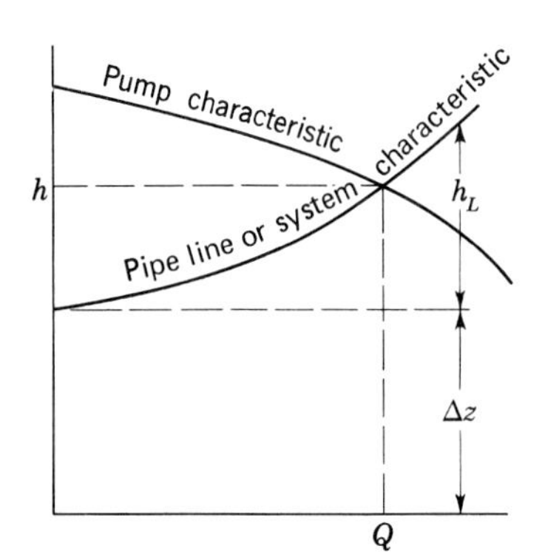

Figure 17.7 Graphical method for finding the operating condition of a pump and pipeline.

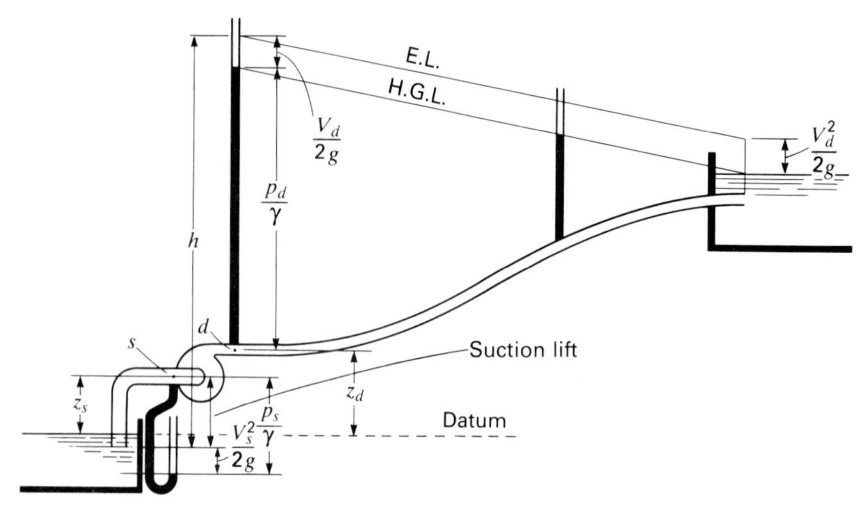

Figure 17.8 Head developed by pump. In this case p_s/γ is negative.

flow conditions at the discharge flange are usually too irregular for accurate pressure measurement, and it is more reliable to measure the pressure at 10 or more pipe diameters away from the pump and to add an estimated pipe friction head for that length of pipe. On the intake side, prerotation sometimes exists in the pipe near the pump, and this will cause the pressure reading on a gage to be higher than the true average pressure at that section.

17.5 SPECIFIC SPEED

For pumps there are two commonly used definitions of specific speed in the English system of units as given by Eqs. (14.24) and (14.25). They are

$$n_s = \frac{n_e\sqrt{Q}}{h^{3/4}} \qquad \text{and} \qquad N_s = \frac{n_e\sqrt{\text{gpm}}}{h^{3/4}} \tag{17.2}$$

where n_e is the rotative speed for maximum efficiency and $N_s = 21.2n_s$. Pumps are customarily rated in gallons per minute; hence N_s is the more commonly used specific speed and will be the one referred to in this text. Computed values of specific speed for a given pump throughout its entire operating range from zero discharge (shutoff head) to maximum discharge (zero head) would give values from zero to infinity, but the only value that has any real significance is that corresponding to the values of head, discharge, and speed at the point of maximum efficiency (i.e., their normal values).

As in the case of the reaction turbine, the numerical value of the specific speed indicates the type of pump. On the gallons-per-minute basis specific speeds for volute centrifugal pumps range from 500 to 5,000, for mixed-flow pumps from

4,000 to 10,000, and for axial-flow pumps from 10,000 to 15,000 as approximate limits.

For a double-suction pump it is customary to base the specific speed on one-half of the total capacity of the pump, on the assumption that a double-suction impeller is the equivalent of two single-suction impellers placed back to back.

Illustrative Example 17.1 (*a*) It is desired to deliver 1,600 gpm at a head of 900 ft with a single-stage pump. What would be the minimum rotative speed that could be used?

Assuming that the minimum practical specific speed is 500, from Eq. (17.2) we get

$$n_e = \frac{N_s h^{3/4}}{\sqrt{\text{gpm}}} = \frac{500(900)^{3/4}}{\sqrt{1{,}600}} = 2{,}050 \text{ rpm}$$

(*b*) For the conditions of (*a*), how many stages must the pump ($N_s = 500$) have if a rotative speed of 600 rpm is to be used?

$$h^{3/4} = \frac{n_e\sqrt{\text{gpm}}}{N_s} = \frac{600\sqrt{1{,}600}}{500} = 48$$

or

$$h = 172 \text{ ft per stage}$$

Hence

$$\tfrac{900}{172} = 5.23 \qquad \text{(6 stages are required)}$$

To meet the exact specifications of head and capacity, either the rotative speed or the specific speed or both could be changed slightly.

Illustrative Example 17.2 (*a*) Determine the specific speed of a pump that is to deliver 2,000 gpm against a head of 150 ft with a rotative speed of 600 rpm.

$$n_s = \frac{n_e\sqrt{\text{gpm}}}{h^{3/4}} = \frac{600\sqrt{2{,}000}}{(150)^{3/4}} = 626$$

(*b*) If the rotative speed were doubled, what would be the flow rate and the head developed by this pump? Assume no change in efficiency.

Eq. (14.16): $Q \propto n$, so $Q = 2 \times 2{,}000 = 4{,}000$ gpm

Eq. (14.15): $h \propto n^2$, so $h = 2^2 \times 150 = 600$ ft

(*c*) Check the specific speed for the conditions given in (*b*).

$$N_s = \frac{1{,}200\sqrt{4{,}000}}{(600)^{3/4}} = 626$$

This result was expected, for the same impeller was involved in (*a*) and (*b*).

(*d*) Find the required operating speed of a two-stage pump ($N_s = 626$) to satisfy the requirements in (*a*).

$$N_s = 626 = \frac{n_e\sqrt{2{,}000}}{(75)^{3/4}}$$

$$n_e = 357 \text{ rpm}$$

17.6 CHARACTERISTICS AT CONSTANT SPEED

Though some centrifugal pumps are driven by variable-speed motors, the usual mode of operation of a pump is at constant speed and typical characteristics of a centrifugal pump for such operation are shown in Fig. 17.9. The head-versus-discharge curve[1] may be transformed into that for some other speed by means of the similarity laws ($Q \propto n$ and $h \propto n^2$); however, the efficiency of the pump drops off as the rotative speed is moved away from the optimum speed. To illustrate this point, the head-versus-discharge curves for a certain centrifugal pump at several different rotative speeds as determined by laboratory test are plotted in Fig. 17.10 together with contours of equal efficiency. Thus, we see that at optimum operating conditions this pump will deliver 700 gpm against a head of 120 ft at a rotative speed of 1,450 rpm. The important feature shown in Fig. 17.10 is that, if a pump is not operating near the optimum point, its efficiency drops off, depending on how far the mode of operation is from optimum.

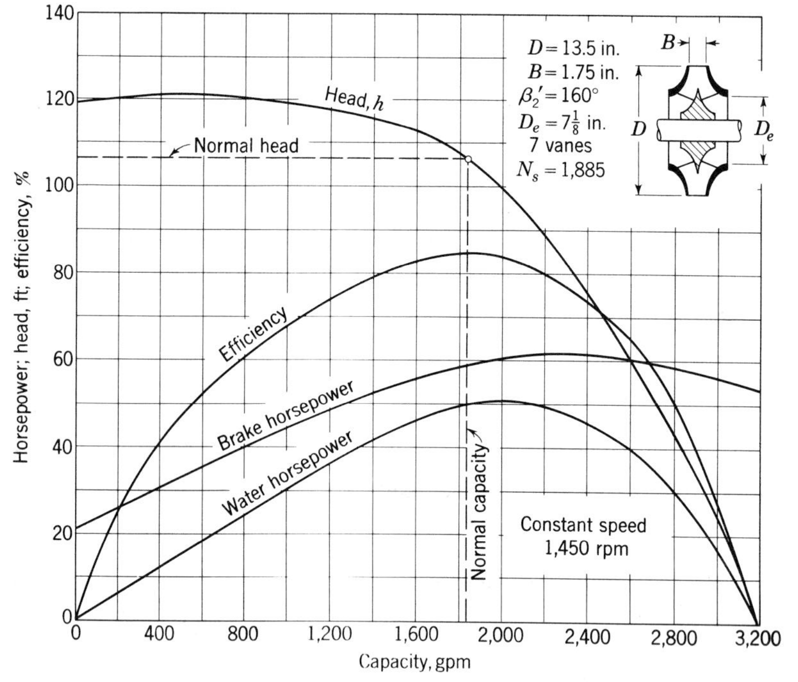

Figure 17.9 Characteristics of a centrifugal pump at constant speed.

[1] A single curve applicable to a family of homologous pumps of different sizes operating at different speeds can be developed by plotting h/n^2D^2 versus Q/nD^3.

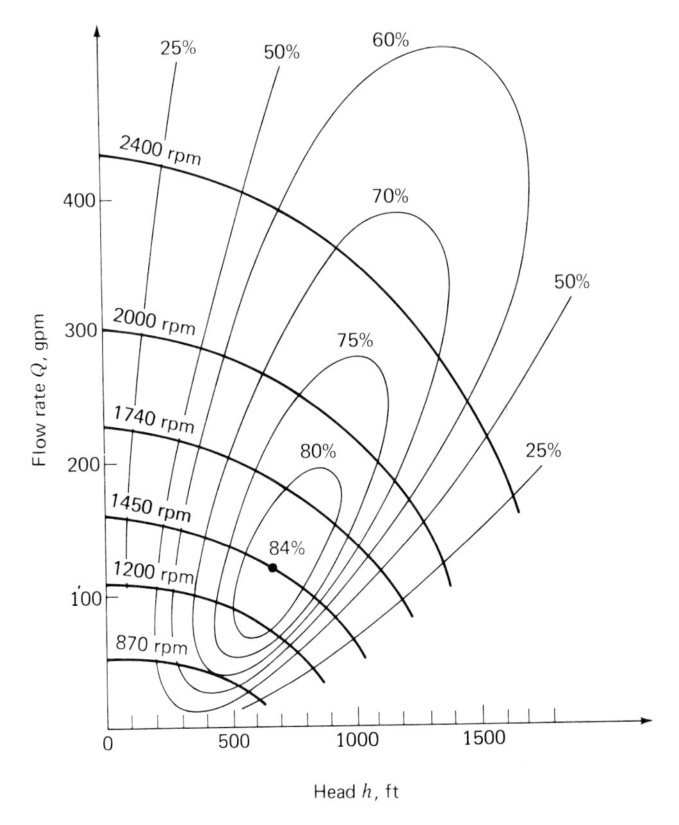

Figure 17.10 Characteristics of a centrifugal pump at various speeds of rotation with contours of equal efficiency.

By different impeller and casing designs it is possible to vary the characteristics, as shown in Fig. 17.11; each one has special advantages for particular conditions. Thus a *flat characteristic* permits a considerable variation in the rate of discharge with but very little change in head, while a *steep characteristic* gives only a small variation in the flow for a relatively large change in head.

The axial-flow pump has a much steeper head-capacity curve than does any centrifugal pump, and instead of the brake power at shutoff being a minimum, as for the centrifugal pump, it not only is a maximum but is very much larger than the power required at the point of maximum efficiency. This is a disadvantage both in starting up and in continued operation at low capacity. The characteristics of the axial-flow pump shown in Fig. 17.11 are for the fixed-blade type. In a few instances this type of pump is made with adjustable blades similar to the runner of the Kaplan turbine, and the blades can be adjusted during operation to suit conditions.

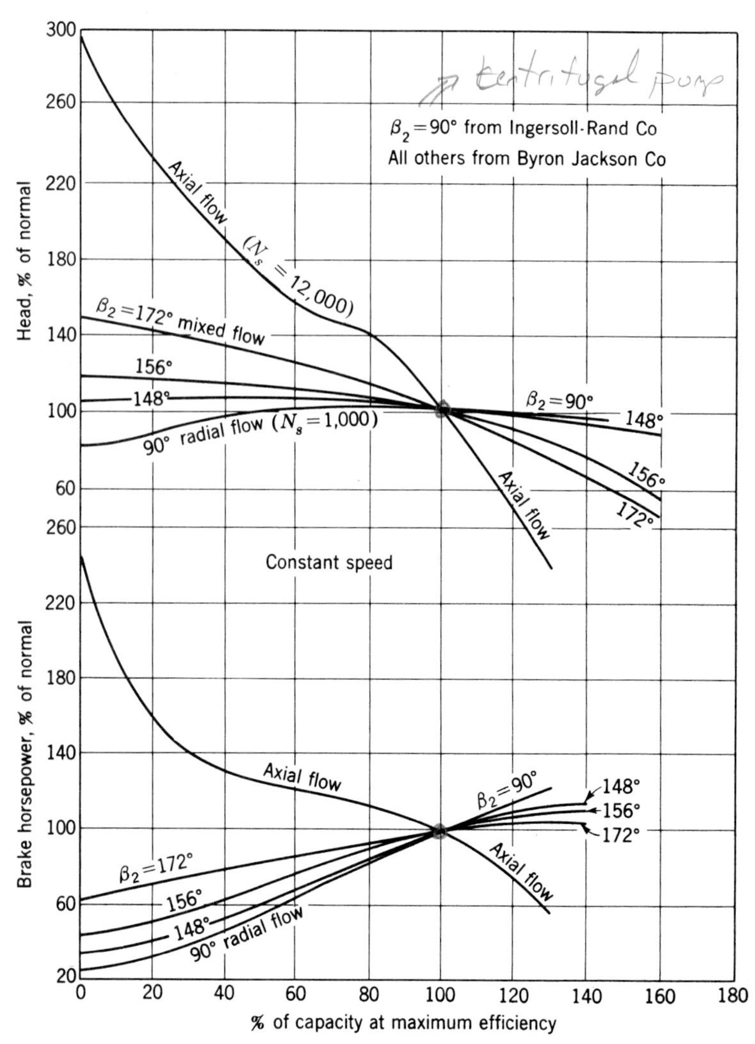

Figure 17.11 Head-capacity characteristics for different types of centrifugal and axial-flow pumps.

17.7 SHUTOFF HEAD

When a pump filled with the fluid to be pumped is operated at normal speed with the discharge valve closed, the head developed is called the *shutoff head.* Ideally, this would appear to be a case of a forced vortex with a pressure-head difference between the eye and the periphery of the impeller of $(u_2^2 - u_1^2)/2g$. However, it is not that simple. Although there is no flow delivered, there is a great deal of circulation within the impeller, which causes a rotation of the fluid in the eye of the impeller and for a distance of several pipe diameters in the intake pipe. The result of this prerotation is that the actual shutoff head is approximately given by $u_2^2/2g$. However, the flow conditions are so complex that any precise theoretical analysis is out of the question. It is known that the actual shutoff head is affected by the value of the impeller-vane angle at exit, by the design of the case, and even by the nature of the intake, since the latter has some influence on the prerotation. Thus, in one instance, where the same impeller was tested at the same speed but with several different types of intakes, the shutoff head ranged from 240 to 282 ft.

17.8 ENERGY LOSSES IN PUMPS

The design of a pump is a specialized field which is beyond the scope of this text.[1] The discussion presented here is to enable one to understand those characteristics of pumps that should be of value to users of pumps. In Sec. 6.9 it was shown that the torque exerted on the fluid by the impeller of a centrifugal pump (Fig. 6.11) is given by

$$T = \rho Q(r_2 V_2 \cos \alpha_2 - r_1 V_1 \cos \alpha_1) \tag{17.3}$$

By setting $T \times \omega = \gamma Q h''$, we find the head h'' imparted to the fluid by the impeller of the pump is given by

$$h'' = \frac{u_2 V_2 \cos \alpha_2 - u_1 V_1 \cos \alpha_1}{g} \tag{17.4}$$

As can be seen from Eq. (17.4), when $\alpha_1 = 90°$ the value of h'' is a maximum. Hence, for maximum efficiency, α_1 will be close to 90°.

The net head h [Eq. (17.1)] may be expressed as

$$h = h'' - h_L \tag{17.5}$$

where h_L represents the hydraulic head loss in the flow through the impeller. An expression [Eq. (6.23)] for h_L was derived in Sec. 6.11. Combining Eq. (6.23) with Eqs. (17.1) and (17.5) gives

$$h'' = \frac{u_2^2 - u_1^2}{2g} + \frac{V_2^2 - V_1^2}{2g} + \frac{v_1^2 - v_2^2}{2g} \tag{17.6}$$

The first term of this expression is the increase in pressure due to centrifugal action, the second term is the increase in kinetic energy, and the third term shows the gain or loss of pressure in flowing through the impeller passages according to whether the areas are such that the relative velocity decreases or increases.

The head loss h_L has several components. First of all, as water enters the vanes of the impeller, there may be a *shock loss* due to turbulence because of an improper relative-velocity angle at vane inlet. This loss is relatively large at low

[1] For detailed methods of design see A. J. Stepanoff, "Centrifugal and Axial Flow Pumps," 2d ed., John Wiley & Sons, Inc., New York, 1957.

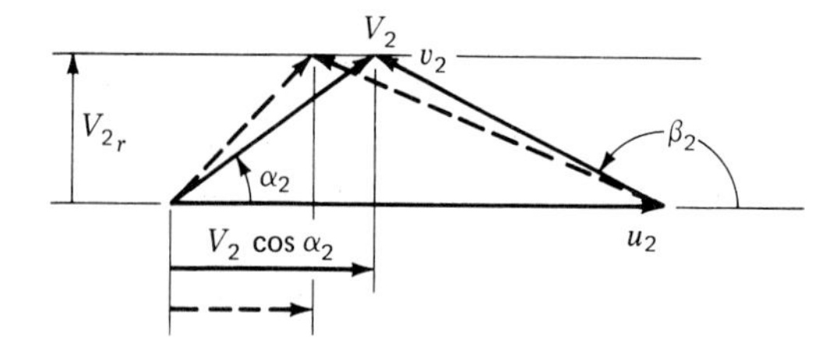

Figure 17.12 Effect of circulatory flow at vane exit. The solid vectors are ideal with angle of efflux β_2 = blade angle β'_2. The dashed vectors show actual values.

and at high flow rates; it grows smaller as optimum operating conditions are approached and is almost nonexistent at optimum conditions. The second loss is that of *fluid friction* in the passages between the vanes. This loss varies approximately as Q^2. The third loss is due to *circulatory flow* at discharge from the impeller created by the difference in pressure on the two sides of each vane. This results in a decrease in the velocity along the working face of the vane and an increase in relative velocity on the back face of the vane. The result of this unequal velocity distribution is that the average angle β_2 of the fluid leaving the impeller is greater than the vane angle β'_2 (Fig. 17.12). Thus the full ideal value of $V_2 \cos \alpha_2$ is not achieved. This component of the hydraulic head loss changes very little with flow rate.

In addition to these hydraulic head losses the efficiency of a pump is reduced by bearing and packing friction and by disk friction as well as by the effect of leakage as described in Sec. 14.1. A typical relationship among these various losses is shown in Fig. 17.13.

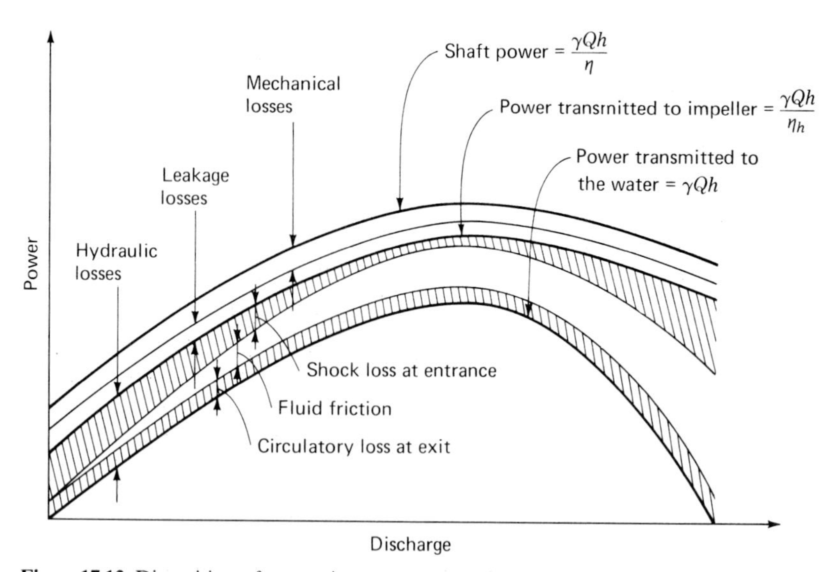

Figure 17.13 Disposition of power in a pump operating at variable head and constant speed.

17.9 CAVITATION

An important factor in satisfactory operation of a pump is the avoidance of cavitation, both for the sake of good efficiency and for the prevention of impeller damage (Secs. 4.8 and 16.13). For pumps a *cavitation parameter* σ has been defined as

$$\sigma = \frac{(p_s)_{\text{abs}}/\gamma + V_s^2/2g - p_v/\gamma}{h} = \frac{\text{NPSH}}{h} \tag{17.7}$$

where subscript s refers to values at the pump intake (i.e., suction side of the pump), h is the head developed by the pump, and p_v is the vapor pressure. As the latter is normally given in absolute units, it follows that p_s must also be absolute pressure. NPSH, the numerator of Eq. (17.7), is referred to as the *net positive suction head.*

With a long straight inlet pipe it may be possible to measure p_s with precision and to compute an accurate value of the mean V_s from the continuity equation. But where prerotation exists or a fitting, such as an elbow, precedes this section by a short distance, neither of these values can be accurately determined. It is then preferable to write the energy equation between the surface of the liquid source and the pump intake (Fig. 17.8). Thus, using absolute pressures,

$$\frac{(p_0)_{\text{abs}}}{\gamma} - h_L = z_s + \frac{(p_s)_{\text{abs}}}{\gamma} + \frac{V_s^2}{2g}$$

where z_s is the elevation of the pump intake above the surface of the liquid, as in Fig. 17.8, and $(p_0)_{\text{abs}}$ is the absolute pressure upon that surface. If the liquid is drawn from a closed tank, this pressure could be either greater or less than the atmospheric pressure. Making this substitution in Eq. (17.7),

$$\sigma = \frac{(p_0)_{\text{abs}}/\gamma - p_v/\gamma - z_s - h_L}{h} = \frac{\text{NPSH}}{h} \tag{17.8}$$

The critical value σ_c is that at which there is an observed change in efficiency or head or some other property indicative of the onset of cavitation. The value will depend not only upon what criterion is used, but also upon the conditions of operation. In Fig. 17.14 is shown an experimental curve where the total head and capacity were kept constant while the intake pressure was decreased, resulting in

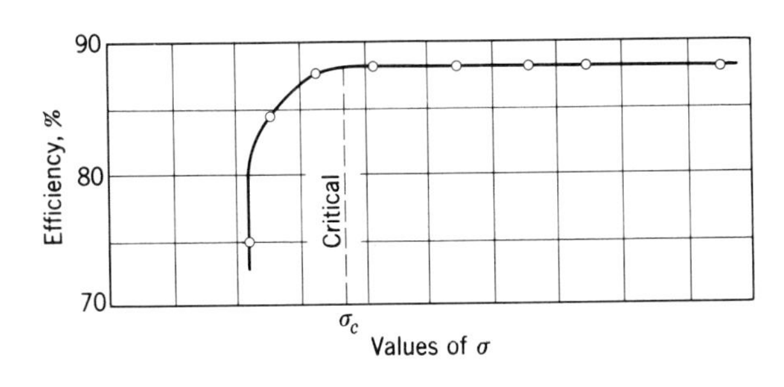

Figure 17.14 Effect of varying the cavitation parameter.

a decrease in σ. The critical value is fixed by the point where the efficiency was found to drop. A different value of σ_c would be found for a different capacity. For safe operation it is desirable to operate at values above the critical σ_c for the capacity involved. Approximate values of σ_c for centrifugal pumps operating under normal conditions near optimum efficiency are shown in Fig. 17.15. The critical value of σ_c for any specified operating condition depends on the design of the particular pump, and in important installations should be determined experimentally in a model study. For modeling purposes, a useful parameter, the *suction specific speed S*, has been defined as

$$S = \frac{n\sqrt{\text{gpm}}}{\text{NPSH}^{3/4}} \tag{17.9}$$

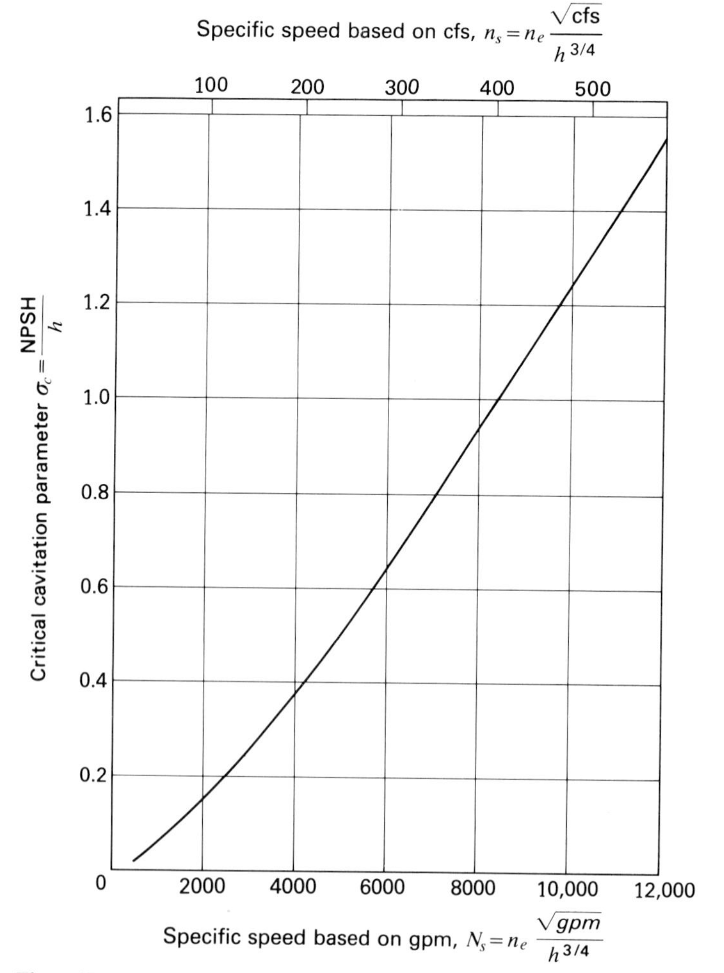

Figure 17.15 Approximate values of critical cavitation parameter σ_c for centrifugal pumps as a function of specific speed.

If model and prototype are operated at identical values of N_s and S, similarity of flow and cavitation is achieved provided the model and prototype are geometrically similar to one another. In dealing with a double-suction pump (Sec. 17.2), the total capacity should be divided by 2 when using Eq. (17.9) to find S.

Introducing the critical value of σ into Eq. (17.8), we obtain

$$(z_s)_{max} = \frac{(p_0)_{abs}}{\gamma} - \frac{p_v}{\gamma} - \sigma_c h - h_L \tag{17.10}$$

which will give the maximum allowable elevation of the pump intake above the surface of the liquid. It is apparent from inspection of Eq. (17.10) that, to ensure freedom from cavitation, the pump should be set lower, particularly if (*a*) it is to be operated at a high elevation above sea level, (*b*) the total head developed is increased, (*c*) the specific speed for a given head is increased, or (*d*) the vapor pressure of the liquid is increased, for example at higher temperatures.

If the value of $(z_s)_{max}$ determined by this equation is negative, it indicates that the pump must be placed below the surface of the liquid. Recommended limiting heads for the prevention of cavitation for single-stage, single-suction pumps as a function of specific speed and suction lift (the elevation difference between the energy line at suction and the eye of the impeller) are given in Fig. 17.16. A positive suction lift indicates that the impeller is above the energy line (Fig. 17.8).

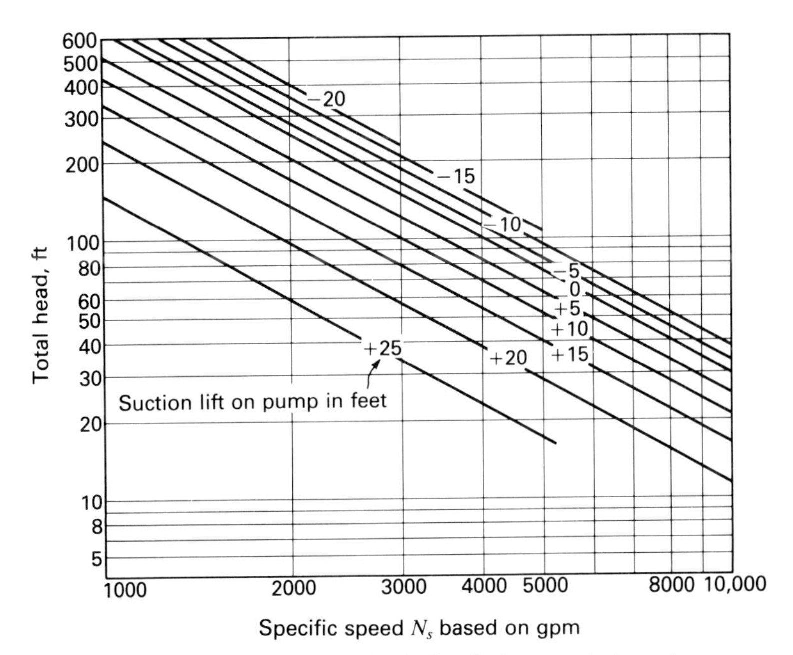

Figure 17.16 Recommended limiting heads for single-stage, single-suction pumps as a function of specific speed and suction lift. At sea level with water temperature of 80°F.

Illustrative Example 17.3 In the accompanying figure is shown the effect of net positive suction heads at intake (also expessed as vacuums) on the operating characteristics of a double-suction centrifugal pump as determined by experimental test at sea level. At the point of maximum efficiency the critical value of NPSH is 10.4 ft. Determine the value of σ_c for this pump, and find where the pump should be set to assure against cavitation for this operating condition. Assume that the friction loss in the intake pipe is 3 ft.

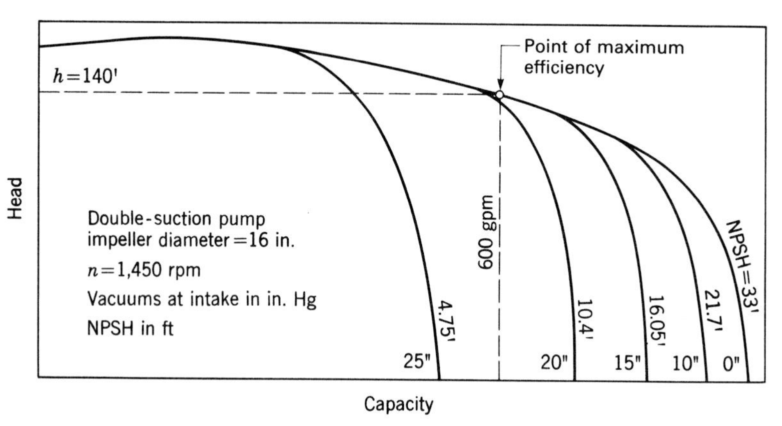

Illustrative Example 17.3

$$N_s = \frac{1{,}450\sqrt{600/2}}{(140)^{3/4}} = 617$$

From Fig. 17.15, $\sigma_c \approx 0.04$ for $N_s = 617$. By direct computation, using Eq. (17.8)

$$\sigma_c = \frac{\text{NPSH}}{h} = \frac{10.4}{140} = 0.0743$$

which is not a close check. The fact that this critical value is larger than the typical value indicated in Fig. 17.15 for such a low specific speed merely emphasizes the fact that variation in design will give values which differ from the norm.

As this particular pump was tested at sea level and with cold water, it may be assumed that $(p_0)_{abs}/\gamma \approx 34$ ft and $p_v/\gamma \approx 1$ ft. Assuming the friction losses in the intake piping to be 3 ft, employing Eq. (17.10), we get $(z_s)_{max} = 34 - 1 - 0.074(140) - 3 = 19.6$ ft, which would be the maximum allowable elevation above the surface to avoid cavitation at this one operating point. If it is desired to avoid cavitation at any point, even for maximum discharge, the test results in the figure show that NPSH should be about 33 ft. In this case $(z_s)_{max} = 34 - 1 - 33 - 3 = -3$ ft, which means the pump should be submerged by that amount to be safe from cavitation. It is doubtful, however, that the pump would be operated close to maximum discharge as its efficiency would be very low at that point.

17.10 VISCOSITY EFFECT

Centrifugal pumps are also used to pump liquids with viscosities different from that of water. Figure 17.17 shows actual test curves of performance for the very extreme range, from water to an oil with a kinematic viscosity 3,200 times that of water. It is seen that, as the viscosity is increased, the head-capacity curve becomes steeper and the power required increases. The dashed line indicates the

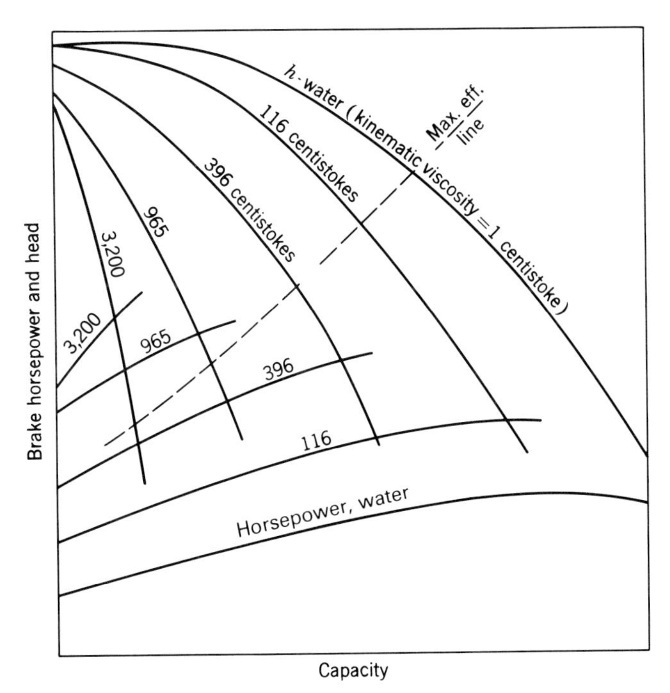

Figure 17.17 Centrifugal pump with viscous oils.

maximum efficiency points for each viscosity curve. It is seen that both the head and the capacity at the point of maximum efficiency decrease with increasing viscosity. As these are accompanied by an increase in the brake horsepower, there is a marked decrease in efficiency. For example, for the situation depicted in Fig. 17.17, if the optimum efficiency of the pump when pumping water is 0.85, its optimum efficiency is only 0.47 when pumping a liquid whose viscosity is 116 times that of water and only 0.18 when pumping a liquid whose viscosity is 396 times that of water.

17.11 EFFICIENCY

Figure 17.18 presents what are believed to be approximate optimum efficiencies of modern water pumps of large capacity. The figure also shows typical runner profiles for a few specific speeds. It is seen that there is a gradual merging of one type into another, and so the dashed line indicates the probable maximum values in these border zones. These curves do not necessarily represent absolute maximum values nor is it to be expected that all pumps will attain efficiencies as high as shown, since these efficiencies apply to pumps of large size whose design and construction has been done with great care. Generally, the larger the pump the higher the attainable efficiency. In Fig. 17.19 are shown typical efficiency curves for normal commercial radial-flow pumps as a function of capacity.

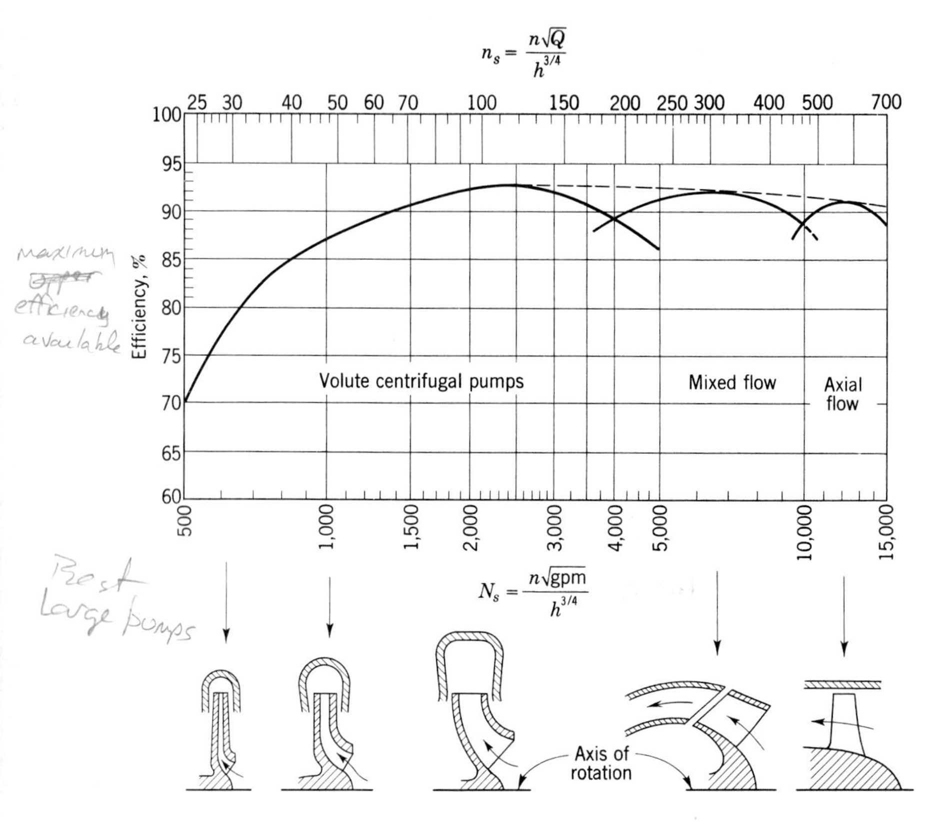

Figure 17.18 Optimum efficiency of water pumps as a function of specific speed.

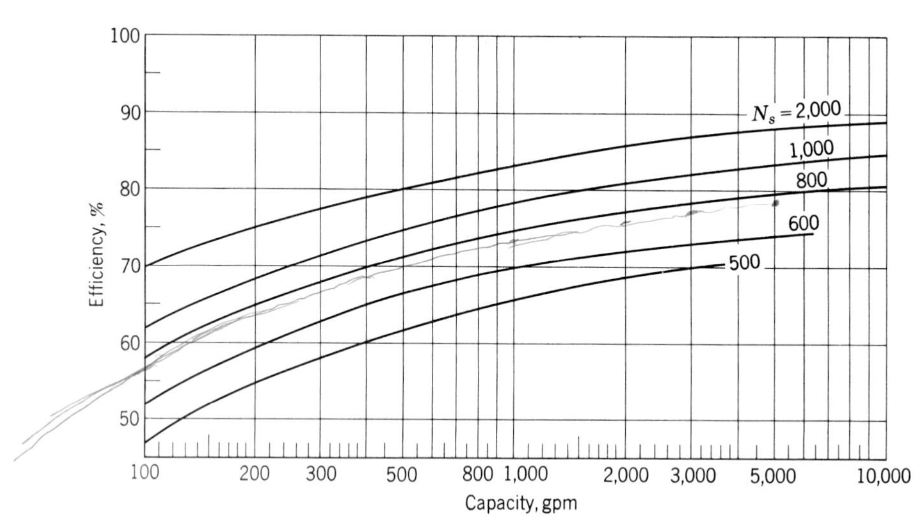

Figure 17.19 Efficiency of commercial radial-flow pumps as a function of capacity.

Best expected efficiencies as a function of Q optimum

For most purposes the specific speed of a double-suction pump is computed by using one-half of the total capacity, and this is especially necessary in considering conditions at entrance to the impeller and with regard to cavitation. But the efficiency of a pump is largely determined by the conditions at exit from the impeller and in the casing and is practically unaffected by subdividing the inlet. Hence, for efficiency diagrams such as Figs. 17.18 and 17.19, the specific speed for double-suction pumps is based on the total capacity.

The relation between the efficiency of a model pump and its prototype can be estimated with reasonable accuracy through application of the Moody formula for turbines [Eq. (16.4)].

17.12 PROPORTIONS AND FACTORS FOR PUMPS

The discharge from any type of impeller (Fig. 17.20) may be found by multiplying the outlet area by the velocity. The area that is most readily computed is the circumferential area for radial-flow impellers and corresponding areas for other types of impellers as shown in Fig. 17.20. The effective flow area is $f\pi DB$ for radial- and mixed-flow impellers and $f\pi D_e^2/4$ for axial-flow impellers where f is a factor (typically about 0.95) to allow for the space taken up by the vanes or hub respectively. The effective flow area is to be multiplied by the component of

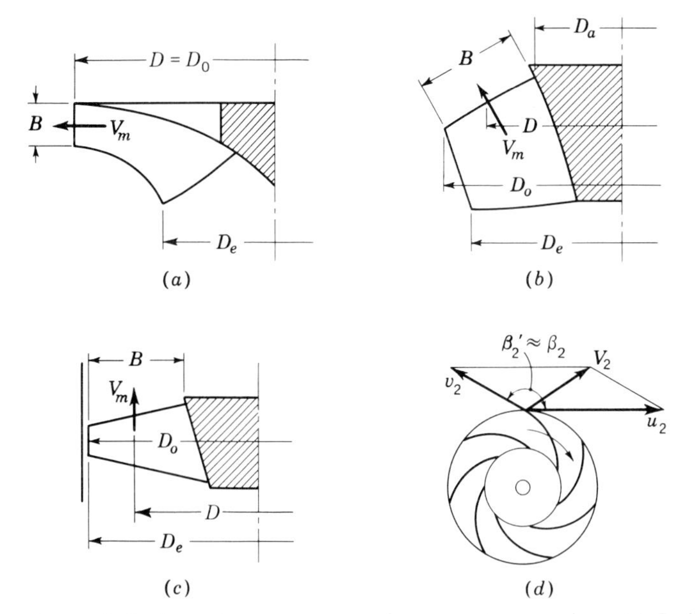

Figure 17.20 Nomenclature for pump factors and proportions. (*a*) Radial flow. (*b*) Mixed flow. (*c*) Axial flow. (*d*) Definition of β'_2. (*Note:* In the radial-flow pump, the extreme diameter D_0 is the same as the mean exit diameter D. In the axial-flow pump, $D = D_0 - B$ and the eye diameter $D_e = D_0$.)

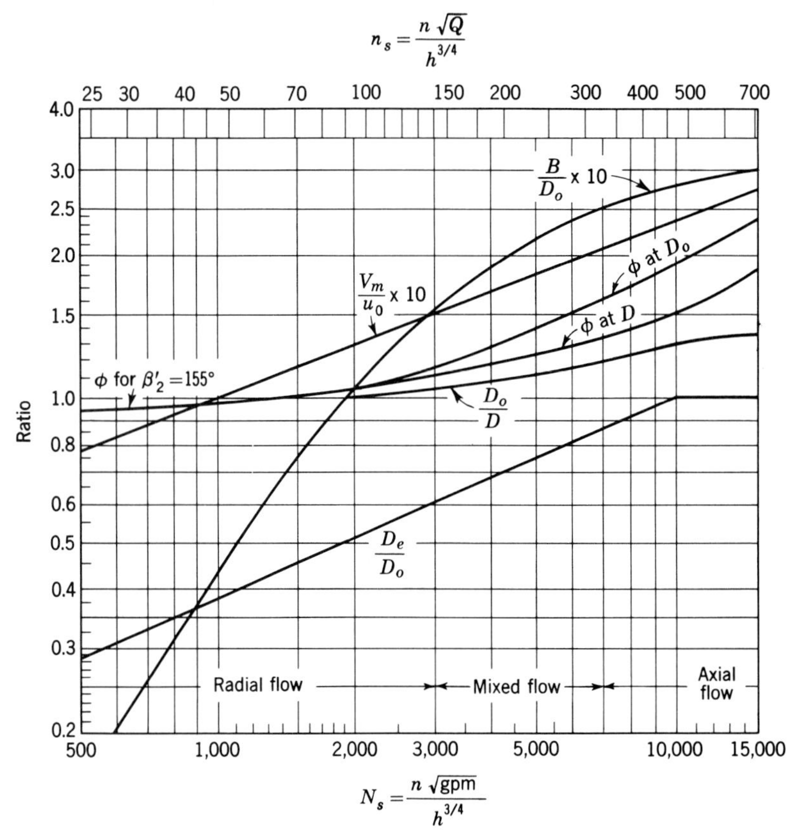

Figure 17.21 Factors and proportions for pumps as a function of specific speed. (u_0 is the peripheral velocity at the extreme diameter D_0 and is referred to as u_2 in the text.)

velocity that is normal to it, which is the radial component for the pure centrifugal impeller, or the axial component for the propeller type, or in general the *meridional* component,[1] which we shall designate by V_m. Then the discharge of radial-flow and mixed-flow pumps is given by $f\pi DBV_m$ and that of axial-flow pumps by $f\pi D_e^2 V_m/4$. The meridional component V_m is proportional to u_2 or to $\sqrt{2gh}$, but the exact relationship must be determined by experience. The ratio of V_m to u_2 depends[2] primarily upon the specific speed, but is also affected by the vane angle at exit, the number of vanes or blades, and the casing design.

Typical values of pump proportions and factors as a function of specific speed are shown in Fig. 17.21. These values are not the only ones that are used, for each manufacturer will have values based upon experience that apply to particular designs.

[1] Meridional velocity V_m for the different types of impellers is defined in Fig. 17.20.

[2] In Fig. 17.21 u_2 is referred to as u_0 as it represents the peripheral velocity at the extreme diameter D_0.

Illustrative Example 17.4 A pump that will deliver 84,500 gpm against a head of 225 ft when operating at 600 rpm is desired. Determine the specific speed of this pump and its approximate dimensions.

$$N_s = \frac{600\sqrt{84{,}500}}{(225)^{3/4}} = 3{,}000$$

Assume $f = 0.95$ and assume that Fig. 17.21 is applicable. From Fig. 17.21, $\phi_e = 1.1$ at diameter D. Hence $u_2 = 1.1\sqrt{2gh} = 132.4$ fps. From Eq. (14.22),

$$D = 153.2 \times 1.1 \times \frac{\sqrt{225}}{600} = 4.21 \text{ ft} = 50.6 \text{ in}$$

From Fig. 17.21,

$$\frac{D_o}{D} = 1.06 \qquad \frac{B}{D_o} = 0.155$$

$$\frac{D_e}{D_o} = 0.6 \qquad \frac{V_m}{u_o} = 0.15$$

Hence

$$D_o = 1.06 \times 50.6 = 53.6 \text{ in}$$

$$B = 0.155 \times 53.6 = 8.31 \text{ in}$$

The eye diameter is

$$D_e = 0.6 \times 53.6 = 32.2 \text{ in}$$

As an exercise it is suggested that the reader make a sketch of this impeller similar to those of Fig. 17.20. The peripheral velocity of the impeller at D_o is

$$u_0 = 1.06 \times 132.4 = 140 \text{ fps}$$

$$(V_m)_2 = 0.15 \times 140 = 21.0 \text{ fps}$$

$$\text{Circumferential area} = 0.95\pi \times \frac{50.6}{144} \times 8.31 = 8.71 \text{ ft}^2$$

$$Q = A_{\text{circum}}(V_m)_2 = 8.71 \times 21.0 = 182.9 \text{ cfs} = 81{,}900 \text{ gpm}$$

which is a close check to the given capacity of 84,500 gpm. A precise check was not expected because the values given in Fig. 17.21 are only approximate.

17.13 PUMP AND SYSTEM CHARACTERISTICS

For a particular situation a pump (or pumps) should be chosen so that under normal conditions of operation the speed and capacity are such that operation is occurring close to peak efficiency. If this is not the case, energy will be wasted and the operation will be uneconomic.

The choice of a pump (or pumps) for a particular situation is complicated by the large number of alternatives that are possible. First of all, there are many different designs of pumps with a variety of specific speeds (Fig. 17.22*a*). By changing the speed of operation of a particular pump (Fig. 17.22*b*) its operating characteristics can be changed. Also, selecting from among different-sized homologous pumps (Fig. 17.22*c*) will provide a variation in characteristics. In

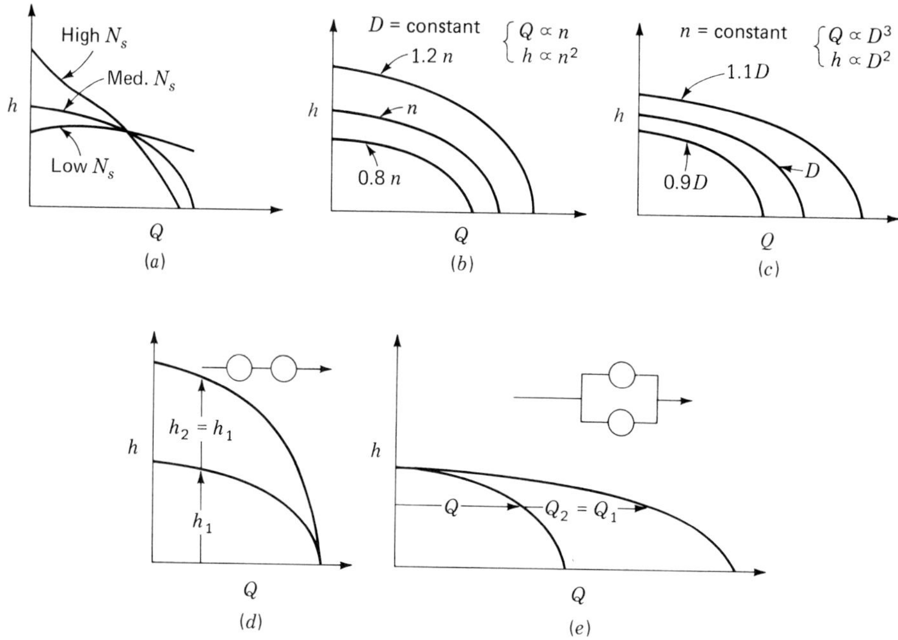

Figure 17.22 Pumping alternatives. (*a*) Different pumps with different characteristics. (*b*) A particular pump at different speeds. (*c*) Homologous pumps of different size. (*d*) Two identical pumps in series. (*e*) Two identical pumps in parallel. (*Note:* In series or parallel the pumps need not be identical, but their operating characteristics should be close to one another.)

addition, different speeds of operation can be used with various sizes of homologous pumps as in Illustrative Example 14.1. Under certain conditions it may be advantageous to install pumps in series (Fig. 17.21*d*) or in parallel (Fig. 17.21*e*). When pumps are installed in series or parallel it is very important that they have reasonably similar head-capacity characteristics throughout their range of operation; otherwise, one pump will carry most of the load and, under certain conditions, all of the load, with the other pump acting as a hindrance rather than a help. In fact, in parallel, if the operating characteristics of the pumps are quite different, a condition of backflow can occur in one of the pumps. Finally, one must always be sure that the selected pump (or pumps) will not encounter cavitation problems over the full range of operating conditions.

The mode of operation is best determined by plotting the pumping characteristics and the pipe system characteristics on the same diagram (Fig. 17.7); the point at which the two curves intersect gives an indication of what will take place. Generally, one can choose between changing the speed of a given pump or selecting a particular size of homologous pump in order to obtain the proper characteristics. The latter is usually preferable because pump efficiency tends to decrease rather rapidly as the speed is changed from the optimum (Fig. 17.10). Several aspects of the relationship between pump and system characteristics are demonstrated in the following example.

Illustrative Example 17.5 Two reservoirs A and B are connected with a long pipe which has characteristics such that the head loss through the pipe is expressible as $h_L = 20Q^2$, where h_L is in feet and Q is the flow rate in 100's of gpm. The water surface elevation in reservoir B is 35 ft above that in reservoir A. Two identical pumps are available for use to pump the water from A to B. The characteristic curve of the pump when operating at 1,800 rpm is given in the following table.

Operation at 1,800 rpm	
Head (ft)	Flow rate (gpm)
100	0
90	110
80	180
60	250
40	300
20	340

At the optimum point of operation the pump delivers 200 gpm at a head of 75 ft. Determine the specific speed N_s of the pump and find the rate of flow under the following conditions: (*a*) A single pump operating at 1,800 rpm; (*b*) two pumps in series, each operating at 1,800 rpm; (*c*) two pumps in parallel, each operating at 1,800 rpm.

The head-capacity curves for the pumping alternatives are plotted and so is the h_L versus Q curve for the pipe system. In this case $h = \Delta z + h_L = 35 + 20Q^2$. The answers are found at the points of intersection of the curves. They are as follows: (*a*) single pump, 156 gpm; (*b*) two pumps in series, 224 gpm; (*c*) two pumps in parallel, 170 gpm.

If Δz had been greater than 100 ft, neither the single pump nor the two pumps in parallel would have delivered any water. If Δz had been -20 ft (i.e., with the water surface elevation in reservoir B 20 feet below that in A), the flows would have been: (*a*) 212 gpm; (*b*) 258 gpm; and (*c*) 232 gpm.

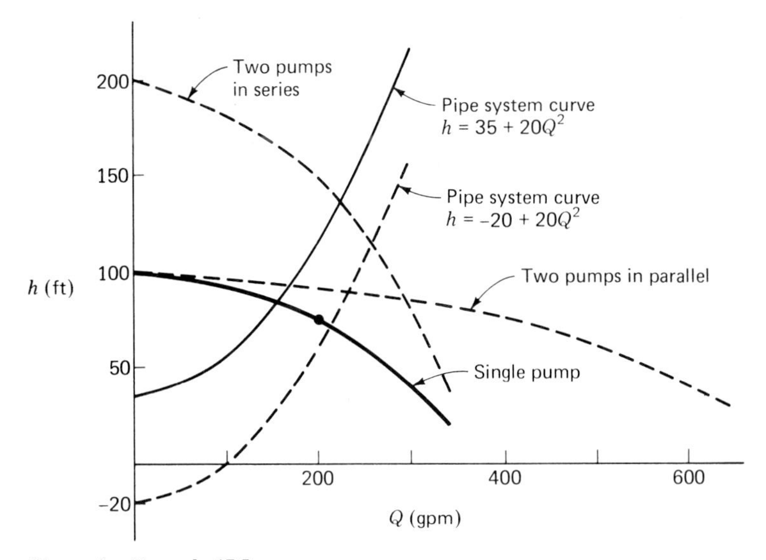

Illustrative Example 17.5

17.14 INSTALLATIONS

A few examples of pump construction and installation are presented as illustrations of modern practice.

The Byron Jackson Company has built pumps with as many as 54 stages. Water has been lifted to heights of several thousand feet by multistage pumps. Ingersoll–Rand produced a 6-in 10-stage pump operating at 3,750 rpm which delivers 1,600 gpm at a head of 6,000 ft, the shutoff head being 7,000 ft. The Worthington Corporation installed a pump at Rocky River to deliver 279.5 cfs at a head of 238.84 ft when running at 327 rpm. The brake horsepower was 8,259, giving an efficiency of 91.7 percent. The impeller diameter was approximately 7.54 ft, and the width at outlet approximately 0.72 ft, with an eye diameter of 4.24 ft.

On the Colorado River Aqueduct the Worthington Corporation built three pumps for the Hayfield plant to deliver 200 cfs each at a head of 444 ft when running at 450 rpm, and three similar pumps for Eagle Mountain with a head of 440 ft. The impeller diameters are approximately 81.6 in, and the eye diameters 34 in. The Byron Jackson Company built three pumps for the Gene plant to deliver 200 cfs each at a head of 310 ft when running at 400 rpm. The impeller diameters are 78 in. At the Intake plant, where the head is 294 ft, the impeller diameters are 76 in. The Allis-Chalmers Company built three pumps for the Iron Mountain plant to deliver 200 cfs each at a head of 146 ft when running at 300 rpm.

A typical moderate-sized large-capacity pumping plant is that at Cartersville, Georgia where the Johnston Pump Company installed a two-stage mixed-flow vertical-shaft pump that delivers 95,000 gpm against a head of 95 ft when operating at 394 rpm. At Marineland of the Pacific in Palos Verdes, California, three-stage Johnston vertical turbine pumps are used to pump salt water to tanks housing marine life. Because of the corrosive action of the sea water the pump bowls are of iron with vitreous enamel coating. All moving parts that come in contact with the sea water are constructed of corrosion-resistant material.

A noteworthy pumping project is that at Grand Coulee on the Columbia River, for which pumps were built by the Byron Jackson Company and the Pelton Water Wheel Company jointly. The head and capacity for the point of maximum efficiency are given in Prob. 17.27, but the pumps may also operate at a head as low as 270 ft and discharge 1,650 cfs, at which point they require approximately 60,000 bhp.

The Allis-Chalmers Company has built a combination reversible pump-turbine for the Hiwasee plant of the TVA, which as a pump will deliver 3,900 cfs at a head of 205 ft at the point of maximum efficiency while requiring approximately 100,000 bhp. The impeller diameter is 266 in, and it runs at 105.9 rpm. The maximum capacity is 5,200 cfs at 135 ft head.

In Italy is a pump built in Switzerland which discharges 250,000 gpm, or 558 cfs, at a head of 787 ft at 450 rpm. It requires 62,000 bhp.

A hot-oil pump to deliver 875 gpm at a head of 8,600 ft with 19 stages has been built by the Byron Jackson Company; this is one of the highest-head pumps in existence.

One of the world's largest pumping installations is the Edmonston Pumping Plant of the State of California water project. This plant lifts water over the Tehachapi mountains. At this plant there are 14 four-stage vertical-shaft centrifugal pumps. Each is capable of delivering 315 cfs against a head of 1,970 ft when rotating at 600 rpm. Their maximum efficiency is about 92 percent. The maximum energy requirements for this plant are approximately 6×10^9 kilowatt-hours per year.

PROBLEMS

17.1 The diameter of the discharge pipe of a pump is 6 in, and that of the intake pipe is 8 in. The pressure gage at discharge reads 30 psi, and the vacuum gage at intake reads 10 in Hg. If $Q = 3.0$ cfs of water and the brake horsepower is 35.0, find the efficiency. The intake and discharge are at the same elevation.

17.2 Compute the specific speed of the pumps at Grand Coulee and of those on the Colorado River Aqueduct at Hayfield, Gene, and Iron Mountain plants. (See Sec. 17.14.)

17.3 Determine the approximate specific speed of the pumps at the Edmonston Pumping Plant in California. (See Sec. 17.14.)

17.4 Suppose 10 stages were to be used for a total head of 900 ft, a capacity of 1,600 gpm, and a pump speed of 600 rpm. What would be the specific speed in both gallons-per-minute and cubic-feet-per-second units?

17.5 A pump is to discharge 10.0 m^3/s at a head of 5.0 m when running at 300 rpm. What type of pump will be required? Suppose the required speed is 450 rpm. What could then be done?

17.6 Under normal operation a centrifugal pump with an impeller diameter of 2.84 in delivers 250 gpm of water at a head of 700 ft with an efficiency of 60 percent at 20,000 rpm. Compute the peripheral velocity, the specific speed and ϕ.

17.7 Under normal operating conditions a centrifugal pump with an impeller diameter of 8.0 cm delivers 12 L/s of water at a head of 262 m with an efficiency of 60 percent at 18,000 rpm. Compute the peripheral velocity, the specific speed and ϕ.

17.8 Select the specific speed of the pump or pumps required to lift 15 cfs of water 375 ft through 10,000 ft of 3-ft-diameter pipe ($f = 0.020$). The pump rotative speed is to be 1,750 rpm. Consider the following cases: single pump, two pumps in series, three pumps in series, two pumps in parallel, three pumps in parallel.

17.9 The pump of Fig. 17.9 is placed in a 10-in-diameter pipe ($f = 0.020$), 1,300 ft long, that is used to lift water from one reservoir to another. The difference in water-surface elevations between the reservoirs fluctuates from 20 to 100 ft. Plot a curve showing delivery rate versus water-surface-elevation difference. Plot also the corresponding efficiencies. The pump is operated at a constant speed of 1,450 rpm. Neglect minor losses.

17.10 Repeat Prob. 17.9 for the case of the same pump operating at 1,200 rpm. Assume efficiency pattern and values remain the same.

17.11 Repeat Prob. 17.9 for the case of a homologous pump whose diameter is 80 percent as large as the pump of Prob. 17.9. Assume efficiency pattern and values remain the same.

17.12 Repeat Prob. 17.9 for the case of a homologous pump with diameter 80 percent as large as the pump of Prob. 17.9 when operating at 1,200 rpm. Assume efficiency pattern and values remain the same.

17.13 A pump homologous to the one whose dimensions and operating characteristics are shown in Figure 17.9 has a diameter $D = 27$ in. When operating at 1,000 rpm this pump delivers 30 cfs through a very long pipeline that connects two reservoirs whose water surface elevations are identical. What will be the flow rate if the pump speed is increased to 1,200 rpm? Assume constant value for the pipe friction factor f and neglect any differences or changes in pump efficiency.

17.14 A pump with a critical value of σ of 0.10 is to pump against a head of 500 ft. The barometric pressure is 14.3 psia, and the vapor pressure of the water is 0.5 psia. Assume the friction losses in the intake piping are 5 ft. Find maximum allowable height of the pump relative to the water surface at intake.

17.15 In a model pump delivering 5.14 cfs with a total head of 400 ft the efficiency started to drop when the gage pressure head plus velocity head at inlet was reduced to 10 ft. What was the value of σ_c if the barometric pressure was 14.3 psia and the water temperature 80°F?

17.16 A boiler feed pump delivers water at 212°F which it draws from an open hot well with a friction loss of 2 ft in the intake pipe between it and the hot well. The barometer pressure is 29 in Hg, and the value of σ_c for the pump is 0.10. What must be the elevation of the water surface in the hot well relative to that of the pump intake? The total pumping head is 240 ft.

17.17 Suppose the pump of Illustrative Example 17.3 were to be operated at its maximum efficiency point at a speed of 3,600 rpm. What would be the minimum allowable value of NPSH and what would be the maximum allowable elevation above the water surface, assuming a barometric pressure of 32 ft of water, a vapor pressure of 1 ft of water, and intake-pipe friction of 3 ft? (*Note:* σ_c is constant at the value found in the illustrative example.)

17.18 Suppose the pump in Illustrative Example 17.3 were to pump gasoline with a vapor pressure of 4.42 psia. Assume the specific gravity of the gasoline to be 0.72. When $h = 140$ ft, $V_s = 10$ fps. Using the same value of σ_c as for water, what is the minimum allowable intake pressure in feet of gasoline and in pounds per square inch? (For gasoline the head-capacity curve is practically the same as that for water, if the head is expressed in feet of gasoline.)

17.19 Suppose a pump were to pump water at a head of 130 ft, the water temperature being 100°F and the barometric pressure being 14.3 psia. At intake the pressure is a vacuum of 17 in Hg and the velocity is 12 fps. What are the values of NPSH and σ?

17.20 The pump of Illustrative Example 17.3 when pumping gasoline delivered 600 gpm at a head of 140 ft of gasoline with an intake pressure of 0 gage. With a vacuum-gage reading of 10 in Hg at the intake, the pump delivered 600 gpm with $h = 94$ ft of gasoline; with a vacuum-gage reading of 15 in Hg, it delivered 250 gpm with $h = 88$ ft of gasoline. These points are neither the points of maximum efficiency nor the points of incipient cavitation. Assume the vapor pressure of the gasoline to be 4.42 psia and the specific weight to be 45 lb/ft^3. If the barometric pressure is 14.7 psia, compute the values of NPSH and of σ for these points, assuming the velocity head to be negligible.

17.21 A pump is delivering 7500 gpm of water at 140°F and the barometric pressure is 13.8 psia. Determine the reading on a pressure gage in inches of mercury vacuum at the suction flange when cavitation is incipient. Assume the suction pipe diameter equals 2 ft and neglect effect of prerotation. Take $\sigma_c = 0.085$.

17.22 A pump delivers 7500 gpm of water at 50°F at an elevation of 5,000 ft. Determine the reading on a pressure gage (inches of mercury vacuum) at the suction flange when cavitation is incipient. Let the diameter of the suction pipe be 2 ft and neglect effect of prerotation. Use $\sigma_c = 0.085$.

17.23 A centrifugal pump, whose operating characteristics at 1,800 rpm are given in Illustrative Example 17.5, is to be placed in a 6-in-diameter pipe line and used to deliver water from reservoir A to reservoir B. The water-surface level in B is 10.0 ft lower than that in A. The pipe line is 1,500 ft long, has a diameter of 6 in and $f = 0.03$. The pump is located very close to reservoir A and head loss between the reservoir and pump should be neglected. If the suction side of the pump is set 5 ft below the water-surface elevation of reservoir A, at what rate could water be reliably pumped? Assume the pump speed can be changed to any value and the pump efficiency remains constant. The water temperature is 50°F, and this installation is in the mountains at elevation 10,000 ft. (a) Solve for the maximum reliable flow rate by computing N_s and using the value of σ_c suggested in the text.

(*b*) Determine the maximum operating speed of the pump below which cavitation will not occur. (*c*) What size homologous pump and rotative speed would you recommend for this situation?

17.24 If the maximum efficiency of the pump of Prob. 17.9 is 82 percent, approximately what would be the maximum efficiency of the pump of Prob. 17.11? Equation (16.4) may be considered applicable.

17.25 This sketch shows the dimensions and angles of the diffuser vanes of a centrifugal pump. The vane passages are 0.80 in wide perpendicular to the plane of the sketch. If the impeller delivers water at the rate of 2.80 cfs under ideal and frictionless conditions, what is the rise in pressure through the diffuser?

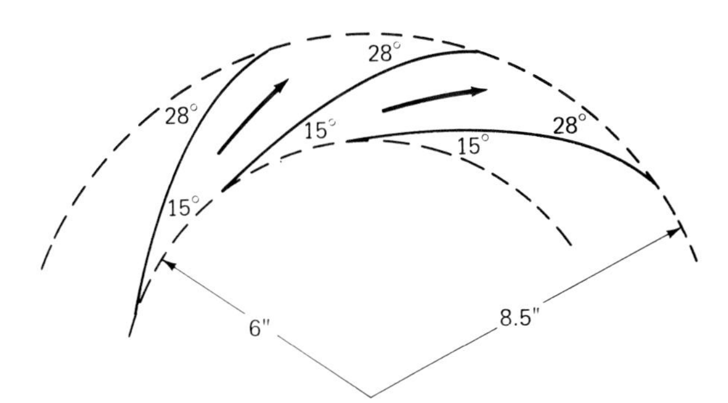

Problem 17.25

17.26 Water leaves the impeller of a centrifugal pump with a velocity of 70 fps at an angle $\alpha_2 = 10°$. It flows through a *whirlpool chamber* consisting of parallel sides before it reaches the volute case. The inner and outer radii of this chamber are 6 and 10 in, respectively. What will be the values of $V \cos \alpha$, V_m, and V for the water as it leaves the chamber and enters the volute? If there were no loss of energy, what would be the gain in pressure head? Assume $f = 1$.

17.27 The Grand Coulee pumps on the Columbia River have impellers with a diameter of $167\frac{3}{8}$ in and a width at exit of 19.5 in. The speed is 200 rpm, and the maximum efficiency 90.8 percent. At the point of maximum efficiency the discharge is 1,250 cfs at a head of 344 ft. The shutoff head is 422 ft. Compute ϕ for maximum efficiency, V_m, and V_m/u_2. Assume that the fractional part of area which is free space is 0.95. (See Fig. 17.20 for definition of V_m.)

17.28 A 54-in pump at Rocky River has an impeller 90 in in diameter, and B is 0.75 ft. It runs at 327 rpm. If $\phi = 1.034$ and $V_m/u_2 = 0.128$, compute the head and capacity. Assume $f = 0.95$.

17.29 Determine the value of ϕ for the pumps of Illustrative Example 14.1. Compare these values with those of Fig. 17.21.

17.30 A pump is required to deliver 2,420 gpm at a head of 150 ft when running at 1,750 rpm. Determine the principal impeller dimensions.

17.31 A pump is required to deliver 9,580 gpm at a head of 36 ft when running at 1,200 rpm. Determine the principal impeller dimensions.

17.32 A 4-stage pump is to be designed to deliver 7,500 gpm against a head of 960 ft at a speed of 1,200 rpm. The four impellers are identical. Using the design factors of Fig. 17.21 determine their approximate dimensions B, D, and D_o. Assume the fractional free space = 0.95.

17.33 A deep-well pump is to deliver 800 gpm against a total head of 200 ft at 1,500 rpm. If the impellers can be no larger than 7 in, how many stages should be used assuming all impellers are identical? As an approximation, assume $f = 1$.

17.34 Consider a pump to deliver 84,500 gpm at a head of 225 ft, as in Illustrative Example 17.4. Determine the rotative speeds and impeller diameters D and D_o for specific speeds of 500, 1,000, 2,000, 5,000, 8,000, 10,000, and 15,000.

17.35 For a constant maximum or outside diameter $D_o = 4$ ft and a constant head of 81 ft, compute the rotative speeds and capacities for specific speeds (N_s) of 500, 1,000, 2,000, 5,000, 10,000, and 15,000.

17.36 Specify the dimensions (D, D_o, D_e, and B) of alternate single-suction pumps to deliver 1,500 gpm against a head of 40 ft. Use motor speeds of 400, 800, 1200, 2,000, and 2,800 rpm. Use design factors of Fig. 17.20 and assume $f = 0.95$. Determine the required motor horsepowers for these pumps using the efficiencies of Figure 17.18. Determine the minimum NPSH for each of these pumps. If the suction-pipe diameters are equal to D_e and $p_v/\gamma = 0.65$ ft, determine the values of the absolute pressure heads at suction below which cavitation will occur.

17.37 Two pumps whose characteristics are given in Illustrative Example 17.5 are to be used in parallel. They must develop a head $h = 35 + 20Q^2$ as in the illustrative example. One pump is to be operated at 1,800 rpm. The speed of the other pump is to be gradually reduced until it no longer delivers water. At approximately what speed will this happen?

17.38 A pump is installed to deliver water from a reservoir of surface elevation zero to another of elevation 300 ft. The 12-in-diameter suction pipe ($f = 0.020$) is 100 ft long and the 10-in-diameter discharge pipe ($f = 0.026$) is 5,000 ft long. The pump characteristic at 1,200 rpm is defined by $h_p = 375 - 24Q^2$ where h_p, the pump head, is in feet and Q is in cubic feet per second. Compute the rate at which this pump will deliver water under these conditions assuming the setting is low enough to avoid cavitation.

17.39 Repeat Prob. 17.38 determining the flow rate if two such pumps were installed in series. Repeat for two pumps in parallel.

17.40 Refer to Illustrative Example 17.5. For the case of the single pump operating at 1,800 rpm plot a curve showing delivery rate versus Δz for Δz values ranging from -20 to $+80$ ft. Repeat for rotative speeds of 1,440 rpm and 2,160 rpm. Assume no problem with cavitation.

17.41 Suppose the pumps of Illustrative Example 17.5 were operated at 1,500 rpm. What then would have been the flow rates for: (*a*) single pump; (*b*) two in series; (*c*) two in parallel. All other data to remain the same.

17.42 Assuming $\phi = 1$ in Eq. (14.22), compute the diameter for the impeller in Illustrative Example 17.1*a*.

17.43 A centrifugal pump with a 12-in-diameter impeller is rated at 600 gpm against a head of 80 ft when rotating at 1,750 rpm. What would be the rating of a pump of identical geometric shape with a 6-in impeller? Assume pump efficiencies and rotative speeds are identical.

17.44 A centrifugal-pump impeller (Fig. 17.20*a*) has dimensions $B = 3.0$ in and $D = 10.0$ in. When operating at optimum conditions the pump delivers 1,600 gpm against a head of 75 ft at 1,450 rpm. The required shaft horsepower is 44.0. Assuming the hydraulic efficiency is 0.83 and water enters the impeller radially, determine the required blade angle at discharge.

APPENDIX

ONE

DIMENSIONS AND UNITS, CONVERSION FACTORS

The systems of dimensions and units used in mechanics are based on Newton's second law of motion, which is force equals mass times acceleration, or $F = ma$, if suitable units are chosen. In the English system, engineers define a pound of force as the force required to accelerate one slug of mass[1] at the rate of one foot per second per second; that is,

$$1 \text{ lb} = 1 \text{ slug} \times 1 \text{ ft/s}^2$$

while in the metric (SI) system,[2] engineers define a newton of force as the force required to accelerate one kilogram of mass at the rate of one meter per second per second; that is,

$$1 \text{ N} = 1 \text{ kg} \times 1 \text{ m/s}^2$$

Physicists, on the other hand, ordinarily use the dyne of force defined as the force required to accelerate one gram of mass at the rate of one centimeter per second per second.

Unfortunately, these different systems tend to create confusion. In many parts of the world engineers use the kilogram for both force and mass units. With universal adoption of SI metric, however, this confusion should gradually disappear.

Any system based on length (L), mass (M), and time (T) is absolute because it is independent of the gravitational acceleration g. A system based on length (L), weight, i.e., force (F), and time (T) is referred to as a gravitational system, since weight depends on the value of g which in turn varies with location (i.e., altitude and latitude). Hence the weight (W) of a certain mass varies with its location. This variation is not generally considered in this text as the variation in the value of g is small as long as we are analyzing a problem

[1] One slug of mass has a weight of approximately 32.2 lb when acted upon by the acceleration of gravity present at the surface of the earth.

[2] The metric (SI) system of units is discussed in the front of this book immediately following the Preface.

on the earth's surface. Fluid problems for other locations, such as the moon where g is quite different than on earth, can be handled by the methods presented in this text if proper consideration is given to the value of g.

On the inside of the front cover of the book a table is presented for converting from the English system of units to SI units. A table for converting from SI units into English units is given on the inside of the back cover of the book.

English system conversions

Area	1 acre = 43,560 ft^2
Energy	1 Btu = 778 ft·lb
Flow rate	1 cfs = 448.83 gpm 1 mgd = 1.55 cfs
Length	1 ft = 12 in, 1 yd = 3 ft, 1 mi = 5,280 ft
Mass	1 slug = 32.2 lb (mass)
Power	1 hp = 550 ft·lb/s = 0.708 Btu/s
Velocity	1 mph = 1.467 fps (30 mph = 44 fps) 1 knot = 1.689 fps = 1.152 mph
Volume	1 ft^3 = 7.48 U.S. gal 1 U.S. gal = 231 in^3 = 0.1337 ft^3 = 8.34 lb of water at 60°F 1 British imperial gal = 1.2 U.S. gal = 10 lb of water
Weight	1 U.S. (short) ton = 2,000 lb 1 metric ton = 2,204 lb = 1000 kg of force 1 British (long) ton = 2,240 lb

Other conversions

Engineering gas constant R	1 ft·lb/(slug·°R) = 0.1672 N·m/(kg·K)
Heat	
Metric	1 cal = 4.187 J (heat required to raise 1.0 g of water 1.0 K)
English	1 Btu = 252 cal (heat required to raise 1.0 lb of water 1.0°R)
Temperature	
Metric	K = 273° + °C
English	°R = 460° + °F
	ΔT of 1°C = ΔT of 1K = ΔT of 1.8°F = ΔT of 1.8°R

Definition of metric quantities

Hectare (ha)	area	10^4 m^2 = (100 m square)
Joule (J)	energy (or work)	newton·meter (N·m)
Liter (L)	volume	10^{-3} m^3
Newton (N)	force (or weight)	1 kg × 1 m/s^2
Pascal (Pa)	pressure	newton/meter^2 (N/m^2)
Poise (P)	viscosity	10^{-1} $N·s/m^2$
Stoke (St)	kinematic viscosity	10^{-4} m^2/s
Watt (W)	power	newton·meter/second (N·m/s)

Commonly used prefixes for SI units

Factor by which unit is multiplied	Prefix	Symbol
10^9	giga	G
10^6	mega	M
10^3	kilo	k
10^{-2}	centi	c
10^{-3}	milli	m
10^{-6}	micro	μ

Relationships between temperatures

[°F (fahrenheit) and °C (celsius)]

°C = (5/9)(°F − 32)

°F = 32 + (9/5) × °C

°F	°C	°C	°F
−20	−28.9	−20	−4
0	−17.8	−10	14
20	−6.7	0	32
32	0.0	10	50
40	4.5	20	68
60	15.5	30	86
80	26.6	40	104
100	37.8	50	122
120	48.9	60	140
140	60.0	70	158
160	71.1	80	176
180	82.2	90	194
212	100.0	100	212

APPENDIX

TWO

VELOCITY OR CELERITY OF PRESSURE (SONIC) WAVES

Consider an elastic fluid at rest in a rigid pipe of cross-sectional area A (Fig. A.1). Suppose a piston at one end is suddenly moved with a velocity V for a time dt. This will produce an increase in pressure which will travel through the fluid with a velocity c. While the piston moves the distance $V\,dt$, the wave front will move the distance $c\,dt$. During this time the piston will displace a mass of fluid $\rho AV\,dt$, and during this same time the increase in pressure dp will increase the density of the portion between 1 and 2 by $d\rho$. Equating the mass displaced by the piston to the gain in mass between 1 and 2 due to increased density, $\rho AV\,dt = Ac\,dt\,d\rho$, from which

$$c = \frac{V\rho}{d\rho} \tag{A.1}$$

From mechanics the impulse of a force equals the resulting increase in momentum. The impulse of the force produced by the piston is $A\,dp\,dt$. The mass $\rho Ac\,dt$ is initially at rest, but as the pressure wave travels through it, each element of it will have its velocity increased to V, so that at the end of the time dt the entire mass up to section 2 will have the velocity V. Hence the increase in momentum is $\rho AcV\,dt$. Thus $A\,dp\,dt = \rho AcV\,dt$, and

$$c = \frac{dp}{V\rho} \tag{A.2}$$

Multiplying Eqs. (A.1) and (A.2), we have

$$c^2 = \frac{dp}{d\rho} \tag{A.3}$$

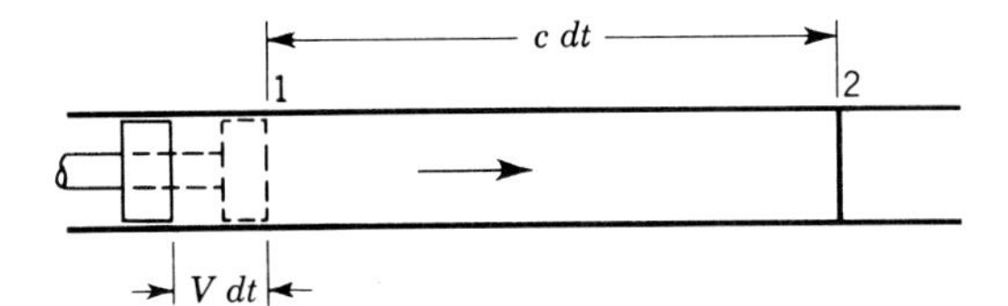

Figure A.1

In Sec. 1.6 the volume modulus of elasticity is defined as $E_v = -(v/dv)\,dp$. Since $\rho = 1/v$, $\rho v = 1.0 =$ constant, and thus $\rho\,dv + v\,d\rho = 0$, and $-v/dv = +\rho/d\rho$. Hence $E_v = \rho\,dp/d\rho$. Substituting this value of $dp/d\rho$ in Eq. (A.3), we have

$$c = \sqrt{\frac{E_v}{\rho}} = \sqrt{\frac{g}{\gamma}E_v} \tag{A.4}$$

This is the velocity or celerity of a pressure (or sound) wave, commonly referred to as the sonic (or acoustic) velocity.

A sonic, or pressure, wave travels through a fluid with such a high velocity that there is no time for any appreciable heat transfer from any heat of compression, moreover, the fluid friction is negligible and thus the process is isentropic. In Sec. 1.9 it is shown that the modulus of elasticity for a perfect gas under isentropic conditions is $E_v = kp$. Inserting this value in Eq. (A.4) gives for a gas,

$$c = \sqrt{\frac{gkp}{\gamma}} = \sqrt{\frac{kp}{\rho}} = \sqrt{kRT} \tag{A.5}$$

This shows that for a gas the sonic velocity is a function of its absolute temperature.

The foregoing analysis has considered the pipe to be rigid. In reality the pipe is elastic, and the stretching of the pipe walls due to the pressure wave makes the modulus of the combination less than that of the fluid alone.

This new modulus will be expressed by K, and we shall let $dv = dv' + dv''$, where dv' is due to compression of the fluid and dv'' is due to stretching of the pipe wall. Thus $K = -v\,dp/(dv' + dv'')$, from which

$$\frac{1}{K} = -\frac{dv'}{v\,dp} - \frac{dv''}{v\,dp} \tag{A.6}$$

The first term on the right is seen to be $1/E_v$ (Sec. 1.6). From the concept of hoop tension the increment of stress in the wall of the pipe is $r\,dp/t$ where r is the radius of the pipe and t is its thickness. If the circumference is stretched an amount dl, the unit deformation is $dl/2\pi r$. Since $dl = 2\pi\,dr$, the increment of unit deformation becomes dr/r. From these relations, recalling that the modulus of elasticity E of a solid = (increment of stress) ÷ (increment of unit deformation), we obtain $dp = Et\,dr/r^2$. For a unit length of pipe and a unit mass of fluid, $v = \pi r^2$, and the increase in volume per unit length of pipe for a unit mass of fluid is equal to the increase in area, so that $dv'' = 2\pi r\,dr$. Substituting these quantities for the three items in the second term of Eq. A.6 gives $dv''/v\,dp = 2r/Et = D/Et$ where D is the pipe diameter. Therefore

$$\frac{1}{K} = \frac{1}{E_v} + \frac{D}{Et} \tag{A.7}$$

Solving for K gives

$$K = \frac{E_v}{1 + (D/t)(E_v/E)} \tag{A.8}$$

The velocity c_p of a pressure wave in an elastic fluid in an elastic pipe is then

$$c_p = \sqrt{\frac{K}{\rho}} = \sqrt{\frac{K}{\rho}\frac{E_v}{E_v}} = c\sqrt{\frac{K}{E_v}} = c\sqrt{\frac{1}{1 + (D/t)(E_v/E)}} \tag{A.9}$$

APPENDIX

THREE

USEFUL TABLES

Table A.1*a* Physical properties of water in English units*

Temperature, °F	Specific weight γ, lb/ft^3	Density ρ, slugs/ft^3	Viscosity $\mu \times 10^5$, lb·s/ft^2	Kinematic viscosity $\nu \times 10^5$, ft^2/s	Surface tension $\sigma \times 10^2$ lb/ft	Vapor pressure p_v, psia	Vapor pressure head p_v/γ, ft	Bulk modulus of elasticity $E_v \times 10^{-3}$, psi
32	62.42	1.940	3.746	1.931	0.518	0.09	0.20	293
40	62.43	1.940	3.229	1.664	0.514	0.12	0.28	294
50	63.41	1.940	2.735	1.410	0.509	0.18	0.41	305
60	62.37	1.938	2.359	1.217	0.504	0.26	0.59	311
70	62.30	1.936	2.050	1.059	0.500	0.36	0.84	320
80	62.22	1.934	1.799	0.930	0.492	0.51	1.17	322
90	62.11	1.931	1.595	0.826	0.486	0.70	1.61	323
100	62.00	1.927	1.424	0.739	0.480	0.95	2.19	327
110	61.86	1.923	1.284	0.667	0.473	1.27	2.95	331
120	61.71	1.918	1.168	0.609	0.465	1.69	3.91	333
130	61.55	1.913	1.069	0.558	0.460	2.22	5.13	334
140	61.38	1.908	0.981	0.514	0.454	2.89	6.67	330
150	61.20	1.902	0.905	0.476	0.447	3.72	8.58	328
160	61.00	1.896	0.838	0.442	0.441	4.74	10.95	326
170	60.80	1.890	0.780	0.413	0.433	5.99	13.83	322
180	60.58	1.883	0.726	0.385	0.426	7.51	17.33	318
190	60.36	1.876	0.678	0.362	0.419	9.34	21.55	313
200	60.12	1.868	0.637	0.341	0.412	11.52	26.59	308
212	59.83	1.860	0.593	0.319	0.404	14.70	33.90	300

* In this table and in the others to follow, if $\mu \times 10^5 = 3.746$ then $\mu = 3.746 \times 10^{-5}$ lb·s/ft^2, etc. For example, at 80°F, $\sigma \times 10^2 = 0.492$ or $\sigma = 0.00492$ lb/ft and $E_v \times 10^{-3} = 322$ or $E_v = 322{,}000$ psi.

Table A.1*b* Physical properties of water in SI units

Temperature, °C	Specific weight γ, kN/m^3	Density ρ, kg/m^3	Viscosity $\mu \times 10^3$, N·s/m^2	Kinematic viscosity $\nu \times 10^6$, m^2/s	Surface tension σ, N/m	Vapor pressure p_v kN/m^2, abs	Vapor pressure head p_v/γ m	Bulk modulus of elasticity $E_v \times 10^{-6}$, kN/m^2
0	9.805	999.8	1.781	1.785	0.0756	0.61	0.06	2.02
5	9.807	1000.0	1.518	1.519	0.0749	0.87	0.09	2.06
10	9.804	999.7	1.307	1.306	0.0742	1.23	0.12	2.10
15	9.798	999.1	1.139	1.139	0.0735	1.70	0.17	2.14
20	9.789	998.2	1.002	1.003	0.0728	2.34	0.25	2.18
25	9.777	997.0	0.890	0.893	0.0720	3.17	0.33	2.22
30	9.764	995.7	0.798	0.800	0.0712	4.24	0.44	2.25
40	9.730	992.2	0.653	0.658	0.0696	7.38	0.76	2.28
50	9.689	988.0	0.547	0.553	0.0679	12.33	1.26	2.29
60	9.642	983.2	0.466	0.474	0.0662	19.92	2.03	2.28
70	9.589	977.8	0.404	0.413	0.0644	31.16	3.20	2.25
80	9.530	971.8	0.354	0.364	0.0626	47.34	4.96	2.20
90	9.466	965.3	0.315	0.326	0.0608	70.10	7.18	2.14
100	9.399	958.4	0.282	0.294	0.0589	101.33	10.33	2.07

Table A.2*a* Physical properties of air at standard atmospheric pressure in English units

Temperature		Density	Specific weight	Viscosity	Kinematic viscosity
T, °F	T, °C	$\rho \times 10^3$, slugs/ft^3	$\gamma \times 10^2$, lb/ft^3	$\mu \times 10^7$, lb·s/ft^2	$\nu \times 10^4$, ft^2/s
−40	−40.0	2.94	9.46	3.12	1.06
−20	−28.9	2.80	9.03	3.25	1.16
0	−17.8	2.68	8.62	3.38	1.26
10	−12.2	2.63	8.46	3.45	1.31
20	−6.7	2.57	8.27	3.50	1.36
30	−1.1	2.52	8.11	3.58	1.42
40	4.4	2.47	7.94	3.62	1.46
50	10.0	2.42	7.79	3.68	1.52
60	15.6	2.37	7.63	3.74	1.58
70	21.1	2.33	7.50	3.82	1.64
80	26.7	2.28	7.35	3.85	1.69
90	32.2	2.24	7.23	3.90	1.74
100	37.8	2.20	7.09	3.96	1.80
120	48.9	2.15	6.84	4.07	1.89
140	60.0	2.06	6.63	4.14	2.01
160	71.1	1.99	6.41	4.22	2.12
180	82.2	1.93	6.21	4.34	2.25
200	93.3	1.87	6.02	4.49	2.40
250	121.1	1.74	5.60	4.87	2.80

Table A.2*b* Physical properties of air at standard atmospheric pressure in SI units

Temperature		Density	Specific weight	Viscosity	Kinematic viscosity
T, °C	T, °F	ρ, kg/m³	γ, N/m³	$\mu \times 10^5$, N·s/m²	$\nu \times 10^5$, m²/s
−40	−40	1.515	14.86	1.49	0.98
−20	−4	1.395	13.68	1.61	1.15
0	32	1.293	12.68	1.71	1.32
10	50	1.248	12.24	1.76	1.41
20	68	1.205	11.82	1.81	1.50
30	86	1.165	11.43	1.86	1.60
40	104	1.128	11.06	1.90	1.68
60	140	1.060	10.40	2.00	1.87
80	176	1.000	9.81	2.09	2.09
100	212	0.946	9.28	2.18	2.31
200	392	0.747	7.33	2.58	3.45

Table A.3*a* The ICAO standard atmosphere in English units

Elevation above sea level, ft	Temp, °F	Absolute pressure, psia	Specific weight γ, lb/ft³	Density ρ, slugs/ft³	Viscosity $\mu \times 10^7$, lb·s/ft²
0	59.0	14.70	0.07648	0.002377	3.737
5,000	41.2	12.24	0.06587	0.002048	3.637
10,000	23.4	10.11	0.05643	0.001756	3.534
15,000	5.6	8.30	0.04807	0.001496	3.430
20,000	−12.3	6.76	0.04070	0.001267	3.325
25,000	−30.1	5.46	0.03422	0.001066	3.217
30,000	−47.8	4.37	0.02858	0.000891	3.107
35,000	−65.6	3.47	0.02367	0.000738	2.995
40,000	−69.7	2.73	0.01882	0.000587	2.969
45,000	−69.7	2.15	0.01481	0.000462	2.969
50,000	−69.7	1.69	0.01165	0.000364	2.969
60,000	−69.7	1.05	0.00722	0.000226	2.969
70,000	−69.7	0.65	0.00447	0.000140	2.969
80,000	−69.7	0.40	0.00277	0.000087	2.969
90,000	−57.2	0.25	0.00168	0.000053	3.048
100,000	−40.9	0.16	0.00102	0.000032	3.150

Table A.3*b* The ICAO standard atmosphere in SI units

Elevation above sea level, km	Temp, °C	Absolute pressure, kN/m², abs	Specific weight γ, N/m³	Density ρ, kg/m³	Viscosity $\mu \times 10^5$, N·s/m²
0	15.0	101.33	12.01	1.225	1.79
2	2.0	79.50	9.86	1.007	1.73
4	−4.5	60.12	8.02	0.909	1.66
6	−24.0	47.22	6.46	0.660	1.60
8	−36.9	35.65	5.14	0.526	1.53
10	−49.9	26.50	4.04	0.414	1.46
12	−56.5	19.40	3.05	0.312	1.42
14	−56.5	14.20	2.22	0.228	1.42
16	−56.5	10.35	1.62	0.166	1.42
18	−56.5	7.57	1.19	0.122	1.42
20	−56.5	5.53	0.87	0.089	1.42
25	−51.6	2.64	0.41	0.042	1.45
30	−40.2	1.20	0.18	0.018	1.51

Table A.4*a* Physical properties of common liquids at standard atmospheric pressure in English units

Liquid	Temperature T, °F	Density ρ, slug/ft³	Specific gravity, s	Viscosity $\mu \times 10^5$, lb·s/ft²	Surface tension σ, lb/ft	Vapor pressure p_v, psia	Modulus of elasticity E_v, psi
Benzene	68	1.74	0.90	1.4	0.002	1.48	150,000
Carbon tetrachloride	68	3.08	1.594	2.0	0.0018	1.76	160,000
Crude oil	68	1.66	0.86	15	0.002		
Gasoline	68	1.32	0.68	0.62	......	8.0	
Glycerin	68	2.44	1.26	3100	0.004	0.000002	630,000
Hydrogen	−430	0.14	0.072	0.043	0.0002	3.1	
Kerosene	68	1.57	0.81	4.0	0.0017	0.46	
Mercury	68	26.3	13.56	3.3	0.032	0.000025	3,800,000
Oxygen	−320	2.34	1.21	0.58	0.001	3.1	
SAE 10 oil	68	1.78	0.92	170	0.0025		
SAE 30 oil	68	1.78	0.92	920	0.0024		
Water	68	1.936	1.00	2.1	0.005	0.34	300,000

Table A.4*b* Physical properties of common liquids at standard atmospheric pressure in SI units

Liquid	Temperature T, °C	Density ρ, kg/m³	Specific gravity, s	Viscosity $\mu \times 10^4$, N·s/m²	Surface tension σ, N/m	Vapor pressure p_v, kN/m², abs	Modulus of elasticity $E_v \times 10^{-6}$, N/m²
Benzene	20	895	0.90	6.5	0.029	10.0	1,030
Carbon tetrachloride	20	1,588	1.59	9.7	0.026	12.1	1,100
Crude oil	20	856	0.86	72	0.03		
Gasoline	20	678	0.68	2.9	.....	55	
Glycerin	20	1,258	1.26	14,900	0.063	0.000014	4,350
Hydrogen	−257	72	0.072	0.21	0.003	21.4	
Kerosene	20	808	0.81	19.2	0.025	3.20	
Mercury	20	13,550	13.56	15.6	0.51	0.00017	26,200
Oxygen	−195	1,206	1.21	2.8	0.015	21.4	
SAE 10 oil	20	918	0.92	820	0.037		
SAE 30 oil	20	918	0.92	4,400	0.036		
Water	20	998	1.00	10.1	0.073	2.34	2,070

Table A.5*a* Physical properties of common gases at standard sea-level atmosphere and 68°F in English units

Gas	Chemical formula	Molecular weight	Specific weight, γ, lb/ft³	Viscosity $\mu \times 10^7$, lb·s/ft²	Gas constant R, ft·lb/(slug·°R) [=ft²/(s²·°R)]	Specific heat, ft·lb/(slug·°R) [=ft²/(s²·°R)] c_p	c_v	Specific heat ratio $k = c_p/c_v$
Air		29.0	0.0753	3.76	1,715	6,000	4,285	1.40
Carbon dioxide	CO_2	44.0	0.114	3.10	1,123	5,132	4,009	1.28
Carbon monoxide	CO	28.0	0.0726	3.80	1,778	6,218	4,440	1.40
Helium	He	4.00	0.0104	4.11	12,420	31,230	18,810	1.66
Hydrogen	H_2	2.02	0.00522	1.89	24,680	86,390	61,710	1.40
Methane	CH_4	16.0	0.0416	2.80	3,100	13,400	10,300	1.30
Nitrogen	N_2	28.0	0.0728	3.68	1,773	6,210	4,437	1.40
Oxygen	O_2	32.0	0.0830	4.18	1,554	5,437	3,883	1.40
Water vapor	H_2O	18.0	0.0467	2.12	2,760	11,110	8,350	1.33

Table A.5*b* Physical properties of common gases at standard sea level and 68°F in SI units

Gas	Chemical formula	Molecular weight	Density ρ, kg/m^3	Viscosity, $\mu \times 10^5$ N·s/m^2	Gas constant R, N·m/(kg·K) [=m^2/(s^2·K)]	Specific heat, N·m/(kg·K) [(=m^2/(s^2·K)] c_p	c_v	Specific heat ratio $k = c_p/c_v$
Air		29.0	1.205	1.80	287	1,003	716	1.40
Carbon dioxide	CO_2	44.0	1.84	1.48	188	858	670	1.28
Carbon monoxide	CO	28.0	1.16	1.82	297	1,040	743	1.40
Helium	He	4.00	0.166	1.97	2,077	5,220	3,143	1.66
Hydrogen	H_2	2.02	0.0839	0.90	4,120	14,450	10,330	1.40
Methane	CH_4	16.0	0.668	1.34	520	2,250	1,730	1.30
Nitrogen	N_2	28.0	1.16	1.76	297	1,040	743	1.40
Oxygen	O_2	32.0	1.33	2.00	260	909	649	1.40
Water vapor	H_2O	18.0	0.747	1.01	462	1,862	1,400	1.33

Table A.6*a* Areas of circles (English units)

Diameter, in	Area in^2	Area ft^2
0.25	0.0491	0.000341
0.5	0.1963	0.001364
1.0	0.785	0.00545
2.0	3.142	0.0218
3.0	7.069	0.0491
4.0	12.57	0.0873
6.0	28.27	0.1963
8.0	50.27	0.349
10.0	78.54	0.545
12.0	113.10	0.785

Table A.6*b* Areas of circles (SI units)

Diameter, cm	Area, m^2
5	0.001963
10	0.00785
15	0.01767
20	0.03142
25	0.04910
30	0.07069
50	0.1963
100	0.7854
150	1.767
200	3.142

Table A.7 Properties of areas

	Sketch	Area	Location of centroid	I or I_c
Rectangle	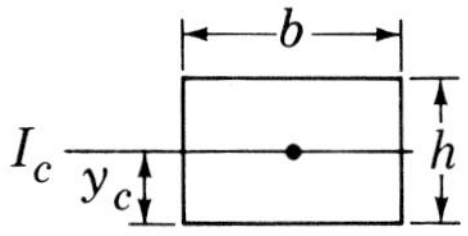	bh	$y_c = \frac{h}{2}$	$I_c = \frac{bh^3}{12}$
Triangle	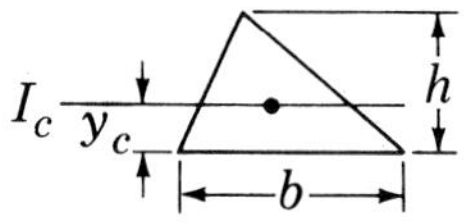	$\frac{bh}{2}$	$y_c = \frac{h}{3}$	$I_c = \frac{bh^3}{36}$
Circle	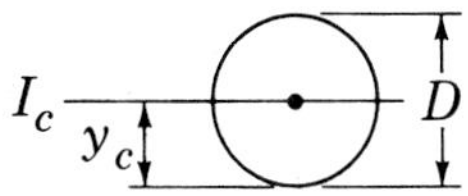	$\frac{\pi D^2}{4}$	$y_c = \frac{D}{2}$	$I_c = \frac{\pi D^4}{64}$
Semicircle	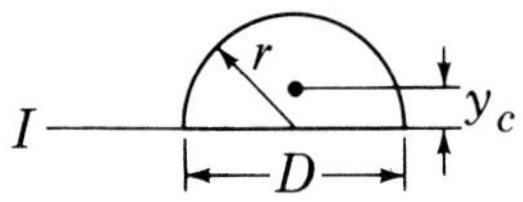	$\frac{\pi D^2}{8}$	$y_c = \frac{4r}{3\pi}$	$I = \frac{\pi D^4}{128}$
Ellipse	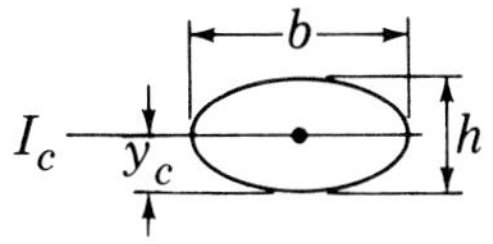	$\frac{\pi bh}{4}$	$y_c = \frac{h}{2}$	$I_c = \frac{\pi bh^3}{64}$
Semiellipse	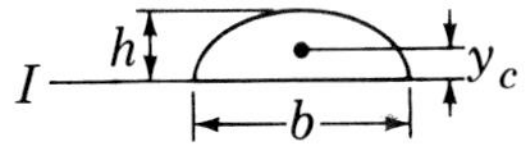	$\frac{\pi bh}{4}$	$y_c = \frac{4h}{3\pi}$	$I = \frac{\pi bh^3}{16}$
Parabola	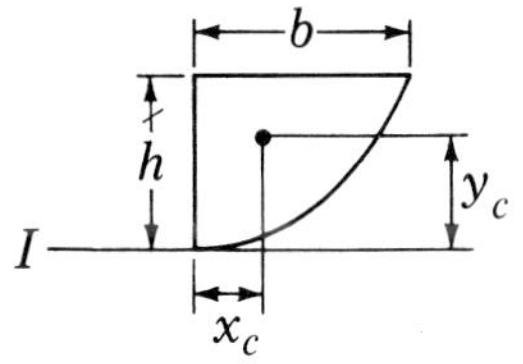	$\frac{2bh}{3}$	$x_c = \frac{3b}{8}$ $y_c = \frac{3b}{5}$	$I = \frac{2bh^3}{7}$

Table A.8 Properties of solid bodies

	Sketch	Volume	Location of center of mass
Cylinder	y_c, h, D	$\frac{\pi D^2 h}{4}$	$y_c = \frac{h}{2}$
Cone	y_c, h, D	$\frac{1}{3}\left(\frac{\pi D^2 h}{4}\right)$	$y_c = \frac{h}{4}$
Sphere	y_c, D	$\frac{\pi D^3}{6}$	$y_c = \frac{D}{2}$
Hemisphere	y_c, r, D	$\frac{\pi D^3}{12}$	$y_c = \frac{3r}{8}$
Paraboloid	y_c, h, D	$\frac{1}{2}\left(\frac{\pi D^2 h}{4}\right)$	$y_c = \frac{h}{3}$

APPENDIX

FOUR

REFERENCES

There is a great volume of literature available on the various aspects of fluid mechanics and hydraulics. The results of original research may be found in papers published in technical journals. A list of books covering various topics of fluid mechanics and its engineering applications is presented here for the convenience of the student. This list by no means includes all the important books that have been written; the intent here is merely to provide a representative list of books. The student is encouraged to "probe deeper" and to widen his horizons by further reading.

Ackers, P., W. R. White, J. A. Perkins, and A. J. M. Harrison: "Weirs and Flumes for Flow Measurement," John Wiley and Sons, Inc., New York, 1978.

Batchelor, G. K.: "An Introduction to Fluid Dynamics," Cambridge University Press, Cambridge, 1967.

Bear, J.: "Hydraulics of Groundwater," McGraw-Hill Book Co., New York, 1979.

Benedict, R. P.: "Fundamentals of Pipe Flow," John Wiley and Sons, Inc., New York, 1980.

Bergeron, L. J. P.: "Water Hammer in Hydraulics and Wave Surges in Electricity," translated by American Society of Mechanical Engineers, John Wiley and Sons, Inc., 1961.

Bird, R. B., W. E. Stewart, and E. N. Lightfoot: "Transport Phenomena," John Wiley and Sons, Inc., 1960.

Brater, E. F., and H. W. King: "Handbook of Applied Hydraulics," 6th ed., McGraw-Hill Book Co., New York, 1976.

Cambel, A. B., and B. H. Jennings: "Gas Dynamics," Dover Publications, Inc., New York, 1968.

Cermak, J. E.: "Wind Engineering—Proceedings of the 5th International Conference," Colorado State University, 2 vols, Pergamon, New York, 1981.

Chow, Van Te: "Open-channel Hydraulics," McGraw-Hill Book Co., New York, 1959.

Croley, T. E.: "Hydrologic and Hydraulic Computations on Small Programmable Calculators," Iowa Institute of Hydraulic Research, The University of Iowa, Iowa City, Iowa, 1977.

Csanady, G. T.: "Theory of Turbomechanics," McGraw-Hill Book Co., New York, 1964.

Daily, J. W., and D. R. F. Harleman: "Fluid Dynamics," Addison-Wesley, Reading, Mass., 1966.

Davis, C. V., and K. E. Sorenson: "Handbook of Applied Hydraulics," 3d ed., McGraw-Hill Book Co., New York, 1969.

Daugherty, R. L.: "Centrifugal Pumps," McGraw-Hill Book Co., New York, 1915.
———: "Hydraulic Turbines," McGraw-Hill Book Co., New York, 1920.
Eckert, E. R. G., and R. M. Drake: "Analysis of Heat and Mass Transfer," McGraw-Hill Book Co., New York, 1972.
Fischer, H.: "Mixing in Inland and Coastal Waters," Academic Press, 1979.
"Fluid Meters: Their Theory and Application," 6th ed., American Society of Mechanical Engineers, New York, 1971.
Freeze, R. A., and J. A. Cherry: "Groundwater," Prentice-Hall, New York, 1979.
Goldstein, S.: "Modern Developments in Fluid Dynamics," 2 vols., Oxford University Press, Fair Lawn, N.J., 1938.
Graf, W. H.: "Hydraulics of Sediment Transport," McGraw-Hill Book Co., New York, 1971.
Henderson, F. M.: "Open Channel Flow," The Macmillan Company, New York, 1966.
Hinze, J. O.: "Turbulence," 2d ed., McGraw-Hill Book Co., New York, 1975.
Hydraulic Institute: "Standards of the Hydraulic Institute," 13th ed., New York, 1975.
Hydraulic Models, ASCE Manual Eng. Practice, 25, American Society of Civil Engineers, New York, 1942.
Jaeger, Charles: "Engineering Fluid Mechanics," Blackie and Son, Ltd., Glasgow, 1956.
Jeppson, R. W.: "Analysis of Flow in Pipe Networks," Ann Arbor Science Pub., Ann Arbor, Mich., 1976.
Kuethe, A. M., and J. D. Schetzer: "Foundations of Aerodynamics," 3d ed., John Wiley and Sons, Inc., New York, 1976.
Lamb, H.: "Hydrodynamics," 6th ed., Dover Publications, New York, 1945.
Langhaar, H. L.: "Dimensional Analysis and Theory of Models," John Wiley and Sons, New York, 1951, reprint by Kreiger, 1979.
Levich, V. G.: "Physiochemical Hydrodynamics," Prentice-Hall, Inc., Englewood Cliffs, N.J., 1962.
Linsley, R. K., and J. B. Franzini: "Water-Resources Engineering," 3d ed., McGraw-Hill Book Co., New York, 1979.
Morris, H. M., and J. M. Wiggert: "Applied Hydraulics in Engineering," 2d ed., John Wiley and Sons, New York, 1972.
Norrie., D. H.: "Incompressible Flow Machines," E. Arnold, London, 1963.
Pai, S. I.: "Viscous Flow Theory," 2 vols., D. Van Nostrand Company, Princeton, N.J., 1955.
Parmakian, John: "Waterhammer Analysis," Prentice-Hall, Inc., Englewood Cliffs, N.J., 1955.
Prandtl, L., and O. G. Tietjens: "Fundamentals of Hydro- and Aeromechanics," McGraw-Hill Book Co., New York, 1934.
Reynolds, W. C., and H. C. Perkins: "Engineering Thermodynamics," 2d ed., McGraw-Hill Book Co., New York, 1977.
Robertson, J. M.: "Hydrodynamics in Theory and Applications," Prentice-Hall, Inc., Englewood Cliffs, N.J., 1965.
——— and G. F. Wislicenus (eds.): "Cavitation—State of Knowledge," American Society of Mechanical Engineers, New York, 1969.
Rouse, H. (ed.): "Engineering Hydraulics," John Wiley and Sons, Inc., New York, 1950.
——— and S. Ince: "History of Hydraulics," Iowa Institute of Hydraulic Research. 1957.
Sabersky, R. H., and A. J. Acosta: "Fluid Flow," 2d ed., The Macmillan Company, New York, 1971.
Schlichting, H.: "Boundary Layer Theory," 7th ed., McGraw-Hill Book Co., New York, 1979.
Shames, I. H.: "Mechanics of Fluids," McGraw-Hill Book Co., New York, 1962.
Shapiro, A. H.: "The Dynamics and Thermodynamics of Compressible Flow," 2 vols., John Wiley and Sons, Inc., New York, 1953.
Smith, P. D., "BASIC Hydraulics," Butterworth Publishers, Stoneham, MA., 1982.
Soo, S. L.: "Fluid Dynamics of Multiphase Systems," Blaisdell, Waltham, Mass., 1967.
Stepanoff, A. J.: "Centrifugal and Axial Flow Pumps," 2d ed., John Wiley and Sons, Inc., New York, 1957.
Stoker, J. J.: "Water Waves," John Wiley and Sons, Inc., New York, 1957.
Streeter, V. L. (ed.): "Handbook of Fluid Dynamics," McGraw-Hill Book Co., New York, 1961.
——— and E. B. Wiley: "Fluid Mechanics," 7th ed., McGraw-Hill Book Co., New York, 1979.

Sutton, G. P., and D. M. Ross: "Rocket Propulsion Elements," 4th ed., John Wiley and Sons, Inc., 1976.

Sutton, O. G.: "The Science of Flight," Penguin Books, New York, 1949.

Thompson, P. A.: "Compressible Fluid Dynamics," McGraw-Hill Book Co., New York, 1971.

Todd, D. K.: "Groundwater Hydrology," 2d ed., John Wiley and Sons, Inc. New York, 1980.

Vallentine, H. R.: "Applied Hydrodynamics," 2d ed., Butterworth and Co. (Publishers), Ltd., London, 1967.

Vennard, J. K., and R. L. Street: "Elementary Fluid Mechanics," 6th ed., John Wiley and Sons, New York, 1982.

von Kármán, T.: "Aerodynamics," Cornell University Press, Ithaca, N.Y., 1954.

Watters, G. Z.: "Analysis and Control of Unsteady Flow in Pipelines," 2d ed., Butterworth Publishers, Stoneham, MA., 1984.

White, F. M.: "Viscous Fluid Flow," McGraw-Hill Book Co., New York, 1974.

Wilkinson, W. L.: "Non-Newtonian Fluids," Pergamon Press, New York, 1960.

Yalin, M. S.: "Theory of Hydraulic Models," The Macmillan Company, London, 1971.

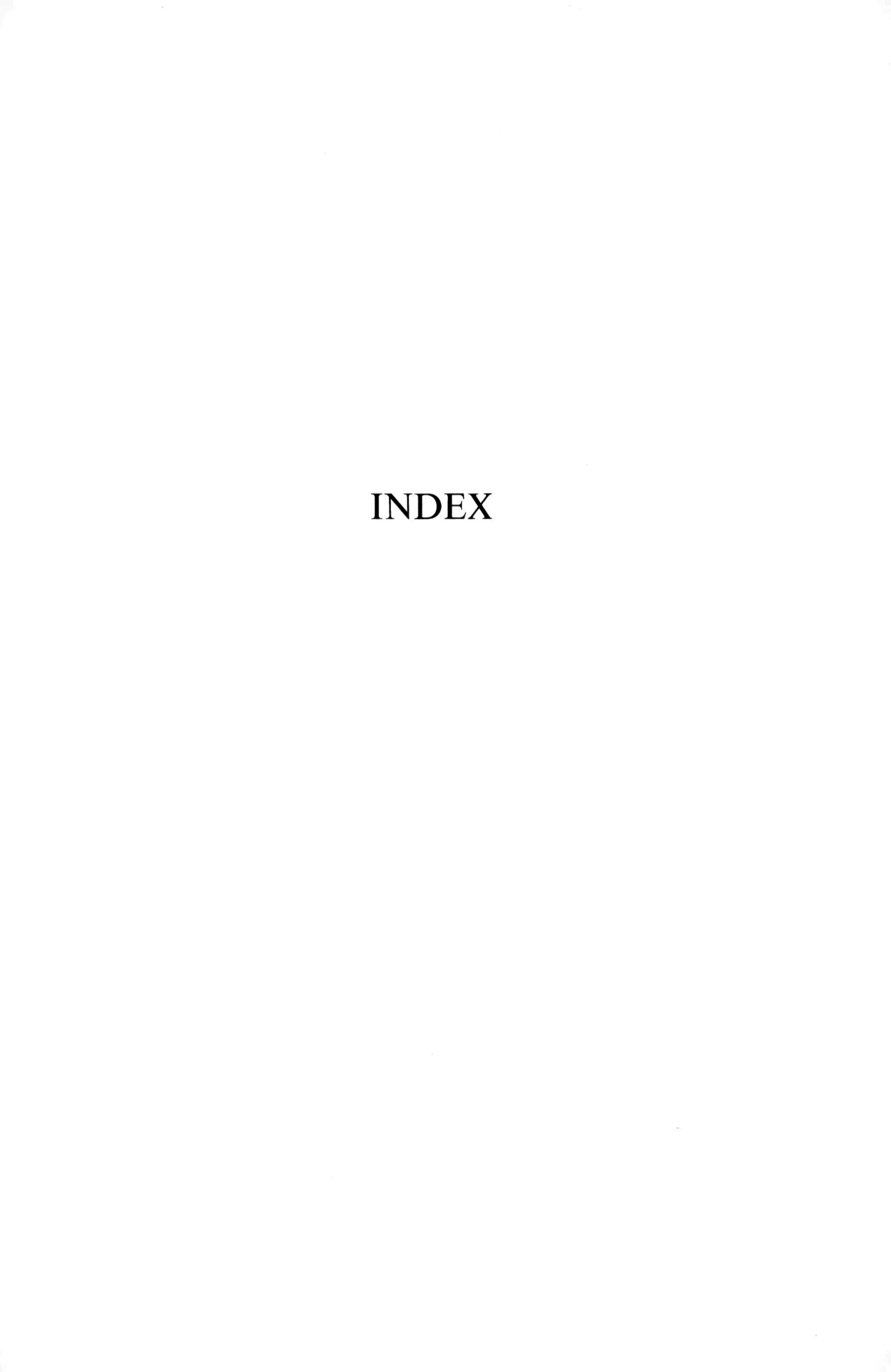

INDEX

INDEX